Advances in Intelligent Systems and Computing

Volume 816

Series editor

Janusz Kacprzyk, Polish Academy of Sciences, Warsaw, Poland
e-mail: kacprzyk@ibspan.waw.pl

The series "Advances in Intelligent Systems and Computing" contains publications on theory, applications, and design methods of Intelligent Systems and Intelligent Computing. Virtually all disciplines such as engineering, natural sciences, computer and information science, ICT, economics, business, e-commerce, environment, healthcare, life science are covered. The list of topics spans all the areas of modern intelligent systems and computing such as: computational intelligence, soft computing including neural networks, fuzzy systems, evolutionary computing and the fusion of these paradigms, social intelligence, ambient intelligence, computational neuroscience, artificial life, virtual worlds and society, cognitive science and systems, Perception and Vision, DNA and immune based systems, self-organizing and adaptive systems, e-Learning and teaching, human-centered and human-centric computing, recommender systems, intelligent control, robotics and mechatronics including human-machine teaming, knowledge-based paradigms, learning paradigms, machine ethics, intelligent data analysis, knowledge management, intelligent agents, intelligent decision making and support, intelligent network security, trust management, interactive entertainment, Web intelligence and multimedia.

The publications within "Advances in Intelligent Systems and Computing" are primarily proceedings of important conferences, symposia and congresses. They cover significant recent developments in the field, both of a foundational and applicable character. An important characteristic feature of the series is the short publication time and world-wide distribution. This permits a rapid and broad dissemination of research results.

More information about this series at http://www.springer.com/series/11156

Jagdish Chand Bansal · Kedar Nath Das
Atulya Nagar · Kusum Deep
Akshay Kumar Ojha
Editors

Soft Computing for Problem Solving

SocProS 2017, Volume 1

Springer

Editors
Jagdish Chand Bansal
Department of Mathematics
South Asian University
New Delhi, India

Kusum Deep
Department of Mathematics
Indian Institute of Technology Roorkee
Roorkee, Uttarakhand, India

Kedar Nath Das
Department of Mathematics
National Institute of Technology Silchar
Silchar, Assam, India

Akshay Kumar Ojha
School of Basic Sciences
Indian Institute of Technology Bhubaneswar
Bhubaneswar, Odisha, India

Atulya Nagar
Department of Mathematics and Computer
  Science, Faculty of Science
Liverpool Hope University
Liverpool, UK

ISSN 2194-5357   ISSN 2194-5365   (electronic)
Advances in Intelligent Systems and Computing
ISBN 978-981-13-1591-6   ISBN 978-981-13-1592-3   (eBook)
https://doi.org/10.1007/978-981-13-1592-3

Library of Congress Control Number: 2018947855

This Springer imprint is published by the registered company Springer Nature Singapore Pte Ltd.
The registered company address is: 152 Beach Road, #21-01/04 Gateway East, Singapore 189721, Singapore

Preface

SocProS, which stands for 'Soft Computing for Problem Solving,' is entering its seventh edition as an established and flagship international conference. This particular annual event is a joint collaboration between a group of faculty members from the institutes of repute like South Asian University, New Delhi; NIT Silchar; Liverpool Hope University, UK; IIT Roorkee; and IIT Bhubaneswar.

The first in the series of SocProS started in 2011 and was held from 20th to 22nd December on the IIT Roorkee Campus with Prof. Deep (IITR) and Prof. Nagar (Liverpool Hope University) as the general chairs. JKLU Jaipur hosted the second SocProS from December 28 to 30, 2012. Coinciding with the Golden Jubilee of the IIT Roorkee's Saharanpur Campus, the third edition of this international conference, which has by now become a brand name, took place at the Greater Noida Extension Centre of IIT Roorkee during December 26–28, 2013. Afterward, in 2014, it has been organized at NIT Silchar, Assam, during December 27–29, 2014. The next conference series was held at Saharanpur Campus of IIT Roorkee during December 18–20, 2015. In the last year, Thapar University, Patiala, has hosted the conference during December 23–24, 2016.

Like earlier SocProS conferences, the focus of SocProS 2017 is on soft computing and its applications to real-life problems arising in diverse areas of medical and health care, supply chain management, signal processing and multimedia, industrial optimization, image processing, cryptanalysis, etc. SocProS 2017 attracted a wide spectrum of thought-provoking articles. A total of 164 high-quality research papers have been selected for publication in the form of this two-volume proceeding.

We hope that the papers contained in this proceeding will prove helpful toward improving the understanding of soft computing at teaching as well as research level and will inspire more and more researchers to work in the field of soft computing.

The editors would like to express their sincere gratitude to SocProS 2017 patron, plenary speakers, invited speakers, reviewers, program committee members, international advisory committee, and local organizing committee; without whose support, the quality and standards of the conference could not be maintained. We express special thanks to Springer and its team for this valuable support in the publication of this proceeding.

Over and above, we would like to express our deepest sense of gratitude to the 'Indian Institute of Technology (IIT) Bhubaneswar' to facilitate the hosting of this conference. Our sincere thanks to all the sponsors of SocProS 2017.

SAU New Delhi, India   Jagdish Chand Bansal
NIT Silchar, India   Kedar Nath Das
LHU, Liverpool, UK   Atulya Nagar
IIT Roorkee, India   Kusum Deep
IIT Bhubaneswar, India   Akshay Kumar Ojha

About the Book

The proceedings of SocProS 2017 will serve as an academic bonanza for scientists and researchers working in the field of soft computing. This book contains theoretical as well as practical aspects using fuzzy logic, neural networks, evolutionary algorithms, swarm intelligence algorithms, etc., with many applications under the umbrella of 'soft computing.' The book will be beneficial for young as well as experienced researchers dealing across complex and intricate real-world problems for which finding a solution by traditional methods is a difficult task.

The different application areas covered in the proceedings are image processing, cryptanalysis, industrial optimization, supply chain management, newly proposed nature-inspired algorithms, signal processing, problems related to medical and health care, networking optimization problems, etc.

Contents

About the Editors

Dr. Jagdish Chand Bansal is Assistant Professor at the South Asian University, New Delhi, India, and Visiting Research Fellow at Liverpool Hope University, Liverpool, UK. He has an excellent academic record and is a leading researcher in the field of swarm intelligence, and he has published numerous research papers in respected international and national journals.

Dr. Kedar Nath Das is Assistant Professor in the Department of Mathematics, National Institute of Technology Silchar, Assam, India. Over the past 10 years, he has made substantial contributions to research on 'soft computing.' He has published several research papers in prominent national and international journals. His chief area of interest is evolutionary and bio-inspired algorithms for optimization.

Prof. Atulya Nagar holds the Foundation Chair as Professor of Mathematical Sciences and is Dean of the Faculty of Science, Liverpool Hope University, UK. He is an internationally respected scholar working at the cutting edge of theoretical computer science, applied mathematical analysis, operations research, and systems engineering.

Prof. Kusum Deep is Professor in the Department of Mathematics, Indian Institute of Technology Roorkee, India. Over the past 25 years, her research has made her a central international figure in the area of nature-inspired optimization techniques, genetic algorithms, and particle swarm optimization.

Dr. Akshay Kumar Ojha is Associate Professor at the School of Basic Sciences, Indian Institute of Technology Bhubaneswar, Odisha, India. He completed his B.Sc., M.Sc., and Ph.D. at Utkal University in 1978, 1980, and 1997, respectively. His research interest areas are geometric programming, artificial neural networks, genetic algorithms, particle swarm optimization, fractional programming, nonlinear optimization, data analysis and optimization, and portfolio optimization. He has 34 years of experience and has published over 30 journal articles and 6 books.

Power Distribution Network Reconfiguration Using an Improved Sine–Cosine Algorithm-Based Meta-Heuristic Search

Usharani Raut and Sivkumar Mishra

Abstract This paper proposes an improved sine–cosine algorithm for solving power distribution network reconfiguration (PDNR) problem. The sine–cosine algorithm is a recently proposed population-based meta-heuristic optimization algorithm which uses the mathematical sine and cosine functions for searching the solution space. The search procedure looks for the best solution by repeatedly making small changes to an initial solution until no further improved solutions are found. To maintain a balance between local and global search, four random variables (r_1, r_2, r_3 and r_4) are integrated into this algorithm. For applying this algorithm to the PDNR problem, some improvements are proposed in this meta-heuristic search algorithm along with a new data structure-based load flow method to minimize power loss as the single objective. The effectiveness of the proposed PDNR algorithm is tested by considering five standard test distribution systems (33, 69, 84, 119 and 136 buses).

Keywords Radial distribution networks · Network reconfiguration · Sine–cosine algorithm

1 Introduction

The power distribution network reconfiguration (PDNR) is an old but quite relevant power optimization problem, which is a process of changing the topology of the power distribution network by altering the open/closed status of the switches. The main objective of the technique is to find a radial operating structure that minimizes the power losses of the distribution system under normal operating conditions. PDNR belongs to a class of complex combinatorial optimization problems where complexity arises from the requirement of radial network topology and nonlinear power flow

U. Raut · S. Mishra (✉)
International Institute of Information Technology, Bhubaneswar, Odisha, India
e-mail: sivkumar@iiit-bh.ac.in

U. Raut
e-mail: usharani@iiit-bh.ac.in

© Springer Nature Singapore Pte Ltd. 2019
J. C. Bansal et al. (eds.), *Soft Computing for Problem Solving*,
Advances in Intelligent Systems and Computing 816,
https://doi.org/10.1007/978-981-13-1592-3_1

1

constraints. Meta-heuristic search-based techniques are the most preferred ones as they always guarantee global minimum results.

Several meta-heuristic-based PDNR methods are available in the literature [1]. In these methods, a population rather than a single agent is considered for searching the solution space. Maintaining a balance between the local search (exploitation phase) and global search (exploration phase), these optimization algorithms converge to a global optimum value without getting trapped in local optimum values. However, these methods are slow in convergence as it stochastically moves in the solution space and it has to check the fitness for every member of the population. So, in a population-based meta-heuristic PDNR method, it is almost necessary to carry out load flow for each member of the population for every generation. Simultaneously, the configurations are also to be checked for radiality and connectivity. In PDNR problem, the most popular objective is the overall active power loss minimization (APLM), although there are other objectives when considered simultaneously makes the optimization problem multiobjective. Bus voltage deviation minimization (BVDM), load balancing index minimization (LBIM), voltage profile improvement (VPI), improvement of voltage stability index (VSI) and number of switching minimization (NSM) are some of the other commonly adopted objectives in PDNR. Similarly, to improve the system reliability, the objectives can be considered as to minimize the reliability-based indices like system average interruption frequency index (SAIFI), system average interruption duration index (SAIDI) and energy not supplied (ENS).

In Table 1, a survey of the recently proposed PDNR methods using meta-heuristic algorithms, such as genetic algorithm (GA), particle swarm optimization (PSO), gravity search algorithm (GSA), symbiotic organism search (SOS), runner root algorithm (RRA), grey wolf algorithm (GWA), harmony search algorithm (HSA), artificial bee colony algorithm (ABCA), flower pollination algorithm (FPA), galaxy-based search algorithm (GbSA), artificial immune algorithm (AIA), teaching–learning-based algorithm (TLBA), invasive weed optimization (IWO), cuckoo search algorithm (CSA), plant growth simulation algorithm (PGSA), and fireworks algorithm (FWA), is presented.

Sine–cosine algorithm (SCA) is a recently proposed meta-heuristic algorithm [17] and has promising features for application to PDNR problem. Several modifications are being proposed during the last one-and-half year for improving the basic SCA. Hafez et al. [18] successfully implemented SCA for feature selection without any change in the original algorithm. Sahlol et al. [19] also used SCA (without any modification) for improving the prediction accuracy of liver enzymes of fish fed by nano-selenite by developing a neural network model based on SCA. Bureerat and Pholdee [20] proposed an adaptive SCA by integrating differential evolution mutation operator for updating the population to detect structural damage. In [21], an interactive SCA is proposed to improve power system security where a micro-SCA is introduced for enhancing the local search. Sindhu et al. [22] proposed an SCA with elitism strategy to update new best solution for feature selection. Kumar et al. combined SCA with Weibull and pareto distribution functions in [23] and Cauchy and Gaussian distribution functions in [24] to enhance the global exploration ability

Table 1 A survey of recent meta-heuristic-based PDNR methods

Author(s) (year)	Meta-heuristics	Objective(s)	Salient features
Faria Jr. et al. (2017) [2]	GA	APLM	Biased random key-based GA
Fathy et al. (2017) [3]	PSO and GSA	APLM, SAIFI, SAIDI and ENS	Hybrid GSA-PSO-based meta-heuristic approach
Sedighizadeh et al. (2017) [4]	SOS and PSO	APLM, BVDM and LBIM	Hybrid SOS-PSO with fuzzified objective approach
Abdelaziz (2017) [5]	GA	APLM	GA with varying population
Nguyen et al. (2017) [6]	RRA	APLM, BVDM, LBIM and NSM	RRA-based fuzzy multiobjective approach
Muthukumar and Jayalalitha (2017) [7]	GWO	APLM and improvement of VSI	GWO-based PDNR with renewable distributed generators (DGs)
Siavash et al. (2017) [8]	PSO, HSA and ABCA	APLM and improvement of VDI	Hybrid HSA-ABCA-based
Namachivayam et al. (2016) [9]	FPA	Minimize cost of energy loss, capacitors and voltage penalty	FPA-based PDNR with capacitor placement
Tolabi et al. (2016) [10]	GbSA	APLM, LBIM and VPI	GbSA-based fuzzy multiobjective approach
Souza et al. (2016) [11]	AIA	Minimize cost of energy	AIA-based PDNR with fixed and variable demands
Lotifour et al. (2016) [12]	TLBA	APLM and VPI	Discrete TLBA-based PDNR
Rani et al. (2015) [13]	IWO	APLM, BVDM, NSM and LBIM	Pareto-based multiobjective PDNR using IWO
Nguyen and Truong (2015) [14]	CSA	APLM and VDIM	CSA-based PDNR
Rajaram et al. (2015) [15]	PGSA	APLM	PGSA-based PDNR in presence of DGs
Imran et al. (2014) [16]	FWA	APLM and VPI	FWA-based PDNR

for maximum power point tracking of a photovoltaic array. Tawhid and Savsani [25] proposed the SCA for multiobjective optimization using elitist non-dominated sorting method and diversity preserving crowding distance approach of NSGA-II.

In the light of above developments, the objective of this paper is to propose an improved SCA-based PDNR. The organization of the paper is as follows: in Sect. 2, the improved SCA (ISCA) is explained. In Sect. 3, a graph theory-based load flow method is presented. The load flow used to implement ISCA works independent of any particular numbering and ordering scheme of nodes and system data. This algorithm first checks the radial nature of the given distribution network and then the connectivity of all the nodes to substation node using the data structure-based technique. In Sect. 4, the algorithm is implemented for several test RDNs and the results are presented and analysed.

2 Improved Sine–Cosine Algorithm

The SCA is a population-based optimization method proposed by Mirjalili [17]. The main success of a population-based meta-heuristic optimization algorithm is achieved by maintaining the balance between global search or exploration and the local search or exploitation. This balance in turn maintains the population diversity which helps to avoid early convergence to local optimum values. In SCA, the cyclic pattern of the sine and cosine function is utilized to explore and exploit the search space. Compared to other methods, SCA method has fewer control parameters and is far simpler to implement. The search procedure for SCA is similar to other meta-heuristic methods which includes three main steps, population initialization, population updating and population selection. The search process of SCA starts with an initial population, and after evaluating the objective function for each member of the population, the best solution is found. The new population for the next generation is then updated by using Eq. (1), and the objective function values of its members are calculated as $f(x_{i,i=1 \text{ to population size}})$.

$$x_{\text{new},i} = \begin{cases} x_{\text{old},i} + r_1 \sin(r_2)\left|r_3 x_{\text{best},i} - x_{\text{old},i}\right| & \text{if } r_4 < 0.5 \\ x_{\text{old},i} + r_1 \cos(r_2)\left|r_3 x_{\text{best},i} - x_{\text{old},i}\right| & \text{otherwise} \end{cases} \quad (1)$$

Here, $x_{\text{best},i}$ is the current best solution of ith iteration. The variables r_2, r_3 and r_4 are random parameters in the ranges of $[0, 2\pi]$, $[0, 2]$ and $[0, 1]$, respectively. The variable r_1 is an iterative adaption parameter (conversion parameter) as calculated by (2), a is a constant parameter, t is the current iteration number and $t_{\max}$ is the maximum number of iterations.

$$r_1 = a\left(1 - \frac{t}{t_{\max}}\right) \quad (2)$$

The current best is compared with the best solution of the newly generated population, and the better one is saved to the next generation. The process is repeated until a termination criterion is met.

While applying SCA to PDNR problem, authors observed that the conversion parameter (r_1) has an important influence on the optimization performance of the algorithm. The performance bettered with a quadratic decreasing parameter (r_1) in place of the linearly decreasing one. So, in this work, SCA with a nonlinear decreasing conversion parameter is suggested as an improvement as per Eq. (3).

$$r_1 = a\left(1 - \frac{t}{t_{max}}\right)^2 \tag{3}$$

In addition to this, a heuristic-based approach [26] is considered to generate the initial population, which also enhances the convergence of the overall search. Moreover, a robust load flow algorithm [27] is used for PDNR, which is explained in the next section.

3 Load Flow Method for Implementing the PDNR

Several repeated load flows need to be executed while solving the PDNR problem for a distribution system using a meta-heuristic approach [27]. Hence, a fast and efficient load flow is very important for the successful implementation of any meta-heuristic algorithm as it promptly checks the fitness of any valid configuration which are the members of a population considered. The load flow must be adaptive for the changing network topologies, and it must simultaneously ensure that the overall network remains radial and all nodes are connected to the root node. In other words, an efficient load flow method not only checks the fitness but can also be used to eliminate the invalid candidates (configurations) by checking the radiality and connectivity. In this paper, a graph theory-based load flow method is used which has all the above features and it is successfully integrated with the ISCA for solving the PDNR problem.

3.1 Basic Load Flow Method

The equivalent current injection (ECI) at bus i for kth iteration of a radial distribution network (RDN) [28] can be written as Eq. (4), where P_i and Q_i are the constant active and reactive power loads connected at bus i. I_i^k and V_i^k are the ECI and bus voltage of ith bus and for kth iteration, respectively.

$$I_i^k = \left(\frac{P_i + jQ_i}{V_i^k}\right)^* \tag{4}$$

The relation between ECIs at all buses and bus voltage deviations from the sub-station bus voltage for the $k+1$th iteration can be expressed using a distribution load flow (DLF) matrix [29] as:

$$\left[\Delta V^{k+1}\right] = [\text{DLF}] \cdot \left[I^k\right] \tag{5}$$

The DLF matrix can be found directly using a *path* array proposed in the next subsection. A diagonal element of DLF is found by adding the impedances of all the lines stored in the *path [ ]* concerning a particular bus, and an off-diagonal element is found by adding all the impedances of the lines which appear in common in the *path[ ]* between two concerned buses. The order of the *DLF* matrix is $(\text{nb} - 1) \times (\text{nb} - 1)$, where nb is the number of buses in the RDN.

3.2   Proposed Arrays

In order to find DLF matrix directly from the system data, the following arrays and matrices are proposed. The proposed matrices are *Adjm* and *G* (the sparse version of *Adjm*). An entry of the adjacent matrix *Adjm*(u, v)= 1 if the bus '*u*' is connected to bus '*v*'; otherwise it is zero. Similarly, the proposed arrays are *path[ ]* and *dist [ ]*. Two pointer arrays *mf [ ]* and *mt[ ]* are also introduced to store the start and end locations in the *path* array. First, the *dist[ ]* array is formed using Dijkstra's algorithm [30], and then, the *path* array is formed using the *dist* array. The *dist[ ]* using Dijkstra's shortest path is presented in Algorithm 1.

Algorithm 1

STEP 1:   (Initialization)

- Assign the distance of root node '*s*' from itself as zero and its status as permanent (*p*), i.e. the state of node '*s*' is $(0, p)$.
- Other nodes are assigned distances of value as ∞ (a very high value) and their status as temporary (*t*), i.e. the states of other nodes are (∞, t).
- Designate the node '*s*' as the current node.

STEP 2:   (Distance values and current node update)
Let '*u*' be the index of current node.

(1)   The set V of nodes with temporary status can be reached from current node '*u*' by a link (u, v), where $v \in V$ from the system data.
(2)   For each $v \in V$, the distance value $dist_v$ of node v is updated as follows: $dist_v = \min\{dist_v, dist_u + Adjm(u, v)\}$
(3)   Change the status of '*v*' to permanent, and designate this node as the current node.

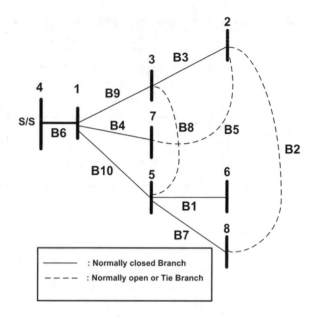

Fig. 1 Eight-bus sample RDN

STEP 3:   (Termination)
           If all the nodes are reached from the root node, then stop. Otherwise, go
           to STEP 2.
STEP 4:   (Output)
           The *dist* array stores the distance values of all the nodes from the root
           node.

Using the above algorithm, the contents of *path* and *dist* and the pointers *mf* and *mt* for a sample 8-bus RDN (Fig. 1) are formed and presented in Table 1. It is to be noted that in the sample RDN considered in Fig. 1, the nodes or buses as well as the branches (including the open switch or tie branches) are numbered and ordered arbitrarily. Even the root node is arbitrarily numbered as '4'. This arbitrariness can also be maintained in the data set of the system as the algorithm based on the proposed arrays is completely independent of any order or numbering scheme to carry out the load flow (Table 2).

3.3   Algorithm for Checking the Radial Configuration

Checking the radiality for a given data is an important feature of this approach. This check promptly sorts out the invalid members (non-radial topologies) from the population in meta-heuristic approaches before the time-consuming fitness evaluation through load flow. The Algorithm 2 describes the radiality check of the network followed by the pseudo-code to implement it.

Table 2 Contents of various arrays formed for the 8-bus sample RDNs

Bus no. [i]	[m]	mf [i]	mf [i]	dist [i]	path [m]
1	1	1	1	1	6
2	2, 3, 4	2	4	3	3, 9, 6
3	5, 6	5	6	2	9, 6
4	7	7	7	0	0
5	8, 9	8	9	2	10, 6
6	10, 11, 12	10	12	3	1, 10, 6
7	13, 14	13	14	2	4, 6
8	15, 16, 17	15	17	3	7, 10, 6

Algorithm 2

STEP 1   Create adjacency matrix (*Adjm*).
STEP 2   Create sparse matrix *G* of *Adjm* (*sparse* is a MATLAB command of the data structure tool box, which stores only non zero entries).
STEP 3   Use depth first search (*DFS*) algorithm to find the order of bus traversal. (*DFS* is a MATLAB command of the data structure tool box).
STEP 4   The path is mapped out between the predecessor node and the discovered node.
STEP 5   If the length of the path is equal to 2, the network is radial. Otherwise loop is present.

3.4   Algorithm for Checking the Connectivity of All Nodes to the Root Node

Checking connectivity of all nodes to the substation bus or root node is another important condition for PDNR which needs to be verified before fitness evaluation for a given topology. Thus, after confirming the radiality of a given network the connectivity of all nodes to the substation node is checked. For this, the traversed path for the given network is found out. If the number of buses in the traversed path is equal to the total number of buses in the network, then the graph representing the system has all nodes connected to the substation node; otherwise, discontinuity is there. Pseudo-code for this is given as Algorithm 3.

```
bus_order = graphtraverse(G, 's','Method','DFS')
if
        length(bus_order) = nb
```

```
                    `The system is connected'
        else
                    `The system is not connected'
  end
```

3.5 Application of the Load Flow for PDNR Problem

This load flow can be effectively used to solve the PDNR problem, particularly with a meta-heuristic approach. In a typical population-based meta-heuristic approach, each member of the population is checked for fitness in every iteration based on the objective to be fulfilled. In PDNR problem, execution of load flows for all the members of the population is almost mandatory for fitness evaluation. However, this has to be carried out only for the valid radial and fully connected topologies. The inclusion of radiality and connectivity checking before load flow execution in this proposed algorithm makes it perfectly suitable for meta-heuristic-based searching for its ability to sort out invalid members from the population before fitness evaluation. For applying the algorithm in any meta-heuristic approach, the step of random generation of open switches can be replaced by the respective local or exploitation and global or exploration-based search procedures. Last but not least that this proposed algorithm considers the set of open branch numbers as the input set for successive load flow execution. The overall flow chart for implementing the proposed ISCA-based PDNR is shown in Fig. 2.

4 Results

The proposed algorithm is implemented in MATLAB (R2011a) following the flow chart in Fig. 2. The reconfiguration results for the five test distribution systems (33, 69, 84, 119 and 136 buses) are presented in the next subsection.

4.1 Reconfiguration Results of 33, 69, 84, 119 and 136 Bus RDNs

As the overall active power loss minimization (APLM) is considered as the objective of PDNR in this paper, accordingly the reconfiguration results for the five standard RDNs (33, 69, 84, 119 and 136 buses) by the proposed ASCA algorithm are presented in Table 3. The system data for all these RDNs are available in [1]. Similarly, the

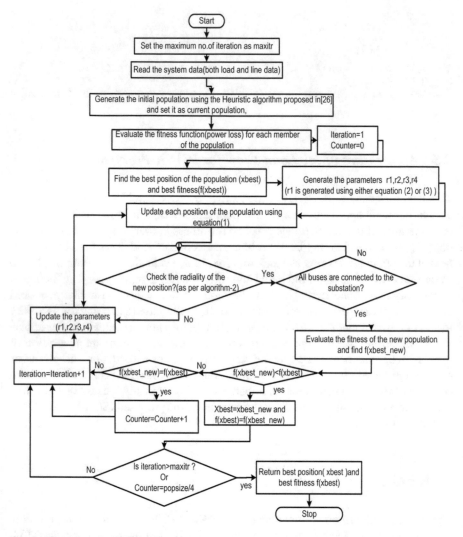

Fig. 2 Flow chart for PDNR with ISCA

optimal results as obtained by various other meta-heuristic methods are presented in Table 4.

Table 3 Reconfiguration results by the proposed method

Test systems	Initial open switches	Initial power loss (kW)	Final open switches	Final power loss (kW)	Exec. time (s)
33-bus	33–37	202.65	7, 9, 14, 32, 37	139.55	5.6
69-bus	69–73	224.97	14, 56, 61, 69, 70	98.605	9.0
84-bus	84–96	530.91	7, 13, 34, 39, 42, 55, 62, 72, 83, 86, 89, 90, 92	463.3	25
119-bus	119–133	1298.1	23, 25, 34, 39, 42, 50, 58, 71, 74, 95, 97, 109, 121, 129, 130	869	34
136-bus	136–156	320.35	7, 35, 51, 90, 96, 106, 118, 126, 135, 137, 138, 141, 142, 144–148, 150, 151, 155	280.2	50

Table 4 Reconfiguration results by other meta-heuristic method

Test systems /method	M–H used	Initial power loss (kW)	Final open switches	Final power loss (kW)	Av. exec. time (s)
33-bus [6]	RRA	202.69	7, 9, 14, 32, 37	139.55	74.7
69 bus [2]	69 73	224.95	14, 57, 61, 69, 70	98.57	26.7
84-bus [13]	84–96	526.97	7, 13, 34, 39, 42, 55, 62, 72, 83, 86, 89, 90, 92	469.1	Not reported
119-bus [16]	119–133	1298	24, 26, 35, 40, 43, 51, 59, 72, 75, 96, 98 110, 122, 130, 131	854.06	7.72
136-bus [31]	136–156	320.35	7, 35, 51, 90, 96, 106, 118, 126, 135, 137, 138, 141, 142, 144–148, 150, 151, 155	280.19	33.98

5 Conclusion

In this paper, an improved SCA has been used to solve the single objective PDNR problem. The initial population has been generated using a heuristic approach. The conversion parameter r_1 has been set to decrease nonlinearly, and a new graph theory-based load flow method has been used for solving PDNR. In most of the test systems, the algorithm has been shown to be faster and capable of obtaining global minimum results. However, the approach still requires fine tuning for avoiding local minimum traps and improving the convergence speed for becoming the best.

References

1. Mishra, S., Das, D., Paul, S.: A comprehensive review on power distribution network reconfiguration. Energy Syst. **8**(2), 227–284 (2017)
2. Faria, Jr., H., Resende, M.G.C., Ernst, D. : A biased random key genetic algorithm applied to the electric distribution network reconfiguration problem. J. Heuristics (2017). https://doi.org/10.1007/s10732-017-9355-8
3. Fathy, A., El-Arini, M., El-Baksawy, O.: An efficient methodology for optimal reconfiguration of electric distribution network considering reliability indices via binary particle swarm gravity search algorithm. Neural Comput. Appl. (2017). https://doi.org/10.1007/s00521-017-2877-z
4. Sedighizadeh, M., Esmaili, M., Moarref, A.E. : Hybrid symbiotic organisms search for optimal fuzzified joint reconfiguration and capacitor placement in electric distribution systems. INAE Lett. (2017). https://doi.org/10.1007/s41403-017-0029-5
5. Abdelaziz, M.: Distribution network reconfiguration using a genetic algorithm with varying population size. Electr. Power Syst. Res. **142**, 9–11 (2017)
6. Nguyen, T.T., Nguyen, T.T., Truong, A.V., Nguyen, Q.T., Phung, T.A.: Multi-objective electric distribution network reconfiguration solution using runner-root algorithm. Appl. Soft Comput. **52**, 93–108 (2017)
7. Muthukumar, K., Jayalalitha, S.: Integrated approach of network reconfiguration with distributed generation and shunt capacitors placement for power loss minimization in radial distribution networks. Appl. Soft Comput. **52**, 1262–1284 (2017)
8. Siavash, M., Pfeifer, C., Rahiminejad, A., Vahidi, B. : Reconfiguration of smart distribution network in the presence of renewable DG's using GWO algorithm. In: 2nd International Conference on Green Energy Technology (ICGET 2017), pp. 1–8 (2017)
9. Namachivayam, G., Sankaralingam, C., Perumal, S.K., Devanathan, S.T.: Reconfiguration and capacitor placement of radial distribution systems by modified flower pollination algorithm. Electr. Power Compon. Syst. **13**, 1492–1502 (2016)
10. Tolabi, H.B., Shakarami, M.R., Hosseini, R., Ayob, S.B.M.: Novel FGbSA: fuzzy-galaxy-based search algorithm for multi-objective reconfiguration of distribution systems. Russ. Electr. Eng. **87**(10), 588–595 (2016)
11. Souza, S.S.F., Romero, R., Pareira, J., Saraiva, J.T.: Artificial immune algorithm applied to distribution system reconfiguration with variable demand. Int. J. Electr. Power Energy Syst. **82**, 561–568 (2016)
12. Lotfipour, A., Afrakhte, H.: A discrete teaching–learning-based optimization algorithm to solve distribution system reconfiguration in presence of distributed generation. Int. J. Electr. Power Energy Syst. **82**, 264–272 (2016)
13. Rani, D.S., Subrahmanyam, M., Sydulu, N.: Multiobjective invasive weed optimization-an application to optimal network reconfiguration in radial reconfiguration systems. Int. J. Electr. Power Energy Syst. **73**, 932–942 (2015)
14. Nguyen, T.T., Truong, A.V.: Distribution network reconfiguration for power loss minimization and voltage profile improvement using cuckoo search algorithm. Int. J. Electr. Power Energy Syst. **68**, 233–242 (2015)
15. Rajaram, R., Kumar, K.S., Rajasekar, N.: Power system reconfiguration in a radial distribution network for reducing losses and to improve voltage profile using modified plant growth simulation algorithm with distributed generation. Energy Rep. **1**, 116–122 (2015)
16. Imran, A.M., Kowsalya, M.: A new power system reconfiguration scheme for power loss minimization and voltage profile enhancement using fireworks algorithm. Int. J. Electr. Power Energy Syst. **62**, 312–322 (2014)
17. Mirjalili, S.: SCA: A sine cosine algorithm for solving optimization problems. Knowl.-Based Syst. **96**, 120–133 (2016)
18. Hafez, A.I., Zawbaa, H.M., Emary, E., Hassanien, A.E.: Sine cosine optimization algorithm for feature selection. In: International Symposium on Innovations in Intelligent Systems and Applications. Sinaia, Romania (2016)

19. Sahlol, A.T., Ewees, A.A., Hemdan, A.M., Hassanien, A.E.: Training feedforward neural networks using sine-cosine algorithm to improve the prediction of lever enzymes on fish farmed on nano-selenite. In: 12th International Computer Engineering Conference. Cairo, Egypt (2016)
20. Bureerat, S., Pholdee, N.: Adaptive sine cosine algorithm integrated with differential evolution for structural damage detection. In: Garvesi, O., et al. (eds.) Computational Science and Its Applications- ICCSA 2017. LNCS, vol. 10404, pp. 71–86. Springer, Cham (2017)
21. Mahdad, B., Srairi, K.: A new interactive sine cosine algorithm for loading margin stability improvement under contingency. Electr. Eng. (2017). https://doi.org/10.1007/s00202-017-0539-x
22. Sindhu, R., Ngadiran, R., Yacob, Y.N., Zahri, N.A.H., Hariharan, M. : Sine–cosine algorithm for feature selection with elitism strategy and new updating mechanism. Neural Comput. Appl. (2017). https://doi.org/10.1007/s00521-017-2837-7
23. Kumar, N., Hussain, I., Singh, B., Panigrahy, B.K. : Single sensor based MPPT of partially shaded PV system for battery charging by using Cauchy and Gaussian sine cosine optimization. IEEE Trans. Energy Convers. (2017). https://doi.org/10.1109/tec.2017.2669518
24. Kumar, N., Hussain, I., Singh, B., Panigrahy, B.K.: Peak power detection of PS solar PV panel by using WPSCO. IET Renew. Power Gener. 11(4), 480–489 (2017)
25. Tawhid, M.A., Savsani, V. : Multi-objective sine-cosine algorithm (MO-SCA) for multi-objective engineering design problems. Neural Comput. Appl. (2017). https://doi.org/10.1007/s00521-017-3049-x
26. Mishra, S., Das, D., Raut, U.: A simple branch exchange based network reconfiguration method for loss minimization with distributed generation. In: 1st IEEE WIE Conference on Electrical and Computer Engineering, pp. 1–5. Dhaka, Bangladesh (2015)
27. Raut, U., Mishra, S.: A robust load flow algorithm to solve power distribution network reconfiguration problem with population based meta heuristic approach. In: 6th IEEE International Conference on Computer Application in Electrical Engineering-Recent Advances (CERA-2017), pp. 74–79. Roorkee, India (2017)
28. Mishra, S.: A simple algorithm for unbalanced radial distribution system load flow. In: IEEE Region 10 Conference (TENCON-2008) Hyderabad, India, pp. 1–6 (2008)
29. Teng, J.H.: A direct approach for distribution load flow solutions. IEEE Trans. Power Delivery 18(3), 882–887 (2003)
30. Hernandez, M., Ramos, G.: Meta-heuristic reconfiguration for future distribution networks operation. In: IEEE/PES Transmission and Distribution Conference and Exposition (2016)
31. Duan, D.L., Ling, X.D., Wu, X.Y., Zhong, B.: Reconfiguration of distribution network for loss reduction and reliability improvement based on an enhanced genetic algorithm. Electr. Power Energy Syst. 64, 88–95 (2015)

Artificial Neural Network for Strength Prediction of Fibers' Self-compacting Concrete

L. V. Prasad Meesaraganda, Prasenjit Saha and Nilanjan Tarafder

Abstract This paper investigates the applicability of artificial neural network model for strength prediction of fibers' self-compacting concrete under compression. The available 99 experimental data samples of fibers self-compacting concrete were used in this research work. In this paper, computational-based research is carried for predicting the strength of concrete under compression and model was developed using ANN with five input nodes and feed-forward three-layer back-propagation neural networks with ten hidden nodes were examined using learning algorithm. ANN model proposed analytically was verified, and it gives more compatible results. Hence, the ANN model is proposed to predict the strength of fibrous self-compacting concrete under compression.

1 Introduction

Artificial neural network (ANN) is a computing tool based on the process of genetic neural networks. The ANN techniques are applicable to civil engineering problems, because of their potentiality of learning straightly from examples. Correct response to deficient work, their ability to extract the results from minimal data, and their generalized results production are the other important properties of ANN [1]. The capabilities mentioned above give rise to ANN a very commanding mechanism to determine solution for several engineering problems, where data is insufficient [2]. The basic idea of ANN-based mathematical model for material performance is, to educate an ANN system using that material for series of experiments including enough information in the results, about the materials behavior, to succeed as a material model [3]. Such trained ANN system replicates the outcome of experiments and is also able to estimate the outcome in other experiments through their simplification potential [4].

L. V. P. Meesaraganda (✉) · P. Saha · N. Tarafder
Civil Engineering Department, NIT Silchar, Silchar, Assam, India
e-mail: prasadsmlv@gmail.com

© Springer Nature Singapore Pte Ltd. 2019   15
J. C. Bansal et al. (eds.), *Soft Computing for Problem Solving*,
Advances in Intelligent Systems and Computing 816,
https://doi.org/10.1007/978-981-13-1592-3_2

Self-compacting concrete is a concrete which is having high workability and high performance; it has the ability to flow into any shape and fill all the voids without any external vibration and passes throw the congested reinforcements under its self-weight without any vibration mechanism externally [5, 6]. As a result, it can be used in the construction where high strength and high performance are the main requirements of the developed concrete. If it is used in mass construction activity, its use in construction gives the economical results [7]. It was first developed in Kochi University of Technology, Japan, in 1980 by Professor Hajime Okamura. The developed new concrete gives a solution to the problem of durability because of its self-consolidation and high segregation resistance [8, 9]. The research is still going on till today to expect the compression value of SCC with optimal mix proportion [10].

Fibers develop the ductile behavior of the concrete members, and inclusion of the small discrete fiber enhances the crack arresting capacity of the concrete member by increasing its tensile strength and modulus of rupture values at a significant percentage enhancement [11, 12]. The tensile strength of concrete is almost considered as zero in the concrete elements design, whereas the compression behavior of concrete is strong. Because of these characteristics, concrete elements cannot sustain such masses that typically receive for the period of their service time. With the make use of fibers in concrete member, the structural integrity of the member can be increased. Fiber-reinforced self-compacting concrete shows the better improvement in mechanical properties as compared to the control SCC specimens [13, 14]. Compression imparted by fibers will result in improved toughness of the members and helpful in avoiding sudden failure under stationary loading, energy assimilation under dynamic loading [15, 16].

The literature review reveals that limited amount of research has been produced in the direction of self-compacting concrete using analytical modeling. So this research work focuses on using ANN model to envisage the strength of concrete under compression so that, reduce the cost of construction, conserve energy and minimizing the wastage of material for achieving the required strength and avoiding the number of trial mixes of concrete to get the target strength etc. can be established.

2 Scope of the Investigation

The ANN technique gives solution to the very difficult problems with the assistance of soft computing network elements. The objective of this study is to predict the strength of SCC under compression by an ANN model and predicting the strength values. In this paper, multilayer feed-forward neural networks with a back-propagation algorithm model were used and arranged in a format of five input parameters and one output parameter. For this use, required data for creating the artificial neural network model is obtained from the experimental results. The investigation also focuses on using analytical model to predict the SCC compression strength value, so that there is a reduction in the cost of construction, conserve energy, and it will lead to a reduction of CO_2 production from cement industries.

3 Methodology and Materials

3.1 Tests on Self-compacting Concrete

The data for the self-compacting concretes (SCCs) were collected from the 99 experimental data samples of the research work carried out by the author and of these data samples were utilized for training and for testing. In which for training the network, 70 data sets are used and the rest has been used for testing the data set. Su et al. (2001) gave a mix design procedure for SCC by following the guideline of the EFNARC specifications, and the same method is followed to arrive the trial mixes [17, 18]. European guidelines give specifications for testing, material selection, mixture designs, and testing methods of SCC in fresh state. The first priority test method for horizontal flow measuring (slum value) with T_{50} was selected for the filling ability of SCC, and second priority test is the V-funnel tests, alternatives to the T_{50} measurement. The passing ability of fresh SCC can be determined by utilizing the U box test or L box test [19]. In this investigation, the behavior of SCC in green and hardened state is confirmed with the EFNARC specifications [17] and reaching the target compressive strength.

3.2 Materials

In this research work, cement used was portland cement (type II) confirming to ASTM C 150/C 150M and the density and specific surface area were 3210 and 310 m^2/kg, correspondingly. Use of fly ash reduced the consumption of cement; hence, locally available fly ash [20] was utilized for replacement of cement partially. The sand used as fine aggregate was river sand as per the standards of ASTM C 29-16, and its bulk density was 1470 kg/m^3. Crushed granite was the coarse aggregate, and it is well-graded aggregate of 20-mm nominal size; density was 1560 kg/m^3 [19]. In this investigation for preparation of concrete mix and curing, potable water was used. To improve the flowability of self-compacting concrete, water-reducing admixture was used and it is polycarboxylic ether-based superplasticizer, confirming to ASTM C494 -2001 [21]. The fibers used were glass and polypropylene fibers, and the diameter of glass fiber was 24 micron with a length of 12 mm, and density was 2600 kg/m^3, and polypropylene fiber (PP) length was 14 mm and density 900 kg/m^3.

4 Network Development for ANN

4.1 Artificial Neural Network model Development

The ANN technique was first developed by McCulloch and Pitts in 1943 [22, 23, 2]. The inspiration behind this mathematical tool is the human brain, and by using this model, different problems in civil engineering are being solved. It perceives the problem as the human brain does like thinking, having ability to learn from the happenings, and getting trained by the consequences. An artificial neuron in Fig. 1 is modeled artificially which is a complete replica of biological neuron. Let us assume that there are N number of inputs (i.e., $I_1, I_2, ..., I_n$) having different weights (i.e., $[W] = [W_{1j}, W_{2j}, ..., W_{nj}]$) that are provided to neuron j. Weighted inputs are summed at the summing junction whose function is same as that of combined dendrites and cell body, whereas the activation function completes the task of axon and synapse.

The output of summing junction is sometimes equal to zero which is undesirable so to avoid this kind of situation, and a biasing factor b_j is added to it. Thus, transfer function f becomes as $U_j = \sum_{k=1}^{n} I_k W_{KJ} + b_j$.

The output of jth neuron, that is O_j, can be achieved as follows:

$$O_j = f(u_j) = f\left(\sum_{k=1}^{n} I_k W_{KJ} + b_j\right)$$

In an ANN, the transfer function highly influences the output of a neuron and they are dependent on it. Most of the transfer functions which are in use are hard limit, linear, log-sigmoid, tan-sigmoid, and others.

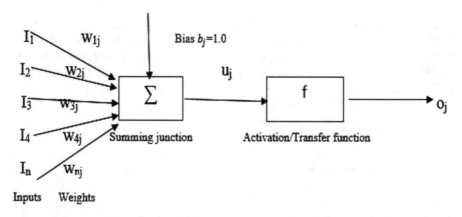

Fig. 1 A schematic diagram showing an artificial neuron

4.2 Feed-Forward Algorithm

Among the different types of algorithms available, in this work the feed-forward ANN algorithm was used. In the feed-forward algorithm, the neuron in each layer is interconnected to a neuron in the next layer and the neurons in the same layer are not connected. Each layer neuron is connected to another layer by using weights. To determine the best configuration, ANN is based on a number of trials and requires some experience to make it [24, 3].

4.3 Back-Propagation Algorithm

The back-propagation technique is a reiterate process to modify the weights from the output layer to input layer till no further betterment is required. The back-propagation technique determines the error, then till the error is going to minimize, it distributes this error in a backward direction from the output node to input node. This method is mainly based on steepest gradient descent principle. The main aim is to minimize the error of actual test data and output of the model [25].

5 Predictive Model Development

5.1 Artificial Neural Network Model

For the sake of conciseness, it is self-possessed to the short discussion of ANN in the present study which can be found in the literature [2, 26]. An ANN-based predicted model has been identified as ANN model for self-compacting concrete (ANN), and the applied ANN model is shown in Fig. 2. The input parameters are coarse aggregate, fine aggregate, fiber, superplasticizer, water powder ratio of ANN, and 28 days compressive strength as an output parameter. The statistical parameters of input and output are shown in Table 1, and it shows the maximum, minimum, mean, and standard deviation. The performance of ANN model is shown in Table 2 in terms of the statistical parameters. Linear correlation coefficient (R), mean square error (MSE), and mean absolute percentage error (MAPE) are the parameters. The ANN model consists of ten number of neuron in a hidden layer with a tan-sigmoid transfer function and output layer with linear function. Levenberg–Marquardt algorithm is used for its better generalization to the training data.

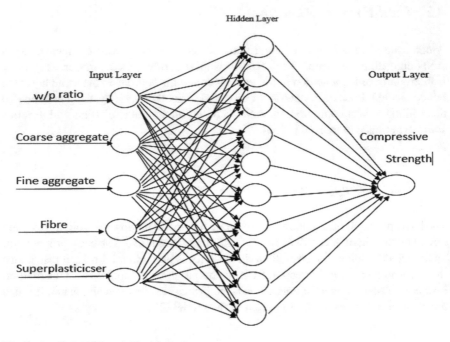

Fig. 2 Applied ANN model in this study

Table 1 Statistical parameters of ANN model

Model variable	Cement	Sand	Coarse aggregate	Fly ash	Fiber	Water	SP	VMA	Compressive strength (MPa)
Mean	385	910.25	778.14	113.72	0.69	191.15	9.75	0.55	49.26
Standard deviation	72.10	60.29	49.54	42.36	0.58	9.85	3.20	0.25	14.28
Maximum	500	1100	870	190.00	1.75	210	17.2	0.90	68.53
Minimum	254	830	660	65.00	0	175	4.50	0.25	27.80

Table 2 Performance of ANN model

Model	Data set	Statistical parameters		
		R^2	MSE	MAPE
ANN	Training data	0.953	0.001872	5.005375
	Testing data	0.938	0.002195	6.404328

5.2 Steps to be followed to Design an Artificial Neural Network

To design a suitable neural network, the following steps are subsequently used

Step 1: Identify input and output variables of the process to be modeled. Input variable is independent in nature.

Step 2: Normalize the variable in the range of 0.0–1.0.

Step 3: Initialize the number of hidden layer and that of neurons in each layer and select appropriate transfer functions for the neurons in each layer.

Step 4: Generate the connections weights in normalized scale, bias values, and coefficient of the transfer functions at random.

Step 5: Update the above parameters iteratively using a back-propagation algorithm implemented through either an incremental or batch mode training.

Step 6: Continue the iterations until the termination criterion is reached.

Step 7: Testing of the optimized neural network.

6 Discussion of Test Results

6.1 Artificial Neural Network Model

MATLAB software is used to perform ANN model using neural network toolbox. Total 99 experimental data sample is used for ANN. In which for training the network, 70 data sets are used and the rest has been used for testing the data set (Figs. 3 and 4). The ANN training with 70 sets of data is executed till the RMS error within permissible level, and training data correlation value is above 95.30% as shown in Fig. 5a. As the data set is tested by ANN, it provides extremely closed value to the experimental results. However, the testing data correlation value is 93.80% as shown in Fig. 5b. On the basis of these high correlation values, the precision of the network is satisfactory, which also validates the success of ANN model in learning this problem and predicting an accurate value.

Statistical performance of developed ANN model is summarized in Table 2, and R, MSE and MAPE values of ANN model for training and testing data were 0.953, 0.001872, 5.005375 and 0.938, 0.002195, 6.404328, respectively. The statistical values in Table 2 ANN model predicted the experimental data very well. All of the statistical values in table represent that ANN model predicts the compressive strength and is very close to the experimental value.

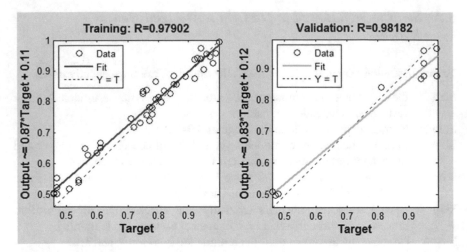

Fig. 3 Training data results of ANN

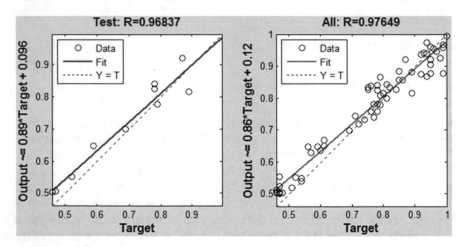

Fig. 4 Testing data results of ANN

7 Conclusion

In this study, ANN model has been realized with some appropriate data for assessment of the strength of self-compacting concrete under compression. A simple feedforward back-propagation technique was applied in this investigation to model different problems concerning nonlinear variables.

The model developed by using ANN technique is compared with the results of empirical methods statistical parameters. The modeling carried for the test data of SCC based on ANN could predict compression value with a correlation coefficient of 0.9380. The results of the analysis indicate that the ANN model performance is good

(a) **(b)**

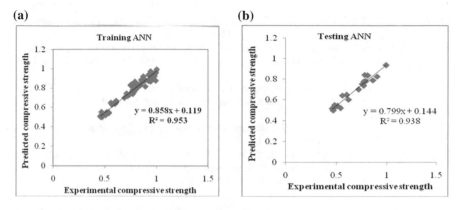

Fig. 5 **a** Training stage of ANN and **b** testing stage of ANN

in predicting the strength value for self-compacting concrete. The model proposed exhibits an advantage of estimating the modulus of elasticity value from concrete within reliable limits. The strength prediction level of ANN technique shows better performance; hence, the model proposed by ANN techniques could be adopted to concrete materials if enough source data is available.

References

1. Saridemir, M.: Prediction of compressive strength of concretes containing metakaolin and silica fume by artificial neural networks. Adv. Eng. Softw. **40**, 350–355 (2009)
2. Kostic, S., Vasovic, D.: Prediction model for compressive strength of basic concrete mixture using artificial neural networks. Neural Comput. Appl. **26**, 1005–1024 (2015)
3. Bilim, C., Atis, C.D., Tanyildizi, H., Karahan, O.: Predicting the compressive strength of ground granulated blast furnace slag concrete using artificial neural network. Adv. Eng. Softw. **40**, 334–340 (2009)
4. Najigivi, A., Khaloo, A., zad, A.I., Rashid, S.A.: An artificial neural networks model for predicting permeability properties of nano silica-rice husk ash ternary blended concrete. Int. J. Concr. Struct. Mater. **7**(3), 225–238 (2013)
5. Dehwah, H.A.F.: Mechanical properties of self-compacting concrete incorporating quarry dust powder silica fume or fly ash. Constr. Build. Mater. **26**, 547–551 (2012)
6. Muciacciaa, G., Cattaneob, S., Rosatia, G., Cangianoc, S.: Properties of lime stone self-compacting concrete at fresh and hardened state. Eur. J. Environ. Civil Eng. **19**(5), 598–613 (2015)
7. Maghsoudi, A.A., Mohamadpour, Sh, Maghsoudi, M.: Mix design and mechanical properties of self compacting light weight concrete. Int. J. Civil Eng. **9**(3), 230–236 (2011)
8. Deeb, R., Karihaloo, B.L.: Mix proportioning of self-compacting normal and high-strength concretes. Mag. Concr. Res. **65**(9), 546–556 (2013)
9. Okamura, H., Ouchi, M.: Self-compacting to achieve durable concrete structures. J. Adv. Concr. Technol. **1**(1), 5–15 (2003)
10. Dinakar, P.: Design of self-compacting concrete with fly ash. Mag. Concr. Res. **64**(5), 401–409 (2012)

11. Acikgenc, M., Ulas, M.: Using an artificial neural network to predict mix compositions of steel fiber-reinforced concrete. Arab. J. Sci. Eng. **40**, 407–419 (2015)
12. Prasad, M.L.V., RathishKumar, P.: Prediction of the moment—curvature relationship of confined fiber reinforced self compacting concrete. Int. J. Cement Wapno Beton **20**, 269–283 (2015)
13. Shafieyzadeh, M.: Prediction of compressive strength of concretes containing silica fume and styrene-butadiene rubber (SBR) with a mathematical model. Int. J. Concr. Struct. Mater. **7**(4), 295–301 (2013)
14. Aslani, F., Nejadi, S.: Mechanical characteristics of self compacting concrete with and without fibres. Mag. Concr. Res. **65**(10), 608–622 (2013)
15. Prasad, M.L.V., Saha, P., Laskar, A.I.: Behaviour of self compacting reinforced concrete beams strengthened with hybrid fiber under static and cyclic loading. Int. J. Civil Eng. Published online on 2nd November 2016. https://doi.org/10.1007/s40999-016-0114-2
16. Uysal, M., Tanyildizi, H.: Estimation of compressive strength of self compacting concrete containing polypropylene fiber and mineral additives exposed to high temperature using artificial neural network. Constr. Build. Mater. **27**, 404–414 (2012)
17. Su, N., Hsu, KCh., Chai, H.W.: A simple mix design method for self compacting concrete. Cement Concr. Res. **31**, 1799–1807 (2001)
18. EFNARC: Specifications and Guidelines for Self-Compacting Concrete. European Federation of Producers and Applicators of Specialist Products for Structures, Association House, 99 West Street, Farnham, UK (2005)
19. Rubaye, A.L., Kulasegaram, S., Karihaloo, B.L.: Simulation of self-compacting concrete in an L-box using smooth particle hydrodynamics. Mag. Concr. Res. (2016). http://dx.doi.org/10.1680/jmacr.16.00408
20. ASTM C618-15: Standard Specification for Coal Fly Ash and Raw or Calcined Natural Pozzolan for Use in Concrete. American Society of Testing and Materials, West Conshohocken, PA (2015)
21. ASTMC 494/C 494M: Standard Specification for Chemical Admixtures for Concrete. Annual Book of ASTM Standards (2001)
22. McCulloch, W.S., Pitts, W.: A logical calculus of the ideas immanent in nervous activity. Bull. Math. Biophys. **5**, 115–133 (1943)
23. McCulloch, W.S., Pitts, W.: A logical calculus of the ideas immanent in nervous activity. Bull. Math. Biol. **52**, 99–115 (1990)
24. Sadrmomtazi, A., Sobhani, J., Mirgozar, M.A.: Modelling compressive strength of EPS lightweight concrete using regression, neural network and ANFIS. Constr. Build. Mater. **42**, 205–216 (2013)
25. Yaprak, H., Karaci, A.: Prediction of the effect of varying cure conditions and *w/c* ratio on the compressive strength of concrete using artificial neural networks. Neural Comput. Appl. **22**, 133–141 (2013)
26. Ghafari, E., Bandarabadi, M., Costa, H., Júlio, E.: Prediction of fresh and hardened state properties of UHPC: comparative study of statistical mixture design and an artificial neural network model. J. Mater. Civ. Eng. (2015). https://doi.org/10.1061/(ASCE)MT.1943-5533.0001270

Using Chaos in Grey Wolf Optimizer and Application to Prime Factorization

Harshit Mehrotra and Saibal K. Pal

Abstract The Grey Wolf Optimizer (GWO) is a swarm intelligence meta-heuristic algorithm inspired by the hunting behaviour and social hierarchy of grey wolves in nature. This paper analyses the use of chaos theory in this algorithm to improve its ability to escape local optima by replacing the key parameters by chaotic variables. The optimal choice of chaotic maps is then used to apply the Chaotic Grey Wolf Optimizer (CGWO) to the problem of factoring a large semi-prime into its prime factors. Assuming the number of digits of the factors to be equal, this is a computationally difficult task upon which the RSA cryptosystem relies. This work proposes the use of a new objective function to solve the problem and uses the CGWO to optimize it and compute the factors. It is shown that this function performs better than its predecessor for large semi-primes, and CGWO is an efficient algorithm to optimize it.

1 Introduction

In the world of mathematical optimization and computational intelligence, solving NP-hard problems is a major challenge as no exact or complete methods exist which can solve these problems in polynomial time. These methods incur huge memory and runtime costs. A solution is to compromise on the chances of getting the correct solution of the problem by using population-based meta-heuristics. They are low on computation time and may also hold lower memory requirements. These meta-heuristics do not need gradient-related information. They begin with a set of candidate

H. Mehrotra (✉)
Department of Computer Science and Engineering, Indian Institute of Technology
(Banaras Hindu University), Varanasi, India
e-mail: harshit.mehrotra.cse15@iitbhu.ac.in

S. K. Pal
Scientific Analysis Group (SAG), Defence Research and Development
Organization (DRDO), New Delhi, India
e-mail: skptech@yahoo.com

© Springer Nature Singapore Pte Ltd. 2019
J. C. Bansal et al. (eds.), *Soft Computing for Problem Solving*,
Advances in Intelligent Systems and Computing 816,
https://doi.org/10.1007/978-981-13-1592-3_3

25

solutions which are then improved over the course of the runtime to achieve the true solution of the problem.

A branch of population-based meta-heuristics that is very popular for mathematical optimization is swarm intelligence (SI). In the words of Bonabeau et al., it is the emergent collective intelligence of groups of simple agents [2]. These algorithms usually draw inspiration from the behaviour of organisms in nature. A number of such nature-inspired SI algorithms have been devised in order to solve problems of continuous optimization as well as combinatorial optimization (like travelling salesman problem, knapsack problem, vehicle routing problem). Some of these are the particle swarm optimization (PSO) [6], firefly algorithm [20], ant colony optimization (ACO) [4], artificial bee colony (ABC) [10], bat algorithm [21] to name a few.

Another such algorithm that was found to be very efficient in solving continuous optimization problems is the Grey Wolf Optimizer (GWO) [13]. It is based on the social hierarchy and hunting behaviour of grey wolves. This meta-heuristic has some tendency to avoid local optimum and is also efficient in moving towards the true solution [22]. However, since it advances towards exploitation, it is not always good for global search. Here, we have tried to eradicate this shortcoming of the GWO by using chaotic variables. Chaotic dynamics find a major application in optimization meta-heuristics to solve the problem of local optimum convergence [19]. This has been successfully proven by application to many SI algorithms like firefly algorithm [8], bat algorithm [9] and PSO [1]. We apply chaos on the key parameters of GWO and compare the performance of the Chaotic GWO (CGWO) with the standard GWO using six benchmark functions.

We have then explored the efficiency of the CGWO to solve discrete optimization problems by applying it to the problem of factorizing a product of two large prime numbers. Such numbers, known as semi-primes are very difficult to factorize when the factors are of almost equal number of digits [3]. No exact methods exist to solve this problem in polynomial time. Hence, some SI-based techniques have been used to solve this problem using certain objective functions [3, 11, 14]. Here, we propose the use of another objective function and compare its performance with a previously used function along with optimization using CGWO.

The rest of the paper goes into the details of each point discussed above. Section 2 discusses the standard GWO meta-heuristic, followed by a detailed discussion of our model of the CGWO in Sect. 3. Various aspects of the prime factorization problem have been discussed in Sect. 4. Results of testing on the benchmark functions and semi-primes are given in Sect. 5. Section 6 gives concluding remarks of this study.

2 The Grey Wolf Optimizer

Grey wolves are among the predators that form the top of the food chain. They have a strong social structure which is followed while hunting. In decreasing order of dominance, wolves in a pack are classified as alpha (α), beta (β), delta (δ) and omega (ω) wolves. The hunting process is also divided into three phases [17]: tracking, chasing the prey; encircling it and attacking it. Inspired by these properties of grey wolves, the Grey Wolf Optimizer was given by Mirjalilli et al. in 2014 [13]. In the mathematical model of the GWO, the fittest solution is labelled as the alpha (α), followed by the beta (β) and the delta (δ) which are the second and third fittest solutions, respectively. All other solutions are omegas (ω) and follow the other three kinds. The process of encircling the prey is modelled by calculating a distance vector and using it to update the position of a wolf. The hunt is usually guided by the alpha and occasionally by the beta and delta. Eliminating this uncertainty for the purpose of mathematical modelling, it is assumed the best three solutions have better knowledge about the optimum and all other solutions are updated according to the positions of the α, β and δ [13]. All the above discussed operations are formulated for an agent with position vector $\mathbf{X}$ as follows:

$$\mathbf{D}_\alpha = |\mathbf{C}_1 \cdot \mathbf{X}_\alpha - \mathbf{X}|, \ \mathbf{D}_\beta = |\mathbf{C}_2 \cdot \mathbf{X}_\beta - \mathbf{X}|, \ \mathbf{D}_\delta = |\mathbf{C}_3 \cdot \mathbf{X}_\delta - \mathbf{X}| \quad (1)$$

$$\mathbf{X}_1 = \mathbf{X}_\alpha - \mathbf{A}_1 \cdot \mathbf{D}_\alpha, \ \mathbf{X}_2 = \mathbf{X}_\beta - \mathbf{A}_2 \cdot \mathbf{D}_\beta, \ \mathbf{X}_3 = \mathbf{X}_\delta - \mathbf{A}_3 \cdot \mathbf{D}_\delta \quad (2)$$

$$\mathbf{X}(t + 1) = \frac{\mathbf{X}_1 + \mathbf{X}_2 + \mathbf{X}_3}{3} \quad (3)$$

The vectors $\mathbf{A}$ and $\mathbf{C}$ are defined as:

$$\mathbf{A} = 2\mathbf{a} \cdot \mathbf{r}_1 - \mathbf{a} \quad (4)$$

$$\mathbf{C} = 2\mathbf{r}_2 \quad (5)$$

The components of $\mathbf{a}$ are uniformly decreased from 2 to 0 over iterations. $\mathbf{r}_1$ and $\mathbf{r}_2$ are random vectors in [0, 1]. A detailed discussion of the nature of search (explorative and exploitative) is presented in the next section where chaotic improvements to the GWO are suggested.

After having been proved effective for continuous optimization and some engineering problems [13], various other versions of the GWO have been proposed to solve various problems like optimizing the control parameters of a DC motor [12], feature extraction [7], training q-Gaussian radial basis functional link nets [16], to name a few.

3 Chaos in Grey Wolf Optimizer

Chaos is a characteristic of any nonlinear system. It is basically a bounded unstable behaviour that occurs in a deterministic nonlinear system. Any chaotic system possesses the property of sensitive dependence on initial conditions, implying that the slightest change in the parameters or initial conditions can lead to a vast difference in the future behavious of the system [5].

3.1 Why Chaos?

As the optimization problem gets tougher with a large number of local optima (like multi-modal functions), the chances that a population-based meta-heuristic will get trapped in one such local optimum increases. Chaos has been used in recent times to solve this problem by developing chaotic optimization algorithms [19]. When used suitably, the pseudo-randomness, ergodicity and irregularity of chaotic variables helps algorithms alternate between exploration and exploitation and hence avoid getting trapped in a local solution. Moreover, chaos is non-repetitive, thus enabling these methods to carry out overall searches at higher speeds than stochastic searches that depend on probabilities [5].

3.2 Chaotic Maps

In order to incorporate chaos in an optimization algorithm, we use one-dimensional functions called chaotic maps which exhibit the property of 'sensitive dependence on initial conditions' and are used in place of key parameters of the algorithm. Some popular chaotic maps which have been used in this paper are [9]:

1. Gauss map:

$$x_{k+1} = \begin{cases} 0 & x_k = 0 \\ (1/x_k) \mod 1 & \text{otherwise} \end{cases} \tag{6}$$

It generates chaotic sequences in (0, 1).
2. Logistic map:

$$x_{k+1} = ax_k(1 - x_k) \tag{7}$$

It generates chaotic sequences in (0, 1) provided that $x_0 \in (0, 1)$ and that $x_0 \notin 0.0, 0.25, 0.75, 0.5, 1.0$. Here, we have used $a = 4$.
3. Chebyshev map:

$$x_{k+1} = \cos(k \cos^{-1} x_k) \tag{8}$$

This map generates a chaotic sequence in (−1, 1).
4. Iterative map:

$$x_{k+1} = \sin\left(\frac{a\pi}{x_k}\right) \tag{9}$$

Here, $a \in (0, 1)$ is a suitable parameter. The chaotic sequence lies in (−1, 1).

5. Singer map:

$$x_{k+1} = \mu \left(7.86x_k - 23.31x_k^2 + 28.75x_k^3 - 13.3x_k^4\right) \tag{10}$$

Here, μ lies between 0.9 and 1.08.

6. Tent map:

$$x_{k+1} = \begin{cases} \frac{x_k}{0.7} & x_k < 0.7 \\ \frac{10}{3}(1 - x_k) & x_k \geq 0.7 \end{cases} \tag{11}$$

It generates a chaotic sequence in (0, 1).

7. Sinusoidal map:

$$x_{k+1} = ax_k^2 \sin(\pi x_k) \tag{12}$$

This map also generates chaotic sequences in (0, 1). When $a = 2.3$ and $x_0 = 0.7$, it simplifies as:

$$x_{k+1} = \sin(\pi x_k) \tag{13}$$

3.3 Adding Chaos to the GWO

A lot of studies have been done in the field of chaotic optimization algorithms by developing chaotic versions of algorithms by replacing control parameters or random variables by chaotic variables [1, 8, 9]. The main motive is to alternate between exploration and exploitation so that the agents do not get caught in a local optimum.

The key parameters in GWO are **A** and **C**. Since $\mathbf{r}_1$ has its elements in [0, 1], the value of **A** can vary in $[-a, a]$. We have $\mathbf{C} = 2\mathbf{r}_2$ where $\mathbf{r}_2$ has its elements in [0, 1], thus implying that **C** has its elements in [0, 2].

Introducing chaos in a

From the position update equation, we get that the new position of a wolf is between the current position and the prey, i.e. attacking it (exploitation) when $|A| < 1$ [13]. As a is uniformly decreased from 2 to 0, once a comes below 1 (after half the number of maximum iterations), the search will be invariably exploitative. In the first half, it changes between explorative and exploitative. To keep such a behaviour intact, we replace a by a chaotic sequence normalized to give values in [1, 2]. This makes sure that the search can have both natures throughout the iterations. Figures 1 and 2 show the variation of A for a standard GWO and a Chaotic GWO (CGWO) with tent map used for a, respectively, when run for 100 iterations. The red line indicates a, $-a$ and the blue line indicates A. The bold black line indicates the values 1 and -1. It can be seen clearly that the CGWO provides exploration at regular intervals throughout

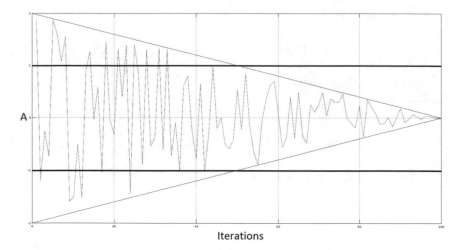

Fig. 1 Variation of A in standard GWO

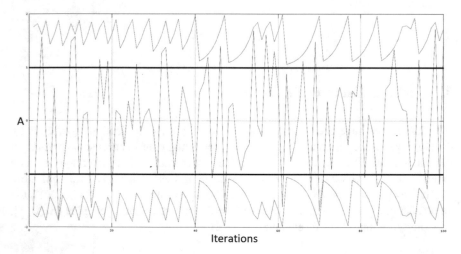

Fig. 2 Variation of A with tent map applied to a

the search, thereby creating better balance. The standard GWO explores the search space for intervals only in the initial half.

Introducing chaos in C

The parameter C assigns weight to the role of the prey in movement of the wolves [13]. When $C > 1$, the role of the prey is emphasized and is given less importance for $C < 1$. So, we redefine C chaotically by replacing the random variable r_2 with a chaotic variable normalized between 0 and 1. Owing to the ergodicity and mixing property of chaos, replacing a random variable by a chaotic one is expected to give better results. Results of comparison of GWO and CGWO on some objective functions are given in Sect. 5.

4 The Prime Factorization Problem

A semi-prime number is a composite number obtained by multiplying two prime numbers. Factorizing such numbers is a difficult task, particularly when the two factors are almost similar sized [3]. This can be understood by knowing that prime factorization is a one-way trapdoor function, which means that given two prime numbers p and q, we can easily compute the corresponding semi-prime number N. However, it is very difficult to obtain p and q if we are given N. The difficulty of this problem is used in the RSA cryptosystem. The problem is NP-complete, so no polynomial time algorithms exist for it for non-quantum computers. The most efficient exact method developed is the general number field sieve method [18] whose asymptotic time complexity is

$$
O\left(\exp\left(\left(\frac{64}{9}\right)^{\frac{1}{3}}(\log b)^{\frac{2}{3}}\right)\right) \tag{14}
$$

Efforts have been made to explore the possibility of solving this problem using meta-heuristics [3, 11, 14]. Various techniques have been applied. Some of them have given promising results as well. We assume that both factors have equal number of digits and propose an improvement to the current methods.

4.1 Choosing an Objective Function

Solving any problem using meta-heuristics involves the optimization of an objective function. Prime factorization follows a similar procedure. Since the problem is very difficult, changes in the objective function can gave largely varying results. In fact, the main challenge of solving this problem is choosing a suitable objective function [15]. Initially, the following two-dimensional function was used for the purpose [11]:

$$
\textbf{minimize } f(x) = (x^2 - y^2) \quad \text{mod } N \qquad x, y \in [2, N-1] \tag{15}
$$

$$
\textbf{constraint } (x \pm y) \quad \text{mod } N \neq 0 \tag{16}
$$

The factors are then calculated as the GCD of $x + y$ and $x - y$ with N. This function is highly chaotic with many local minima even for a small N. Moreover, the search space is very large. Thus, a one-dimensional function was introduced [3] :

$$
\textbf{minimize } f(x) = N \quad \text{mod } x \qquad x \in (2, \sqrt{N}) \tag{17}
$$

This function attains 0 when x is a factor of N. It is the smaller factor as the upper bound is $\sqrt{N}$. The other factor is obtained by dividing N by the obtained x. The lower bound becomes 10^{d-1} under the assumption that both factors have same number of digits (d is number of digits in $\sqrt{N}$). The probability that a randomly generated solution is correct in Eq. (15) is $\frac{1}{(N-1)^2}$ and in Eq. (17) is more than $\frac{1}{(N-1)}$. Equation (17) also has several local minima, but being one-dimensional, its performance is better as seen in [3]. This motivates us to explore possibilities of a better objective function to solve the problem. One aim should be to narrow down the search space as well.

Objective function used

Consider a semi-prime N with prime factors $p(<\sqrt{N})$ and $q(>\sqrt{N})$. Both will be greater than 2, hence their sum is an even integer.

$$p + q = 2n \tag{18}$$

Using $N = pq$ to eliminate q, we get:

$$p^2 - 2pn + N = 0 \tag{19}$$

Solving Eq. (19), we get two solutions which are the values of p and q.

$$p = n - \sqrt{n^2 - N} \quad \text{and} \quad q = n + \sqrt{n^2 - N} \tag{20}$$

Since both the factors are integers, we need to find a positive integer n such that the radical $\sqrt{n^2 - N}$ is also an integer. Next, we need to fix the bounds within which such an n has to be searched.

We continue with our assumption that both factors have equal number of digits, d. This gives us:

$$10^{d-1} < p < \sqrt{N} \quad \text{and} \quad \sqrt{N} < q < 10^d - 1 \tag{21}$$

Writing p and q as their difference from $\sqrt{N}$:

$$p = \sqrt{N} - x \qquad 0 < x < \sqrt{N} - 10^{d-1} \tag{22}$$
$$q = \sqrt{N} + y \qquad 0 < y < 10^d - 1 - \sqrt{N} \tag{23}$$

Using the above two equations and $N = pq$, we obtain the following relation:

$$xy = 2n\sqrt{N} - 2N \tag{24}$$

The mean n can now be written as:

$$n = \sqrt{N} + \frac{y - x}{2} \tag{25}$$

We now use Eq. (24) to get a quadratic equation in x by substituting for y in Eq. (25). This gives us the following value of n:

$$n = \frac{x^2 - 2x\sqrt{N} + 2N}{2(\sqrt{N} - x)} \tag{26}$$

Using differentiation techniques, it can be calculated that the above expression will be increasing for the domain of x given in Eq. (22). Hence, the domain of n comes out to be:

$$n \in \left(\sqrt{N}, \frac{10^{d-1} + \frac{N}{10^{d-1}}}{2}\right) \tag{27}$$

Using the domains of x, y in Eqs. (22) and (23) and the fact that $n > \sqrt{N}$ (AM-GM inequality), we obtain another range for n:

$$n \in \left(\sqrt{N}, \frac{\sqrt{N} + 10^d - 1}{2}\right) \tag{28}$$

The upper bound for searching n will be the least of the ones obtained in Eqs. (27) and (28) depending upon the N used. So, the final objective function becomes:

$$\textbf{minimize } \left\{\sqrt{x^2 - N}\right\} \qquad x \in \left(\sqrt{N}, \min\left(\frac{10^{d-1} + \frac{N}{10^{d-1}}}{2}, \frac{\sqrt{N} + 10^d - 1}{2}\right)\right) \tag{29}$$

where $\{.\}$ refers to the fractional part function.

Results of optimizing this function using CGWO have been given and compared with those using with the function in Eq. (17) in the next section.

5   Experiments and Results

For the first part of the experiments, the CGWO with chaos in a (Sect. 3.3) and C (Sect. 3.3) is compared with the standard GWO for performance on five minimization continuous benchmark functions. These are the Rastrigin function (f_1), Ackley function (f_2), Sphere function (f_3), Goldstein–Price function (f_4) and Griewank function (f_5). Four of these being multi-modal, test the explorative ability and the Sphere function tests the efficiency of exploitation. Parameters for these functions are given in Table 1, and their definitions are given below. All tests are carried out in GNU Octave 4.0.2 on a 8GB, 2.20 GHz CPU laptop computer running Windows 8.

Table 1 Parameters for the benchmark functions used

Definition	Dim	Domain	f_{min}
$f_1(\mathbf{x})$	30	$[-5.12, 5.12]$	0
$f_2(\mathbf{x})$	30	$[-32, 32]$	0
$f_3(\mathbf{x})$	30	$[-100, 100]$	0
$f_4(x, y)$	2	$[-2, 2]$	3
$f_5(\mathbf{x})$	30	$[-600, 600]$	0

$$f_1(\mathbf{x}) = f(x_1, \ldots, x_n) = 10n + \sum_{i=1}^{n}(x_i^2 - 10\cos(2\pi x_i)) \tag{30}$$

$$f_2(\mathbf{x}) = -20.\exp\left(-0.2\sqrt{\frac{1}{n}\sum_{i=1}^{n}x_i^2}\right) - \exp\left(\frac{1}{d}\sum_{i=1}^{n}\cos(2\pi x_i)\right) + 20 + e \tag{31}$$

$$f_3(\mathbf{x}) = \sum_{i=1}^{n}x_i^2 \tag{32}$$

$$f_4(x, y) = [1 + (x + y + 1)^2(19 - 14x + 3x^2 - 14y + 6xy + 3y^2)][30 + (2x - 3y)^2(18 - 32x + 12x^2 + 4y - 36xy + 27y^2)] \tag{33}$$

$$f_5(\mathbf{x}) = 1 + \sum_{i=1}^{n}\frac{x_i^2}{4000} - \prod_{i=1}^{n}\cos\left(\frac{x_i}{\sqrt{i}}\right) \tag{34}$$

The number of search agents and maximum iterations is fixed at 30 and 500, respectively. Mean value and standard deviation are calculated for 30 runs of each type of chaotic map. The initial value of the chaotic variable is taken to be a random number between 0 and 1 to get an unbiased idea of the efficiency of the map. The chaotic maps used are the ones described in Sect. 3.2. The GWO was proven to be better (or competitive, at least) for continuous optimization when compared to other popular meta-heuristics in [13]. Here, we have compared the CGWO with GWO and have evaluated the results using the mean, standard deviation (SD) and success ratio (SR). The success ratio is defined as the percentage of runs for which the solution is found successfully, i.e. [9]

$$\sum_{d=1}^{D}(X_i^{obt} - X_i^*)^2 \le (UB - LB) \times 10^{-4} \tag{35}$$

Table 2 Evaluation results for chaos in a

Map	f_1		f_2		f_3		f_4		f_5	
	Mean	SD	Mean	SD	Mean	SD	Mean	SD	Mean	SD
Gauss	55.041	33.024	2.18E−14	3.01E−15	3.23E−50	8.74E−50	3	4.77E−05	0.0076	0.0084
Logistic	51.037	29.315	2.18E−14	2.72E−15	1.56E−51	4.31E−51	3	3.81E−05	0.0081	0.0081
Sinusoidal	11.0952	14.9742	1.47E−14	2.82E−15	5.72E−56	1.41E−55	3	2.54E−05	0.0026	0.0062
Tent	42.413	36.275	2.18E−14	2.59E−15	9.82E−51	4.54E−50	3	2.63E−05	0.0041	0.0044
Singer	24.487	29.45	1.47E−14	3.02E−15	7.74E−52	1.71E−51	3	3.64E−05	0.0027	0.0071
Chebychev	47.73	36.838	2.18E−14	3.53E−15	7.93E−52	1.56E−51	3	2.47E−05	0.0051	0.0084
Iterative	47.377	31.994	2.18E−14	3.15E−15	1.62E−47	5.16E−47	3	2.51E−05	0.0047	0.0076
Standard	13.955	10.079	8.57E−14	7.31E−15	3.94E−31	6.63E−31	3	2.90E−05	0.0046	0.0088

Table 3 Success ratios with chaos in a

Map	f_1	f_2	f_3	f_4	f_5	Total SR
Gauss	0	100	100	100	50	350
Logistic	0	100	100	100	50	350
Sinusoidal	52	100	100	100	84	436
Tent	0	100	100	100	90	390
Singer	40	100	100	100	87	427
Chebychev	0	100	100	100	70	370
Iterative	0	100	100	100	70	370
Standard	20	100	100	100	75	395

5.1   Results with Chaos Applied in a

As described in Sect. 3.3, a is replaced by a chaotic variable normalized in $[0, 1]$ and the results are given in Tables 2 and 3. Moreover to give final exploitation, a is given a value of 0.2 for the final 50 iterations, so as to give A small values. It can be seen that in functions where the SR was already 100 (f_2, f_3), chaos helped in improving the mean value of the solution found. This improvement is of several orders in the case of f_3, a purely exploitation-intensive function. Performance for f_4 is the same, which was already perfect. Whereas, in f_1 and f_5, where the standard GWO failed to give correct answers every time, chaos improved the SR significantly with some maps and gave mean values comparable or better in those cases. Looking at both the tables, we can easily say that CGWO with the sinusoidal map is the best chaotic improvement in a.

5.2   Results with Chaos Applied in C

Next chaos is applied to C as described in Sect. 3.3. The results are provided in Tables 4 and 5. The performance in case of f_3 is better in a few maps, but only by few orders. This is in contrast to the CGWO with chaos in a. This shows that a major part of exploitation is played by a and not C. For f_2, most maps give improved means as compared to the standard GWO. Once again, the results are same for f_4. A number of maps give better success ratios and mean values of solutions for f_1 and f_5. From both the tables, it is inferred that CGWO with chaos in C gives best performance with sinusoidal and iterative maps (almost the same between them).

Table 4 Evaluation results for chaos in C

Map	f_1		f_2		f_3		f_4		f_5	
	Mean	SD	Mean	SD	Mean	SD	Mean	SD	Mean	SD
Gauss	31.541	16.724	7.46E−14	1.59E−14	1.78E−28	5.57E−28	3	1.42E−05	0.0085	0.0156
Logistic	12.138	10.181	3.64E−14	6.10E−15	7.89E−33	2.75E−32	3	1.22E−05	0.0022	0.0071
Sinusoidal	10.24	11.67	7.20E−14	9.12E−15	1.01E−29	1.039E−29	3	2.17E−06	0.0025	0.0068
Tent	35.124	15.031	1.02E−13	3.46E−14	6.54E−27	1.90E−26	3	1.05E−05	0.0067	0.0106
Singer	17.293	9.3137	1.59E−12	1.80E−12	1.05E−22	1.68E−22	3	2.72E−05	0.005	0.0083
Chebychev	20.801	21.197	3.50E−14	6.66E−15	6.67E−33	1.92E−32	3	1.12E−05	0.0037	0.008
Iterative	14.48	23.236	3.38E−14	5.08E−15	9.60E−34	2.30E−33	3	1.08E−05	0.0028	0.0076
Standard	13.955	10.079	8.57E−14	7.31E−15	3.94E−31	6.63E−31	3	2.90E−05	0.0046	0.00878

Table 5 Success ratios with chaos in C

Map	f_1	f_2	f_3	f_4	f_5	Total SR
Gauss	4	100	100	100	64	368
Logistic	17	100	100	100	90	407
Sinusoidal	33	100	100	100	87	420
Tent	0	100	100	100	67	367
Singer	10	100	100	100	67	377
Chebychev	20	100	100	100	80	400
Iterative	30	100	100	100	90	420
Standard	20	100	100	100	75	395

Table 6 Results for the maximization problem

Maps	Chaos in a			Chaos in C		
	Mean	SD	SR	Mean	SD	SR
Gauss	58941	291.79	0	59999.05	0.7377	60
Logistic	58927.5	276.327	0	59998.829	0.8612	53
Sinusoidal	58854.6	327.887	0	59999.322	0.6797	60
Tent	59999.386	0.75433	84	59999.007	1.6546	77
Singer	58708.8	369.906	0	59998.476	1.7212	50
Chebychev	59059	15.55	0	59999.031	0.8901	67
Iterative	59059	15.55	0	59998.633	1.5617	50
Standard	59998.128	1.6351	37	59998.128	1.6351	37

5.3   A Different Maximization Problem

In order to test the efficiency of the CGWO, another function, quite a different one
was also used. This function is quite smooth but there are plenty of local optima that
yield values very close to the actual solution.

$$f(x) = f(s_1, x_2, x_3, x_4, x_5) = \prod_{i=1}^{5} x_i \quad \mathrm{mod}\ 60{,}000 \qquad x_i \in (1, 10) \qquad (36)$$

The problem is a maximization one with the true maximum value tending towards
60000. Keeping the parameters same, the results obtained are provided in Table 6. It
is observed that only the tent map gives very good results for chaotic a, whereas all
maps give average to good results for a chaotic C. Still, chaos has evidently shown
great improvement to the standard performance.

5.4 Results on the Prime Factorization Problem

The one-dimensional objective function suggested in Eq. (29) (F2) is now tested for its performance with the most successful function used till now (Eq. 17), i.e. F1. We carry on with our assumption that both factors have equal number of digits. The challenge for optimizing using CGWO is choosing a proper combination of maps. From the experiments done above, it can be inferred that sinusoidal, tent map are efficient for a and sinusoidal, iterative map are efficient for C. For the functions now in concern, it was found from performing a few runs that a combination of sinusoidal (a), iterative (C) map works well for Eq. (17) and that of sinusoidal map for both a and C works well for Eq. (29). We went ahead with these choices and the optimization results are presented in Table 7. SR is the number of times, out of 30, the correct factors were computed within the maximum iterations. MI and SD are measured for the successful runs only.

In this problem of prime factorization, accuracy plays a bigger role than the time taken, provided that the difference in the latter is no very large. From the results obtained, it can be observed that our suggested function outperforms the currently used modular function in comparable number of iterations. The difference can be seen in a more clear way for the larger numbers. Following these results, this function can perhaps be used for further research even when number of digits of the factors are not same by performing calculations to restrict the search space.

6 Conclusion and Future Research Challenges

This work has shown the effect of chaos in improving the performance of the Grey Wolf Optimizer for continuous optimization. A new one-dimensional function has also been proposed which has shown better performance than the one being used so far. However, there still remain challenges to solving the prime factorization problem using meta-heuristics. The standard deviation in the number of iterations is of the same order as the mean iterations (in case of both functions). This shows that the performance is quite erratic within a range. This does raise concerns regarding the scalability and reliability of the methods. Secondly, a function needs to be developed that is smooth and has fewer local minima in order to increase the success ratios, specially for large semi-primes. A function was proposed in [15], which is smooth leading to better selection pressure.

$$\textbf{minimize } f(x) = |\log(N) - \log(x) - \log(y)| \qquad x \in [10^{d-1}, \sqrt{N}], y \in [\sqrt{N}, 10^d - 1] \tag{37}$$

However, the actual solution is located in a region having the shape of a two-dimensional curve. This region has a very large number of points yielding values

Table 7 Comparison of objective functions for prime factorization

Max Iter	Bits	N	Agents	F1			F2		
				MI	SD	SR	MI	SD	SR
100	15	50,759	30	15.57	13.772	30	16.96	17.909	30
			40	10.35	8.9017	30	12.44	11.012	30
			50	8.99	6.9143	30	9.5	9.2731	30
200	19	370,627	30	53.08	59.67	30	46.83	42.6	30
			40	37.91	47.675	30	35.03	35.295	30
			50	29.35	32.444	30	24.89	24.006	30
500	24	10,909,343	80	74.45	67.096	30	106.44	105.031	30
			120	51.72	49.33	30	66.47	72.03	30
			160	35.67	35.261	30	50.8	49.549	30
500	25	29,835,457	80	133.93	114.68	29	112.467	88.834	30
			120	88.6	74.573	30	80.667	104.975	30
			160	67.8	49.353	30	77.167	71.852	30
1000	29	392,913,607	100	238.32	204.46	28	319.03	253.86	29
			300	138.1	102.036	30	171.333	144.276	30
			500	69.567	77.716	30	121.933	157.9	30
2000	33	5,325,280,633	100	824.48	539.73	25	801.57	480.27	21
			300	485.9	489	29	465	463.13	29
			500	305.2	258.093	30	264.133	321.435	30
3000	35	42,336,478,013	300	1065.1	875.46	28	948.25	814.95	28
			500	594.2	419.345	30	852.37	783.42	30
			700	348.4	241.267	30	518.633	404.516	30

(continued)

Table 7 (continued)

Max Iter	Bits	N	Agents	F1				F2			
				MI	SD	SR		MI	SD	SR	
4000	38	2,72,903,119,607	300	2038.63	988.36	19		1889.375	1208.1	24	
			500	1318	901.62	23		1410.65	1076.82	26	
			700	1020.4	823.96	26		1190.83	1008.43	29	
12000	44	11,683,458,677,563	700	5048.42	3343.37	20		5477.52	3712.87	22	
			1000	4994.3	3597.93	24		4322.84	3174.13	26	
			1300	4346.96	3845.87	26		4189	2637.23	29	

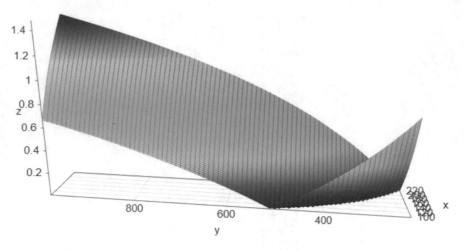

Fig. 3 Graph of Eq. (37) for $N = 50{,}759$

very close to the true minimum value, i.e. 0. This is evident from the graph of
the function for $N = 50{,}759$ in Fig. 3. This function was also tested but yielded
poor success ratios even for small numbers. Meta-heuristics are a promising method
for factorizing semi-primes. Overcoming these challenges can bring about major
improvements.

References

1. Alatas, B., Akin, E., Ozer, A.B.: Chaos embedded particle swarm optimization algorithms.
   Chaos, Solitons Fractals **40**(4), 1715–1734 (2009)
2. Bonabeau, E., Dorigo, M., Theraulaz, G.: Swarm Intelligence: From Natural to Artificial Systems, vol. 1. Oxford University Press, Oxford (1999)
3. Dass, P., Sharma, H., Bansal, J.C., Nygard, K.E.: Meta heuristics for prime factorization problem. In: 2013 World Congress on Nature and Biologically Inspired Computing (NaBIC), pp.
   126–131. IEEE, New York (2013)
4. Dorigo, M., Birattari, M., Stutzle, T.: Ant colony optimization. IEEE Comput. Intell. Mag.
   1(4), 28–39 (2006)
5. dos Santos Coelho, L., Mariani, V.C.: Use of chaotic sequences in a biologically inspired
   algorithm for engineering design optimization. Expert Syst. Appl. **34**(3), 1905–1913 (2008)
6. Eberhart, R., Kennedy, J.: A new optimizer using particle swarm theory. In: Proceedings of the
   Sixth International Symposium on Micro Machine and Human Science, 1995. MHS'95, pp.
   39–43. IEEE, New York (1995)
7. Emary, E., Zawbaa, H.M., Hassanien, A.E.: Binary grey wolf optimization approaches for
   feature selection. Neurocomputing **172**, 371–381 (2016)
8. Gandomi, A.H., Yang, X.-S., Talatahari, S., Alavi, A.H.: Firefly algorithm with chaos. Commun.
   Nonlinear Sci. Numer. Simul. **18**(1), 89–98 (2013)
9. Gandomi, A.H., Yang, X.-S.: Chaotic bat algorithm. J. Comput. Sci. **5**(2), 224–232 (2014)
10. Karaboga, D., Basturk, B.: A powerful and efficient algorithm for numerical function optimization: Artificial Bee Colony (ABC) algorithm. J. Global Optim. **39**(3), 459–471 (2007)

11. Laskari, E.C., Meletiou, G.C., Vrahatis, M.N.: Problems of cryptography as discrete optimization tasks. Nonlinear Anal.: Theor. Methods Appl. **63**(5), e831–e837 (2005)
12. Madadi, A., Motlagh, M.M.: Optimal control of DC motor using grey wolf optimizer algorithm. TJEAS J. **4**(4), 373–79 (2014)
13. Mirjalili, S., Mirjalili, S.M., Lewis, A.: Grey wolf optimizer. Adv. Eng. Softw. **69**, 46–61 (2014)
14. Mishra, M., Chaturvedi, U., Pal, S.K.: A multithreaded bound varying chaotic firefly algorithm for prime factorization. In: 2014 IEEE International Advance Computing Conference (IACC), pp. 1322–1325. IEEE, New York (2014)
15. Mishra, M., Gupta, V., Chaturvedi, U., Shukla, K.K., Yampolskiy, R.V.: A study on the limitations of evolutionary computation and other bio-inspired approaches for integer factorization. Procedia Comput. Sci. **62**, 603–610 (2015)
16. Muangkote, N., Sunat, K., Chiewchanwattana, S.: An improved grey wolf optimizer for training q-Gaussian radial basis functional-link nets. In: 2014 International Computer Science and Engineering Conference (ICSEC), pp. 209–214. IEEE, New York (2014)
17. Muro, C., Escobedo, R., Spector, L., Coppinger, R.P.: Wolf-pack (Canis lupus) hunting strategies emerge from simple rules in computational simulations. Behav. Process. **88**(3), 192–197 (2011)
18. Pomerance, C.: A tale of two sieves. Biscuits Number Theor. **85**, 175 (2008)
19. Yang, D., Li, G., Cheng, G.: On the efficiency of chaos optimization algorithms for global optimization. Chaos, Solitons Fractals **34**(4), 1366–1375 (2007)
20. Yang, X.-S.: Firefly algorithms for multimodal optimization. In: International Symposium on Stochastic Algorithms, pp. 169–178. Springer, Berlin (2009)
21. Yang, X.-S.: Bat algorithm for multi-objective optimisation. Int. J. Bio-Inspired Comput. **3**(5), 267–274 (2011)
22. Yang, X.-S., Gandomi, A.H., Talatahari, S., Alavi, A.H.: Metaheuristics in Water, Geotechnical and Transport Engineering. Newnes, London (2012)

On the Consecutive Customer Loss Probabilities in a Finite-Buffer Renewal Batch Input Queue with Different Batch Acceptance/Rejection Strategies Under Non-renewal Service

A. D. Banik, Souvik Ghosh and M. L. Chaudhry

Abstract This paper deals with a finite-buffer renewal input queueing system, where the arrivals occur in batches of random size and the server serves the customers singly. The successive service times are correlated and its representation is expressed through the continuous-time Markovian service process (C-MSP). As the buffer capacity is finite, the partial batch rejection policy and the total batch rejection policy are considered in this paper. The blocking probabilities and mean waiting time of the first, last, and an arbitrary customer of a batch are determined using the steady-state system-length distribution at pre-arrival epoch. Further, the probability of k or more consecutive customer loss (i.e., k-CCL) during a busy period is computed. The results are illustrated by some tables and graphs for different inter-batch-arrival distributions as well as different C-MSP representations.

Keywords Renewal batch-arrival · Finite-buffer · Continuous-time Markovian service process (C-MSP) · Blocking probabilities · Consecutive customer losses (CCL)

A. D. Banik · S. Ghosh (✉)
School of Basic Sciences, Indian Institute of Technology Bhubaneswar,
Argul Campus, Khurda 752050, Odisha, India
e-mail: souvikghosh589@gmail.com

A. D. Banik
e-mail: banikad@gmail.com

M. L. Chaudhry
Department of Mathematics and Computer Science,
Royal Military College of Canada, P.O. Box 17000,
STN Forces, Kingston, ON K7K 7B4, Canada
e-mail: chaudhry-ml@rmc.ca

© Springer Nature Singapore Pte Ltd. 2019
J. C. Bansal et al. (eds.), *Soft Computing for Problem Solving*,
Advances in Intelligent Systems and Computing 816,
https://doi.org/10.1007/978-981-13-1592-3_4

1 Introduction

The applicability of different queueing models in various areas such as manufacturing systems, production systems, communication networks makes the queueing theory interesting. Queueing models with different service and arrival processes have been studied by many researchers. In early researches, the service and arrival processes were assumed to be Poisson which preserves the memoryless property. As the technology develops, the correlated nature of arrival and service processes needs to be introduced in queueing models. The correlation among the inter-arrival can be explained by continuous- and discrete-time Markovian arrival process (*C-MAP* and *D-MAP*). Considerable work on queueing models with *C-MAP* arrivals includes Lucantoni et al. [13]. It is to be mentioned here that continuous-time Markovian arrival process (*C-MAP*) can suitably represent the versatile Markovian point process (*VMPP*); see Neuts [15] and Ramaswami [19]. Further, the discrete-time batch Markovian arrival process (*D-BMAP*) has been considered by Samanta et al. [21]. Recent study considering correlated arrival includes Yu and Alfa [22] and Banerjee et al. [3]. Analogous to the correlated inter-arrival times, the correlation among the service times can be explained by continuous-time Markovian service process (*C-MSP*) which is a generalization of many other service processes including Markov-modulated Poisson process (*MMPP*) and *PH*-type renewal process. The process *C-MSP* and its bibliographic details have been discussed by Gupta and Banik [10], where they have used a combination of embedded Markov chain process, supplementary variable technique, and matrix-geometric method. Applying perturbation theory, the asymptotic nature of $GI/C\text{-}MSP/1$ queueing system has been explained by Alfa et al. [1]. Allowing the server to take vacations, Machihara [14] has investigated the $G/SM/1/\infty$ queueing model. Later, Banik and Gupta [4] have investigated the batch-arrival queueing system with finite-buffer capacity under total and partial batch acceptance policy. Using the roots of the characteristic equation, Chaudhry et al. [5] have investigated the $GI^{[X]}/C\text{-}MSP/1/\infty$ model. Queueing systems with finite-buffers are more realistic than the corresponding infinite-buffer models.

In this study, we assume that the arrival occurs in batches of random sizes, the inter-batch-arrival times are generally distributed, and the service is governed by *C-MSP*. This model can be denoted by $GI^{[X]}/C\text{-}MSP/1/N$ where N denotes the finite capacity of the buffer including the one who is with the server. As the system capacity is considered to be finite, it may happen that an arriving batch can't be accommodated fully in the system. If an arriving batch contains more customer than the available space in the system, then only the required number of customers are allowed to join the system; i.e., a portion of the arriving batch is accepted, this is defined as partial batch rejection (PBR) policy. The system can also adopt the total batch rejection (TBR) policy; i.e., an entire batch is rejected if the batch is loaded with more customer than the available system space. The analysis of $GI^{[X]}/C\text{-}MSP/1/N$ and $GI/C\text{-}$

$BMSP/1/N$ queueing systems under PBR policy using RG-factorization technique have been presented by Banik, Ghosh and Chaudhry in [2]. The steady-state system-length distribution at pre-arrival epoch in the $GI^{[X]}/C\text{-}MSP/1/N$ model with PBR and TBR strategies are determined here. It may be remarked here that considering an initial phase distribution at the time when the service process resumes from an idle period, the $GI^{[X]}/C\text{-}MSP/1/N$ model has been analyzed by Banik and Gupta [4]. They obtain the steady-state system-length distribution at pre-arrival epoch by applying GTH algorithm to the one-step transition probability matrix of embedded Markov chain. This paper also analyzes the $GI^{[X]}/C\text{-}MSP/1/N$ queueing model with slightly different assumptions, i.e., the busy period starts with the service phase which was last phase of the server in the previous busy period.

In this paper, we compute some important performance measures, such as the blocking probabilities and mean waiting time of the customers in the $GI^{[X]}/C\text{-}MSP/1/N$ queueing system. The blocking probability in the $GI^{[X]}/M\,(n)//N$ queueing system has been addressed by Ferreira and Pacheco [8]. Moreover, the consecutive customer losses (CCLs) in communication networks with finite-buffer capacity is also an important characteristic with respect to the quality of service (QoS); see Chydzinski [6]. The well-known k-CCL probability in the $M/G/1/N$ and $G/M/c/N$ queueing systems have been explored by De Boer [7]. Pacheco and Ribeiro [16] have computed the k-CCL probabilities for the simple $M/G/1/N$ and $GI/M/1/N$ queueing system. Further, Pacheco and Ribeiro [17] have derived the k-CCL probabilities for the oscillating $GI^{[X]}/M//N$ queueing systems with phase-dependent service rates. The consecutive loss probabilities in the regular and oscillating $M^X/G/1/N$ queueing system is also studied by Pacheco and Rebeiro [18]. This paper determines the k-CCL probabilities in a busy period of the $GI^{[X]}/C\text{-}MSP/1/N$ queueing system. Generally, a busy period starts when the idle server becomes busy and that busy period ends when the server becomes idle again after that. As defined by Pacheco and Ribeiro [17], we use a modified definition of the busy period (may be initiated with more than one customers) for determining the k-CCL probabilities in the $GI^{[X]}/C\text{-}MSP/1/N$ queueing system.

2   Description of the Model

A single server finite-buffer queue with batch-arrivals is discussed here. The batch size of the arriving batches is assumed to be of random size, and the random variable is symbolized by X with $P(X = i) = g_i, i \geq 1$. The average batch size is determined as $E(X) = \sum_{i=1}^{\infty} ig_i = \bar{g}$ and further g_i' it is defined as $g_i' = \sum_{r=i}^{\infty} g_r, i \geq 1$. The time interval between the occurrence of two successive batches; i.e., the batch inter-arrival times are considered to be independent and identically distributed (iid) random variables (rv's) A with mean $1/\lambda$. Hence, the mean arrival rate is calculated as λ.

The arrival process is also considered to be independent of the service process. Let the cumulative distribution function (cdf), probability density function (pdf), and the Laplace–Stieltjes transform (LST) of the pdf of a random variable Y be symbolized by $F_Y(y)$, $f_Y(y)$, and $f_Y^*(s)$, respectively. Further, it is assumed that including the customer in service the buffer capacity of the system is finite, say N.

Two matrices L_0 and L_1 of order m describe the m-state C-MSP. L_1 (L_0) represents the state transition for a (no) service. Then the generator matrix of the C-MSP is $L = L_0 + L_1$. It has been considered that one customer can be served by the server at a time. Further, details on C-MSP can be found in Chaudhry et al. [5]. $N(t)$ and $J(t)$ represent the number of customers served and state of the underlying continuous-time Markov chain (CTMC) at time t, respectively. Let us define the Markov process $X(t) = \{N(t), J(t)\}$ with state space $E = \{(n, r) : 0 \le n \le N, 1 \le r \le m\}$, where n is the level variable and r be the phase variable of the Markov chain. If $\overline{\pi}_j$ denotes the probability that a customer is getting service in steady state with the service phase j $(1 \le j \le m)$, then $\overline{\pi} = [\overline{\pi}_1, \overline{\pi}_2, \ldots, \overline{\pi}_m]$ denotes the stationary probability vector and can be determined from the relations $\overline{\pi} L = 0$ and $\overline{\pi} e = 1$, where e and 0 are the column vector of suitable size with all the entries 1 and row vector of suitable size with all the entries 0, respectively. Hence, the fundamental service rate or the mean service rate of customers is calculated by $\mu^* = \overline{\pi} L_1 e$. The traffic intensity or offered load is given by $\rho = \lambda \bar{g} / \mu^*$.

For $0 \le n \le N$, $t \ge 0$ and $1 \le i, j \le m$, if we define the conditional probability as

$$P_{i,j}(n, t) = Pr\{N(t) = n, J(t) = j | N(0) = 0, J(0) = i\},$$

then $P(n, t)$ represent an $m \times m$ matrix whose (i, j)th component is $P_{i,j}(n, t)$. Hence, using matrix notation the system may be expressed as

$$\frac{d}{dt} P(0, t) = P(0, t) L_0, \tag{1}$$

$$\frac{d}{dt} P(n, t) = P(n, t) L_0 + P(n - 1, t) L_1, \ 1 \le n \le N - 1, \tag{2}$$

$$\frac{d}{dt} P(N, t) = P(N, t) L + P(N - 1, t) L_1, \tag{3}$$

with $P(0, 0) = I_{m \times m}$ and $P(n, 0) = \overline{0}_{m \times m}$ $(n \ge 1)$, where $I_{m \times m}$ and $\overline{0}_{m \times m}$ are the identity matrix and null matrix of dimension $m \times m$, respectively. Throughout the paper, we do not mention the dimension as subscript of the identity matrix and the null matrix and mentioned those matrices as I and $\overline{0}$, respectively. However, the dimension of the identity matrix and the null matrix is mentioned as subscript wherever it is required. Let us define two matrices S_n and S_n^* of order $m \times m$ as follows: For $0 \le n \le N$, the (i, j)th $(1 \le i, j \le m)$ element of the matrix S_n represents the conditional probability of n service completion in a inter-batch-arrival time span provided that at the previous batch-arrival instant there were at least n customers in the system, where i and j indicate the phase of the service process at the beginning of the inter-batch-arrival time and at the ending of the inter-batch-arrival time. For $1 \leq n \leq N$, the (i, j)-th $(1 \leq i, j \leq m)$ element of the matrix $\mathbf{S}_n^*$ denotes the probability that exactly n customers are in the system at a batch-arrival instant with phase of the service is i and then at least n service has completed during the next inter-batch-arrival interval while the phase of the service changes to j at nth service completion epoch.

Then the matrices $\mathbf{S}_n$ and $\mathbf{S}_n^*$ can be expressed as

$$\mathbf{S}_n = \int_0^\infty \mathbf{P}(n, t)\, dF_A(t), \quad 0 \leq n \leq N, \quad \text{and} \quad \mathbf{S}_n^* = \frac{1}{\lambda}\boldsymbol{\Omega}_{n-1}\mathbf{L}_1, \ 1 \leq n \leq N, \quad (4)$$

where $\boldsymbol{\Omega}_{n-1} = \lambda \int_0^\infty \mathbf{P}(n, x)(1 - F_A(x))\, dx$ and can be computed as

$$\boldsymbol{\Omega}_0 = \lambda\left(\mathbf{I} - \mathbf{S}_0\right)(-\mathbf{L}_0)^{-1}, \quad \boldsymbol{\Omega}_N = \left(\boldsymbol{\Omega}_{N-1}\mathbf{L}_1 - \lambda\mathbf{S}_N\right)(-\mathbf{L})^{-1} \quad \text{and} \quad (5)$$

$$\boldsymbol{\Omega}_n = \left(\boldsymbol{\Omega}_{n-1}\mathbf{L}_1 - \lambda\mathbf{S}_n\right)(-\mathbf{L}_0)^{-1}, \quad 1 \leq n \leq N - 1. \quad (6)$$

Interested readers are referred to Chaudhry et al. [5] for a detailed discussion on the matrices $\mathbf{S}_n$, $\mathbf{S}_n^*$, and $\boldsymbol{\Omega}_n$. Let us consider the embedded points as the time epochs just before the batches arrive in the system and denote the embedded points as t_i^- $(i = 0, 1, 2, \ldots)$, where t_i's are the time instants at which a batch-arrival is about to occur. Denoting the $\mathcal{N}(t_i^-)$ and $\mathcal{J}(t_i^-)$ as the number of customers in the system and the phase of the service process at time t_i^- $(i = 0, 1, 2, \ldots)$, respectively. Then $\mathcal{X}(t_i^-)_{t_i^- \geq 0} = (\mathcal{N}(t_i^-), \mathcal{J}(t_i^-))$ constitutes a discrete-time Markov chain (DTMC) with state space E as defined earlier. Let $\pi_{j,n}^-$ denote the limiting case that at pre-arrival instant of a batch-arrival n customers are present in the system and the state of the service is j, that is

$$\pi_j^-(n) = \lim_{i \to \infty} P(\mathcal{N}(t_i^-) = n, \ \mathcal{J}(t_i^-) = j), \ 0 \leq n \leq N, \ 1 \leq j \leq m.$$

For $0 \leq n \leq N$, we denote $\boldsymbol{\pi}^-(n) = [\pi_1^-(n), \pi_2^-(n), \pi_3^-(n), \ldots, \pi_m^-(n)]$. Now if the corresponding one-step transition probability matrix (TPM) of the PBR and TBR model embedded at pre-arrival epoch is denoted by $\mathcal{P}_{\text{PBR}}$ and $\mathcal{P}_{\text{TBR}}$, respectively. Then the matrices are given as follows:

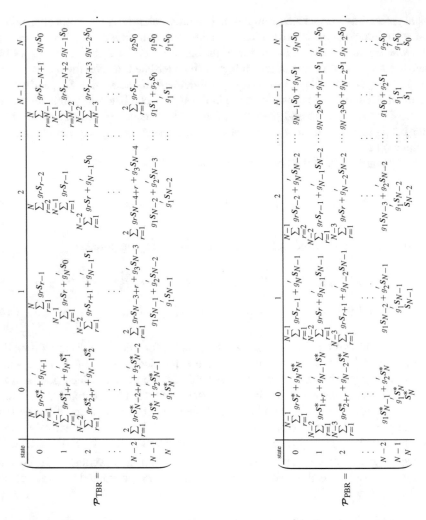

Let us denote the one-step TPM of either model as $\mathcal{P}$ and $\pi^- = [\pi^-(0), \pi^-(1), \pi^-(2), \ldots, \pi^-(N)]$. Then the stationary pre-arrival epoch probabilities $\pi^-(n)$ $(0 \le n \le N)$ can be determined by solving the equation $\pi^-\mathcal{P} = \pi^-$. One may use GTH algorithm (see Grassmann et al. [9]) or the RG-factorization technique (see Latouche and Ramaswamy [11] and Li [12]) for computation of the probabilities $\pi^-(n)$ $(0 \le n \le N)$.

3 k-CCL Probabilities in the $GI^{[X]}/C$-$MSP/1/N$ Queueing System

As defined by Pacheco and Ribeiro [17], we define a c^*-busy period which begins at batch-arrival epoch, where c^* denotes a pair (n, i) with n and i denotes the system length and the state of the MSP just after the initiation of the busy period, respectively. Further, we define (c, k)-CCL probability as the probability of losing k (or more) consecutive customer in a c^*-busy period, where $c = (c^*, j)$ ($j \in \{1, 2, \ldots, m\}$) and j indicates the phase of the MSP just after the occurrence of the k-CCL probability. With the state space E, we define $D = (D(t))_{t \geq 0}$ as a CTMC between two consecutive batch-arrivals. Let T is the random variable with distribution of the inter-batch-arrival time. Then, as discussed by Pacheco and Ribeiro [17], the TPM of D is given as $\mathcal{Q} = \sum_{l=0}^{\infty} v_l \mathcal{P}^l$, where $v_l = \int_0^{\infty} e^{-\theta t} \frac{(\theta t)^l}{l!} dF_A(t)$ and $\theta = \max\{-[L_0]_{ii}, 1 \leq i \leq m\}$.

For computation of (c, k)-CCL probability, it is acceptable to extend the state space of the DTMC D to construct a DTMC $Y = (Y_l)_{l \in \mathbb{N} \cup \{0\}}$ which is embedded at batch-arrival instant until either the system becomes empty or a k-CCL event takes place. The DTMC Y can be constructed by adding the accumulated number of consecutive customer losses at the embedded epochs to the state space E. Hence, the state space of the DTMC Y is given by $E^* = E \cup \{(N_r, i) : 1 \leq r \leq k$ and $1 \leq i \leq m\}$, where N_r indicates the event that N customers are in the system immediately after a batch-arrival instant accumulating r-CCL. Further, for $l \in \mathbb{N}$ and $(n, i) \subset E^*$, $\{(Y_l) = (n, i)\}$ can be interpreted as follows:

- For $n \in \{1, 2, \ldots, N\}$ and $1 \leq i \leq m$, $\{(Y_0) = (n, i)\}$ indicates that initially n number of customers are present in the system and the state of the service is i.
- For $n = N_r$ with $r \in \{1, 2, \ldots, k\}$ and $1 \leq i \leq m$, $\{(Y_0) = (N_r, i)\}$ indicates that initially there are N number of customers are present with r-CCL accumulating in the system and the phase of the service process is i.
- For $1 \leq i \leq m$, $\{(Y_l)_{l>0} = (0, i)\}$ indicates that before the arrival of the l-th batch arrival, the phase of the MSP is i and the system is emptied without the occurrence of any k-CCL event.
- For $n \in \{1, 2, \ldots, N\}$ and $1 \leq i \leq m$, $\{(Y_l)_{l>0} = (n, i)\}$ indicates that before the arrival of the l-th batch, there are n number of customers in the system with the service process is in phase i and neither the system becomes empty nor a k-CCL event takes place.
- For $n = N_r$ ($r \in \{1, 2, \ldots, k - 1\}$) and $1 \leq i \leq m$, $\{(Y_l)_{l>0} = (n, i)\}$ indicates that before the arrival of the l-th batch, there are N number of customers in the system with r-CCL accumulating and the service process is in phase i while neither the system becomes empty nor a k-CCL event takes place.
- For $1 \leq i \leq m$, $\{(Y_l)_{l>0} = (N_k, i)\}$ indicates that before the l-th batch-arrival instant, there are N number of customers in the system and the state of the service is i while a k-CCL takes place without emptying the system.

Let $\mathcal{R}$ denote the TPM of the DTMC Y, and from the definition, it can be given as

level	0	1	2	3	$\cdots$	$N-1$	N
0	I	$\mathbf{0}$	$\mathbf{0}$	$\mathbf{0}$	$\cdots$	$\mathbf{0}$	$\mathbf{0}$
1	$\mathcal{Q}_{1,0}$	$\bar{\mathbf{0}}$	$\sum_{l=1}^{1}\mathcal{Q}_{1,l}g_{2-l}$	$\sum_{l=1}^{2}\mathcal{Q}_{1,l}g_{3-l}$	$\cdots$	$\sum_{l=1}^{N-2}\mathcal{Q}_{1,l}g_{N-1-l}$	$\sum_{l=1}^{N-1}\mathcal{Q}_{1,l}g_{N-l}$
2	$\mathcal{Q}_{2,0}$	$\bar{\mathbf{0}}$	$\sum_{l=1}^{1}\mathcal{Q}_{2,l}g_{2-l}$	$\sum_{l=1}^{2}\mathcal{Q}_{2,l}g_{3-l}$	$\cdots$	$\sum_{l=1}^{N-2}\mathcal{Q}_{2,l}g_{N-1-l}$	$\sum_{l=1}^{N-1}\mathcal{Q}_{2,l}g_{N-l}$
3	$\mathcal{Q}_{3,0}$	$\bar{\mathbf{0}}$	$\sum_{l=1}^{1}\mathcal{Q}_{3,l}g_{2-l}$	$\sum_{l=1}^{2}\mathcal{Q}_{3,l}g_{3-l}$	$\cdots$	$\sum_{l=1}^{N-2}\mathcal{Q}_{3,l}g_{N-1-l}$	$\sum_{l=1}^{N-1}\mathcal{Q}_{3,l}g_{N-l}$
$\vdots$	$\vdots$	$\vdots$	$\vdots$	$\vdots$	$\ddots$	$\vdots$	$\vdots$
$N-1$	$\mathcal{Q}_{N-1,0}$	$\bar{\mathbf{0}}$	$\sum_{l=1}^{1}\mathcal{Q}_{N-1,l}g_{2-l}$	$\sum_{l=1}^{2}\mathcal{Q}_{N-1,l}g_{3-l}$	$\cdots$	$\sum_{l=1}^{N-2}\mathcal{Q}_{N-1,l}g_{N-1-l}$	$\sum_{l=1}^{N-1}\mathcal{Q}_{N-1,l}g_{N-l}$
N	$\mathcal{Q}_{N,0}$	$\bar{\mathbf{0}}$	$\sum_{l=1}^{1}\mathcal{Q}_{N,l}g_{2-l}$	$\sum_{l=1}^{2}\mathcal{Q}_{N,l}g_{3-l}$	$\cdots$	$\sum_{l=1}^{N-2}\mathcal{Q}_{N,l}g_{N-1-l}$	$\sum_{l=1}^{N-1}\mathcal{Q}_{N,l}g_{N-l}$
N_1	$\mathcal{Q}_{N,0}$	$\bar{\mathbf{0}}$	$\sum_{l=1}^{1}\mathcal{Q}_{N,l}g_{2-l}$	$\sum_{l=1}^{2}\mathcal{Q}_{N,l}g_{3-l}$	$\cdots$	$\sum_{l=1}^{N-2}\mathcal{Q}_{N,l}g_{N-1-l}$	$\sum_{l=1}^{N-1}\mathcal{Q}_{N,l}g_{N-l}$
$\vdots$	$\vdots$	$\vdots$	$\vdots$	$\vdots$	$\ddots$	$\vdots$	$\vdots$
N_{k-1}	$\mathcal{Q}_{N,0}$	$\bar{\mathbf{0}}$	$\sum_{l=1}^{1}\mathcal{Q}_{N,l}g_{2-l}$	$\sum_{l=1}^{2}\mathcal{Q}_{N,l}g_{3-l}$	$\cdots$	$\sum_{l=1}^{N-2}\mathcal{Q}_{N,l}g_{N-1-l}$	$\sum_{l=1}^{N-1}\mathcal{Q}_{N,l}g_{N-l}$
N_k	$\bar{\mathbf{0}}$	$\bar{\mathbf{0}}$	$\bar{\mathbf{0}}$	$\bar{\mathbf{0}}$	$\cdots$	$\bar{\mathbf{0}}$	$\bar{\mathbf{0}}$

level	N_1	$\cdots$	N_{k-1}	N_k
0	$\mathbf{0}$	$\cdots$	$\mathbf{0}$	$\mathbf{0}$
1	$\sum_{l=1}^{N}\mathcal{Q}_{1,l}g_{N+1-l}$	$\cdots$	$\sum_{l=1}^{N}\mathcal{Q}_{1,l}g_{N+n-1-l}$	$\sum_{l=1}^{N}\mathcal{Q}_{1,l}g'_{N+n-l}$
2	$\sum_{l=1}^{N}\mathcal{Q}_{2,l}g_{N+1-l}$	$\cdots$	$\sum_{l=1}^{N}\mathcal{Q}_{2,l}g_{N+n-1-l}$	$\sum_{l=1}^{N}\mathcal{Q}_{2,l}g'_{N+n-l}$
3	$\sum_{l=1}^{N}\mathcal{Q}_{3,l}g_{N+1-l}$	$\cdots$	$\sum_{l=1}^{N}\mathcal{Q}_{3,l}g_{N+n-1-l}$	$\sum_{l=1}^{N}\mathcal{Q}_{3,l}g'_{N+n-l}$
$\vdots$	$\vdots$	$\ddots$	$\vdots$	$\vdots$
$N-1$	$\sum_{l=1}^{N}\mathcal{Q}_{N-1,l}g_{N+1-l}$	$\cdots$	$\sum_{l=1}^{N}\mathcal{Q}_{N-1,l}g_{N+n-1-l}$	$\sum_{l=1}^{N}\mathcal{Q}_{N-1,l}g'_{N+n-l}$
N	$\sum_{l=1}^{N-1}\mathcal{Q}_{N,l}g_{N+1-l}+\mathcal{Q}_{N,N}g_1$	$\cdots$	$\sum_{l=1}^{N-1}\mathcal{Q}_{N,l}g_{N+n-1-l}+\mathcal{Q}_{N,N}g_{n-1}$	$\sum_{l=1}^{N-1}\mathcal{Q}_{N,l}g'_{N+n-l}+\mathcal{Q}_{N,N}g'_n$
N_1	$\sum_{l=1}^{N-1}\mathcal{Q}_{N,l}g_{N+1-l}$	$\cdots$	$\sum_{l=1}^{N-1}\mathcal{Q}_{N,l}g_{N+n-1-l}+\mathcal{Q}_{N,N}g_{n-2}$	$\sum_{l=1}^{N-1}\mathcal{Q}_{N,l}g'_{N+n-l}+\mathcal{Q}_{N,N}g'_{n-1}$
$\vdots$	$\vdots$	$\ddots$	$\vdots$	$\vdots$
N_{k-1}	$\sum_{l=1}^{N-1}\mathcal{Q}_{N,l}g_{N+1-l}$	$\cdots$	$\sum_{l=1}^{N-1}\mathcal{Q}_{N,l}g_{N+n-1-l}$	$\sum_{l=1}^{N-1}\mathcal{Q}_{N,l}g'_{N+n-l}+\mathcal{Q}_{N,N}g'_1$
N_k	$\bar{\mathbf{0}}$	$\cdots$	$\bar{\mathbf{0}}$	I

It may be noted that the dimension of the TPM $\mathcal{R}$ is $m(N+k+1)\times m(N+k+1)$ and for $1\le i\le m$, $(0,i)$ and (N_k,i) are the absorbing states of the DTMC Y while all other states are transient. Moreover, the state space of the DTMC Y can be partitioned as $E^*=\{(0,i):1\le i\le m\}\cup E_t^*\cup\{(N_k,i):1\le i\le m\}$, where $E_t^*=\{(n,r):1\le n\le N,1\le r\le m\}\cup\{(N_r,i):1\le r\le k-1\text{ and }1\le i\le m\}$. Hence, the TPM $\mathcal{R}$ of the DTMC Y can be block partitioned as follows:

$$\mathcal{R}=\begin{bmatrix} I_{m\times m} & \mathbf{0}_{m\times m(N+k-1)} & \mathbf{0}_{m\times m} \\ \varsigma_{m(N+k-1)\times m} & \mathcal{B}_{m(N+k-1)\times m(N+k-1)} & \vartheta_{m(N+k-1)\times m} \\ \mathbf{0}_{m\times m} & \mathbf{0}_{m\times m(N+k-1)} & I_{m\times m} \end{bmatrix}. \tag{7}$$

Let us denote the matrices $\varsigma_{m(N+k-1)\times m}$, $\mathcal{B}_{m(N+k-1)\times m(N+k-1)}$ and $\vartheta_{m(N+k-1)\times m}$ as ς, $\mathcal{B}$ and ϑ, respectively, and the entries of those matrices can be determined directly from the TPM $\mathcal{R}$ of the DTMC Y.

For $c^* \in E^*$ and $j \in \{1, 2, \ldots, m\}$, let us define $\gamma_{(c^*,j)}$ as a probability that a k-CCL event takes place in a c^*-busy period and just after the occurrence of the k-CCL event the server will be in phase j. Thus, $\gamma_{(c^*,j)}$ can be interpreted as follows:

<u>Case 1:</u> $n \in \{1, 2, \ldots, N\}$ and $i, j \in \{1, 2, \ldots, m\}$, then $\gamma_{(c^*,j)}$ denotes the probability that a k-CCL event takes place in the c^*-busy period, where $c^* = (n, i)$ and the state of the service process will be j just after occurrence of the k-CCL event.

<u>Case 2:</u> $n = N_r$ ($r \in \{1, 2, \ldots, k\}$) and $i, j \in \{1, 2, \ldots, m\}$, then $\gamma_{(c^*,j)}$ denotes the probability that a k-CCL event takes place in the c^*-busy period with r-CCL accumulating at the starting of the busy period, where $c^* = (N, i)$ and the phase of the service process will be j just after occurrence of the k-CCL event.

For $c^* \in E^*$, we define a row vector as $\gamma_{c^*} = [\gamma_{(c^*,1)}, \gamma_{(c^*,2)}, \ldots, \gamma_{(c^*,m)}]$. Further, according to the block structure of the matrix $\mathcal{R}$, we define $\gamma = [\gamma_{c^*}]_{c^* \in E^*}$ as a matrix with dimension $m(N + k - 1) \times m$. One may note that for $c^* \in E^*$, the (c, k)-CCL probability can be computed as the probability that the DTMC Y reaches to the state (N_k, j) ($j \in \{1, 2, \ldots, m\}$) when starting from the state c^*, i.e., the probability of absorption to the absorbing state (N_k, j) from the transient state c^*. Hence, from the calculus of absorption probabilities (see, e.g., Resnick [20]), the (c, k)-CCL probabilities in the $GI^{[X]}/C\text{-}MSP/1/N$ queueing system can be computed as

$$\gamma = [I - \mathcal{B}]^{-1}\vartheta. \tag{8}$$

For $n \in \{1, 2, \ldots, N\} \cup \{N_1, N_2, \ldots, N_{k-1}\}$ and $i \in \{1, 2, \ldots, k\}$, let us denote $\hat{\gamma}_{c^*}$ as the $((n, i), k)$-CCL probability, i.e, the probability of occurring k-CCL event in c^*-busy period, where $c^* = (n, i)$. Further, according to the block structure of the matrix $\mathcal{R}$, we define a column vector $\overline{\gamma} = [\hat{\gamma}_{c^*}]_{c^* \in E^*}$. Thus we can compute $\hat{\gamma}_{c^*} = \gamma_{c^*}e$ and $\overline{\gamma} = \gamma e$.

Remark 1 The accurate computation of the matrix γ needs precise determination of the matrices $\mathcal{B}$ and ϑ, which in turn needs the precise computation of the matrix $\mathcal{Q}$, as the matrices $\mathcal{B}$ and ϑ completely depend on the matrix $\mathcal{Q}$. Thus, if one can compute the $\mathcal{Q}$ precisely, then the matrix ϑ as well as the (c, k)-CCL probabilities in the $GI^{[X]}/C\text{-}MSP/1/N$ queueing system can be computed accurately using the Eq. (8). However, being an infinite sum, the matrix $\mathcal{Q}$ cannot be computed accurately but only be approximated. The approximation can be done in the following way.

(i) Choose $\epsilon > 0$ as the desired accuracy of the approximation.

(ii) Choose a large number $\mathcal{L}$ and compute $\mathcal{Q}_{\mathcal{L}} = \sum_{l=0}^{\mathcal{L}} v_l \mathcal{P}^l$.

(iii) Determine the spectral radius of the matrices $\mathcal{Q}_{\mathcal{L}}$ and $\mathcal{Q}_{\mathcal{L}+1}$ as $\eta_{\mathcal{L}}$ and $\eta_{\mathcal{L}+1}$, respectively.

(iv) If $\eta_{\mathcal{L}+1} - \eta_{\mathcal{L}} < \epsilon$, then approximate $\mathcal{Q} \simeq \mathcal{Q}_{\mathcal{L}}$, otherwise repeat step (ii) – (iv) with a higher value of $\mathcal{L}$.

4 Some Useful Performance Measures of the $GI^{[X]}/C\text{-}MSP/1/N$ Queueing System

Using the pre-arrival epoch probabilities, we have computed some important performance measures which are discussed in this section. Let P_{BF}, P_{BL}, and P_{BA} denote the blocking probabilities of the first, last, and an arbitrary customer in a batch. Further, we denote the average waiting time in the system of the first, last, and an arbitrary customer by w_F, w_L, and w_A. Now, using those notations, we derive the blocking probabilities and waiting times as follows.

4.1 Partial Batch Rejection Strategy

Following Banik and Gupta [4], the average waiting times in the system of the first, last, and an arbitrary customer for the PBR model can be obtained as follows:

$$
w_F = \frac{1}{1 - P_{BF}} \Big[\sum_{n=0}^{N-1} \boldsymbol{\pi}^-(n) \sum_{j=0}^{n} (-\boldsymbol{L}_0^{-1}\boldsymbol{L}_1)^j (-\boldsymbol{L}_0^{-1})e \Big], \tag{9}
$$

$$
w_A = \frac{1}{1 - P_{BA}} \Big[\sum_{n=0}^{N-1} \boldsymbol{\pi}^-(n) \sum_{j=0}^{N-n-1} \sum_{k=0}^{n+j} g_j^- (-\boldsymbol{L}_0^{-1}\boldsymbol{L}_1)^k (-\boldsymbol{L}_0^{-1})e \Big], \tag{10}
$$

$$
w_L = \frac{1}{1 - P_{BL}} \Big[\sum_{n=0}^{N-1} \boldsymbol{\pi}^-(n) \sum_{j=1}^{N-n} \sum_{k=0}^{n+j-1} g_j (-\boldsymbol{L}_0^{-1}\boldsymbol{L}_1)^k (-\boldsymbol{L}_0^{-1})e \Big], \tag{11}
$$

where

$$
P_{BF} = \boldsymbol{\pi}^-(N)e, \quad P_{BA} = \sum_{i=0}^{N} \sum_{j=N-i}^{\infty} \boldsymbol{\pi}^-(i) g_j^- e, \quad P_{BL} = \sum_{i=0}^{N} \sum_{j=N-i+1}^{\infty} \boldsymbol{\pi}^-(i) g_j e, \tag{12}
$$

with $g_r^- = \frac{1}{\bar{g}} \sum_{i=r+1}^{\infty} g_i$, $r \geq 0$.

4.2 Total Batch Rejection Strategy

The mean system and queue length of the system in case of the TBR model is the same as PBR model. But the blocking probabilities in the TBR model is different from the PBR model as an entire batch is rejected if the whole batch is not accommodated in the system. Therefore, the blocking probability of the first and last customer under total rejection policy is the same. For the TBR strategy, the expressions of the blocking probability and the mean waiting time of the first (last) and an arbitrary customer are derived by Banik and Gupta [4] and are given as follows:

$$P_{BF} = P_{BL} = \sum_{i=0}^{N} \sum_{j=N-i+1}^{\infty} \pi^{-}(i)g_j e, \quad P_{BA} = \sum_{i=0}^{N} \sum_{j=N-i+1}^{\infty} \pi^{-}(i)\frac{jg_j}{\bar{g}}e, \quad (13)$$

$$w_F = w_L = \frac{1}{1 - P_{BL}} \Big[\sum_{n=0}^{N-1} \pi^{-}(n) \sum_{j=1}^{N-n} \sum_{k=0}^{n} g_j(-L_0^{-1}L_1)^k(-L_0^{-1})e \Big], \quad (14)$$

$$w_A = \frac{1}{1 - P_{BA}} \Big[\sum_{n=0}^{N-1} \pi^{-}(n) \sum_{j=1}^{N-n} \sum_{r=0}^{j-1} \sum_{k=0}^{n+r} \frac{g_j}{\bar{g}}(-L_0^{-1}L_1)^k(-L_0^{-1})e \Big]. \quad (15)$$

5 Numerical Illustrations

Using the results obtained in the previous sections, several numerical experiments have been done for the $GI^{[X]}/C\text{-}MSP/1/N$ queueing system. Among those experiments, few numerical results are presented in this section. Although numerical calculations were carried out with high precision, due to lack of place the results have been reported to 6 decimal places. All the numerical calculations are performed in MAPLE 2015.

5.1 Comparative Study of Blocking Probabilities and Mean Waiting Times for the Case of Partial Batch Rejection and Total Batch Rejection Policy in a $GI^{[X]}/C\text{-}MSP/1/N$ Queueing System

We obtain the blocking probabilities and mean waiting times of a $PH^{[X]}/C\text{-}MSP/1/10$ queueing system under PBR policy and TBR policy. For this queueing system, the g_n's ($n \geq 1$) are considered as the coefficient of z^n in the expansion of $4z/(3-z)^2$ with $\bar{g} = 2.0$. The representation of two different three-phase PH-type density function of the inter-batch-arrival time is taken as

$$\alpha_1 = \alpha_2 = [0.2, 0.3, 0.5], \; T_1 = \begin{pmatrix} -2.5 & 0.5 & 0.2 \\ 0.1 & -2.0 & 0.1 \\ 0.2 & 0.4 & -3.7 \end{pmatrix} \text{ and } T_2 = \begin{pmatrix} -1.5 & 0.5 & 0.2 \\ 0.1 & -1.0 & 0.1 \\ 0.2 & 0.4 & -1.7 \end{pmatrix}.$$

The four-phase *C-MSP* representation is taken as

$$L_0 = \begin{pmatrix} -6.5 & 0.1 & 0.6 & 0.3 \\ 0.3 & -5.4 & 0.8 & 0.2 \\ 0.2 & 0.7 & -4.3 & 0.1 \\ 0.6 & 0.4 & 0.1 & -3.9 \end{pmatrix}, \quad L_1 = \begin{pmatrix} 1.1 & 1.7 & 1.5 & 1.2 \\ 1.3 & 0.9 & 0.7 & 1.2 \\ 1.4 & 0.9 & 0.3 & 0.7 \\ 0.9 & 0.5 & 1.1 & 0.3 \end{pmatrix}.$$

For this representation of *C-MSP*, it is found that $\mu^* = 3.862496$ and $\bar{\pi} = [0.224993, 0.240930, 0.283610, 0.250467]$. For the first (second) inter-arrival time distribution, the mean arrival rate (λ) of the customers and the traffic intensity (ρ) of the system are calculated as 2.221854 (0.894436) and 1.150476 (0.463139), respectively. For different system capacity (N), the two models under both the PBR and TBR policy are investigated and the blocking probability and the mean waiting time of an arbitrary customer are presented in Figs. 1 and 2, respectively.

Fig. 1 Blocking probability of an arbitrary customer versus buffer capacity of the system

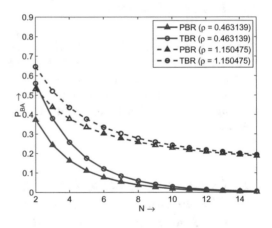

Fig. 2 Mean waiting time of an arbitrary customer versus buffer capacity of the system

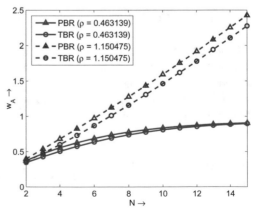

From Fig. 1, it is seen that as the system capacity increases the blocking probability of an arbitrary customer under both rejection policies approaches to a certain value. It is also observed that P_{BA} converges to zero as N takes large value. Figure 2 demonstrates that with increasing system capacity (N), the waiting time of an arbitrary customer under both rejection policies approaches to a certain value if ρ is less than one. But when ρ is greater than one, the waiting time of an arbitrary customer under both the policies grows linearly as the system capacity increases.

5.2 The Effect of the Correlation in the C-MSP on the CCL Probabilities in the $GI^{[X]}/C\text{-}MSP/1/N$ Queueing System

We study the behavior of the CCL probabilities in the $GI^{[X]}/C\text{-}MSP/1/N$-PBR queueing system for different lag-1 correlation coefficient of the $C\text{-}MSP$. This study considers the following $C\text{-}MSP$ representations with mean service rate (μ^*) 3.4750.

Representation 1: $L_0 = \begin{pmatrix} -6.95 & 3.475 \\ 3.475 & -6.95 \end{pmatrix}$ and $L_1 = \begin{pmatrix} 1.7375 & 1.7375 \\ 1.7375 & 1.7375 \end{pmatrix}$ with lag-1 correlation coefficient 0.0; i.e., the $C\text{-}MSP$ is uncorrelated.

Representation 2: $L_0 = \begin{pmatrix} -6.5 & 0.025 \\ 0.025 & -2.0 \end{pmatrix}$ and $L_1 = \begin{pmatrix} 6.4 & 0.075 \\ 0.025 & 1.95 \end{pmatrix}$ with lag-1 correlation coefficient 0.195379.

Representation 3: $L_0 = \begin{pmatrix} -9.5 & 0.025 \\ 0.025 & -0.5 \end{pmatrix}$ and $L_1 = \begin{pmatrix} 9.4 & 0.075 \\ 0.025 & 0.45 \end{pmatrix}$ with lag-1 correlation coefficient 0.393053.

In Table 1, we have given the 5-CCL probabilities in a $PH^{[X]}/C\text{-}MSP/1/5$-PBR queueing system for different lag-1 correlation coefficient of the $C\text{-}MSP$ as given before. For this investigation, we have considered $g_n = (1-p)^{n-1}p$, where $p = 1/3.4750$ and $n \geq 1$, and the PH-type inter-batch-arrival time distribution is taken as $\alpha = [1.0, 0.0]$, $T = \begin{pmatrix} -1.2 & 0.0 \\ 0.0 & -1.2 \end{pmatrix}$. Hence, one can compute $\bar{g} = 3.4750$, $\lambda = 1.2$ and thus $\rho = 1.2$.

From Table 1, we can conclude if the service time is uncorrelated then the state of the service in the starting of the busy period has no effect on the k-CCL probabilities in the $GI^{[X]}/C\text{-}MSP/1/N$ queueing system. But if the service time is correlated, then the state of the service in the beginning of the busy period has a certain impact on the k-CCL probabilities in the $GI^{[X]}/C\text{-}MSP/1/N$ queueing system. Further, from Table 1, one can observe that with higher correlation in $C\text{-}MSP$ the CCL probabilities are higher.

Table 1 5-CCL probabilities in the $PH^{[X]}/C\text{-}MSP/1/5$-PBR queueing system for different lag-1 correlation coefficient of the $C\text{-}MSP$

n	lag-1 correlation coefficient 0.0		lag-1 correlation coefficient 0.195379		lag-1 correlation coefficient 0.393053	
	$\hat{\gamma}_{((n,1),5)}$	$\hat{\gamma}_{((n,2),5)}$	$\hat{\gamma}_{((n,1),5)}$	$\hat{\gamma}_{((n,2),5)}$	$\hat{\gamma}_{((n,1),5)}$	$\hat{\gamma}_{((n,2),5)}$
1	0.327972	0.327972	0.238532	0.479303	0.213297	0.605946
2	0.339452	0.339452	0.249512	0.491783	0.220956	0.611830
3	0.349890	0.349890	0.260217	0.502010	0.228516	0.616901
4	0.359379	0.359379	0.270601	0.510610	0.235963	0.621717
5	0.368006	0.368006	0.280634	0.518028	0.243288	0.626435
5_1	0.389408	0.389408	0.300502	0.540374	0.259979	0.655462
5_2	0.416298	0.416298	0.325814	0.567549	0.281082	0.688235
5_3	0.450084	0.450084	0.358079	0.600594	0.307893	0.725229
5_4	0.492536	0.492536	0.399228	0.640770	0.342112	0.766975

5.3 Effect of the Batch Size, Service Time, and Batch Inter-Arrival Time Distributions on the CCL Probabilities in the $GI^{[X]}/C\text{-}MSP/1/N$ Queueing System

Considering different batch size, service time, and inter-batch-arrival time distribution, we study the k-CCL probabilities in the $GI^{[X]}/C\text{-}MSP/1/N$ queueing system. For this study, we assume that either $g_n = (1-p)^{n-1}p$ with $n \geq 1$ and $0 < p < 1$, i.e., geometric distribution or $g_1 = 0.4$, $g_2 = 0.3$, $g_3 = 0.2$, $g_4 = 0.1$, and $g_n = 0.0$ ($n \geq 5$). For the first batch, size distribution the $\bar{g}$ is $1/p$, and for the second distribution, it can be calculated as 2.0. The $C\text{-}MSP$ is taken as

$$
L_0 = \begin{pmatrix} -(\sigma + \sigma^*) & 0.5\sigma & 0.5\sigma \\ 0.5\sigma & -(\sigma + \sigma^*) & 0.5\sigma \\ 0.5\sigma & 0.5\sigma & -(\sigma + \sigma^*) \end{pmatrix} \text{ and } L_1 = \begin{pmatrix} 0.3\sigma^* & 0.3\sigma^* & 0.4\sigma^* \\ 0.3\sigma^* & 0.3\sigma^* & 0.4\sigma^* \\ 0.3\sigma^* & 0.3\sigma^* & 0.4\sigma^* \end{pmatrix},
$$

where σ is a positive number and $\sigma^* = 0.1\sigma$. Hence, one can calculate $\mu^* = \sigma^*$ and the lag-1 correlation coefficient is 0.0; i.e., the $C\text{-}MSP$ is uncorrelated. Further, we consider four different types of inter-batch-arrival time distributions with equal mean $1/\lambda$ as (i) PH-type with representation $\alpha = [1.0, 0.0]$, $T_1 = \begin{pmatrix} -\lambda & 0.0 \\ 0.0 & -\lambda \end{pmatrix}$, (ii) deterministic with mean $1/\lambda$, i.e., $D(\lambda)$, (iii) Erlang 3-distribution, i.c., $E_3(3\lambda)$, and (iv) Weibull with distribution function $1 - e^{(t/b)^a}$, i.e., $W(a, b)$ and we set $a = 2.4$ and $b = 1/\lambda\Gamma(1 + 1/a)$.

Considering the batch is geometrically distributed with different value of p, the effect of batch size distribution on k-CCL probabilities (particularly $((1, 1), 3)$-CCL probability) in a $GI^{[X]}/C\text{-}MSP/1/10$ queueing system is presented in Fig. 3. For this study, we set $\lambda = 0.5$, $\sigma = 10$ and $p = 1/p^*$, where $1.0 \leq p^* \leq 9.0$. The effect of the inter-batch-arrival time distribution on k-CCL probabilities (particularly $((1, 1), 5)$-CCL probability) in a $GI^{[X]}/C\text{-}MSP/1/5$ queueing system is shown in Fig. 4. This study is carried out by considering $g_1 = 0.4$, $g_2 = 0.3$, $g_3 = 0.2$, $g_4 = 0.1$, and $g_n = 0.0$ ($n \geq 5$), $\sigma = 20.0$ and $0.5 \leq \lambda \leq 3.5$. The impact of the batch service time distribution on k-CCL probabilities (particularly $((1, 1), 1)$-CCL probability) in a $GI^{[X]}/C\text{-}MSP/1/5$ queueing system is illustrated in Fig. 5. For this study, the parameters are set as $g_1 = 0.4$, $g_2 = 0.3$, $g_3 = 0.2$, $g_4 = 0.1$, and $g_n = 0.0$ ($n \geq 5$), $\lambda = 0.5$ and $1.0 \leq \sigma \leq 30.0$.

From Fig. 3, it can be observed that starting from 0.0 the $((1, 1), 3)$-CCL probability approaches to 1.0 as the mean batch size increases. From Fig. 4, it can be observed that starting from 0.0 the $((1, 1), 5)$-CCL probability approaches to 1.0 as the mean arrival rate increases. From Fig. 5, one can observe that starting from 1.0 the $((1, 1), 1)$-CCL probability approaches to 0.0 as the mean service rate increases.

Fig. 3 Effect of the batch size distribution on the 3-CCL probabilities in the $GI^{[X]}/C\text{-}MSP/1/10$-PBR queueing system

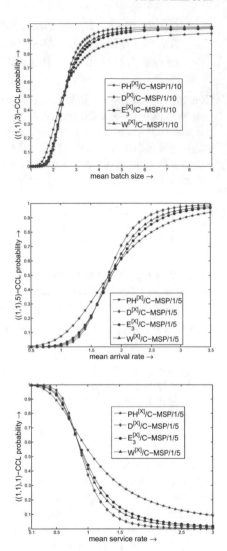

Fig. 4 Effect of the inter-batch-arrival time distribution on the 5-CCL probabilities in the $GI^{[X]}/C\text{-}MSP/1/5$-PBR queueing system

Fig. 5 Effect of the service time distribution on the 1-CCL probabilities in the $GI^{[X]}/C\text{-}MSP/1/5$-PBR queueing system

6 Conclusion

In this paper, we discuss the $GI^{[X]}/C\text{-}MSP/1/N$ queueing system and obtain steady-state probability distributions at pre-arrival and arbitrary epochs. We study the so-called k-CCL probabilities in the $GI^{[X]}/C\text{-}MSP/1/N$ queueing system. We also obtain stationary performance measures such as mean waiting time and blocking probabilities for the first customer, an arbitrary customer, and the last customer in an accepted batch. This study can be extended for renewal service and correlated batch service, i.e., the $GI^X/G/1/N$ and $GI^X/C\text{-}BMSP/1/N$ queueing system. One may compute the k-CCL probabilities in vacation queueing systems with exhaustive and non-exhaustive service discipline. However, computation of k-CCL probabilities in a queueing system with correlated inter-batch-arrival times and renewal/non-renewal service, i.e., $C\text{-}BMAP/G/1/N$ and $C\text{-}BMAP/C\text{-}MSP/1/N$, is also an interesting problem which is left as our future research plan.

Acknowledgements The third author was supported partially by NSERC under research grant number RGPIN-2014-06604.

References

1. Alfa, A.S., Xue, J., Ye, Q.: Perturbation theory for the asymptotic decay rates in the queues with Markovian arrival process and/or Markovian service process. Queueing Syst. **36**(4), 287–301 (2000)
2. Ayesta, U., Boon, M., Prabhu, B., Righter, R., Verloop, M.: European Conference on Queueing Theory 2016 (2016)
3. Banerjee, A., Gupta, U.C., Chakravarthy, S.R.: Analysis of a finite-buffer bulk-service queue under Markovian arrival process with batch-size-dependent service. Comput. Oper. Res. **60**, 138–149 (2015)
4. Banik, A.D., Gupta, U.C.: Analyzing the finite buffer batch arrival queue under Markovian service process: $GI^X/MSP/1/N$. Top **15**(1), 146–160 (2007)
5. Chaudhry, M.L., Banik, A.D., Pacheco, A.: A simple analysis of the batch arrival queue with infinite-buffer and Markovian service process using roots method: $GI^{[X]}/C\text{-}MSP/1/\infty$. Ann. Oper. Res. **252**, 135–173 (2017)
6. Chydzinski, A.: On the remaining service time upon reaching a given level in $M/G/1$ queues. Queueing Syst. **47**(1–2), 71–80 (2004)
7. De Boer, P.T.: Analysis and efficient simulation of queueing models of telecommunication systems. Ph.D. thesis, Centre for Telematics and Information Technology University of Twente (2000)
8. Ferreira, F., Pacheco, A.: Analysis of $GI^X/M(n)//N$ systems with stochastic customer acceptance policy. Queueing Syst. **58**(1), 29–55 (2008)
9. Grassmann, W.K., Taksar, M.I., Heyman, D.P.: Regenerative analysis and steady state distributions for Markov chains. Oper. Res. **33**(5), 1107–1116 (1985)
10. Gupta, U.C., Banik, A.D.: Complete analysis of finite and infinite buffer $GI/MSP/1$ queue - a computational approach. Oper. Res. Lett. **35**(2), 273–280 (2007)
11. Latouche, G., Ramaswami, V.: Introduction to Matrix Analytic Methods in Stochastic Modeling, vol. 5. SIAM, Philadelphia, PA (1999)
12. Li, Q.L.: Constructive Computation in Stochastic Models with Applications: the RG-Factorizations. Springer Science & Business Media, Berlin (2011)
13. Lucantoni, D.M., Meier-Hellstern, K.S., Neuts, M.F.: A single-server queue with server vacations and a class of non-renewal arrival processes. Adv. Appl. Probab. **22**, 676–705 (1990)
14. Machihara, F.: A $G/SM/1$ queue with vacations depending on service times. Stoch. Models **11**(4), 671–690 (1995)
15. Neuts, M.F.: A versatile Markovian point process. J. Appl. Probab. **16**, 764–779 (1979)
16. Pacheco, A., Ribeiro, H.: Consecutive customer loss probabilities in $M/G/1/n$ and $GI/M(m)//n$. In: n systems. In: Proceedings from the 2006 Workshop on Tools for Solving Structured Markov Chains, Pisa, Italy (2006)
17. Pacheco, A., Ribeiro, H.: Consecutive customer losses in oscillating $GI^X/M//n$ systems with state dependent services rates. Ann. Oper. Res. **162**(1), 143–158 (2008)

18. Pacheco, A., Ribeiro, H.: Consecutive customer losses in regular and oscillating $M^X/G/1/n$ systems. Queueing Syst. **58**(2), 121–136 (2008)
19. Ramaswami, V.: The $N/G/1$ queue and its detailed analysis. Adv. Appl. Probab. **12**, 222–261 (1980)
20. Resnick, S.I.: Adventures in Stochastic Processes. Springer Science & Business Media, Berlin (2013)
21. Samanta, S.K., Gupta, U.C., Sharma, R.K.: Analyzing discrete-time D-BMAP/$G/1/N$ queue with single and multiple vacations. Eur. J. Oper. Res. **182**(1), 321–339 (2007)
22. Yu, M., Alfa, A.S.: Algorithm for computing the queue length distribution at various time epochs in DMAP/$G^{(1,a,b)}/1/N$ queue with batch-size-dependent service time. Eur. J. Oper. Res. **244**(1), 227–239 (2015)

Fuzzy Enhancement for Efficient Emotion Detection from Facial Images

Payal Bhattacherjee and M. M. Ramya

Abstract Human–computer interaction is one of the emerging fields facilitating computers to better understand human behavior. One such mode is through emotions displayed in facial images. Emotion detection requires real-time data acquisition. More often, the images are degraded due to poor illumination. Image enhancement becomes mandatory for precise emotion detection from facial images. Fuzzy logic works well for contrast enhancement as it deals with uncertainties in image acquisition well. A new methodology for contrast enhancement of facial images based on new improved fuzzy set theory is proposed. The proposed approach is carried out in three phases. Firstly, the overall brightness is adjusted using a trigonometric function to change the dynamic range of the image. Secondly, two different membership functions are established based on the histogram to adjust the local contrast of image details. The performance metrics like average information count (AIC) and natural image quality evaluator (NIQE) were used to evaluate the proposed method which generated on an average 89% decrease in NIQE value and 10% increase in AIC value.

1 Introduction

Image processing of facial images in recent times has garnered a lot of attention. Facial images have long being used as biometric. Off, lately emotion detection from facial images has been an active area of research for image processing-based technologies. Emotion detection from facial images cannot be done with degraded images. Most of the applications of emotion detection require real-time image acquisition. There exits non-uniform illumination, use of inadequate shutter speed and

P. Bhattacherjee (✉) · M. M. Ramya
Center for Automation and Robotics, Hindustan Institute of Technology and Science,
Padur, Chennai, Tamil Nadu, India
e-mail: payal.bhattacherjee708@gmail.com

M. M. Ramya
e-mail: mmramya@hindustanuniv.ac.in

© Springer Nature Singapore Pte Ltd. 2019
J. C. Bansal et al. (eds.), *Soft Computing for Problem Solving*,
Advances in Intelligent Systems and Computing 816,
https://doi.org/10.1007/978-981-13-1592-3_5

time, use of improper lens, which leads to low contrast images. Hence, image enhancement is the pivotal method in image processing of facial images.

Image enhancement techniques can be broadly classified into gray-level transformation, gradient-based methods, reflectance field estimation [1]. Gray-level transformation techniques like histogram equalization, logarithmic transformation, gamma intensity correction (GIC) are global enhancement techniques. These techniques tend to stretch the intensities, and this approach to contrast enhancement does not work very well as amplitude of image histogram is high at one or several ends and lower at others [1]. These techniques cannot handle side lighting and give a washed out effect to the images. Moreover, these techniques tend to amplify existing noise. Gradient-based methods like grayscale derivative (DGD), Laplacian of Gaussian (LoG) extract edge-based information that is immune to lighting conditioning but cannot overcome shadows [2].

Gray-level grouping (GLG) technique focuses on grouping the grayscale values into various bins based on some predefined threshold. These bins are redistributed to meet certain enhancement criterion. The underlying objective is to get an equally spread histogram [1].

Reflectance field estimation methods include Single Scale retinex (SSR), Gaussian High Pass (GHP), Logarithmic cosine transform (LDCT) enhance both the well-illuminated regions of the image as well as the region with shadow [2]. Light Random Sprays Retinex (LrsRetinex) is an improved version of Random Sprays Retinex which is based on the original retinex theory. LrsRetinex is computational inexpensive compared to RSR and successful in removal of noise [3].

Fuzzy-based enhancement is computationally inexpensive when compared to transform domain techniques. It is also an effective methodology for compensating shadows [4].

1.1 Fuzzy Enhancement

Fuzzy set theory is proposed as a middle path between classical sets and inherent ambiguity of real time. Fuzzy enhancement is found to solve contrast issues due to non-uniform illumination [5]. Data acquisition and conversion to digital images give raise to umpteen uncertainties and ambiguities. These ambiguities are handled well by human reasoning, but quantifying them is a task for conventional algorithms. This has led to many researches toward fuzzy enhancement [6–10].

Contrast intensification operator (INT) is calculated, and membership of the gray levels is modified on the basis of the INT operator. However, the INT operator solely depends on a fixed threshold and defined membership function. The threshold is fixed at 0.5 that does not work very well for over contrast images [11, 12].

There has been extensive work on fuzzy enhancement using various membership functions [13] have used modified Gaussian membership function as it makes smoother transition from darker region to bright region [14]. Have also used fuzzy logic for contrast enhancement, histogram analysis was carried out for fixing the threshold.

There exists other fuzzy rule-based contrast enhancement, which utilizes human domain knowledge to form the membership function [15–17]. These algorithms also tend to have a predefined threshold for each condition, and hence choosing an optimal threshold could become challenging. The enhancement factor is fixed at $k = 128$ statically, which yields good results only for very low contrast images. Hence, there is a requirement for an improved fuzzy-based enhancement with a modified membership function and image restoration factor.

This paper aims at adaptive fuzzy enhancement of color facial images for pain estimation. Each emotion is associated with certain combination of action units. Every action unit can be decimated into its respective muscle movements. This analysis can be used for building intelligent systems, which can be used for emotion recognition-based decision making. The prima facie of emotion recognition is face detection. The facial images for emotion detection are captured at natural environment. These images suffer glare halo shadows. Hence, there is a distinct requirement for image enhancement. The overall image brightness is improved by trigonometric function, and then modified membership function is established for enhancing local contrast of the image details. This paper is organized as follows. Sect. 2 discusses details on materials and methods. Section 3 gives results and comparisons of various techniques based on performance metrics. Section 4 discusses on evaluations of proposed method. Section 5 concludes the paper.

2 Materials and Methods

2.1 Materials

The main objective of this paper is to enhance contrast of color facial images using fuzzy theory for emotion detection, in particular pain.

The data for emotion recognition was acquired from Affect Analysis Group Research Lab, University of Pittsburg [18]. It is made publically available for research work in pain detection and monitoring. The dataset contains 48398 images of size (320 * 240) of patients suffering shoulder pain when they are dealt with active and passive motion test. Each of the 48,398 images is taken as frames from video sequences. The level of pain experienced at each frame level was annotated by the patient's self-report and also by the observer. The images are facial action coding system (FACS) coded.

The images are well illuminated but suffer from glare and shadows. These images when used as it is for face detection do not give satisfactory results. The algorithm was tested with two images from the dataset. The contrast of the images was reduced by 20, 40, and 60% using Photoshop. The algorithm was tested on the images from dataset and also on the contrast-reduced images using opencv-3.1.0 on Mac operating system.

2.2 Methods

Fuzzy-based image enhancement works well for degraded images, wherein edges are uncertain and inaccurate [5, 10]. A new approach to fuzzy rule-based enhancement is proposed, wherein two different thresholds are calculated from the input image and membership function is applied on the dynamic thresholds.

2.2.1 Proposed Methodology

Mapping the intensities values from gray level to fuzzy domain does fuzzy-based image enhancement; the enhancement is carried out in three stages: fuzzification, modification of membership (fuzzy enhancement), and defuzzification [11, 12].

An image of size $i * j$ with intensities values 0–255 can be represented as a union of fuzzy singletons.

$$I = \cup\{\mu I_V((i, j))\}$$

where $\mu(I_V(i, j)$ is the grade of membership of intensity $I_V(i, j)$ to a given class.

A fuzzy enhancement method for efficient emotion detection from color images is proposed. In this method, the input image is normalized as shown in Eq. 1, to convert the image to fuzzy plane [0, 1].

$$I_v = 0; \quad I(i, j) \leq 0;$$

$$I_v = I(i, j) - \min(I(i, j)/\max(I(i, j) - \min(I(i, j))) \tag{1}$$

where I_v = Normalized image min, $I(i, j)$=minimum intensity in $I(i, j)$, and max $I(i, j)$=maximum intensity in $I(i, j)$.

Overall Brightness Adjustment

Global brightness is adjusted using a trigonometric function as shown in Eq. 2, which possesses parabolic function properties. This function has a distinct advantage; that is, the brightness in the dark areas will not be enhanced too fast and brightness in light areas will not be compressed too much [19, 20].

$$I_0 = 1 - (1 - \rho \sin(I_V)) \tag{2}$$

$$I_V \in [0, 1]$$

$I_V =$ Intensity value of each channel

$I_0 =$ overall brightess adjusted image

$\rho =$ brightness adjustment factor

$$\rho = \left\{ \frac{\pi}{2} \quad I_v(i, j) \leq 0.23529; \right.$$

$$1.4 \quad 60 < I_v(i, j) \leq 0.70588;$$

$$1.2 \quad I_v(i, j) > 0.70588 \}$$

If 30% of the intensities are below 0.23529, i.e., the image is darker, then $\rho = \frac{\pi}{2}$. In images where 80% of the image is on brighter side, then $\rho = 1.2$; otherwise $\rho = 1.4$. This mapping of intensities will increase the overall brightness of low-illuminated images, but local contrast is affected.

Local Contrast Adjustment

For local contrast enhancement, fuzzy set theory is applied. Membership function is applied with modified thresholds for obtaining uniform enhancement results over various images with varying levels of contrast. Initial threshold is established from the histogram [21]. The weighted sum is calculated as:

$$s = \sum_{i,j=0}^{i,j=m,n} I_{V(i,j)} \, he(i, j) / \sum_{i,j=0}^{i,j=m,n} he(i, j) \tag{3}$$

where

$he(i, j) =$ histogram of the image and

$s =$ the weighted average intensity value in the image.

s divides the image into two regions $a1$ $(0 \geq s - 1)$ and $a2$ $(s \geq$ max intensity). $I_0(i, j)$ is modified according to the membership function values which is established using the following rules.

1. If the intensity values $I_0(i, j)$ are away from s, it will be enhanced more, and the intensity values near to s will be stretched less as shown in Eq. 4. An average between the s and the minimum intensity value is calculated. This value is set as the threshold for the first class.

$$a = (s + \text{min intensity})/2$$

$$\mu_{a1} = (1 - (a - I_0(i, j) * 255)/a \quad 0 \leq I_0(i, j) \leq a - 1 \tag{4}$$

where $a =$ new threshold for class 1

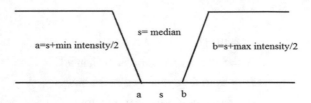

Fig. 1 Membership function

$$\mu_{a1} = \text{degree of membership of intensity } I(i, j) \text{ to class 1}$$
$$I_o(i, j) = \text{globally enhanced image}$$

2. If the intensity values are closer to 255, it will be stretched less and intensities away from 255 will be enhanced more as shown in Eq. 5. An average between the s and the maximum intensity value is calculated. This value is set as the threshold for the second class.

$$b = (s + \text{max intensity})/2$$
$$\mu_{a2} = (255 - I_0(i, j) * 255)/255 - b, \quad b \le I_0(i, j) \le \text{max intensity} \quad (5)$$

where b = new threshold for class 2

$$\mu_{a2} = \text{degree of membership of intensity to class 2}$$
$$I_o(i, j) = \text{globally enhanced image}$$

The intermediate frequencies between a (threshold for dark region) and b (threshold for bright region) remain unmodified (Figs. 1 and 2).

Defuzzification

Image defuzzification is done to change to intensity plane from fuzzy plane.

$$I_{el1} = I_V * 255 + \mu_{a1} * t \quad 0 < I_0(i, j) \le a \quad (6)$$
$$I_{el2} = I_V * \mu_{a2} * 255 + (255 - \mu_{a2} * t) \quad b \le (i, j) \le \text{max intensity} \quad (7)$$

where μ_{a1} = degree of membership of intensity $I(i, j)$ to class 1

$$\mu_{a2} = \text{degree of membership of intensity } I(i, j) \text{ to class 2}$$

The parameter t determines the stretching intensity, which is calculated from analyzing the peak and valley of the histogram, which makes the algorithm adaptive.

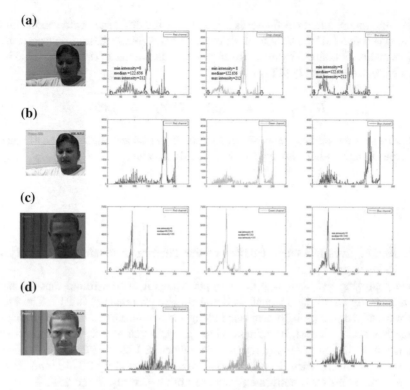

Fig. 2 Comparison of histogram of original and enhanced images. **a** First original image, **b** enhanced image of (**a**), **c** second original image, **d** enhanced image of (**c**)

From the above analysis, the stretching intensity t was found to be varying in between 8 and 15 for UNBC Pain Archive.

2.3 Performance Metrics

The empirical analysis of proposed enhancement technique was carried out using performance metrics like average information count (AIC), natural image quality evaluator (NIQE) [22]. AIC is the measure of information in an image, and higher values indicate greater information.

$$\text{AIC} = -\sum_{i=0}^{i=L-1} p(i) \log p(i) \tag{8}$$

Where $p(i)$ is the probability density function of the ith intensity.

Natural image quality evaluator is calculated as the difference between the multivariate Gaussian (MVG) fit of natural scene statistics (NSS) features extracted from corpus and NSS features extracted from low contrast images. The lower value indicates better image quality [22–24].

$$D = \sqrt{(u1 - u2)^T((\varepsilon1 + \varepsilon2)/2)^{-1}(u1 - u2)} \qquad (9)$$

where $u1, u2, \varepsilon1, \varepsilon2$ are the mean vectors and covariance a matrix of MVG model of corpus image and test image and D is the NIQE value.

3 Results

This section gives a detailed analysis of the experiment conducted to verify the proposed methodology.

On analyzing the histogram of the input image a, it was found that the minimum intensity in all the channels is 8 and maximum intensity is 212. The image was normalized and globally enhanced using the trigonometric function. Since the image has good contrast, brightness factor ρ was taken as 1.4. For local contrast enhancement, the weighted average was calculated at 122.656. This threshold does not work very well when the image contrast was reduced by 20, 40, and 60% as the minimum and maximum intensities are not necessarily 0 and 255. Therefore, new thresholds were calculated as the average between the minimum intensity and weighted average for dark regions and average between maximum intensity and weighted average for bright regions. The threshold for dark region of image a was found to be 65.328 for all the three channels. Threshold for the bright region was found to be 167.328. On these thresholds, the membership function was applied and 98% enhanced image was obtained. This can be validated from the histogram of the enhanced image b. The minimum intensity of the enhanced image was 0 and maximum intensity 251.

Similar observation can be made on image c; minimum intensity of the image is 8, and maximum intensity is 161 with median 80.5283. The new thresholds were calculated as $a = 44.2929$ and $b = 120.793$, and on applying the membership function the enhanced image d was obtained with minimum and maximum intensity of 0 and 251, respectively. Defuzzification was done to convert the image from fuzzy domain to gray scale by using a stretching factor 8 for both the images. The stretching factor was obtained from peak and valley analysis of histogram.

4 Discussion

The proposed algorithm was tested on images with varying contrast.

Figures 3 and 4 show the processing results of various enhancement techniques. Images are labeled as a, b, c, d, 1, 2, 3, 4; a, 1 are the original image; b, c, d, 2, 3, 4 are 20, 40, 60% contrast-reduced images, respectively. The images were processed with contrast limited adaptive histogram equalization (CLAHE), contrast stretching, pals-king algorithm, GLG, retinex and proposed algorithm.

The images a (ii), b (ii), c (ii), d (ii) processed with adaptive histogram technique like contrast limited adaptive histogram equalization technique (CLAHE) tend to show artifacts at strong edges. Contrast stretching causes uneven enhancement in the images a (iii), b (iii), c (iii), d (iii). The Pals-king algorithm enhanced images a (iv), b (iv), c (iv), d (iv) and shows that darker regions are made more darker and brighter regions are made more brighter. Hence, intermediate intensities are lost. Gray-level grouping (GLG) causes excessive enhancement in the blue channel of the images a (v), b (v), c (v), d (v) and loses in color information. Light Random Sprays Retinex (LrsRetinex) works fine, but overall brightness of the images a (vi), b (vi), c (vi), d (vi) is reduced. Similar results can be seen on images 1, 2, 3, 4 with different methodologies.

The proposed algorithm facilitates calculation of dynamic thresholds and modifies the membership function by calculating the average between weighted average and minimum intensity in dark areas. Average between weighted average intensity and maximum gray level value in bright region of the histogram. Image normalization is done before applying the membership function. Image normalization changes the range of intensity values to [0–1], stretches the amplitude of the histogram to cover the entire gray scale, and hence improves the contrast. The enhanced images do

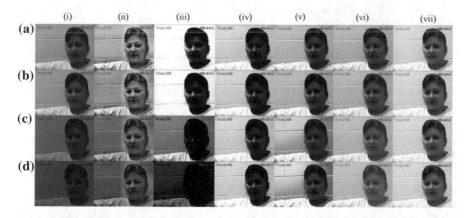

Fig. 3 Comparison of various methods on first image from the dataset with various levels of contrast: (i) input image, (ii) CLAHE, (iii) contrast stretching, (iv) pals, (v) GLG, (vi) LrsRetinex, (vii) proposed algorithm: **a** original image, **b** 20% contrast-reduced image, **c** 40% contrast-reduced image, **d** 60% contrast-reduced image

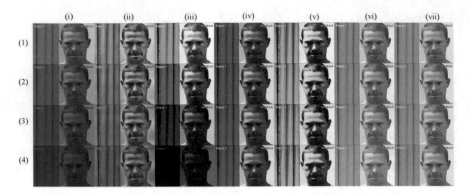

Fig. 4 Comparison of various methods on second image from the dataset with various levels of contrast: (i) input image, (ii) CLAHE, (iii) contrast stretching, (iv) pals, (v) GLG, (vi) LrsRetinex, (vii) proposed algorithm: (1) original image, (2) 20% contrast-reduced image, (3) 40% contrast-reduced image, (4) 60% contrast-reduced image

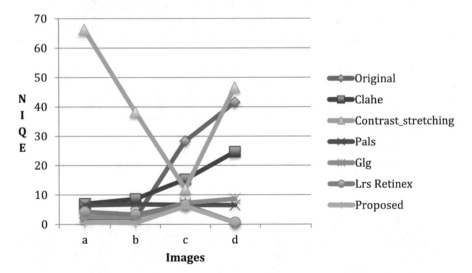

Fig. 5 Comparison of various NIQE values for the first image from the dataset with varying contrast

not introduce any artifacts, there is no amplification of noise, and there exists even enhancement in all the regions of the images.

Figures 5 and 6 show the comparison of NIQE values of various contrast enhancement methods. NIQE value is measured on a scale of 0–100 (0 represents best quality, and 100 represents worst). The images were processed with various existing methodologies, and on comparing the various NIQE values the proposed methodology gives a NIQE value of 0.65 and 0.56 for the original images which is the least value. For contrast-reduced images, also the proposed methodology generates the least values.

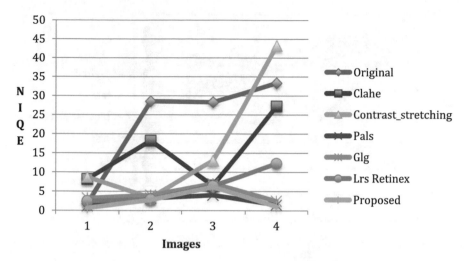

Fig. 6 Comparison of various NIQE values for the second image from the dataset with varying contrast

Figures 7 and 8 show the average information count in an image. NIQE and AIC values of various methods were compared, and it reflects the fact that the proposed method outperforms the traditional methods. Figure 9 depicts the error rate calculated as difference between average skin pixels in a skin patch in original image and average skin pixels in differently enhanced images of various contrasts. Figure 10 depicts the error rate calculated as difference between average background pixels in a background patch in original image and average background pixels in differently enhanced images of various contrast. The above error rate establishes the fact that the proposed algorithm enhances skin pixels and background in a steady uniform manner. This kind of improvement in intensity values enhances the subtle details in areas of constant texture, which is valuable toward image classification on texture features. The proposed method performance is well suited for commercial and industrial application. The analysis results indicate that proposed method not only improves the contrast but generates visually enhanced images.

The objective of the work was enhancing the contrast of facial images for accurate detection of pain. Though there exist traditional algorithms [25], it fails to enhance low contrast images. The membership threshold is fixed at the weighted average intensity. This thresholding does not work in low contrast images where the intensity values condense in a narrow range.

The proposed algorithm is successful in enhancing low contrast images by normalizing the value to [0–1]; this spreads the otherwise concentrated intensities. Moreover, the membership function is applied with different thresholds for different regions. For darker region, average between the lowest intensity and weighted average is taken as threshold; for brighter regions, average between weighted average and maximum

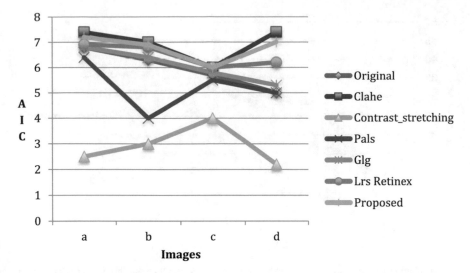

Fig. 7 Comparison of AIC values for the first image from the dataset with varying contrast

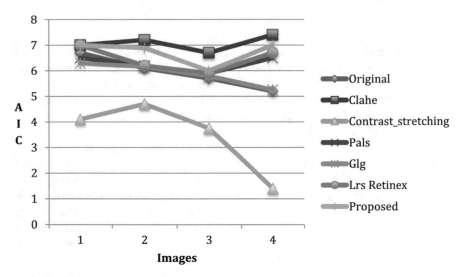

Fig. 8 Comparison of AIC values for the second image from the dataset with varying contrast

value is taken as threshold. This kind of flexible threshold decreases the fuzziness and hence makes the algorithm suitable for images with varied levels of contrast. Hence, the algorithm is truly adaptive in nature.

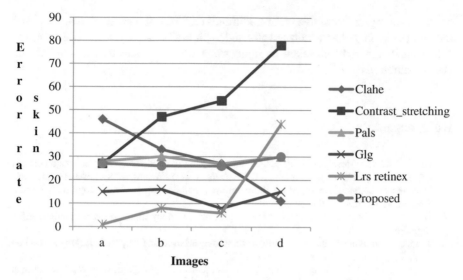

Fig. 9 Error rate for skin

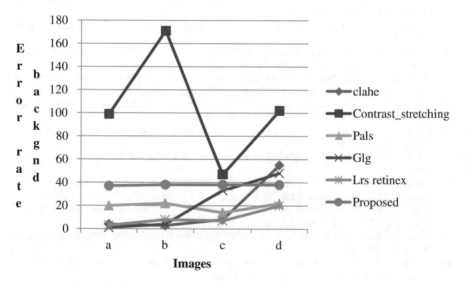

Fig. 10 Error rate background

5 Conclusion

In this work, we proposed an algorithm for adaptive enhancement of contrast. The global brightness is modified using a sine function, and fine details are enhanced using fuzzy logic. The algorithm shows great adaptability by enhancing low contrast, good contrast images with shadows, glare, and halo equally efficiently. The proposed method is compared with five other traditional methods. It was found that proposed method performs better which is validated with NIQE, AIC values. The contrast-enhanced images will be taken as input for action unit detection to predict and classify emotions.

References

1. Chen, Z.A.: Gray-level grouping (GLG): an automatic method for optimized image contrast enhancement-part I: the basic method. IEEE Trans. Image Process. **15**(8), 2290–2302 (2006)
2. Han, H.: A comparative study on illumination preprocessing in face recognition. Pattern Recogn. **46**(6), 1691–1699 (2013)
3. Banić, N.: Light random sprays Retinex: exploiting the noisy illumination estimation. IEEE Sign. Process. Lett. **20**(12), 1240–1243 (2013)
4. Tang, L.: Removing shadows from urban aerial images based on fuzzy Retinex. Dianzi Xuebao (Acta Electronica Sinica), pp. 500–503 (2005)
5. Gopikakumari, V.L.: Fuzzy rule based enhancement in the SMRT domain for low contrast images. Procedia Comput. Sci. **46**, 1747–1753 (2015)
6. Hanmandlu, M.: An optimal fuzzy system for color image enhancement. IEEE Trans. Image Process. **15**(10), 2956–2966 (2006)
7. Hanmandlu, M.D.: Color image enhancement by fuzzy intensification. Pattern Recogn. Lett. **24**(1), 81–87 (2003)
8. Lu, S.Z.: Neuro-fuzzy synergism to the intelligent system for edge detection and enhancement. Pattern Recogn. **36**(10), 2395–2409 (2003)
9. Nachtegael, M.: Classical and fuzzy approaches towards mathematical morphology. Fuzzy techniques in image processing. Physica-Verlag HD, pp. 3–57 (2000)
10. Zadeh, L.A.: Outline of a new approach to analysis of complex systems and decision processes. IEEE Trans. Syst. Man Cybern. **1**, I28–I44 (1973)
11. Pal, S.K.: Image enhancement using smoothing with fuzzy sets. IEEE Trans. Syst. Man Cyber. **11**(7), 494–500 (1981)
12. Pal, S.K.-M.: Fuzzy Mathematical Approach to Pattern Recognition. Halsted Press, New York (1986)
13. Hasikin, K.: Adaptive fuzzy intensity measure enhancement technique for non-uniform illumination and low-contrast images. Sign. Image Video Process. **9**(6), 1419–1442 (2015)
14. Raju, G.: A fast and efficient color image enhancement method based on fuzzy-logic and histogram. AEU-Int. J. Electron. Commun. **68**(3), 237–243 (2014)
15. Bhutani, K.R.: An application of fuzzy relations to image enhancement. Pattern Recogn. Lett. **16**(9), 901–909 (1995)
16. Young Sik, C., Krishnapuram, R.: A robust approach to image enhancement based on fuzzy logic. IEEE Trans. Image Process **6**, 808–825 (1997)
17. Choi, Y.: A fuzzy-rule-based image enhancement method for medical applications. In: Proceedings of the Eighth IEEE Symposium on Computer-Based Medical Systems. IEEE, New York (1995)
18. Lucey, P.: Painful data: the UNBC-McMaster shoulder pain expression archive database. In: IEEE International Conference on . IEEE, New York (2011)
19. Guo, P.: An adaptive enhancement algorithm for low-illumination image based on hue reserving. In: Cross Strait Quad-Regional Radio Science and Wireless Technology Conference (CSQRWC), vol. 2. IEEE, New York (2011)
20. Zheng, J.-Y.J.-L.: Color image enhancement based on RGB gray value scaling. Comput. Eng. **38**(2), 226–228 (2012)
21. Magudeeswaran, V.: Fuzzy logic-based histogram equalization for image contrast enhancement. Math. Probl. Eng. (2013)

22. Moorthy, A.K.: Blind image quality assessment: from natural scene statistics to perceptual quality. IEEE Trans. Image Process. **20**(12), 3350–3364 (2011)
23. Gonzalez, R.C.: Morphological image processing (2008)
24. Sheikh, H.R.: LIVE image quality assessment database release **2**, 2007 (17 July 2005). http://live.ece.utexas.Edu/research/quality
25. Yun, H.-J.: A novel enhancement algorithm combined with improved fuzzy set theory for low illumination images. Math. Prob. Eng. (2016)
26. Mittal, A.R.: Making a completely blind image quality analyzer. IEEE Signal Process. Lett. **20**(3), 209–212 (2013)

Multi-objective Linear Fractional Programming Problem with Fuzzy Parameters

Suvasis Nayak and Akshay Kumar Ojha

Abstract In this paper, a method is developed to derive the acceptable ranges of objective values for a multi-objective linear fractional programming problem (MOLFPP) with fuzzy parameters both in objectives and constraints. α- and β-cuts are respectively used in the objectives and constraints to specify the degrees of satisfaction and transform the fuzzy parameters into closed intervals. Using variable transformation and Taylor series expansion, the interval-valued fractional objectives are approximated by intervals of linear functions. The objective functions are assigned proper weights using analytic hierarchy process. Weighting sum method is used to transform the interval-valued multiple objectives into single objective. MOLFPP in interval-valued form is equivalently formulated as two linear problems which derive the acceptable ranges of objective values. Two numerical examples are illustrated to demonstrate the proposed method.

Keywords Multi-objective optimization · Linear fractional programming
Fuzzy parameters · Analytic hierarchy process · Weighting sum method

1 Introduction

Mathematical formulation of numerous real-world problems comprises multiple conflicted and inter-related objectives in rational form of physical and/or economical quantities which are optimized simultaneously being restricted to a common set of constraints. Such mathematically modeled problems are interpreted as multi-criteria fractional programming or ratio optimization. Some practical instances of fractional objectives are profit/cost, inventory/sale, output/employee, risk-assets/capital, debit/equity, etc. It has gained intensive research interest because of its wide range of

S. Nayak (✉)
School of Basic Sciences, Indian Institute of Technology, Bhubaneswar, India
e-mail: sn14@iitbbs.ac.in

A. K. Ojha
e-mail: akojha@iitbbs.ac.in

© Springer Nature Singapore Pte Ltd. 2019
J. C. Bansal et al. (eds.), *Soft Computing for Problem Solving*,
Advances in Intelligent Systems and Computing 816,
https://doi.org/10.1007/978-981-13-1592-3_6

application in numerous important fields like engineering, economics, information theory, finance, business, management science, transportation, water resources, corporate planning. Stancu-Minasian [12] discussed various methods, applications, and theoretical concepts on fractional programming. Charnes and Cooper [2] developed variable transformation technique to transform the fractional objective into linear form with an additional constraint. Mehra et al. [8] proposed a methodology to solve linear fractional programming problem(LFPP) with fuzzy coefficients. Chinnadurai and Muthukumar [3] solved LFPP in fuzzy environment. Liu [7] discussed geometric programming with fuzzy parameters. Bogdana and Milan [14] developed a method to solve LFPP with fuzzy coefficients in the objective function. Stancu-Minasian and Pop [13] modified the fuzzy approach of Dutta et al. [4] to obtain efficient solution of MOLFPP. Toksari [15] used Taylor series approach to solve MOLFPP. Chakraborty and Gupta [1] formulated MOLFPP into an equivalent linear form and used fuzzy approach to develop a method of solution. Mishra [10] implemented weighting sum method to derive a set of non-dominated solutions of a bi-level LFPP. In this paper, MOLFPP is studied with fuzzy parameters since many practical problems which are mathematically modeled, decision maker(DM) does not know the precise values of the required data. In such situations, fuzzy numbers are appropriate to be considered instead of fixed values.

The paper is organized as: Following introduction, Sect. 2 interprets some definitions of fuzzy theory and operations on fuzzy numbers. Section 3 describes analytic hierarchy process to determine proper weights and Sect. 4 explains variable transformation method to solve FPP. The mathematical formulation of the MOLFPP with fuzzy parameters and the proposed solution procedure are incorporated in Sect. 5 and Sect. 6, respectively. Two numerical examples are discussed in Sect. 7. Finally Sect. 8 contains the conclusion part.

2   Preliminaries

The concept of fuzzy set theory was first developed by Zadeh to mathematically model the imprecise data of the real-world problems. Some useful fuzzy concepts [6] implemented to develop the proposed work are discussed below.

Definition 1 If X is the universal set and $A \subset X$ then a fuzzy set $\tilde{A}$ in X is defined as $\tilde{A} = \{(x, \mu_{\tilde{A}}) | x \in X\}$.

Definition 2 $\mu_{\tilde{A}} : X \to [0, 1]$ is defined as the degree of membership or the membership function of x in A.

Definition 3 The α-cut of a fuzzy set A is defined as $\tilde{A}_{\alpha} = \{x | \mu_{\tilde{A}}(x) \geq \alpha\}$.

Definition 4 The α-cuts of the triangular and trapezoidal fuzzy numbers $\tilde{a} = (a_1, a_2, a_3)$ and $\tilde{b} = (b_1, b_2, b_3, b_4)$ are defined as $\tilde{a}_{\alpha} = [a_1 + \alpha(a_2 - a_1), a_3 - \alpha(a_3 - a_2)]$ and $\tilde{b}_{\alpha} = [b_1 + \alpha(b_2 - b_1), b_4 - \alpha(b_4 - b_3)]$, respectively.

Since a fuzzy number can be equivalently expressed as a closed interval by using α-cuts, the arithmetic operations on the fuzzy numbers $\tilde{a}$ can be achieved through their corresponding intervals $\tilde{a}_\alpha = [a^L, a^U]$. Using interval analysis [11], the following operations are defined.

- $\tilde{a} + \tilde{b} = [a^L + b^L, a^U + b^U]$
- $c\tilde{a} = \begin{cases} [ca^L, ca^U], \text{ if } c > 0 \\ [ca^U, ca^L], \text{ if } c < 0 \end{cases}$, for $c \in \mathbb{R}$
- $\tilde{a}\tilde{b} = [\min S_1, \max S_1]$
  where $S_1 = \{a^L b^L, a^L b^U, a^U b^L, a^U b^U\}$
- $\frac{\tilde{a}}{\tilde{b}} = [\min S_2, \max S_2]$ where $S_2 = \{\frac{a^L}{b^L}, \frac{a^L}{b^U}, \frac{a^U}{b^L}, \frac{a^U}{b^U}\}$
- $\tilde{a} \preceq \tilde{b}$ iff $a^L \leq b^L$ and $a^U \leq b^U$.

3 Analytic Hierarchy Process (AHP)

AHP [5, 10] is originally developed by Saaty to determine the proper weights for the objectives of multiple criteria decision-making problems. The importance of the objectives by the decision maker(DM) is reflected in the entries of the pairwise comparison matrix. The weights are ascertained for a n-objective optimization problem using the following process.

- Construct the pairwise comparison matrix,
  $C = (c_{ij})$ where $i, j = n$
- Construct the normalized pairwise comparison matrix,
  $N = (n_{ij})$ where $n_{ij} = (c_{ij})/(\sum_{i=1}^n c_{ij})$
- The weights are obtained as,
  $w_i = (\sum_{j=1}^n n_{ij})/(n), i = 1, 2, \ldots, n$

After getting the weighting vector(w_i), check the consistency of the method [5] by deriving the consistency ratio(CR) = CI/RI where CI and RI represent the consistency and random index, respectively. If CR $\leq$ 0.1, then the process is consistent otherwise redefine the pairwise comparison matrix.

4 Variable Transformation Method (VTM)

VTM was developed by Charnes and Cooper [2] to derive the optimal solution of single objective linear fractional programming problem by introducing an extra variable. Consider the following linear fractional programming problem (LFPP), (P_1) and linear programming problem (LPP), (P_2).

$$(P_1) : \max f(x) = \frac{cx + q}{dx + r}$$
$$\text{subject to} \tag{1}$$
$$S_1 = \{Ax \leq b, x \geq 0\}$$

$$(P_2) : \max \ g(y, z) = cy + qz$$
$$\text{subject to} \tag{2}$$
$$S_2 = \{dy + rz = 1, Ay - bz \leq 0, \quad y \geq 0, z > 0\}$$

where the transformations $z = \frac{1}{dx+r}$ and $y = xz$ derive (P_2) from (P_1).

Lemma 1 ([12]) *For any feasible solution* $(y, z) \in S_2$, *z is positive.*

Theorem 1 ([12]) *If* (y^*, z^*) *is an optimal solution of* (P_2) *then* $x^* = \frac{y^*}{z^*}$ *is the optimal solution of* (P_1).

5 Problem Formulation

In context of decision-making problems, DM does not always know the exact numeric values of the available data which causes the use of fuzzy numbers to fit the model with the practical problem. Multi-objective LFPP with fuzzy parameters can be mathematically formulated as follows.

$$\max \ f(x) = (f_1(x), f_2(x), \ldots, f_k(x))$$
$$\text{subject to} \tag{3}$$
$$S = \{x \in \mathbb{R}^n | Ax \leq b, x \geq 0\}$$

where $f_i(x) = \dfrac{\tilde{c}_i x + \tilde{q}_i}{\tilde{d}_i x + \tilde{r}_i}$, $i = 1, 2, \ldots, n$; $A = (\tilde{a}_{tj})_{m \times n}$ and $b = (\tilde{b}_t)_{m \times 1}$, $t = 1, 2, \ldots, m$.

Assume that, $\tilde{c}_i, \tilde{d}_i, \tilde{q}_i, \tilde{r}_i, \tilde{a}_{tj}, \tilde{b}_t$ are positive and triangular fuzzy numbers defined over the universal set of the real numbers $\mathbb{R}$ as:

$$\tilde{c}_i = (c_i^1, c_i^2, c_i^3), \tilde{d}_i = (d_i^1, d_i^2, d_i^3), \tilde{q}_i = (q_i^1, q_i^2, q_i^3), \tilde{r}_i = (r_i^1, r_i^2, r_i^3),$$
$$\tilde{a}_{tj} = (a_{tj}^1, a_{tj}^2, a_{tj}^3), \tilde{b}_t = (b_t^1, b_t^2, b_t^3).$$

6 Solution Procedure

The fuzzy parameters of the objectives and constraints can be transformed into intervals using α- and β-cuts, respectively.

Objectives:

$$
\begin{aligned}
f_i(x) &= \frac{[c_i^1 + \alpha(c_i^2 - c_i^1), c_i^3 - \alpha(c_i^3 - c_i^2)]x + [q_i^1 + \alpha(q_i^2 - q_i^1), q_i^3 - \alpha(q_i^3 - q_i^2)]}{[d_i^1 + \alpha(d_i^2 - d_i^1), d_i^3 - \alpha(d_i^3 - d_i^2)]x + [r_i^1 + \alpha(r_i^2 - r_i^1), r_i^3 - \alpha(r_i^3 - r_j^2)]} \\
&= \frac{[(c_i^1 + \alpha(c_i^2 - c_i^1))x + q_i^1 + \alpha(q_i^2 - q_i^1), (c_i^3 - \alpha(c_i^3 - c_i^2))x + q_i^3 - \alpha(q_i^3 - q_i^2)]}{[(d_i^1 + \alpha(d_i^2 - d_i^1))x + r_i^1 + \alpha(r_i^2 - r_i^1), (d_i^3 - \alpha(d_i^3 - d_i^2))x + r_i^3 - \alpha(r_i^3 - r_i^2)]} \\
&= [\frac{(c_i^1 + \alpha(c_i^2 - c_i^1))x + q_i^1 + \alpha(q_i^2 - q_i^1)}{(d_i^3 - \alpha(d_i^3 - d_i^2))x + r_i^3 - \alpha(r_i^3 - r_i^2)}, \frac{(c_i^3 - \alpha(c_i^3 - c_i^2))x + q_i^3 - \alpha(q_i^3 - q_i^2)}{(d_i^1 + \alpha(d_i^2 - d_i^1))x + r_i^1 + \alpha(r_i^2 - r_i^1)}] \\
&= [f_i^L(x), f_i^U(x)], \quad i = 1, 2, \ldots, k
\end{aligned}
$$

Constraints:

$$
\sum_{j=1}^{n} \tilde{a}_{tj} x_j \le \tilde{b}_t, \quad t = 1, 2, \ldots, m
$$

$$
\sum_{j=1}^{n} [a_{tj}^1 + \beta(a_{tj}^2 - a_{tj}^1), a_{tj}^3 - \beta(a_{tj}^3 - a_{tj}^2)]x_j
$$
$$
\le [b_t^1 + \beta(b_t^2 - b_t^1), b_t^3 - \beta(b_t^3 - b_t^2)]
$$

$$
\sum_{j=1}^{n} [a_{tj}^L, a_{tj}^U]x_j \le [b_t^L, b_t^U], t = 1, 2, \ldots, m
$$

This can be splitted into the following inequalities.

$$
\sum_{j=1}^{n} (a_{tj}^1 + \beta(a_{tj}^2 - a_{tj}^1))x_j \le b_t^1 + \beta(b_t^2 - b_t^1), \quad t = 1, 2, \ldots, m
$$

$$
\sum_{j=1}^{n} (a_{tj}^3 - \beta(a_{tj}^3 - a_{tj}^2))x_j \le b_t^3 - \beta(b_t^3 - b_t^2), \quad t = 1, 2, \ldots, m
$$

The MOLFPP (3) can be stated as:

$$
\begin{aligned}
\max \quad & f_i(x) = [f_i^L(x), f_i^U(x)], \quad i = 1, 2, \ldots, k \\
& \text{subject to} \\
& \sum_{j=1}^{n} a_{tj}^L x_j \le b_t^L, t = 1, 2, \ldots, m \\
& \sum_{j=1}^{n} a_{tj}^U x_j \le b_t^U, t = 1, 2, \ldots, m \\
& x \ge 0
\end{aligned}
\tag{4}
$$

The fractional objectives of (4) can be approximated by linear functions using first order Taylor series expansion of the objectives about their individual optimal solutions(IOS). VTM as defined in Sect. 4 is used to derive the IOS of $f_i^L(x)$ and $f_i^U(x)$ for $i = 1, 2, \ldots, k$. Particularly using VTM, the mathematical formulation for the lower bound of the ith objective, i.e., $f_i^L(x)$ can be stated as:

$$\max \quad (c_i^1 + \alpha(c_i^2 - c_i^1))y + (q_i^1 + \alpha(q_i^2 - q_i^1))z$$
$$\text{subject to}$$
$$(d_i^3 - \alpha(d_i^3 - d_i^2))y + (r_i^3 - \alpha(r_i^3 - r_i^2))z = 1$$
$$\sum_{j=1}^{n}(a_{tj}^1 + \beta(a_{tj}^2 - a_{tj}^1))y_j \le (b_t^1 + \beta(b_t^2 - b_t^1))z, \ t = 1, 2, \dots, m \quad (5)$$
$$\sum_{j=1}^{n}(a_{tj}^3 - \beta(a_{tj}^3 - a_{tj}^2))y_j \le (b_t^3 - \beta(b_t^3 - b_t^2))z, \ t = 1, 2, \dots, m$$
$$y_j \ge 0, z > 0$$

where $y_j = x_j z, j = 1, 2, \dots, n$.

Solving (5), the IOS of $f_i^L(x)$ can be obtained as $x = (x_j = \frac{y_j}{z}, j = 1, 2, \dots, n)$. Let, $X_i^{L*} = (X_{i1}^{L*}, X_{i2}^{L*}, \dots, X_{in}^{L*})$ and $X_i^{U*} = (X_{i1}^{U*}, X_{i2}^{U*}, \dots, X_{in}^{U*}), (i = 1, 2, \dots, k)$ be the IOS of $f_i^L(x)$ and $f_i^U(x)$ respectively. Using Taylor series expansion, $f_i^L(x)$ and $f_i^U(x)$ can be approximated as:

$$f_i^L(x) \cong \bar{f}_i^L(x) = f_i^L(X_i^{L*}) + \sum_{k=1}^{n}(x_k - X_{ik}^{L*})\frac{\partial f_i^L(X_i^{L*})}{\partial x_k}, i = 1, 2, \dots, n$$

$$f_i^U(x) \cong \bar{f}_i^U(x) = f_i^U(X_i^{U*}) + \sum_{k=1}^{n}(x_k - X_{ik}^{U*})\frac{\partial f_i^U(X_i^{U*})}{\partial x_k}, i = 1, 2, \dots, n$$

The objectives of (4) can be reformulated as: $\max f_i(x) = [\bar{f}_i^L(x), \bar{f}_i^U(x)]$, $i = 1, 2, \dots, k$

Using weighting sum method [9, 10], the MOLFPP (4) can be stated as:

$$\max \ f(x) = \left[\sum_{i=1}^{n} w_i \bar{f}_i^L(x), \sum_{i=1}^{n} w_i \bar{f}_i^U(x) \right]$$
$$\text{subject to}$$
$$\sum_{j=1}^{n} a_{tj}^L x_j \le b_t^L, t = 1, 2, \dots, m \quad (6)$$
$$\sum_{j=1}^{n} a_{tj}^U x_j \le b_t^U, t = 1, 2, \dots, m$$
$$x \ge 0$$

where w_i represent the weights assigned to the objectives $f_i(x)$ in order of their importance and are determined by the decision maker(DM) using AHP as defined in Sect. 3. Problem (6) can be equivalently stated as two LPP:

$$\textbf{(LFP)} : \max \ f^L(x) = \sum_{i=1}^{n} w_i \bar{f}_i^L(x)$$

subject to

$$\sum_{j=1}^{n} a_{tj}^L x_j \leq b_t^L, t = 1, 2, \ldots, m \tag{7}$$

$$\sum_{j=1}^{n} a_{tj}^U x_j \leq b_t^U, t = 1, 2, \ldots, m$$

$$x \geq 0$$

$$\textbf{(UFP)} : \max \ f^U(x) = \sum_{i=1}^{n} w_i \bar{f}_i^U(x)$$

subject to

$$\sum_{j=1}^{n} a_{tj}^L x_j \leq b_t^L, t = 1, 2, \ldots, m \tag{8}$$

$$\sum_{j=1}^{n} a_{tj}^U x_j \leq b_t^U, t = 1, 2, \ldots, m$$

$$x \geq 0$$

Let x^{L*} and x^{U*} be the solutions of (7) and (8) respectively. The solution x^{U*} can be considered as the α-β acceptable solution of the MOLFPP (3) by the DM since it provides larger intervals of objective values as compared to the solution x^{L*}. Thus, it generates the acceptable ranges of objective values.

7 Numerical Examples

To illustrate the proposed solution procedure, the following MOLFPP with triangular fuzzy parameters are solved.

7.1 Example 1

max
$$f_1(x) = \frac{(7, 9, 11)x_1 + (12, 13, 15)x_2 + (6, 7, 9)x_3 + (4, 5, 6)}{(1, 3, 5)x_1 + (1, 2, 5)x_2 + (3, 5, 7)x_3 + (1, 2, 3)}$$
$$f_2(x) = \frac{(14, 15, 16)x_1 + (10, 12, 14)x_2 + (7, 8, 9)x_3 + (3, 4, 5)}{(11, 12, 13)x_1 + (7, 9, 11)x_2 + (6, 7, 8)x_3 + (3, 5, 7)}$$

subject to
$$(1, 2, 3)x_1 + (2, 3, 5)x_2 + (1, 1, 4)x_3 \preceq (5, 7, 9)$$
$$x = (x_1, x_2, x_3) \geq 0$$

Specifying the degrees of satisfaction $\alpha = \beta = 0.5$, the problem can be formulated as:

max
$$f_1(x) = \left[\frac{8x_1 + 12.5x_2 + 6.5x_3 + 4.5}{4x_1 + 3.5x_2 + 6x_3 + 2.5}, \frac{10x_1 + 14x_2 + 8x_3 + 5.5}{2x_1 + 1.5x_2 + 4x_3 + 1.5} \right]$$
$$f_2(x) = \left[\frac{14.5x_1 + 11x_2 + 7.5x_3 + 3.5}{12.5x_1 + 10x_2 + 7.5x_3 + 6}, \frac{15.5x_1 + 13x_2 + 8.5x_3 + 4.5}{11.5x_1 + 8x_2 + 6.5x_3 + 4} \right]$$

subject to
$$1.5x_1 + 2.5x_2 + x_3 \le 6$$
$$2.5x_1 + 4x_2 + 2.5x_3 \le 8$$
$$x_1, x_2, x_3 \ge 0$$

Using VTM and Taylor series expansion, the lower and upper bounds of the interval-valued fractional objectives are approximated as:
$$f_1^L(x) \approx \tilde{f}_1^L(x) = -0.4654x_1 + 0.1717x_2 - 1.2770x_3 + 2.7619$$
$$f_1^U(x) \approx \tilde{f}_1^U(x) = -1.0864x_1 + 0.6296x_2 - 4.8395x_3 + 6.1852$$
$$f_2^L(x) \approx \tilde{f}_2^L(x) = 0.0204x_1 + 0.0033x_2 - 0.0138x_3 + 1.0195$$
$$f_2^U(x) \approx \tilde{f}_2^U(x) = -0.1019x_1 + 0.0400x_2 - 0.0706x_3 + 1.4450$$

The pairwise comparison matrix for the objectives $f_1(x)$ and $f_2(x)$ is supposed to be constructed by the DM as:

$$C = \begin{bmatrix} 1 & 3 \\ 1/3 & 1 \end{bmatrix}$$

Using AHP, the weights are determined as: $w_1 = 0.75$ and $w_2 = 0.25$. Using weighting method, the MOLFPP is transformed into the following problems.

LFP: max $f^L(x) = 0.75\tilde{f}_1^L(x) + 0.25\tilde{f}_2^L(x)$

subject to
$$1.5x_1 + 2.5x_2 + x_3 \le 6$$
$$2.5x_1 + 4x_2 + 2.5x_3 \le 8$$
$$x_1, x_2, x_3 \ge 0$$

UFP: max $f^U(x) = 0.75\tilde{f}_1^U(x) + 0.25\tilde{f}_2^U(x)$

subject to
$$1.5x_1 + 2.5x_2 + x_3 \le 6$$
$$2.5x_1 + 4x_2 + 2.5x_3 \le 8$$
$$x_1, x_2, x_3 \ge 0$$

The ranges of the optimal objective values at the solution of the UFP are obtained as $f_1(x) = [3.1053, 7.4444], f_2(x) = [0.9808, 1.5250]$ for $\alpha = \beta = 0.5$ and the weights $w_1 = 0.75, w_2 = 0.25$. At various degrees of satisfaction $\alpha, \beta \in [0.5, 1]$, the ranges of acceptable objective values are listed in Table 1.

Table 1 The ranges of the acceptable objective values

$\alpha = \beta$	$f_1(x) = \bar{f}_1(x)$	$f_2(x)$	$\bar{f}_2(x)$
0.5	[3.1053, 7.4444]	[0.9808, 1.5250]	[1.0261, 1.5250]
0.6	[3.3967, 7.5211]	[1.0260, 1.4585]	[1.0617, 1.4585]
0.7	[3.7417, 6.4307]	[1.0735, 1.3961]	[1.1325, 1.3961]
0.8	[4.1557, 6.0103]	[1.1233, 1.3375]	[1.1367, 1.3375]
0.9	[4.6632, 5.6365]	[1.1757, 1.2825]	[1.1767, 1.2825]

7.2 Example 2

max

$$f_1(x) = \frac{(4, 7, 11)x_1 + (6, 9, 15)x_2 + (3, 5, 9)x_3 + (2, 5, 7)}{(2, 5, 6)x_1 + (4, 5, 7)x_2 + (3, 6, 8)x_3 + (2, 5, 8)}$$

$$f_2(x) = \frac{(6, 9, 11)x_1 + (5, 7, 9)x_2 + (8, 10, 12)x_3 + (3, 4, 5)}{(5, 8, 12)x_1 + (3, 5, 7)x_2 + (8, 12, 14)x_3 + (4, 5, 8)}$$

$$f_3(x) = \frac{(3, 5, 12)x_1 + (5, 7, 8)x_2 + (8, 11, 15)x_3 + (2, 5, 6)}{(1, 3, 4)x_1 + (2, 8, 14)x_2 + (3, 9, 12)x_3 + (1, 3, 5)}$$

subject to

$$(1, 3, 5)x_1 + (2, 5, 7)x_2 + (3, 5, 6)x_3 \preceq (6, 8, 12)$$
$$x = (x_1, x_2, x_3) \geq 0$$

Specifying the degrees of satisfaction $\alpha = \beta = 0.5$, the problem can be formulated as:

max

$$f_1(x) = \left[\frac{5.5x_1 + 7.5x_2 + 4x_3 + 3.5}{5.5x_1 + 6x_2 + 7x_3 + 6.5}, \frac{9x_1 + 12x_2 + 7x_3 + 6}{3.5x_1 + 4.5x_2 + 4.5x_3 + 3.5} \right]$$

$$f_2(x) = \left[\frac{7.5x_1 + 6x_2 + 9x_3 + 3.5}{10x_1 + 6x_2 + 13x_3 + 6.5}, \frac{10x_1 + 8x_2 + 11x_3 + 4.5}{6.5x_1 + 4x_2 + 10x_3 + 4.5} \right]$$

$$f_3(x) = \left[\frac{4x_1 + 6x_2 + 9.5x_3 + 3.5}{3.5x_1 + 11x_2 + 10.5x_3 + 4}, \frac{8.5x_1 + 7.5x_2 + 13x_3 + 5.5}{2x_1 + 5x_2 + 6x_3 + 2} \right]$$

subject to

$$2x_1 + 3.5x_2 + 4x_3 \leq 7$$
$$4x_1 + 6x_2 + 5.5x_3 \leq 10$$
$$x_1, x_2, x_3 \geq 0$$

Using VTM and Taylor series expansion, the lower and upper bounds of the interval-valued fractional objectives are approximated as:

$$f_1^L(x) \approx \bar{f}_1^L(x) = 0.0101x_1 + 0.1019x_2 - 0.1690x_3 + 0.7999$$
$$f_1^U(x) \approx \bar{f}_1^U(x) = 0.0661x_1 + 0.1240x_2 - 0.3306x_3 + 2.1569$$
$$f_2^L(x) \approx \bar{f}_2^L(x) = -0.0413x_1 + 0.0661x_2 - 0.0992x_3 + 0.7080$$
$$f_2^U(x) \approx \bar{f}_2^U(x) = -0.0341x_1 + 0.1443x_2 - 0.4451x_3 + 1.3565$$

$f_3^L(x) \approx \bar{f}_3^L(x) = 0.0231x_1 - 0.4429x_2 - 0.1269x_3 + 1.0011$
$f_3^U(x) \approx \bar{f}_3^U(x) = 0.1224x_1 - 1.6582x_2 - 1.4184x_3 + 3.5154$

The pairwise comparison matrix for the objectives $f_1(x)$, $f_2(x)$, and $f_3(x)$ is supposed to be constructed by the DM as:

$$C = \begin{bmatrix} 1 & 3 & 5 \\ 1/3 & 1 & 3 \\ 1/5 & 1/3 & 1 \end{bmatrix}$$

Using AHP, the weights are determined as: $w_1 = 0.6333$, $w_2 = 0.2605$ and $w_3 = 0.1062$. Using weighting method, the MOLFPP is transformed into the following problems.

LFP:
max $f^L(x) = 0.6333\bar{f}_1^L(x) + 0.2605\bar{f}_2^L(x) + 0.1062\bar{f}_3^L(x)$

subject to

$2x_1 + 3.5x_2 + 4x_3 \le 7$
$4x_1 + 6x_2 + 5.5x_3 \le 10$
$x_1, x_2, x_3 \ge 0$

UFP:
max $f^U(x) = 0.6333\bar{f}_1^U(x) + 0.2605\bar{f}_2^U(x) + 0.1062\bar{f}_3^U(x)$

subject to

$2x_1 + 3.5x_2 + 4x_3 \le 7$
$4x_1 + 6x_2 + 5.5x_3 \le 10$
$x_1, x_2, x_3 \ge 0$

The ranges of the optimal objective values at the solution of the UFP are obtained as $f_1(x) = [0.8519, 2.3265]$, $f_2(x) = [0.7063, 1.4217]$, $f_3(x) = [1.0588, 3.8214]$ for $\alpha=\beta=0.5$ and the weights $w_1 = 0.6333$, $w_2 = 0.2605$ and $w_3 = 0.1062$. At various degrees of satisfaction $\alpha, \beta \in [0.5, 1]$, the ranges of acceptable objective values are tabulated below.

Tables 2 and 3 represent the ranges of acceptable objective values of the fractional objectives and the approximated linear objectives respectively.

Remark 1 The same values of the degrees of satisfaction for the objectives and constraints, i.e., $\alpha = \beta$ are considered in the above examples for the sake of convenience but their different values $\alpha \ne \beta$ can also be used to derive different ranges of the acceptable objective values.

Table 2 The ranges of the acceptable objective values

$\alpha = \beta$	$f_1(x)$	$f_2(x)$	$f_3(x)$
0.5	[0.8519, 2.3265]	[0.7063, 1.4217]	[1.0588, 3.8214]
0.6	[0.9300, 2.0542]	[0.7653, 1.3388]	[1.1631, 3.2361]
0.7	[1.0126, 1.8216]	[0.8296, 1.2622]	[1.2752, 2.7474]
0.8	[1.2562, 1.8018]	[1.0108, 1.3223]	[0.8240, 1.2377]
0.9	[1.3689, 1.6415]	[1.0865, 1.2432]	[0.9176, 1.1230]

Table 3 The ranges of the acceptable approximated objective values

$\alpha = \beta$	$\bar{f}_1(x)$	$\bar{f}_2(x)$	$\bar{f}_3(x)$
0.5	[0.8252, 2.3221]	[0.6047, 1.2713]	[1.0589, 3.8214]
0.6	[0.8972, 2.0355]	[0.6614, 1.1973]	[1.1631, 3.2361]
0.7	[0.9721, 1.7860]	[0.7238, 1.1294]	[1.2752, 2.7474]
0.8	[1.2562, 1.8018]	[1.0108, 1.3223]	[0.4985, 0.7268]
0.9	[1.3689, 1.6415]	[1.0865, 1.2432]	[0.6090, 0.7358]

8 Conclusions

This paper studies MOLFPP with fuzzy parameters both in the objectives and constraints. The acceptable ranges of objective values can be determined using the proposed solution approach. The method is described using triangular fuzzy numbers but linear/ trapezoidal fuzzy numbers and interval parameters can also be considered. The use of different values of α and β can derive a set of alternatives for the DM to choose the ranges of objective values. Numerical examples are discussed to illustrate the solution approach. Multi-level MOLFPP with fuzzy parameters is a further scope of research in this field.

Acknowledgements Authors are thankful to the Editors and anonymous referees for their valuable comments and suggestions to improve the quality of the paper.

References

1. Chakraborty, M., Gupta, S.: Fuzzy mathematical programming for multi objective linear fractional programming problem. Fuzzy sets and syst. **125**, 335–342 (2002)
2. Charnes, A., Cooper, W.W.: Management models and industrial applications of linear programming. Mang. Sci. **4**, 38–91 (1957)
3. Chinnadurai, V., Muthukumar, S.: Solving the linear fractional programming problem in a fuzzy environment: numerical approach. Appl. Math. Model. **40**, 6148–6164 (2016)
4. Dutta, D., Tiwari, R.N., Rao, J.R.: Multiple objective linear fractional programminga fuzzy set theoretic approach. Fuzzy Sets and Syst. **52**, 39–45 (1992)

5. Golden, B.L., Wasil, E.A., Harker, P.T.: The Analytic Hierarchy Process. Springier, New York (1989)
6. Zimmermann, H.-J.: Fuzzy Set Theory and Its Applications. Kluwer Academic Publishers, Dordrecht (1985)
7. Liu, S.-T.: Geometric programming with fuzzy parameters in engineering optimization. Int. J. Approx. Reason. **46**, 484–498 (2007)
8. Mehra, A., Chandra, S., Bector, C.R.: Acceptable optimality in linear fractional programming with fuzzy coefficients. Fuzzy Optim. Decis. Making **6**, 5–16 (2007)
9. Miettinen, K.: Nonlinear Multiobjective Optimization. Springer Science & Business Media, Berlin (2012)
10. Mishra, S.: Weighting method for bi-level linear fractional programming problems. Eur. J. Oper. Res. **183**, 296–302 (2007)
11. Moore, R.E.: Interval Analysis. Prince-Hall, Englewood Cliffs, NJ (1966)
12. Stancu-Minasian, I.M.: Fractional Programming: Theory, Methods and Applications. Kluwer Academic Publishers, Dordrecht (1997)
13. Stancu-Minasian, I.M., Pop, B.: On a fuzzy set approach to solving multiple objective linear fractional programming problem. Fuzzy Sets Syst. **134**, 397–405 (2003)
14. Stanojević, B., Stanojević, M.: Solving method for linear fractional optimization problem with fuzzy coefficients in the objective function. Int. J. Comp. Commun. Control. **8**, 146–152 (2012)
15. Toksari, M.D.: Taylor series approach to fuzzy multiobjective linear fractional programming. Inf. Sci. **178**, 1189–1204 (2008)

Hindi Speech Synthesis Using Paralinguistic Content Expression

T. V. Prasad

Abstract A long-standing problem of monotonicity in naturalness has been solved using a well-founded model, namely the Speech Hierarchy Model. This model is based on the fact that all natural speech signals have infinite variations. For example, red light is present in an infinite number of frequencies in nature, whereas a computer has only a few numbers within a finite range to create red color. Paralinguistic content, which is a part of a speech signal, also varies infinitely. Using the concept of paralinguistic content expression, which can be used to express any form of variation onto a speech signal, the present methods of synthesizing speech are enhanced and will lead to technology which is more natural in the human sense. This paper implements the method and results in a tool for synthesizing Hindi speech which gave high intelligibility in 81% of input text samples.

Keywords Speech synthesis · Variability · Paralinguistic content · Paralinguistic content expression

1 Introduction

Speech synthesis consists of converting text-to-speech by a human being or a machine. Various tools have been developed over the years for machine-based speech synthesis, and these include Android eSpeak and Microsoft Speech Application Programming Interface (MS SAPI). eSpeak is a tool which was developed by applying Klatt's formant synthesizing method [1].

Speech synthesis may be categorized as restricted (for synthesizing messages) and unrestricted (text-to-speech) syntheses. The former is suitable for announcing menus or messages and as part of information systems. The latter is needed, for example, in applications (such as ATMs) for the visually impaired. The text-to-speech procedure

T. V. Prasad (✉)
Godavari Institute of Engineering and Technology, Rajahmundry 533296,
Andhra Pradesh, India
e-mail: tvprasad2002@yahoo.com

© Springer Nature Singapore Pte Ltd. 2019
J. C. Bansal et al. (eds.), *Soft Computing for Problem Solving*,
Advances in Intelligent Systems and Computing 816,
https://doi.org/10.1007/978-981-13-1592-3_7

consists of two main phases, usually called high-level and low-level syntheses. In high-level synthesis, the input text is converted into such form that the low-level synthesizer can create the output speech signal. The three basic methods for low-level synthesis are the formant, concatenative, and articulatory synthesis [2].

Naturalness consists of expressing gender variations, speaker variations, and other forms of paralinguistic content variations. Deep learning is a recent technology which is based on neural networks and is used to synthesize speech [3].

Section 1 introduces the paper. Section 2 including Sects. 2.1, 2.2, 2.3 and 2.4 describes the methodology. Sections 3 and 4 present the results and discussion, respectively. Section 5 concludes the paper.

2 Methodology

Speech synthesis is the process of converting written text into speech by a machine. Speech synthesis is fairly advanced as of today and very accurate. However, what is lacking in speech synthesis is the naturalness usually attributed to human speech.

Human speech never repeats. It has infinite variation. Based on this observation, this research work consisted of mimicking natural human speech by introducing variations in speech the way it is observed in human speech. A major task was to overcome the problem of naturalness by removing robot-like monotonicity in speech synthesis. The present research successfully overcomes this monotonicity from the viewpoint of introducing variations and presents a tool which produces non-repeating speech. This is based on the Speech Hierarchy Model [6]. The Speech Hierarchy Model was implemented as the Speech Hierarchy-based Soft Computing System (SpHiSoCS).

2.1 Paralinguistic Content Expression

Constant variations are the truth of nature, and since speech signals are natural signals, they portray infinite variations. However, the human mind can fathom and handle these variations easily. Variations also lead to uniqueness in nature, and the human mind can recognize each individual variation very precisely. The mind does not forget a person's face and can recognize it even after 20–30 years. Not just this, the mind can recognize patterns and classify the variations very accurately.

Speech synthesis is the reverse process of speech-to-text conversion. The problem of speech synthesis becomes the problem of introducing variability into speech [4] while speaking a set of words. Variability exists in the form of paralinguistic content in natural speech. Therefore, expressing variability in the form of different paralinguistic contents in each utterance will introduce greater "naturalness" into synthesized speech. This process is termed as paralinguistic content expression (PCX) and is the reverse of paralinguistic content elimination (PCE).

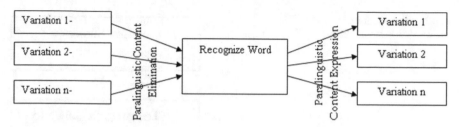

Fig. 1 Overall process of variability elimination by eliminating paralinguistic content to obtain the underlying word, followed by synthesizing speech variations of the underlying word, using paralinguistic content expression

Thus, characterizing patterns in a set of variations of speech signals consists of eliminating paralinguistic content and identifying the underlying words as a pattern. This is how Speech Recognition (SR) works. On the other hand, creating variations from a set of words involves PCX, which results in speech synthesis, as depicted in Fig. 1.

A block diagram of the process of speech synthesis using PCX is presented in Fig. 2.

The applicability of fuzzy elements to the problem of speech processing was found to be suitable in experiments conducted as part of this research work since they mimic the way human beings carry out speech processing functions. The underlying concept of paralinguistic content expression is that each variation of each phoneme is representable by a membership value of the phoneme's spoken version in a fuzzy set mapped to that phoneme as discussed [4].

2.2 A Brief Description of the Method of Expressing Infinite Variability

Figure 3 depicts the three variations of the Hindi sentence "MAIN THEEK HOON" (which means "I am fine" in English) as uttered by three different speakers. This represents the variations in speech spoken by human beings.

The method implemented as part of this research work consists of synthesizing speech signals by concatenating phonemes of a speaker and then expressing paralinguistic content in the form of speaker variations, that is, the variations of the same spoken text created by the same speaker's voice or voices of different speakers. Thus, paralinguistic content is expressed to synthesize more natural speech.

Table 1 shows the stages in this research work for speech synthesis.

Fig. 2 Block diagram of
speech synthesis with PCX

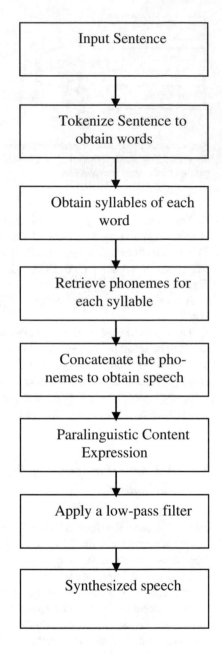

The generalized algorithm for speech synthesis using PCX is presented in Algorithm 1. This version of the algorithm is applicable for synthesizing speech by including other forms of speech variability which exist in human speech, such as variability in emotions.

(a)

(b)

(c)

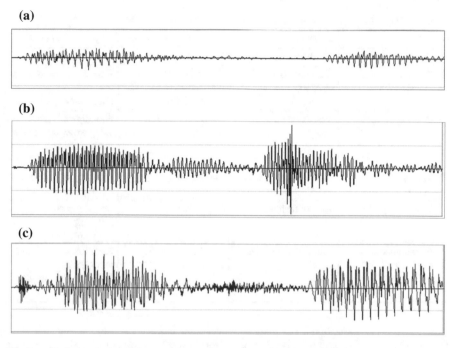

Fig. 3 **a** Variation 1 of the Hindi sentence "MAIN THEEK HOON" from Speaker 1. **b** Variation 2 of the Hindi sentence "MAIN THEEK HOON" from Speaker 2. **c** Variation 3 of the Hindi sentence "MAIN THEEK HOON" from Speaker 3

Algorithm 1 Generalized algorithm for Speech Synthesis using Paralinguistic Content Expression.

1. Input sentence ST in the form of a set of words W_i, where $i = 1, \ldots, I$, and I is the number of words in ST.
2. Retrieve the phonetized form H_i of each word W_i from the dictionary. Each H_i is a set of phonemes P_{ij}, where $j = 1, \ldots, K$, and K is the number of phonemes in H_i.
3. For each phoneme P_{ij} retrieve the spoken version V_{ij} for the speaker r.
4. For an unvoiced phoneme, filter and remove the last half of the speech segment of the phoneme, For a voiced phoneme, filter and remove the first and the last one-thirds of the speech segment of the phoneme. This is required since the energy of a voiced phoneme is concentrated in the middle portion of the phoneme signal. Create a shaping curve and introduce variability into the shaping curve by generating a different shaping curve each time randomly and superimposing each shaping curve with the voiced phoneme speech sample V_{ij}.
5. Concatenate the speech segments V_{ij} of all the phonemes into a single speech signal VZ.

Table 1 Stages in the process of speech synthesis and approaches applied at each stage

Stage	Output(s)	Problem(s) handled	Approach(es) applied
1. Preprocessing	Words in a sentence	Word extraction	Tokenization
2. Word-to-phoneme conversion	Phoneme string of sentence	Phoneme generation with emotion-specific pronunciation	Pronunciation rule bases Dictionary
3. Prosody analysis	Annotated phoneme string	Segment duration Intonation Energy	Phoneme marking using fuzzy degrees of membership PSOLA, TDSOLA
4. Speech nit recognition	Annotated speech unit recognition string	Intelligibility	Mapping of phonemes to stored speech units
5. Acoustic synthesis	Spoken response	Naturalness	Waveform shaping Filtering Pitch correction
6. Paralinguistic content expression	Variations of spoken response	Speaker variability expression	Amplitude variability expression Emphasis variability expression

6. Modify the pitch and the phoneme duration of the phonemes in the speech signal VZ using the Pitch Synchronous OverLap and Add (PSOLA) and Time Domain Synchronous OverLap and Add (TDSOLA) algorithms.
7. Post-processing steps for restoring naturalness include

   a. Filter VZ using a lowpass FIR filter to suppress low frequencies,
   b. Filter the result of step 8(a) using a bandpass IIR Chebyshev filter to enhance voice frequencies,
   c. Smooth the result of step 8(b) to remove harshness to obtain the final output signal VZ.

8. Output the synthesized speech signal VZ.

Current Speech Synthesis Tools have a high degree of accuracy but lack in the synthesis of speech which exhibits naturalness. The speech synthesized sounds monotonous and machinelike.

When a human being speaks the same sentence repetitively, each repetition is different from all the others in terms of paralinguistic content such as energy and emphasis in vowels. This is a form of variability which lends the "naturalness" inherent in human speech.

Thus, the problem of incorporating naturalness in synthesized speech reduces to the introduction of variability in synthesized speech by applying PCX keeping in view the observation that no two human beings or even one human being does not utter the same sentence in exactly the same manner if he speaks it twice or more.

This corrected the monotonous nature of speech synthesized by currently existing tools. The voice synthesized using the SpHiSoCS Speech Synthesis Tool was intelligible and had improved naturalness as compared to existing tools.

2.3 Description of Problems Handled

In the Speech Synthesis Tool, two forms of variability have been introduced to demonstrate the above ideas. The two forms of variability that have been expressed include emphasis and speaker variability. The initial steps of speech synthesis were implemented as is with an additional step of PCX. Incorporating this step leads to the synthesis of variations in the speech signal which mimics the way human beings speak. Human beings never create the same speech signal twice even if they repeat the same words. This concept has led to an enhancement in the naturalness of speech.

2.4 Application of Soft Computing Techniques: Paralinguistic Content Expression with Fuzzy Theory

The algorithm for speech synthesis consists of recording all the phonemes of Hindi from a single speaker. The pronunciation of each word is also stored in the dictionary table of the database. This mimics the way a dictionary for a natural language store the pronunciation of each word. The user then enters a sentence comprising of words already in the dictionary. This sentence is converted into phonemes using the pronunciation stored in the dictionary, and then the spoken phonemes are picked up from the recordings. These recordings are then concatenated and then enhanced using filters and the PSOLA and TDSOLA. Finally, the paralinguistic content is expressed for speaker variations and emphasis variations. Algorithm 2 presents the text-to-speech synthesis using PCX. The algorithm is specifically applicable to synthesizing speech by incorporating amplitude variability and emphasis variability only.

Algorithm 2 Text-to-speech synthesis with Paralinguistic Content Expression

1. Input Sentence SZ in the form of a set of words W_i, where $i = 1, ..., I$, and I is the number of words in SZ.
2. Select speaker r, either, male or female.
3. Retrieve the phonetized form H_i of each word W_i from the Dictionary. Each H_i is a set of phonemes P_{ij}, where $j = 1, ..., J$, and J is the number of phonemes in H_i.
4. For each phoneme P_{ij} retrieve the spoken version V_{ij} for the speaker r.
5. For an unvoiced phoneme, filter and remove the last half of the speech segment of the phoneme, For a voiced phoneme, filter and remove the first and the last one-thirds of the speech segment of the phoneme. This step is important since the

energy of a voiced phoneme lies in the central one-third portion of the phoneme. Create a shaping curve and introduce variability into the shaping curve by generating a different shaping curve each time randomly and combining it with the voiced phoneme speech sample V_{ij}.

6. Concatenate the speech segments V_{ij} of all the phonemes into a single speech signal VZ.
7. Modify the pitch and the phoneme duration of the phonemes in the speech signal VZ using the PSOLA and TDSOLA algorithms. The selection of the pitch and phoneme duration is to be done based on the specified gender by selecting the pitch on the basis of membership functions representing the variability present in the fuzzy set of pitch values.
8. Post-processing steps for restoring naturalness include

   a. Filter VZ using a lowpass FIR filter to suppress low frequencies,
   b. Filter the result of step 9(a) using a low pass filter to enhance voice frequencies,
   c. Smoothen the result of step 9(b) to remove harshness to obtain the final output signal VZ.

9. Output the synthesized speech signal VZ.

A snapshot of the GUI implementing the above algorithm is shown in Fig. 4.

3   Experimental Results

The data set for the Speech Synthesis Tool is depicted in Table 2. Data Set #1 had 14 question–answer pairs, and Data Set #2 included 36 question–answer or assertion–response pairs in Hindi. Table 2a gives a segment of the Data Set #1 in Hindi, and Table 2b gives the same translated into English. Table 2c, d presents a segment of the Data Set #2 in Hindi and English, respectively. The input to this tool is a textual sentence. The output of the tool is a set of possible variations of the synthesized speech of the textual input.

3.1   Results

The description of the test set for the tool is depicted in Table 3.

Sample results are as depicted in Fig. 6a, b. The results for the rest sets are shown in Table 4 and Fig. 5.

Table 2 Segment of the Hindi version of the dialog used as part of the data set

Sentence no.	Sentence	No. of words
(a) The topic of the dialog was the "Indian Education System" (Data Set #1)		
Sentence 1	Namaskar, aap kaisae hain?	4
Sentence 2	Namaste, main thheek hoon	4
Sentence 3	Aapkaa naam kyaa hai?	4
Sentence 4	Meraa naam Vishnu hai	4
Sentence 5	Aapkii aayu kyaa hai?	4
Sentence 6	Main 25 varsh kaa hoon	5
Sentence 7	Main Faridaabaad mein rehtaa hoon	5
Sentence 8	Main tumhaari aatmaa mein rehtaa hoon	6
Sentence 9	Bhaarat ki shiksha pranaali kaa aadhaar kyaa hai?	8
Sentence 10	Bhaarat ki shiksha pranaali kaa aadhaar naitiktaa hai	8
(b) English translation of Data Set #1		
Sentence 1	Hello, how are you?	4
Sentence 2	Hello, I am fine	4
Sentence 3	What is your name?	4
Sentence 4	My name is Vishnu	4
Sentence 5	How old are you?	4
Sentence 6	I am 25 years old	5
Sentence 7	I live in Faridabad	4
Sentence 8	I reside in your soul	5
Sentence 9	What was the Indian Education System based on?	8
Sentence 10	The Indian Education System was based on morality	8
(c) Data Set #2—A segment of the data set of Hindi sentences		
1	KYAA AAP PYAALAA RAKHAENGAE?	4
2	HAAN LO RAKH DIYAA	4
3	KYAA MAIN KHAANAA DAAL DOON?	5
4	HAAN DAAL DO	3
5	MAIN KAAGAZ KAISAE KAATOON?	4
6	KAINCHEE SAE KAATO	3
7	TUMKO KISNAE MAARAA?	3
8	MUJHAE KISMAT NAE MAARAA	4
9	TUMKO KISNAE PIITAA?	3
10	ANKAL NAE	2

(continued)

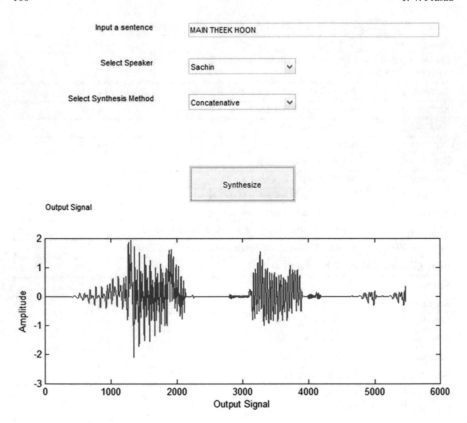

Fig. 4 A snapshot of the interface for the Speech Synthesis Tool

Table 2 (continued)

(d) English translation of Data Set #2		
1	WILL YOU PUT DOWN THE CUP?	6
2	YES, I HAVE PUT IT DOWN	6
3	SHOULD I SERVE THE FOOD?	5
4	YES SERVE IT	3
5	HOW SHOULD I CUT THE PAPER?	6
6	USING SCISSORS	2
7	WHO BEAT YOU UP?	4
8	MY LUCK HAS BEATEN ME	5
9	WHO HAS BEATEN YOU?	4
10	MY UNCLE	2

Table 3 Description of test set and approach for the tool on speech synthesis

Parameter	Value
Input sentence	MAIN THEEK HOON
Language	Hindi
Gender of speaker	Male
Output format	.wav file
Objective	To introduce variability in synthesized speech by applying PCX to achieve naturalness keeping in view the observation that no two human beings or even one human being utters the same sentence in exactly the same manner if he speaks it twice or more times. This will remove the monotonous nature of speech synthesized by currently existing tools
Approach followed	Concatenative synthesis from pre-recorded phonemes based on input sentence followed by pitch synchronous overlap and add (PSOLA) algorithm and time domain synchronous overlap and add (TDSOLA) algorithm being applied for pitch correction to restore naturalness Post-processing includes (a) Filtering using a low-pass FIR filter to suppress low frequencies (b) Filtering using a band-pass IIR Chebyshev filter to enhance voice frequencies (c) Smoothening of final signal to remove harshness
Intelligibility	High and clear
Types of variability introduced	(a) Amplitude/energy variability in vowels (b) Emphasis variability by using a shaping curve that is different for different vowels with each variation of the same sentence produced by the tool
Number of sentences in test set	Test Set 1 had 28 Hindi sentences, and Test Set 2 had 72 Hindi sentences

Table 4 Results of experiments for the test set for the Speech Synthesis Tool

Parameter		Test set 1 (%)	Test set 2 (%)
Intelligibility	Intelligible	57	22
	Partially intelligible	39	51
	Unintelligible	0	24
	Error	4	3

3.2   Comparison with Existing Tools

The tool for speech synthesis was compared with other tools in terms of features and their approach. The other tools were Android IVONA, Android eSpeak, and Microsoft SAPI. The comparison is depicted in Table 5.

A new step was added in SpHiSoCS TTS, i.e., paralinguistic content expression, to the process of concatenative speech synthesis. The new step led to enhancement in naturalness of previously monotonous synthesized speech as was the case in concatenative speech synthesis. The addition opened a new direction in signal synthesis, whereby natural-like synthesis has become possible. Till now, synthesis was possible in a rather artificial voice, whereas this method allows variations in synthesized speech which is much more humanlike.

4   Discussion

The fundamental problem in text-to-speech conversion according to Dutoit [5] is how to produce natural sounding intonation and rhythm without having access to the higher levels of linguistic information, namely syntax, semantics, and pragmatics? According to him, the answer is that it is not possible. However, the Speech Hierarchy Model provides a solution to this problem. The solution is to use Intraword, Intrasentence (or Interword), Response, and Intersentence Fuzzy Rules for representing this information [6]. Then, this paralinguistic content information is expressed

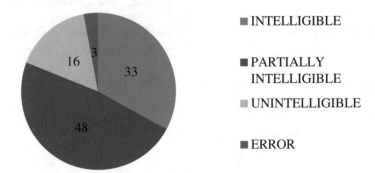

Fig. 5 Pie chart showing the results of testing the SpHiSoCS STT tool. The degree of intelligibility is subjectively measured in terms of four levels, namely intelligible, partially intelligible, unintelligible, and error. The first two levels are considered as correct synthesis, whereas the latter two are considered as incorrect synthesis

(a)

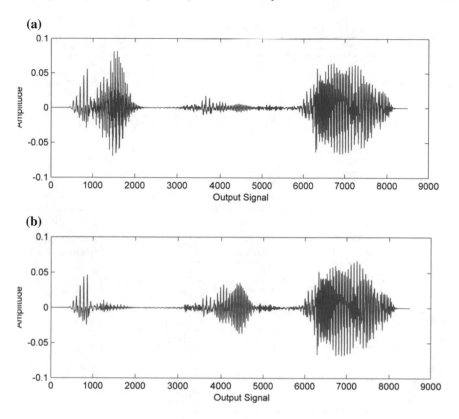

(b)

Fig. 6 **a** Variation 1 of the synthesis of the sentence "MAIN THEEK HOON." **b** Variation 2 of the synthesis of the sentence "MAIN THEEK HOON"

onto speech synthesized making it sound non-monotonic. This non-monotonicity allows infinite variations to be created lending it the naturalness required for speech.

The patterns in the neural networks of the brain may be split up due to there being certain patterns which are noisy and certain patterns which are noiseless. Noisy patterns may arise from digital devices. Each value in a digital signal is discrete, and as such the signal resembles a discontinuous noise signal. However, noiseless signals may be analog signals. Natural sounds are all analog and continuous since they are three-dimensional and follow a spiral pattern. Thus, Indian musical instruments such as the *veena*, the *sitar*, the non-electric guitar, the *tabla*, the flute produce natural sounds and are melodious to the ears. Essentially, they are composed of natural materials and therefore produce natural continuous sounds.

A limitation of the tool is that the pronunciation in the form of syllables has to be stored in the dictionary for every word. This is similar to a dictionary of words where the pronunciation of each word is printed along with the meanings of the

Table 5 Comparison of Speech Synthesis Tool with existing tools

Parameter	Android IVONA	Android eSpeak	Microsoft SAPI	SpHiSoCS TTS
Open source	Yes	Yes	No	Yes
Speech output	Spoken	As.wav file	An API for supporting other software	Spoken and as .wav file
Command line inter-face/graphical user interface	GUI	CLI	CLI	GUI
Algorithm	Not available	Converts text to phonemes with pitch and length information	API black box functions	Training phase using ANFIS Testing phase using ANFIS
Technique used	Not available	Formant synthesis	Not available	Concatenative synthesis
Support for Hindi	No	Yes	No	Yes
Platform	BrightVoice technology	Shared library in C++	.Net	MATLAB
Intelligibility	Best	Poor	Good	Good
Naturalness	Good	Poor	Poor	Good
Number of languages supported	13	51	Not applicable	1(since at prototype stage)
Scalability	Yes	No	Yes	At prototype stage
Paralinguistic content expression	No	No	No	Yes

word. The results of this implementation were found to be fairly accurate with a high intelligibility of the synthesized speech.

5 Conclusions

The paper gives an insight into the way nature creates signals. Further, it emphasizes that natural signals rather than digital signals are more amenable to the human brain which does not use digital signals. Digital signals may be accurate and precise but they are monotonous and cannot produce natural signals found in nature. Further, since digital signals are not amenable to the brain, they create a form of pollution in the memory of the brain since digital signals are discrete and more like a noise signal rather than a three-dimensional natural sound signal.

6   Future Work

The future work for this tool will consist of incorporating emotions using PCX to create speech which is more natural. PCX is a promising technology for creating natural synthesized speech.

References

1. Klatt, D.H.: Software for a cascade/parallel formant synthesizer. J. Acoust. Soc. Am. **67**(3), 971–995 (1980)
2. Lemmetty, S.: Review of Speech Synthesis Technology, Master's Thesis, Helsinki University of Technology, Helsinki, Finland (1999)
3. Bengio, Y.: Learning deep architectures for AI. Found. Trends Mach. Learn. **2**(1), 1–127 (2009)
4. Lakra, S., Prasad, T.V., Ramakrishna, G.: Using Fuzzy sets to model paralinguistic content in speech as a generic solution for current problems in speech recognition and speech synthesis. J. Theoret. Appl. Info. Tech. **78**(3), 441–446 (2015)
5. Dutoit, T.: High Quality Text-to-Speech Synthesis of the French Language. Ph.D. thesis, The Austrian Research Institute for Artificial Intelligence (OFAI), Vienna, Austria (1993)
6. Lakra, S., Prasad, T.V., Ramakrishna, G.: Modelling and Simulating response generation by a computer using a rule-based approach. Int. J. Soft Comput. **11**(5), 299–304 (2016)

An Effective Parameter Tuning for a Bi-objective Genetic Algorithm to Solve the Sum Coloring Problem

Olfa Harrabi and Jouhaina Chaouachi Siala

Abstract To tackle several combinatorial optimization problems, genetic algorithms were frequently used as powerful evolutionary methods. However, performance of these algorithms is heavily dependent on finding appropriate parameters and operators. This phenomenon is known as the parameter tuning problem. The latter falls into the category of hard problems. Therefore, we develop a heuristic approach for effective parameter tuning of genetic algorithms. We also focus on determining the most appropriate genetic operators. Our approach detects the promising available set of parameters. For the practical demonstration, we tackle the sum coloring of graphs. Results show the practical performance of the proposed method to enhance the genetic algorithm.

Keywords Genetic algorithms · Parameter tuning · Sum coloring problem

1 Introduction

Evolutionary algorithms (EA) include several heuristics, able to solve optimization tasks using some aspects of natural evolution. These stochastic search methods incrementally improve the quality of a given solution by means of many operators of variation (e.g., crossover and mutation) and other selection operators (e.g., parent selection and survivor selection). For a given problem, EA used a population of different possible solutions. Among the different EA, genetic algorithms (GAs) have been considered as the most powerful method to generate diversified solutions in significant ways to better exploring the search space of solution [1]. To achieve these issues, GA employs the selection, crossover, mutation, and replacement during

O. Harrabi (✉)
Higher Institute of Management of Tunis, ISG Tunis, Tunis, Tunisia
e-mail: olfa-harrabi@hotmail.fr

J. C. Siala
Institute of Advanced Business Studies of Carthage, IHEC Carthage, Ben Arous, Tunisia
e-mail: siala.jouhaina@gmail.com

© Springer Nature Singapore Pte Ltd. 2019
J. C. Bansal et al. (eds.), *Soft Computing for Problem Solving*,
Advances in Intelligent Systems and Computing 816,
https://doi.org/10.1007/978-981-13-1592-3_8

107

the search process. Interestingly, genetic algorithms show a wealth of pertinence to several practical and theoretical applications. In fact, GA is effective for solving hard theoretical problems, such as traveling salesman [2], graph bisection [3], graph coloring [4], and sum coloring of graphs [5]. On the other hand, GA has also many real-world applications: routing [6], scheduling [7].

However, different parameters related to GA as the crossover operator, the crossover rate, the mutation operator, and the mutation rate can considerably influence the performance of genetic algorithms. As Sivanandam and Deepa state [1], choosing the appropriate parameters for genetic algorithms is one of the most persisting challenge tasks. Unfortunately, there are few works studying the effects of different GA parameters and tuning them. In practice, most researches adopt fixed parameter with reference to previous similar literature or after some preliminary experiments. However, the significant way to determine the well adapted parameters is iterating the search process form different starting values. Such issue, consisting on designing the parameters for a specific algorithm, is called the parameter tuning problem (PTP for short). In the literature, this problem was proven to be hard [8].

In this research paper, we focus on:

- Exposing the conducted research in the field of the PTP related to genetic algorithms.
- Designing an efficient method for tuning parameters of a proposed bi-objective genetic algorithm.
- Looking for the appropriate combination of genetic operators (crossover operator and mutation operator) and parameter values (crossover rate, tournament selection size and mutation rate) that could enhance the performance our proposed bi-objective genetic algorithm.
- Showing the advantages of using a parameter tuning procedure in improving the efficiency of GA. As a showcase, we adopt the resolution of the NP-hard theoretical problem: the sum coloring of graphs [9].

This paper is structured as follows. Section 2 describes the sum coloring of graphs considered in this work. Section 3 presents the proposed solution approach with the different tested operators. Section 4 is devoted for the elaborated parameter tuning method. We report simulation results and comparisons in Sect. 5. The last section sums up the paper.

2   The Sum Coloring of Graphs

2.1   Preliminary Definitions and Properties

In [9], Kubicka has introduced the sum coloring problem (SCP for short) and proved its NP hardness. The problem aims at coloring a graph using the minimum sum of assigned colors. Formally, the SCP could be defined using a simple undirected graph $G = (V, E)$ (V is a set of n vertices and E is a set of m edges). A **proper coloring** of the graph G consists on assigning colors to all the vertices with no adjacent nodes colored similarly. Otherwise, the coloring is **non-proper**, the nodes are **conflict vertices** and the edge linking the vertices is an **edge in conflict**. The minimal sum necessary to properly color the graph is called **chromatic sum** and denoted by $\sum(G)$. The number of colors needed to obtain the chromatic sum is called **strength** of the graph and denoted by $s(G)$.

2.2 Bi-objective Mathematical Formulation

During the evolutionary process, we evaluate the goodness of configurations using two fitness functions: the sum of used colors and the number of conflicting edges.

We present in what follows the bi-objective mathematical model of the sum coloring of graphs. We list bellow the necessary notation:

- V: are the vertices of the graph.
- E: are the edges of the graph.
- K: are the used colors.
- c_k: is the cost of color k.
- $x_{ik} = \begin{cases} 1 \text{ if vertex i is colored with color } k, \\ \forall k \in K, \forall i \in V \\ 0 \text{ otherwise} \end{cases}$
- $y_{ij} = \begin{cases} 1 \text{ if edge } (i, j) \text{ violated the proper} \\ \text{coloring, } \forall (i, j) \in E \\ 0 \text{ otherwise} \end{cases}$

Hence, the final specification of the proposed ILP model is as follows:

$$\text{Minimize} \quad \sum_{i \in V} \sum_{k \in K} c_k x_{ik} \tag{1}$$

$$\text{Minimize} \quad \sum_{(i,j) \in E} y_{ij} \tag{2}$$

$$\sum_{k \in K} x_{ik} = 1 \quad \forall i \in V \tag{3}$$

$$y_{ij} \geq x_{ik} + x_{jk} - 1 \quad \forall k \in K, \forall (i, j) \in E \tag{4}$$

$$x_{ik} \in \{0, 1\} \quad \forall k \in K, \forall i \in V \tag{5}$$

$$y_{ij} \in \{0, 1\} \quad \forall (i, j) \in E \tag{6}$$

Equation (1) looks for the chromatic sum and Eq. (2) is about looking for proper colorings.

Constraints (3) force each vertex to be colored only once. Feasibility of the solution is indicated by constraints (4). The last constraints (5) and (6) state that the decision variables x_{ik} and y_{ij} are binary-valued.

2.3   Resolution Techniques

Given the NP hardness of the SCP [9], there exist exact approaches for coloring only specific graphs [9–11]. Therefore, much effort has been devoted to elaborate approximate approaches. These methods have been categorized into three classes: the greedy algorithms, the evolutionary algorithms, and the local search heuristics. For more details about the mentioned algorithms, one can refer to [12].

3   Proposed Solution Approach: Vector-Evaluated Genetic Algorithm (VEGA)

3.1   The VEGA Algorithm

From an algorithm perspective, our proposed genetic algorithm follows the VEGA framework [13]. The latter is the first practical algorithm proposed to operate with population having multiple objectives. Basically, VEGA decomposes the total population into n sub-populations according to the number of objectives (n presents the number of objective functions). Therefore, our initial population is decomposed into sub-populations of equal sizes. A such decomposition is straightforward optimization technique to treat separately each objective in turn. Then, each sub-population is evaluated using its related objective functions. Subsequently, proportionate selection operator is used to create the mating pool. The proposed VEGA approach is detailed in the following Algorithm 1.

The main goal of our evolutionary algorithm is minimizing the sum of used colors given by a set of k-colorings. The solution is evaluated using the objective functions presented in (Eqs. 1 and 2).

During the optimization process, we opted for different genetic operators in order to reach proposer solution with a minimum sum of colors: selection, crossover, and mutation operators. At a final step, we decide whether an improved offspring will be

---

Algorithm 1 Pseudo-code of the VEGA algorithm procedure

1: **Input**:A graph G, population of size N
2: **Output**: Best-Indiv: the best found coloring
3: **Begin**
4: **While** a maximum number of iterations is not reached **do**
5:    Random population initialization (P, N);
6:    $(Pop_1, Pop_2) \longleftarrow$ Divide P into two
sub-populations of sizes $\frac{N}{2}$;
7:    **For** all Individual in the population **do**
8:       (parent1, parent2) $\longleftarrow$ **Selection** (Pop_1, Pop_2);
9:       Offspring $\longleftarrow$ **Crossover** (parent1, parent2);
10:      Offspring' $\longleftarrow$ **Mutation** (Offspring);
11:      Evaluate (Offspring');
12:      $X \longleftarrow$ Get-Worst(P) ;
13:      $\Delta \longleftarrow$ F(offspring') - F(X);
14:      **If** $\Delta < 0$ **then**
15:         Insert (Offspring', P, X);
16:      **end if**
17:    **End for**
18:**End while**
19: Best-Indiv $\longleftarrow$ Best-Individual (P);
20: Return Best-Indiv
21: **End**

---

inserted into the population or not according to a population updating rule.

Since crossover and mutation are the main genetic operators that considerably influence the quality of obtained offsprings, we attempt to vary several operators to determine the most appropriate for our problem. We also propose a parameter tuning algorithm to find the best parameters values such as tournament selection size, crossover rate, and mutation rate.

3.2 Selection Operator

Selection or reproduction within a genetic algorithm involves selection of chromosomes for the next generation. For our proposed VEGA algorithm, we adopt the Tournament Selection operator in the following way. Firstly, we select randomly n individuals from the population. Then, we choose the best two individuals based on the objective functions Eqs. (1) and (2).

3.3 Crossover Operators

In [14], authors study several crossover operators in order to predict its efficiency when dealing with the sum coloring of graphs. In this context, five operators have been tested: one-point crossover, two-point crossover, uniform crossover, Greedy Partition Crossover (GPX), and Sum Partition Crossover (SPX). Authors investigate a Pareto front analysis of the different tested operators using a set of graph instances. Extensive results showed a robust performance of the SPX operator since it almost leads to a non-dominated Pareto fronts. This insights us to adopt the SPX as a crossover operator during our evolutionary search.

3.4 Mutation Operators

In our study, we sought to determine the best neighborhood operators that we could use during the mutation step. In this context, we developed three neighborhood operators: the exchange mutation, the displacement mutation, and the insertion mutation. For all the developed operators, we assume that each coloring is encoded using a bit string of colors.

- Exchange mutation: It chooses at random two genes in the chromosome and swaps them.
- Displacement mutation: It selects at random two position in the chromosome and reinserts the group of genes between these two positions at a random chosen position.
- Insertion mutation: It randomly selects one gene in the chromosome and displaces it at a random position.

4 The Parameter Tuning Problem: PTP

4.1 Definitions and Survey

Tuning the parameters of a given algorithm is necessary to guarantee good algorithm performance. However, optimizing parameters is a complex optimization task beyond the huge number of parameters and a lack of analytic solvers. Therefore, as Agoston cited in [8], the PTP is known as a hard problem.

Generally, a good design of an algorithm includes all decisions relative to implementation details and some other specifications to tackle a determined problem. In this work, we investigate a parameter tuning approach as successful key for designing the VEGA algorithm. In other words, we use the PTP as a dedicated search method to get the suitable parameters in the VEGA algorithm. Generally, tuning parameters is devised into two cases:

- A parameter tuning: The values of the parameter are fixed at the beginning of the algorithm. Values could be not changed during the algorithm run.

- A parameter control: The values of the parameter are also fixed at the beginning of the algorithm. However, changes of value are allowed during the algorithm run.

In fact, several works have been proposed to study parameter control. In this context, we note several parameter control methods applied to metaheuristics: GA, evolution strategies, particle swarm optimization. For more details, interested reader can refer to [8]. Technically, tuning parameters seek finding the appropriate parameters' values related to a given algorithm. Initially, the method starts from a vector of parameters and searches for the best configuration based on the considered objective functions. For instance, one distinguishes both of structural and parametric tuning [15]. In practice, structural tuning are composed of a set of *qualitative* parameters (symbolic, e.g., crossover operator, mutation operator). However, parametric tuning aims at defining the best values of *quantitative* parameters (numerical, e.g., crossover rate, mutation rate).

4.2 A Parameter Tuning Proposal

In this section, we detail our proposed algorithm to tune parameters of the VEGA solution approach. As previously mentioned, the parameter tuning methods try to efficiently guide the search process in order to generate the appropriate parameter vectors. During our search process, we try to find the appropriate:

- Qualitative parameters: choosing the best mutation operator.
- Quantitative parameters: choosing the best tournament size, crossover rate, and mutation rate.

In this context, we propose to tune three versions of the VEGA algorithm:

- VERSION I: is a VEGA algorithm that performs with tournament selection, an SPX crossover, and an exchange mutation operators.
- VERSION II: is a VEGA algorithm that performs with tournament selection, an SPX crossover, and a displacement mutation operators.
- VERSION III: is a VEGA algorithm that performs with tournament selection, an SPX crossover, and an insertion mutation operators.

The parameter tuning method is fully shown in Fig. 1.

Basically, we choose for our initial population different parameter settings. Then, we randomly change the values of these quantitative parameters. Subsequently, we test the obtained parameter vector and evaluate its performance regarding its fitness value and the consumed computational time.

Fig. 1 Proposed parameter
tuning algorithm for the
VEGA solution approach

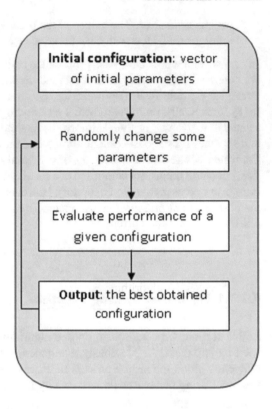

5   Preliminaries Computational Results

5.1   Tested Instances and Experimental Protocol

This section has two main objectives. First, we show the well-suited genetic operators through three tested versions of the VEGA algorithm. Moreover, we assess
the benefits of our adaptive parameter tuning method through a comparison with a
static scheme of the algorithm (does not refer to a tuning strategy). Our proposed
algorithm was programmed in C++ using visual studio 2008.[1] We conducted tests
on a set of benchmarks reputed to be hard when reporting computational results for
the sum coloring problem: DIMACS graphs.[2] For the same instance, our proposed
algorithm obtained feasible and non-feasible solutions. We retain the best proper
solutions with the minimum sum of colorings based on Eq. (1). Qualities of these
solutions are evaluated in term of Gap metric between the produced solution denoted

---

[1]The experiments were implemented on a i3 processor with 2.53 GHz of clock speed and 4 GB of
available memory.

[2]http://dimacs.rutgers.edu/Challenges/.

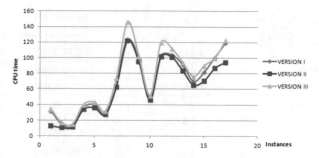

Fig. 2 CPU consumption of three versions of VEGA algorithm

by $Sol_{heuristic}$ and the current best lower bound LB [16]. The percentage deviation Gap is calculated as follows:

- Gap $= \frac{(Sol_{heuristic} - LB)}{LB} \times 100$

We turned VEGA algorithm adopting the tuning procedure for 50 generations.

5.2 Comparison Between Different Genetic Operators

In a first phase of experiments, we propose to compare between the three versions of our VEGA solution approach: VERSION I, VERSION II, and VERSION III. We report for comparisons only proper solutions. Tests were conducted in order to show the best-performing mutation operator. It is significantly relevant to compare the versions providing the same computational efforts. In this context, we propose to run the different versions with the same parameters adopted in [5].[3] Table 1 displays for each algorithm version the average of Gap (%) value and CPU computational running time (in minutes).

Looking at Table 1, one can note that VERSION II outperforms all the other versions and successfully gets improved solutions with an average Gap equal to 0.694%. Furthermore, in terms of CPU consumption, VERSION II is interestingly the fastest. This is clearly illustrated in Fig. 2. In fact, the latter requires an average time of 62.682 min compared 69.235 min for VERSION I and 74.747 min for VERSION III. Empirical simulations favor the utility of using the displacement operator during the mutation step.

---

[3]Tournament size=120, crossover rate=0.9, and mutation rate=0.4.

Table 1 Impact of different mutation operators

Instances	VERSION I		VERSION II		VERSION III	
	Gap	t1	Gap	t2	Gap	t3
DSJC125.1	0.361	32	0.272	13	0.392	34.8
DSJC125.5	0.719	15	0.694	11	0.709	17
DSJC125.9	0.508	12	0.498	11.5	0.514	15
DSJC250.1	0.662	36	0.652	34	0.661	39.9
DSJC250.5	1.421	40	1.379	36	1.411	43
DSJC250.9	0.930	28	0.926	27.5	0.930	32
DSJC500.1	1.295	72	1.187	63	1.187	74
DSJC500.5	2.846	123.5	2.762	122	2.843	146
DSJC500.9	1.745	99	1.741	95.4	1.748	102
Le450-15a	0.158	48	0.133	46	0.162	52
Le450-15b	0.172	103	0.132	102	0.145	120
Le450-15c	0.529	104	0.505	101.4	0.543	112
Le450-15d	0.485	91.5	0.481	84	0.483	94
Le450-25a	0.036	70	0.035	65.7	0.038	76
Le450-25b	0.045	82	0.021	71	0.041	90
Le450-25c	0.180	101	0.180	87.3	0.184	100
Le450-25d	0.211	120	0.205	94.8	0.221	123
Average	0.724	69.235	0.694	62.682	0.718	74.747

Table 2 Quantitative parameters to be tuned and their ranges

Parameter	Min	Max
Tournament size	1	200
Crossover rate	0	1
Mutation rate	0	1

5.3   Benefits of Parameter Tuning Method

During our second step of experiments, we focus on determining the best configuration for the quantitative parameters: the best tournament size, crossover rate, and mutation rate. Authors in [17] state the ranges of the cited parameters are described in Table 2.

To validate the performance of our parameters tuning method, we compare the output results against a static scheme of VEGA algorithm[4] and a recent hybrid memetic algorithm (MA) [18]. Table 3 shows the results of experiments where we report for each tested approach the average of Gap (%) value and the consumed

---

[4]Static scheme adopting the same parameters in [5].

Table 3 Comparative results of VEGA, static scheme of VEGA, and MA algorithms

Instances	VEGA			Static scheme		MA
	Gap	t1	Parameters	Gap	t2	Gap
DSJC125.1	0.268	11	[50, 0.8, 0.3]	0.272	13	0.268
DSJC125.5	0.686	10	[52, 0.8, 0.3]	0.694	11	0.682
DSJC125.9	0.475	10	[50, 0.8, 0.3]	0.498	11.5	0.470
DSJC250.1	0.644	30.4	[51, 0.8, 0.3]	0.652	34	0.641
DSJC250.5	1.374	32	[50, 0.8, 0.3]	1.379	36	1.370
DSJC250.9	0.924	20.9	[57, 0.8, 0.3]	0.926	27.5	0.924
DSJC500.1	1.184	60	[58, 0.8, 0.3]	1.187	63	1.231
DSJC500.5	2.748	120.7	[50, 0.8, 0.3]	2.762	122	2.814
DSJC500.9	1.739	90.9	[56, 0.8, 0.3]	1.741	95.4	1.750
Le450-15a	0.112	42	[100, 0.9, 0.3]	0.133	46	0.133
Le450-15b	0.114	100	[100, 0.9, 0.3]	0.132	102	0.131
Le450-15c	0.499	98.9	[102, 0.9, 0.4]	0.505	101.4	0.529
Le450-15d	0.407	80	[104, 0.9, 0.5]	0.481	84	0.459
Le450-25a	0.027	60.4	[100, 0.9, 0.4]	0.035	65.7	0.035
Le450-25b	0.020	68	[110, 0.9, 0.4]	0.021	71	0.015
Le450-25c	0.167	80.4	[108, 0.9, 0.4]	0.180	87.3	0.167
Le450-25d	0.172	91	[100, 0.9, 0.4]	0.205	94.8	0.205
Average	0.680	59.211	–	0.694	62.682	0.695

CPU time.[5] For the VEGA algorithm, we mention in the last column the obtained parameter values for each optimal solutions.

It is noteworthy from Table 3 that VEGA algorithm achieves optimal solutions with different parameter values even for the same category of benchmarks. We also observe from the same table that VEGA algorithm, adopting the parameter tuning method, presents the best performance on all the tested benchmark graphs. It outperforms the static scheme and MA algorithm in all cases of graphs. In fact, VEGA achieves the best results in term of Gap and computational times. Interestingly, it only requires 59.211 min to get a Gap value of 0.680%. This reflects the ability of our proposed parameter tuning method in guiding the search process to find the best exploration of the search space. We also note that static scheme outperforms the MA algorithm which approves the robustness of the VEGA algorithm with all its main components.

---

[5]Computational times are not available for MA algorithm.

6 Conclusion

In this paper, we illustrate the feasibility of tuning parameters for a bi-objective genetic algorithm: VEGA. During the search process, we attempt to determine both quantitative and qualitative parameters. The adaptive parameter tuning algorithm was performed since it provided us the best configuration of the different parameters. Currently, we are trying to perform a more elaborated parameter tuning method that could calibrate other evolutionary algorithms. Another future work consists on the integration of some statistical tests in order to confirm our findings and return the best parameter values for the considered algorithm.

References

1. Sivanandam, S.N., Deepa, S.N.: Introduction to Genetic Algorithms. Springer Science & Business Media, Berlin (2007)
2. Grefenstette, J., Gopal, R., Rosmaita, B., Van Gucht, D.: Genetic algorithms for the traveling salesman problem. In: Proceedings of the first International Conference on Genetic Algorithms and their Applications, pp. 160–168. Lawrence Erlbaum, New Jersy (July 1985)
3. Reeves, C.R.: Genetic algorithms for the operations researcher. INFORMS J. Comput. 9(3), 231–250 (1997)
4. Fleurent, C., Ferland, J.A.: Genetic and hybrid algorithms for graph coloring. Ann. Oper. Res. 63(3), 437–461 (1996)
5. Harrabi, O., Fatnassi, E., Bouziri, H., Chaouachi, J.: A bi-objective memetic algorithm proposal for solving the minimum sum coloring problem. In: Proceedings of the Genetic and Evolutionary Computation Conference Companion, pp. 27–28. ACM, New York (July 2017)
6. Tasan, A.S., Gen, M.: A genetic algorithm based approach to vehicle routing problem with simultaneous pick-up and deliveries. Comput. Ind. Eng. 62(3), 755–761 (2012)
7. Prez, E., Posada, M., Herrera, F.: Analysis of new niching genetic algorithms for finding multiple solutions in the job shop scheduling. J. Intell. Manuf. 23(3), 341–356 (2012)
8. Eiben, A.E., Smit, S.K.: Parameter tuning for configuring and analyzing evolutionary algorithms. Swarm Evol. Comput. 1(1), 19–31 (2011)
9. Kubicka, E., Schwenk, A.J.: An introduction to chromatic sums. In: Proceedings of the 17th Conference on ACM Annual Computer Science Conference, pp. 39–45. ACM, New York (February 1989)
10. Bar-Noy, A., Kortsarz, G.: Minimum color sum of bipartite graphs. J. Algorithms 28(2), 339–365 (1998)
11. Kroon, L.G., Sen, A., Deng, H., Roy, A.: The optimal cost chromatic partition problem for trees and interval graphs. In: International Workshop on Graph-Theoretic Concepts in Computer Science, pp. 279–292. Springer, Berlin (June 1996)
12. Jin, Y., Hamiez, J.P., Hao, J.K.: Algorithms for the minimum sum coloring problem: a review. Artif. Intell. Rev. 47(3), 367–394 (2017)
13. Schaffer, J.D.: Multiple objective optimization with vector evaluated genetic algorithm. In: Proceeding of the First International Conference of Genetic Algorithms and Their Application, pp. 93–100 (1985)
14. Bouziri, H., Harrabi, O.: Behavior study of genetic operators for the minimum sum coloring problem. In: 2013 5th International Conference on Modeling, Simulation and Applied Optimization (ICMSAO), pp. 1–6. IEEE, New York (April 2013)
15. Birattari, M., Kacprzyk, J.: Tuning Metaheuristics: A Machine Learning Perspective, vol. 197. Springer, Berlin (2009)

16. Jin, Y., Hao, J.K.: Hybrid evolutionary search for the minimum sum coloring problem of graphs. Inf. Sci. **352**, 15–34 (2016)
17. Smit, S.K., Eiben, A.E.: Parameter tuning of evolutionary algorithms: generalist versus specialist. In: European Conference on the Applications of Evolutionary Computation, pp. 542–551. Springer, Berlin (April 2010)
18. Moukrim, A., Sghiouer, K., Lucet, C., Li, Y.: Upper and lower bounds for the minimum sum coloring problem, submitted for publication (2014)

Optimal Combined Overcurrent and Distance Relay Coordination Using TLBO Algorithm

Saptarshi Roy, P. Suresh Babu and N. V. Phanendra Babu

Abstract Relay coordination is an important aspect to maintain proper power system operation and control. Relays must be organized in such a way that every main relay should have a backup relay and coordination time interval (CTI) between primary and backup and different zones of the relays should be maintained to achieve proper fault identification and fault clearance sequences. The relays should operate in minimum desirable time satisfying all the coordination constraints. So, relay coordination is nothing but a highly constraint problem. Heuristic techniques are often used to get optimal solution of this kind of problem. In this paper, this constraint problem is solved by teaching learning-based optimization (TLBO) on a IEEE 14 bus system. Proper desirable time-setting multiplier (TSM) with minimum operating time of relays is calculated. We also incorporated intelligent overcurrent relay characteristics selection to get the desired results in this work. The results seem to be satisfactory as it is working better when compared with contemporary other techniques like genetic algorithm (GA) or particle swarm optimization (PSO).

Keywords Coordination of relays · Coordination time interval · Teaching learning-based optimization · Plug setting · Time-setting multiplier · Overcurrent relay characteristics

1 Introduction

Relays must be organized in such a way that every main relay should have a backup relay and CTI between primary and backup and different zones of the relay should be

S. Roy (✉) · P. S. Babu · N. V. P. Babu
National Institute of Technology Warangal, Warangal 506004, Telangana, India
e-mail: saptarshi.roy.ju@gmail.com

P. S. Babu
e-mail: drsureshperli@nitw.ac.in

N. V. P. Babu
e-mail: phanendra229@gmail.com

© Springer Nature Singapore Pte Ltd. 2019
J. C. Bansal et al. (eds.), *Soft Computing for Problem Solving*,
Advances in Intelligent Systems and Computing 816,
https://doi.org/10.1007/978-981-13-1592-3_9

maintained. For achieving proper fault identification and fault clearance sequences, the proper relay coordination is necessary. These relays should be able to differentiate between the normal operating currents and overcurrents due to fault conditions. During faulty conditions, these relays must respond quickly and isolate the faulty part of the circuit from healthy part and thus allow normal operation of the healthy part of the circuit. If primary relay meant for the clearance of the fault fails, backup relay must operate after providing sufficient time gap. So, backup relay must properly coordinate with the main relay and the flexible settings of the relays (e.g., plug setting, time multiplier setting and suitable relay operating characteristics) must be set to achieve the desired objectives.

Overcurrent and distance relays are often used for protection of power system. Nowadays, this scheme is used in almost all sub-transmission system. To achieve better co-ordination, a distance with a distance, an over current with a over current relay and an over current relay with a distance relay must be coordinated. One of them will act as main relay and another one as backup. Proper coordination time interval should be maintained between them.

The study of coordination of relays was first done among overcurrent relays. Initially, it is done by using linear programming method including simplex, two-phase simplex, and dual simplex methods [1–4]. But the problem regarding using these methods is that the solution will not come unless all the constraints are satisfied. So, people gradually started to use intelligent and meta-heuristic approaches which gives optimal solution instead of exact solution meeting all the constraints criteria. In Ref. [5], optimal coordination is done by genetic algorithm. Reference [6] shows optimal coordination by using particle swarm optimization and Ref. [7] shows the time coordination by using evolutionary algorithm. But these schemes are having two types of problems. First one is mis-coordination and other one is lack of solution for relays with both discrete and continuous time-setting multipliers (TSMs). The problems are resolved in [8] by adding a new expression with the objective function. All the above-discussed methodologies are done by using overcurrent relays and the relay characteristics are assumed to be fixed. While in digital relays different overcurrent relay characteristics can be selected. So, the algorithm for relay coordination should be capable of selecting the best-fitting characteristics of overcurrent relays to have optimal coordination.

Reference [9] shows relay coordination with an hybrid GA algorithm which is helpful in relay coordination of overcurrent and distance relays. Reference [10] shows relay coordination using GA and intelligent relay characteristics selection. TLBO came into picture in 2011. TLBO can be used to solve both constraint and unconstraint problems [11, 12]. References [13–15] shows relay coordination using TLBO for small systems but all of them used fixed characteristics (standard IDMT). None of them used different intelligent characteristics available in digital relays.

In this paper, we are using teaching learning-based optimization (TLBO) for distance and overcurrent relay coordination with intelligent overcurrent relay characteristics selection. Distance and overcurrent relays are used as pairs to protect transmission lines. Relay coordination using TLBO and with intelligent overcurrent relay characteristics is a novel contribution in this paper. The method is more simple and reliable than previous methods used. We have taken IEEE 14 bus system and implement the discussed method. We obtained the results in terms of optimal TMS settings, convergence characteristics, operating time of relays, etc. The results seem to be satisfactory as it is performing better when compared with other contemporary techniques like GA and PSO.

2 Teaching Learning-Based Optimization (TLBO)

TLBO is an algorithm inspired by teaching learning process. It is proposed by Rao and Wagmare [11]. The learning process will be done through two stages such as teacher stage and learner stage. While modeling the algorithm, the group of learners was modeled as population; subjects opted by learners were modeled as design variables. Here, learners' result becomes the fitness value. After iteration, the best solution inside the population becomes teacher and the constraints of optimization problem become design variables [16–18].

2.1 Teacher Stage

The teacher stage is the first stage of this algorithm. As all of us know teacher teaches students and increases the mean of their result depending upon their capability. Suppose that there are 'a' number of subjects or design variables and 'b' number of learners exist and the mean result of the learners is $m_{c,d}$ in a particular subject 'c'. The best overall result considering all the subjects together obtained in the entire population of learners can be considered the result of the best learner, Kbest. The best learner will be considered as teacher. The difference between the existing mean result of each subject and the teacher for each subject is given by Eq. (1)

$$\text{difference\_mean}_{c,k,d} = r_d(x_{c,\text{Kbest},d} - t_f \, m_{c,d})k = 1 \text{ to } b \tag{1}$$

where $x_{c,\text{Kbest},d}$, is the result of the best learner (i.e., teacher) in subject c. t_f is the teaching factor and r_d is the random number in the range [0, 1]. The value of t_f can be either 1 or 2. The value of t_f is decided randomly with equal probability as follows:

$$t_f = \text{round}[1 + \text{rand}(0, 1)\{2 - 1\}] \tag{2}$$

The value of t_f is randomly decided by the algorithm using Eq. (2). Based on the difference_mean$_{c,K,d}$, the existing solution is updated in the teacher phase according to the following expression [Ref. Eq. (3)]:

$$x'_{c,K,d} = x_{c,K,d} + \text{difference\_mean}_{c,K,d} \tag{3}$$

$x'_{c,K,d}$ is the new value of $x_{c,K,d}$. $x'_{c,K,d}$ should be accepted if it improves the value of the function. After teacher stage, all fitted values will be given as input to the learner stage. So, it means the learner stage depends on teacher stage.

2.2 Learner Stage

The second part of the algorithm is learner phase. Learners boost up their knowledge by interactions among themselves. Consider a population size of 'b', and the learning phenomenon of this phase can be expressed below.

Randomly select two learners e and f such that $x'_{\text{total-}e,d} \neq x'_{\text{total-}f,d}$ (where $x'_{\text{total-}e,d}$ and $x'_{\text{total-}f,d}$ are the updated values of $x_{\text{total-}e,d}$ and $x_{\text{total-}f,d}$, respectively, at the end of teacher phase) [Ref. Eqs. (4) and (5)]:

$$x''_{c,e,d} = x'_{c,e,d} + r_d(x'_{c,e,d} - x'_{c,e,d}), \quad \text{if } x'_{\text{total}-e,d} < x'_{\text{total}-f,d} \tag{4}$$

$$x''_{c,e,d} = x'_{c,e,d} + r_d(x'_{c,f,d} - x'_{c,e,d}), \quad \text{if } x'_{\text{total}-e,d} < x'_{\text{total}-f,d} \tag{5}$$

Accept $x''_{c,e,d}$ if it gives a better function value.

3   Problem Statement

For achieving better protection, both distance and overcurrent relays are used as main and backup relays, respectively, in protection schemes. In this situation, it is necessary to coordinate these two types of relays simultaneous. With an intention to find a solution of the above problem, we have chosen a fitness function similar to Eq. (6) and all its parameters are described in Eqs. (7)–(9). All the constraints are described in Eqs. (10)–(15).

$$\text{Fitness function} = \min(\alpha \sum_{i=1}^{n} t_i + \beta \sum_{i=1}^{n} |T_{\text{DIOC}i} - |T_{\text{DIOC}i}||$$

$$+ \lambda \sum_{i=1}^{n} |T_{\text{OCD}i} - |T_{\text{OCD}i}|| + \delta \sum_{i=1}^{n} |T_{\text{OC}i} - |T_{\text{OC}i}||) \tag{6}$$

where

$$T_{\text{OC}i} = T_{\text{oc backup}\,i} - T_{\text{oc main}\,i} - \text{CTI}' \tag{7}$$

$$T_{\text{DIOC}i} = T_{\text{oc}\,i} - T_{z2\,i} - \text{CTI}' \tag{8}$$

$$T_{\text{OCD}i} = T_{z2\,i} - T_{\text{oc}\,i} - \text{CTI}' \tag{9}$$

T_{oc} is the operating time of overcurrent relay, and T_{z2} is the operating time of second zone of the distance relay, and α, β, λ, δ are penalty factors.

4 Constraints

The several constraints need to be satisfied to obtain optimal coordination.

4.1 Coordination Constraints

$$T_{z2\,backup} - T_{oc\,main} \geq CTI' \tag{10}$$

$$T_{oc\,main} - T_{z2\,backup} \geq CTI' \tag{11}$$

CTI is coordination time interval whose typical value is between 0.2 and 0.3 s.

4.2 Relay Characteristics

The overcurrent relay characteristics are typically of below nature:

$$t = TSM\left(\frac{K}{M^\alpha - 1} + L\right) \tag{12}$$

t time of operation of the relay
TSM   Time-setting multiplier.

K, L, and α are constants. It varies characteristics to characteristics.

M is the ratio between short-circuit current I_{sc} and pickup current I_p. TSM is supposed to be continuous and can take any value between 0.05 and 1.1. Coordinating time interval in each cases is supposed to be 0.25 s. Reference [10] shows eight types of intelligent characteristics available in digital overcurrent relay.

4.3 Pickup Current Constraints

Pickup current is having a limit. The relay coordination problem is highly dependent on the value of the pickup current of the relays. For sensing a small amount of fault current, the pickup current should be less than minimum fault current. On the other hand, the minimum pickup current may be doubled under small overloaded condition to avoid any maloperation. The limits of pick up current can be expressed as below [15]:

$$Ip_{\min} \leq Ip \leq Ip_{\max} \tag{13}$$

4.4 TSM Constraints

TSM is supposed to be continuous and can take any value between 0.05 and 1.1. Mathematically, it can be expressed as below:

$$\text{TSM}_{\min} \leq \text{TSM} \leq \text{TSM}_{\max} \tag{14}$$

4.5 Constrains on Relay Operating Time

For minimizing or mitigating maloperation due to transient, overshoot, or any other critical condition of the network, relays should operate after a minimum time. Limits on time of operation of relay (t_{op}) can be expressed as:

$$t_{\text{op}\min} \leq t_{\text{op}} \leq t_{\text{op}\max} \tag{15}$$

Minimum operation time of relay is 0.1 s, and maximum depends on the requirement of the user.

5   Test Results

To test the methodology, an IEEE-14 bus system has been selected. Test system data are obtained from Ref. [19]. The relay arrangement is shown for this power system as per Fig. 1. The mho directional relays are used here.

Overcurrent relays are arranged using time-graded protection scheme with inverse definite minimum time (IDMT) characteristics. Different types of intelligent overcurrent characteristics available in digital relays are obtained from Ref. [10].

The main and backup relay pairs with short-circuit current data are shown in Table 1.

From Table 1, it is found that relay R_4, R_5, R_6, R_{12} having more protective reliability as they have two backups. The typical operating time of first, second, and third zones of all distance relays are 20 ms, 0.3 s (or more), and 0.6 s (or more). Second zone starts from 80% length of the lines for all lines. The short-circuit currents of the main and backup overcurrent relays must be calculated from close in bus fault cases (critical fault locations). The information regarding pickup current settings is shown

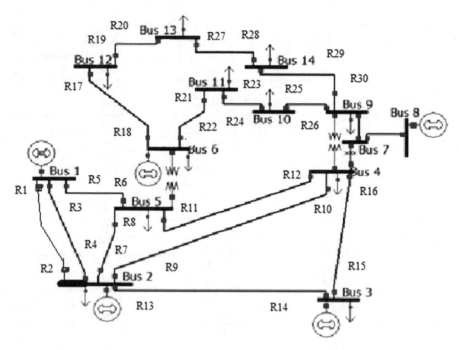

Fig. 1 Relay arrangement in IEEE 14 bus system

in Table 2. The value of pickup current of each overcurrent relay is assumed roughly 1.25 times of the relevant maximum load in approximated integer form. All the

Table 1 Main and backup relay pairs with short-circuit current data

Main relay (R_i)	Backup relay (R_i)	Main relay short-circuit current (amps)	Backup relay short-circuit current (amps)
R_1	R_6	1494	125
R_2	R_{14}	621	268
R_3	R_6	1494	125
R_4	R_{10}	621	260
R_4	R_8	621	281
R_5	R_2	3393	338
R_5	R_4	3393	338
R_6	R_{12}	1716	880
R_6	R_7	1716	678
R_7	R_3	2732	1816
R_8	R_5	703	788

(continued)

Table 1 (continued)

Main relay (R_i)	Backup relay (R_i)	Main relay short-circuit current (amps)	Backup relay short-circuit current (amps)
R_9	R_3	2753	132
R_{10}	R_{11}	1766	1088
R_{11}	R_5	1109	795
R_{12}	R_{15}	1380	464
R_{12}	R_9	1380	699
R_{13}	R_1	2774	1820
R_{14}	R_{16}	1118	795
R_{15}	R_{13}	1249	925
R_{16}	R_{11}	2004	1092
R_{17}	R_{20}	415	415
R_{18}	R_{21}	1419	270
R_{19}	R_{18}	565	565
R_{20}	R_{28}	996	267
R_{21}	R_{24}	528	528
R_{22}	R_{17}	1189	35
R_{23}	R_{22}	641	641
R_{24}	R_{26}	921	921
R_{25}	R_{23}	377	377
R_{26}	R_{29}	1239	168
R_{27}	R_{19}	913	186
R_{28}	R_{30}	608	608
R_{29}	R_{27}	406	406
R_{30}	R_{25}	1327	264

relevant data (regarding load current, pickup current, and short-circuit current) are obtained from Power World simulation. Complex calculations are obtained by using MATLAB coding.

The process of finding objective function is trial and error. The ultimate objective of choosing the objective function is to reduce the time of operation of relay, same as in the case of overcurrent to overcurrent relay coordination case. The only difference here is some additional terms are coming due to the presence of distance relay. When $|T_{\mathrm{DIOC}i}|$ is positive, then the second term of objective function is becoming zero, but when $|T_{\mathrm{DIOC}i}|$ is negative, then the second term is additive with the objective function and increasing its value. Since it is a minimization problem, the chance of survival of such fitness value is mitigated by this approach. As per coordination constraints, $|T_{DIOCi}|$ value should be always greater than equals to zero. Its value can be negative only in case of mis-coordination. So, with such approach the chance

Table 2 Pickup current values of the relay

Relay number (R_i)	Load current (amps)	Pickup current (amps)	Relay number (R_i)	Load current (amps)	Pickup current (amps)
1	312	390	16	100	125
2	310	388	17	32	40
3	312	390	18	32	40
4	310	388	19	8	10
5	298	373	20	8	10
6	298	373	21	32	40
7	166	208	22	32	40
8	166	208	23	16	20
9	224	280	24	16	20
10	224	280	25	26	33
11	260	325	26	26	33
12	260	325	27	24	30
13	294	368	28	24	30
14	294	368	29	40	50
15	100	125	30	40	50

of mis-coordination problem is almost nullified. The same kind of explanation can be given for choosing the third and fourth term of the fitness function also.

By applying teaching learning-based optimization (TLBO) in the network of Fig. 1, the output results are obtained. TSMs and overcurrent relay characteristics selected by TLBO are shown in Table 3. In all cases, TSMs are considered to be continuous (0.05–1.1). The time of operation of relays in each case is also shown in the table (Table 3). The various outputs from this work are shown pictorially from Figs. 2, 3, 4, 5 and 6.

From Tables 2 and 3, it is observed that in case of relays with less pickup current, characteristic 2 (standard inverse) is more suitable.

6 Comparative Study with Other Methodologies

The results obtained by using TLBO is compared with the results obtained by using genetic algorithm (GA) and particle swarm optimization (PSO) for the same problem. Table 4 lists the comparative study. Figures 7 and 8 show the convergence curve obtained by other algorithms.

From Table 4, it is seen that TLBO is converging in fastest time with compare to GA and PSO. TLBO does not use any algorithm-specific parameter like GA (uses mutation rate, selection rate, and crossover probability) and PSO (uses inertia weight, social and cognitive parameters). Because of this property, it converges very fast and

Table 3 Output table

Relay (R_i)	Second zone operation time (for distance relay) (T_{z2}) (s)	TSM (for overcurrent relay)	No. of selected characteristic
1	0.42825	0.0898	3
2	0.27816	0.05	6
3	0.5388	0.113	3
4	0.27816	0.05	6
5	0.4546	0.1465	2
6	0.5249	0.14	3
7	0.5353	0.2822	2
8	0.44928	0.0792	3
9	0.4505	0.1505	2
10	0.3067	0.12057	3
11	0.44658	0.0798	3
12	0.4662	0.1121	3
13	0.2736	0.1325	3
14	0.45	0.07225	2
15	0.45	0.15142	2
16	0.38253	0.18996	8
17	0.4537	0.15513	2
18	0.4514	0.23845	2
19	0.4569	0.2743	2
20	0.4178	0.3434	5
21	0.4501	0.1704	2
22	0.4502	0.2256	2
23	0.45	0.2308	2
24	0.4521	0.25697	2
25	0.4566	0.16278	2
26	0.4569	0.24553	2
27	0.4569	0.23077	2
28	0.45	0.19935	2
29	0.2625	0.13844	3
30	0.4518	0.21868	2
Average value	0.427682	0.16822	–
Fitness value	211.4055	–	–

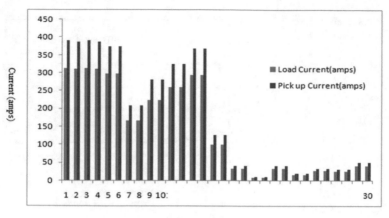

Fig. 2 Comparison of load current and pickup current of various relays

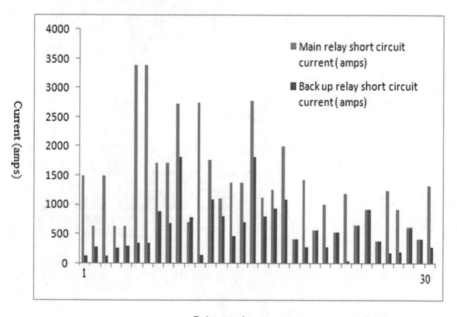

Fig. 3 Comparison of short-circuit currents of main and backup relays

faster than other contemporary algorithms like GA and PSO. Here, the simulation is done on a Dell Laptop, processor: Intel® core (TM) i3-2350M CPU@2.30 GHZ, RAM: 8.00 GB, Simulator : MATLAB R 2014b.

Fig. 4 Convergence curve
of the TLBO algorithm

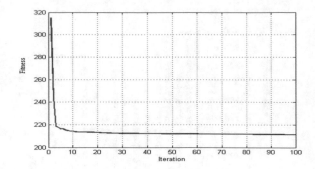

Fig. 5 Comparison of
operating time of second
zone of relays

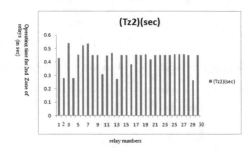

Fig. 6 Comparison of
optimum TSM of relays

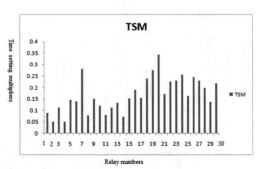

Table 4 Comparative study with GA and PSO

Attributes	TLBO	GA	PSO
Number of iterations to converge	12	10	34
Average time per iteration (sec) taken	0.0078	0.0139	0.0054
Total time taken to converge (CPU elapsed time) (sec)	0.0936	0.139	0.1836

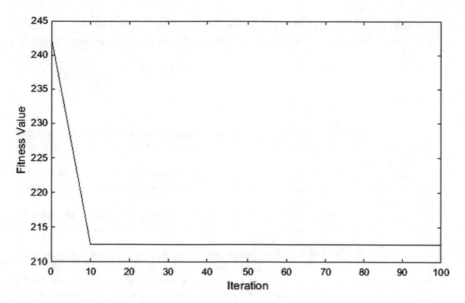

Fig. 7 Convergence curve using GA

Fig. 8 Convergence curve using PSO

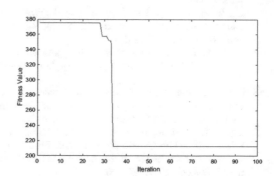

7 Conclusions

This paper focused on optimal coordination of directional and overcurrent relays. The problem statement and various constraints which are to be satisfied are already presented in the paper. Teaching learning-based optimization (TLBO), which is a modern meta-heuristic technique, is applied to solve the problem. The optimum time of operation, TSM, pickup currents of relays are calculated for an IEEE 14 bus system. All the constraints are found within desirable range. Which intelligent over current characteristics are required to get the desired result are also selected. Coordination time interval is taken 0.25 s for each cases. The protection settings seem to be satisfactory for the discussed power network as it is performing better when it is compared with other contemporary algorithms like GA and PSO. As a future scope of this work, relay coordination on higher test systems can be implemented.

References

1. Urdaneta, A.J., Nadira, R., Perez, L.G.: Optimal coordination of directional overcurrent relays in interconnected power system. IEEE Trans. Power Delivery 3(3) (1998). https://doi.org/10. 1109/61.193867
2. Urdaneta, A.J., Resterpo, H., Fajardo, J., Sanchez, J.: Coordination of directional over current relays timing using linear programming. IEEE Trans. Power Deivery 11(1), 122–129 (1996). https://doi.org/10.1109/61.484008
3. Chattopadhyay, B., Sidhu, T.S., Sachdev, M.S.: An on-line relay coordination algorithm for adaptive protection using linear programming techniques. IEEE Trans. Power Deivery 11(1), 165–173 (1996). https://doi.org/10.1109/61.484013
4. Askarian Abyaneh, H., Keyhani, R.: Optimal co-ordination of over current relays in power system by dual simplex method. In: The AUPEC Conference, Perth, Australia (1995)
5. So, C.W., Lai, K.T., Li, K.K., Fung, K.Y.: Application of genetic algorithm for over current relay coordination. In: Proceedings of Instrumentation Electrical Engineering Conference Developments in Power System Protection. March 25–27, pp. 66–69 (1997). https://doi.org/10.1049/c p:19970030
6. Zeineldin, H., El-Saadany, E., Salama, M.A.: Optimal coordination of over current relays using a modified particle swarm optimization. Electr. Power Syst. Res. 76(11) (2006). https://doi.or g/10.1016/j.epsr.2005.12.001
7. So, C.W., Li, K.K.: Time coordination method for power system protection by evolutionary algorithm. IEEE Trans. Ind. Appl. 36(5), 1235–1240 (2000). https://doi.org/10.1109/28.8712 69
8. Razavi, F., Askaria, H., Abyaneh, Al-Dabbagh, M., Mohammadi, R., Torkaman, H.: A new comprehensive genetic algorithm method for over current relays coordination. Electric Power Syst. Res. 92(9) (2008). https://doi.org/10.1016/j.epsr.2007.05.013
9. Sadeh, J., Aminotojari, V., Bashir, M.: Optimal coordination of over current and distance relays with hybrid genetic algorithm. In: Proceedings of 10th International Conference on Environment and Electrical Engineering. IEEE, Rome, Italy, 8–11 May (2011). https://doi.or g/10.1109/eeeic.2011.5874690
10. Chabanloo, R.M., Abyanch, H.A., Kamangar, S.S.H., Razavi, F.: Optimal combined overcurrent and distance relays coordination incorporating intelligent overcurrent relay characteristics selection. IEEE Trans. Power Delivery 26(3), 1381–1391 (2011). https://doi.org/10.1109/TP WRD.2010.2082574
11. Rao, R.V., Wagmare, G.G.: A comparative study of a teaching–learning-based optimization algorithm on multi-objective unconstrained and constrained functions. J. King Saud University-Comput. Inf. Sci. 26, 332–346 (2012). https://doi.org/10.1016/j.jksuci.2013.12.004
12. Rao, R.V., Patel, V.: An elitist teaching-learning-based optimization algorithm for solving complex constrained optimization problems. Int. J. Ind. Eng. Comput. 3, 535–560 (2013). https://doi.org/10.5267/j.ijiec.2012.03.007
13. Singh, M., Panigrahi, B.K., Abhyankar, A.R.: Optimal coordination of directional over-current relays using teaching learning-based optimization (TLBO) algorithm. Electr. Power Energy Syst. 50, 33–41 (2013). https://doi.org/10.1016/j.ijepes.2013.02.011
14. Kalaage, A.A., Ghawghawe, N.D.: Optimum coordination of directional overcurrent relays using modified adaptive teaching learning based optimization algorithm. Intell. Ind. Syst. 2(1), 55–71 (2016). https://doi.org/10.1007/s40903-016-0038-9
15. Saha, D., Dutta, A., Saha Roy, B.K., Das, P.: Optimal coordination of DOCR in interconnected power systems. In: IEEE 2nd International Conference on Control, Instrumentation, Energy

and Communication, 28–30 January, Kolkata, India (2016). https://doi.org/10.1109/ciec.2016.7513834

16. Ojaghi, M., Ghahremani, R.: Piece–wise linear characteristic for coordinating numerical over current relays. IEEE Trans. Power Delivery **32**(1), 145–151 (2017)

17. Rao, R.V., Balic, J., Savsani, V.V.J.: Teaching–learning-based optimization algorithm for unconstrained and constrained real-parameter optimization problems. Eng. Optim. **44**(12), 1447–1462 (2012). https://doi.org/10.1109/tpwrd.2016.2578324

18. Rao, R.V.: Review of applications of TLBO algorithm and a tutorial for beginners to solve the unconstrained and constrained optimization problems. Decis. Sci. Lett. **5**(1), 1–30 (2016). https://doi.org/10.5267/j.dsl.2015.9.003

19. Power flow data for IEEE 14 bus test case, University of British Columbia. http://www.ece.ubc.ca/~hameda/download_files/case14.m

CORO-LABs: Complexity Reduction of Layered Approach in Codifying Business Solutions Using Tuxedo

Ankit Shrivastava, Ashish Kumar and Pradeep Kumar Tiwari

Abstract The aim of paper is to reduce the complexity of layered approach in codifying the business solutions by providing the cross-layer integration. In order to migrate from legacy systems to SAP modules, we adopt a 3 tier service-oriented architecture (SOA) using Tuxedo as a middleware, for centralizing all the business flow. This paper consists of an architecture by which the clients place their request through the frontend and the generated request is forwarded to scheduling team which takes care of stock availability. If the stock is available, it prepares a backlog based on it and the logistics part mainly deals with warehouse, logistics and shipment which consist of information about the stock. The challenges of this work lie with the migrating issues like communicating with legacy system E1 with proposed system E2 and the SAP modules as data transfer has to take place in a way which the system understands. We also implement solution which helps the customers in easing the performance of the logistics movements of warehouses.

Keywords Cross-layer integration · Tuxedo · Legacy system

1 Introduction

In order to migrate from legacy systems to SAP modules, we adopt a three-tier SOA-based architecture [1] using Tuxedo as a middleware, for centralizing all the business flow and providing transactional mechanism-based reliable services. This

A. Shrivastava (✉) · A. Kumar
Department of Computer Science & Engineering, Manipal University Jaipur,
Jaipur, Rajasthan, India
e-mail: ankitshrivastavaieee@gmail.com

A. Kumar
e-mail: aishshub@gmail.com

P. K. Tiwari
Department of Computer Science & Engineering, Vindhya Institute of Technology,
Satna, Madhya Pradesh, India
e-mail: pradeeptiwari.mca@gmail.com

© Springer Nature Singapore Pte Ltd. 2019
J. C. Bansal et al. (eds.), *Soft Computing for Problem Solving*,
Advances in Intelligent Systems and Computing 816,
https://doi.org/10.1007/978-981-13-1592-3_10

application consists of modules from where the clients place their request through the frontend and the generated request is forwarded to scheduling team which takes care of stock availability. If the stock is available, it prepares a backlog based on it and the logistics part mainly deals with warehouse, logistics and shipment which consist of information about the stock.

The challenges of this project lie with the migrating issues like communicating with legacy system E1 with proposed system E2 and the SAP modules as data transfer has to take place in a way which the system understands. We also implement solutions which helps the warehouse managers in easing the analysis parameters of the logistics movements of warehouses. Most of the applications have frontend designed as desktop applications but due to many viabilities of the desktop applications, we have tried to implement the Web application version of SGA 06 Applications, which overcomes most of the limitations of traditional desktop applications.

We look for a futuristic approach, taking in account cross-browser compatibility issues, and move towards ExtJS, a mature JavaScript framework for developing rich Internet applications, without making any changes to the middleware and the backend.

The paper is divided into different sections. Literature review is given in Sect. 2. Section 3 discussed the proposed architecture. Tuxedo framework and its key ideas are described in Sect. 4. Section 5 described the methodology. Testing of the developed system is discussed in Sect. 6, and at last the conclusion is written in Sect. 7.

2 Literature Review

Wen and Jingsha [2] demonstrate applications in banking system having TUXEDO as middleware. This paper exhibits the BEA Tuxedo and presents the utilization of Tuxedo in a keeping money framework. This paper clarifies the promotion vantages of Tuxedo as takes after. The three-layer structure display in the light of Tuxedo gives multi-layer structure, correspondence, exchanges, security, adaptation to internal failure and numerous other essential administrations for building a huge scale, superior disseminated applications. Tuxedo as an exchange middleware of a circulated OLTP (online exchange process) gives efficient instrument that can be utilized to fabricate the OLTP framework with three-layer outline. Tuxedo permits the expansion of assets in an arranged framework without changing the current application.

Sun and Zhao [3] demonstrate paper on small bank intermediary business system for social insurance based on Tuxedo. This paper presents the overview of three-tier architecture containing Tuxedo as middleware.

Chu et al. [4] explain transaction middleware model for SCA programming. This paper exhibits an exchange middleware show for SCA (Administration Part Design) programming in view of the Tuxedo outline work, presents the ATMI official and gives the techniques for sending SCA segments in a Tuxedo domain and summoning Tuxedo benefit from SCA customer.

Andrade et al. [5] emphasize on open online transaction processing with Tuxedo system. This paper shows the necessities driving open OLTP today's showcase and an outline of the Tuxedo items intended to meet those prerequisites. The Tuxedo middleware provides many services to meet the requirements for open online transaction process.

Felt [6] explains distributed transaction processing in Tuxedo system. This paper presents the advantages and the functionality of Tuxedo middleware. He also explains various trends in the Web application development and the latest technologies adopted in Internet, the evolution of Web browsers. He discusses how a Web browser can be used as an application platform and address various issues on cross-browser compatibility.

Lihua and Jianan [7] try to implement Web applications based on ExtJS framework using the MVC design patterns and address the compatibility issues with different browsers and integration with different frameworks.

3 Proposed Architecture

The legacy applications in Cobol running on HP 3K systems, since not catering to the business requirements, cause overhead in implementing the requirements, which were not centralized and a single change requirement has overhead of changing the code base in all the places where the application is running. Around 25 decentralized logistic sites connected and providing online visibility of worldwide logistic business within.

- About 200,000 transactions exchanged daily.
- 250,000 line items considered in the extraction for picking daily.
- 6000 shipments created daily.
- 70,000 parcels moved between two sites/customers/sub-contractors daily.

As we can see, the robustness of the system and the functionality, which system has to offer, we adopt a three-tier-based SOA architecture which we will be discussing in next module. The legacy applications in E1 are now migrated to SAP modules, but in order for the modules to communicate to each other in a way that the transferred data is understood, we have come up with E2, the architecture is explained in Fig. 1, E2 now holds all the business logic comprising of sales and marketing team, PRIS

Fig. 1 Proposed architecture

Fig. 2 Central pick
technical architecture

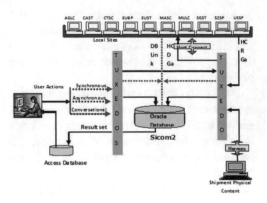

team, central logistics team, SAP team etc. Figure 2 shows the technical architecture
of Tuxedo in relation to the business plan.

There are some monitors which continuously poll some tables for particular status
and do the intended work when find some rows in the table. HCDGATE is used
when central pick application wants to access E1 programs. HCRGATE is used
when E1 wants to access central pick Tuxedo services. The Tuxedo services have
all the business rules. It does all the intended work that application needs updating,
deletion, insertion of data in the database as shown in Fig. 2.

4   Tuxedo

The Prophet Tuxedo framework is a middleware item that circulates applications
over numerous stages, databases and working frameworks utilizing message-based
interchanges and, if wanted, conveyed exchange handling [8]. Middleware is utilized
as a part of conveyed preparing among different servers, oversee circulated exchanges
and incorporate numerous database stages. Tuxedo was planned from the earliest
starting point for high accessibility and to give greatly adaptable applications to
help applications requiring a large number of exchanges every second on ordinarily
accessible conveyed frameworks [9]. The key ideas in Tuxedo framework are:

4.1   Messaging Core

Tuxedo is at its centre a message directing and lining framework. Solicitations are
sent to named administrations and Tuxedo utilizes memory based between process
correspondence offices to line the solicitations to servers. The requester is uninformed
of where the server that really forms the demand is found or how it is actualized [10].

Tuxedo can utilize the substance of the message to figure out what servers ought to be used to get the demand by methods for information subordinate steering.

4.2 Clustering

The core of the Tuxedo framework is the Notice Board (BB). This is a common memory section that contains the condition of a Tuxedo area [11]. Servers, administrations, exchanges and customers are altogether enrolled in the BB giving a worldwide perspective of their state over the machines inside a space. Release Board Contact (BBL) will coordinate all BBs. Another procedure on each machine called the Extension is in charge of passing solicitations starting with one machine then onto the next [12]. This enables Tuxedo to spread load over the different machines inside an area and enables servers and administrations to keep running on various machines.

4.3 Flexible Buffer Formats

Tuxedo applications can use an assortment of message designs relying on the kind of information that will be passed. FML buffers can contain a self-assertive number of named fields of self-assertive sort. Fields can be rehashed and settled. VIEW buffers are basically records like C structures. A VIEW buffer has an outside depiction which enables Tuxedo to get to the fields inside it if important for things like information subordinate directing. Other buffer designs incorporate XML, Auto Beam, STRING and MBSTRING. Tuxedo can consequently and straightforwardly change over FML buffers to and from XML buffers.

4.4 Gateways

To encourage the sharing of administrations crosswise over areas, Tuxedo gives space doors. A space passage permits bringing in and trading administrations from re-bit areas. This enables the nearby area to see benefits on remote areas just as they were neighbourhood administrations. The space portals are in charge of engendering security and exchange setting to the remote area.

4.5 Transaction Monitoring and Coordination

Tuxedo applications can ask for that all administration summons and their related updates to any assets controlled by asset administrators be controlled by an exchange.

Once the application starts an exchange, all resulting administration summons and settled summons are incorporated as a component of that exchange, even those administrations that were executed on remote spaces. Tuxedo at that point arranges the confer preparing with the asset supervisors to guarantee nuclear updates to all affected assets [13].

4.6   Queuing Subsystem

Tuxedo gives a lining subsystem called /Q. This office gives transient and tenacious lines that enable application to expressly line solicitations to named lines. Lines can be requested by message accessibility time, lapse time, need, LIFO, FIFO, or a mix. The Prophet Tuxedo framework offers many highlights to suit the structural angles.

- **Distributed Administrations**—enables straightforward access to application as well as framework administrations situated on different equipment stages.
- **Fast, Connectionless Interchanges**—customer conveys to notice board as opposed to servers, which enhances framework execution.
- **Server Straight Forwardness**—the index of administrations on the release board maps benefit names to servers. Customers don't should know about server character.

4.7   Tuxedo's Approach to Client/Server

A customer is a program that gathers or ask for information from a client which passes that demand to a server fit for satisfying it. It can dwell on a PC or a workstation as a component of the frontend of an application.

To be a customer, a program must have the capacity to conjure the Prophet Tuxedo libraries of capacities and strategies referred to aggregately as the Application to Exchange Screen Interface or ATMI. A customer joins Prophet Tuxedo application by calling the ATMI customer instatement schedule. When it has joined an application, a customer would be able to exchange limits and call ATMI capacities that empower it to speak with different projects in the application. A customer leaves the application by ATMI end work. As shown in Fig. 3, Application Exchange Screen Interface (ATMI) is the Application Programming Interface to BEA Tuxedo. It incorporates Exchange Schedules, Message Taking Care of Schedules, Administration Interface Schedules and Buffer Administration Schedules.

Announcement Board is an accumulation of shared information structures intended to monitor a running Tuxedo Framework/T Application. It contains data about servers, administrations, customers and exchanges relating to an application. A Prophet Tuxedo server is a procedure that directs an arrangement of administrations, dispatching them consequently for customers that demand them.

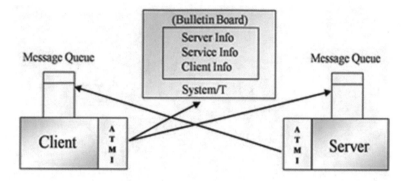

Fig. 3 Tuxedo client server architecture

An administration, thusly, is a capacity inside a server program that plays out a specific assignment required by a business. Administration capacities execute business rationale through calls to database interfaces, for example, SQL and, potentially, calls to the ATMI to get to extra administrations, lines and different assets. The servers which these administrations live at that point answer to the customers or send the customer solicitations to another administration.

5 Methodology

It covers the points of interest clarification of philosophy that is being utilized to influence this work to finish and functioning admirably. The strategy is used to accomplish the target of the work that will fulfil an impeccable outcome. With a specific end goal to assess this work, the technique used in view of Framework Advancement Life Cycle (FALC) for the most part of noteworthy advancement are Arranging, Outline, Actualizing, Testing and Upkeep (Fig. 4).

Inputs: Plant code for which the report is to be generated, Report Type, Flow Type, Date from, Date to find the duration between which to generate report, Mailing list to send performance report.

Fig. 4 Planning activities

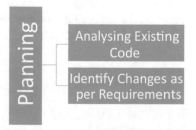

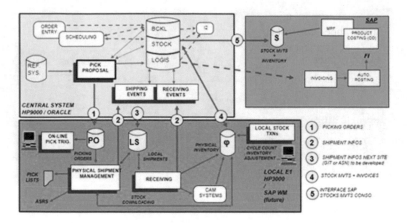

Fig. 5 Business flow between different modules in migration

Output: An excel file is generated containing the performance analysis of the plant by classifying the shipment type with proposed and achieved count of shipments.

The block diagram in Fig. 5 dissipates the flow of data from the legacy applications to SAP module via E2. The block in blue can be considered as E2, in yellow as E1 and in purple as SAP module.

The client sends their through the frontend like the VB.Net screens and depending on the requirements and the inputs, a particular service written in Tuxedo is called which has access to the Oracle database, retrieving particular record sets and are populated. We use FML buffers, i.e. data existing in key-value pairs as XML to transfer or maintain the data flow between the modules. For easy flow of data from E2 to the SAP system, we use Web Methods as the middleware, as it is easy for SAP to communicate with SAP, where further processing takes place and the reply is sent back to the E2. E2 acts as central repository for all the transactions and has updated information about the entire system. E2 again consists of several modules which have predefined functionality and rules and communicate with each other.

The major modules are sales order creation, PRIS, central logistics. The module implemented mainly deals with pick–pack activities of the warehouses maintained by ST. This module mainly deals with shipments generated by the customers and has track of various activities that happens from creation of shipment to time from release to check shipment. The main motivation behind this module is to automate and ease the analysis of performance of the warehouses, i.e. the response time they take to serve the client.

6   Testing

It incorporates location of mistakes in the progressions done in the application. The testing procedure begins with a test arrange for that perceives test-related exercises, for example, experiment age, testing criteria, and asset portion for testing. The code is tried and mapped against the plan report made in the outline stage. The yield of the testing stage is a test report containing mistakes that happened while testing the application.

Testing is divided into three phases as follows:

– Unit testing
– Integration testing
– Functional testing.

6.1   Unit Testing

Unit testing refers to the testing of individual software units or related units, where a unit is the smallest functional part of an application. Unit testing makes heavy use of White box testing techniques along with Black box techniques.

In our environment, a typical screen and its associated components make a unit. White box testing measures like code-walkthroughs, control flow graphing are used extensively at this level apart from functional testing efforts like messages, boundary values etc.

Unit testing centres check exertion around the littlest unit of programming outline—the part. The units are recognized at the point-by-point configuration period of the product advancement life cycle, and the unit testing can be led parallel for different units. Five viewpoints are tried under Unit testing contemplations:

1. The part interface is tried to guarantee that data legitimately streams into and out of the program unit under test.
2. The nearby information structure is analysed to guarantee that information put away incidentally keeps up its uprightness amid all means in a calculation's execution.
3. Limit conditions are tried to guarantee that the module works appropriately at limits set up to constrain or confine handling.
4. Every single autonomous way (premise ways) through the control structure are practised to guarantee that all announcements in a module have been executed at any rate once.
5. Lastly, all mistakes taking care of ways are tried.

6.2   Integration Testing

Integration testing refers to the testing in which software units of an application are combined and tested for evaluating the interaction between them. Black box test case design is most prevalent during integration, though white box testing techniques like control flow graphing and execution tracing are also carried out. Inter-module and inter-product integration issues are the prime focus areas here. We concentrate on the application's business rules and ensure they are validated across different modules.

6.3 Functional Testing

Testing led on a total, incorporated framework to assess the framework's consistence with its predefined necessities. Programming once approved for meeting utilitarian necessities must be confirmed for legitimate interface with other framework components like equipment, databases and individuals. Practical testing confirms that all these framework components work legitimately and the product accomplishes general capacity/execution.

7 Conclusion

Achievement of an information stockroom usage can be measured when the counselling manager and the customer have executed the task as per the concurred necessities and undertaking extension. In an information distribution centre condition, it is far-fetched that partners can indicate all their detailing and investigation needs in advance. Thusly, getting out some example information from the information distribution centre right off the bat in the undertaking will give the clients an essence of the kind of data they can escape the information stockroom. This iterative way of "show and tell" will produce significantly more precise comprehension of the necessities along these lines guaranteeing that the information stockroom execution is following as needs be.

References

1. Rosen, M., Lublinsky, B., Smith, K.T., Balcer, M.J.: Applied SOA: Service-Oriented Architecture and Design Strategies. Wiley, New York (2012)
2. Wen, Y., Jingsha, H.: The application of tuxedo middleware in the banking system. In: 2011 IEEE 3rd International Conference on Communication Software and Networks (ICCSN), pp. 594–597. IEEE, New York (2011)
3. Sun, G., Zhao, D.: A small bank intermediary business system for social insurance based on tuxedo. In: 2011 International Conference on Internet Computing & Information Services (ICICIS), pp. 41–44. IEEE, New York (2011)
4. Chu, Q., Shen, Y., Jiang, Z.: A transaction middleware model for SCA programming. In: First International Workshop on Education Technology and Computer Science, 2009. ETCS'09, vol. 3, pp. 568–571. IEEE, New York (2009)
5. Andrade, J.M., Carges, M.T., MacBlane, M.R.: The Tuxedo tm system: An open on-line transaction processing environment. Data Eng. **51**, 33 (1994)

6. Felt, E.P.: Distributed transaction processing in the Tuxedo system. In: Proceedings of the Second International Conference on Parallel and Distributed Information Systems, pp. 266–267. IEEE, New York (1993)
7. Lihua, L., Jianan, X.: Research and implementation of e-mail direct marketing management system based on ExtJS. In: 2013 Third International Conference on Intelligent System Design and Engineering Applications (ISDEA), pp. 1106–1109. IEEE, New York (2013)
8. Chattopadhayay, B.: Getting Started with Oracle Tuxedo. Packt Publishing Ltd, Birmingham (2013)
9. Hall, C.L.: Building Client/Server Applications Using Tuxedo. Wiley, New York (1996)
10. Celar, S., Mudnic, E., Seremet, Z.: State-of-the-art of messaging for distributed computing systems. Ann. DAAAM Proc. 27 (2016)
11. Tyde, F.: Oracle cloud computing strategy. Program Executive, High Performance, Grid, Cloud Computing Oracle ASEAN (2009)
12. Böhm, K., Grabs, T., Röhm, U., Schek, H.J.: Evaluating the coordination overhead of replica maintenance in a cluster of databases. In: Euro-Par 2000 Parallel Processing, pp. 435–444. Springer, Berlin (2000)
13. Andrade, J.M., Carges, M.T., MacBlane, M.R.: Open online transaction processing with the tuxedo system. In: Compcon Spring'92. Thirty-Seventh IEEE Computer Society International Conference, Digest of Papers, pp. 366–371. IEEE, New York (1992)

Chaotic Spider Monkey Optimization Algorithm with Enhanced Learning

Nirmala Sharma, Avinash Kaur, Harish Sharma, Ajay Sharma and Jagdish Chand Bansal

Abstract Spider monkey Optimization (SMO) algorithm is a category of swarm intelligence-based algorithms, which mimics the fission–fusion social system (FFSS) comportment of spider monkeys. Although, SMO is proven to be a balanced algorithm, i.e., it balances the exploration and exploitation phenomena, sometimes the performance of SMO is degraded due to slow convergence in the search process. This article presents an efficient modified SMO algorithm which is capable of suppressing these inadequacies and is named as chaotic spider monkey optimization with enhanced learning (CSMO) algorithm. In this proposed algorithm, a chaotic factor is introduced in the Global Leader stage for providing appropriate stochastic nature and enhanced learning methods are commenced over the Local Leader stage and the Local Leader Learning stage in the form of learning method and exploring method, respectively. These changes help to enhance the exploration and exploitation proficiencies of SMO algorithm. Moreover, this proposed strategy is analysed on 12 different benchmark functions, and the results are being contrasted with original SMO and two of its recent variants, namely power law-based local search in SMO (PLSMO) and Lévy flight SMO (LFSMO).

Keywords Meta-heuristic optimization techniques · Spider monkey optimization algorithm · Swarm intelligence · Chaotic factor

N. Sharma · A. Kaur · H. Sharma (✉)
Rajasthan Technical University, Kota, Rajasthan, India
e-mail: harish.sharma0107@gmail.com

N. Sharma
e-mail: nirmala_rtu@yahoo.com

A. Kaur
e-mail: kauravinash05@gmail.com

A. Sharma
Government Engineering College, Jhalawar, Rajasthan, India
e-mail: ajay_2406@yahoo.com

J. C. Bansal
South Asian University, New Delhi, India
e-mail: jcbansal@gmail.com

© Springer Nature Singapore Pte Ltd. 2019
J. C. Bansal et al. (eds.), *Soft Computing for Problem Solving*,
Advances in Intelligent Systems and Computing 816,
https://doi.org/10.1007/978-981-13-1592-3_11

1 Introduction

Swarm intelligence (SI) is one of the efficient meta-heuristic optimization techniques to solve the optimization problems. Swarming nature of animals, insects, birds, etc. can solve complex optimization engineering problems without centralized controls [7].

Researchers found that swarm intelligence algorithms have great potential to solve combinatorial and numerical optimization problems [4, 8–11, 13]. Spider monkey optimization (SMO) algorithm is one of the known algorithms under this swarm intelligence category, which is developed by conceiving motivation from the cognitive behaviour and proficiency of communication among spider monkeys. They reside in group of atmost 40–50 monkeys in a single group and forage the food for whole group in the leadership of a female leader. The food searching behaviour of spider monkeys is regarded as fission–fusion social system (FFSS). The same phenomenon is applied in SMO algorithm for searching the appropriate solutions, but SMO [2, 10] also experiences the issue of slow convergence and stagnation. Hence, it is needed to sustain an appropriate balance amid intensification and diversification properties of the solutions during the solution search process.

This paper introduces a modified version of SMO, named as chaotic SMO with enhance learning (CSMO), to enhance the algorithm's performance. In this proposed technique, three different modifications are incorporated with three stages of SMO, namely the learning method, a chaotic factor, and the exploring method in the Local Leader stage, Global Leader stage, and the Local Leader Learning stage, respectively. The learning method always assures to produce a better solution after each iteration, while a chaotic factor provides a good randomization in order to capture solutions uniformly from the whole search space. Further, the Exploring method, which considers the impact of the worst solution of the swarm in the solution search space. Further, this strategy is being tested over 12 benchmark functions, and the outcomes are being contrasted with the original SMO and two of its recent variants, namely power law-based local search in SMO (PLSMO) and Lévy flight SMO (LFSMO).

This paper is organized as follows: Section 2 describes the original SMO algorithm. The proposed CSMO is explained in Sect. 3. Section 4 contains the results and experiments. Finally, Sect. 5 encompasses the surmised work.

2 Spider Monkey Optimization (SMO) Algorithm

There are six stages in SMO algorithm, namely Local Leader stage, Global Leader stage, Global Leader Learning stage, Local Leader Learning stage, Local Leader Decision stage, and Global Leader Decision stage [2].

Firstly, the population is initialized, and then, these stages are executed one by one, respectively. The initialization of N solutions of dimension D in the search space is done by the following equation:

$$SM_{ij} = SM_{\min j} + U(0, 1) \times (SM_{\max j} - SM_{\min j}) \tag{1}$$

where SM_{ij} is the ith spider monkey in jth dimension and $SM_{\min j}$ and $SM_{\max j}$ are the lower and upper bounds. $U(0, 1)$ is a uniformly distributed random function in the range $[0, 1]$.

1. **Local Leader stage**: This is the first stage in which the spider monkeys update their positions according to the Local Leader position as well as from the randomly selected monkey of the whole group. So for ith spider monkey, the position update equation will be:

$$\begin{aligned} SMnew_{ij} &= SM_{ij} + U(0, 1) \times (LL_{kj} - SM_{ij}) \\ &+ U(-1, 1) \times (SM_{rj} - SM_{ij}) \end{aligned} \tag{2}$$

where $SMnew_{ij}$ is the updated solution, LL_{kj} is the Local Leader in jth dimension of kth group, and SM_{rj} is the randomly selected spider monkey of jth dimension.

2. **Global Leader stage**: Now in the Global Leader stage, the solutions make the updations according to the position of the Global Leader as well as according to the position of any randomly selected monkey of the whole group. So, for ith spider monkey, the position update equation for this stage will be:

$$\begin{aligned} SMnew_{ij} &= SM_{ij} + U(0, 1) \times (GL_j - SM_{ij}) \\ &+ U(-1, 1) \times (SM_{rj} - SM_{ij}) \end{aligned} \tag{3}$$

where GL_{kj} is the Global Leader in jth dimension of kth group.

3. **Global Leader Learning stage**: The Global Leader Learning stage helps to select the Global Leader in the swarm by employing the greedy selection process based upon the fitness values. Also, there is a counter, namely *GlobalLimitCount*, which is incremented by one, if the Global Leader is not updating its position in each iteration.

4. **Local Leader Learning stage**: Same in the Local Leader Learning stage, the Local Leader is selected who is having the best fitness in the particular local group of the swarm. If that position of the Local Leader is not updated in each iteration, then the associated counter, namely *LocalLimitCount*, is incremented by one.

5. **Local Leader Decision stage**: In this stage, the Local Leader is regarded as stuck in local optima, if the associated counter, namely *LocalLimitCount*, crosses the predefined threshold *LocalLeaderLimit*. Hence, to avoid this condition, the local

group members are updated either by random initialization or by the following
equation as per the perturbation rate (pr):

$$\text{SMnew}_{ij} = \text{SM}_{ij} + U(0, 1) \times (\text{GL}_j - \text{SM}_{ij})$$
$$+ U(0, 1) \times (\text{SM}_{ij} - \text{LL}_{ij}) \tag{4}$$

It is clear from above Eq. 4 that the solutions tend to attract towards the Global
Leader and move away from the Local Leader so as to come out from the local
value.

6. **Global Leader Decision stage**: Here, in the Global Leader Decision stage, the
   GlobalLimitCount is regulated over the threshold value, namely *GlobalLeader
   Limit*. If the associated counter crosses that limit, the Global Leader divides its
   group into subgroups, but that subgroups should not exceed the maximum group
   (MG) limit which is calculated by total number of monkeys divided by ten, i.e.,
   $N/10$.

2.1 Spider Monkey Optimization (SMO) Algorithm

The complete steps of SMO algorithm is shown in Algorithm 1:

Algorithm 1 Spider Monkey Optimization (SMO)

Initialization of solutions using equation 1.1, also initialize *LocalLeaderLimit*,
GlobalLeaderLimit and *pr*.

Calculation of fitness of the solutions based upon the distance of the food source from the spider
monkeys.

Selection of Global Leader and Local Leader.

while until termination condition is not reached **do**

  Update the solutions in local groups using equation 1.2.

  Apply the greedy selection process according to the fitness values of all solutions.

  Calculate the probability ($prob_i$) for all group members.

  Now according to the selected $prob_i$, global leader, other group members and by self-experience,
  update the positions of the solutions using equation 1.3.

  Again apply that greedy selection process on all group members and update the Global and
  Local Leaders.

  cycle = cycle + 1

  Whenever the condition happens for Local Leader, which is described in STEP 5 of SMO
  algorithm, equation 1.4 is applied.

  Similarly, when Global Leader is stagnated, the solution which is described in STEP 6, is
  followed.

end while Result the best solution detected so far.

Here, $pr \, \varepsilon [0.1, 0.8]$, and N is the swarm size.

3   Chaotic Spider Monkey Optimization Algorithm with Enhanced Learning (CSMO)

To enhance the exploration and exploitation capabilities, three modifications, namely learning method, a chaotic factor, and exploring method are introduced in the SMO as follows:

The Learning method: In the Local Leader stage, the position update process is modified as shown in Algorithm 2. The third term of the position update equation presents, the case in which the randomly selected solution (nth) is more fit than the current solution (ith), then the nth solution is attracted towards the ith solution, otherwise the ith solution will be attracted towards the nth solution [6]. In the position update equation, SM_{nj} is the *neighbour* of ith solution of jth dimension, and fit_i and fit_n are the fitness values of ith and nth solutions, respectively. It is clear from the Algorithm 2 that the less fit solutions are attracted towards higher fit solutions in the solution search process. This change will make the modified algorithm exploiting more efficiently and help producing the better solutions.

Algorithm 2 Improved position update process in Local Leader stage:

---

  for each $k \in \{1, \ldots, MG\}$ **do**
    for each member $SM_i \in k^{th}$ group **do**
      for each $j \in \{1, \ldots, D\}$ **do**
        if $U(0, 1) \geq pr$ **then**
          if $fit_i > fit_n$ **then**
            $SMnew_{ij} = SM_{ij} + U(0, 1) \times (LL_{kj} - SM_{ij}) + U(-1, 1) \times (SM_{ij} - SM_{nj})$
          else
            $SMnew_{ij} = SM_{ij} + U(0, 1) \times (LL_{kj} - SM_{ij}) + U(-1, 1) \times (SM_{nj} - SM_{ij})$
          end if
        else
          $SMnew_{ij} = SM_{ij}$
        end if
      end for
    end for
  end for

A Chaotic factor: In this modification, the position update process of Global Leader stage is shown in Algorithm 3. Chaos [3] is actually a universal phenomena of nonlinear dynamics, and the chaotic optimization mechanism means the better stochastic search and hence more randomicity, ergodicity, and regularity to find the better results.

The Exploring method: In the Local Leader Decision stage, the solutions when get stuck somewhere at local optima, and could not update themselves, a disturbance is introduced in the solution search process to divert the solutions at other locations. In the modified Local Leader Decision stage, a new component $U(0, 1)$ $(Gworst_j - SM_{ij})$ is incorporated with the existing position update equation as shown

Algorithm 3 Improved position update process in Global Leader stage:

---

for $k = 1$ to MG **do**
   $count = 1$;
   $GS = k^{th}$ group size;
   while $count < GS$ **do**
     for $i = 1$ to GS **do**
       if $U(0, 1) < prob_i$ **then**
         $count = count + 1.$
         Randomly select $j \in \{1 \ldots D\}$.
         Randomly select SM_r from k^{th} group s.t. $r \neq i$.
         Calculate a Chaotic factor as follows:
         $w = 0.5 \times U(0, 1) + 0.5 \times (4 \times U(0, 1) \times (1 - U(0, 1))).$
         $SMnew_{ij} = SM_{ij} + w \times (GL_j - SM_{ij}) + U(-1, 1) \times (SM_{rj} - SM_{ij}).$
       end if
     end for
     if i is equal to GS **then**
       $i = 1$;
     end if
   end while
end for

---

in Algorithm 4. This introduced component will divert the stuck solutions towards the worst solution found so far in the swarm and thus enhances the exploration ability of the swarm [5] as shown in Algorithm 4.

Algorithm 4 Improved Local Leader Decision stage:

---

for $k = \{1 \ldots MG\}$ **do**
   if $LocalLimitCount_k > LocalLeaderLimit$ **then**
     $LocalLimitCount_k = 0.$
     $GS = k^{th}$ group size;
     for $i \in \{1 \ldots GS\}$ **do**
       for each $j \in \{1 \ldots D\}$ **do**
         if $U(0, 1) \geq pr$ **then**
           $SMnew_{ij} = SM_{minj} + U(0, 1) \times (SM_{maxj} - SM_{minj})$
         else
           $SMnew_{ij} = SM_{ij} + U(0, 1) \times (GL_j - SM_{ij}) + U(0, 1) \times (SM_{ij} - LL_{ij}) +$
           $U(0, 1) \times (Gworst_j - SM_{ij})$
         end if
       end for
     end for
   end if
end for

The pseudocode of modified SMO is depicted in Algorithm 5.

Algorithm 5 Chaotic Spider Monkey Optimization Algorithm with Enhanced Learning (CSMO)

Initialize the parameters: *MI* (Maximum number of iterations), *D* (Dimension of the problem), and *N* (Swarm Size)
Initialize the swarms, SM_i where (i=1,2,...,*N*) using equation (1.1)
iter = 1
while iter <> MI **do**
   Local Leader stage using Algorithm 2
   Global Leader stage using Algorithm 3
   Global Leader Learning stage
   Local Leader Learning stage
   Local Leader Decision stage using Algorithm 4
   Global Leader Decision stage
   iter=iter+1
end while Result the ideal solution detected so far.

---

4 Experimental Results

The proposed CSMO algorithm is being analysed on 12 different global optimization problems (f_1 to f_{12}) [1, 14] which are presented in Table 1.

A correlative comparison is also taken among CSMO, SMO, and its significant variants, namely PLSMO and LFSO). Experimental settings, which are adopted to test CSMO, SMO, PLSMO, and LFSMO over the regarded test problems, are as follows:

- The number of simulations/run=100,
- Swarm size $N = 50$ and maximum group MG = $N/10$,
- Rest of the parameter settings for the algorithms SMO, PLSMO, and LFSMO are similar to their original research paper.

Table 2 shows the comparison among CSMO, SMO, PLSMO, and LFSMO in terms of standard deviation (SD), mean error (ME), average number of function evaluations (AFE), and success rate (SR). And it is clearly demonstrated that CSMO is better in terms of reliability, efficiency, and accuracy as contrasted to the other regarded algorithms. Moreover, the convergence speeds of the regarded algorithms are contrasted by means of average number function evaluations (AFEs). Higher the convergence speed, smaller the AFE, so far as reflected in CSMO. To suppress the impact of stochastic nature of the algorithms, function evaluations for every problem are taken as average over 100 runs. Also, the acceleration rate (AR) [12] is calculated in order to contrast the convergence speeds, which basically depends upon the AFEs of the algorithms as shown below:

$$AR = \frac{AFE_{ALGO}}{AFE_{CSMO}}, \tag{5}$$

Here, ALGO $\in$ { SMO, PLSMO, and LFSMO } and AR > 1 depicts that CSMO is comparatively fast.

Table 1 Test problems

S. No.	Test problem	Objective function	Search range	D	Acc error
1	Sphere	$f_1(x) = \sum_{i=1}^{D} x_i^2$	[−5.12, 5.12]	30	1.00E−05
2	De Jong f4	$f_2(x) = \sum_{i=1}^{D} i.(x_i)^4$	[−5.12, 5.12]	30	1.00E−05
3	Griewank	$f_3(x) = 1 + \frac{1}{4000}\sum_{i=1}^{D} x_i^2 - \prod_{i=1}^{D}\cos(\frac{x_i}{\sqrt{i}})$	[−600, 600]	30	1.00E−05
4	Zakharov	$f_4(x) = \sum_{i=1}^{D} x_i^2 + (\sum_{i=1}^{D}\frac{ix_i}{2})^2 + (\sum_{i=1}^{D}\frac{ix_i}{2})^4$	[−5.12, 5.12]	30	1.00E−02
5	Salomon problem	$f_5(x) = 1 - \cos(2\pi\sqrt{\sum_{i=1}^{D}x_i^2}) + 0.1(\sqrt{\sum_{i=1}^{D}x_i^2})$	[−100, 100]	30	1.00E−01
6	Step function	$f_6(x) = \sum_{i=1}^{D} (\lfloor x_i + 0.5 \rfloor)^2$	[−100, 100]	30	1.00E−05
7	Colville function	$f_7(x) = 100[x_2 - x_1^2]^2 + (1 - x_1)^2 + 90(x_4 - x_3^2)^2 + (1 - x_3)^2 + 10.1[(x_2 - 1)^2 + (x_4 - 1)^2] + 19.8(x_2 - 1)(x_4 - 1)$	[−10, 10]	4	1.00E−05
8	Kowalik	$f_8(x) = \sum_{i=1}^{11}[a_i - \frac{x_1(b_i^2 + b_i x_2)}{b_i^2 + b_i x_3 + x_4}]^2$	[−5, 5]	4	1.00E−04
9	2D Tripod function	$f_9(x) = p(x_2)(1 + p(x_1)) + |(x_1 + 50p(x_2)(1 - 2p(x_1)))| + |(x_2 + 50(1 - 2p(x_2)))|$	[−100, 100]	2	1.00E−04
10	Goldstein-Price	$f_{10}(x) = (1 + (x_1 + x_2 + 1)^2 \cdot (19 - 14x_1 + 3x_1^2 - 14x_2 + 6x_1x_2 + 3x_2^2)) \cdot (30 + (2x_1 - 3x_2)^2 \cdot (18 - 32x_1 + 12x_1^2 + 48x_2 - 36x_1x_2 + 27x_2^2))$	[−2, 2]	2	1.00E−14
11	Hosaki problem	$f_{11} = (1 - 8x_1 + 7x_1^2 - 7/3x_1^3 + 1/4x_1^4)x_2^2 \exp(-x_2)$, subject to $0 \le x_1 \le 5, 0 \le x_2 \le 6$	[0, 5][0, 6]	2	1.00E−06
12	Meyer and Roth problem	$f_{12}(x) = \sum_{i=1}^{5}(\frac{x_1 x_3 t_i}{1 + x_1 t_i + x_2 v_i} - y_i)^2$	[−10, 10]	3	1.95E−03

Table 2 Test problems comparison of the results

Test Problem	Algorithm	SD	ME	AFE	SR
f_1	CSMO	8.15E−07	8.88E−06	13117.49	100
	SMO	8.37E−07	8.87E−06	13642.30	100
	PLSMO	8.15E−07	8.96E−06	13174.79	100
	LFSMO	9.02E−07	9.00E−06	13182.96	100
f_2	CSMO	1.25E−06	8.20E−06	11102.84	100
	SMO	1.20E−06	8.49E−06	12725.66	100
	PLSMO	9.49E−07	8.68E−06	11198.07	100
	LFSMO	1.26E−06	8.54E−06	12325.96	100
f_3	CSMO	4.94E−03	2.12E−03	69888.60	80
	SMO	5.48E−03	2.96E−03	77028.55	72
	PLSMO	4.25E−03	2.13E−03	74710.82	77
	LFSMO	4.65E−03	2.45E−04	85957.91	73
f_4	CSMO	6.41E−04	9.38E−03	111585.85	100
	SMO	6.18E−04	9.41E−03	141818.34	100
	PLSMO	4.95E−04	9.65E−03	121187.10	100
	LFSMO	1.11E−03	9.44E−03	132136.90	100
f_5	CSMO	3.76E−02	1.83E−02	192815.75	17
	SMO	3.13E−02	1.89E−01	190609.68	11
	PLSMO	2.82E−02	1.98E−01	197091.37	15
	LFSMO	1.71E−02	1.97E−02	195537.39	13
f_6	CSMO	0.00E+00	0.00E+00	12214.54	100
	SMO	9.95E−02	1.00E−02	13986.92	99
	PLSMO	0.00E+00	0.00E+00	17812.30	100
	LFSMO	2.18E−02	5.00E−02	13305.71	99
f_7	CSMO	2.13E−04	7.90E−04	14456.54	100
	SMO	2.23E−04	7.86E−04	54591.31	100
	PLSMO	1.70E−04	8.58E−04	18452.62	100
	LFSMO	5.87E−05	9.72E−04	16728.67	100
f_8	CSMO	1.84E−05	8.85E−05	44247.84	100
	SMO	9.66E−05	1.03E−04	40350.34	97
	PLSMO	1.40E−04	1.19E−04	34295.95	97
	LFSMO	1.17E−04	1.17E−04	34870.39	97
f_9	CSMO	2.45E−05	6.24E−05	12144.14	100
	SMO	2.55E−05	6.08E−05	14448.76	100
	PLSMO	2.37E−05	6.65E−05	13709.36	100
	LFSMO	2.40E−05	6.64E−05	13508.80	100
f_{10}	CSMO	4.79E−14	4.96E−14	103704.76	51
	SMO	4.81E−14	5.70E−14	115459.67	43
	PLSMO	4.73E−14	6.16E−14	125301.11	44
	LFSMO	4.88E−14	5.37E−14	107600.93	47

(continued)

Table 2 (continued)

Test Problem	Algorithm	SD	ME	AFE	SR
f_{11}	CSMO	4.01E−06	1.00E−05	179296.02	24
	SMO	3.18E−06	1.07E−05	174092.01	10
	PLSMO	3.63E−06	1.04E−05	178104.34	11
	LFSMO	4.92E−06	9.30E−06	158185.78	21
f_{12}	CSMO	2.69E−06	1.95E−03	1680.97	100
	SMO	2.77E−06	1.95E−03	2026.62	100
	PLSMO	3.01E−06	1.95E−03	1755.23	100
	LFSMO	3.05E−06	1.95E−04	1733.32	100

Table 3 Comparison of the basic SMO, PLSMO and LFSMO based on Acceleration Rate (AR) of CSMO

Test problems	SMO	PLSMO	LFSMO
f_1	1.040008416	1.004368214	1.004368214
f_2	1.146162603	1.008577085	1.008577085
f_3	1.102161869	1.068998664	1.068998664
f_4	1.2709348	1.086043616	1.086043616
f_5	0.988558663	1.022174641	1.022174641
f_6	1.145104114	1.458286599	1.458286599
f_7	3.776236222	1.276420222	1.276420222
f_8	0.911916604	0.775087552	0.775087552
f_9	1.189772186	1.128886854	1.128886854
f_{10}	1.113349763	1.208248397	1.208248397
f_{11}	0.970975318	0.993353561	0.993353561
f_{12}	1.205625323	1.044176874	1.044176874

Using Eq. (5), *AR* is calculated betwixt CSMO and SMO, CSMO and PLSMO, and CSMO and LFSMO. This comparison is tabulated in Table 3 which clarifies that CSMO is fastest among all the regarded algorithms for most of the test functions. Additionally, boxplots evaluation [8] for average number of function evaluations is carried out for all the algorithms CSMO, SMO, PLSMO, and LFSMO to show the empirical distribution of data pictorially. The boxplots diagram is shown in Fig. 1. It is divulged easily from the figure that interquartile range and medians of CSMO are contrastingly less.

Therefore, it is easily concluded from boxplots that CSMO is cost efficient than SMO, PLSMO, and LFSMO. Apart from this, an alternative statistical test is also performed in order to examine that whether these results are with a considerable difference or not. As the average number of function evaluations for regarded algorithms is not scattered uniformly, hence a nonparametric statistical test is needed to distinguish the results of the algorithms.

Fig. 1 Boxplots graphs for average number of function evaluation

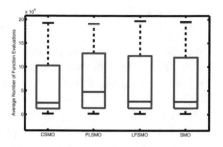

A well-substantiated and nonparametric test is available, namely Mann–Whitney U rank sum [10], which is used to do comparison among non-Gaussian data. The level of consequence is taken as 5% in this paper, i.e., $\alpha = 0.05$ for comparing betwixt CSMO–SMO, and CSMO–PLSMO, CSMO–LFSMO.

The results of the Mann–Whitney U rank sum test for the average number function evaluations of 100 simulations are being shown in the Table 4. Firstly, we will inspect the resulted dissimilarity evaluated through Mann–Whitney U rank sum test; i.e., we will check whether there is a significant difference betwixt two data sets or not. If that difference is not seen, then that means there is a null hypothesis; hence, '=' sign appears for that. Other than that, we can contrast the average number of function evaluations. Furthermore, for the conditions where CSMO is performed over less or more average number of function evaluations than the other algorithms, the signs '+' and '−' are used, respectively. It is all shown in Table 4, which contains 29 '+' signs out of 36 comparisons. Thus, it can be concluded that the results of CSMO are considerably cost-effective than SMO, PLSMO, and LFSMO over regarded test problems.

Table 4 Depending upon the mean function evaluations and the Mann-Whitney U rank sum test, comparison is done by taking $\alpha = 0.05$. at noteworthy level ('+' shows CSMO is consequently better, '−' shows CSMO is worse and '=' shows that there is no consequent distinction), TP: Test Problem

Test problems	CSMO versus SMO	CSMO versus PLSMO	CSMO versus LFSMO
f_1	+	+	+
f_2	+	+	+
f_3	+	+	+
f_4	+	+	+
f_5	−	+	+
f_6	+	+	+
f_7	+	+	+
f_8	−	−	−
f_9	+	+	+
f_{10}	+	+	+
f_{11}	−	−	−
f_{12}	+	+	+

5 Conclusion

To improve the intensification and diversification properties of spider monkey optimization (SMO) algorithm, three major changes have been performed in this paper, and the resulting algorithm is named as chaotic spider monkey optimization algorithm with enhanced learning (CSMO). Firstly, the learning method is introduced in the Local Leader stage to explore the search space fully. After that, a chaotic factor is applied in the Global Leader stage to provide better stochastic effect. And finally, the exploration method is introduced in Local Leader Decision stage in order to avoid the premature condition and exploring more by considering a worst solution of the swarm. This proposed strategy is experimentally evaluated through various tests, and it is being analysed that for solving real-world continuous optimization problems, CSMO is a very useful strategy. Also for resolving the discrete optimization problems, CSMO may be regarded and its proficiency may be examined.

References

1. Ali, M.M., Khompatraporn, C., Zabinsky, Z.B.: A numerical evaluation of several stochastic algorithms on selected continuous global optimization test problems. J. Global Optim. **31**(4), 635–672 (2005)
2. Bansal, J.C., Sharma, H., Jadon, S.S., Clerc, M.: Spider monkey optimization algorithm for numerical optimization. Memetic Comput. **6**(1), 31–47 (2014)
3. Feng, Y., Teng, G.-F., Wang, A.-X., Yao, Y.-M.: Chaotic inertia weight in particle swarm optimization. In: Second International Conference on Innovative Computing, Information and Control, 2007. ICICIC'07, pp. 475–475. IEEE, New York (2007)
4. Gupta, K., Deep, K., Bansal, J.C.: Improving the local search ability of spider monkey optimization algorithm using quadratic approximation for unconstrained optimization. Comput. Intell. (2016)
5. Liu, H., Xu, G., Ding, G., Sun, Y.: Human behavior-based particle swarm optimization. Sci. World J. 2014 (2014)
6. Rao, R.V., Savsani, V.J., Vakharia, D.P.: Teaching-learning-based optimization: a novel method for constrained mechanical design optimization problems. Comp. Aided Des. **43**(3), 303–315 (2011)
7. Rauff, J., et al.: Nature-inspired optimization algorithms. Math. Comput. Educ. **49**(3), 208 (2015)
8. Sharma, A., Sharma, H., Bhargava, A., Sharma, N.: Optimal design of PIDA controller for induction motor using spider monkey optimization algorithm. Int. J. Metaheuristics **5**(3–4), 278–290 (2016)
9. Sharma, A., Sharma, H., Bhargava, A., Sharma, N.: Power law-based local search in spider monkey optimisation for lower order system modelling. Int. J. Syst. Sci. pp. 1–11 (2016)
10. Sharma, A., Sharma, H., Bhargava, A., Sharma, N., Bansal, J.C.: Optimal placement and sizing of capacitor using Limaçon inspired spider monkey optimization algorithm. Memetic Comput. pp. 1–21 (2016)
11. Sharma, A., Sharma, H., Bhargava, A., Sharma, N., Bansal, J.C.: Optimal power flow analysis using Lévy flight spider monkey optimisation algorithm. Int. J. Artif. Intell. Soft Comput. **5**(4), 320–352 (2016)

12. Sharma, N., Sharma, H., Sharma, A., Bansal, J.C.: Modified artificial bee colony algorithm based on disruption operator. In: Proceedings of Fifth International Conference on Soft Computing for Problem Solving, pp. 889–900. Springer, Berlin (2016)
13. Singh, U., Salgotra, R., Rattan, M.: A novel binary spider monkey optimization algorithm for thinning of concentric circular antenna arrays. IETE J. Res. pp. 1–9 (2016)
14. Suganthan, P.N., Hansen, N., Liang, J.J., Deb, K., Chen, Y.-P., Auger, A., Tiwari, S.: Problem definitions and evaluation criteria for the CEC: special session on real-parameter optimization. KanGAL Report **2005005** (2005)

Analysis of Liver Cancer Using Data Mining SVM Algorithm in MATLAB

Srinivas Vadali, G. V. S. R. Deekshitulu and J. V. R. Murthy

Abstract Liver cancer is one among the normal types of cancer. Detection and determination of liver tumor at early stage are vital. The vast majority of the cancer passings can be anticipated by early detection, determination, and compelling treatment. It is required to fragment the liver tumor from the medical images for tumor analysis. A robotized framework is proposed for segmentation and classification of liver tumor which is an effective and simple to utilize technique. The proposed framework comprises of PreAprocessing, segmentation, postAprocessing, and a last classification as benign and malignant. Amid the preAprocessing stage, the image is resized to 256×256. In the segmentation stage, level set strategy is connected for sectioning the suspicious area. In postAprocessing stage, the district of intrigue is acquired from the first image. At long last the Pseudo Zenerike minute and GLDM is utilized for highlight extraction from CT image. These components are given as contribution to the SVM for classification of tumor as benign or malignant. The SVM is prepared utilizing four images. The proposed framework can accomplish precision rate of 86.7%.

Keywords Medical diagnosis · Lung cancer · MATLAB · SVM classification

S. Vadali (✉)
Department of CSE, Jawaharlal Nehru Technological University Kakinada (JNTUK), Kakinada, India
e-mail: vadalisrinivas16@gmail.com

G. V. S. R. Deekshitulu
Department of Mathematics, University College of Engineering Kakinada (UCEK),
Jawaharlal Nehru Technological University Kakinada (JNTUK), Kakinada, India
e-mail: dixitgvsr@hotmail.com

J. V. R. Murthy
Department of CSE, University College of Engineering Kakinada (UCEK),
Jawaharlal Nehru Technological University Kakinada (JNTUK), Kakinada, India
e-mail: mjonnalagedda@gmail.com

© Springer Nature Singapore Pte Ltd. 2019
J. C. Bansal et al. (eds.), *Soft Computing for Problem Solving*,
Advances in Intelligent Systems and Computing 816,
https://doi.org/10.1007/978-981-13-1592-3_12

163

1   Introduction

A cancer is a development of cells in irregular way, more often than not started from a specific cell which is unusual . Control components of the cells are lost, coming about the ceaseless irregular development, contaminate the close-by tissues, spread to other part of the body, and animate the fresh recruits vessels development. Cancerous cells can develop from any part of the tissue inside the body. Cancerous cells create and increment a mass of cancer-causing tissue called a tumor. That assaults and destroys commonplace touching tissues. Tumors can be malignant or noncancerous. Cancerous cells from the essential (introductory) site can spread all through the body (metastasize). Typically, the cancer is produced by the outside viewpoints, for example, irresistible infection, utilization of tobacco, and inside perspectives, for example, from heredity, hormones, and resistance limit. Treatment of ailment is conceivable up to some connect through surgery, radiation, chemotherapy, hormone treatment, safe treatment, and treatment. Normal cancer sorts melanoma, lung cancer, melanoma, leukemia, pancreatic cancer, thyroid cancer, breast cancer, kidney cancer. All tumors cannot be cured. Malignant tumor cannot be cured, but benign tumor can be cured. A man's threat peril can be decreased with strong choices like keeping up a key separation from tobacco, confining alcohol utilize, protecting your skin from the sun and avoiding indoor tanning, eating a regimen rich in items like vegetables and natural products, keeping a sound weight, and being physically progressive.

2   Related Work

R. Rajagopal et al. [1] proposed a strategy for segmentation of liver sores. Numerical morphology utilized as a middle separating and Otsu's thresholding technique. Also, morphological sifting is utilized to remove the sores shape and edges. Gabar change channel for edge detection. For various sorts of liver tumors, an exact outcome is created with no manual cooperation. The system can be enhanced by incorporating a fuzzy and neural system algorithm.

Pack Chen et al. [2] proposed a strategy where multistep LSM with different introduction is utilized. Bend advancement is initially developed independently by quick walking technique and afterward LSM. At that point, both are joined with the curved corridor algorithm for getting an unpleasant liver counter. FMM and LSM are executed in light of the underlying bends. Halfway segmentation yield is utilized for CH algorithm. Liver form is smoothened by LSM. LSM with different instatements is much speedier which covers more liver locales. Yet at the same time, over-spillage and over-segmentation issue exists. To consequently finish up perfusion bends, a robotized perfusion analysis strategy is proposed at the same time, still under segmentation issue on lower sharp corner injuries display because of a low inclination meaning of the lower half liver sores.

Chen Zhaoxue et al. [3] proposed a strategy for liver sores segmentation which incorporates a basic line seek technique and a plane segmentation to get a twofold image which is made out of groups of segregated white pixels from the liver segment. Extraction of liver injury in light of the spatial attributes and histogram. For associating the secluded pixel groups, Gaussian obscuring procedure is utilized. To obscured image, a thresholding is connected after the post-preparing venture for retouching gaps and size channels.

S. Luo et al. [4] proposed a technique which comprises of three stages for liver segmentation. For extricating pixel level components, surface examination is utilized. Wavelet coefficient and Haralick surface portrayal likewise utilized. What's more, classification is done utilizing SVM. To expel commotion an outline liver, morphological strategy is utilized. By joining the elements delivers a best outcome then utilizing just haralick surface portrayal when SVM is utilized to order.

Shraddha Sangewar et al. [5] proposed a technique for liver sores segmentation which comprise the altered k-implies strategy with confined molding algorithm, and to isolate a image into various injuries, isolate sores are distinguished and segmentation is accomplished. This technique gives an exact liver segmentation.

O. FekryAbd-Elaziz et al. [6] proposed a programmed segmentation technique which comprises of a mix of district developing and force examination with pre-preparing step. A strategy for programmed liver segmentation. This technique lessens the time many-sided quality by barring alternate locales in the image. Generally, liver part is sectioned by beginning from the seed point in locale developing technique. Here, seed point is chosen naturally.

Wenhan Wang et al. [7] a novel technique is proposed; area developing algorithm is utilized for evacuating non-liver tissues. Furthermore, morphological component is utilized with various level settings. Discrete purposes of liver injuries are gotten. Edges are adjusted by slope data, and three-dimensional reclamations are utilized to enhance the recouped image which has a lower time unpredictability yet there is an over-segmentation issue exists.

Ina Singh et al. [8] discussed the k-means algorithm, and they investigated the weakness of the standard k-implies algorithm, similar to estimation of separation between the every data protest and the bunches in every emphasess. Consequently, effectiveness of the every group is very little. This paper delivered an enhanced k-implies algorithm to beat these issues with standard k-implies. In any case, requires a straightforward data structure for putting away data of every emphases, utilized as a part of next cycle. This strategy does not process the separation of every data protest group over and over. An outcome demonstrates that this strategy enhances the speed of running time, bunching, and precision.

Gambino, O. what's more, et al. [9] proposed a technique where a programmed surface based district developing for liver segmentation. Three-dimensional-seeded district becoming in light of surface element is utilized, and a seed voxel is naturally chosen inside the liver organ, and algorithm of the limit esteem for area developing ceasing condition is additionally programmed. Elements are removed for CT stomach volumes and the seed locale developing relied on upon highlight space insights.

Xinjian Chen et al. [10] proposed a strategy which depends on the dynamic appearance which is deliberately consolidated model (AAM), live wire, and diagram cuts for segmentation which comprises of three principle parts they are model building, object acknowledgment, and depiction. In model building, they have developed an AAM and afterward prepared the LW capacity and GC parameters. In the second part they proposed a algorithm for enhancing the ordinary AAM, AAM and LW are consolidated viably, which brings about situated AAM. For protest instatement, a multi-question system is adjusted. Pseudo 3D introduction and segmentation of organs cuts with the assistance of OAAM. For outline section, 3D-shaped obliged GC is proposed.

B. BalaKumar and A. Mohamed Syed Ali proposed a completely programmed framework for liver segmentation which utilized a fuzzy c-implies bunching alongside the level set. By enhancing the distinction of one of a kind image, the limits are made clearer. At that point for removing the liver sores, the anatomical data and spatial fuzzy c-means are utilized.

Laura Manuel proposed a technique for liver segmentation from CT checks, by utilizing the dynamic surface strategy. By joining within and outside powers, it finds a surface which minimizes the vitality work. This technique is productive in liver sore extraction. This lessens time unpredictability and builds the precision.

Sajith A.G. and Hariharan S. proposed a straightforward strategy liver tumor segmentation from CT check, which is valuable for centers. Level set is generally utilized for segmentation as a part of image preparing. Here, firstly fuzzy c-implies algorithm is utilized, and for fine outline, level set is utilized. The liver locale and the segmentation performed by this technique are extremely very much characterized.

N. UmaDevi1 and R. Poongodi In this strategy the spatial fuzzy c-implies alongside the level set is utilized for element variety limits for segmentation. The fuzzy level set is developed from the spatial fuzzy bunching segmentation. This is helpful for medical image segmentation.

Xuechen Li et al. presented an approach for programmed segmentation of liver CT examines by utilizing the fuzzy c-implies bunching and the level set. The framework gives a proficiency of 0.99 precision with the specificity of 0.9989. At the point when this framework is contrasted and the standard level set, the over-segmentation issue is diminished in this framework.

3   Data Mining

Data mining is a vital procedure where keen techniques are connected to discover data designs. It is the way toward finding intriguing example and vital data or data from a lot of data. It is mainstream because of the fruitful applications in media transmission, promoting, and tourism. Nowadays, the convenience of the techniques has been demonstrated likewise in medical field. Data mining is otherwise called learning mining from data, data extraction, data/design analysis, data archaic exploration, and data digging. Notwithstanding, there are different facilities to consider while picking the pertinent data mining strategy to be utilized as a part of a specific application. The "best" model is regularly found by trial or hit and miss: attempting distinctive advancements and algorithms. One of the algorithms is data classification is the way toward finding a model (or capacity) that clarify and distinctive data classes. Data mining innovation gives a simple to utilize approach i.e. client situated way to deal with decide the concealed examples in data. Classification procedure has been connected in different regions of issue like drug, social administration, and designing fields. Different sorts of issue like sicknesses conclusion, image acknowledgment and credit advancement utilizing SVM algorithms or methods [2].

4   Overview of Liver Cancer

Liver cancer is a difficult issue. As of not long ago, in the most medical research, the explanations behind affliction from liver cancer are indistinct. It is hard to identify the liver cancer at beginning stage, and it is working extremely well or regularly when it is partially harmed [3]. Liver cancer is one of the more hazardous and debilitating infections at worldwide level with more than one million cases analyzed every year [4]. It is the fourth regular cancer in world and third driving reason for cancer mortality. Liver cancer is hard to recognize at early stage because of the absence of indications [3]. A few sorts of hazard variable like cirrhosis, stoutness, smoking, hepatitis B and hepatitis C or liquor abuse are exceptionally connected to the liver cancer [4, 5]. The medical term that is utilized for the liver cancer is hepatocellular carcinoma. It is most destructive to life, and it is more regular for men than ladies [6]. Medical finding is imperative and more convoluted assignment. It should be executed precisely or entirely and productively. As, there is unnerve of assets, for example, skill to handle these sorts of life undermining sicknesses, a robotized framework would be taken as better choice to put all assets people and mastery all together to make esteem and advantages for society at least cost and greatest proficiency [1]. The usage of a computerized framework needs a point by point analysis of different procedures accessible.

5   Problem Statement

Liver segmentation and classification of tumor is pivotal amid treatment stage. In any case, segmentation of liver is troublesome because of comparable powers of the neighboring tissues. What's more, classification of tumor physically requires part of time, and it is expensive. Thus, there is requirement for a framework which recognizes and group liver tumor consequently.

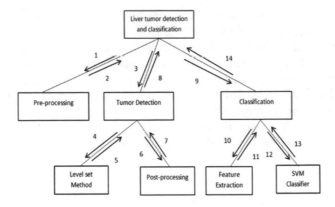

Fig. 1 Proposed system architecture

6 Proposed System

A framework to distinguish and characterize liver tumor consequently is planned and created. Proposed framework depends on the CT images. For segmentation, level set technique is utilized, and SVM classifier is utilized for characterizing tumor. SVM is prepared with different elements from the liver tumor images (Fig. 1).

- Image pre-processing stage: The CT image is resized into 256×256 and changed over into dim-level image.
- Level set strategy stage: The pre-prepared image is given as contribution to the level set technique. The yield of this module is the tumor fragmented image.
- Post-preparing stage: The fragmented image is given as contribution for post-handling module for getting the tumor area of intrigue image.
- Feature extraction: Features are extricated from the tumor image for classification.
- SVM classifier: The extricated elements are taken as info, and the yield of this phase is tumor classification as malignant and benign.

7 Results Analysis

Analysis of system result

Analysis of the proposed framework is fundamental since it gives a substance of frameworks productivity. For analysis of the outcome, 15 CT images of liver are considered which comprise of both malignant and benign tumor. The underneath table demonstrates the dataimage given to framework, and the level set connected to image and the divided image yield and the classification of the tumor is done, i.e., malignant or benign class which is simply in view of the parameters. Tables 1 and 2 show the input and output of the system.

Table 1 Level set output

Sl. no.	Image ID	Input	Level set output	Classification result
1	L_1	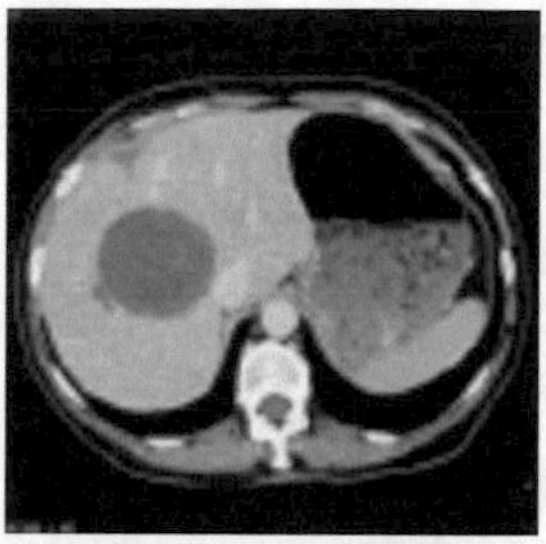	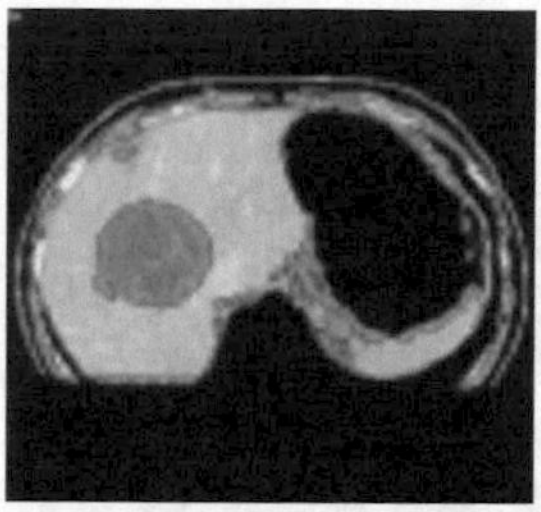	Malignant
2	L_2	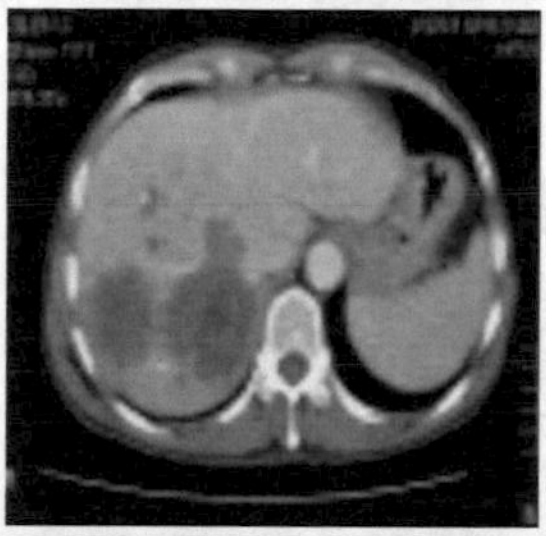	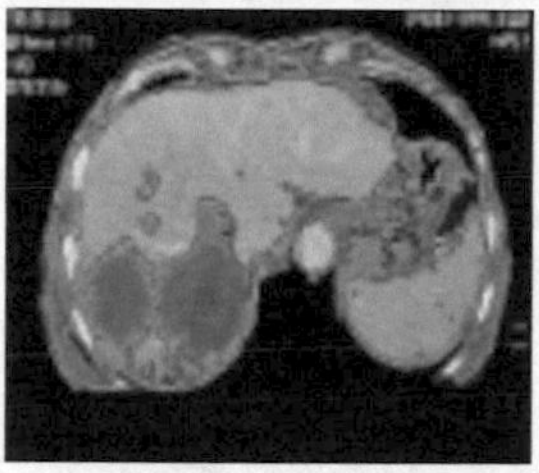	Malignant
3	L_3	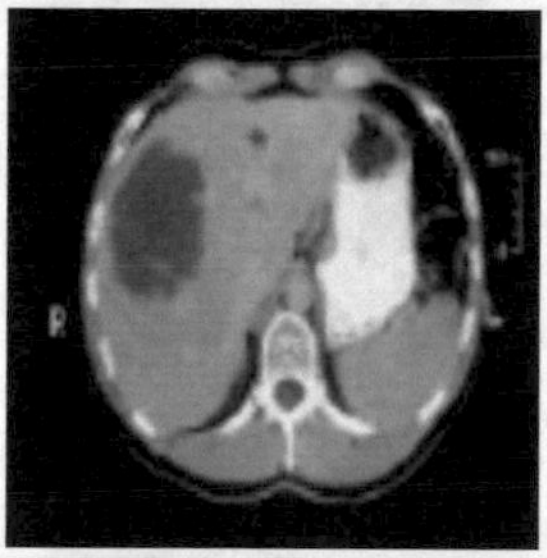	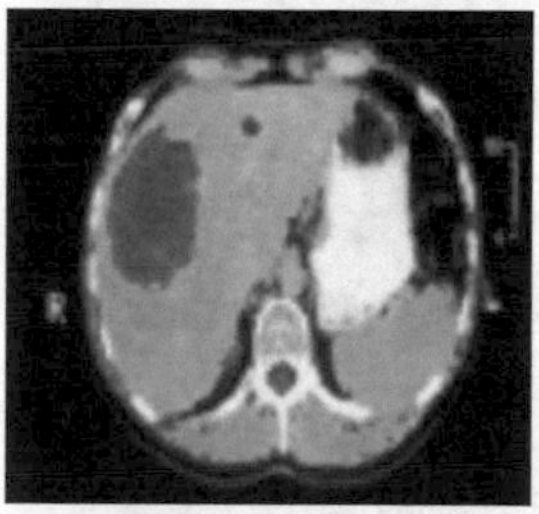	Benign
4	L_4	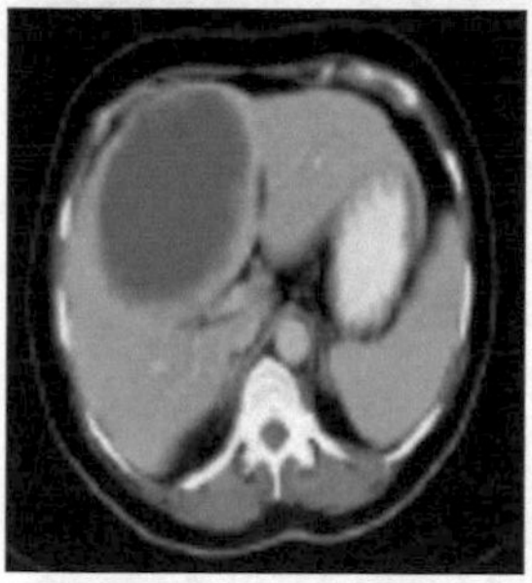	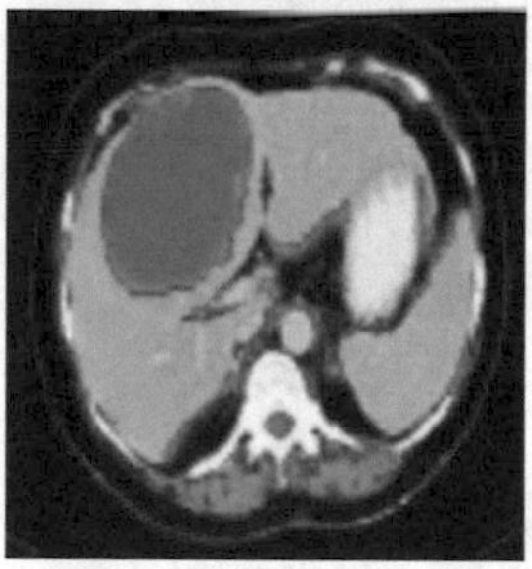	Malignant

(continued)

Table 1 (continued)

Sl. no.	Image_ID	Input	Level set output	Classification result
5	L_5			Benign
6	L_6			Malignant
7	L_7			Malignant
8	L_8			Benign

(continued)

Table 1 (continued)

Sl. no.	Image_ID	Input	Level set output	Classification result
9	L_9			Benign
10	L_10			Malignant
11	L_11			Malignant
12	L_12			Benign

(continued)

Table 1 (continued)

Sl. no.	Image_ID	Input	Level set output	Classification result
13	L_13			Benign
14	L_14			Malignant
15	L_15			Malignant

Image Source The UCI machine learning repository [11]

The consistency, reliability, and efficiency of the system are known, when the performance measure is done on the proposed system. The parameters such as "sensitivity," "accuracy," and the "specificity" are used for the classifier performance checking. The equation for computing these parameters:

$$\text{Precision} = (TP)/(TP + FP) \quad \text{Recall} = (TP)/(TP + FN).$$

$$\text{Specificity} = (TN)/(TN + FP) \quad \text{Accuracy} = (TP + TN)/(TP + FP + TN + FN)$$

where

TP: Abnormal liver identify as abnormal.

Table 2 Proposed system input output

Input Image	Grey scale Image	Segmented Image	Region of Interest	Classificatio n
				Benign
				Malignant

Image Source The UCI machine learning repository [11]

TN: Normal liver identify as normal.
FP: Abnormal liver wrongly identify as normal.

The parameters such as specificity, precision, recall, and accuracy described below for checking the performance of the proposed system.

- **Specificity**: It describes the correctly identified the negative tumor as negative,

$$\text{Specificity} = (TN)/(TN + FP)$$

- **Accuracy**: It describes correctly identified tumor out of total tumor.

$$\text{Accuracy} = (TP + TN)/(TP + FP + TN + FN)$$

- **Precision**: It describes the correct classification of the tumor out of total number of +Ve tumor.

$$\text{Precision} = (TP)/(TP + FP).$$

- **Recall**: It describes the total number of correctly classified +Ve tumor out of entire +Ve tumor. Recall $= (TP)/(TP+FN)$.

Here, we took 15 images for the performance check of the proposed system and the parameter values that are found as shown below.

$$TP = 7\,FP = 2\,FN = 0\,TN = 6$$

$$\text{Total no. of images} = 15$$

Table 3 Performance of the system

Sl. no.	Measure of performance	Percentage (%)
1	Specificity	75
2	Accuracy	86.7
3	Precision	77
4	Recall	100

The obtained parameter values for recall, precision, specificity, and accuracy are given in Table 3.

8 Conclusion

In this paper, we designed and actualized an algorithm which will identify liver as benign or malignant. The venture was tried on real CT liver image. Testing of every module is completed and discovered fruitful. SVM is prepared with the known data and tried with irregular obscure data to characterize the tumor. The algorithm was planned and coded in MATLAB. The aftereffect of our proposed strategy demonstrates that it recognizes and arranged the tumor as benign or malignant precisely. The precision rate of the proposed framework is 86.7%.

Future work

- Features utilized as a part of proposed framework are less for classification. In future, accuracy of the framework can be expanded by expanding the quantity of components utilized.
- Proposed framework works just with 2D CT images. CT image contains less data contrasted with 3D image. In 3D CT image subtle elements like area, profundity, region, and so forth are clearer. Thus, proposed strategy can be progressed to deal with 3D image. As there are loads of data in 3D images, expanded precision can be accomplished.

References

1. Soni, J., Ansari, U., Sharma, D., Soni, S.: Predictive data mining for medical diagnosis: overview of heart diseases prediction. IJCST (0975–8887) **17**(8), 43–48 (2011)
2. Karabatak, M., Ince, M.C.: An expert system for detection of breast cancer based on association rules and neural networks. Expert Syst. Appl. **36**(2), 3465–3469 (2009) (Elsevier Science)
3. Ubaidillah, S.H.A., Sallehuddin, R., Mustaffa, N.H.: Classification of liver cancer using artificial neural network and support vector machine. In: Proceedings of International Conference on Advance in Communication Network, and Computing, CNC, pp. 488–493. Elsevier Science (2014)

4. Lam, Y.H.B.: Proteomic Classification of Liver Cancer using Artificial Neural Network (2005)
5. Oh, J.H., Gao, J.: Fast Kernel discriminant analysis for classification of liver cancer mass spectra. IEEE/ACM Trans. Comput. Biol. Bioinf. **8**(6), 1522–1534 (2011)
6. Thangaraju, P., Mehala, R.: Novel classification based approaches over cancer diseases. IJAR-CCE **4**(3), 294–297 (2015)
7. Neshat, M., Yaghobi,M., Naghibi, M.B., Esmaelzadeh, A.: Fuzzy expert system design for diagnosis of liver disorders. In: IEEE International Symposium on Knowledge Acquisition and Modeling, pp. 252–256 (2008)
8. Beheshti, Z., Shamsuddin, S.M.H., Beheshti, E., Yuhaniz, S.S.: Enhancement of artificial neural network learning using centripetal accelerated particle swarm optimization for medical diseases diagnosis, vol. 18, pp. 2253–2270. Springer Science (2013)
9. Fakoor, R., Ladhak, F., Nazi, A., Huber, M.: Using deep learning to enhance cancer diagnosis and classification. In: Proceeding of the 30th International Conference on Machine Learning, Atlanta, Georgia USA, vol. 28 (2013)
10. Chou, S., Hsu, C.-L.: MMDT: a multivalued and multi-labeled decision tree classifier for data mining. Expert Syst. Appl. **28**, 799–812 (2005) (Elsevier Science)
11. The UCI Machine Learning Repository

ACOPF-Based Transmission Network Expansion Planning Using Grey Wolf Optimization Algorithm

Ashish Khandelwal, Annapurna Bhargava, Ajay Sharma
and Harish Sharma

Abstract Transmission network expansion planning problem (TNEP) is extensive, complicated, nonlinear programmable issue. A cost-effective result of TNEP is required to get the power generation pattern with load demand in economical and reliable manner. In this paper, AC optimal power flow-based TNEP is solved using a recent nature-inspired algorithm, namely grey wolf optimization algorithm (GWO). The GWO is tested to this TNEP for solution of six bus Graver's system and 24 bus IEEE system. The reported outcomes reveal that GWO is cost-effective and more accurate strategy to solve this TNEP.

Keywords Transmission network expansion planning problem · Grey wolf optimization · Six bus Graver's system · 24 bus IEEE system

1 Introduction

Transmission network expansion planning problem (TNEP) describes cost-effective and optimal expansion of new lines to meet the load increment in planning horizon. The solutions of TNEP also satisfy the system operational constraints. In view of the solutions of TNEP, several analytical formulations like as hybrid, transportation, AC, DC model are designed [1]. In this work, AC optimal power flow-based TNEP formulation is adopted.

A. Khandelwal (✉) · A. Bhargava · H. Sharma
Rajasthan Technical University, Kota, India
e-mail: ashish_khandelwal042@yahoo.com

A. Bhargava
e-mail: abhargava@rtu.ac.in

H. Sharma
e-mail: hsharma@rtu.ac.in

A. Sharma
Government Engineering College, Jhalawar, India
e-mail: ajay_2406@yahoo.com

© Springer Nature Singapore Pte Ltd. 2019   177
J. C. Bansal et al. (eds.), *Soft Computing for Problem Solving*,
Advances in Intelligent Systems and Computing 816,
https://doi.org/10.1007/978-981-13-1592-3_13

Researchers are performing continuous endeavours to achieve the economical and optimal solution of TNEP. Various solution techniques like mathematical optimization, heuristic, and metaheuristic have been applied to solve TNEP [2]. Mathematical optimization methods like linear programming [3], bender decomposition [4], nonlinear programming [5], mixed-integer programming [6–8] have been applied to solve TNEP.

Engineer experience-based technique known as heuristic method is also applied to solve TNEP [9, 10]. Nowadays, metaheuristic method, a combination of heuristic and optimization method are applied to solve optimization problems due to their easiness, derivation-free component, flexibility and limitation of local optima [11–13]. Various meta-heuristic approaches have been applied to achieve TNEP result like genetic algorithm (GA) [14, 15], simulated annealing (SA) [16], ant colony optimization (ACO) [17], harmony search (HS) [18].

Recently, a meta-heuristic approach, namely grey wolf optimization algorithm (GWO) is attracting researcher to achieve result of real-world issue like power system load frequency control [19], solar system maximum power point tracker (Mppt) modelling [20], multi-input multi-output (MIMO) system [21], selection of features problem [22], etc. The achieved results by GWO are promising. In this work, the GWO is tested to TNEP for six bus Graver's and 24 bus IEEE framework.

The remaining paper is structured in the following manner: mathematical formulation of ACOPF-based TNEP is illustrated in Part 2. In Sect. 3, GWO strategy is elaborated. Section 4 presents the outcomes of TNEP using GWO. In the end, the work conclusion is reported in Sect. 5.

2 Transmission Network Expansion Planning Problem (TNEP)

In this work, ACOPF-based TNEP formulation is selected and the additional number of circuits to be achieved such that least expansion cost without any over burdens is provided within planning period [16, 23].

The analytical modelling without any security constraints is as follows:

$$\text{cost} = \sum_{(i,j) \in \alpha} c_{ij} k_{ij} \tag{1}$$

Subject to

$$P(V, \theta, k) + P_d - P_g = 0 \tag{2}$$

$$Q(V, \theta, k) + Q_d - Q_g = 0 \tag{3}$$

$$I_{ij} \leq (k_{ij}^0 + k_{ij}) I_{ij}^{\max} \tag{4}$$

$$0 \leq k_{ij} \leq k_{ij}^{\text{max}} \tag{5}$$

$$P_g^{\text{max}} \leq P_g \leq P_g^{\text{min}} \tag{6}$$

$$Q_g^{\text{max}} \leq Q_g \leq Q_g^{\text{min}} \tag{7}$$

fitness function (FF)

$$\text{cost} = \sum_{(i,j) \in \alpha} c_{ij} k_{ij} + M_1 \sum_{ol} |(I_{ij} - I_{ij}^{\text{max}})| + M_2(k_{ij} - k_{ij}^{\text{max}}) \tag{8}$$

In the above equations, abbreviations are as follows.

cost = cost of expansion
c_{ij} = additional branch cost
k_{ij} = additional branch between i to j
P_g = gross real power of generator
P_d = gross real power of load
Q_g = gross reactive power of generator
Q_d = gross reactive power of load
I = vector with elements I_{ij}
k_{ij}^0 = branches from i to j in initialization
α = set of all candidate circuits
k_{ij} = No. of lines in branch ij respectively
k_{ij}^{max} = maximum no. of lines in branch ij respectively
I_{ij} = gross power in line ij
I_{ij}^{max} = maximum power of line ij.

The constraints details are as follows.

Equation 2 $\Rightarrow$ Real power node balance constraint
Equation 3 $\Rightarrow$ Reactive power node balance constraint
Equation 4 $\Rightarrow$ Transmission line power flow limit constraint
Equation 5 $\Rightarrow$ Right of way of transmission lines constraint.

3 Grey Wolf Optimization Algorithm

GWO has been developed by Seyedali Mirjalili et al. in the year 2013, encouraged by collective behaviour of grey wolves (GW) [24]. The GWO is depended on grouping along with food searching procedure of GW. The step-by-step procedure of this algorithm are as follows.

3.1 Mechanism of Grouping

The GW are divided into four categorizies which are denoted as alpha (α), beta (β), delta (δ) and omega (ω). Most superior outcome is α. Subsequently, superiority of outcomes are denoted as β and δ, and rest of results refers to ω.

3.2 Mechanism of Circling

Before hunting, GW circles the prey. Mechanism of encircling are represented by thefollowing equations:

$$\vec{\nabla} = |\vec{\Upsilon}.\vec{\kappa_p}(T) - \vec{\kappa}(T)|, \tag{9}$$

$$\vec{\kappa}(T+1) = \vec{\kappa_p}(T) - \vec{\varsigma}.\vec{\nabla} \tag{10}$$

Here, T denotes the present iteration, $\vec{\varsigma}$ $\vec{\Upsilon}$ are vectors of coefficient, $\vec{\kappa_p}$ the direction of location for the prey, and $\vec{\kappa}$ shows the direction of location for GW. The vectors $\vec{\varsigma}$ and $\vec{\Upsilon}$ are computed as described:

$$\vec{\varsigma} = 2.\lambda.\vec{\theta_1} - \vec{\lambda_1} \tag{11}$$

$$\vec{\Upsilon} = 2.\vec{\theta_2} \tag{12}$$

Here, the values of $\vec{\lambda}$ are gradually decreased from 2 to 0 during the computation. $\vec{\theta_1}$ and $\vec{\theta_2}$ are arbitrary vectors between [0 , 1].

3.3 Mechanism of Hunting

The GW finds the location of food and then circles them. α mentors the hunt. The β and δ seldom take interest in hunting. Locations of other search agents are modified in reference to α position. The hunting mechanism is described by the following equation.

$$\vec{\nabla_\alpha} = |\vec{\Upsilon_1}.\vec{\kappa_\alpha} - \vec{\kappa}|, \vec{\nabla_\beta} = |\vec{\Upsilon_2}.\vec{\kappa_\beta} - \vec{\kappa}|, \vec{\nabla_\delta} = |\vec{\Upsilon_3}.\vec{\kappa_\delta} - \vec{\kappa}| \tag{13}$$

$$\vec{\kappa_1} = \vec{\kappa_\alpha} - \vec{\varsigma_1}.(\vec{\nabla_\alpha}), \vec{\kappa_2} = \vec{\kappa_\beta} - \vec{\varsigma_2}.(\vec{\nabla_\beta}), \vec{\kappa_3} = \vec{\kappa_\delta} - \vec{\varsigma_3}.(\vec{\nabla_\delta}) \tag{14}$$

$$\vec{\kappa}(T+1) = \frac{\vec{\kappa_1} + \vec{\kappa_2} + \vec{\kappa_3}}{3} \tag{15}$$

3.4   Mechanism of Attacking and Food Searching

The ability of the GW to achieve global optima is represented as exploitation feature. The range of λ lies between 2 and 0, ς is also diminished. In view of that, ς is an arbitrary number in the interim $[-2\lambda, 2\lambda]$. In condition of $|\varsigma| < 1$, the GW is compelled to strike the prey. This GW feature to find the prey in search space is known as exploration feature. The arbitrary values of ς accomplish the search agent to discard from the prey. For $|\varsigma| > 1$, the GW is compelled to discard from the prey. The GWO algorithm is described as follows 1.

Algorithm 1 GWO

Start with initial estimate of GW population κ_i (i = 1, 2, ..., n)
Start λ, ς, and Υ
Define the MX = number of maximum iterations
Evaluate fitness value of each candidate solution
κ_α = Location of superior candidate solution
κ_β = Location of second superior candidate solution
κ_δ = Location of third superior candidate solution
while t < MX **do**
  for each candidate solution **do**
    Modify the location of the existing candidate solution by equation 15
  end for
  Modify λ, ς, and Υ
  compute the FF for all candidate solutions
  update κ_α, κ_β, and κ_δ
  T = T + 1
end while
return κ_α

---

4   Transmission Network Expansion Planning Using GWO

TNEP is solved by GWO for two test systems described as follows. Initial parameter of algorithm is also given.

4.1   Algorithm Setting

- Number of agents in search space = 30,
- Function evaluations (FEs) maximum number = 200,000,
- Numbers of computation = 100,
- Braking condition = either maximum number of FE is reached or TNEP tolerable error has been achieved.

The ACOPF-based TNEP is evaluated using GWO on the following test system.

4.2 Six Bus Graver's System

Gravers system is consist of six nodes, load demand of 760 MW, and 15 new candidate circuit [1, 3]. The TNEP is tested on this system through GWO algorithm. The algorithm is computed for hundred runs for the ACOPF-based TNEP. The utmost repeated outcome is considered. The outcomes of GWO for TNEP are compared with specialized branch and bound technique (SP.BB) [25, 26], artificial bee colony (ABC) algorithm [27] and hierarchical decomposition approach [28] as presented in Table 1.

In Table 1, outcomes are computed in terms of expansion cost. The operational cost is least for GWO algorithm which replicates that GWO is better than other algorithms to solve TNEP.

4.3 24 Bus IEEE System

In 24 bus IEEE system, there are 24 buses, 38 lines connecting them, 41 new candidate circuit and load of 8550 MW [10, 29]. The TNEP is tested on this system through GWO algorithm. The algorithm is computed for hundred runs for the ACOPF-based TNEP. The utmost repeated outcome is considered. The outcomes of GWO for TNEP are compared with branch and bound (BB) algorithm [30], (ABC) algorithm [27], and GA [14, 31], as presented in Table 2. The results show that GWO is better candidate among other techniques to solve TNEP.

Table 1 Comparison of results for Graver's Six Bus system

Strategy	Optimal cost $\times 10^3$(US$)	Bus topology
GWO	190	$n_{4-6} = 2, n_{6-2} = 3, n_{3-5} = 1, n_{2-3} = 1$
ABC	200	$n_{4-6} = 2, n_{6-2} = 4, n_{3-5} = 1$
SP.BB	190	$n_{4-6} = 2, n_{6-2} = 1, n_{3-5} = 2, n_{2-3} = 2, n_{1-5} = 1$
HD	200	$n_{4-6} = 2, n_{6-2} = 4, n_{3-5} = 1$

Table 2 Comparison of results for 24 Bus IEEE system

Strategy	Optimal cost $\times 10^3$(US$)	Bus topology
GWO	150	$n_{7-8} = 2, n_{6-10} = 1, n_{12-13} = 1, n_{16-17} = 1$
ABC	152	$n_{7-8} = 2, n_{6-10} = 1, n_{10-12} = 1, n_{14-16} = 1$
BB	152	$n_{7-8} = 2, n_{6-10} = 1, n_{10-12} = 1, n_{14-16} = 1$
GA	152	$n_{7-8} = 2, n_{6-10} = 1, n_{10-12} = 1, n_{14-16} = 1$

5   Conclusion

In this paper, AC optimal power flow (ACOPF) modelled transmission network expansion planning problem (TNEP) is computed using GWO algorithm. The algorithm is tested for TNEP on six bus Graver's and 24 bus IEEE framework. The comparison between outcomes achieved by GWO and available algorithms in the literature are performed. Through the examination of results, it is discovered that GWO is a proficient and financially savvy method to solve TNEP. Further in future, the work might be stretched out for solution of dynamic TNEP.

References

1. Romero, R., Monticelli, A., Garcia, A., Haffner, S.: Test systems and mathematical models for transmission network expansion planning. IEE Proc. Gener. Transm. Distrib. **149**(1), 27–36 (2002)
2. Latorre, G., Cruz, R.D., Areiza, J.M., Villegas, A.: Classification of publications and models on transmission expansion planning. IEEE Trans. Power Syst. **18**(2), 938–946 (2003)
3. Garver, L.L.: Transmission network estimation using linear programming. IEEE Trans. Power Apparatus Syst. **7**, 1688–1697 (1970)
4. Binato, S., Pereira, M.V.F., Granville, S.: A new benders decomposition approach to solve power transmission network design problems. IEEE Trans. Power Syst. **16**(2), 235–240 (2001)
5. Al-Hamouz, Z.M., Al-Faraj, A.S.: Transmission expansion planning using nonlinear programming. In: Transmission and Distribution Conference and Exhibition 2002: Asia Pacific. IEEE/PES, vol. 1, pp. 50–55. IEEE (2002)
6. Alguacil, N., Motto, A.L., Conejo, A.J.: Transmission expansion planning: a mixed-integer LP approach. IEEE Trans. Power Syst. **18**(3), 1070–1077 (2003)
7. Bahiense, L., Oliveira, G.C., Pereira, M., Granville, S.: A mixed integer disjunctive model for transmission network expansion. IEEE Trans. Power Syst. **16**(3), 560–565 (2001)
8. Zhang, H., Vittal, V., Heydt, G.T., Quintero, J.: A mixed-integer linear programming approach for multi-stage security-constrained transmission expansion planning. IEEE Trans. Power Syst. **27**(2), 1125–1133 (2012)
9. Lee, C.W., Ng, S.K.K., Zhong, J., Wu, F.F.: Transmission expansion planning from past to future. In: Power Systems Conference and Exposition, 2006. PSCE'06. 2006 IEEE PES, pp. 257–265. IEEE (2006)
10. Romero, R., Rocha, C., Mantovani, J.R.S., Sanchez, I.G.: Constructive heuristic algorithm for the dc model in network transmission expansion planning. IEE Proc. Gener. Transm. Distrib. **152**(2), 277–282 (2005)
11. Sharma, A., Sharma, H., Bhargava, A., Sharma, N., Bansal, Jagdish Chand: Optimal power flow analysis using lévy flight spider monkey optimisation algorithm. Int. J. Artif. Intell. Soft Comput. **5**(4), 320–352 (2016)
12. Sharma, A., Sharma, H., Bhargava, A., Sharma, N., Bansal, J.C.: Optimal placement and sizing of capacitor using limaçon inspired spider monkey optimization algorithm. Memetic Comput. 1–21 (2016)
13. Sharma, A., Sharma, H., Bhargava, A., Sharma, N.: Power law-based local search in spider monkey optimisation for lower order system modelling. Int. J. Syst. Sci. **48**(1), 150–160 (2017)
14. Silva, I.D.J., Rider, M.J., Romero, R., Garcia, A.V., Murari, C.A.: Transmission network expansion planning with security constraints. IEE Proc. Gener. Transm. Distrib. **152**(6), 828–836 (2005)

15. Shivaie, M., Sepasian, M.S., Sheikh-El-Eslami, M.K.: Multi-objective transmission expansion planning using fuzzy-genetic algorithm. Iran. J. Sci. Technol. Trans. Electr. Eng. **35**(E2), 141–159 (2011)
16. Romero, R., Gallego, R.A., Monticelli, A.: Transmission system expansion planning by simulated annealing. IEEE Trans. Power Syst. **11**(1), 364–369 (1996)
17. da Silva, A.M.L., Rezende, L.S., da Fonseca Manso, L.A., de Resende, L.C.: Reliability worth applied to transmission expansion planning based on ant colony system. Int. J. Electr. Power Energy Syst. **32**(10):1077–1084 (2010)
18. Verma, A., Panigrahi, B.K., Bijwe, P.R.: Harmony search algorithm for transmission network expansion planning. Gener. Transm. Distrib. IET **4**(6), 663–673 (2010)
19. Guha, D., Roy, P.K., Banerjee, S.: Load frequency control of interconnected power system using grey wolf optimization. Swarm Evol. Comput. **27**, 97–115 (2016)
20. Mohanty, S., Subudhi, B., Ray, P.K.: A new mppt design using grey wolf optimization technique for photovoltaic system under partial shading conditions. IEEE Trans. Sustain. Energy **7**(1), 181–188 (2016)
21. El-Gaafary, A.A.M., Mohamed, Y.S., Hemeida, A.M., Mohamed, A.A.A.: Grey wolf optimization for multi input multi output system. Univ. J. Commun. Netw. **3**(1), 1–6 (2015)
22. Emary, E., Zawbaa, H.M., Hassanien, Aboul Ella: Binary grey wolf optimization approaches for feature selection. Neurocomputing **172**, 371–381 (2016)
23. Romero, R., Mantovani, J.R.S., Rider, M.J. et al.: Transmission-expansion planning using the dc model and nonlinear-programming technique. IEE Proc. Gener. Transm. Distrib. **152**(6), 763–769. IET (2005)
24. Mirjalili, S., Mirjalili, S.M., Lewis, Andrew: Grey wolf optimizer. Adv. Eng. Softw. **69**, 46–61 (2014)
25. Sérgio Haffner, A., Monticelli, A.Garcia, Romero, R.: Specialised branch-and-bound algorithm for transmission network expansion planning. IEE Proc. Gener. Transm. Distrib. **148**(5), 482–488 (2001)
26. Haffner, S., Monticelli, A., Garcia, A., Mantovani, J., Romero, R.: Branch and bound algorithm for transmission system expansion planning using a transportation model. IEE Proc. Gener. Transm. Distrib. **147**(3), 149–156 (2000)
27. Rathore, C., Roy, R., Sharma, U., Patel, J.: Artificial bee colony algorithm based static transmission expansion planning. In: 2013 International Conference on Energy Efficient Technologies for Sustainability (ICEETS), pp. 1126–1131. IEEE (2013)
28. Romero, R., Monticelli, A.: A hierarchical decomposition approach for transmission network expansion planning. IEEE Trans. Power Syst. **9**(1), 373–380 (1994)
29. Fang, R., Hill, D.J.: A new strategy for transmission expansion in competitive electricity markets. IEEE Trans. Power Syst. **18**(1), 374–380 (2003)
30. Rider, M.J., Garcia, A.V., Romero, R.: Transmission system expansion planning by a branch-and-bound algorithm. IET Gener. Transm. Distrib. **2**(1), 90–99 (2008)
31. Romero, R., Rider, M.J., Silva, I.D.L.: A metaheuristic to solve the transmission expansion planning. IEEE Trans. Power Syst. **22**(4), 2289–2291 (2007)

Use of Improved Gravitational Search Algorithm for 3D Reconstruction of Space Curves Using NURBS

Amarjeet Singh and Kusum Deep

Abstract Gravitational Search Algorithm (GSA) is a memory-less, nature-inspired algorithm for nonlinear continuous optimization problems. In Singh et al. (a new Improved Gravitational Search Algorithm for function optimization using a novel "best-so-far" update mechanism. IEEE, pp. 35–39 (2015) [21]), Singh and Deep proposed an Improved GSA using best-so-far mechanism. In this paper, the problem of 3D reconstruction is modelled as a nonlinear optimization problem. GSA and Improved GSA are used to solve three reconstruction problems. Based on the several computational experiments and analysis, it is concluded that the performance of improved GSA is better than original GSA in terms of convergence and solution quality.

Keywords Gravitational search algorithm · Continuous function optimization
Heuristic technique · Reconstruction · NURBS

1 Introduction

Most of the nonlinear continuous optimization problems are arised from different branch of engineering, and it is a great challenge to find its optimum solution. The available literature for solving optimization problem can be studied into probabilistic methods and deterministic methods. The applicable area of deterministic methods is restricted in comparison to probabilistic methods. Mostly, probabilistic algorithms are developed from the natural behaviour of different species present on the earth or

A. Singh (✉)
Department of Mathematics, Janki Devi Memorial College, University of Delhi,
New Delhi 110060, India
e-mail: amarjeetiitr@gmail.com

K. Deep
Department of Mathematics, Indian Institute of Technology Roorkee, Roorkee 247667,
Uttrakhand, India
e-mail: kusumfma@iitr.ac.in

© Springer Nature Singapore Pte Ltd. 2019
J. C. Bansal et al. (eds.), *Soft Computing for Problem Solving*,
Advances in Intelligent Systems and Computing 816,
https://doi.org/10.1007/978-981-13-1592-3_14

natural laws. Genetic Algorithm (GA) [1], particle swarm optimization (PSO) [2], Artificial Bee Colony (ABC) [3], glow-worm swarm optimization (GSO) [4, 5], etc., are some popular methods.

Gravitational Search Algorithm (GSA), a nature-inspired optimization, is proposed by Rasedi et al. [6] in 2009 to solve continuous nonlinear optimization problems. From the past studies on GSA [7], it is observed that GSA has some demerits. It is found that GSA could not provide the better solution for highly complex problems and have a chance to get trapped in local optima. To get rid from these demerits, GSA has been hybridized with many heuristic algorithms and operators. Sarafrazi et al. [8] introduced "Disruption" operator and Doraghinejad et al. [9] introduced "Black Hole" operator for the hybridization of GSA. Xu et al. [10] proposed Improved GSA. Singh and Deep [11] hybridized GSA with Laplace Crossover operator and Power Mutation operator and proposed LXGSA, PMGSA and LXPMGSA algorithms. In [12, 13], these algorithms are extended for constrained optimization and mixed integer constraint problems.

Reconstruction of three-dimensional (3D) space curves is a vital and challenging problem in computer vision. The objective is to develop a model such that an undetermined curve/surface approximates a given set of data points. In general, these data points are obtained from a pair of image point taken from two different views. Reconstruction is also related to camera calibration and orientation where the results depend on the exactness of the picture division. The single-view reconstruction from multi-view images has several optical and geometric constraints. Also, multi-view reconstruction has feature matching among various perspectives. A variety of reconstruction methods are available in the literature. These methods can be studied in two diverse methodologies: model-based and constraint-based. In model-based approaches, the reconstruction is obtained by the accumulation of essential shapes, for example lines, circles, ovals, rectangles, parallelograms, crystals. For complex objects, these methods may require the decomposition of the scene or more basic shape.

From the last few decades, nature-inspired optimization algorithms are playing an important role to solve various application problems such as clustering, games, designing, chemical engineering problems. These algorithms are also applied in various branches of computer vision. Saini et al. [14] presented a survey on particle swarm optimization and its variants in human motion tracking problem. Voisin et al. [15] proposed a Genetic Algorithm strategies to reconstruct a point cloud using numerous 3D shapes and a fitness function defined to integrate a tolerance threshold to handle noisy data. Ning et al. [16] used simulated annealing for 3D reconstruction from 2D orthogonal projections. Simulated annealing [17] is also used in structure determining problem, where a 3D structure is reconstructed from the projected images. Siddique and Zakaria [18] proposed a 3D reconstructing system based on single 2D image using Genetic Algorithm. Wong and Ng [19] used particle swarm optimization for 3D reconstruction from multiple views. Koch and Dipanda [20] composed uncalibrated stereovision system (USS) by locating five cameras on an arc of a circle around the object and proposed an evolutionary-based method for 3D panoramic reconstruction from USS.

The rest of paper is organized as follows: in Sect. 2, Gravitational Search Algorithm and Improved Gravitational Search Algorithm are explained. In Sect. 3, non-uniform random B-spline and objective function for reconstruction problems are defined. In Sect. 4, three reconstruction problems are explained. Experimental results on reconstruction problems are discussed in Sect. 5, whereas in Sect. 6, conclusions are drawn.

2   Gravitational Search Algorithm

Gravitational Search Algorithm is a recently proposed algorithm in view of the analogy of gravity and mass communications. In this algorithm, the position of the particle represents the solution of the problem at the specified dimension and particle's mass represents the quality of the solution; greater mass corresponds to better solution. Every iteration of GSA goes through three phases: (i) initialization population, (ii) force calculation and (iii) motion.

In initialization phase, a population of N_P particles is initialized randomly and the velocity of each particle is set to zero. The fitness value of ith particle is denoted $f_i(t)$. Let the position of ith particle is represented by

$$x_i = \left(x_i^1, x_i^2, \ldots, x_i^d, \ldots, x_i^m \right) \quad \text{for} \quad i = 1, 2, \ldots, N_P \tag{1}$$

where x_i^d is the position of ith particle in dth direction.

In force calculation phase, the fitness of particle is evaluated using a fitness function then gravitational mass and inertia mass of each particle is evaluated as follows:

$$M_{ai}(t) = M_{pi}(t) = M_{ii}(t) = M_i(t) \tag{2}$$

$$m_i = \frac{f_i(t) - \text{worstt}(t)}{\text{bestt}(t) - \text{worstt}(t)}, \quad i = 1, 2, \ldots, N_P. \tag{3}$$

$$M_i(t) = \frac{m_i(t)}{\sum_{j=1}^{N_P} m_j(t)} \tag{4}$$

where $M_{ai}(t)$, $M_{pi}(t)$ and $M_{ii}(t)$ are the active gravitational mass, passive gravitational mass, inertia mass of particle i, respectively, and bestt(t) and worstt(t) represent best and worst at time t.

For minimization problem bestt$(t) = \underset{j \in \{1,2,\ldots,N_P\}}{\text{Min}} f_j(t)$ and worstt$(t) = \underset{j \in \{1,2,\ldots,N_P\}}{\text{Max}} f_j(t)$

Then the force acting on mass 'i' from 'j' is evaluated by

$$F_{ij}^d(t) = G(t) \frac{M_{pi}(t) \times M_{aj}(t)}{R_{ij}(t) + \varepsilon} \left(x_j^d(t) - x_i^d(t) \right) \tag{5}$$

here ε is a small value and gravitational constant $G(t)$ is evaluated by

$$G(t) = G(0) \exp\left(\frac{-\alpha t}{\text{max\_itr}}\right) \tag{6}$$

$R_{ij}(t)$ represents Euclidean distance of particles i and j. The force acting on particle i in dimension d is evaluated by

$$F_i^d(t) = \sum_{j \in Kb, j \neq i} rd_j F_{ij}^d(t) \tag{7}$$

where rd_j is uniformly distributed random number in $(0, 1]$, and Kb is the collection of first k best particles and it decreases as iteration increases.

In motion phase, acceleration of every particle is evaluated by

$$a_i^d(t) = \frac{F_i^d(t)}{M_i(t)} \tag{8}$$

where $a_i^d(t)$ is the dth dimension acceleration of particle i at time t. Then the velocities and next position of particles i in the dth dimension are updated by

$$v_i^d(t+1) = rd_i \times v_i^d(t) + a_i^d(t) \tag{9}$$

$$x_i^d(t+1) = x_i^d(t) + v_i^d(t+1) \tag{10}$$

where rd_i is randomly distributed number in the interval $(0, 1]$.

The fitness of best particle is set to Gbest in the initial population, and in the next iterations, the fitness of best particle is compared to Gbest. Gbest is updated, if best particle has better fitness, otherwise Gbest remains same.

Since Gbest remains unchanged when Gbest has better fitness in comparison to the fitness of best particle, the current population may not have global best solution searched by the algorithm. Due to this inherent drawback, a valuable information stored in Gbest may be lost. In order to overcome this drawback, Singh and Deep [21] proposed Improved Gravitational Search Algorithm (IGSA). In IGSA, if best particle of current population has better fitness than the fitness of "best-so-far", i.e. Gbest, then Gbest is replaced by best particle, otherwise the worst particle of the population is set as Gbest. By applying the above process, IGSA population never lose the global best particle. Figure 1 shows the pseudo-code of IGSA for minimization problems.

3   Reconstruction of 3D Curves and Surfaces

In this section, the problem of 3D reconstruction curves and surfaces is modelled as a nonlinear optimization problem in which control points are to be determined by minimizing the objective function which is the error function between given data

Set population size= N_p
Set dimension = m
Set $G(0), \alpha$
Initialize population
Let $x_i(t) = \left(x_i^1(t), \ldots, x_i^d(t), \ldots, x_i^m(t) \right)$ represents the position of i particle.
Evaluate fitness f
$bestt(t) = \underset{j \in \{1, \ldots, N_P\}}{\text{Min}} f_j(t), worstt(t) = \underset{j \in \{1, \ldots, N_P\}}{\text{Max}} f_j(t),$
Gbest=bestt(t),
Set maximum iteration = m_itr
t=0
while (t $\leq$ m_itr)
$\quad$ { $\quad G(t) = G(0) exp(-\alpha t / \text{m\_itr})$
$\qquad bestt(t) = \underset{j \in \{1, \ldots, N_P\}}{\text{Min}} f_j(t), worstt(t) = \underset{j \in \{1, \ldots, N_P\}}{\text{Max}} f_j(t),$
$\qquad m_i(t) = \dfrac{f_i(t) - worstt(t)}{bestt(t) - worstt(t)}, i = 1, \ldots, N_P$
$\qquad M_i(t) = m_i(t) / \sum_{j=1}^{N} m_j(t), \, i = 1, \ldots, N_P$
$\quad$ **for** $i = 1$ to N_p
$\qquad$ { **for** d = 1 to m
$\qquad\quad$ { $\quad F_i^d(t) = \underset{j \in kb, j \neq i}{\sum} rd_j \, G(t) \dfrac{M_{pi}(t) \times M_{aj}(t)}{R_{ij}(t) + \varepsilon} \left(x_j^d(t) - x_i^d(t) \right)$
$\qquad\qquad a_i^d(t) = F_i^d(t) / M_{ii}(t)$
$\qquad\qquad v_i^d(t+1) = rd_i \, v_i^d(t) + a_i^d(t)$
$\qquad\qquad x_i^d(t+1) = x_i^d(t) + v_i^d(t+1)$
$\qquad\quad$ }
$\qquad$ }
$\quad$ Evaluate fitness f
$\quad bestt(t) = \underset{j \in \{1, \ldots, N_P\}}{\text{Min}} f_j(t), worstt(t) = \underset{j \in \{1, \ldots, N_P\}}{\text{Max}} f_j(t),$
$\quad$ If f(bestt(t)) <f(Gbest)
$\qquad\qquad$ Gbest = bestt(t);
$\quad$ else
$\qquad\qquad$ worstt(t)= Gbest;
$\qquad$ t=t+1
}

Fig. 1 Pseudo-code of IGSA

points and data points on the generated curve/surface. GSA and IGSA are used to solve 3D reconstruction problem.

NURBS Fitting: A rational B-spline curve $P(t) : R^3 \to R$ is given by

$$P(t) = \sum_{i=1}^{n+1} B_i R_{i,k}(t) \tag{11}$$

where B_i represents the ith control point of the space curve $P(t)$. $R_{i,k}(t)$ represents the ith rational B-spline basis function and evaluated by

$$R_{i,k}(t) = \frac{w_i N_{i,k}(t)}{\sum_{i=1}^{n+1} w_i N_{i,k}(t)}, \ i = 1, 2, \ldots, n + 1. \tag{12}$$

where w_i represents the weight related to the control point B_i and $N_{i,k}(t)$ is the normalized B-spline basis function of order k (degree k-1) and evaluated by Cox deBor recursive relation as follows:

$$N_{i,1}(t) = \begin{cases} 1, \ if \ x_i \le t < x_{i+1} \\ 0, \ \text{otherwise} \end{cases} \tag{13}$$

and

$$N_{i,k}(t) = \frac{(t - x_i)N_{i,k-1}(t)}{x_{i+k-1} - x_i} + \frac{(x_{i+k} - t)N_{i+1,k-1}(t)}{x_{i+k} - x_{i+1}} \tag{14}$$

where t is independent variable which varies from t_{min} to t_{max}. x_i is the element of knot vector $X = \{x_1, x_2, \ldots, x_{n+k+1}\}$. If knot vector is not uniformly spaced, then these curves are known as non-uniform random B-spline (NURBS). All conic sections can be represented by NURBS and variety of shapes can be assigned using NURBS.

For a given set of control points, the shape of the curve/surface can be determined through NURBS fitting and the data points lie on the curve/surface can be determined from Eq. (11) for different parametric values of t, but in the present study, the aim is the other way round, i.e. given a set of unorganized 3D data points, determine the 3D control points/NURBS curve which approximate the set of data points correctly. Hence, the objective function of the problem is defined as sum of the square of the difference between given unorganized data points and the point on the NURBS curve/surface, i.e.

$$\text{error} = \sum_{j} \left[\left(\sum_{i=1}^{n+1} B_i^x R_{i,k}(t_j) - P^x(t_j) \right)^2 + \left(\sum_{i=1}^{n+1} B_i^y R_{i,k}(t_j) - P^y(t_j) \right)^2 + \left(\sum_{i=1}^{n+1} B_i^z R_{i,k}(t_j) - P^z(t_j) \right)^2 \right] \tag{15}$$

where $P(t_j) = \left(P^x(t_j), P^y(t_j), P^z(t_j) \right)$ is the jth unorganized data points.

4 Reconstruction Problems

Problem 1 Helix

A set of 101 data points is considered and shown in Fig. 2a. The data points are obtained from the curve which is generated by the NURBS fitting of 13 control points. The control points and the NURBS curve are shown in Fig. 2b.

Problem 2 Ruled Surface

A set of 861 data points is considered and shown in Fig. 3a. These data points are obtained from the ruled surface, constructed from two NURBS curves. The control points of first and second curve are shown by 'o' and '*', respectively, in Fig. 3a. The ruled surface is shown in Fig. 3b.

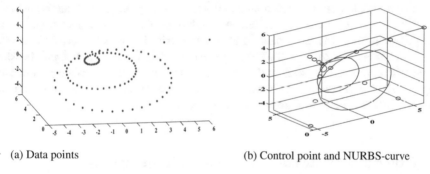

(a) Data points   (b) Control point and NURBS-curve

Fig. 2 Helix **a** data points. **b** Control point and NURBS curve

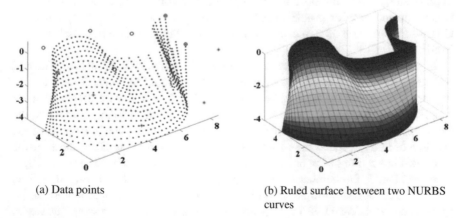

(a) Data points   (b) Ruled surface between two NURBS curves

Fig. 3 Ruled surface between two NURBS curves

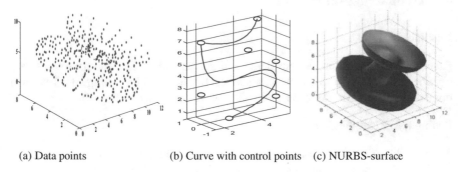

(a) Data points   (b) Curve with control points   (c) NURBS-surface

Fig. 4 Hollow dumbbell

Problem 3 Hollow Dumbbell

A set of 400 data points is considered and shown in Fig. 4a. These data points are obtained from the surface, generated by revolving a curve about z-axis, and then the generated 3D surface is rotated 20° about x-axis, 15° about y-axis, 25° about z-axis and translated to (5, 5, 0). The curve is obtained by seven control points. The curve with control points is shown in Fig. 4b, and the resultant 3D surface is shown in Fig. 4c. The control points are shown by 'o'. These curves and surfaces are generated by MATLAB NURBS Toolbox available online.

5   Experimental Results on Reconstruction Problems

In problem 1, since the number of control points is 13 (three-dimensional), the dimension of the optimization problem becomes 39, the population size for implementing GSA and IGSA is set 58 (which is 1.5 times of population size). The search space is fixed to $[-10, 10]$.

In problem 2, the ruled surface is constructed from two NURBS curves. The first NURBS curve is generated by seven two-dimensional control points (shown by o in Fig. 3a), and second NURBS curve is generated by four control points (shown by * in Fig. 3a). The control points of first curve are generated by algorithm and the control points of second curve kept fix. Therefore, the dimension of the optimization problem becomes 14, and the population size for implementing GSA and IGSA is 21 (which is 1.5 times of population size). The search space is fixed to $[-10, 10]$.

In problem 3, the number of control points is 7 (three-dimensional); therefore, the dimension of the optimization problem becomes 21, and the population size for implementing GSA and IGSA is 31 (which is 1.5 times of population size). The search space is fixed to $[-10, 10]$. The algorithms generate the control points, and through NURBS fitting, a 3D curve/surface is fitted and points are calculated for different parametric values. The error is evaluated between data points and evaluated points

Table 1 Best, worst, average, median and standard deviation of error over 30 runs

Problem	Algo.	Best	Worst	Average	Median	STD
Helix	GSA	1.27E−07	0.498241	0.026792	0.001086	0.091449
	IGSA	1.27E−07	0.006825	0.000232	1.27E−07	0.001245
Ruled surface	GSA	5.11E−16	132.139	11.08459	2.46E−15	34.13193
	IGSA	4.77E−16	28.96157	0.990616	2.13E−15	5.284652
Hollow dumbbell	GSA	4.74E−16	8.12E−15	2.58E−15	2.34E−15	1.58E−15
	IGSA	4.26E−16	7.6E−15	1.98E−15	1.74E−15	1.48E−15

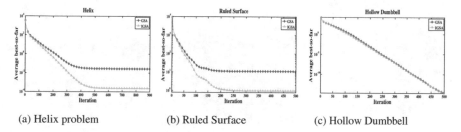

(a) Helix problem   (b) Ruled Surface   (c) Hollow Dumbbell

Fig. 5 Convergence behaviour of GSA and IGSA

from reconstructed curve/surface. This error is set as the objective function for the algorithms. The aim is to minimize error. $G(0)$, α and population size are algorithm parameters, and during experiments, their values are set $G(0) = 100$, $\alpha = 20$, population size $= 50$.

The maximum iteration for helix problem is set 900 and for others it is 500. Both the algorithms are run 30 times. Statistical results are calculated over 30 runs and shown in Table 1.

It is observed from the Table 1 that the performance of IGSA is significantly better than GSA on helix and ruled surface problem but on dumbbell problem, IGSA is marginally better than GSA. The convergence plot for these problems is also plotted and shown in Fig. 5. The convergence plots of helix and ruled surface problem show that the convergence of IGSA is significantly better than GSA. The convergence plot of dumbbell problem shows that IGSA has slightly better convergence.

From the above, it is clear that the performance of IGSA is better than others. Hence, IGSA is run for a typical run and resultant shape is plotted and shown in Fig. 6 for helix, Fig. 7 ruled surface and Fig. 8 for hollow dumbbell. In Fig. 6a, the 3D curve generated by the algorithm at iteration = is shown. Similarly, in Fig. 6b–f curve generated by the algorithm at iteration = 0, 50, 100, 450 and 900 is shown.

Figure 7 shows the evolution of ruled surface generated by the algorithm for a typical run. In Fig. 7a, the shape of ruled surface generated by the algorithm

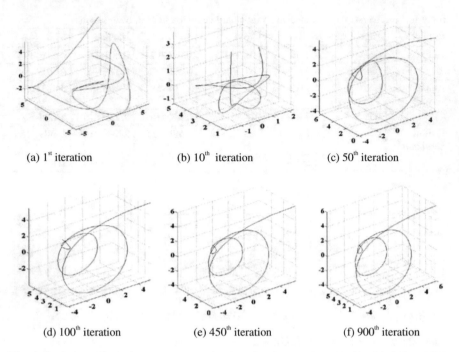

(a) 1st iteration   (b) 10th iteration   (c) 50th iteration

(d) 100th iteration   (e) 450th iteration   (f) 900th iteration

Fig. 6 Evolution of helix for a typical run of IGSA when iteration is 1, 10, 50, 100, 450 and 900

at iteration = 1 is shown. Similarly, in Fig. 7b–f, ruled surface generated by the algorithm at iteration = 10, 50, 100, 250 and 500 is shown.

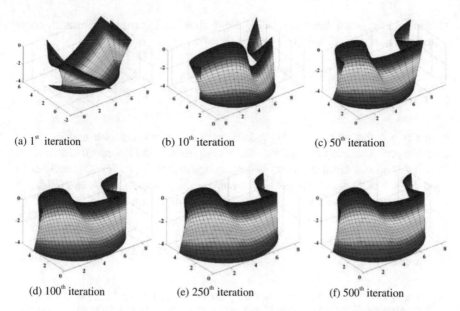

(a) 1st iteration (b) 10th iteration (c) 50th iteration

(d) 100th iteration (e) 250th iteration (f) 500th iteration

Fig. 7 Evolution of ruled surface for a typical run of IGSA when iteration is 1, 10, 50, 100, 250 and 500

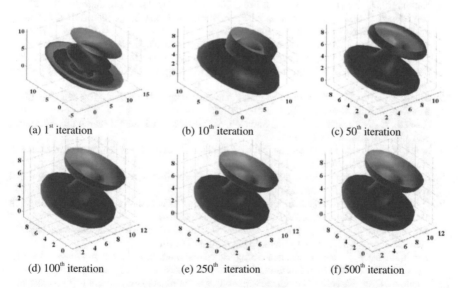

(a) 1st iteration (b) 10th iteration (c) 50th iteration

(d) 100th iteration (e) 250th iteration (f) 500th iteration

Fig. 8 Evolution of hollow dumbbell for a typical run of IGSA, when iteration is 1, 10, 50, 100, 250 and 500

Figure 8 shows the evolution of hollow dumbbell generated by the algorithm for a typical run. In Fig. 8a, the shape of dumbbell generated by the algorithm at iteration = 1 is shown. Similarly, in Fig. 8b–f, dumbbell generated by the algorithm at iteration = 10, 50, 100, 250 and 500 is shown.

6 Conclusions

In this paper, the problem of 3D reconstruction is modelled as a nonlinear optimization problem, and Gravitational Search Algorithm and Improved Gravitational Search Algorithm are used to solve three reconstruction problems, namely helix, ruled surface and hollow dumbbell. The convergence plots of IGSA and GSA indicate that the performance of IGSA is the best in comparison to GSA.

References

1. Back, T.: Evolutionary Algorithms in Theory and Practice: Evolution Strategies, Evolutionary Programming, Genetic Algorithms. Oxford Univ. Press, New York, USA (1996)
2. Kennedy, J.: Particle swarm optimization. In: Encyclopedia of Machine Learning, pp. 760–766. Springer, US, (2010)
3. Karaboga, D., Basturk, B.: Artificial bee colony (ABC) optimization algorithm for solving constrained optimization problems. In: Foundations of Fuzzy Logic and Soft Computing, pp. 789–798. Springer, Berlin Heidelberg (2007)
4. Singh, A., Deep, K.: How improvements in glowworm swarm optimization can solve real-Life problems. In: Proceedings of Fourth International Conference on Soft Computing for Problem Solving, Advances in Intelligent Systems and Computing, vol. 336, pp. 279–291. Springer, India (2015)
5. Singh, A., Deep, K.: New variants of glowworm swarm optimization based on step size. Int. J. Syst. Assur. Eng. Manage. 6(3), 286–296 (2015)
6. Rashedi, E., Nezamabadi-Pour, H., Saryazdi, S.: GSA: a gravitational search algorithm. Inf. Sci. 179(13), 2232–2248 (2009)
7. Sabri, N.M., Puteh, M., Mahmood, M.R.: A review of gravitational search algorithm. Int. J. Adv. Soft Comput. Appl. 5(3), 1–39 (2013)
8. Sarafrazi, S., Nezamabadi-pour, H., Saryazdi, S.: Disruption: a new operator in gravitational search algorithm. Sci. Iranica 18(3), 539–548 (2011)
9. Doraghinejad, M., Nezamabadi-pour, H.: Black hole: a new operator for gravitational search algorithm. Int. J. Comput. Intell. Syst. 7(5), 809–826 (2014)
10. Xu, B.C., Zhang, Y.Y.: An improved gravitational search algorithm for dynamic neural network identification. Int. J. Autom. Comput. 11(4), 434–440 (2014)
11. Singh, A., Deep, K.: Real coded genetic algorithm operators embedded in gravitational search algorithm for continuous optimization. Int. J. Intell. Syst. Appl. 7(12), 1–22 (2015)
12. Singh, A., Deep, K.: Novel Hybridized variants of gravitational search algorithm for constrained optimization. Int. J. Swarm Intell. 3(1), 1–22 (2017)
13. Singh, A., Deep, K.: Hybridized gravitational search algorithms with real coded genetic algorithms for integer and mixed integer optimization problems. In: Proceedings of Sixth International Conference on Soft Computing for Problem Solving, pp. 84–112. Springer, Singapore (2017)
14. Saini, S., Rambli, B.A., Rohaya, D., Zakaria, M.N.B., Bt Sulaiman, S.: A review on particle swarm optimization algorithm and its variants to human motion tracking. Math. Prob. Eng. 2014, 1–16 (2014)

15. Voisin, S., Abidi, M.A., Foufou, S., Truchetet, F.: Genetic algorithms for 3D reconstruction with supershapes. In: 16th International Conference on Image Processing, pp. 529–532. IEEE (2009)
16. Ning, J., McClean, S., Cranley, K.: 3D reconstruction from two orthogonal views using simulated annealing approach. In: Third International Conference on 3-D Digital Imaging and Modeling, pp. 309–313 (2001)
17. Ogura, T., Sato, C.: A fully automatic 3D reconstruction method using simulated annealing enables accurate posterioric angular assignment of protein projections. J. Struct. Biol. **156**(3), 371–386 (2006)
18. Siddique, M. T., Zakaria, M. N.: 3D Reconstruction of geometry from 2D image using Genetic Algorithm. In: 2010 International Symposium in Information Technology, vol. 1, pp. 1–5 (2010)
19. Wong, Y.P., Ng, B.Y.: 3D reconstruction from multiple views using Particle Swarm Optimization. In: Congress on Evolutionary Computation, pp. 1-8. IEEE (2010)
20. Koch, A., Dipanda, A.: Evolutionary-based 3D reconstruction using an uncalibrated stereovision system: application of building a panoramic object view. Multimedia Tools Appl. **57**(3), 565–586 (2012)
21. Singh, A., Deep, K., Nagar, A.: A new improved gravitational search algorithm for function optimization using a novel "best-so-far" update mechanism. In: Second International Conference on Soft Computing and Machine Intelligence, IEEE 2015, pp. 35–39 (2015)

Approaches to Question Answering Using LSTM and Memory Networks

G. Rohit, Ekta Gautam Dharamshi and Natarajan Subramanyam

Abstract Question answering (QA) is a field of Natural Language Processing that deals with generating answers automatically to questions asked to a system. It can be categorized into two types—open-domain and closed-domain QA. Open-domain QA can deal with questions about anything, whereas closed-domain QA deals with questions in a specific domain. In our work, we use the architectures of LSTM and memory networks to perform closed-domain question answering and compare the performances of the two. LSTMs are specialized RNNs that can remember necessary data and forget the irrelevant bits. Since data in QA consist of stories and questions based on them, this model seems appropriate, with the ability to handle long sequences. On the other hand, memory networks provide an architecture where there is a provision to store the information learnt by the system in an explicit memory component, rather than just as weight matrices. This also seems like an architecture well-suited to question answering. We implement each model and train it on the Facebook bAbi dataset. This dataset is specifically generated for the purpose of evaluating QA systems on the twenty prerequisite toy bAbi tasks. Each dataset corresponds to one task and checks whether the model is able to perform chaining, counting, answer with single and multiple supporting facts, understand relations, directions, etc. Based on the performances of each model on the bAbi tasks, we perform a comparative study of the two.

Keywords Memory networks · Long short-term memory · Question answering

G. Rohit · E. G. Dharamshi (✉)
PES Institute of Technology, Bangalore, India
e-mail: ektagd410@gmail.com

G. Rohit
e-mail: grohitpo@gmail.com

N. Subramanyam
PES University, Bangalore, India
e-mail: natarajan@pes.edu

© Springer Nature Singapore Pte Ltd. 2019
J. C. Bansal et al. (eds.), *Soft Computing for Problem Solving*,
Advances in Intelligent Systems and Computing 816,
https://doi.org/10.1007/978-981-13-1592-3_15

199

1 Introduction

One of the main challenges of computer science is to build systems that are more intelligent and can understand human beings without being explicitly told so. A major breakthrough in this is in the form of question answering (QA) systems. Question answering system is, as the name describes, a system that can answer questions. The quest for knowledge is deeply human, and so with the advent of computers and that of NLP, question answering took off.

A comparison can be drawn between a question answering system and the way human beings perform the same task. Humans would either "know" the answer from before or attempt to "discover" the answer. If a human knew the answer, it can be assumed that he/she has been taught about it. It could be learning by example, previous experience, explicit tutoring, etc. On the other hand, if the human being attempts to discover the answer, it could be by reading up from an information source, like a book or the Internet, understanding it, and then identifying the most relevant answer for the question.

The first case corresponds to closed-domain QA, in which the system is explicitly trained for the task by giving it examples of questions and right answers. This is similar to reading comprehension [1]. The second case is similar to closed-domain QA, in which a question is posed to the system and the system looks up multiple knowledge bases and returns a list of answers, ranked in order of relevance. In our work, we make an attempt to compare two leading techniques of solving the problem of closed-domain QA.

2 Literature Survey

Question answering is a field in the domain of Natural Language Processing (NLP) and artificial intelligence (AI) that deal with the ability of computers to be able to answer questions posed to it in natural language. NLP techniques sometimes make use of ML techniques to simulate the working of a brain and make the model learn as per the requirements. In this section, we begin by introducing the problem of QA and how the fields of Natural Language Processing and Machine Learning could hold potential solutions. Next, we get into neural networks and their customizations to meet our needs, and then into memory networks. Lastly, we discuss how we plan to evaluate and draw comparisons between the models we will build.

2.1 Question Answering

As envisaged by Das in [2], we have always wanted computers to act intelligent. The major challenge in this is to the inability of computers to communicate in natural language. It is a branch of artificial intelligence (AI). Question answering is one quintessential application of NLP. The task performed by a question answering system is that when given a question and small set of relevant data, it finds the exact answer for the question. It has practical applications in various domains like education, health care, and personal assistance.

2.2 Conventional Techniques for Question Answering

Before the advent of neural networks, problems in the field of Natural Language Processing were solved using traditional techniques like Information Retrieval (IR), Information Extraction (IE), Stemming, Lemmatization, Part-of-Speech (POS) Tagging, Named Entity Recognition (NER).

The overall steps in a generic QA system can be summarized as follows. First, the question type is determined, and based on this, the expected answer type is identified. Then, some NLP tasks, like the ones mentioned above, are performed to identify relations between words and interpret semantic meanings. As more NLP tasks are applied, the system narrows down on the parts of the corpus most likely to contain the answer. As a result of this, the exact words can be singled out and the answer is generated.

2.3 Machine Learning

Machine learning is a field in AI that gives computers the ability to learn without having any explicit knowledge in the field of operation. It is an evolution of pattern recognition and computational learning. We think that machine learning seems to be the best approach to go about solving the problem of question answering as it is the only technique that allows the system to interpret natural language in a way similar to the human brain, and hence this method seems to have the most potential.

2.4 Long Short-Term Memory

Theoretically, recurrent neural networks (RNNs) can make use of information in arbitrarily long sequences, but in practice, they are limited to looking back only a few steps. This is due to the problem of Vanishing Gradients. The activation functions produce an output between 0 and 1 or −1 and 1. Since gradients are calculated by chain rule, which involves multiplication, it is possible that the gradients may be multiplied by very small values each time, having little or no effect on the weight matrices. Due to this, and the fact that the weight matrices are randomly initialized and are to be "learnt" based on the data, it can happen that the weight matrices do not

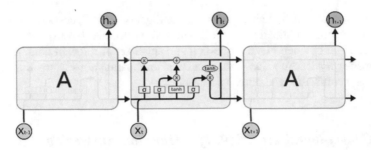

Fig. 1 Architecture of LSTM

change much at all. This problem was first identified by Hochreiter in his diploma thesis [3]. The solution provided by him is known as long short-term memory (LSTM) [4].

An LSTM is an improvised RNN which can remember longer sequences more efficiently by providing a feature to "remember" the relevant and "forget" the irrelevant parts of the data [5]. LSTMs provide finer control over what is needed by making changes to the internal structure of a neuron, as can be seen in Fig. 1. Here, x_t corresponds to the input vector at time step t (or the tth word), h_{t-1} is the hidden state vector from the previous time step, and the calculations for the various internal states such as "forget gate" (controls which part of the information is propagated to the next time step), "input gate" (controls which part of the input is considered as important information), and cell state are also represented in the figure.

The advantages of the LSTM network are so significant that Hermann et al. [6] argue for using supervised learning as opposed to classical NLP techniques for the task of question answering. They describe a model that focuses on the aspects of the document that it believes will help it answer the question. This can encode the statements in the story of the dataset. From this, when questions are posed to it, it predicts the answer. This paper also talks about the dearth of good-quality datasets for question answering and how there is a need to synthetically generate datasets for employing supervised learning to the task of question answering.

In our work, the story is vectorized using Google's word2vec. The question corresponding to each story is appended to it and fed to the LSTM, one word at each time step. Based on the query and the story, errors in the output predictions are computed and propagated back through the LSTM model for each time step.

2.5 Memory Networks

Memory networks are a technique that proposes to extend the machine learning models by adding a memory component to it. In the absence of the memory component, the model just consists of the weight matrices to represent all the data features it has

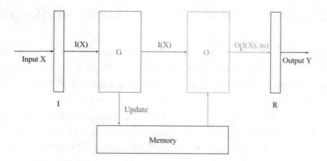

Fig. 2 Architecture of memory networks

learnt. But the addition of this memory component obviously increases the scope and performance of the model.

Weston et al. [7] introduce memory networks which have a separate dedicated memory component for storing information and accessing relevant information as and when needed. They argue that the memory in a typical recurrent neural network is too less to accommodate all data and answer any question accurately.

As shown in Fig. 2, a memory network consists of a memory m and four (potentially learned) components I, G, O, and R as follows:

I: (input feature map)—This module converts the incoming input to the internal feature representation, for example, an array of vectors or an array of strings.
G: (generalization)—This module updates old memories given the new input. This is called generalization as there is an opportunity for the network to compress and generalize its memories at this stage for some intended future use.
O: (output feature map)—This module produces a new output (in the feature representation space), given the new input and the current memory state.
R: (response)—This module converts the output into the response format desired, for example, a textual response or an action.

Given an input X (e.g., an input character, word, or sentence depending on the granularity chosen, an image, or an audio signal), the flow of the model is as follows:

1. X is converted to an internal feature representation $I(X)$.
2. Memory mi is updated, given the new input: $mi = G (mi, I(X), m)$, $\forall i$.
3. Output features o are computed, given the new input and the memory: $o = O (I(X), m)$.
4. Output features o are decoded to give the final response: $r = R (o)$.

This process is applied at both train and test time and memories are also stored at test time, but the model parameters of $I, G, O,$ and R are not updated. Memory networks cover a wide class of possible implementations. One particular instantiation of a memory network is where the components are neural networks. This is the model used in our work.

2.6   Evaluation Metrics–Toy Tasks in Question Answering

To determine which of the aforementioned two techniques is more suitable, we require a sound basis for comparison. Various kinds of questions may be present, and we need to identify which model is able to gauge the answer to what kind of questions better (or worse) than the other.

The main tasks that could help evaluate reading comprehension via question answering have been well-explored in [8]. They list out about 20 "leaf" tasks that form the basis for testing a question answering system. These tasks measure understanding in several ways: whether a system is able to answer questions via chaining facts, simple induction, deduction and many more. The tasks are designed to be prerequisites for any system that aims to be capable of conversing with a human. The examples for each of the tasks listed in this paper are summarized in Table 1.

In our work, we intend to build our own RNN and memory network models and compare them on the basis of the tasks listed out by this paper. We also intend to analyze what effect, if any, language has on the performance of either models. The bAbi dataset is suitable for this because it provides synthetically generated data, as recommended in [5].

3   Implementation

3.1   LSTM

We wrote our code in Python, making use of the Numpy and Keras modules to build the model. This gave us the advantage of not having to code each layer and each cell of the network from scratch. We could easily define layers with their activation functions using just a few statements. We experimented with various values of vector lengths and layer sizes in the neural network to arrive at optimum values to get the best possible results.

3.2   Memory Network

We used the LSTM itself as the central component of the memory network model, essentially creating a memory neural network. Our choice to use this was based on two reasons—the amount of generalization needed by the memory component could be learned by the model, providing the best value; and having the same benchmark to compare both models. Essentially, it was the use of LSTM for question answering against the use of LSTM in the memory component of memory network-based question answering system.

Table 1 Toy tasks

Task	Sample question	Answer
Single supporting fact	Mary went to the bathroom John moved to the hallway Mary travelled to the office Where is Mary?	Office
Two supporting facts	John is in the playground John picked up the football Bob went to the kitchen Where is the football?	Playground
Three supporting facts	John picked up the apple John went to the office John went to the kitchen John dropped the apple Where was the apple before the kitchen?	Office
Two argument relations	The office is north of the bedroom The bedroom is north of the bathroom The kitchen is west of the garden What is north of the bedroom? What is the bedroom north of?	Office Bathroom
Three argument relations	Mary gave the cake to Fred Fred gave the cake to Bill Jeff was given the milk by Bill Who gave the cake to Fred? Who did Fred give the cake to?	Mary Bill
Yes/No questions	John moved to the playground Daniel went to the bathroom John went back to the hallway Is John in the playground? Is Daniel in the bathroom?	No Yes
Counting	Daniel picked up the football Daniel dropped the football Daniel got the milk Daniel took the apple How many objects is Daniel holding?	Two
Lists/Sets	Daniel picks up the football Daniel drops the newspaper Daniel picks up the milk John took the apple What is Daniel holding?	Milk, Football
Simple negation	Sandra travelled to the office Fred is no longer in the office Is Fred in the office? Is Sandra in the office?	No Yes
Indefinite knowledge	John is either in the classroom or the playground Sandra is in the garden Is John in the classroom? Is John in the office?	Maybe No

(continued)

Table 1 (continued)

Task	Sample question	Answer
Basic co-reference	Daniel was in the kitchen Then he went to the studio Sandra was in the office Where is Daniel?	Studio
Conjunction	Mary and Jeff went to the kitchen Then Jeff went to the park Where is Mary? Where is Jeff?	Kitchen Park
Compound co-reference	Daniel and Sandra journeyed to the office Then they went to the garden Sandra and John travelled to the kitchen After that, they moved to the hallway Where is Daniel?	Garden
Time reasoning	In the afternoon Julie went to the park Yesterday Julie was at school Julie went to the cinema in the evening Where did Julie go after the park? Where was Julie before the park?	Cinema School
Basic deduction	Sheep are afraid of wolves Cats are afraid of dogs Mice are afraid of cats Gertrude is a sheep What is Gertrude afraid of?	Wolves
Basic induction	Lily is a swan Lily is white Bernhard is green Greg is a swan What color is Greg?	White
Positional reasoning	The triangle is to the right of the blue square The red square is on top of the blue square The red sphere is to the right of the blue square Is the red sphere to the right of the blue square? Is the red square to the left of the triangle?	Yes Yes
Size reasoning	The football fits in the suitcase. The suitcase fits in the cupboard. The box is smaller than the football Will the box fit in the suitcase? Will the cupboard fit in the box?	Yes No

(continued)

Table 1 (continued)

Task	Sample question	Answer
Path finding	The kitchen is north of the hallway The bathroom is west of the bedroom The den is east of the hallway The office is south of the bedroom How do you go from den to kitchen? How do you go from office to bathroom?	West, North North, West
Agent's Motivations	John is hungry John goes to the kitchen John grabbed the apple there Daniel is hungry Where does Daniel go? Why did John go to the kitchen?	Kitchen Hungry

4 Results and Discussions

We trained and tested both our models on the Facebook bAbi dataset. We made use of the toy tasks listed in [8] to benchmark the performance of each of our models. We used them also as a means of comparison between the two models. We also intend to infer whether language has any role to play in the degree of understanding a model has, i.e., if it is easier to interpret one language over the other.

The results of this work are shown in Table 2.

The first thing that struck us is that in some tasks accuracies were as high as 98% and more, while in others, they were lower than 30%. The tasks that had the least accuracies were tasks 3 and 19. Task 3 is the task of three supporting facts, and task 19 is path finding based on directions. It is surprising that task 2, which is based on two supporting facts, had a good performance but the models had a hard time comprehending three supporting facts. Path finding is a task about locating a path between two places when the relative position between each pair of places is known. The explanation for the performance of the models in this task could be that the relative positions could be understood in a one-dimensional space, but path finding requires two dimensions, which the models were not able to do.

Overall, we noticed that memory networks perform significantly better than LSTMs. This can be seen in Fig. 3. The most striking fact is that memory networks ace the first ten tasks by a significant margin. The same is true also for tasks 11, 12, 14, and 17. In the rest of the tasks, their difference is at most 2–3%, with the advantage swinging either way.

When we analyzed how language affected the performance of either model, we found some surprising behavior. In most tasks, English seems to be easier for the models to comprehend. However, in tasks 3, 4, 6, and 7, the LSTM performed significantly better on the Hindi dataset. On the other hand, the accuracy of memory networks on the English dataset is much higher than on the Hindi dataset.

Table 2 Results of the work

Heading level	LSTM		Memory networks	
	English	Hindi	English	Hindi
Task 1	93.8	91.3	99.9	100
Task 2	33.2	29.3	44.8	44.9
Task 3	13.3	18.2	23.5	19.3
Task 4	76.3	81.2	99.7	83.5
Task 5	58.1	55.8	91.1	87.6
Task 6	50.3	57.9	90.2	81.4
Task 7	53.6	79.2	85.5	86.2
Task 8	77.2	76.2	77.5	81.4
Task 9	84.1	50.6	88.3	67.3
Task 10	75.9	58.0	96.2	70.7
Task 11	99.6	83.4	98.5	99.9
Task 12	89.4	72.6	97.6	99.5
Task 13	92.3	92.3	96.6	91.4
Task 14	41.8	40.6	46.2	45.3
Task 15	58.3	47.7	56.1	51.9
Task 16	47.4	45.5	47.8	48.3
Task 17	52.0	50.7	70.2	58.8
Task 18	90.3	93.4	92.8	94.1
Task 19	10.5	11.5	11.2	12.3
Task 20	97.9	98.1	98.2	97.4

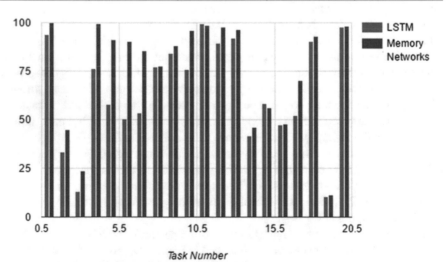

Fig. 3 LSTMs versus memory networks on the English dataset

References

1. Kapashi, D., Shah, P.: Answering Reading Comprehension Using Memory Networks. Report for Stanford University Course cs224d (2015)
2. Das, B.: A Survey on Question Answering System, Survey for Indian Institute of Technology, Bombay (2014)
3. Hochreiter, S.: Untersuchungen zu dynamischen neuronalen Netzen [in German]. Diploma Thesis, TU München (1991)
4. Hochreiter, S., Schmidhuber, S.: Long short term memory. Neural Comput. **9**(8), 1735–1780 (1997)
5. Understanding LSTM Networks: http://colah.github.io/posts/2015-08-Understanding-LSTMs/. Last accessed 4 Mar 2017
6. Hermann, K.M., Kocisky, T., Grefenstette, E., Espeholt, L., Kay, W., Suleyman, M., Blunsom, R.: Teaching machines to read and comprehend. In: 28th International Proceedings on Advances in Neural Information Processing Systems, pp. 1693–1701. MIT Press, Montreal (2015)
7. Weston, J., Chopra, S., Bordes, A.: Memory Networks, arXiv preprint, arXiv:1410.3916 (2014)
8. Weston, J., Bordes, A., Chopra, S., Rush, A.M., van Merriënboer, B., Joulin, A., Mikolov, T.: Towards AI-Complete Question Answering: A Set of Prerequisite Toy Tasks, arXiv preprint, arXiv: 1502.05968 (2015)

Data Extraction from Traffic Videos Using Machine Learning Approach

Anshul Mittal, Mridul Gupta and Indrajit Ghosh

Abstract Traffic safety has become one of the major concerns in most of the countries with extensive road networks. With the ever-increasing traffic and its various types, it has become increasingly difficult to check if the road network can sustain the surge. To evaluate the efficiency and safety of a network, several factors such as speed, vehicular composition, traffic volume are required. Data collection for calculating each of these factors is time-consuming. Most of the current activities in the area of intelligent transportation systems involve the collection of data through various sources such as surveillance cameras, but the collection alone is not sufficient. It requires a lot of time to process this data and determine the safety level of the road network which becomes manifold for a country with a vast network like India. It is a necessity to expand the use of intelligent transportation for the processing of the data. To achieve this initially, it is required to have traffic flow data. Therefore, high-resolution video cameras were placed at vantage points approximately 100–150 m away from the center of intersection locations. Two such intersections were selected from the National Capital Region (NCR) of India. Traffic flow-related data was recorded from 10 am to 4 pm during good weather condition. The obtained videos were then processed to segregate different types of vehicles. The proposed algorithm deals with the vehicles which are up to 70% occluded. A CNN-LSTM (Krizhevsky et al. in Advances in neural information processing systems, pp 1097–1105, 2012 [6]) model is trained for the recognition of a vehicle. Following this, a minimal cover volume algorithm is developed using bi-grid mapping for classifying vehicles and evaluating various parameters such as base center, orientation, and minimizing error due to occlusion. The proposed algorithm is based on machine learning, and it can estimate the required parameters with minimal human assistance and accuracy of 95.6% on test video and 87.6% on cifar-100 for object detection.

A. Mittal (✉) · M. Gupta · I. Ghosh
Department of Civil Engineering, Indian Institute of Technology Roorkee, Roorkee, India
e-mail: anshulmittal71@gmail.com

M. Gupta
e-mail: mridulgupta9@gmail.com

I. Ghosh
e-mail: indrafce@iitr.ac.in

© Springer Nature Singapore Pte Ltd. 2019
J. C. Bansal et al. (eds.), *Soft Computing for Problem Solving*,
Advances in Intelligent Systems and Computing 816,
https://doi.org/10.1007/978-981-13-1592-3_16

Keywords Machine learning · Intelligent transportation system · ITS
Computer vision · Safety performance indicators

1 Introduction

Road transport system is one the fastest growing sectors increasing from 200 billion tonne km in 1980 to 700 billion tonne km in 2012 in India. This hike in the vehicles, on the one hand, represents the development of a country, and on the other hand, it leads to an increase in the road accidents. Over 1.2 million people worldwide are killed in road accidents each year, and many more are injured. These estimates are expected to increase by about 65% over the next 20 years unless we commit ourselves to its prevention. Safety performance indicators which are causally related to the number of crashes or to the injury consequences of a crash are increasing used. However, confirming the safety of a road network on the basis of crashes and injuries only is insufficient, and more information about the safety performance indicators (SPIs) such as speed, vehicular composition, maximum flow is needed. With rapidly expanding road networks and increasing vehicles, it is necessary to monitor these roads for safety standards at a similar pace. Thus, it is not preferable to waste time on manually extracting the SPIs from the data, which could very well be used in designing better road network. What we need is a software that can analyze the video of the traffic flow and supply us with all the required SPIs.

In this paper, we have presented an algorithm that can detect and accurately categorize the vehicles in three different weight classes, namely light, medium, and heavy to track the movement of the vehicles with reasonable accuracy. The categorization of vehicles' accuracy has been calculated on cifar-10 dataset and also on our own video (due to the absence of standard dataset). The remainder of the paper is structured as follows. In Sect. 3, we have explained the proposed algorithm in detail and the corresponding results are then presented in Sect. 4. Finally, we conclude the paper in Sect. 5.

2 Related Work

Recently, several papers have proposed methods of using deep networks for classification of different objects. In the VGG method [1], deep networks (CNN) are trained with increasing depth using an architecture with small convolutional filter. With increasing depth, the accuracy of classification also improves and achieves maxima at 16–19 weight layers. The Maxout method [2] is the same as convolutional neural networks (CNN) or multi-layer perceptron, i.e., a feed-forward architecture, but explores the effect on accuracy by using a new activation function, the maxout unit. In the OverFeat method [3], a single object is assumed and a fully connected layer is trained to predict a box around it. In the multi-box algorithm [4, 5], regions are predicted on the basis of a fully connected network that predicts multiple boxes which are used for regional convolutional neural network object detection. A comparison of our algorithm with VGG and Maxout methods has been provided in Table 5.

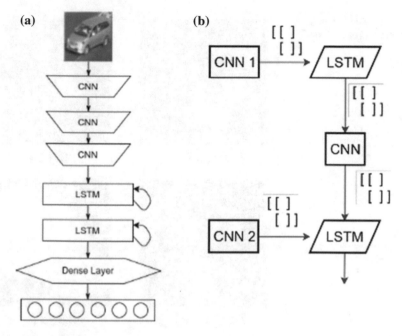

Fig. 1 **a** CNN-LSTM architecture for detecting class of vehicle and road surface. **b** Modified LSTM used in the network

3 Methodology

A. Vehicle and road detection

In this paper, we have used a deep CNN-LSTM architecture for recognizing vehicles and their classes, namely light, medium, and heavy vehicles as described by Krizhevsky et al. [6] and Sainath et al. [7]. We have used three CNN layers and two LSTM layers with a single dense layer in the end followed by softmax classification at the learning rate of 0.0001 with cross-validation. Memory cell comprises of CNN unit and a storage unit which convolves memory each time a new input arrives at input gate, and as a result, we get a convolving memory LSTM (see Fig. 1b) (CNN i for $i = 1$ to N where N is equal to number of features). Another contribution of this paper is to develop a mathematical model for detection of vehicles which are as much as 70% occluded. Results are compiled in the results section.

We segment the vehicle from its background using algorithm 1. Output sequence at each step of algorithm is shown in Fig. 2. Coordinates of the bounding box so obtained (Fig. 2d) are used for calculating minimum area rectangle to cover the segmented vehicle.

Fig. 2 Showing output of algorithm 1 as each step

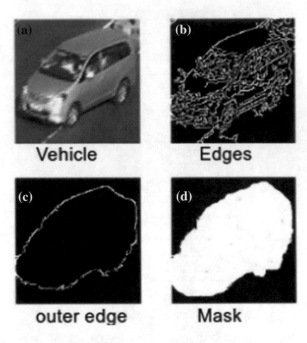

(a) Vehicle	(b) Edges
(c) outer edge	(d) Mask

Algorithm 1: Vehicle Extraction
//F(x,y) ← RGB Image
Edges ← edgeDetection(F(x,y))//Canny edge
outline ← outerMostCountour(Edges)
mask ← fillPolygon(outline)
return boundingBox(mask)

$F(x, y)$ is the section of RGB image of road scene where detection of vehicle is positive Fig. 2a. The function edgeDetection accepts image as an input and returns edges as a vector and their hierarchy in the image Fig. 2b. The function outMostContour accepts the edges vector and returns the contour of which every other contour is a part of Fig. 2c. Function fillPolygon accepts the outermost contour and creates the binary image by filling inside the contour Fig. 2d. In the end, the algorithm returns the coordinates of the bounding rectangle of the mask Fig. 2d.

Fig. 3 Showing marking of line l_1, l_2, l_3, l_4, l_5, and l_6

B. Scene parameters and cube fitting

Perspective images of road scene captured from the camera consist of information regarding depth (Y) as well its placement (X) with respect to the surface of road. For calculating parameters of perspective projection of the system, we follow the following procedure:

- We mark two line segments on the road scene l_1 and l_2 to meet at a point. For l_1, we have two coordinates (x_{1a}, y_{1a}) and (x_{1b}, y_{1b}) similarly, for line l_2, we have two coordinates (x_{2a}, y_{2a}) and (x_{2b}, y_{2b}) (Fig. 3).
- Next, we mark four vertical line segments for calculating boundary condition denoting projection of vertical objects l_3, l_4, l_5, and l_6 whose coordinates are $[(x_{3a}, y_{3a}), (x_{3b}, y_{3b})]$, $[(x_{4a}, y_{4a}), (x_{4b}, y_{4b})]$, $[(x_{5a}, y_{5a}), (x_{5b}, y_{5b})]$, and $[(x_{6a}, y_{6a}), (x_{6b}, y_{6b})]$ for 2 m height (Fig. 3b). **Note**: for this algorithm to work always choose $(x_{3a}, y_{3a}) = (x_{1a}, y_{1a})$, $(x_{4a}, y_{4a}) = (x_{1b}, y_{1b})$, $(x_{5a}, y_{5a}) = (x_{2a}, y_{2a})$, and $(x_{6a}, y_{6a}) = (x_{2b}, y_{2b})$ (Fig. 3).
- Using minimal volume analysis as explained in the following section (Fig. 4), we calculate the orientation for the detected vehicle.

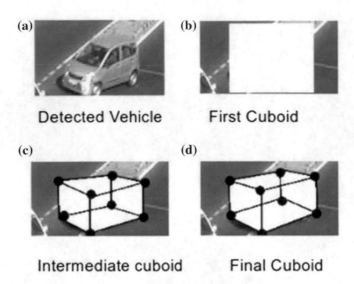

Fig. 4 Showing the time snaps for calculating intermediate cuboid

Table 1 Boundary condition to evaluate A and B

$[X', Y']$	$[X, Y]$
$[0, 0]$	$[x_{1a}, y_{1a}]$
$[1, 0]$	$[x_{1b}, y_{1b}]$
$[0, 1]$	$[x_{2a}, y_{2a}]$
$[1]$	$[x_{3a}, y_{3a}]$

C.   Minimal volume cover

To evaluate the minimal cover, we have divided vehicles into three major classes, i.e., light, medium, and heavy. For each class, we have standardized the height of the vehicle to evaluate the minimal cover (Table 1). Using Eq. 1, we transform coordinates from perspective grid system to rectangular grid system, and using Eq. 2, we transform base of cuboid from rectangular grid system to perspective grid system, and using Eq. 3, we project height of cuboid from rectangular grid system to perspective grid system. For a given system, x, y represents coordinates in perspective distance and x', y' represents coordinates in transformed rectangular region.

$$X'_0 = A * X_0 \tag{1}$$

$X_0' = [w'^*x', w'^*y', w]$ $w' = w$ for $w! = 0$ else infinity, $A' =$ is a 3×3 matrix, $X_0 = [x, y, 1]$. Using value from Table 2, we evaluate A.

$$X_1 = B * X'_1 \tag{2}$$

Table 2 Boundary condition to evaluate C

$[X', Y', Z']$	$[X, Y]$
[0,0,0]	$[x_{3a}, y_{3a}]$
[0,0,2]	$[x_{3b}, y_{3b}]$
[1,0,0]	$[x_{4a}, y_{4a}]$
[1,0,2]	$[x_{4b}, y_{4b}]$
[0,1,0]	$[x_{5a}, y_{5a}]$
[0,1,2]	$[x_{5b}, y_{5b}]$
[1,1,0]	$[x_{6a}, y_{6a}]$
[1,1,2]	$[x_{6b}, y_{6b}]$

Table 3 Showing vehicle parameters for different classes

Class	Width	Length	Height
Light	0.64	1.87	1.5
Medium	1.6	4	2
Heavy	2.43	10	3.5

$X_1' = [x', y', 1]$, $X_1 = [w'*x, w'*y, w]$ $w' = w$ for $w! = 0$ else infinity, $B =$ is a 3×3 matrix. Using value from Table 2, we evaluate B.

$$X_2 = C * X_2' \tag{3}$$

$X_2' = [x', y', z', 1]$, $X_2 = [w'*x, w'*y, w]$ $w' = w$ for $w! = 0$ else infinity, $C =$ is a 4×3 matrix using value from Table 2, we evaluate C.

Parameters A, B, C are calculated by solving system of linear equations.

Using Algorithm 2, we evaluate cuboid having minimal area (minCbids) covering the detected vehicle(s). Persp2Rect transforms bounding box of max evaluated at the end to algorithm 1 (Box) to rectangular grid system using Eq. 1 and stores in RectBox. MinX and MinY returns the minimum value of x', y' of the RectBox, respectively. Similarly, MaxX and MaxY returns the maximum value of x', y' of the RectBox, respectively. For a given class of vehicle, we use the value of width (w), length (l), and height (h) (Table 3, [8]) which is used by the function createBox to calculate coordinates of the cuboid rotated at the angle α along z' axis in clockwise direction. rect2Persp uses Eqs. 2 and 3 to project cube on the perspective grid. calcProjArea uses the projected area of the cuboid on the perspective grid and evaluates the vehicle region covered.

Algorithm 2: Minimal Cover
\\Box: Stores coordinates of bounding box of mask
Procedure MinimalCover(k,box) \\k stores the number of cubes required to fill the box
minCbids \\Stores k cuboids initialized with maximum volume
RectBox ← Presp2Rect(Box);
x`min ← minX (RectBox)
x`max ← maxX (RectBox)
y`min ← minY (RectBox)
y`max ← maxY (RectBox)
for **x`** ∈ (x`min , x`max):
for **y`** ∈ (y`min , y`max):
for α ∈ (0 , 90):
cbids ← createBox(x`,y`, α, w, l, h,k); \\returns the tensor of k cuboids
for each (cbid,minCbid) in (cbids, minCbids):
projCube ← rect2Presp(cbid);
Area ← calcProjArea(projCube);
if Area< minCbid.area:
minCbid ← cbid;
return minCbids;

After evaluating minimum cover cuboid, we can store its parameters, i.e., x', y', α, class of the vehicle for further data analysis. This procedure is repeated for every vehicle detected in the frame, and using mean shift algorithm, the vehicle's path is traced in a given video sequence.

D.   Occlusion correction

Algorithm 3 deals with the occlusion present in the video of road. Procedure bounding cubes take unique classes detected in a box as input as well as box (section of image which is being analyzed). This procedure returns the tensor of cuboids which is used to segment box and each segment is again passed to the neural network for vehicle class prediction.

Table 4 Accuracy assessment for different models

S. no.	Model	Cifar-10 (%)	Scene specific (%)
1	1C-0L	75.1	72.3
2	1C-1L	85.7	76.8
3	2C-1L	90.4	82.5
4	2C-2L	97.3	85.6
5	3C-1L	96.1	84.9
6	3C-2L	99.5	95.6
7	3C-3L	99.4	95.2

Note xC-yL implies that x layers of CNN followed by y layers of LSTM network

Algorithm 3: Maximum number of bounding cubes
Procedure boundingCubes(classes,box):
Flag = True
Volume = 0
k = classes
while(Flag):
tempVolume = 0
cuboids = MinimalCover(k,box)
for each cuboid in cuboids:
tempVolume += cuboid.volume
if tempVolume <= 0.8*Volume:
Volume = tempVolume
k+=1
else :
return cuboids

4 Results

E. Accuracy of CNN-LSTM architecture for detecting class of vehicle and road surface.

We have experimented with multiple CNN-LSTM model, out of which we have chosen the one which is giving maximum accuracy. In Cifar-10 and scene-specific data, we have created ground truth manually for classification into light, medium, and heavy. For accuracy, we have given 1 for each correct and 0 for incorrect prediction. Results are summarized in Table 4. Results are evaluated by taking average of accuracy over 1000 images.

Table 5 Table for accuracy comparison for recognition on different datasets with recent papers

Dataset	VGG [1] (%)	Maxout [2] (%)	Our method (%)
Cifar-10	98.68	92.3	99.5
Image net	91.57	94.6	98.3

Table 6 Accuracy for different classes of vehicles

	2 Wheelers (Light) (%)	4 Wheelers (Medium) (%)	Trucks (Heavy) (%)
Our method	78.2	94.6	93.2

F. Comparative Studies

We have summarized results of comparison of our method with few current methods in Table 5. Our method outperforms all the mentioned neural networks.

G. Vehicle category recognition accuracy

In Table 6, we have summarized the accuracy of our algorithm on different classes of vehicles.

5 Conclusion

In this paper, we have presented a machine learning approach for faster vehicle data extraction with the overall accuracy of 95.6%. After detection using neural networks, minimal cover cuboid method was used to evaluate the orientation of the on-road vehicle. Since we have used standard widths and lengths of the vehicles from [8] and have estimated the height from the average height of the vehicles, we can estimate orientation of a vehicle even if it is occluded. Data collection and analysis which takes the major portion of the time of designing of a road network can now be done in considerably lesser time using this algorithm. It took only 2 h to analyze a 24 h video of an intersection and preparing a table for percentage occlusion, orientation, and vehicle classes. Therefore, the algorithm presented in this paper is faster, robust, and free from human intervention, thus saving large amount of money.

References

1. Simonyan, K., Zisserman, A.: Very deep convolutional networks for large-scale image recognition. In: ICLR (2015)
2. Goodfellow, I.J., Warde-Farley, D., Mirza, M., Courville, A., Bengio, Y.: Maxout networks. arXiv:1302.4389 (2013)
3. Sermanet, P., Eigen, D., Zhang, X., Mathieu, M., Fergus, R., LeCun, Y.: Overfeat: integrated recognition, localization and detection using convolutional networks. In: ICLR (2014)
4. Erhan, D., Szegedy, C., Toshev, A., Anguelov, D.: Scalable object detection using deep neural networks. In: CVPR (2014)
5. Szegedy, C., Reed, S., Erhan, D., Anguelov, D.: Scalable, high-quality object detection. arXiv: 1412.1441v2 (2015)
6. Krizhevsky, A., Sutskever, I., Hinton, G.E.: Imagenet classification with deep convolutional neural networks. In: Advances in Neural Information Processing Systems, pp. 1097–1105 (2012)
7. Sainath, T.N., Vinyals, O., Senior, A., Sak, H.: Convolutional, long short-term memory, fully connected deep neural networks. In: 2015 IEEE International Conference on Acoustics, Speech and Signal Processing (ICASSP), pp. 4580–4584. IEEE (2015)
8. Nokandeh, M.M., Ghosh, I., Chandra, S.: Determination of passenger-car units on two-lane intercity highways under heterogeneous traffic conditions. J. Transp. Eng. **142**(2), 04015040 (2015)
9. World Health Organization (WHO): World Report on Road Traffic Injury Prevention, WHO, Geneva, http://www.who.int/violence_injury_prevention/publications/road_traffic/world_report/en/ (2004)
10. NTDPC ~ Vol-02 Part1 ~ Ch02.indd26 page30, http://planningcommission.nic.in/sectors/NTDPC/volume2_p1/trends_v2_p1.pdf

Stability Analysis and Controller Design for Unstable Systems Using Relay Feedback Approach

D. Kishore, K. Anand Kishore and R. C. Panda

Abstract In this paper, a novel approach to identify the parameters of the unstable systems is considered and the stability analysis of the unstable system has been carried out. The relay feedback approach has been chosen to obtain the sustained oscillations, and single-relay feedback approach is considered for different bench examples for the analysis purpose unlike the conventional Z–N methods. The simplest method to generate sustained oscillation is used. An attempt is made to stabilize the oscillations of the unstable process considered, and the stability checking criterion such as Nyquist plot was also used. Simple relay identification with hysteresis is used to identify the unstable system parameters. In this analysis, a few transfer functions are considered with ratio of time delay to time constant less than 0.693 for analysis purpose. A new method of control is proposed for different unstable systems.

Keywords Unstable systems · Relay feedback · Nyquist stability · Limit cycles
Control system

1 Introduction

A system is said to be unstable if it produces an unbounded output for bounded input or precisely if any one root of the characteristic equation lies on the left half plane of the s-plane. In process industries, it is necessary to operate some processes at unstable states for achieving safety and maximizing the product. Control of unstable systems is a challenging task due to the operation at multiple steady states and nonlinear behavior. For effective control of unstable systems, it is very important to identify the system. Identification is the process in which a system can be identified by using the input and output data. The identification of the unstable systems for the purpose

D. Kishore (✉) · K. Anand Kishore
Department of Chemical Engineering, NIT Warangal, Warangal, India
e-mail: Kishorenit.trichy@gmail.com

R. C. Panda
Central Leather Research Institute, Chennai, India

© Springer Nature Singapore Pte Ltd. 2019
J. C. Bansal et al. (eds.), *Soft Computing for Problem Solving*,
Advances in Intelligent Systems and Computing 816,
https://doi.org/10.1007/978-981-13-1592-3_17

of controller design is as important as that of the stable systems. Chinta Shankar rao [1] has extensively carried out studies on unstable systems and reported that unstable systems exhibit multiple steady states due to certain nonlinearity of the systems. Some of the steady states may be unstable and need to operate at unstable steady state for economic and/or safety reasons. Chinta Shankar Rao [1] has reported the identification of an unstable subsystem operating in stable closed loop in the presence of noise. Chinta Shankar Rao [1] presented jacketed CSTR. Chidambaram reviewed on unstable jacketed CSTR. Chinta Shankar Rao [1] has presented Identification using equation error input covariance the method can be implemented online, except for the case where the identified model has poles and zeros on imaginary axis. Chinta Shankar Rao [1] has reported the output error and Box–Jenkins model structures for identification of the unstable systems. Jacobsen [1] has identified transfer function model between the composition of the distillate and the recycle ratio where an unstable second-order model with one unstable and one unstable zero. Chinta Shankar Rao [1] has discussed a reliable automatic PID tuning method for the open-loop unstable processes. Identification is carried out with low-order models by means of two relay tests, one with an additional delay, which does not require a priori knowledge about the process. Chinta Shankar Rao [1] discussed the CSTR. Linearized model results for unstable operating point model with two unstable poles and a negative zero. Presented several unstable systems: gas-phase polyolefin reactor, jacketed CSTR, isothermal reactor, cart-and-pole problem, and helicopter. Chinta Shankar Rao [1] has reported new direct closed-loop identification method based on output intersampling scheme. The above all methods are discussed the conventional identification methods, and it is very difficult and requires in-depth knowledge and extensive simulations for identification of the different process parameters. In the current work describing function method was used for the identification of two key parameters, and a simplest method was proposed for the identification and control of different unstable systems. The basic block diagram is shown in Fig. 1.

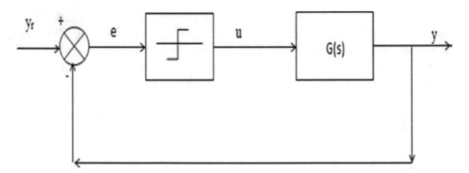

Fig. 1 Relay feedback control system

2 Materials and Methods

The DF method was among the pioneering work for relay identification and has been continuously developed up to the present [2]. The key idea behind this approach is the use of the ultimate amplitude and phase conditions for steady oscillations under a relay test to establish fitting conditions for model identification. A detailed review on the various methods that are developed for the analysis of relay feedback methods was given in detail [3]. In the Current work describing function method was used to estimate other parameters from the relay feedback method and the graphical results are shown in the following figures for the identified process using relay feedback approach stability analysis is carried out with given parameters which show the system is stable for the range of the operating point. The block diagram of the conventional relay feedback approach is shown in Fig. 1.

1. A single-relay feedback experiment is performed on the given unstable process with proposed method, and dynamics of the relay can be varied to achieve sustained oscillations.
2. From the sustained oscillations we got from the relay identification method by making using of the describing function method, ultimate gain and ultimate period are calculated.
3. From the two parameters such as K_u and P_u, controllers are designed based on pole placement method.
4. Based on the designed controller and respective process, closed stability test is conducted.
5. By using Nyquist stability criteria, the analysis is carried out.
6. A simple MATLAB/SIMULINK approach was adopted for identification and control of the various unstable systems.
7. A new control structure is adopted for the different unstable processes.

The describing function was (DF) was given by

$$K_u = \frac{4h}{\pi a} \tag{1}$$

$$p_u = \frac{2\pi}{\omega} \tag{2}$$

where

h Height of the relay
a Peak amplitude of the relay response
p_u Ultimate period
K_u Ultimate gain

Case study 1: Let us consider the transfer function with first order with time delay FOPTD of the process given below

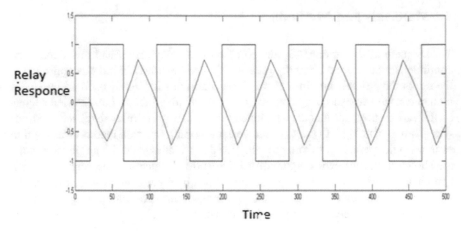

Fig. 2 Sustained oscillations for case study 1 using relay feedback approach

$$G(s) = \frac{3.3226e^{-20s}}{99.69s - 1} \tag{3}$$

The relay feedback test was conducted, and the parameters such as ultimate gain (K_u) and ultimate period P_u are obtained as shown in Fig. 2

where

K_u Ultimate gain of the process
P_u Ultimate period of the process

where

$\theta=$ Time delay
τ Time constant of the process, and the during relay feedback test, the parameters obtained are
K_u 1.689,
P_u 0.01117.

3   Controller Design Aspects

Controller design for unstable systems has drawn a great attention for researchers. Most of the chemical systems are unstable in nature such as bioreactor, distillation column, CSTR, and moreover, the unstable systems will exhibit the multiple steady states and be complex in nature. Controller design for unstable systems was not that much easy as compared to stable systems. In the literature, there are different control methods that are proposed for different unstable systems. Padma Sree and

Chidambaram have proposed various methods for unstable systems such as synthesis method, pole placement method, equation coefficient method, optimization methods, but all these methods require extensive calculations and complex nonlinear equation solvers, which makes it difficult. In order to overcome these problems, a new SIMULINK approach has been proposed for the controlling of various processes such as FOPTD, FOPTD with one zero, SOPTD, SOPTD with one unstable process, and simulation studies have been performed for various processes using the proposed control structure. In this work, a new control approach has been designed based on several simulation studies using the MATLAB/SIMULINK approach.

4 Proposed Control Structure

The proposed control structure for the design of the controllers is shown in Fig. 3. The proposed control structure introduced here was the simplest approach which did not require any computational efforts and rigorous tuning for unstable systems like other methods in the existing literature. The conventional sustained oscillations for the given transfer function was shown in the Fig. 4. The controller for the FOPTD process was designed, and the response was plotted for servo response as shown in below Fig. 5. For stabilizing the unstable systems was very difficult compared with other methods in the literature, but the proposed method gives very simplest and satisfactory servo response for the unstable FOPTD process (Fig. 4).

Case study 2: Let us consider the transfer function of the unstable process with one zero and one unstable pole given by the process in Eq. (4)

$$G(s) = \frac{(24.6s + 2.21)e^{-20s}}{98.3S - 1} \tag{4}$$

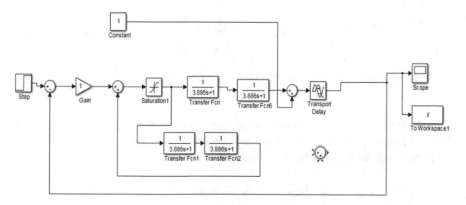

Fig. 3 Proposed controlled structure for various unstable processes

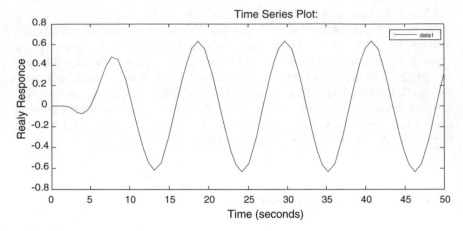

Fig. 4 Sustained oscillations of case study 2 using relay feedback approach

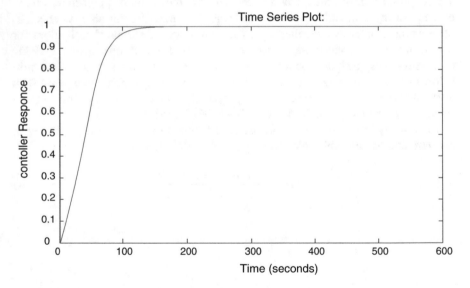

Fig. 5 Servo response of the unstable FOPTD process for case study 1

The relay feedback test was conducted, and the parameters such as ultimate gain (k_u) and ultimate period P_u are obtained as given below; the sustained oscillations obtained using the single-relay feedback test are shown in Fig. 4

$$K_u = 1.689, \; P_u = 0.01117$$

5 Controller Design

The controller design for the above transfer function is carried out by the method proposed in the above procedure, and the response is shown in the following figure. Figure 7 shows control response of the proposed method for case study 2 with the same analysis, and the ultimate period and ultimate gain are obtained; the graphical results are obtained as shown in Fig. 6, and also the input and output oscillations. For the transfer function given in case study 2, the control response was obtained and the response is shown in Fig. 7.

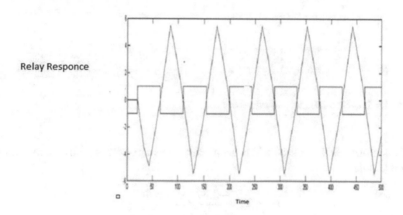

Relay Responce

Time

Fig. 6 Generated limit cycles of case study 2 using relay feedback approach

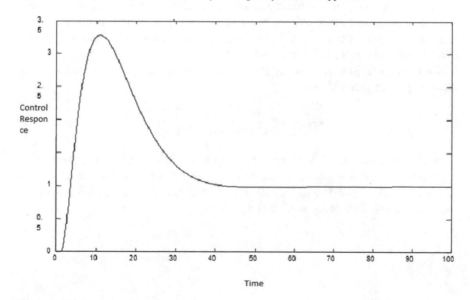

Control Respon ce

Time

Fig. 7 Control response of case study 2

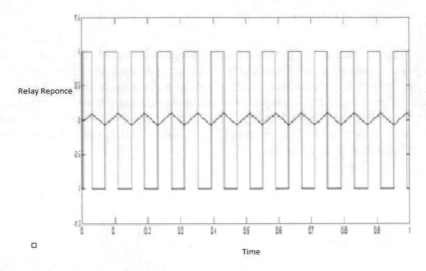

Fig. 8 Limit cycle oscillations of case study 3

Case study 3: Let us consider an unstable system with negative zero and one unstable pole

$$G(s) = \frac{-0.8714s + 6.963}{s^2 + 2.848s - 1.132} \tag{5}$$

The ultimate period and ultimate gain are obtained; the values are $K_u = 13.545$ and $p_u = 12.5$; respective graphical results are shown in Fig. 8; and the control response was obtained which is less oscillatory like in other methods in the literature for SOPTD process which is shown in Fig. 9.

Case study 4: Let us consider the process having one positive zero and one unstable pole as given below

$$G(s) = \frac{2.044s + 3.8687}{0.4769s^2 - 0.348s + 1} \tag{6}$$

And similar procedure is carried out to get sustained oscillations, and control response is obtained as per the procedure adopted for case study 4; corresponding graphical results are shown in Fig. 10, and the parameters obtained using the relay feedback approach were $K_u = 4.2$ and $p_u = 0.3$.

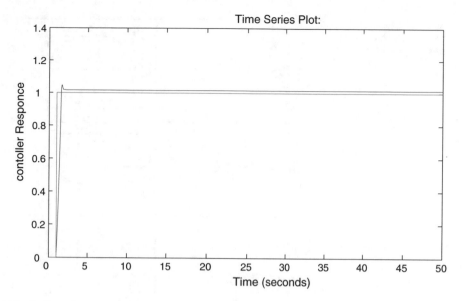

Fig. 9 Control response of the proposed method for case study 3

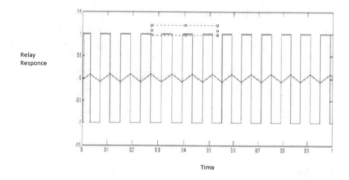

Fig. 10 Sustained oscillations of case study 4 using relay feedback approach

6 Stability Analysis

For the identified system, using relay feedback approach Nyquist stability analysis was carried out, and corresponding graphical results are given in Table 1.

Table 1 Stability analysis of different unstable systems using Nyquist diagram

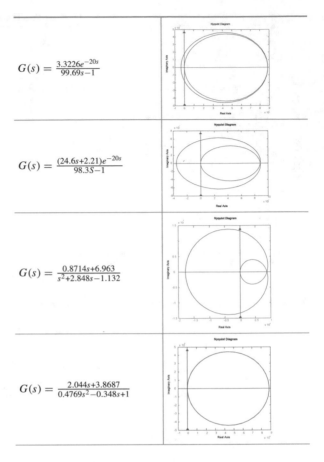

$$G(s) = \frac{3.3226e^{-20s}}{99.69s-1}$$

$$G(s) = \frac{(24.6s+2.21)e^{-20s}}{98.3S-1}$$

$$G(s) = \frac{0.8714s+6.963}{s^2+2.848s-1.132}$$

$$G(s) = \frac{2.044s+3.8687}{0.4769s^2-0.348s+1}$$

7 Conclusion

In this work a relay feedback approach was introduced to identify the unstable process parameters using describing function method and from the identified parameters a stability analysis was carried out using Nyquist stability criteria, in addition to the identification of two key parameters such as ultimate gain and ultimate period a new control method was proposed for the control of various unstable processes using the simplest SIMULINK approach which was not carried out by previous researchers even though there exist lot of methods available for controller design for unstable systems they require much computational parameters and involve the extensive simulations, the method yields satisfactory results.

References

1. A Thesis on Subspace Identification of Unstable Systems, submitted to IIT Madras by Chinta Sankar Rao
2. Liu, T., Wang, Q.-G., Huang, H.P.: A tutorial review on process identification from step or relay feedback tests. J. Process Control
3. Majhi, S.: Advanced control Theory: A Relay Feedback approach, 2nd edn
4. Åström, K.J., Hägglund, T.: Automatic tuning of simple regulators with specification on phase angle and amplitude margins. Automatica **20**, 645–651 (1984)

Ranking Alternatives Using QUALIFLEX Method by Computing All Spanning Trees from Pairwise Judgements

Debasmita Banerjee, Debashree Guha and Fateme Kouchakinejad

Abstract *QUALIFLEX*, a well-known outranking method based on Jacquet-Lagreze's permutation method, investigates all possible permutations of alternatives in order to find the final decision result. It is worthy to mention that to choose the best ranking order, it is not required to always consider total $m!$ permutations of the ranking order of alternatives if the number of alternatives is m. This drawback of *QUALIFLEX* of implanting the huge computations for all permutations of alternatives motivates us to develop a new ranking method. In our proposed ranking method, the algorithmic complexity of *QUALIFLEX* method is overcome by generating only the possible preference orders of alternatives from the set of judgements, which are in the form of real data, using graph theoretic approach. Then we calculate the concordance and discordance indices in the desired sets of preferences and finally choose the ultimate alternative. Furthermore, an example is provided to illustrate the application of the proposed method, together with comparison analysis.

Keywords Multi-criteria decision-making · *QUALIFLEX* method · *EAST* method · Preference relation

1 Introduction

In the domain of multi-criteria decision-making (MCDM), different ranking methods have been developed to choose the most suitable alternative from a set of predefined alternatives and those methods have been applied in many fields of

D. Banerjee (✉) · D. Guha
Department of Mathematics, Indian Institute of Technology, Patna 801103, India
e-mail: debasmitabanerjee12@gmail.com

D. Guha
e-mail: deb1711@yahoo.co.in

F. Kouchakinejad
Department of Pure Mathematics, Shahid Bahonar University of Kerman, Kerman, Iran
e-mail: kouchakinezhad@gmail.com

© Springer Nature Singapore Pte Ltd. 2019
J. C. Bansal et al. (eds.), *Soft Computing for Problem Solving*,
Advances in Intelligent Systems and Computing 816,
https://doi.org/10.1007/978-981-13-1592-3_18

real life. Among them, *QUALIFLEX* (i.e. QUALItative FLEXible multiple criteria method) is a well-known method, which was first developed by Paelinck [14–16] as a generalization of Jacquet-Lagreze's permutation method. After that several useful extensions have been made to enhance the *QUALIFLEX* method [2–4, 8, 11, 22, 24] and have been applied to MCDM [3, 11, 22], medical decision-making [4], risk evaluation problem [24], etc. Due to its flexibility with respect to cardinal and ordinal information, *QUALIFLEX* method is a very convenient ranking method for decision-making. This method evaluates all possible rankings or permutations of alternatives. For each pair of alternatives of the said permutations, concordance and discordance indices are calculated which reflects the concordance and discordance of their ranks. By this method, we try to find out the best order of the alternatives based on the level of concordance and select the most preferred one from the set of alternatives depending on different criteria. Although the usefulness and applicability of the *QUALIFLEX* method in decision-making field have been thoroughly studied in previous years, it is important to note that the number of permutations increases drastically with an increase in the number of alternatives. Moreover, to choose the best ranking order, it is not required to always consider all $m!$ permutations of the ranking order of alternatives if number of alternatives is m. This difficulty of considering all possible permutations can be easily overcome by generating only the possible preference order of the alternatives based on the decision data. With this analysis, in the present study we first generate all possible preferences from the set of judgements using graph theoretic approach. These sets of preferences are induced by enumerating all spanning trees from pairwise comparison matrix. Then we calculate the concordance and discordance indices in the desired sets of preferences and finally choose the ultimate alternative.

This paper is organized as follows: In Sect. 2, basic concepts related to the above-mentioned methods are reviewed. In Sect. 3, we propose an updated *QUALIFLEX* method to rank alternatives integrated with the graph theoretic approach of generating all spanning trees from decision data and applied it to an MCDM problem. To implement the proposed method, a numerical example is presented in Sect. 4 along with a comparative analysis. Finally, some conclusions and future works are presented in Sect. 5.

2 Background

In decision-making problems, decision maker gives his/her opinion in the form of real values for each alternative which indicates the performance of that alternative according to his/her point of view. These preferences are represented in different ways [5, 6]. One of such representations is using a utility function.

Definition 1 Consider a set of alternatives $\{X_1, X_2, ..., X_m\}$. Preferences are given on the basis of a positive ratio scale $U = \{u_i : i = 1, 2, ..., m\}$, $u_i \in [0, 1]$. These real numbers u_i for each alternative X_i are called utility values.

The relation between utility value and satisfaction of decision maker is proportional [13, 21]. Decision maker's preferences over the set of alternatives are also represented by multiplicative preference relation.

2.1 Multiplicative Preference Relation

Definition 2 [12] $M = (a_{ij})_{m \times m}$ is called a multiplicative preference relation, if the elements of M satisfy

$$a_{ij} = \frac{1}{a_{ji}}, \ a_{ii} = 1, \ a_{ij} \in R^+, \ \forall \, i, j = 1, 2, ..., m \tag{1}$$

where a_{ij} represents a multiplicative preference degree of alternative X_i over X_j.

If the preference relation satisfies transitivity property, i.e. $a_{ij} = a_{ik}a_{kj}, \forall \, i, j, k = 1, 2, ..., m$, then M is called a consistent multiplicative preference relation.

Remark 1 Here the intensity of preferences a_{ij} is measured in R^+. However, this intensity of preference was first proposed by Saaty. According to Saaty [17], the a_{ij}s are measured in $1 - 9$ scale where $a_{ij} = 1$ indicates indifference between X_i and X_j, $a_{ij} = 9$ indicates that X_i is absolutely preferred than X_j, and $a_{ij} \in 2, 3, ..., 8$ indicates intermediate evaluations.

Herrera in his paper [9] describes that, to derive a multiplicative preference relation from a set of utility values, there exits a transformation function $h:[0, 1]^2 \to R^+$ such that,

$$a_{ij} = h(u_i, u_j). \tag{2}$$

This transformation function h must follow that the higher value of u_i results higher a_{ij} and the higher value of u_j results lower a_{ij}, i.e. function h must be increasing in the first argument and decreasing in the second argument. On the other hand, if the pair of values (u_i, u_j) change slightly, the preference between that pair of alternatives should change slightly too, i.e. the function h must be a continuous function. According to Chiclana [6], this transformation function acts coherently because it does not change the informative content of the multiplicative preference relations when we make the information uniform in the decision model. It can be written in another way as,

$$h(u_i, u_j) = l\left(\frac{u_i}{u_j}\right). \tag{3}$$

where l is an increasing function. The detail of this type function and its analysation are discussed in [5, 13]. The simplest form of this transformation function to find out the intensity of preference was proposed and used by Saaty in his AHP method [18, 19] as,

$$a_{ij} = h(u_i, u_j) = l\left(\frac{u_i}{u_j}\right) = \frac{u_i}{u_j}. \tag{4}$$

We can interpret, this ratio $\frac{u_i}{u_j}$ as preference intensity of X_i over X_j, which is given by the decision maker on the basis of a specific criterion. In [9], it is also discussed that preferences between alternatives do not change when the utility values of the implied alternatives change in the same proportion. So,

$$a_{ij} = \left(\frac{u_i}{u_j}\right)^c, \quad c > 0. \tag{5}$$

can also be used for the evaluation of intensity of preference.

In the following section, we will discuss our proposed ranking method.

3   The Proposed Method

3.1   Construction of Pairwise Comparison Matrix Based on the Decision Matrix

Consider a decision-making problem, where a decision maker gives his/her opinion for m alternatives $X_1, X_2, ..., X_m$ based on n criteria $C_1, C_2, ..., C_n$. He/she provides his/her opinion in the form of real valuation, known as utility value, $A_{ij} \in R^+$ for ith alternative with respect to jth criterion. The decision maker's opinion can be condensed in the following decision matrix:

$$D = \begin{matrix} & \begin{matrix} C_1 & C_2 & \cdots & C_n \end{matrix} \\ \begin{matrix} X_1 \\ X_2 \\ \vdots \\ X_m \end{matrix} & \begin{pmatrix} A_{11} & A_{12} & \cdots & A_{1n} \\ A_{21} & A_{22} & \cdots & A_{2n} \\ \vdots & \vdots & \cdots & \vdots \\ A_{m1} & A_{m2} & \cdots & A_{mn} \end{pmatrix} \end{matrix}. \tag{6}$$

For each criterion to obtain the preference value of alternative X_i over X_j, we derive multiplicative preference relation from above given set of utility values, using Eq. (4), i.e. the simplest form of transformation function h.

So, for kth criterion, where $k = 1, 2, ..., n$, multiplicative preference relation can be evaluated as,

$$D_{C_k} = \begin{matrix} & \begin{matrix} X_1 & X_2 & \cdots & X_m \end{matrix} \\ \begin{matrix} X_1 \\ X_2 \\ \vdots \\ X_m \end{matrix} & \begin{pmatrix} \frac{A_{1k}}{A_{1k}} & \frac{A_{1k}}{A_{2k}} & \cdots & \frac{A_{1k}}{A_{mk}} \\ \frac{A_{2k}}{A_{1k}} & \frac{A_{2k}}{A_{2k}} & \cdots & \frac{A_{2k}}{A_{mk}} \\ \vdots & \vdots & \cdots & \vdots \\ \frac{A_{mk}}{A_{1k}} & \frac{A_{mk}}{A_{2k}} & \cdots & \frac{A_{mk}}{A_{mk}} \end{pmatrix} \end{matrix}. \tag{7}$$

The importance weight of each criterion needs to be taken into account in practical applications, as different criteria must have different levels of importance. Let $w = (w_1, w_2, ..., w_n)$ be the weight vector of n criteria where $w_k > 0, k = 1, 2, ..., n, \sum_{k=1}^{n} w_k = 1$. To aggregate individual judgements into collective judgement with respect to all criteria, we use weighted geometric averaging operator,

$$a_{ij} = \prod_{k=1}^{n} \left(\frac{A_{ik}}{A_{jk}} \right)^{w_k}. \tag{8}$$

Thus, the judgement matrix can be evaluated as,

$$M = \begin{matrix} & X_1 & X_2 & \cdots & X_m \\ X_1 \\ X_2 \\ \vdots \\ X_m \end{matrix} \begin{pmatrix} \prod_{k=1}^{n} \left(\frac{A_{1k}}{A_{1k}} \right)^{w_k} & \prod_{k=1}^{n} \left(\frac{A_{1k}}{A_{2k}} \right)^{w_k} & \cdots & \prod_{k=1}^{n} \left(\frac{A_{1k}}{A_{mk}} \right)^{w_k} \\ \prod_{k=1}^{n} \left(\frac{A_{2k}}{A_{1k}} \right)^{w_k} & \prod_{k=1}^{n} \left(\frac{A_{2k}}{A_{2k}} \right)^{w_k} & \cdots & \prod_{k=1}^{n} \left(\frac{A_{2k}}{A_{mk}} \right)^{w_k} \\ \vdots & \vdots & \vdots & \vdots \\ \prod_{k=1}^{n} \left(\frac{A_{mk}}{A_{1k}} \right)^{w_k} & \prod_{k=1}^{n} \left(\frac{A_{mk}}{A_{2k}} \right)^{w_k} & \cdots & \prod_{k=1}^{n} \left(\frac{A_{mk}}{A_{mk}} \right)^{w_k} \end{pmatrix}. \tag{9}$$

which is represented as,

$$M = \begin{matrix} & X_1 & X_2 & \cdots & X_m \\ X_1 \\ X_2 \\ \vdots \\ X_m \end{matrix} \begin{pmatrix} a_{11} & a_{12} & \cdots & a_{1m} \\ a_{21} & a_{22} & \cdots & a_{2m} \\ \vdots & \vdots & \vdots & \vdots \\ a_{m1} & a_{m2} & \cdots & a_{mm} \end{pmatrix}. \tag{10}$$

where a_{ij} is the ratio between respective utility values of the alternatives. Using of weighted geometric averaging operator ensures that the matrix M is also a multiplicative preference relation. Thus, utilizing weighted geometric averaging operator, pairwise comparison matrix M can be generated from the decision matrix D.

3.2 Enumerating All Possible Preference Orders from Pairwise Comparison Matrix Using Graph Theoretic Approach

Let $J = \{a_{ij}\}$ be the pairwise comparison judgement set containing maximum $\frac{m(m-1)}{2}$ different independent judgements due to reciprocal property provided by the decision maker based on different criteria. Now inspired by the $EAST$ method [20] of enumeration all spanning trees from a set of pairwise judgements, we take the set of m alternatives as vertices of a complete directed graph $G = (X, J)$ and edges connecting them as preference degree between each pair of alternatives where J represents the weighted edges. Then we can generate a forest $\Gamma = \{S_1, S_2, ..., S_\eta\}$ containing η number of spanning trees where each tree presents an independent judgement. Back in 1889, Cayley developed the well-known formula $\eta = m^{(m-2)}$ for the number of spanning trees in the complete graph with m vertices. If we draw all those spanning trees in a figure, then this displays every possible combination of minimum edges connecting all those m vertices together. Each spanning tree $S_i = \{a_{ij}\} \subset J$, $i = 1, 2, ..., \eta$ represents a set of $(m - 1)$ independent judgements. These judgements are said to be consistent if $a_{ij} = a_{ik}a_{kj}, \forall\, i, j, k$. According to Saaty [19], if this m-dimensional comparison matrix M is a consistent matrix, then its maximal eigenvalue ($\lambda_{\max}$) must be equal to m. But in reality, due to different environmental effects, human judgements get affected, which generates inconsistencies in the judgements of decision makers. In that case, maximal eigenvalue exceeds the value m. By consistency ratio, we can judge whether a comparison matrix is consistent or not and this consistency ratio (CR), i.e. deviation of $\lambda_{\max}$ from m can be calculated [23] from,

$$CR = \frac{CI}{RI} \tag{11}$$

where CI is consistency index, RI is random consistency index. When CR ≤ 0.1 (i.e. 10%), the matrix can be acceptable as consistent matrix. Consistency index can be computed as

$$CI = \frac{\lambda_{\max} - m}{m - 1}. \tag{12}$$

RI with respect to different size matrix is shown in Table 1.

By preference symbol $\rightarrow$ ordinal consistency can be expressed as, $X_i \rightarrow X_j \rightarrow X_k$ implies $X_i \rightarrow X_k$. The preference judgements are ordinally inconsistent if $X_k \rightarrow X_i$ when $X_i \rightarrow X_j \rightarrow X_k$, i.e. it will form a three-way cycle in G as $X_i \rightarrow X_j \rightarrow X_k \rightarrow X_i$. For a pairwise comparison matrix, the total number of three-way cycles implies total number of ordinal inconsistencies in J [1, 7, 10]. So, while choosing J, there should not exist any k permitting the relation $a_{ij} \Leftrightarrow a_{ik}a_{kj}$ otherwise the graph will be disconnected. From each independent judgement, weight vector of the alternatives $W = (W_1, W_2, ..., W_m)$, where $W_i > 0$, can be calculated by using the formulas,

Table 1 RI for matrices with different size

Number of elements	3	4	5	6	7	8	9	10	11	12	13
RI		0.52	0.89	1.11	1.25	1.35	1.40	1.45	1.49	1.51	1.54	1.56

$$a_{ij} = \frac{W_i}{W_j}, \quad \sum W_i = 1. \tag{13}$$

Depending on η different weight vectors generated from each spanning tree, we can rank alternatives in all possible ways. So by this process, we obtain maximum η different rankings of alternatives.

3.3 Computation of Concordance/ Discordance Index for Every Possible Preference Ordering

We can construct a set $P = \{P_1, P_2, ..., P_\eta\}$ of all possible rankings of alternatives by $EAST$ method from the pairwise comparison judgements. Let $P_l = (..., X_\rho, ..., X_\beta, ...)$ denote the lth ranking order where $l = 1, 2, ..., \eta$ and in P_l alternative X_ρ is ranked higher than or equal to X_β. Now we calculate the distance $d(A_{ij}, 0)$ for each A_{ij} in decision matrix D using Euclidean distance formula,

$$d(A_{ij}, 0) = \sqrt{(A_{ij} - 0)^2} = |A_{ij} - 0|. \tag{14}$$

Next, compute the concordance/discordance index [4] $I_j^l(X_\rho, X_\beta)$, for each pair of alternatives (X_ρ, X_β) in the preference ranking P_l with respect to the criterion C_j using,

$$I_j^l(X_\rho, X_\beta) = d(A_{\rho j}, 0) - d(A_{\beta j}, 0). \tag{15}$$

If $I_j^l(X_\rho, X_\beta) > 0$, $I_j^l(X_\rho, X_\beta) = 0$, $I_j^l(X_\rho, X_\beta) < 0$, then it is said to be concordance, ex-aequo, discordance index, respectively.

Consider the weight w_j of each criteria to compute the value $I_j^l(X_\rho, X_\beta).w_j$. Next, to determine the weighted concordance/discordance index $I^l(X_\rho, X_\beta)$ for each pair of (X_ρ, X_β) in P_l apply

$$I^l(X_\rho, X_\beta) = \sum_{j=1}^{n} I_j^l(X_\rho, X_\beta).w_j = \sum_{j=1}^{n} \left(d(A_{\rho j}, 0) - d(A_{\beta j}, 0) \right).w_j. \tag{16}$$

Derive the comprehensive concordance/discordance index I^l for each P_l using

$$I^l = \sum_{X_\rho, X_\beta \in X} \sum_{j=1}^{n} I_j^l(X_\rho, X_\beta).w_j = \sum_{X_\rho, X_\beta \in X} \sum_{j=1}^{n} \left(d(A_{\rho j}, 0) - d(A_{\beta j}, 0) \right).w_j. \tag{17}$$

The evaluation criterion of the chosen hypothesis for ranking of the alternatives is the arithmetic sum of all weighted differences of distances corresponding to the element-by-element consistency. The highest value of comprehensive concordance/discordance index indicates the better ranking of that alternative.

The steps of this algorithm are summarized as follows:

3.4 Algorithm

Step 1. Convert decision matrix $D = (A_{ij})_{m \times n}$ into pairwise comparison matrix $M = (a_{ij})_{m \times m}$ using Eq. (8).

Step 2. From the above obtained preference matrix (9), generate all possible spanning trees.

Step 3. Then based on all preference vectors for judgement J, find out the preference weights of the alternatives by enumerating all spanning trees from the pairwise judgements by using Eq. (13).

Step 4. Analysing all weights, rank the alternatives based on the highest weight assigned with respect to the alternatives.

Step 5. Calculate the distance $d(A_{ij}, 0)$ for each A_{ij} using Formula (14).

Step 6. Calculate concordance and discordance index for each pair of alternatives for all possible ranking order using Eqs. (15) and (16).

Step 7. Finally computing comprehensive concordance/discordance index for each permutation using Eq. (17), optimal ranking can be obtained.

In the *QUALIFLEX* method, we consider all $m!$ possible rankings/ permutations of alternatives from which we find out the optimal ranking order of the alternatives. This number of permutations rapidly increases with an increasing number of alternatives. On the other hand by generating all spanning trees from pairwise comparison judgements, we take into account maximum $m^{(m-2)}$ number of possible ranking of alternatives in which some of these possible orderings may repeat. Also another important issue is that as some of the preference orderings are obtained same in *EAST* method, thus to compute concordance and discordance indices, it is not required to consider all $m^{(m-2)}$ number of orderings, while in *QUALIFLEX* all $m!$ permutations are needed to be considered. On the other hand, the method used in *EAST* to choose the final ordering is not accurate and equitable as it generates the ranking based on the majority of the generated trees. But in any case, if equal majority for any two alternatives are obtained then how to rank them was not clearly discussed in [20]. These drawbacks of the existing methods are overcome by our new ranking method. The following numerical example shows the procedure.

4 Numerical Example

In this section, a real-life-based example is presented to illustrate the working procedure of proposed algorithm. Suppose an engineering corporation intend to choose a sub-contractor from four candidates, namely X_1, X_2, X_3, X_4. The evaluation of the best sub-contractor is based upon the following five criteria:

1. Reliability (C_1),
2. Labour quality (C_2),
3. Experience (C_3),
4. Business management (C_4),
5. Tender quality (C_5).

All the criteria are not equally important. The set of weighting coefficient of each criteria is $w = [0.3, 0.25, 0.2, 0.1, 0.15]$. Suppose a decision maker has made strict evaluation for these four sub-contactors from those five aspects. This evaluation is summarized in the following decision matrix:

$$D = \begin{array}{c} \\ X_1 \\ X_2 \\ X_3 \\ X_4 \end{array} \begin{pmatrix} C_1 & C_2 & C_3 & C_4 & C_5 \\ 0.91 & 0.3 & 0.02 & 0.8 & 0.2 \\ 0.9 & 0.04 & 0.21 & 0.67 & 0.34 \\ 0.11 & 0.8 & 0.05 & 0.3 & 0.2 \\ 0.87 & 0.17 & 0.51 & 0.04 & 0.26 \end{pmatrix}.$$

From the given decision matrix D, pairwise comparison matrix A can be generated by using weighted geometric averaging operator Eq. (8). Pairwise comparison matrix is as follows:

$$M = \begin{array}{c} \\ X_1 \\ X_2 \\ X_3 \\ X_4 \end{array} \begin{pmatrix} X_1 & X_2 & X_3 & X_4 \\ 1.0000 & 0.9235 & 1.0000 & 0.9614 \\ 1.0828 & 1.0000 & 1.0828 & 1.0411 \\ 1.0000 & 0.9235 & 1.0000 & 0.9614 \\ 1.0401 & 0.9606 & 1.0401 & 1.0000 \end{pmatrix}.$$

The graph for four alternatives shown in Fig. 1 confirms that there does not exist any three-way cycle, which ensures the ordinally consistent set of judgements. Using Cayley's theorem, from these four vertices we can generate 16 spanning trees, i.e. in this case the value of η is 16. The forest of trees generated by the given set of judgements is shown in Fig. 2. This figure displays every combination of all the vertices with minimum possible number of edges. From each tree, a preference

Fig. 1 Graph for J

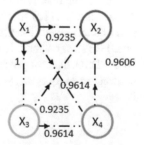

Fig. 2 A forest containing all 16 spanning trees for J

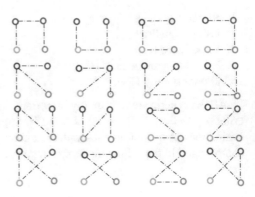

Table 2 Possible rankings of alternatives generated from J

S_η	$\{a_{ij}\} \subset J$	Ranking	S_η	$\{a_{ij}\} \subset J$	Ranking
S_1	$\{a_{12}, a_{14}, a_{23}\}$	$X_3 \prec X_1 \prec X_4 \prec X_2$	S_9	$\{a_{13}, a_{14}, a_{24}\}$	$X_1 \prec X_3 \prec X_4 \prec X_2$
S_2	$\{a_{12}, a_{24}, a_{34}\}$	$X_1 \prec X_3 \prec X_4 \prec X_2$	S_{10}	$\{a_{12}, a_{23}, a_{24}\}$	$X_3 \prec X_1 \prec X_4 \prec X_2$
S_3	$\{a_{13}, a_{23}, a_{24}\}$	$X_1 \prec X_3 \prec X_4 \prec X_2$	S_{11}	$\{a_{12}, a_{23}, a_{34}\}$	$X_3 \prec X_1 \prec X_4 \prec X_2$
S_4	$\{a_{13}, a_{23}, a_{34}\}$	$X_1 \prec X_3 \prec X_4 \prec X_2$	S_{12}	$\{a_{13}, a_{24}, a_{34}\}$	$X_1 \prec X_3 \prec X_4 \prec X_2$
S_5	$\{a_{13}, a_{14}, a_{23}\}$	$X_1 \prec X_3 \prec X_4 \prec X_2$	S_{13}	$\{a_{12}, a_{13}, a_{14}\}$	$X_1 \prec X_3 \prec X_4 \prec X_2$
S_6	$\{a_{14}, a_{23}, a_{34}\}$	$X_1 \prec X_3 \prec X_4 \prec X_2$	S_{14}	$\{a_{12}, a_{14}, a_{34}\}$	$X_1 \prec X_3 \prec X_4 \prec X_2$
S_7	$\{a_{14}, a_{23}, a_{24}\}$	$X_3 \prec X_1 \prec X_4 \prec X_2$	S_{15}	$\{a_{12}, a_{13}, a_{24}\}$	$X_1 \prec X_3 \prec X_4 \prec X_2$
S_8	$\{a_{14}, a_{24}, a_{34}\}$	$X_1 \prec X_3 \prec X_4 \prec X_2$	S_{16}	$\{a_{12}, a_{13}, a_{34}\}$	$X_1 \prec X_3 \prec X_4 \prec X_2$

weight vector W can be obtained from which we can rank the alternatives. All possible rankings are presented in Table 2.

Among all 16 rankings produced from each spanning tree, only two different rankings are obtained as $X_3 \prec X_1 \prec X_4 \prec X_2$ and $X_1 \prec X_3 \prec X_4 \prec X_2$. Between these two rankings, to find out the optimal one, we now calculate concordance/discordance index. Denoting these two rankings as $P_1 = (X_2, X_4, X_1, X_3)$ and $P_2 = (X_2, X_4, X_3, X_1)$, we compute the distance $d(A_{ij}, 0)$ for each A_{ij} in D using Eq. (14). Now for each pair of alternatives (X_ρ, X_β) in the permutation P_l with respect to each criterion C_j, the results of the concordance/discordance index are presented in Table 3. Then using Eq. (16) we calculate weighted concordance/discordance index for each pair of alternatives. Finally, the comprehensive concordance/discordance index computed for each permutation is $I^1 = 0.3875$ and $I^2 = 0.0795$. So, I^1 gives the highest value. Thus, the best permutation is $P_1 = (X_2, X_4, X_1, X_3)$ and the best order of subcontractors is $X_2 \succ X_4 \succ X_1 \succ X_3$. Therefore, the best choice for the sub-contractor is X_2.

Table 3 The results of concordance/discordance index

P_1	$I_j^1(X_2, X_4)$	$I_j^1(X_2, X_1)$	$I_j^1(X_2, X_3)$	$I_j^1(X_4, X_1)$	$I_j^1(X_4, X_3)$	$I_j^1(X_1, X_3)$
C_1	0.03	−0.01	0.79	−0.04	0.76	0.8
C_2	−0.13	−0.26	−0.76	−0.13	−0.63	−0.5
C_3	−0.3	0.19	0.16	0.49	0.46	−0.03
C_4	0.63	−0.13	0.37	−0.76	−0.26	0.5
C_5	0.08	0.14	0.14	0.06	0.06	0
P_2	$I_j^2(X_2, X_4)$	$I_j^2(X_2, X_3)$	$I_j^2(X_2, X_1)$	$I_j^2(X_4, X_3)$	$I_j^2(X_4, X_1)$	$I_j^2(X_3, X_1)$
C_1	0.03	0.79	−0.01	0.76	−0.04	−0.8
C_2	−0.13	−0.76	−0.26	−0.63	−0.13	0.5
C_3	−0.3	0.16	0.19	0.46	0.49	0.03
C_4	0.63	0.37	−0.13	−0.26	−0.76	−0.5
C_5	0.08	0.14	0.14	0.06	0.06	0

4.1 Comparison Analysis

In this section, we compare the results of our numerical example derived from proposed method with other existing processes. As $EAST$ method generates the ranking based on the majority of derived weights of alternatives from pairwise comparison matrix, so maximum number of highest weight will proceed to better ranking of the alternatives. From Table 2, it is observed that 12 out of 16 spanning trees give the ranking as, $X_1 \prec X_3 \prec X_4 \prec X_2$. While four of these spanning trees generate ranking as $X_3 \prec X_1 \prec X_4 \prec X_2$. So by majority, $X_1 \prec X_3 \prec X_4 \prec X_2$ will be optimal ranking obtained by $EAST$ method. However, the majority-based ranking method is not very appropriate. As if in any case, equal majority for any two alternatives are obtained, and then how to rank them was not clearly discussed there. On the other hand in case of $QUALIFLEX$ method, for a set of four alternatives it generates all possible $4! = 24$ permutations of the alternatives. For each permutation P_l if we calculate comprehensive concordance/discordance index, then we will get the maximal value for the ranking $X_3 \prec X_2 \prec X_4 \prec X_1$. So, if we consider four alternatives to rank based on different criterion, by $QUALIFLEX$ method we need to compute the concordance and discordance indices for all $4! = 24$ rankings which are time consuming. In our proposed ranking method, we first compute the maximum 16 number of possible preference orders of the alternatives by applying $EAST$ method on pairwise comparison matrix derived from the decision data. Then we observe that among the 16 rankings produced from each spanning tree, only two different rankings are obtained as $X_3 \prec X_1 \prec X_4 \prec X_2$ and $X_1 \prec X_3 \prec X_4 \prec X_2$. So finally we calculate concordance/discordance index based on only two rankings, and this reduces computational complexity, which is the advantages of our proposed model. The ranking order of alternatives in each method is summarized in Table 4.

Table 4 Ranking order of alternatives in different methods

Methods	Ranking
EAST	$X_1 \prec X_3 \prec X_4 \prec X_2$
QUALIFLEX	$X_3 \prec X_2 \prec X_4 \prec X_1$
Proposed method	$X_3 \prec X_1 \prec X_4 \prec X_2$

5 Conclusion

In this paper, we have developed a new ranking method to handle MCDM problems based on real data. By developing this new ranking method, we have tried to overcome the drawbacks of the well-known outranking method, *QUALIFLEX*, by integrating it with the graph theoretic approach of generating all spanning trees from pairwise comparison matrix. Our proposed method is divided into three stages. First, we have converted the decision matrix obtained from decision makers opinion for each alternative based on different criteria to a pairwise comparison matrix where preference of each alternative with respect to another one is presented. In the second part, from the obtained pairwise comparison matrix we have generated all independent judgements as spanning tree by taking all alternatives as nodes and weighted edges as preference degree of each pair of alternatives and then derived the weights of each alternative from which different rankings of alternatives are generated. And in the third part, from those different rankings by calculating concordance and discordance indices, we have obtained the optimal rank of alternatives. Finally, a practical example has been presented to illustrate the applicability of the proposed method and further analysis has been also conducted to compare the experimental result with the results of the other existing methods. In future, we can extend this method for imprecise environment where decision maker cannot give his/her opinion very precisely due to the complexity of the situation.

References

1. Ali, I., Cook, W.D., Kress, M.: On the minimum violations ranking of a tournament. Manag. Sci. **32**, 660–672 (1986)
2. Chen, T.-Y.: Data construction process and qualiex-based method for multiple criteria group decision making with interval-valued intuitionistic fuzzy sets. Int. J. Inf. Tech. Decis. **12**, 425–467 (2013)
3. Chen, T.-Y.: Interval-valued intuitionistic fuzzy qualiflex method with a likelihood-based comparison approach for multiple criteria decision analysis. Inf. Sci. **261**, 149–169 (2014)
4. Chen, T.-Y., Chang, C.-H., Lu, J.R.: The extended qualiflex method for multiple criteria decision analysis based on interval type-2 fuzzy sets and applications to medical decision making. Eur. J. Oper. Res. **226**, 615–625 (2013)
5. Chiclana, F., Herrera, F., Herrera-Viedma, E.: Integrating three representation models in fuzzy multipurpose decision making based on fuzzy preference relations. Fuzzy

Set Syst. **97**, 33–48 (1998)

6. Chiclana, F., Herrera, F., Herrera-Viedma, E.: Integrating multiplicative preference relations in a multipurpose decision-making model based on fuzzy preference relations. Fuzzy Set Syst. **122**, 277–291 (2001)

7. Gass, S.I.: Tournaments, transitivity and pairwise comparison matrices. J. Oper. Res. Soc. **49**(6), 616–624 (1998)

8. Griffith, D.A., Paelinck, J.H.P.: Qualireg, a qualitative regression method. In: Non-standard Spatial Statistics and Spatial Econometrics, pp. 227–233 . Springer (2011)

9. Herrera, F., Herrera-Viedma, E., Chiclana, F.: Multiperson decision-making based on multiplicative preference relations. Eur. J. Oper. Res. **129**, 372–385 (2001)

10. Jensen, R.E., Hicks, T.E.: Ordinal data AHP analysis: a proposed coefficient of consistency and a nonparametric test. Math. Comput. Model. **17**, 135–150 (1993)

11. Li, J., Wang, J.-Q.: An extended qualiflex method under probability hesitant fuzzy environment for selecting green suppliers. Int. J. Fuzzy Syst. 1–14 (2017)

12. Liu, F., Zhang, W.-G., Wang, Z.-X.: A goal programming model for incomplete interval multiplicative preference relations and its application in group decision-making. Eur. J. Oper. Res. **218**, 747–754 (2012)

13. Luce, R.D., Bush, R.R., Eugene, G. (ed.): Handbook of Mathematical Psychology (1963)

14. Paelinck, J.H.P.: Qualitative multiple criteria analysis, environmental protection and multiregional development. Pap. Reg. Sci. **36**, 59–76 (1976)

15. Paelinck, J.H.P.: Qualitative multicriteria analysis: an application to airport location. Environ. Plan A. **9**, 883–895 (1977)

16. Paelinck, J.H.P.: Qualiflex: a flexible multiple-criteria method. Econ. Lett. **1**, 193–197 (1978)

17. Saaty, T.L.: A scaling method for priorities in hierarchical structures. J. Math. Psychol. **15**, 234–281 (1977)

18. Saaty, T.L.: Exploring the interface between hierarchies, multiple objectives and fuzzy sets. Fuzzy Set Syst. **1**, 57–68 (1978)

19. Saaty, T.L.: The Analytic Hierarchy Process. McGraw-Hill (1980)

20. Siraj, S., Mikhailov, L., Keane, J.A.: Enumerating all spanning trees for pairwise comparisons. Comput. Oper. Res. **39**, 191–199 (2012)

21. Tanino, T.: On group decision making under fuzzy preferences. In: Multiperson Decision Making Using Fuzzy Sets and Possibility Theory, pp. 172–185 (1990)

22. Tian, Z.-P., Wang, J., Wang, J.-Q., Zhang, H.-Y.: A likelihood-based qualitative flexible approach with hesitant fuzzy linguistic information. Cogn. Comp. **8**(4), 670–683 (2016)

23. Tzeng, G.-H., Huang, J.-J.: Multiple attribute decision making: methods and applications. CRC Press, Boca Raton (2011)

24. Zhang, X.: Multicriteria pythagorean fuzzy decision analysis: a hierarchical qualiflex approach with the closeness index-based ranking methods. Inf. Sci. **330**, 104–124 (2016)

Association Rule Hiding Using Chemical Reaction Optimization

N. P. Gopalan and T. Satyanarayana Murthy

Abstract In recent days, enormous data are generated from departmental stores, hospitals, social media, banks, etc. These datasets are associated with different association rules for monitoring the business operations. During this process, to avoid leaking of sensitive information leads to development of association rule hiding algorithms. Many heuristic algorithms are developed but they are limited to optimal solutions. In this paper, an efficient meta-heuristic algorithm has been developed for association rule hiding based on chemical reaction optimization algorithm. The results of the proposed approach are compared with the genetic algorithm, particle swarm optimization, and cuckoo-based algorithms. The experimental results of the proposed algorithm are tested on the benchmark datasets.

Keywords Association rules · Association rule hiding · Sensitive data
Chemical reaction optimization

1 Introduction

Huge amount of data collected from the hospitals, banks, and social media may contain sensitive information in the form of sensitive rules. These are extracted by association rule mining technique in data mining [1–3] which may lead to hiding the sensitive information. The hiding of sensitive information without leaking the business secrets leads to rule hiding. The main goal is to preserve the disclosure of the sensitive association rules. Many approaches are developed for association rule hiding like heuristic, border, reconstruction, and exact-based algorithms. Currently, meta-heuristic algorithms play vital role in hiding the sensitive association rules but limit to optimal results. In this paper, an efficient meta-heuristic algorithm has been

N. P. Gopalan · T. S. Murthy (✉)
Department of Computer Applications, National Institute of Technology, Trichy 620015, India
e-mail: murthyteki@gmail.com

N. P. Gopalan
e-mail: npgopalan@nitt.edu

© Springer Nature Singapore Pte Ltd. 2019
J. C. Bansal et al. (eds.), *Soft Computing for Problem Solving*,
Advances in Intelligent Systems and Computing 816,
https://doi.org/10.1007/978-981-13-1592-3_19

developed based on chemical reaction optimization (CRO). In this hiding process, datasets are modified and lead to several side effects like hiding failure, lost rule, ghost rule. In this paper, main objective is reducing the ghost rules and avoiding hiding failure. This paper was organized into different sections: Sect. 2 contains literature and introduction of association rule hiding. Section 3 describes the problem statement and terminology used. Section 4 explains the proposed algorithm of association rules hiding. Sections 5 and 6 analyze the results.

2   Literature Survey

Chemical reaction optimization [4, 5], a population-based meta-heuristic algorithm was inspired by the molecular reactions taken place in a chemical process. It is a molecule-based meta-heuristic algorithm, where each molecule contains two types of energies like potential energy and kinetic energy. Energy generated by the molecule because of its motion is coined as kinetic energy, and the energy generated by the molecule because of its virtue of its position is coined as potential energy. In a chemical reaction process, a product has been prepared by changing the molecular structures. Molecule structures are based on on-wall ineffective collision, decomposition, synthesis, and inter-molecular ineffective collision. The parameters used in this process are molecular structure, potential energy, kinetic energy, number of collisions, minimum structure, minimum potential energy, and minimum collisions. Molecular structure represents the solution in a chemical reaction optimization. It contains number, matrix, etc. A potential energy in chemical reaction optimization acts as an objective function. A kinetic energy is represented as a nonnegative value. It indicates the worst solution. The number of collisions is an integer number which represents number of molecular collisions. A minimum structure is a molecular structure that represents the minimum potential energy. A minimum potential energy represents the minimum structure of a molecule. A minimum collision determines the number of molecule collisions. Association rule hiding techniques are broadly categorized into heuristic, border, exact, reconstruction, and cryptographic techniques. Atalla et al. [6] used heuristic approach for hiding the sensitive association rules based on cyclic approach. Dasseni et al. [7] proposed three techniques for hiding the sensitive association rules based on support and confidence values. Five algorithms are proposed by Verykios et al. [8] among them, three algorithms are rule-oriented and remaining are item-oriented. These approaches can hide only one single rule. Olivera et al. [9] proposed a balanced factor for hiding the multiple sensitive rules. Wang et al. [10] proposed ISL and DSR algorithms, but they suffer with hiding failure and lost rules. Shah et al. [11] proposed a ADSRCC and RRLR algorithm for hiding the sensitive association rules. Damandiya et al. [12] proposed a support reduction technique, and it performs better than ADSRCC. Hong et al. [13] proposed a SIF-IDF algorithm for rule hiding but the running time is very high. To overcome the drawback with SIF-IDF algorithm, Lin et al. [14] proposed an HMAU algorithm for hiding the sensitive rules based on the support. Ghalesefidi et al. [15] proposed a hybrid algorithm for association rule hiding. Belwal et al. [16] proposed a novel method for hiding the association rules based on support and confidence. A clustering-based algorithm was proposed by Modi et al. [17, 18] to reduce the effects on the perturbed dataset. Keshava Murthy et al. [19] proposed a multi-rule association rule hiding algorithm using genetic algorithm. Lin et al. [20] proposed a sanitization approach for hiding sensitive items based on PSO. Afshari et al. [21] proposed an association rule mining algorithm based on cuckoo optimization algorithm. Satyanarayana Murthy et al. [22–25] proposed a novel algorithm for privacy-preserving data mining.

3 Problem Statement

Huge number of transactions $T_1, T_2, T_3, T_4 \ldots T_n$ collects from social media, hospitals, banks, etc. These transactions contain collection of items $i_1, i_2, \ldots i_n$. Generate the association rules by using apriori algorithm. These rules are categorized into private rules and public rules. ARH4CRO was developed for hiding the private rules, and also, it avoids hiding failure and reduces lost rules.

4 Proposed Algorithm

Input: Dataset D, minimum support threshold (MST), minimum confidence threshold (MCT), support(S), confidence (C), transactional array A, MAX_SUP, MAX_CONF.
Output: Sanitized dataset D.

1. Apply apriori algorithm for generating the rules R.
2. Calculate the support S, confidence C for all the rules where s>MST and c>MCT.
3. A support set S = {s1,s2...sn} and confidence set C = {c1,c2,...cn}.
4. Determine the MAX_SUP and MAX_CONF constant values.
5. Determine the sensitive rules based on MAX_SUP and MAX_CONF.
   for(rules R, i variable)
   do
   Let   an   array   B[i]   =   if(Rule(i).S(i)   >MAX_SUP   and Rule(i).C(i) > MAX_CONF)
   until..No Rules.

6. B[i] = {b1,b2,b3,b4...bn} are the sensitive rules.
7. Once rules are generated, Dataset D given as an input to the ARH4CRO algorithm generates disturbed dataset D^1.
8. Call ARH4CRO Algorithm().

9. Perturbed dataset D^1 given as an input to the apriori algorithm generates rules that contain PUR with sensitivity.

5 Analysis of the Proposed Algorithm

Given a dataset D made up of transactions t_1, t_2, $t_3 \ldots t_n$. These transactions contain collection of items i_1, $i_2, \ldots i_n$. Generate the association rules using apriori algorithm. This algorithm produces rules r_1, r_2, $r_3 \ldots r_n$. Set the values of MST = 40% and MCT = 40% to choose the best rule from the set of association rules. The ARH4CRO algorithm hides the sensitive items from disclosure. In this paper, sensitive items are involved in OnWallIneffectiveCollision. The molecular structure represents the solution of the problem. Potential energy determines the objective function. In this algorithm, PE determines the objective function Objf1 = Hiding Failure, Objf2 = Lost Rules and Objf3 = Ghost Rules. The aim of the algorithm can achieve these objective functions. Kinetic energy determines the worst solution which represents that increase of hiding failure and lost rules. A number of hits of the molecular wall represent that how many sensitive items are there in a transactional dataset. The parameters used are hiding failure, misses cost, lost rules, artificial patterns, dissimilarity, and side effect factor. Hiding failure defined as how many sensitive rules are leaked in the process of hiding. Lost rule represents how many rules are lost during the mining process with original rules. Misses cost represents the percentage of non-critical rules that are hidden during the sanitization process. *Ghost Rules* represents that how many rules are additionally generated during the mining process with original rules. *Dissimilarity* measures the dissimilarity among the original and sanitized datasets. *Side Effect Factor* determines the percentage of non-sensitive rules removed during the sanitization process. *Lost Rule Recovery* determines the percentages of rules recovered with overall sensitive rules. *Ghost Rule Recovery* determines the percentages of rules reduced with overall non-sensitive rules. The proposed algorithm compares with genetic, particle swarm optimization, and cuckoo optimization techniques, and it achieves better performance. OnWallIneffectiveCollision deals with the molecule in bombarding with the borders. Similarly, the sensitive rules bombard with cell walls and lose its support like a molecule. These sensitive pattern, S_p in a transactional dataset D can be represented as s_1, s_2, $s_3 \ldots s_n$ then

$$P \ E \ S_p \ where \ support(S_p) > MST \ and \ confidence(S_p) > MCT \ 1 <= S_p <= n$$

On-Wall Ineffective Collision:

On-wall ineffective collision states that a molecule collides with boundaries in a unit. During this collision, the molecule w converts into w_1. During the process of association rule hiding, a critical transaction consists of one or more sensitive association rules, which have a single or multiple consequent rules. Single rules are undergone for an on-wall ineffective collision like, given W a sensitive item converted into **W->W1** by achieving an objective function PE(W) = HF. W be a sensitive item information privacy preserved by replacing a specialized value like 0 or ? or null.

Input: Transactional dataset D, public rule PUR, private rule PTR, sensitive item W

Algorithm 1: *Sensitive Rules are hidden by the OnWallIneffectiveCollision technique*
PE(W) = f(w1)   //objective function
    obf1 = HF
end if
Algorithm 2: *Lost Rules*
PE(W) = f(w1)   //objective function
    obf1 = LR
end if
Algorithm 3: *Ghost Rules*
PE(W) = f(w1)   //objective function
    obf1 = GR
end if

6 Results and Discussion

The experiments are conducted on a desktop computer with Intel Core i5 Processor with 16 GB of RAM uses Windows operating system. Our experiment uses standard benchmark datasets like the mushroom, chess. This transactional dataset given as an input to the apriori algorithm; it produces the best association rules. The following are the benchmark datasets used to conduct the experiments and their results are shown in (Figs. 1 and 2, Tables 1 and 2).

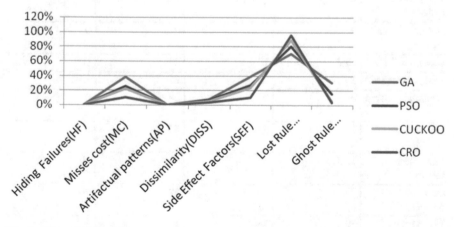

Fig. 1 Chess dataset results

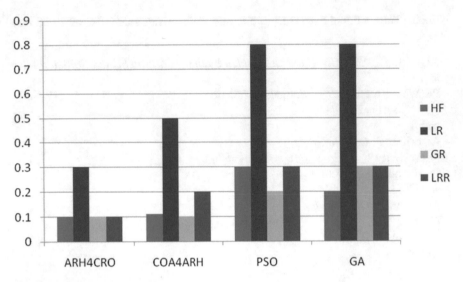

Fig. 2 Mushroom dataset results

Table 1 Benchmark datasets

Dataset	Instances	Attributes
Mushroom	8124	22
Chess	3196	36

Table 2 Experimental results for chess dataset

Comparison parameters	GA (%)	PSO (%)	CUCKOO (%)	CRO (%)
Hiding failures (HF)	0	0	0	0
Lost rules (LR)	15	15	20	10
Misses cost (MC)	38	25	22	10
Artificial patterns (AP)	0	0	0	0
Dissimilarity (DISS)	7.4	6.4	5.3	3.3
Side effect factors (SEF)	38.5	27	24	10
Lost rule recovery (LRR)	70	80	89	96
Ghost rule recovery (GRG)	30	15	5	3

References

1. Agrawal, R., Imielinski, T., Swami, A.: Mining association rules between sets of items in large databases. In: ACM SIGMOD International Conference on Management of Data SIGMOD, p. 207 (1993)
2. Hahsler, M., Grun, B., Hornik, K.: Arules—A computational environment for mining association rules and frequent itemsets. J. Stat. Softw. (2005)
3. Tan, P.-N., Steinbach, M., Kumar, V.: Association analysis: basic concepts and algorithms. In: Introduction to Data Mining (2005)
4. Lam, A.Y.S., Li, V.O.K.: Chemical reaction optimization (2012)
5. Lam, A.Y.S., Li, V.O.K.: Real Coded Chemical reaction optimization. IEEE Trans. Evol. Comput. **16**(3), 339–353 (2012)
6. Atallah, M., Bertino, E., Elmagarmid, A., Ibrahim, M., Verykios, V.: Disclosure limitation of sensitive rules, pp. 45–52 (1999)
7. Dasseni, E., Verkios, V.S., Elmagarmid, A.K., Bernito, E.: Hiding association rules by using confidence and support, pp. 369–383 (2000)
8. Verykios, V.S., Elmagarmid, A.K., Bertino, E., Saygin, Y., Dasseni, E.: Association rule hiding. Trans. Knowl. Data Eng. **16**(4), 434–447 (2004)
9. Oliveira, S.R.M., Zaiane, O.R.: Privacy preserving frequent itemset mining (2002)
10. Wang: Hiding informative association rules sets. Expert Syst. Appl. **33**(2), 316–323 (2007)
11. Shah, K., Thakkar, A., Ganatra, A.: Association rule hiding by heuristic approach to reduce side effects and hide multiple R.H.S. Int. J. Comput. Appl. **45**(1), 1–7 (2012)
12. Damandiya, N.H., Rao, U.P.: Hiding sensitive association rules to maintain privacy and data quality in database, pp. 1306–1310 (2013)
13. Hong, T.-P., Lin, C.-W., Yang, K.-T., Wang, S.L.: A heuristic data-sanitization approach based on SIF-IDF (2011)
14. Lin, C.-W., Hong, T.-P., Hsu, H.C.: Reducing side effects of hiding sensitive item sets in privacy preserving data mining. Sci. World J. (2014)
15. Ghalehsefidi, N.J., Dehkordi, M.N.: A hybrid algorithm based on heuristic method to preserve privacy in association rule mining. Indian J. Sci. Technol. **9**(27) (2016)
16. Belwal, R., Varshney, J., Khan, S.: Hiding sensitive association rules efficiently by introducing new variable hiding counter (2013)
17. Modi, C.N., Rao, U.P., Patel, D.R.: Maintaining privacy and data quality in privacy preserving association rule mining (2010)
18. Modi, C.N., Rao, U.P., Patel, D.R.: An efficient solution for privacy preserving association rule mining. Int. J. Comput. Netw. **2**(5), 79–85 (2010)
19. Kesava Murthy, B., Khan, A.M.: Privacy preserving association rule mining over distributed databases using genetic algorithm. Neural Comput. Appl. **22**(1), S351–S364 (2013)
20. Lin, J.C.W,: A sanitization approach for hiding sensitive Itemsets based on particle swarm optimization. Eng. Appl. Artif. Intell. **53**, 1–18 (2016)
21. Afshari, M.H.: Association rule hiding using cuckoo optimization algorithm. Expert Syst. Appl. **64**,340–351 (2016)
22. Satyanarayana Murthy, T.: Pine Apple Expert System Using Improved C4.5 Algorithm, pp. 1264–1266 (2013)
23. Satyanarayana Murthy, T.: Privacy Preserving for Expertise Data using K-anonymity Technique to Advise the Farmers (2013)
24. Gopalan, N.P., Satyanarayana Murthy, T., Venkateswarlu, Y.: Hiding Critical Transaction using Unrealization approach (2017)
25. SatyanarayanaMurthy, T., Gopalan, N.P., Venkateswarlu, Y.: An Efficient Method for Hiding Association Rules with Additional Parameter Metrics (2017)

Inspection–Maintenance-Based Availability Optimization of Feeder Section Using Particle Swarm Optimization

Aditya Tiwary

Abstract Proper maintenance is important for reliable and efficient operation of the distribution system. The time duration after which the maintenance is required in the distribution system is important to identify. A methodology on PSO for evaluating the optimum value of inspection and maintenance is developed. Optimum value of duration between two inspections is obtained. The cost function and optimization constraints have also been considered. Two sample power systems were used for implementation of this problem. The output received is compared with other variants of particle swarm optimization (PSO) such as bare bones PSO (BBPSO), coordinated aggregation-based PSO (CAPSO), enhanced leader PSO (ELPSO).

Keywords Reliability function · Unavailability · Particle swarm technique CAPSO · ELPSO

Symbols

T_c	Cycle time
t_{ins}	Inspection duration
t_{er}	Repair time expected
τ	Period between inspections
c_i	Cost coefficient
τ_i	Interval between inspection and repair
NC	Number of sections
NLP	Load points
U_k	Unavailability
U_{d-k}	Unavailability threshold value
A_K	Availability

A. Tiwary (✉)
Electrical & Electronics Engineering Department, Institute of Engineering & Science,
IPS Academy, Indore, India
e-mail: raditya2002@gmail.com

© Springer Nature Singapore Pte Ltd. 2019
J. C. Bansal et al. (eds.), *Soft Computing for Problem Solving*,
Advances in Intelligent Systems and Computing 816,
https://doi.org/10.1007/978-981-13-1592-3_20

257

w	Inertia weight
$rand_1$, $rand_2$	Random digits
c_1, c_2	Coefficients of acceleration
$iter_{max}$	Maximum number of iteration
w_{max}	Maximum value specified for inertia weight
w_{min}	Minimum value specified for inertia weight

1 Introduction

Studies on failure have concluded that failure of power system is mainly due to failure in the distribution subsystem of power system. Therefore, to improve the system availability improvement in the system is necessary. Proper maintenance at regular interval is required to improve the availability of the supply system. New methods for power system reliability are discussed in the literature [1]. Various techniques for preventive maintenance are discussed [2]. Gangel et al. [3] discussed methods for distribution system reliability performance. Various methods for reliability calculations are provided [4–10]. Pereira et al. [11] presented a tool based on computing application for evaluation of reliability. Su et al. [12] proposed reliability design of distribution system. Chang et al. [13] used polynomial-time algorithm for optimal reliability design. For operational planning optimization, a method was described by Popov et al. [14]. Ant colony algorithm under performance was proposed by Meziane et al. [15]. Value-based distribution system optimization was proposed by Sohn et al. [16]. Bakkiyaraj et al. [17] developed a methodology for optimum reliability planning for power system. Louit et al. [18] proposed a methodology for obtaining optimum interval action. Dehghanian et al. [19] proposed a comprehensive scheme for reliability centred maintenance in distribution power system. Parallel Monte Carlo approach for distribution system assessment is discussed [20]. Huda et al. [21] developed accelerated distribution system reliability evaluation by multi-level MCS. Adefarati et al. [22] proposed distributed generation integration with power system. Reliability indices improvement method was proposed by Lopez et al. [23] and Ray et al. [24]. Tiwary et al. [25] developed a methodology for optimizing availability using teaching–learning algorithm. Arya et al. [26] proposed method for evaluating reliability indices. Tiwary et al. [27] proposed a method for optimum period. Arya et al. [28] proposed SBS method to calculate reliability. Tiwary et al. [29] determined reliability using state transition sampling technique.

Kerdphol et al. [30] proposed optimum sizing of battery using PSO. Bhattacharya and Raj [31] have developed PSO-based bio-inspired algorithm for reactive power planning. The optimization of energy storage using PSO was proposed by Kerdphol et al. [32]. Banerjee et al. [33] discussed a method based on PSO technique for scheduling. A method on PSO for optimal solution for the capacitor allocation in distribution system was developed by Ramadan et al. [34]. The optimization for optimal placement of capacitor is proposed [35]. Hashemi et al. [36] proposed an algorithm concerning load reduction based DR. Huo et al. [37] proposed a methodology for optimal operation of interconnected energy hubs by using decomposed hybrid particle swarm and interior-point approach. PSO for economic power dispatch was proposed by Kumar et al. [38]. Jadoun et al. [39] developed a modulated PSO for economic emission. Basu [40] developed a modified PSO for nonconvex economic dispatch. Suganthan et al. [41] proposed particle swarm optimiser with neighbourhood operator. Bansal et al. [42] developed inertia weight strategies in PSO. Bansal and Deep [43] proposed modified binary PSO algorithm. Deep and Bansal [44] proposed a new mean PSO for function optimization. Bansal et al. [45] proposed a method for optimization of directional over current relay times by PSO. Jadon et al. [46] proposed a self-adaptive acceleration factor in PSO. Bakkiyaraj et al. [47] proposed optimal reliability planning of power system.

PSO-based methodology is proposed identifying the optimum period between two maintenance by minimizing the cost function and satisfying availability constraints. PSO is a swarm intelligence optimization method tested on many optimization problems and has given better result in literature [30–40], and therefore, PSO is used in obtaining the optimum solution for the availability problem. Results obtained using PSO have been compared with other variants of PSO such as BBPSO, CAPSO and ELPSO, and statistical inference has been provided. Proposed method is implemented on radial and meshed distribution system.

2 Availability Model for Distribution System and Problem Formulation

The complete cycle for the availability model consists of inspection duration, expected repair time and duration between two inspections and can be expressed as follows [25].

$$T_c = \tau + t_{\text{ins}} + t_{\text{er}} \tag{1}$$

The downtime can be represented as

$$t_{\text{dn}} = t_{\text{ins}} + t_r \tag{2}$$

The repair time expected after the inspection is complete can be obtained as follows [25]

$$t_{\text{er}} = [1 - R(\tau)]t_r \tag{3}$$

Or

$$t_{\text{er}} = [1 - R(\tau)].t_{\text{dn}} \tag{4}$$

Availability can be obtained by the following equation

$$A(\tau) = \frac{\int_0^\tau R(t)dt}{T_c} \tag{5}$$

Or

$$A(\tau) = \frac{1 - \exp(-\lambda\tau)}{\lambda T_c} \tag{6}$$

Assume failure rate as λ in above expression.

The following objective function is to be minimized [25]

$$J = \sum_{i=1}^{NC} c_i/\tau_i \tag{7}$$

Constraints on objective function based on the unavailability are provided as:

$$U_k \leq U_{d-k} \tag{8}$$

The unavailability is calculated as:

$$U_K = 1 - A_K \tag{9}$$

The availability can be calculated from above equation.

3 Optimization Techniques: An Overview

3.1 Particle Swarm Optimization (PSO) in Brief

PSO is a swarm intelligence method proposed by Kennedy and Eberhart [48]. PSO is a population-dependant algorithm and is developed from social behaviour of bird flocking. Each and every particle searches for the best solution. The particle approaches towards better result by guidance provided by velocity and position.

Velocity and position of particle are modified as:

$$\rho_i^{iter} = w\rho_i^{iter-1} + c_1 rand_1(P_{best}^{iter-1} - S_i^{iter-1}) + c_2 rand_2(G_{best}^{iter-1} - S_i^{iter-1}) \tag{10}$$

$$S_i^{iter} = S_i^{iter-1} + \rho_i^{iter} \tag{11}$$

Inertia weight w for the updation of the velocity is obtained as:

$$w^{\text{iter}} = w_{\max} - \frac{w_{\max} - w_{\min}}{\text{iter}_{\max}} * \text{iter} \tag{12}$$

The process is terminated due to occurrence of maximum number of iterations.

3.2 Bare Bones Particle Swarm Optimization (BBPSO) in Brief

In 2003, Kennedy [49] proposed BBPSO. In BBPSO, velocity vector is eliminated and only position vector is available for obtaining the optimum solution. The updated position can be obtained by using the following relation.

$$x_i^{(t+1)} = N\left(\frac{G_{\text{best}} + P_{\text{best}}}{2}, |G_{\text{best}} - P_{\text{best}}|\right) \tag{13}$$

where N represents a GD with mean $\mu = \left(\frac{G_{\text{best}} + P_{\text{best}}}{2}\right)$ and SD $\sigma = |G_{\text{best}} - P_{\text{best}}|$ for each particle.

3.3 Coordinated Aggregation-Based Particle Swarm Optimization (CAPSO) in Brief

This method of optimization is applied in the literature [50]. Velocity relation for particle except the best achievement is as follows.

$$\rho_i^{\text{iter}} = w^{\text{iter}-1}.\rho_i^{\text{iter}-1} + \sum \text{rand}.AF^{\text{iter}-1}[S_j^{\text{iter}-1} - S_i^{\text{iter}-1}] \tag{14}$$

Velocity of the best particle

$$\rho_b^{\text{iter}} = w^{\text{iter}-1}.\rho_i^{\text{iter}-1} + \text{rand}.[S_p^{\text{iter}-1} - S_b^{\text{iter}-1}] \tag{15}$$

where

$$AF^{\text{iter}-1} = \frac{J(S_j^{\text{iter}-1}) - J(S_i^{\text{iter}-1})}{\sum [J(S_j^{\text{iter}-1}) - J(S_i^{\text{iter}-1})]} \tag{16}$$

3.4 Enhanced Leader Particle Swarm Optimization (ELPSO) in Brief

ELPSO is a variant of PSO and was proposed by Jordehi et al. [51].

Successive mutation strategy stage	Mutation applied
1st	Gaussian mutation
2nd	Cauchy mutation
3rd	Opposition-based mutation
4th	Opposition-based mutation is applied to whole group
5th	DE-based mutation operator is applied

By applying the above process, the optimum value is achieved.

4 PSO-Based Availability Algorithm

The following are the steps of the proposed algorithm.

1: Initial swarm generalization:

$$\rho_i^0 = \{XV_{i1}^0, XV_{i2}^0, \ldots \ldots XV_{iNC}^0\}$$

$$S_i^0 = \{X_{i1}^0, X_{i2}^0, \ldots \ldots X_{iNC}^0\}$$

2: Say iter $=1$
3: Evaluate A_k, U_k at each load point.
4: Obtain the feasible and not feasible solutions. Unavailability threshold value (U_{d-k}) is taken as 0.99.
5: Calculate cost function $J(X_i^{(0)})$, and obtain $P_{best,i}$ and G_{best}.
6: Calculate inertia weight by using Eq. (12).
7: Update position and velocity by Eqs. (10) and (11).
8: Apply inequality constraints. Apply bounce back technique [52].
9: Obtain optimum solution.

$$iter = iter + 1$$

10: $iter > iter_{max}$.

Stop, otherwise go to step 5.

5 Results and Discussions

PSO technique and its variants, BBPSO, CAPSO, ELPSO are discussed in Fig. 1.

5.1 Radial Distribution System

The initial values of the radial distribution system and control parameters are shown in Tables 1 and 2, respectively. In Table 3, optimum duration between inspection-repair and optimum availability is given. Table 4 provides optimum availability. Table 5 provides different statistics. Figure 2 shows the evolution of best fitness value. Table 6 provides the CPU times required for convergence on Intel Core 2 duo Processor, 2.10 GHz. CPU time by BBPSO has reduced by 17% than in comparison to PSO algorithm. 18.12% reduction in CPU time by CAPSO with comparison with PSO algorithm. PSO variants, ELPSO and CAPSO, have given better results (Fig. 3).

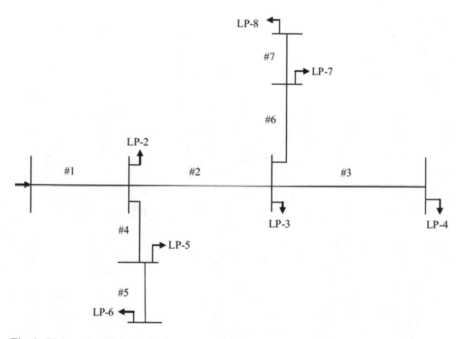

Fig. 1 Eight-node radial distribution system [26]

Table 1 Initial system data [25]

Distribution segment	#1	#2	#3	#4	#5	#6	#7
λ/year $\times$ 10^{-5}	0.04	0.02	0.03	0.05	0.02	0.01	0.01
C_i (Rs.) $\times$ 10^7	2.00	3.00	4.00	3.50	4.50	5.00	1.00
τ_j, $\min(h)$	500	500	500	500	500	500	500
τ_j, $\max(h)$	6000	6000	6000	6000	6000	6000	6000
$t_{ins}(h)$	12	10	15	20	10	18	18
$t_{ri}(h)$	36	20	25	30	20	24	24

Table 2 Control parameters

S. no.	Parameters	Values			
		ELPSO	CAPSO	BBPSO	PSO
1	Population size	20	20	20	20
2	c_1	NA	NA	NA	1.2
3	c_2	NA	NA	NA	0.12
4	(k_{max})	500	500	500	500

Table 3 Optimal interval between inspection and repair

Distribution segments	ELPSO		CAPSO		BBPSO		PSO	
	Optimal duration, τ_i (h)	Optimum availability $A_i(\tau_i)$	Optimal duration, τ_i (h)	Optimum availability $A_i(\tau_i)$	Optimal duration, τ_i (h)	Optimum availability $A_i(\tau_i)$	Optimal duration, τ_i (h)	Optimum availability $A_i(\tau_i)$
#1	4705	0.9965	4703	0.9965	4658	0.9965	4485	0.9964
#2	4735	0.9974	4733	0.9974	4728	0.9974	4545	0.9974
#3	4690	0.9962	4688	0.9961	4670	0.9961	4501	0.9960
#4	5150	0.9948	5145	0.9948	5105	0.9948	4925	0.9947
#5	4693	0.9975	4691	0.9974	4689	0.9974	4418	0.9973
#6	4610	0.9959	4608	0.9959	4598	0.9959	4410	0.9957
#7	4628	0.9959	4625	0.9959	4619	0.9959	4488	0.9958

Table 4 Optimal availability

LP	Optimum availability (ELPSO)	Optimum availability (CAPSO)	Optimum availability (BBPSO)	Optimum availability (PSO)
2	0.9966	0.9965	0.9965	0.9964
3	0.9939	0.9939	0.9939	0.9938
4	0.9901	0.9901	0.9900	0.9898
5	0.9914	0.9913	0.9913	0.9911
6	0.9889	0.9888	0.9887	0.9884
7	0.9899	0.9898	0.9898	0.9895
8	0.9860	0.9859	0.9858	0.9854

Table 5 Statistics for minimum value of objective function with ELPSO, CAPSO, BBPSO, PSO

Parameters	Values of parameters (ELPSO)	Values of parameters (CAPSO)	Values of parameters (BBPSO)	Values of parameters (PSO)
Average value (Rs.) $\times 10^4$	6.39	6.40	6.46	6.89
Standard deviation (Rs.) $\times 10^3$	15.08	15.12	15.30	17.80
Coefficient of variation	0.2360	0.2364	0.2368	0.2583
Median value (Rs.) $\times 10^4$	5.97	5.98	6.01	6.12
Minimum value (Rs.) $\times 10^4$	4.37	4.39	4.40	4.41
Maximum value (Rs.) $\times 10^4$	9.83	9.84	9.85	9.99
Length of confidence interval (Rs.) $\times 10^3$	5.94	5.95	6.02	7.01
Upper bound CI (Rs.) $\times 10^4$	6.68	6.69	6.76	7.24
Lower bound CI (Rs.) $\times 10^4$	6.09	6.10	6.16	6.54
Frequency of convergence (f)	60	59	58	51

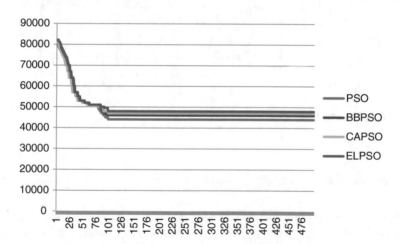

Fig. 2 Variation of best value

Table 6 CPU time

DS	CPU time (s) ELPSO	CPU time (s) CAPSO	CPU time (s) BBPSO	CPU time (s) PSO
Radial	8.98	8.99	9.02	10.98

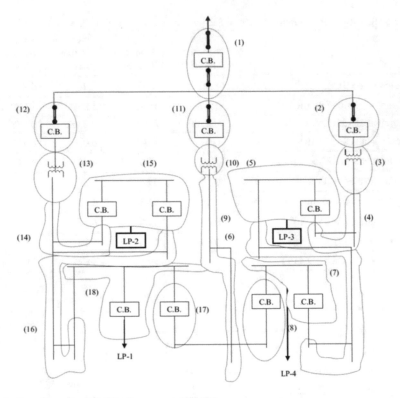

Fig. 3 Sample-meshed distribution system [28, 53]

5.2 Meshed Distribution System

The initial values of the meshed distribution system and control parameters are shown in Tables 7 and 8, respectively. In Table 9, optimum duration and optimum availability are given. Table 10 provides the optimum value of availability. Table 11 provides different statistics. Figure 4 shows the evolution of best fitness value. Table 12 provides the CPU time required. CPU time by BBPSO has reduced by 12% than to PSO. 13.23% reduction is seen in CPU time for ELPSO algorithm in comparison with PSO algorithm.

Table 7 Initial system data [25]

Distribution section	λ_i^0 failure/year $\times$ 10^{-5}	C_i (Rs) $\times 10^7$	τ_j, min(h)	τ_j, max(h)	$t_{ins}(h)$	$t_{ri}(h)$
#1	0.03	2.00	500	6000	10	30
#2	0.01	3.00	500	6000	15	20
#3	0.05	3.50	500	6000	12	23
#4	0.01	1.00	500	6000	20	28
#5	0.03	4.00	500	6000	18	40
#6	0.05	4.50	500	6000	16	38
#7	0.04	2.50	500	6000	13	36
#8	0.03	1.00	500	6000	10	33
#9	0.01	2.00	500	6000	17	26
#10	0.05	5.00	500	6000	18	28
#11	0.01	1.50	500	6000	20	31
#12	0.05	4.30	500	6000	11	21
#13	0.02	3.20	500	6000	16	38
#14	0.03	4.80	500	6000	18	39
#15	0.04	3.00	500	6000	20	34
#16	0.03	4.30	500	6000	10	32
#17	0.05	4.40	500	6000	15	38
#18	0.03	4.60	500	6000	10	29

Table 8 Control parameters

S. no.	Parameters	Values			
		ELPSO	CAPSO	BBPSO	PSO
1	Population size	20	20	20	20
2	c_1	NA	NA	NA	1.2
3	c_2	NA	NA	NA	0.12
4	(k_{max})	500	500	500	500

6 Conclusion

A technique is proposed for obtaining optimum interval between preventive maintenance using particle swarm optimization. The duration after which the maintenance is required for the distribution system is important to obtain. If the optimum duration for the maintenance is known, then measures can be taken in advance in order to minimize the failure of the system. If during inspection there arises a need for the repair then repair can be carried out. Formulated problem has also been solved using different variants of PSO, such as ELPSO, CAPSO and BBPSO algorithm. The

Table 9 Optimal interval

DS	ELPSO		CAPSO		BBPSO		PSO	
	Optimal duration, τ_i (h)	Optimum availability $A_i(\tau_i)$	Optimal duration, τ_i (h)	Optimum availability $A_i(\tau_i)$	Optimal duration, τ_i (h)	Optimum availability $A_i(\tau_i)$	Optimal duration, τ_i (h)	Optimum availability $A_i(\tau_i)$
#1	2499	0.9956	2499	0.9956	2498	0.9956	2313	0.9953
#2	2797	0.9946	2795	0.9945	2782	0.9945	2598	0.9941
#3	4885	0.9964	4878	0.9964	4875	0.9963	4619	0.9962
#4	5298	0.9960	5294	0.9960	5292	0.9960	5048	0.9958
#5	5234	0.9958	5233	0.9958	5220	0.9958	5098	0.9957
#6	5096	0.9956	5095	0.9956	5089	0.9956	4797	0.9955
#7	4599	0.9962	4598	0.9962	4595	0.9962	4385	0.9962
#8	4591	0.9971	4590	0.9971	4583	0.9971	4378	0.9971
#9	2890	0.9940	2890	0.9940	2885	0.9940	2738	0.9937
#10	4979	0.9951	4970	0.9951	4957	0.9951	4761	0.9950
#11	4999	0.9957	4994	0.9957	4971	0.9957	4792	0.9956
#12	4585	0.9964	4571	0.9964	4551	0.9964	4412	0.9964
#13	4659	0.9961	4654	0.9961	4648	0.9961	4502	0.9960
#14	4772	0.9955	4765	0.9955	4738	0.9955	4598	0.9954
#15	4546	0.9947	4538	0.9947	4524	0.9947	4375	0.9946
#16	4797	0.9972	4795	0.9972	4789	0.9972	4512	0.9971
#17	4512	0.9955	4505	0.9955	4499	0.9955	4223	0.9954
#18	4549	0.9971	4541	0.9971	4528	0.9971	4339	0.9970

Table 10 Optimal availability

LP	Optimum availability (ELPSO)	Optimum availability (CAPSO)	Optimum availability (BBPSO)	Optimum availability (PSO)
1	0.9927	0.9927	0.9926	0.9922
2	0.9905	0.9904	0.9902	0.9898
3	0.9918	0.9918	0.9913	0.9909
4	0.9922	0.9921	0.9917	0.9914

results have been obtained for radial and meshed distribution system and compared with statistical inferences. The results obtained from the three variants of the PSO are in close argument.

Table 11 Statistics for minimum value of objective function

Parameters	Values of parameters (ELPSO)	Values of parameters (CAPSO)	Values of parameters (BBPSO)	Values of parameters (PSO)
Average value (Rs.) $\times 10^4$	1.45	1.47	1.52	1.71
Standard deviation (Rs.) $\times 10^3$	1.98	2.01	2.09	2.55
Coefficient of variation	0.1370	0.1371	0.1375	0.1491
Median value (Rs.) $\times 10^4$	1.45	1.46	1.50	1.55
Minimum value (Rs.) $\times 10^4$	1.18	1.19	1.22	1.30
Maximum value (Rs.) $\times 10^4$	1.85	1.87	1.90	2.10
Length of confidence interval (Rs.) $\times 10^2$	7.80	7.91	8.23	10.04
Upper bound CI (Rs.) $\times 10^4$	1.48	1.50	1.56	1.76
Lower bound CI (Rs.) $\times 10^4$	1.43	1.45	1.47	1.65
Frequency of convergence (f)	63	62	60	49

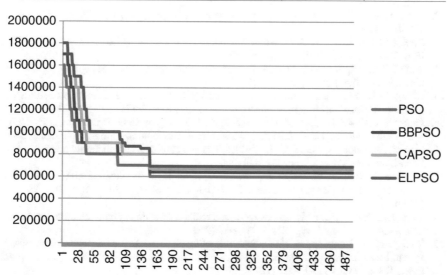

Fig. 4 Variation of best value

Table 12 CPU time

DS	CPU time (s) ELPSO	CPU time (s) CAPSO	CPU time (s) BBPSO	CPU time (s) PSO
Meshed	10.42	10.46	10.56	12.01

References

1. Elmakias, D.: New Computational Methods in Power System Reliability. Springer International Edition (2000)
2. Gertsbakh, I.: Reliability Theory with Applications to Preventive Maintenance. Springer International Edition (2000)
3. Gangel, M.W., Ringlee, R.J.: Distribution system reliability performance. IEEE Trans. PAS-1968 **87**(7), 1657–1665 (1968)
4. IEEE Committee Report, Bibliography on the application of probability methods in power system reliability evaluation. IEEE Trans. Power Appar. Syst. **PAS-91**, 649–660 (1972)
5. IEEE Committee Report, Bibliography on the application of probability methods in power system reliability evaluation, 1971–1977. IEEE Trans. Power Appar. Syst. **PAS-97**, 2235–2242, (1978)
6. Allan, R.N., Dialynas, E.N., Homer, I.R.: Modelling and evaluating the reliability of distribution systems. IEEE Trans. Power Appar. Syst. **PAS-98**, 2181–2189 (1979)
7. Billinton, R., Allan, R.N.: Reliability Evaluation of Engineering Systems. Springer International Edition (1992)
8. Pereira, M.V.F., Pinto, L.M.V.G.: A new computational tool for composite reliability evaluation. IEEE Trans. PAS **7**(1), 258–263 (1992)
9. Billinton, R., Allan, R.N.: Reliability Evaluation of Power System. Springer International Edition (1996)
10. Allan, R.N., Billinton, R.: Probabilistic assessment of power systems. Proc. IEEE **88**(2), 140–162 (2000)
11. Pereira, M.V.F., Pinto, L.M.V.G.: A new computational tool for composite reliability evaluation. IEEE Trans. Power Syst. **7**(1), 258–264 (1992)
12. Su, C.T., Lii, G.R.: Reliability design of distribution systems using modified genetic algorithms. Electr. Power Syst. Res. **60**, 201–206 (2002)
13. Chang, W.F., Wu, Y.C.: Optimal reliability design in an electrical distribution system via polynomial-time algorithm. Int. J. Electr. Power Energy Syst. **25**(8), 659–666 (2003)
14. Popov, V.A., Canha, L.N., Farret, F.A., Abaide, A.R., Rodrigues, M.G., Bernardon, D.P., Konig, A.L., Comassetto, L., Licht, A.P.: Algorithm of reliability optimization for operational planning of distribution systems. In: IEEE/PES Transmission and Distribution Conference and Exposition: Latin America, pp. 523–528 (2004)
15. Meziane, R., Massim, Y., Zeblah, A., Ghoraf, A., Rahli, R.: Reliability optimization using ant colony algorithm under performance and cost constraints. Electr. Power Syst. Res. **76**, 1–8 (2005)
16. Sohn, J.M., Nam, S.R., Park, J.K.: Value-based radial distribution system reliability optimization. IEEE Trans. Power Syst. **21**(2), 941–947 (2006)
17. Bakkiyaraj, A., Kumarappan, N.: Optimal reliability planning for a composite electric power system based on Monte Carlo simulation using particle swarm optimization. Int. J. Electr. Power Energy Syst. **47**, 109–116 (2013)
18. Louit, D., Pascual, R., Banjevic, D.: Optimal interval for major maintenance actions in electricity distribution networks. Int. J. Electr. Power Energy Syst. **31**(7–8), 396–401 (2009)
19. Dehghanian, Payman, Fotuhi-Firuzabad, Mahmud, Aminifar, Farrokh, Billinton, Roy: A comprehensive scheme for reliability centered maintenance in power distribution systems—part I: methodology. IEEE Trans. Power Deliv. **28**, 761–770 (2013)

20. Martinez-Velasco, Juan A., Guerra, Gerardo: Parallel Monte Carlo approach for distribution reliability assessment. IET Gener. Transm. Distrib. **8**, 1810–1819 (2014)
21. Huda, A.S.N., Živanović, R.: Accelerated distribution systems reliability evaluation by multilevel Monte Carlo simulation: implementation of two discretisation schemes. IET Gener. Transm. Distrib. **11**, 3397–3405 (2017)
22. Adefarati, T., Bansal, R.C.: Reliability assessment of distribution system with the integration of renewable distributed generation. Appl. Energy **185**, 158–171 (2017)
23. López, J.C., Lavorato, M., Rider, M.J.: Optimal reconfiguration of electrical distribution systems considering reliability indices improvement. Int. J. Electr. Power Energy Syst. **78**, 837–845 (2016)
24. Ray, S., Bhattacharya, A., Bhattacharjee, S.: Optimal placement of switches in a radial distribution network for reliability improvement. Int. J. Electr. Power Energy Syst. **76**, 53–68 (2016)
25. Tiwary, A., Arya, L.D., Arya, R., Choube, S.C.: Inspection repair based availability optimization of distribution systems using teaching learning based optimization. J. Inst. Eng. (India) Series B, **97**(3), 355–365 (2016)
26. Arya, L.D., Choube, S.C., Arya, R., Tiwary, Aditya: Evaluation of reliability indices accounting omission of random repair time for distribution systems using Monte Carlo simulation. Int. J. Electr. Power Energy Syst. **42**, 533–541 (2012)
27. Tiwary, A., Arya, R., Choube, S.C., Arya, L.D.: Determination of optimum period between inspections for distribution system based on availability accounting uncertainties in inspection time and repair time. J. Inst. Eng. (India) Series B (Springer) **93**(2), 67–72 (2012)
28. Arya, R., Tiwary, A., Choube, S.C., Arya, L.D.: A smooth bootstrapping based technique for evaluating distribution system reliability indices neglecting random interruption duration. Int. J. Electr. Power Energy Syst. **51**, 307–310 (2013)
29. Tiwary, A., Arya, R., Choube, S.C., Arya, L.D.: Determination of reliability indices for distribution system using a state transition sampling technique accounting random down time omission. J. Inst. Eng. (India) Series B (Springer) **94**(1), 71–83 (2013)
30. Kerdphol, T., Qudaih, Y., Mitani, Y.: Optimum battery energy storage system using particle swarm optimization considering dynamic demand response for microgrids. Int. J. Electr. Power Energy Syst. **83**, 58–66 (2016)
31. Bhattacharyya, B., Raj, S.: PSO based bio inspired algorithms for reactive power planning. Int. J. Electr. Power Energy Syst. **74**, 396–402 (2016)
32. Kerdphol, T., Fuji, K., Mitani, Y., Watanabe, M., Qudaih, Y.: Optimization of a battery energy storage system using PSO for stand-alone microgrids. Int. J. Electr. Power Energy Syst. **81**, 32–39 (2016)
33. Banerjee, S., Dasgupta, K., Chanda, C.K.: Short term hydro-wind-thermal scheduling based on particle swarm optimization technique. Int. J. Electr. Power Energy Syst. **81**, 275–288 (2016)
34. Ramadan, H.S., Bendary, A.F., Nagy, S.: Particle swarm optimization algorithm for capacitor allocation problem in distribution system with wind turbine generators. Int. J. Electr. Power Energy Syst. **84**, 143–152 (2017)
35. Su, X., Masoum, A.S., Wolfs, P.J.: PSO and Improved BSFS based sequential comprehensive placement and real-time multi-objective control of delta-connected switched capacitors in unbalanced radial MV distribution networks. Int. J. Electr. Power Energy Syst. **31**, 612–622 (2016)
36. Hashemi, S., Aghamohammadi, M.R., Sangrody, H.: Restoring desired voltage security margin based on demand response using load to source impedance ratio index and PSO. Int. J. Electr. Power Energy Syst. **96**, 143–151 (2018)
37. Huo, D., Blond, S.L., Gu, C., Wei, W., Yu, D.: Optimal operation of interconnected energy hubs by using decomposed hybrid particle swarm and interior-point approach. Int. J. Electr. Power Energy Syst. **95**, 36–46 (2018)
38. Kumar, R., Sharma, D., Sadu, A.: A hybrid multi-agent based particle swarm optimization algorithm for economic power dispatch. Int. J. Electr. Power Energy Syst. **33**, 115–123 (2011)

39. Jadoun, V.K., Gupta, N., Niazi, K.R., Swarnkar, A.: Modulated particle swarm optimization for economic emission dispatch. Int. J. Electr. Power Energy Syst. **73**, 80–88 (2015)

40. Basu, M.: Modified particle swarm optimization for nonconvex economic dispatch problems. Int. J. Electr. Power Energy Syst. **69**, 304–312 (2015)

41. Suganthan, P.N.: Particle swarm optimizer with neighbourhood operator. In: Proceeding of Congress on Evolutionary Computation, pp. 1958–1962 (1999)

42. Bansal, J.C., Singh P.K., Saraswat, M., Verma, A., Jadon, S. S., Abraham, A.: Inertia weight strategies in particle swarm optimization. In: Third World Congress on Nature and Biologically Inspired Computing, pp. 633–640 (2011)

43. Bansal, J.C., Deep, K.: A modified binary particle swarm optimization for knapsack problems. Appl. Math. Comput. **218**(22), 11042–11061 (2012)

44. Deep, K., Bansal, J.C.: Mean particle swarm optimisation for function optimisation. Int. J. Comput. Intell. Stud. **1**(1), 72–92 (2009)

45. Bansal, J.C., Deep, K.: Optimization of directional overcurrent relay times by particle swarm optimization. In: Swarm Intelligence Symposium, pp. 1–7 (2008)

46. Jadon, S.S., Sharma, H., Bansal, J.C., Tiwari, R.: Self adaptive acceleration factor in particle swarm optimization. In: Proceeding of International Conference on Bio Inspired Computing: Theories and applications, pp. 325–340 (2013)

47. Bakkiyaraj, R.A., Kumarappan, N.: Optimal reliability planning for a composite electric power system based on Monte Carlo simulation using Particle swarm optimization. Int. J. Electr. Power Energy Syst. **47**, 109–116 (2013)

48. Kennedy, J., Eberhart, R.: Particle Swarm optimization. In: Proceeding of IEEE international Conference on Neural Networks, Piscataway, NJ, pp. 1942–1948 (1995)

49. Kennedy, J.: Bare bones particle swarms. In: Proceeding of IEEE Swarm intelligence symposium, pp. 80–87 (2003)

50. Vlachogiannis, J.G., Lee, K.Y.: A comparative study on particle swarm optimization for optimal steady State performance of power systems. IEEE Trans. Power Syst. **21**(4), 1318–1728 (2006)

51. Jordehi, A.R., Jasni, J., Wahab, N.A., Kadir, M.Z., Javadi, M.S.: Enhanced leader PSO (ELPSO): an new algorithm for allocating distributed TCSC's in power system. Int. J. Electr. Power Energy Syst. **64**, 771–784 (2015)

52. Price, K., Storn, R., Lampinen, J.: Differential Evolution: A Practical Approach to Global Optimization. Springer (2005)

53. Su, C.T., Lii, G.R.: Reliability design of distribution systems using modified genetic algorithms. Electr. Power Syst. Res. **60**, 201–206 (2002)

Formulating an Economic Order Quantity Model for Items with Variable Rate of Deterioration and Two-Component Demand

Trailokyanath Singh, Chittaranjan Mallick and Rahul Kumar Singh

Abstract In this paper, an economic order quantity (EOQ) model to minimize the total cost for deteriorating items has developed by incorporating items with two-component demand rate. Firstly, from starting to a certain period of the cycle, the demand rate is considered as constant, and secondly, the demand rate is assumed as a linear function of time for the last part of the cycle. The distribution for the deterioration rate of each item is considered as a three-parameter Weibull distribution function. A solution procedure is provided to determine the EOQ and the optimum average total cost. Finally, results are analyzed and demonstrated with an illustrative example.

Keywords Decay · Economic order quantity · Two-component demand
Three-parameter Weibull distribution

1 Introduction

In practice, certain commodities deteriorate in their lifespan while kept in storage such as food items, crops, fruits, chemicals, volatile liquids, blood stored in blood banks, radioactive substances, electronic components, and others. Thus, deterioration is a natural phenomenon indicating the change, damage, decay, or spoilage of items that cannot be used in original purposes. Therefore, it is very important for

T. Singh (✉) · R. K. Singh
Department of Mathematics, C.V. Raman College of Engineering,
Bhubaneswar 752054, Odisha, India
e-mail: trailokyanaths108@gmail.com

R. K. Singh
e-mail: rahul94377@gmail.com

C. Mallick
Department of Mathematics, Parala Maharaja Engineering College,
Berhampur 761003, Odisha, India
e-mail: cmallick75@gmail.com

© Springer Nature Singapore Pte Ltd. 2019
J. C. Bansal et al. (eds.), *Soft Computing for Problem Solving*,
Advances in Intelligent Systems and Computing 816,
https://doi.org/10.1007/978-981-13-1592-3_21

the researchers for studying the inventory problems in order to control and maintain inventories. From the beginning of the research on this concept, the decaying effect on items was observed by Ghare and Schrader [1] with help of calculus techniques. Dave and Patel [2] proposed a no shortage deteriorating inventory model with varying demand and the instantaneous replenishment, and further, it was extended by Sachan [3] by including the concept of shortages. Besides time-dependent demand patterns, some models with varying deterioration rate were studied. This outcome of varying deterioration rate is noticed in the drugs after its date expires. Also, breakdown of batteries takes place in the process. Covert and Philip [4] and Singh and Pattnayak [5] presented their model with varying deterioration rate as two-parameter Weibull distribution. In reality, some items deteriorate after some time; in that case, the distribution like three-parameter Weibull is appropriate to use in the model. Philip [6] established a three-parameter Weibull distributed model by extending the model of Covert and Philip's. Singh et al. [7] introduced the three-parameter Weibull distribution deterioration in the inventory model for deteriorating items with ramp-type demand rate and shortages. Goyal and Giri [8], Li et al. [9], and Janssen et al. [10] made the extensive surveys of the literature regarding deteriorating items.

In most of the inventory models, the time-dependent demand is linear, exponential, or quadratic. The rapidly increasing or decreasing and accelerated rise or retarded fall in demand rate is characteristics of exponential and quadratic demand pattern, respectively. The optimal policies for deteriorating items developed by Hollier and Mak [11] and Hariga and Benkherouf [12] were based on rapidly decreasing and increasing exponential demand pattern, respectively. Goswami and Chaudhuri [13] and Chakrabarti and Chaudhuri [14] studied replenishment policies for a deteriorating item with shortages by using the assumption of linear increasing function of demand pattern. A heuristic approach for replenishment policy for deteriorating items with linear trended demand was introduced by Kim [15]. Ghosh and Chaudhuri [16] proposed an order-level policy Weibull distribution deterioration, quadratic demand, and shortages. As deterioration, the natural phenomenon increases with respect to time. Therefore, a three-parameter Weibull density function is fit to represent the shelf-life of items which is applicable for the newly launched fashion goods, electronic devices, and automobiles, etc. whose shelf-life deteriorates with respect to time. A deteriorating inventory model with quadratic demand pattern, three-parameter Weibull distribution deterioration, and shortages was established by Sanni and Chukwu [17]. However, the linear demand pattern indicates the uniform change in demand rate. Recently, Srivastava and Gupta [18] worked on the inventory model for deteriorating items having constant deterioration rate and both constant and linear demand pattern as two-component demand rate.

In this paper, an economic order quantity model for deteriorating items with two-component demand pattern is developed under the assumption of the three-parameter Weibull distribution deterioration rate. The concept of two-component demand applies here for constant demand in its first part and a linear demand pattern in its rest part of the cycle. In the 1st part (time 0 to time μ), item remains unchanged with its quality and quantity; but in the 2nd part (time μ to time T), item deteriorates with its quality and/or quantity. The model is applicable for the items which are launched recently into the market. Initially, demand for such items remains constant for a certain period of time, and then, it increases linearly with time when they become popular. Finally, all the results are illustrated numerically.

2 Assumptions and Notations

The model is based on the following assumptions:

(i) The demand and deterioration rate follow two-component form and a three-parameter Weibull distribution deterioration rate, respectively. Demand is constant and a linear function of time in first and second component, respectively.
(ii) Lead time is zero, and replenishment rate is infinite.
(iii) Shortages are not considered.
(iv) There is no repair or replenishment of deteriorated units in the inventory system.
(v) All costs remain unaltered during the cycle under consideration.

3 Notations

The following notations are used for the development and results of the model:

$\theta(t)$:   Deterioration rate exhibits a three-parameter Weibull distribution, i.e., $\theta(t) = \alpha\beta(t - \gamma)^{\beta-1}$, $0 < \alpha \ll 1$, $\beta > 1$ & $\gamma > 0$ where α, β & γ are three-parameter Weibull density function.

$R(t)$:   Demand rate follows a two-component form, i.e., $R(t) = \begin{cases} D, & 0 \le t \le \mu, \\ D + D_0(t - \mu), & \mu < t \le T \end{cases}$. It shows a constant and a linearly increasing function of time in its first and second component, respectively.

$I(t)$:   On-hand inventory level.

T: Order cycle.

I_0:   Order quantity.

c_h: Holding cost/unit/unit of time.

c_o: Ordering cost/order.

c_d: Deterioration cost/unit/unit of time.

μ: Time point where the decay starts as well as demand rises linearly.

$\text{VSC}(T)$:   The variable system cost per unit time

4 Model Development

The inventory system starts initially with a lot size I_0 at time $t = 0$. The inventory level shrinks up to time $t = \mu$ due to constant demand pattern, and then, it ends at time $t = T$ due to both depletion and linearly increasing demand pattern.

During the time period $[0, \mu]$, the total demand (TD) is given by

$$TD = D\mu, \tag{1}$$

where D units per unit time are the constant demand rate.

The rest part (RP) of the inventory reserved for the period $[\mu, T]$ is given by

$$RP = I_0 - D\mu. \tag{2}$$

From Fig. 1, the area (A) of the trapezoidal region with height μ and two parallel sides I_0 and $I_0 - D\mu$ is given by

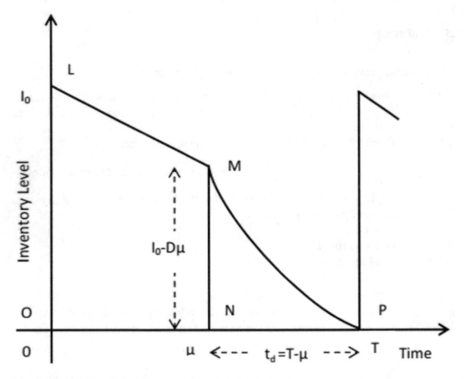

Fig. 1 Graphical representation of the inventory level with time

$$A = (I_0 - D\mu)\left(\mu + \frac{t_d}{2}\right) + \frac{D\mu^2}{2}, \tag{3}$$

with the help of $t_d = (T - \mu)$ as origin has been shifted for calculation.

The inventory level $I(t)$ at time t during the period $[\mu, T]$ is represented by the following first order linear differential equation:

$$\frac{dI(t)}{dt} + \theta(t)I(t) = -[D + D_0(t - \mu)], \quad 0 \leq t \leq t_d, \tag{4}$$

with the boundary condition $I(0) = I_0 - D\mu$ where $\theta(t) = \alpha\beta(t - \gamma)^{\beta-1}$, $0 < \alpha \ll 1$, $\beta > 1$ & $\gamma > 0$.

The solution of above differential equation with its boundary condition is given by

$$\begin{aligned}
I(t) &= (I_0 - D\mu)e^{\alpha[(-\gamma)^\beta - (t-\gamma)^\beta]} \\
&\quad + (D - D_0\mu)\left[\frac{\alpha}{\beta+1}\left((-\gamma)^{\beta+1} - (t-\gamma)^{\beta+1}\right) - t\right]e^{[-\alpha(t-\gamma)^\beta]} \\
&\quad + D_0(D - D_0\mu)\left[\alpha\left(\frac{\gamma(-\gamma)^{\beta+1}}{\beta+1} + \frac{(-\gamma)^{\beta+2}}{\beta+2}\right) - \frac{t^2}{2}\right]e^{[-\alpha(t-\gamma)^\beta]} \\
&\quad - \alpha D_0(D - D_0\mu)\left(\frac{\gamma(t-\gamma)^{\beta+1}}{\beta+1} + \frac{(t-\gamma)^{\beta+2}}{\beta+2}\right)e^{[-\alpha(t-\gamma)^\beta]}, 0 \leq t \leq t_d,
\end{aligned} \tag{5}$$

by ignoring the terms containing the higher power of α, $0 < \alpha \ll 1$.

The initial order quantity is obtained by putting $I(t_1) = 0$ in Eq. (5), i.e.,

$$\begin{aligned}
I_0 &= D\mu + (D - D_0\mu)\left[t_d + \frac{\alpha}{\beta+1}\left((t_1 - \gamma)^{\beta+1} - \alpha(-\gamma)^{\beta+1}\right)\right]e^{[-\alpha(-\gamma)^\beta]} \\
&\quad + D_0\left[\frac{t_d^2}{2} + \alpha\left(\frac{\gamma(t_1 - \gamma)^{\beta+1}}{\beta+1} + \frac{(t_1 - \gamma)^{\beta+2}}{\beta+2}\right)\right]e^{[-\alpha(-\gamma)^\beta]} \\
&\quad - \alpha D_0\left(\frac{\gamma(-\gamma)^{\beta+1}}{\beta+1} + \frac{(-\gamma)^{\beta+2}}{\beta+2}\right)e^{[-\alpha(-\gamma)^\beta]}.
\end{aligned} \tag{6}$$

Individual costs are evaluated before they are grouped together for the determination of variable system cost for the cycle $[0, T]$.

Ordering cost, ICO:

$$\text{ICO} = c_o. \tag{7}$$

Holding cost, ICH:

$$\text{ICH} = c_h \left[(I_0 - D\mu)\left(\mu + \frac{t_d}{2}\right) + \frac{D\mu^2}{2} \right], \tag{8}$$

[from Eq. (4)].

Deterioration cost, ICD:

$$\text{ICD} = c_d \left[(I_0 - D\mu) - \int_0^{t_d} [D + D_0(t - \mu)]dt \right]$$

$$= c_d \left[(I_0 - D\mu) - (D - D_0\mu)t_d - \frac{D_0 t_d^2}{2} \right]. \tag{9}$$

The variable system cost per unit time, VSC, in the period $[0, \ T]$ is comprised of the sum of the ordering cost, the holding cost, and the cost deterioration of items per unit time, i.e.,

$$\text{VSC}(T) = \frac{1}{T}[\text{ICO} + \text{ICH} + \text{ICD}] = \frac{c_o}{T}$$

$$+ \frac{(D - D_0\mu)}{T}\left[T - \mu + \frac{\alpha}{\beta+1}\left((T - \mu - \gamma)^{\beta+1} - \alpha(-\gamma)^{\beta+1}\right)\right]\left[\frac{c_h(T + \mu)}{2} + c_d\right]e^{\left[-\alpha(-\gamma)^\beta\right]}$$

$$+ \frac{D_0}{T}\left[\frac{(T - \mu)^2}{2} + \alpha\left(\frac{\gamma(T - \mu - \gamma)^{\beta+1}}{\beta+1} + \frac{(T - \mu - \gamma)^{\beta+2}}{\beta+2}\right)\right]\left[\frac{c_h(T + \mu)}{2} + c_d\right]e^{\left[-\alpha(-\gamma)^\beta\right]}$$

$$- \frac{\alpha D_0}{T}\left(\frac{\gamma(-\gamma)^{\beta+1}}{\beta+1} + \frac{(-\gamma)^{\beta+2}}{\beta+2}\right)\left[\frac{c_h(T + \mu)}{2} + c_d\right]e^{\left[-\alpha(-\gamma)^\beta\right]}$$

$$+ \frac{1}{2T}\left[c_h D\mu^2 - c_d(T - \mu)(2D - 3D_0\mu + D_0T)\right], \tag{10}$$

[from (6)].

5 Solution Procedures

Using calculus technique, the variable system cost per unit of time can be obtained from

$$\frac{d\text{VSC}(T)}{dT} = 0 \tag{11}$$

satisfying

$$\frac{d^2\text{VSC}(T)}{dT^2} > 0. \tag{12}$$

From Eqs. (11) and (12), we obtain

$$
\frac{dVSC(T)}{dT} = \frac{(D + D_0 T - 2D_0\mu)}{T}\left[1 + \alpha(T - \mu - \gamma)^\beta\right]\left[\frac{c_h(T + \mu)}{2} + c_d\right]e^{\left[-\alpha(-\gamma)^\beta\right]}
$$
$$
+ \frac{c_h(D - D_0\mu)}{2T}\left[T - \mu + \frac{\alpha}{\beta + 1}\left((T - \mu - \gamma)^{\beta+1} - \alpha(-\gamma)^{\beta+1}\right)\right]e^{\left[-\alpha(-\gamma)^\beta\right]}
$$
$$
+ \frac{c_h D_0}{2T}\left[\frac{(T - \mu)^2}{2} + \alpha\left(\frac{\gamma(T - \mu - \gamma)^{\beta+1}}{\beta + 1} + \frac{(T - \mu - \gamma)^{\beta+2}}{\beta + 2}\right)\right]e^{\left[-\alpha(-\gamma)^\beta\right]}
$$
$$
- \frac{\alpha c_h D_0}{2T}\left(\frac{\gamma(-\gamma)^{\beta+1}}{\beta + 1} + \frac{(-\gamma)^{\beta+2}}{\beta + 2}\right)e^{\left[-\alpha(-\gamma)^\beta\right]}
$$
$$
- \frac{1}{T}\left[c_d(D + D_0 T - 2D_0\mu) + VSC(T)\right] = 0, \tag{13}
$$

and

$$
\frac{d^2VSC(T)}{dT^2} = \frac{D_0}{T}\left[1 + \alpha(T - \mu - \gamma)^\beta\right]\left[\frac{c_h(T + \mu)}{2} + c_d\right]e^{\left[-\alpha(-\gamma)^\beta\right]}
$$
$$
+ \frac{\alpha\beta(D + D_0 T - 2D_0\mu)}{T}(T - \mu - \gamma)^{\beta-1}\left[\frac{c_h(T + \mu)}{2} + c_d\right]e^{\left[-\alpha(-\gamma)^\beta\right]}
$$
$$
+ \frac{c_h(D + D_0 T - 2D_0\mu)}{T}\left[1 + \alpha(T - \mu - \gamma)^\beta\right]e^{\left[-\alpha(-\gamma)^\beta\right]}
$$
$$
- \frac{2}{T^2}(D + D_0 T - 2D_0\mu)\left[1 + \alpha(T - \mu - \gamma)^\beta\right]\left[\frac{c_h(T + \mu)}{2} + c_d\right]e^{\left[-\alpha(-\gamma)^\beta\right]}
$$
$$
+ \frac{c_h(D - D_0\mu)}{T^2}\left[T - \mu + \frac{\alpha}{\beta + 1}\left((T - \mu - \gamma)^{\beta+1} - \alpha(-\gamma)^{\beta+1}\right)\right]e^{\left[-\alpha(-\gamma)^\beta\right]}
$$
$$
+ \frac{c_h D_0}{T^2}\left[\frac{(T - \mu)^2}{2} + \alpha\left(\frac{\gamma(T - \mu - \gamma)^{\beta+1}}{\beta + 1} + \frac{(T - \mu - \gamma)^{\beta+2}}{\beta + 2}\right)\right]e^{\left[-\alpha(-\gamma)^\beta\right]}
$$
$$
- \frac{\alpha c_h D_0}{T^2}\left(\frac{\gamma(-\gamma)^{\beta+1}}{\beta + 1} + \frac{(-\gamma)^{\beta+2}}{\beta + 2}\right)e^{\left[-\alpha(-\gamma)^\beta\right]}
$$
$$
+ \frac{1}{T^2}\left[2c_d(D + D_0 T - 2D_0\mu) + 2VSC(T) - D_0 c_d T\right]. \tag{14}
$$

6 Data and Application

Example 1 Consider the parametric values in appropriate units:
$[c_h, c_o, c_d, D, D_0, \mu, \alpha, \beta, \gamma] = [0.50, 18, 80, 20, 0.2, 0.4, 0.02, 2, 0.08]$.
In the numerical example and tables given hereafter, $(*)$ denotes optimal solutions. Solving Eq. (13), $T^* = 2.40772$ months satisfying Eq. (14) as $\frac{d^2VSC(T^*)}{dT^2} = 16.8766 > 0$. Substituting $T^* = 2.40772$ in Eqs. (6) and (10), the respective values of optimum order quantity and optimum system cost per month are $I_0^* = 49.3576$ units and $VSC^*(T) = Rs.\,52.7879$.

It is observed from Table 1 that (i) for increase in c_h and c_d, the values of both optimum cycle time and optimum order quantity increase, whereas optimum total cost decreases; (ii) for increase in c_o, the values of optimum cycle time, optimum total cost, and optimum order quantity increase; (iii) for increase in a, the values of optimum cycle time decreases, whereas both optimum total cost and optimum order

Table 1 Sensitivity analysis

Parameter	Change in parameter	T^*	% Change in T^*	% Change in VSC(T^*)	% Change in I_0^*
c_h	0.750	2.25673	−6.27108	+11.33000	−6.63768
	0.625	2.32926	−3.25868	+05.76293	−3.45904
	0.375	2.49287	+3.53654	−05.97639	+3.77895
	0.250	2.58546	+7.38209	−12.18610	+7.91935
c_o	120	2.76141	+14.68980	+29.2658	+15.88080
	100	2.59677	+07.85183	+15.1311	+08.42727
	60	2.18287	−09.33871	−16.4903	−09.85644
	40	1.89883	−21.13580	−35.0219	−22.07640
c_d	27.0	2.20938	−08.23767	+6.00271	−08.70322
	22.5	2.29823	−04.54746	+3.18271	−04.82114
	13.5	2.54835	+05.84080	−3.67118	+06.25598
	9.0	2.74055	+13.82350	−8.04503	+14.93020
D	30	2.10000	−12.78060	+16.32870	+29.6696
	25	2.23394	−07.21762	+08.62290	+15.3593
	15	2.64804	+09.98123	−09.94038	−16.7737
	10	3.01734	+25.31940	−21.98540	−35.6348
D_0	0.30	2.40064	−0.294054	+0.1448920	−0.0591601
	0.25	2.40416	−0.147858	+0.0727303	−0.0299853
	0.15	2.41132	+0.149519	−0.0731091	+0.0301879
	0.10	2.41495	+0.300284	−0.1465970	+0.0607809
μ	0.6	2.50021	+3.84139	−4.17801	+3.21247
	0.5	2.45346	+1.89972	−2.16676	+1.57666
	0.3	2.36307	−1.85445	+2.33305	−1.51304
	0.2	2.31956	−3.66156	+4.84376	−2.95861
α	0.030	2.20192	−8.54751	+6.19382	−8.33651
	0.025	2.29350	−4.74391	+3.29351	−4.62401
	0.015	2.55716	+6.20670	−3.83463	+6.04061
	0.010	2.76804	+14.9652	−8.47384	+14.5491
β	3.0	2.11874	−12.0022	+3.80357	−12.3511
	2.5	Complex no.	…	…	…
	1.5	Complex no.	…	…	…
	1.0	2.97440	+23.5360	−4.50568	+24.1750
γ	0.12	2.42596	+0.757563	−0.965002	+0.672642
	0.10	2.41680	+0.377120	−0.481081	+0.334903
	0.06	2.39873	−0.373382	+0.478240	−0.332066
	0.04	2.38983	−0.743027	+0.953828	−0.661094

Here "…" denotes the infeasible solution

quantity increase; (iv) for increase in a_0, α, and β, the values of both optimum cycle time and optimum order quantity decrease, whereas optimum total cost increases; (v) for increase in μ and γ, the values of both optimum cycle time and optimum order quantity increase, whereas optimum total cost decreases.

7 Conclusions

In the present paper, an economic order quantity (EOQ) inventory model for deteriorating items with two-component demand rate and a three-parameter Weibull distribution deterioration rate is studied. Shortages are not considered in this problem. A mathematical model is derived to determine the cycle time as the decision variable in order to calculate the optimum order quantity and optimum variable system cost. Finally, a numerical example is provided to verify the theoretical results.

There are numerous scopes in extending present model as the future work. The derived model can be extended in determining time-dependent demand patterns, stochastic demand pattern, product quantity, stock, and others. It can be extended form Weibull distribution deterioration rate to time-proportional deterioration rate. Finally, the present idea can be extended to allow for shortages, partial backlogging, quantity discounts, and others.

Acknowledgements The authors thank the editor and anonymous reviewers for their valuable and constructive comments, which have led to a significant improvement in the manuscript.

References

1. Ghare, P.M., Schrader, G.F.: A model for exponentially decaying inventory. J. Ind. Eng. **14**(5), 238–243 (1963)
2. Dave, U., Patel, L.K.: (T, Si) policy inventory model for deteriorating items with time proportional demand. J. Oper. Res. Soc. **32**(2), 137–142 (1981)
3. Sachan, R.S.: (T, Si) policy inventory model for deteriorating items with time proportional demand. J. Oper. Res. Soc. **35**(11), 1013–1019 (1984)
4. Covert, R.P., Philip, G.C.: An EOQ model for items with Weibull distribution deterioration. AIIE Trans. **5**(4), 323–326 (1973)
5. Singh, T., Pattnayak, H.: An EOQ inventory model for deteriorating items with varying trapezoidal type demand rate and Weibull distribution deterioration. J. Inf. Optim. Sci. **34**(6), 341–360 (2013)
6. Philip, G.C.: A generalized EOQ model for items with Weibull distribution deterioration. AIIE Trans. **6**(2), 159–162 (1974)
7. Singh, T., Mishra, P.J., Pattanayak, H.: An EOQ inventory model for deteriorating items with timedependent deterioration rate, ramp type demand rate and shortages. Int. J. Math. Oper. Res. (In press)
8. Goyal, S.K., Giri, B.C.: Recent trends in modeling of deteriorating inventory. Eur. J. Oper. Res. **134**(1), 1–16 (2001)
9. Li, R., Lan, H., Mawhinney, J.R.: A review on deteriorating inventory study. J. Serv. Sci. Manage. **3**(01), 117–129 (2010)

10. Janssen, L., Claus, T., Sauer, J.: Literature review of deteriorating inventory models by key topics from 2012 to 2015. Int. J. Prod. Econ. **182**, 86–112 (2016)
11. Hollter, R.H., Mak, K.L.: Inventory replenishment policies for deteriorating items in a declining market. Int. J. Prod. Res. **21**(6), 813–836 (1983)
12. Hariga, M.A., Benkherouf, L.: Optimal and heuristic inventory replenishment models for deteriorating items with exponential time-varying demand. Eur. J. Oper. Res. **79**(1), 123–137 (1994)
13. Goswami, A., Chaudhuri, K.S.: An EOQ model for deteriorating items with shortages and a linear trend in demand. J. Oper. Res. Soc. **42**(12), 1105–1110 (1991)
14. Chakrabarti, T., Chaudhuri, K.S.: An EOQ model for deteriorating items with a linear trend in demand and shortages in all cycles. Int. J. Prod. Econ. **49**(3), 205–213 (1997)
15. Kim, D.H.: A heuristic for replenishment of deteriorating items with linear trend in demand. Int. J. Prod. Econ. **39**(1), 265–270 (1995)
16. Ghosh, S.K., Chaudhuri, K.S.: An order-level inventory model for a deteriorating item with Weibull distribution deterioration, time-quadratic demand and shortages. Adv. Model. Optim. **6**(1), 21–35 (2004)
17. Sanni, S.S., Chukwu, W.I.: An inventory model with three-parameter Weibull deterioration, quadratic demand rate and shortages. Am. J. Math. Manage. Sci. **35**(2), 159–170 (2016)
18. Srivastava, M., Gupta, R.: EOQ Model for deteriorating items having constant and time-dependent demand rate. Opsearch **44**(3), 251–260 (2007)

Computational Study of Fluid Flow in Wavy Channels Using Immersed Boundary Method

Mithun Kanchan and Ranjith Maniyeri

Abstract Accurate control and handling of fluids in microfluidic-based bio-medical devices is very important in diverse range of applications such as laboratory-on-chip (LOC), drug delivery, and bio-technology. Flow through medical devices such as kidney dialyzer and membrane oxygenator can be considered as laminar due to low Reynolds number and narrow channel geometry, thus requiring efficient utilization of passive modulation systems to improve fluid mixing in these devices. In the present work, numerical investigation of fluid flow and passive mixing effects is carried out for wavy-walled channel configurations. A two-dimensional computational model based on an immersed boundary finite volume method is developed to perform numerical simulation on a staggered Cartesian grid system. Further, pressure–velocity coupling of governing continuity and Navier–Stokes equations describing the fluid flow is done by SIMPLE algorithm. Fluid variables are described by Eulerian coordinates and solid boundary by Lagrangian coordinates. Linking of these coordinate variables is done using Dirac delta function. A momentum-forcing term is added to the Navier–Stokes equation in order to impose the no-slip boundary condition on the wavy wall. Parametric study is carried out to analyze the fluid flow characteristics by varying wave geometry factor (WG Factor) of crest–crest (CC Model) wavy wall configurations for Reynolds number ranging from 10 to 50. From this work, it is evident that incorporating wavy-walled passive modulators prove to be good and robust method for enhancing mixing in biomedical devices.

Keywords Dirac delta function · Fractional step method · Immersed boundary method · Momentum-forcing · Passive mixing

M. Kanchan (✉) · R. Maniyeri
Department of Mechanical Engineering, National Institute of Technology Karnataka, Surathkal, Mangalore 575025, India
e-mail: mranji1@nitk.edu.in

© Springer Nature Singapore Pte Ltd. 2019
J. C. Bansal et al. (eds.), *Soft Computing for Problem Solving*,
Advances in Intelligent Systems and Computing 816,
https://doi.org/10.1007/978-981-13-1592-3_22

1 Introduction

Mass transport enhancement by mixing is an important aspect in modern-day biomedical devices. Hydrodynamic instabilities can be produced by introducing eddy promoters or passive modulation systems to improve mixing [1]. Inducing chaotic mixing effects in such devices is difficult. Since mixing is achieved primarily through diffusion, long micro-channels are required in order to achieve satisfactory mixing results. This also leads to slower blending of fluids [2]. Maintaining a diminutive scale of these devices and improving its mixing efficiency is a major challenge researchers are facing today. Active perturbations in fluid flow created by employing mechanical components or passive chaotic mixing induced by modifying the geometry of the mixer devices are some of the techniques used to overcome this problem. Recently, wavy channels have been introduced in low Reynolds number and creeping flow conditions which are pertaining to microfluidics and application of fluid mixing such as laboratory-on-chip and micro-evaporators [3]. For moderate Reynolds number (20–300), Tatsuo et al. [4] conducted experimental investigation on flow and mass transfer in symmetric wavy wall channels. They demonstrated that flow separation phenomenon was observed in wavy- and arc-shaped channels, thus exhibiting early transition to turbulence at low Reynolds number. Wang et al. [5] numerically studied converging–diverging channels based on experimental work of Tatsuo et al. [4] for steady and unsteady flow condition. Compared to straight channels, wavy channels produce self-sustaining oscillations which destabilize laminar boundary layers, thus improving heat transfer and mixing capabilities. Bahaidarah et al. [6] also performed numerical simulation of wavy-walled channels for Reynolds number ranging from 25 to 400. The work focuses on improving computational effort by demonstrating the adequacy of using a single wavy module as computational domain. Periodic boundary conditions which use stream-wise constant pressure drop were used in these works. The effects of phase shift of wavy-walled channels on the hydraulic performance were investigated by Ahmed et al. [7]. The sinusoidal wavy channel with 0° phase shift provided the best thermo-hydraulic performance for low Reynolds number. Finite volume simulations were performed by Aslan et al. [8] to study comparison of experimental and numerical results of wavy-walled channels. It is found that the performance of channels with rounded wavy peaks is better in comparison with sharp peaks. The work is concluded with the observation that finite volume-based numerical predictions are close to experimental results. Both the works highlight the importance of using sinusoidal shaped wavy-walled channels with no phase shifts for the present numerical study. Based on the examination of all experimental and numerical works, it can be inferred that introducing wavy-walled channels improve mixing performance at moderate Reynolds number. However, simulating flow behavior of complex geometries such as wavy channels is a computationally intensive procedure. This paper emphasis on applying a computational technique called immersed boundary method to perform numerical simulations at relatively less computational effort and time.

Charles Peskin in 1972 developed immersed boundary method (IBM) to study cardio-mechanism [9]. A unique feature about this method is that the technique is based on non-conforming mesh methods and uses a monolithic fluid structure solver. IBM is a single Cartesian grid-based method and does not require body-fitted grid generation for complex structures, thus resulting in the absence of grid transformation terms. This significantly reduces the per-grid-point operation count [10]. Also in IBM, the grid does not conform to the solid boundary, and hence, the presence of boundary or imposition of boundary conditions is done by modifying the governing equation in the vicinity of solid boundary. IBM uses Eulerian coordinates to describe fluid flow and Lagrangian coordinates to describe solid boundary. Since the development of this method by Peskin, numerous modifications and refinements have been proposed based on different application requirements. IBM can be broadly classified as continuous and discrete forcing approaches. In the former, a force term is added to the continuous Navier–Stokes equation before discretization, whereas in the latter, the forcing is applied either implicitly or explicitly to the discretized equation. Elastic forces, bending resistance, or any other type of structural behavior can be built into the forcing term. Continuous forcing approach as originally introduced by Peskin is suitable for elastic boundaries [11]. However, another approach called the virtual boundary method was developed by Goldstein et al. [12] to simulate rigid body flows. A feedback-forcing scheme is employed to impose the no-slip boundary condition on the rigid boundary in the fluid. A specific version of this technique which considers the structure to be elastic but extremely stiff was developed by Lai and Peskin [13]. The numerical interpretation of this method is that boundary points are secured to a fixed equilibrium position by very stiff springs having constant stiffness. This is referred to as tether forcing and has been successfully used to simulate flow past cylinders at moderate Reynolds number. The present work primarily focuses on this technique.

In this paper, we perform numerical simulation to study Poiseuille flow in a wavy channel with crest-crest (CC Model) wave configuration using continuous forcing-based immersed boundary finite volume method. SIMPLE algorithm on a staggered Cartesian grid system is used to solve continuity and Navier–Stokes equation governing the fluid flow. A FORTRAN code is developed to capture the flow behavior for different Reynolds number ranging from 10 to 50.

2 Mathematical Modeling and Numerical Procedure

The schematic representation of CC Model, i.e., crest and crest of the two wavy walls lying on the same unit length section, is shown in Fig. 1. The length of the channel is taken as L and height H. The amplitude of wavy wall is given as A and wavelength λ.

The immersed boundary is derived from variation principle called "Principle of least action." Two distinct set of grids are required. One is a Cartesian grid which covers the entire fluid domain and other immersed boundary points which discretize

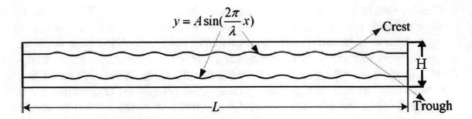

Fig. 1 Schematic illustration of CC Model

the structure. The boundary force is described on these immersed boundary points. The Eulerian grid points are fixed but the Lagrangian boundary points move with respect to fluid flow and they usually do not coincide with each other. The continuity and Navier–Stokes equations in dimensionless form are given by Eqs. (1) and (2).

$$\nabla.\mathbf{u} = 0 \tag{1}$$

$$\frac{\partial \mathbf{u}}{\partial t} + \mathbf{u}.\nabla \mathbf{u} = -\nabla p + \frac{1}{Re}\nabla^2 \mathbf{u} + \mathbf{f}(\mathbf{x}) \tag{2}$$

where $\mathbf{u}$ is the velocity of fluid, p the pressure of fluid, and Re the Reynolds number. The flow-governing equations are non-dimensionalized by using channel height as characteristic length and inlet fluid velocity as characteristic velocity. Periodic boundary condition is applied in the positive x-direction. The flow is driven by a constant pressure gradient of $\frac{12}{Re}$. No-slip boundary conditions are used at the bottom and top walls. The boundary force $\mathbf{F^n}(\mathbf{s})$ is computed from Eq. (3) and applied to the fluid in Eq. (4).

$$\mathbf{F^n}(\mathbf{s}) = S\big(X_e(s) - X^n(s)\big) \tag{3}$$

$$\mathbf{f^n}(\mathbf{x}) = \sum_s \mathbf{F^n}(\mathbf{s})\delta\big(x - X^n(s)\big)\Delta s \tag{4}$$

where discrete delta function is given by Eq. (5), $X_e(s)$ is the equilibrium position, and S is the stiffness constant.

$$\delta(x) = d_h(x)d_h(y) \tag{5}$$

and

$$
d_h(r) = \begin{cases} \frac{1}{8h}\left(3 - \frac{2|r|}{h} + \sqrt{1 + \frac{4|r|}{h} - 4\left(\frac{|r|}{h}\right)^2}\right), & |r| \leq h \\[4mm] \frac{1}{8h}\left(5 - \frac{2|r|}{h} - \sqrt{-7 + \frac{12|r|}{h} - 4\left(\frac{|r|}{h}\right)^2}\right), & h \leq |r| \leq 2h \\[4mm] 0, & \text{otherwise} \end{cases}
$$

$$
\frac{u^* - u^n}{\Delta t} = -\nabla p^* - u^* \nabla u^* + \frac{1}{Re}\nabla^2\left(u^*\right) + f^n(x) \tag{6}
$$

where h is Eulerian grid size, and $u^*(x)$ is the intermediate velocity. The implicit temporal form of Navier–Stokes equation is given in Eq. (6). The pressure–velocity coupling is done using SIMPLE algorithm. The scheme is fully implicit in nature and pressure of the fluid is initially assumed. A second-order differencing technique called the method of deferred correction is used to evaluate nonlinear convection terms and linear diffusion terms. The algebraic equations generated after discretization are solved using line-by-line Thomas algorithm. In order to satisfy the continuity equation given in Eq. (1), a pseudo-pressure term is used to correct the velocity field at each computational step. The new velocity from the Eulerian grid is interpolated into the immersed boundary point given by Eq. (7) and boundary point is moved to new position $X^{n+1}(s)$ given by Eq. (8).

$$
\mathbf{U^{n+1}}(\mathbf{s}) = \sum_x \mathbf{u^{n+1}}(\mathbf{x})\delta\left(x - X^n(s)\right)\Delta x^2 \tag{7}
$$

$$
\frac{X^{n+1} - X^n}{\Delta t} = \mathbf{U^{n+1}}(\mathbf{s}) \tag{8}
$$

A more elaborate discussion regarding IBM and its implementation can be found in [14–16].

3 Results and Discussion

The governing equations are solved on a two-dimensional rectangular dimensionless domain of 4×1. Numerical results related to flow over wavy channels are obtained by developing a FORTRAN code. The Eulerian grid size is taken as 251×101 after conducting an extensive grid refinement study. A total of 501 Lagrangian points (IB points) are considered for the discretization of wavy wall structures. The validation of the present code is done in two parts. Firstly, two virtual rigid walls are placed in the channel at a distance of 0.75 and 0.25. Immersed boundary method enforces the no-slip boundary condition on the walls. Simulations are carried out until steady-state solution is reached for Reynolds number 20. Figure 2 shows the velocity vector plot for this case. From the figure, two smaller and one larger flow field is observed, thus

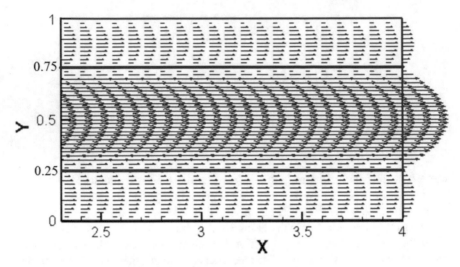

Fig. 2 Velocity vector plot for straight wall in channel at *Re* 20

Fig. 3 Velocity profile comparison between experimental work of Gong et al. [17] and present numerical work

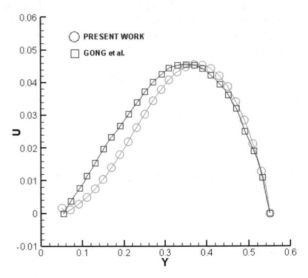

clearly depicting flow separation. This serves as first part of validation for the developed code. Second part of the validation consists of comparing the results obtained from the IBM code with the experimental work of Gong et al. [17]. Results are compared for a test condition of amplitude $A = 0.2$ and wavelength $\lambda = 2.0$ for Reynolds number 20. Figure 3 shows the comparison of velocity profile between experimental and numerical values. A good agreement is established between experimental and numerical results.

Simulations are carried for four different Reynolds number, i.e., $Re = 10, 20, 30,$ and 50. In this paper, the numerical simulations are mainly presented for $Re = 50$.

Table 1 Parametric study variables

Model number	Configuration	WG factor or (A/λ) ratio
Straight line	$A=0, \lambda=0.0$	–
1	$A=0.1, \lambda=1.0$	0.1
2	$A=0.1, \lambda=2.0$	0.05
3	$A=0.2, \lambda=1.0$	0.2
4	$A=0.2, \lambda=2.0$	0.1

A geometry linked parameter called WG Factor is introduced which is defined as the ratio of wave amplitude to wavelength. For the case of Reynolds number 50, velocity contour and vector plots for different models are given by Fig. 4. Also, Table 1 depicts the various wavy configurations used in the present study and WG Factors (wave geometry parameter) generated from these models.

From Fig. 4, it is observed that the flow is streamlined and contoured along the wavy structures. A fully developed Poiseuille-type flow behavior is exhibited. Any significant form of vortices or regeneration pattern is not seen since the flow is dominated by viscous forces. Velocity vector contours from the present work are found to be identical with the results obtained by Gong et al. [3], Ahmed et al. [18], and Zontul et al. [19].

Figure 5a, b show the velocity profile for Models 1–4 for Reynolds number 50. A biased shift of centerline velocity toward the wave peaks is observed. This is an important characteristic feature found in the study of wavy-walled channels for low and moderate Reynolds numbers. Such observations are seen in the works of Husain et al. [20] and Mills et al. [21]. Maximum velocity deviation from the centerline is observed for Model 1 and Model 3 owing to larger amplitudes with respect to wavelength. Centerline velocity inclined toward wave peaks of the wall with parabolic-type profile resemblance. The main factor governing this behavior is the geometry of the wavy channel. This results in thinning of hydrodynamic and thermal boundary layers which significantly enhance thermal and mixing performance. Another important factor is the effect of maximum velocity on mixing enhancement. Narrow contracting sections produce larger pressure drops which lead to increase in centerline velocity and eventually hampering mixing capabilities of the model taken into consideration [22]. Models 1 and 4 have WG Factor of 0.1 and thus produce lower centerline velocity, whereas Model 2 and Model 3 having WG Factor 0.05 and 0.2 produce large velocities as shown in Fig. 5a, b. Thus, an optimum value of WG Factor exists for which the best possible mixing performance can be obtained. It also provides us with a simple yet powerful technique to evaluate geometries of wavy channels when considering its suitability for experiments.

Figure 6 shows the variation of maximum velocity for increasing Reynolds number. Maximum velocity for Model 2 and Model 3 keeps increasing with Reynolds number depicting larger pressure drops. Such kind of pressure drops is not observed for Model 1 and Model 4. A straight wall in channel is compared with various wavy-

Fig. 4 Velocity contour and
vector plots for various wavy
channel configurations at Re
50

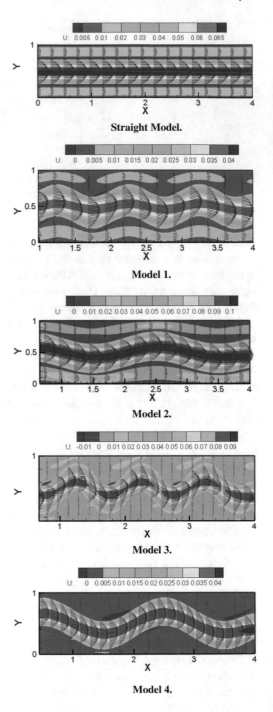

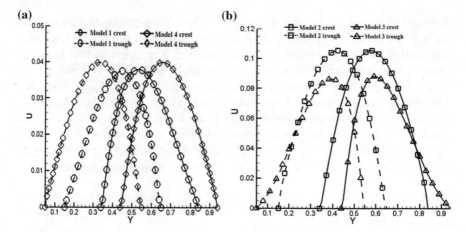

Fig. 5 **a** Velocity distribution along *y*-axis for CC Model for Model 1 and Model 4. **b** Velocity distribution along *y*-axis for CC Model for Model 2 and Model 3

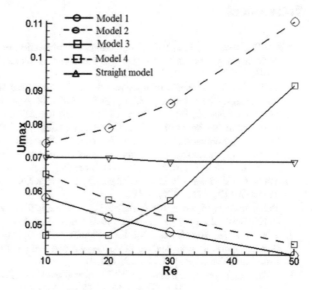

Fig. 6 Variation of maximum velocity with Reynolds number

walled geometric models. It shows almost constant velocity variation with respect to Reynolds number.

4 Conclusion

The present work focuses on developing a two-dimensional computational model based on an immersed boundary finite volume method to perform numerical simulation of wavy-walled channels in order to study passive mixing effects and flow behavior. Numerical simulation of such complex wavy wall geometries is successfully carried out using immersed boundary method. The effect of WG Factor on mixing performance is studied in this work. Configurations having WG Factor in the range of 0.1 are preferred. From our analysis, we conclude that CC Model-type wavy wall channel having geometry $A = 0.1$ and $\lambda = 1.0$ perform the best in terms of mixing enhancement in channel flow with wavy walls. Wavy walls are probably the best possible alternatives to straight micro-channels with respect to mixing performance improvements. The results obtained will be helpful in developing or optimizing the wavy wall channel-based microfluidic devices for biomedical applications.

Acknowledgements This research was supported by Science and Engineering Research Board, a statutory body of Department of Science and Technology (DST), Government of India through the funded project ECR/2016/001501.

References

1. Sui, Y., Teo, C.J., Lee, P.S.: Direct numerical simulation of fluid flow and heat transfer in periodic wavy channels with rectangular cross-sections. Int. J. Heat Mass Transf. **55**(1), 73–88 (2012)
2. Cho, C.C.: A combined active/passive scheme for enhancing the mixing efficiency of microfluidic devices. Chem. Eng. Sci. **63**(12), 3081–3087 (2008)
3. Gong, L., Kota, K., Tao, W., Joshi, Y.: Parametric numerical study of flow and heat transfer in microchannels with wavy walls. J. Heat Transf. **133**(5), 051702 (2011)
4. Tatsuo, N., Shinichiro, M., Shingho, A., Yuji, K.: Flow observations and mass transfer characteristics in symmetrical wavy-walled channels at moderate Reynolds numbers for steady flow. Int. J. Heat Mass Transf. **33**(5), 835–845 (1990)
5. Wang, G.V., Vanka, S.P.: Convective heat transfer in periodic wavy passages. Int. J. Heat Mass Transf. **38**(17), 3219–3230 (1995)
6. Bahaidarah, H.M., Anand, N.K., Chen, H.C.: Numerical study of heat and momentum transfer in channels with wavy walls. Numer. Heat Transf. Part A **47**(5), 417–439 (2005)
7. Ahmed, M.A., Yusoff, M.Z., Ng, K.C., Shuaib, N.H.: The effects of wavy-wall phase shift on thermal-hydraulic performance of Al_2O_3–water nanofluid flow in sinusoidal-wavy channel. Case Stud. Therm. Eng. **4**, 153–165 (2014)
8. Aslan, E., Taymaz, I., Islamoglu, Y.: Finite volume simulation for convective heat transfer in wavy channels. Heat Mass Transf. **52**(3), 483–497 (2016)
9. Peskin, C.S.: Flow patterns around heart valves: a digital computer method for solving the equations of motion. IEEE Trans. Biomed. Eng. **BME-20**(4), 316–317 (1973)
10. Mittal, R., Iaccarino, G.: Immersed boundary methods. Annu. Rev. Fluid Mech. **37**, 239–261 (2005)
11. Peskin, C.S.: The immersed boundary method. Acta numerica **11**, 479–517 (2002)
12. Goldstein, D., Handler, R., Sirovich, L.: Modeling a no-slip flow boundary with an external force field. J. Comput. Phys. **105**(2), 354–366 (1993)
13. Lai, M.C., Peskin, C.S.: An immersed boundary method with formal second-order accuracy and reduced numerical viscosity. J. Comput. Phys. **160**(2), 705–719 (2000)
14. Maniyeri, R., Kang, S.: Numerical study on bacterial flagellar bundling and tumbling in a viscous fluid using an immersed boundary method. Appl. Math. Model. **38**(14), 3567–3590 (2014)

15. Maniyeri, R., Kang, S.: Numerical study on the rotation of an elastic rod in a viscous fluid using an immersed boundary method. J. Mech. Sci. Technol. **26**(5), 1515–1522 (2012)
16. Maniyeri, R., Suh, Y.K., Kang, S., Kim, M.J.: Numerical study on the propulsion of a bacterial flagellum in a viscous fluid using an immersed boundary method. Comput. Fluids **62**, 13–24 (2012)
17. Gong, L.J., Kota, K., Tao, W., Joshi, Y.: Thermal performance of microchannels with wavy walls for electronics cooling. IEEE Trans. Compon. Packag. Manuf. Technol. **1**(7), 1029–1035 (2011)
18. Ahmed, M.A., Yusoff, M.Z., Shuaib, N.H.: Numerical investigation on the nanofluid flow and heat transfer in a wavy channel. In: Engineering Applications of Computational Fluid Dynamics. Springer International Publishing, pp. 145–167 (2015)
19. Zontul, H., Kurtulmuş, N., Şahin, B.: Pulsating flow and heat transfer in wavy channel with zero degree phase shift. Eur. Mech. Sci. **1**(1), 31–38 (2017)
20. Husain, A., Kim, K.Y.: Thermal transport and performance analysis of pressure-and electroosmotically-driven liquid flow microchannel heat sink with wavy wall. Heat Mass Transf. **47**(1), 93–105 (2011)
21. Grant Mills, Z., Shah, T., Warey, A., Balestrino, S., Alexeev, A.: Onset of unsteady flow in wavy walled channels at low Reynolds number. Phys. Fluids **26**(8), 084104 (2014)
22. Ramgadia, A.G., Saha, A.K.: Numerical study of fully developed flow and heat transfer in a wavy passage. Int. J. Therm. Sci. **67**, 152–166 (2013)

A Novel Approach to Handle Forecasting Problems Based on Moving Average Two-Factor Fuzzy Time Series

Abhishekh, S. K. Bharati and S. R. Singh

Abstract In this paper, we present a novel approach to handling forecasting problems based on moving average in two-factor fuzzy time series. The proposed method defines a new technique to partition the universe of discourse into number of intervals based on the number of observations available in the historical time series data. Partition of interval depends on the transformed moving average time series data rather than actual time series data sets. Further, triangular fuzzy set is defined for transformed moving average data set and obtained membership grades of each moving average datum to their corresponding triangular fuzzy sets. Also, variation data set is calculated from transformed moving average data sets to define second factor data set. Further, frequency occurrence of fuzzy logical relationships is used in defuzzification process. The proposed method of moving average forecasting is verified and certified with three different fuzzy time series models. The robustness of proposed method is implemented in forecasting of Bombay Stock Exchange (BSE) Sensex historical data and compared in terms of different statistical error which indicates that the proposed method can provide more accurate forecasted values over with existing fuzzy time series models.

Keywords Fuzzy time series · Triangular fuzzy number (TFN)
Moving average · Variation · Two-factor FLRGs · BSE Sensex data sets

Abhishekh (✉) · S. R. Singh
Department of Mathematics, Institute of Science, Banaras Hindu University,
Varanasi 221005, India
e-mail: abhibhu1989@gmail.com

S. R. Singh
e-mail: srsingh@bhu.ac.in

S. K. Bharati
Department of Mathematics, Kamala Nehru College,
University of Delhi, New Delhi 110049, India
e-mail: skmaths.bhu@gmail.com

© Springer Nature Singapore Pte Ltd. 2019
J. C. Bansal et al. (eds.), *Soft Computing for Problem Solving*,
Advances in Intelligent Systems and Computing 816,
https://doi.org/10.1007/978-981-13-1592-3_23

295

1 Introduction

In the literature, various forecasting models have been implemented on university enrollments, temperature prediction, weather forecasting, and stock index forecasting that are based on fuzzy time series. The major objective of many researchers is to provide better accuracy rates in the forecasted outputs in the field of fuzzy time series. The fuzzy set theory introduced by Zadeh [1] was first applied by Song and Chissom [2, 3] to handle imprecise data sets, and his forecasting model was implemented on university enrollments of Alabama. Fuzzy time series has been further studied by Chen [4] who presented a simplified forecasting model based on centroid method to enhance the accuracy in forecasted outputs. Further, many researchers [5, 6] developed time series forecasting models in fuzzy environment and tried to provide better forecasting accuracy rates.

During the last few decades, some authors like Huarng [7] introduced that different length and number of intervals may affect the accuracy in forecasted outputs and implemented his model to forecast the university enrollment of Alabama. Yu [8] presented a weighted fuzzy time series model for TAIEX forecasting. However, Huarng and Yu [9] proposed a type 2 fuzzy time series model, and Huarng and Yu [10] developed a method of ratio-based length of intervals to enhance in forecasted outputs. Further, Yolcu et al. [11] gave a new approach for determining the length of intervals in fuzzy time series. After generating the length of intervals, the major objective is how to fuzzify the historical real-valued time series data, because it affects to create fuzzy logical relationships (FLRs) of fuzzified data sets. Many researchers used triangular shape membership grades as 0, 0.5, and 1 without giving difference in degree. To resolve this problem of data fuzzification, we define a different membership grade for each observation corresponding to their defined triangular fuzzy sets. In this study, the data fuzzification process is based upon their maximum membership grades of that observation in corresponding triangular fuzzy sets.

In order to make the fuzzy time series forecasting problem more practical, Lee et al. [12] introduced a forecasting method based on two-factor high-order fuzzy time series to enhance the forecasted accuracy rates. Chen and Tanuwijaya [13] presented a novel approach of automatic clustering techniques for high-order fuzzy time series forecasting. The time series models [12, 13] are based on concept of factors fuzzy time series model. Generally, moving average is used as a statistical technique in various methods. We shall now apply this moving average technique in a forecasting model. In order to improve the forecasting accuracy, Sulandari and Yudhanto [14] introduced a forecasting method using a hybrid simple moving average weighted fuzzy time series model. Recently, Abhishekh and Kumar [15] proposed a computational method for rice production high-order fuzzy time series forecasting to enhance the forecasted outputs.

The motivation of this paper is to develop a novel approach to handling forecasting problems based on moving average in two-factor fuzzy time series data. To prove the performance of the developed method, we use the historical time series Sensex data set of Bombay Stock Exchange. The proposed method defines a new technique to partition the universe of discourse into number of overlapping intervals based on the number of observation available in transformed moving average time series data set. In this study, actual Sensex data is not used to define the universe of discourse while transformed moving average data sets are used in partition observed from actual data set. These moving average data sets are examined as a main factor historical time series data sets. After this, variation data set is calculated from moving average data set. These variation data sets are examined as a second factor historical time series data set. Furthermore, we generate a fuzzy logical relationship groups (FLRGs) based on two-factor fuzzy time series. After generating the two-factor FLRGs, we measure the frequency occurrence of FLR in FLRGs, and to overcome the difficulty, we presented two heuristic rule for defuzzification process providing a crisp output of higher forecasting accuracy rates over the existing models of fuzzy time series in the literature.

The major objective of this study is to propose a novel approach to handle forecasting problems based on moving average in two-factor fuzzy time series to improve in forecasted accuracy rates. An outline of the rest of the paper is organized as follows: Sect. 2 reviews the basic concept of fuzzy time series and moving average. Section 3 proposes a novel approach for moving average forecasting based on two-factor FLRGs. Section 4 consists of verification and comparison of the proposed method and has been implemented to forecast the BSE Sensex historical data sets. The conclusion is discussed in Sect. 5.

2 Basic Preliminaries

This section briefly summarizes the basic review of fuzzy time series and moving average concepts. Fuzzy time series is another approach to solve forecasting problems in which the historical real-valued data sets are given in the form of linguistic variables. The fuzzy time series was firstly interpreted by Song and Chissom [2, 3] based on the fuzzy set theory [1], and some definitions are given below.

Definition 1 If X is a collection of objects denoted generically by x, then a fuzzy set $\tilde{A}$ in X is a set of ordered pairs

$$\tilde{A} = \left\{ \left(x, \mu_{\tilde{A}}(x) \right) \middle| x \in X \right\} \tag{1}$$

where $\mu_{\tilde{A}}(x)$ is called the membership function or grade of membership (also degree of compatibility or degree of truth) of x in $\tilde{A}$ that maps X to the membership space M (M is generally taken to be [0, 1]).

Definition 2 A fuzzy set $\tilde{A}$ on real line $\mathbb{R}$ with membership function $\mu_{\tilde{A}}(x) : \mathbb{R} \to$ [0, 1] is called a fuzzy number, and a fuzzy number $\tilde{A} = [a, b, c]$ determined by the triplet is said to be a triangular fuzzy number (TFN), if its membership function is defined as (in Fig. 1).

Fig. 1 Membership functions of a TFN

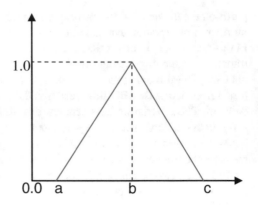

$$\mu_A(x) = \begin{cases} \frac{x-a}{b-a} & \text{if } a \leq x \leq b \\ \frac{c-x}{c-b} & \text{if } b \leq x \leq c \\ 0 & \text{otherwise} \end{cases} \tag{2}$$

Definition 3 Assume that $Y(t)$ ($t = \ldots, 0, 1, 2, \ldots$), a subset of real numbers, be the universe of discourse on which fuzzy sets $f_i(t)$ ($i = 1, 2, \ldots$) are defined. If $F(t)$ is a collection of $f_i(t)$ ($i = 1, 2, \ldots$), then $F(t)$ is called a fuzzy time series on $Y(t)$ ($t = \ldots, 0, 1, 2, \ldots$).

Definition 4 If there exists a fuzzy relation $R(t - 1, t)$ between $F(t - 1)$ and $F(t)$, both $F(t - 1)$ and $F(t)$ are fuzzy sets, such that the relationship can be shown as $F(t) = F(t - 1) \circ R(t - 1, t)$, then $F(t)$ is said to be derived from $F(t - 1)$, represented as $F(t - 1) \rightarrow F(t)$, where '$\circ$' represents the max–min composition operator of fuzzy sets.

Definition 5 Assume that $F(t)$ is derived from $F(t - 1)$ only and is denoted by $F(t - 1) \rightarrow F(t)$, then there is a fuzzy relationship between $F(t)$ and $F(t - 1)$ which is expressed as

$$F(t) = F(t - 1) \circ R(t - 1, t). \tag{3}$$

This relation R is referred to as a first-order model of $F(t)$. If $R(t - 1, t)$ is independent of time t, that is for different time t_1 and t_2, $R(t_1 - 1, t_1) = R(t_2 - 1, t_2)$, then $F(t)$ is called a time-invariant fuzzy time series. Otherwise, it is called a time-variant fuzzy time series.

Definition 6 Assume that $F(t - 1) = \tilde{A}_i$ and $F(t) = \tilde{A}_j$, then fuzzy logical relationship (FLR) can be defined as $\tilde{A}_i \rightarrow \tilde{A}_j$ where $\tilde{A}_i$ is called the current state and $\tilde{A}_j$ is called the next state of fuzzy logical relationship (FLR), respectively. If $F(t)$ is derived by more fuzzy sets $F(t - n), F(t - n + 1), \ldots, F(t - 1)$

then the fuzzy relationship can be defined as $\tilde{A}_{i1}, \tilde{A}_{i2}, \ldots, \tilde{A}_{in} \rightarrow \tilde{A}_j$, where $F(t-n) = \tilde{A}_{i1}, F(t-n+1) = \tilde{A}_{i2}, \ldots, F(t-n) = \tilde{A}_{in}$. The relationship is called nth-order fuzzy time series model.

Definition 7 Assume that $F(t)$ be a fuzzy time series $(t = \ldots, 0, 1, 2, \ldots)$, where the value of $F(t)$ is represented by a fuzzy set. If $F(t)$ is derived by $(F_1(t-1), F_2(t-1))$, then fuzzy logical relationship between them can be represented by $(F_1(t-1), F_2(t-1)) \rightarrow F(t)$, and it is called the two-factor first-order fuzzy time series forecasting model. Here $(F_1(t-1), F_2(t-1))$ and $F(t)$ are called the current state and the next state of FLR, respectively.

Definition 8 Assume that $F(t)$ be a fuzzy time series $(t = \ldots, 0, 1, 2, \ldots)$, where the value of $F(t)$ is represented by a fuzzy set. If $F(t)$ is derived by $(F_1(t-n), F_2(t-n)), \ldots, (F_1(t-2), F_2(t-2)), (F_1(t-1), F_2(t-1))$, then fuzzy logical relationship between them can be represented by $(F_1(t-n), F_2(t-n)), \ldots, (F_1(t-2), F_2(t-2)), (F_1(t-1), F_2(t-1)) \rightarrow F(t)$ and it is called the two-factor nth-order fuzzy time series forecasting model. Here $(F_1(t-n), F_2(t-n)), \ldots, (F_1(t-2), F_2(t-2)), (F_1(t-1), F_2(t-1))$ and $F(t)$ are called the current state and the next state of the nth-order FLR, respectively.

Definition 9 A moving average (MA) is a computation to analyze data points by making a series of averages of different subsets of the total data sets. Then, the subset is modified by 'shifting forward' that is excluding the first number of the series. This creates a new subset of numbers, which is averaged. This process is repeated over the entire data series. Every type of moving average (MA) is a mathematical result that is calculated by averaging a number of past data observations. The plot line connecting all the (fixed) averages is the moving average. In this study for applying the method of moving average, the period of moving average has to be selected. The period may be a two-yearly moving average, three-yearly moving average, four-yearly moving average, etc. For example, two-yearly moving average can be calculated from the data: a, b, c, d, e, f can be computed as

$$\frac{a+b}{2}, \frac{b+c}{2}, \frac{c+d}{2}, \frac{d+e}{2}, \frac{e+f}{2}$$

A moving average of period m can be mathematically expressed as

$$M_t = \frac{x_{t-1} + x_{t-2} + x_{t-3} + \cdots + x_{t-n}}{n} \tag{4}$$

where M_t, is a moving average for the coming period at the point t, n is the number of period for which the moving average is calculated, $x_{t-1}, x_{t-2}, x_{t-3}, \ldots, x_{t-n}$ are the actual data observations in the past period, two periods ago, three periods ago, and so on n periods ago, respectively.

3   Development of Proposed Method for Forecasting Based on Moving Average Two-Factor Fuzzy Time Series

This section presents the stepwise procedure of the proposed method of forecasting using moving average approach for fuzzy time series based on real-valued historical time series data set.

Step 1   First, collect the historical time series data. Further, calculate two-period moving average data sets using moving average approach (as defined in Definition 9) from actual observations of data set. So we have a collection of two-period moving average historical data sets observed from actual observation of data set.

Step 2   Assume transformed moving average data sets as a main factor historical data set. Now define the universe of discourse U of the main factor based on the range of available moving average historical data set as $U = [D_{\min} - D_1, D_{\max} + D_2]$, where $D_{\min}$ and $D_{\max}$ are minimum and maximum value of the moving average historical data set, and D_1 and D_2 are two proper positive assigned real number for easy partitioning the universe of discourse U. In order to partition the universe of discourse U into length of equal intervals, the length of interval is determined as

$$u = \frac{D_{\max} - D_{\min}}{k} \tag{5}$$

where k is the number of observations available in the main factor (moving average data set) historical time series data and u is the length of each partitioned intervals. As the length of intervals is determined, U can be partitioned into $k+1$ equal length of intervals $u_i, i = 1, 2, \ldots, k, k+1$ based on the number of observation available in the moving average historical data set.

Step 3   Now construct the triangular fuzzy sets $\tilde{A}_i, i = 1, 2, \ldots, k, k + 1$ corresponding to the interval $u_i, i = 1, 2, \ldots, k, k + 1$ defined in step 2 and obtained triangular membership grades to each moving average datum in their corresponding triangular fuzzy sets.

Step 4   To fuzzify each historical time series datum of the main factor (moving average data set), we choose maximum degree of membership grades (as defined in step 3) of each datum to $\tilde{A}_i, (i = 1, 2, \ldots, k, k + 1)$, and then we fuzzify that moving average datum to $\tilde{A}_i$ for maximum membership grades available in triangular fuzzy sets $\left(\tilde{A}_1, \tilde{A}_2, \ldots, \tilde{A}_{k+1} \right)$ of x_k for fixed k. x_k is the kth-observations datum of the main factor historical time series data.

Step 5   Find the variation as $\Delta(t - 1) = \{ f(t - 1) - f(t - 2) \}$ for all time t available in the main factor historical data set, so we have a collection of variation data set observed from moving average data set. Here, we assume that set of all variation data as a second factor historical time series data set.

Step 6 Now, define the universe of discourse V of the second factor based on the range of available variations data set as $V = [E_{\min} - E_1, E_{\max} + E_2]$ where $E_{\min}$ and $E_{\max}$ are the minimum and maximum value of the second factor historical data set, and E_1 and E_2 are two proper assigned positive real numbers for easy partitioning the universe of discourse V. In order to partition the universe of discourse V into length of equal intervals, the length of interval is determined as

$$v = \frac{E_{\max} - E_{\min}}{k} \tag{6}$$

where k is the number of observations available in the second factor (variation data sets) historical time series data and v is the length of each partitioned intervals. As the length of interval is determined, V can be partitioned into $k + 1$ equal length of intervals $v_i, i = 1, 2, \ldots, k, k + 1$ based on the number of observation available in the variation historical data set.

Step 7 Construct the triangular fuzzy sets $\tilde{B}_i, i = 1, 2, \ldots, k, k+1$ corresponding to the interval $v_i, i = 1, 2, \ldots, k, k+1$ defined in step 6 and obtained triangular membership grades to each variation data sets in their corresponding triangular fuzzy sets.

Step 8 To fuzzify the each historical time series datum of the second factor (variation data sets), we choose maximum degree of membership grades (as defined in step 7) of each datum to $\tilde{B}_i, (i = 1, 2, \ldots, k, k + 1)$ then we fuzzify that variation datum to $\tilde{B}_i$ for maximum membership grades available in triangular fuzzy sets $\left(\tilde{B}_1, \tilde{B}_2, \ldots, \tilde{B}_{k+1}\right)$ of x_k for fixed k. x_k is the kth-observations datum of the second factor historical time series data.

Step 9 Based on these fuzzified historical data sets of the main factor (moving average data set) and the fuzzified historical data sets of the second factor (variation data set) obtained in step 4 and step 8, respectively, we construct fuzzy logical relationships based on two factors. If the fuzzified historical data of the main factor at time $t - 1$ and t are $\tilde{A}_{k+1}$ and $\tilde{A}_k$, respectively, $\tilde{A}_{k+1}$ and $\tilde{A}_k$ are triangular fuzzy sets and if the fuzzified historical data of the second factor at time $t - 1$ is $\tilde{B}_{k+1}$, where $\tilde{B}_{k+1}$ is triangular fuzzy set then construct the two-factor fuzzy logical relationships (FLR) by the rule

$$\left(\tilde{A}_{k+1}, \tilde{B}_{k+1}\right) \rightarrow \tilde{A}_k$$

where $\left(\tilde{A}_{k+1}, \tilde{B}_{k+1}\right)$ is called the current state and $\tilde{A}_k$ is the next state of two-factor fuzzy logical relationships, respectively. If the constructed FLR having the same current state, then they lie in a same group and form a fuzzy logical relationship groups (FLRGs) of two-factor fuzzy time series historical data. To check the efficiency of the proposed method, we

Table 1 FLRGs and their frequency of occurrence

Groups	FLRGs	Frequency of occurrence
1	$\left(\tilde{A}_r, \tilde{B}_r\right) \rightarrow \tilde{A}_{r_1}$	3
2	$\left(\tilde{A}_s, \tilde{B}_s\right) \rightarrow \tilde{A}_{r_3}$	1
3	$\left(\tilde{A}_r, \tilde{B}_r\right) \rightarrow \tilde{A}_{r_2}$	1

measure the frequency occurrence of repeated fuzzy logical relationships; this affects in forecasted outputs and provides more accurate results. For explanatory purpose, we assume that the fuzzy logical relationships are computed from any historical fuzzy time series data based on two-factor moving average fuzzy time series are obtained as follows:

$$\left(\tilde{A}_r, \tilde{B}_r\right) \rightarrow \tilde{A}_{r_1}, \left(\tilde{A}_s, \tilde{B}_s\right) \rightarrow \tilde{A}_{r_3}, \left(\tilde{A}_r, \tilde{B}_r\right) \rightarrow \tilde{A}_{r_1}, \left(\tilde{A}_r, \tilde{B}_r\right) \rightarrow \tilde{A}_{r_1},$$
$$\left(\tilde{A}_r, \tilde{B}_r\right) \rightarrow \tilde{A}_{r_2}$$

Here, fuzzy set $\tilde{A}_{r_1}$ shown in relationships $\left(\tilde{A}_r, \tilde{B}_r\right) \rightarrow \tilde{A}_{r_1}$ occurs three times; hence, they can be placed in the same groups, and their frequency of occurrence is defined as three. The above fuzzy logical relationships and their frequency of occurrence in FLRGs which is based on two-factor fuzzy time series data can be summarized in the following Table 1.

Step 10 To defuzzify the fuzzified time series data sets which are based on the constructed two-factor FLRGs, we give two heuristic rules to determine the forecasted outputs at time t.

Rule 1 If the two-factor historical fuzzy time series data are used and FLRGs contain in the next state is one or more fuzzified values at time $t - 1$ with frequency of occurrence is ≥ 1 in FLR, assume fuzzy logical relationship groups at time $t - 1$ is $\left(\tilde{A}_{k+1}, \tilde{B}_{k+1}\right) \rightarrow \tilde{A}_1, \tilde{A}_2, \ldots, \tilde{A}_k, \tilde{A}_{k+1}$ with more than one fuzzified value in the next state, further we define the frequency occurrence of two-factor fuzzy logical relationships in FLRGs which are listed in Table 2.

Then, the forecasted value at time t is computed as follows

$$\frac{f_1 \times m_1 + f_2 \times m_2 + \ldots + f_k \times m_k + f_{k+1} \times m_{k+1}}{f_1 + f_2 + \ldots + f_k + f_{k+1}}$$

where f_i $(i = 1, 2, \ldots, k, k + 1)$ denotes the number of frequency occurrence of the fuzzy logical relationship in the FLRGs $\left(\tilde{A}_{k+1}, \tilde{B}_{k+1}\right) \rightarrow \tilde{A}_1, \tilde{A}_2, \ldots, \tilde{A}_k, \tilde{A}_{k+1}$ and k is the number of observations available in the main factor (moving average data set) of the historical time series data sets. $m_1, m_2, \ldots, m_k, m_{k+1}$ are the mid-

Table 2 FLRGs and their frequency occurrence

FLRGs	Frequency of occurrence
$\left(\tilde{A}_{k+1}, \tilde{B}_{k+1} \right) \to \tilde{A}_1$	f_1
$\left(\tilde{A}_{k+1}, \tilde{B}_{k+1} \right) \to \tilde{A}_2$	f_2
…	…
$\left(\tilde{A}_{k+1}, \tilde{B}_{k+1} \right) \to \tilde{A}_k$	f_k
$\left(\tilde{A}_{k+1}, \tilde{B}_{k+1} \right) \to \tilde{A}_{k+1}$	f_{k+1}

points of the intervals $u_1, u_2, \ldots, u_k, u_{k+1}$ in accordance to the main factor historical time series data corresponding to triangular fuzzy sets $\tilde{A}_1, \tilde{A}_2, \ldots, \tilde{A}_k, \tilde{A}_{k+1}$. $\tilde{A}_i$ and $\tilde{B}_i (i = 1, 2, \ldots, k, k + 1)$ are fuzzified values represented by a triangular fuzzy sets of the main factor (moving average data set) and second factor (variations data set) historical fuzzy time series data sets, respectively.

Rule 2 If the two-factor historical fuzzy time series data is used and assume fuzzy logical relationship groups at time $t - 1$ is $\left(\tilde{A}_{k+1}, \tilde{B}_{k+1} \right) \to \#$.

Where the symbol "#" represents an unknown parameter, then the forecasted value at time t is m_{k+1} where m_{k+1} is midpoint of the interval u_{k+1} in accordance to the main factor historical time series data corresponding to triangular fuzzy set $\tilde{A}_{k+1}$. $\tilde{A}_{k+1}$ and $\tilde{B}_{k+1}$ are triangular fuzzy sets of the main factor (moving average data set) and second factor (variation data set) historical fuzzy time series data sets, respectively.

Step 11   In time series forecasting, root mean square error (RMSE) and average forecasting error (AFE) are the common tools to measure the forecasting accuracy rates which are defined as follows:

1. The root mean square error (RMSE) can be defined as

$$\text{RMSE} = \sqrt{\frac{\sum_{i=1}^{k} (F_i - A_i)^2}{k}} \tag{7}$$

2. The average forecasting error (AFE) can be defined as

$$\text{Forecasting error (\%)} = \frac{|F_i - A_i|}{A_i} \times 100 \tag{8}$$

$$\text{AFE(\%)} = \frac{\text{sum of forecasting error}}{k} \tag{9}$$

Here F_i and A_i are forecasted value and actual value (open observations) of historical time series data sets, respectively. k is the number of observations available in the main factor (moving average data set) historical data set. Least the values of RMSE and AFE indicates that the better forecasting accuracy rates.

Table 3 Historical data sets of the BSE Sensex from period January 1, 2012, to February 29, 2012

Date	Open observations (in rupees)	Moving average data sets	Fuzzy moving average	Variation data sets	Fuzzy variation
01/Jan/2012	15,534.67	–	–	–	–
03/Jan/2012	15,640.56	–	–	–	–
04/Jan/2012	15,967.49	15,587.61	$\tilde{A}_1$	0	$\tilde{B}_{21}$
05/Jan/2012	15,893.07	15,804.02	$\tilde{A}_4$	216.41	$\tilde{B}_{37}$
...	...	...	...	...	...
28/Feb/2012	17,545.11	18,027.10	$\tilde{A}_{36}$	−76.08	$\tilde{B}_{15}$
29/Feb/2012	17,919.93	17,760.15	$\tilde{A}_{32}$	−266.95	$\tilde{B}_1$

The second column consists of open observations (actual data set), and third and fourth columns consist of moving average data (main factor data set) and with their corresponding fuzzified values. The last fifth and sixth columns represent variation data sets (named as second factor data set) and with their corresponding fuzzified values

4 Implementation of Proposed Method

In this section, we present the stepwise computations and comparison in terms of root mean square error (RMSE) and average forecasting error (AFE) of proposed forecasting method based on moving average two-factor fuzzy time series. To verify the performance of the proposed method, it is implemented on the Bombay Stock Exchange (BSE) Sensex historical time series data from period January 1, 2012, to February 29, 2012, which is recorded from daily news paper The Times of India, and it is observed that the forecasting accuracy rates can be further improved by applying a proposed method based on moving average forecasting compared with other fuzzy time series forecasting models. The stepwise computation is as follows:

Step 1 The historical time series data from period January 1, 2012, to February 29, 2012, of BSE Sensex and calculated two-period moving average time series data which are obtained from open observations (actual data set) of historical data sets are placed in Table 3.

Step 2 Define the universe of discourse U of the main factor as the moving average historical data sets by taking two proper positive real number as $D_1 = 37.61$ and $D_2 = 14.2$, then $U = [15,550, 18,379]$, where $D_{\max} = 18,364.80$ and $D_{\min} = 15,587.61$ are observed from Table 3. Now, we define the length of interval by which we can make a partition of the universe of discourse U, and then from Eq. (5),

$$\text{Length } u = \frac{18,364.80 - 15,587.61}{40} \approx 69$$

$k = 40$ is the number of observations available in the main factor (moving average data set) historical time series data sets. So, we partition the universe of discourse U into $k + 1$ intervals of equal length $u = 69$ as follows:

$$u_1 = [15{,}550, 15{,}619], \quad u_2 = [15{,}619, 15{,}688], \quad u_3 = [15{,}688, 15{,}757], \ldots$$

Step 3 Triangular fuzzy sets $\tilde{A}_1, \tilde{A}_2, \ldots, \tilde{A}_{40}, \tilde{A}_{41}$ are defined with corresponding intervals $u_1, u_2, \ldots, u_{40}, u_{41}$ defined in step 2 as follows:

$$\tilde{A}_1 = [15{,}550, 15{,}619, 15{,}688], \quad \tilde{A}_2 = [15{,}619, 15{,}688, 15{,}757], \quad \tilde{A}_3 = [15{,}688, 15{,}757, 15{,}826],$$

... ...

The triangular membership grades of each datum available in the main factor (moving average datum) Sensex data sets to corresponding triangular fuzzy sets $\tilde{A}_i$ ($i = 1, 2, \ldots, 41$) are defined as follows:

$$\tilde{A}_1 = \{15{,}587.61, 0.54\}, \quad \tilde{A}_2 = \emptyset(\text{null fuzzy set}), \quad \tilde{A}_3 = \{15{,}804.02, 0.31\}, \ldots$$

Step 4 Now implement the fuzzification process of the proposed method to main factor (moving average data set) BSE Sensex historical data sets is based on their maximum membership grades of each observation available in the main factor historical time series data with their corresponding triangular fuzzy set. The membership grades of each observation are defined in step 3. So all fuzzified values of the main factor historical observations are placed in Table 3.

Step 5 Calculate the variation of the main factor (moving average data set) historical data sets. For example, the variation of date 10/Feb/2012 is calculated as follows.

$$\text{The variation } \Delta(10/\text{Feb}/2012) = f(10/\text{Feb}/2012) - f(09/\text{Feb}/2012)$$
$$= 17{,}639.74 - 17{,}722.71$$
$$= -82.97$$

Similarly, the variations of all historical data available in the main factor are calculated and shown in Table 3.

Step 6 Define the universe of discourse V of the second factor as the variation historical data sets by taking two proper positive real number, $E_1 = 3.05$ and $E_2 = 26.76$ as $V = [-270, 263]$, where $E_{max} = 236.24$ and $E_{min} = -266.95$ are observed from second factor data sets by Table 3. Now we define length of interval by which we can make a partition of the universe of discourse V; then, from Eq. (6)

$$\text{Length } v = \frac{236.24 - (-266.95)}{40} \approx 13$$

$k = 40$ is the number of observations available in the second factor (variations data sets) historical time series data sets. So, we partition the universe of discourse V into $k + 1$ intervals of equal length $v = 13$ as follows.

$v_1 = [-270, -257]$	$v_2 = [-257, -244]$	$v_3 = [-244, -231]$
$v_4 = [-231, -218]$	$v_5 = [-218, -205]$	$v_6 = [-205, -192]$
...	...	...
$v_{36} = [185, 198]$	$v_{37} = [198, 211]$	$v_{38} = [211, 224]$
$v_{39} = [224, 237]$	$v_{40} = [237, 250]$	$v_{41} = [250, 263]$

Step 7 Triangular fuzzy sets $\tilde{B}_1, \tilde{B}_2, \ldots, \tilde{B}_{40}, \tilde{B}_{41}$ are defined with corresponding intervals $v_1, v_2, \ldots, v_{40}, v_{41}$ defined in step 6 as follows:

$\tilde{B}_1 = [-270, -257, -244]$	$\tilde{B}_2 = [-257, -244, -231]$	$\tilde{B}_3 = [-244, -231, -218]$
$\tilde{B}_4 = [-231, -218, -205]$	$\tilde{B}_5 = [-218, -205, -192]$	$\tilde{B}_6 = [-205, -192, -179]$
...	...	...
$\tilde{B}_{36} = [185, 198, 211]$	$\tilde{B}_{37} = [198, 211, 224]$	$\tilde{B}_{38} = [211, 224, 237]$
$\tilde{B}_{39} = [224, 237, 250]$	$\tilde{B}_{40} = [237, 250, 263]$	$\tilde{B}_{41} = [250, 263, 263]$

The triangular membership grades of each datum available in the second factor (variation datum) Sensex data set to corresponding triangular fuzzy sets $\tilde{B}_i$ ($i = 1, 2, \ldots, 41$) are defined as follows:

$\tilde{B}_1 = \{-266.95, 0.23\}, \ \tilde{B}_2 = \phi \,(\text{null fuzzy set}), \ \tilde{B}_3 = \phi \,(\text{null fuzzy set}), \ \tilde{B}_4 = \phi \,(\text{null fuzzy set})$

$\tilde{B}_5 = \phi \,(\text{null fuzzy set}), \ \tilde{B}_6 = \phi \,(\text{null fuzzy set}), \ \tilde{B}_7 = \phi \,(\text{null fuzzy set}), \ \tilde{B}_8 = \{-173.43, 0.42\}$

...   .

$\tilde{B}_{39} = \{236.24, 0.94\}, \ \tilde{B}_{40} = \phi \,(\text{null fuzzy set}), \ \tilde{B}_{41} = \phi \,(\text{null fuzzy set}).$

Step 8 Now implement the fuzzification process of the proposed method based on their maximum membership grades of each observation available in the second factor historical time series data with their corresponding triangular fuzzy set. The membership grades of each observation are defined in step 7. So all fuzzified value of the second factor historical observations is placed in Table 3.

Step 9 With these fuzzified value of main factor data and second factor data set in Table 3, we established a fuzzy logical relationships (FLRs) and consequently fuzzy logical relationship groups (FLRGs) accordingly as defined in step 9 of proposed method in Sect. 3 of forecasting based on moving average two-factor fuzzy time series are placed in Table 4.

Table 4 Two-factor FLRGs based on moving average data sets

Groups	FLRGs	Groups	FLRGs
Group 1:	$\left(\tilde{A}_1, \tilde{B}_{21}\right) \rightarrow \tilde{A}_4$	Group 21:	$\left(\tilde{A}_{22}, \tilde{B}_{11}\right) \rightarrow \tilde{A}_{22}$
Group 2:	$\left(\tilde{A}_4, \tilde{B}_{37}\right) \rightarrow \tilde{A}_6$	Group 22:	$\left(\tilde{A}_{22}, \tilde{B}_{22}\right) \rightarrow \tilde{A}_{25}$
Group 3:	$\left(\tilde{A}_6, \tilde{B}_{30}\right) \rightarrow \tilde{A}_4$	Group 23:	$\left(\tilde{A}_{25}, \tilde{B}_{39}\right) \rightarrow \tilde{A}_{27}$
...	...	...	...
Group 19:	$\left(\tilde{A}_{23}, \tilde{B}_{36}\right) \rightarrow \tilde{A}_{23}$	Group 39:	$\left(\tilde{A}_{36}, \tilde{B}_{15}\right) \rightarrow \tilde{A}_{32}$
Group 20:	$\left(\tilde{A}_{23}, \tilde{B}_{23}\right) \rightarrow \tilde{A}_{22}$	Group 40:	$\left(\tilde{A}_{32}, \tilde{B}_1\right) \rightarrow \#$

Table 5 Forecasted outputs obtained by proposed and other existing models

Date	Open observations	Proposed method	Chen method [4]	Hui-Kuang Yu model [8]	Huarng and Yu type 2 model [9]
01/Jan/2012	15,534.67	–	–	–	16,000
03/Jan/2012	15,640.56	–	16,000	15,499.96	16,000
04/Jan/2012	15,967.49	–	16,000	15,499.96	16,500
05/Jan/2012	15,893.07	15,791.50	16,000	15,499.96	16,000
06/Jan/2012	15,789.08	15,929.50	16,000	15,499.96	16,500
07/Jan/2012	15,893.30	15,791.50	16,000	15,499.96	16,000
...	...	...	...	...	...
27/Feb/2012	17,975.19	18,068.50	18,000	18,249.97	18,000
28/Feb/2012	17,545.11	17,999.50	18,000	17,508.29	18,000
29/Feb/2012	17,919.93	17,723.50	18,000	17,508.29	18,000

Step 10   Fuzzified historical data sets are defuzzified to get the crisp output by using the proposed method of moving average forecasting based on two-factor fuzzy logical relationship groups along with three different forecasted values obtained by Chen method [4], weighted model of Yu [8], and Huarng and Yu [9] model for fuzzy type 2 (Table 5).

Step 11   Forecasting accuracy rates obtained by proposed method and compare it with other models in terms of RMSE and AFE.

From the based performance comparison Table 6, it is observed that the comparative study of RMSE and AFE obtained by the proposed method is of higher forecasting accuracy rates than Chen [4], Yu [8] and Huarng and Yu [9] of fuzzy type 2 model. Also, from Fig. 2, it is cleared that the forecasted output obtained by the proposed method of moving average forecasting based on two-factor FLRGs is significantly in closed accordance to the open observations (actual) Sensex data set.

Table 6 Performance comparison table of RMSE and AFE in proposed method and compares it with other models in terms of RMSE and AFE

Models/parameters	Proposed method	Chen method [4]	Hui-Kuang Yu model [8]	Huarng and Yu type 2 model [9]
RMSE	217.50	445.18	354.09	484.27
AFE (%)	1.04	2.06	1.76	2.43

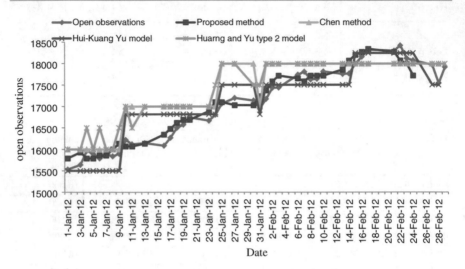

Fig. 2 Graph of the historical Sensex data set

5 Conclusion

In this paper, we have presented a novel approach to handle forecasting problems based on moving average in two-factor fuzzy time series. Proposed method uses a new technique to define the universe of discourse on transformed moving average data set and make partition into intervals based on the number of observations available in transformed moving average data set. Further, variation data set are calculated from moving average data set and construct different triangular fuzzy sets for both moving average and variation data set. Additionally, we obtained membership grades of each observation available in both time series data sets with their corresponding triangular fuzzy set. This study also generates two-factor fuzzy logical relationship groups, and defuzzification rule is based on their frequency occurrence of FLR in FLRGs and obtained a crisp output in a more efficient way. We apply the proposed method on BSE Sensex historical time series data set. From the experimental results as shown in Table 6, we observe that our proposed method can get higher forecasting accuracy rates than the existing models of Chen [4], Hui-Kuang Yu [8], and Huarng and Yu [9].

References

1. Zadeh, L.A.: Fuzzy set. Inf. Control **8**, 338–353 (1965)
2. Song, Q., Chissom, B.: Forecasting enrollments with fuzzy time series-Part I. Fuzzy Sets Syst. **54**, 1–9 (1993)
3. Song, Q., Chissom, B.: Forecasting enrollments with fuzzy time series-Part II. Fuzzy Sets Syst. **64**, 1–8 (1994)
4. Chen, S.M.: Forecasting enrollments based on fuzzy time series. Fuzzy Sets Syst. **81**, 311–319 (1996)
5. Lee, H.S., Chou, M.T.: Fuzzy forecasting based on fuzzy time series. Int. J. Comput. Math. **81**, 781–789 (2004)
6. Singh, S.R.: A computational method of forecasting based on high-order fuzzy time series. Expert Syst. Appl. **36**, 10551–10559 (2009)
7. Huarng, K.: Effectives length of intervals to improve forecasting in fuzzy time series. Fuzzy Sets Syst. **123**, 387–394 (2001)
8. Yu, H.K.: Weighted fuzzy time series models for TAIEX forecasting. Phys. A **349**, 609–624 (2005)
9. Huarng, K., Yu, H.K.: A type 2 fuzzy time series model for stock index forecasting. Phys. A **353**, 445–462 (2005)
10. Huarng, K., Yu, T.H.K.: Ratio-based lengths of intervals to improve fuzzy time series forecasting. IEEE Trans. Syst. Man Cybern. Part B Cybern. **36**, 328–340 (2006)
11. Yolcu, U., Egrioglu, E., Uslu, V.R., Basaran, M.A., Aladag, C.H.: A new approach for determining the length of intervals of fuzzy time series. Appl. Soft Comput. **9**, 647–651 (2009)
12. Lee, L.W., Wang, L.H., Chen, S.M., Leu, Y.H.: Handling forecasting problems based on two-factors high-order fuzzy time series. IEEE Trans. Fuzzy Syst. **14**, 468–477 (2006)
13. Chen, S.M., Tanuwijaya, K.: Fuzzy forecasting based on high-order fuzzy logical relationships and automatic clustering techniques. Expert Syst. Appl. **38**, 15425–15437 (2011)
14. Sulandari, W., Yudhanto, Y.: Forecasting trend data using a hybrid simple moving average-weighted fuzzy time series model. In: IEEE International Conference on Science in Information Technology, Yogyakarta, Indonesia, 27–28 Oct 2015
15. Abhishekh, Kumar, S.: A computational method for rice production forecasting based on high-order fuzzy time series. Int. J. Fuzzy Math. Arch. **13**, 145–157 (2017)

A Multi-objective Optimization Study of Parameters for Low-Altitude Seat Ejections

R. Naveen Raj and K. Shankar

Abstract Flight ejection process exposes the pilot to large accelerations causing spinal injuries at various levels of the spine. The successful ejections in low-altitude seat ejections remain very low when compared to higher altitude ejections, since the low-level ejections are time critical events demanding optimal process parameters to improve chances of successful ejection. The spinal injury depends on variety of factors such as rate of impulse applied, peak acceleration. Dynamic Response Index (DRI) is used as an injury parameter to predict injury of spine. Hitherto a multi-objective optimization-based study has not been conducted, and it can impart various important insights which can be useful in optimizing the low-altitude ejection processes. In this paper, a novel multi-objective optimization approach to maximize the height reached by the pilot to obtain a life-saving height and to minimize DRI value to reduce injury is used to study various process parameters such as fore-aft angle of the aircraft, rate of impulse applied, flight speed. Two different algorithms, NSGA-II and MOPSO, were used in this study.

Keywords Seat ejection · Multi-objective optimization · Low altitude

1 Introduction

From the times of World War Two, the need for ejection seats has been crucial, and various technologies have been investigated to carry out the ejection sequence. The early ejection seats catered to the need of high energy boost mechanism for the aviator to escape safely, by using ballistic catapults which produced sufficient energy for the pilot to get out of the cockpit. Over the years, with the increase in understanding

R. Naveen Raj (✉) · K. Shankar
Department of Mechanical Engineering, Indian Institute of Technology Madras,
Chennai 600036, India
e-mail: navraj94@gmail.com

K. Shankar
e-mail: skris@iitm.ac.in

311

biomechanical tolerances of the human body and mechanics of flying objects, the ejection seats have focused on higher level safety for the aviator. The latest aircrafts employ H-shaped rocket motors attached to the bottom of the seat, which utilizes the latest technology to monitor several factors like altitude, velocity of the aircraft using sensors and decide the accurate ejection parameters as presented by Specker, Plaga and Santi [1]. The increased focus of safety is evident from the improvement in recent fatality rates. High altitude ejections (above 500 ft) have a success rate of 91.4%. However, low-altitude ejections (ejections happening below 500 ft) had a success rate of 51.2%, and the remaining were fatal [2]. This necessitates the need for more research and improvement in ejection seat performances at lower altitudes.

The low rate of successful ejection can be attributed to various factors such as very little or no time to prepare for the ejection, the ejection seat may impact the ground before parachute opening or the parachute may have insufficient time to fully deploy, high bank angle at lower altitudes [2]. These factors make the seat ejection at low altitudes a time critical events, thus optimal conditions available at the instant must be chosen to increase the rate of successful ejections at low altitudes. The ejection seats performance at higher altitudes don't face many complex problems, the only problem is clearing the tail of the aircraft. Thus, the minimum impulse force required to clear the tail is the optimal impulse. Whereas in case of low-altitude seat ejections, a larger amount of force would be useful as it would give the pilot some life-saving height. Although it comes with increased risk of vertebral injuries, most times the minimum forces required would be well within the human tolerable limits. Also, increasing the impulse applied would be the lesser of two evils as reasonable risk of injury is acceptable when the other end is death.

The ejection process involves a complex sequence of events that subjects the human body to large forces from all directions. The injuries associated with various phases of the ejection process [3] are ejection gun firing and rocket motor lead to spinal compression fractures (thoracic and lumbar). The Drogue parachute deployment and the main parachute deployment give raise to spinal injuries (cervical). The parachute landing phase creates lower limb injuries which can be largely avoided by disconnecting the seat before landing and by prior training to the pilots. The spinal compression injuries are the most predominant injuries as observed from the victims of unsuccessful seat ejections happened. The study mainly focuses on the spinal compression fractures which are mainly caused due to the axial impulse travelling from the pelvis to head.

From very early designs [4], improper posture of the pilot to have a curved spine was recognized as one of the major contributing factors to spinal injuries during ejection and thus the proper alignment of the spine along the line of thrust is important to avoid injuries. But these injuries can be avoided with a full-functioning restraint system. This study is focused with such type of scenarios where the pilot is restrained to the seat at all times.

Despite the complexity of human body and the variation between individuals, the response of the human body to accelerations environments can be approximately characterized by an analogous dynamical model [5]. Many injury parameters have been used to characterize the injury attained during flight ejection. The Ruff tolerance curves, developed as early as 1950 [6], were the first attempt to quantify the injuries in spine. These curves were obtained from catapult-based experiments, and it was found that for exposure periods less than 5 ms, the tolerance depends on the strength of the most susceptible vertebrae, and for exposure periods ranging between 5 ms and 1 s the tolerance of the human spine depends upon static compressive strength of the human spine. The Ruff tolerance curves were followed by the Eiband curves [7], and the specialty of these curves were they were available for accelerations along different directions. But the main limitations of the Eiband curves were the line dividing moderate and severe injury which was obtained from animal tests made on pigs and chimpanzees.

The Dynamic Response Index (DRI) which was proposed in 1971 [8] is the most adopted injury parameter up to date [9]. This study also uses the DRI to calculate the injury involved during the application of the impulse.

Multi-objective optimizations have been useful tool for various engineering problem to optimize the parameters involved. In general, a multi-objective optimization algorithm yields a set of non-dominated solutions which are all optimal with respect to the objective function values [10, 11].

In the present paper, an attempt is made to understand the effects of various parameters on the ejection process by obtaining Pareto-optimal fronts for various combinations of the various flight parameters involved. The performance of the multi-objective problem is studied with two different algorithms, (a) real-coded NSGA-II [14] based on evolutionary principle and (b) MOPSO [15] which is a behaviourally inspired algorithm. These are compared, and the best one is chosen for the analysis.

2 Mathematical Modelling of the Objective Functions

2.1 Dynamic Response Index

The dynamic response index is a single degree of freedom spring-mass damper model (Fig. 1) to simulate the biomechanical response of the human upper body/vertebral column/pelvis which tries to identify the spinal compression and relate it to the probability of the injury. Latham first introduced the concept of a mechanical spring-mass model under vertical loading [4] to represent the pilot and the seat. This idea was later incorporated by Payne [5, 12] to develop the spring-mass damper dynamical system approximations of the human body. They derived the stiffness, damping and mass values for the model from the experimental data obtained by Ruff used for developing Ruff tolerance curves. Later Brinkely and Schaffer [8] collected experimental data from real-life pilot ejections to validate the DRI model. Assuming that the DRI values are normally distributed and the relationship between vertebral breaking strength and body weight is random the relationship between the DRI and the probability of injury.

Fig. 1 DRI model

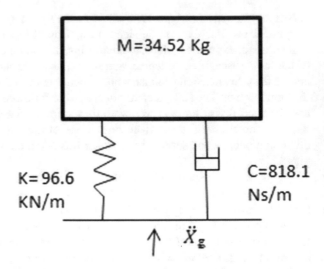

The governing equation for the DRI model can be written as

$$m\frac{\mathrm{d}^2 x}{\mathrm{d}t^2} + c\frac{\mathrm{d}x}{\mathrm{d}t} + kx = -m\ddot{x}_g \tag{1}$$

Writing the equation in terms of 'δ', $\delta = x - x_g$, $\dot{\delta} = \dot{x} - \dot{x}_g$

$$\ddot{x}_g(t) = \ddot{\delta} + 2\zeta\omega_n\dot{\delta} + \omega_n^2\delta \tag{2}$$

where

δ = Compression of the spring (lumbar compression)
$\zeta = 0.224$
$\omega_n = 52.2$ rad/s

The dynamic response index is written as

$$\mathrm{DRI} = \frac{\omega_n^2\delta\mathrm{max}}{g} \tag{3}$$

Thus, the DRI relates to the deflection of the model or compression of the spine and is a way to describe the probability of injury on an ejecting pilot in terms of spinal compression due to an input acceleration. The DRI is also denoted as DRI_z to quantify the injury in the vertical direction. DRI for other three axes was also developed later.

2.2   Pilot Trajectory After Ejection

The pilot trajectory can be modelled with the help of the governing equations of the pilot along with their corresponding initial conditions, the whole ejection process involves pilot moving up the guide rails before a sudden burst of rocket thrust propels it to clear the tail and then free-fall flight under the action of gravity alone. Robinson et al. [13] had developed a trajectory model for a catapult-based seat ejection system which was modified to get a rocket assisted seat ejection system.

The origin (x', y') is fixed at the seat, and the x' axis runs through the body of the aircraft as shown in Fig. 2.

θ_p is the angle of climb of the aircraft
Φ_e is the initial angle of ejection of the pilot relative to the aircraft (75°)
X_s is the aircraft length, as measured from the ejection seat to the vertical stabilizer (15 m)
Y_s is the stabilizer height (5 m)
V_p is the velocity of the plane
Φ is the angle of velocity vector after some time t

The forces acting on the pilot during the process are drag force due to air resistance, gravitational force and impulse force acting for the short span. The governing equation for the motion of the pilot can be split into two phases, first where the impulse force is applied and the free-fall phase (Fig. 3).

The drag force opposing the motion of the pilot is varying as square of the velocity (v^2) assuming that the drag force is of the form,

$$F_D = \frac{1}{2}\rho c_d s v^2 \tag{4}$$

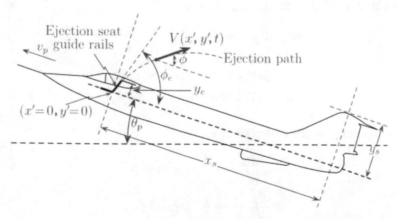

Fig. 2 Schematic diagram showing relevant important dimensions of an aircraft [13]

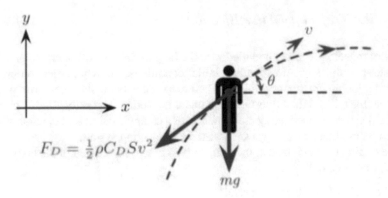

Fig. 3 Forces acting on pilot after ejection during free-fall phase [13]

where,

ρ = air density (1.2 kg/m^3)
c_d = drag coefficient (0.45)
s = surface area of the pilot (2.0 m^2)
v = velocity of the pilot

The equations for the first phase are (impulse is applied for $t = 0.3$ s)

$$m\frac{d^2y}{dt^2} = F_v \sin\theta * g - \frac{1}{2}\rho c_d s v^2 \cos\theta \tag{5}$$

$$m\frac{d^2y}{dt^2} = F_v \cos\theta * g - \frac{1}{2}\rho c_d s v^2 \sin\theta \tag{6}$$

where
    F_v is the impulsive acceleration applied in g units in vertical direction.
The initial conditions for the first phase are

$$x(0) = y_e.\cos(\Phi)$$
$$y(0) = y_e.\sin(\Phi)$$
$$v_x(0) = V_e \cos(\Phi_e - \theta_p) - V_p \cos(\theta_p)$$
$$v_y(0) = V_e \sin(\Phi_e - \theta_p) + V_p \sin(\theta_p)$$

The equations for the second phase are

$$m\frac{d^2x}{dt^2} = -\frac{1}{2}\rho c_d s v^2 \cos\theta \tag{7}$$

$$m\frac{d^2y}{dt^2} = -g - \frac{1}{2}\rho c_d s v^2 \sin\theta \tag{8}$$

The initial conditions for the second phase are obtained from the solution of the Eqs. (5) and (6).

The above second-order differential equations were solved using "MATLAB" using the inbuilt ODE45 solver.

The vertical height reached by the pilot which is one of the objectives of the optimization problem is obtained by solving the Eqs. (5–8).

3   Multi-objective Optimization

3.1   Optimization Problem Formulations

The multi-objective optimization problem is proposed to maximize the height reached and minimize the injury parameter value. The input values required for the optimization problem are as described above in the equations. The upward impulse is applied till $t = 0.3$ s from the point in time when the pilot reaches the top of the guide rails (Table 1).

The objective functions used for the problem are

1. **Minimize**:

$$Z_1 = \text{DRI}$$

where,

DRI—Dynamic Response Index

2. **Maximize**:

$$Z_2 = h$$

where,

h—Height reached by the pilot after ejection

Table 1 Design variables with lower and upper bounds

Sl. No.	Design variables $\{x\}$	Lower limit	Upper limit	Bounds
1	Peak acceleration (F_{max})	5 g	30 g	$5\,g \leq X_1 \leq 30\,g$
2	Angle of ejection from the seat back angle (Φ)	$-10°$	$10°$	$-10° \leq X_2 \leq 10°$

The two objectives are conflicting in nature; a higher ejection force would give an higher height but also result in higher chances injury to the spine, a lower ejection force would have lesser chance of injury but the height reached by the pilot will also be less giving raise to other fatal problems in low-altitude (below 500 ft) ejections. This conflicting nature gives raise to Pareto fronts as no one solution can be considered better than any other with respect to the other. The Pareto-optimal fronts were obtained for these objective functions. The values of parameters such as blank angle (θ_p), the rate of impulse applied (F), aircraft velocity (v_p) were modified according to the needs, to study their effects on the ejection process. The other input parameters were kept constant.

3.2 Algorithm Selection

The main challenge of a multi-objective heuristic algorithm is to obtain a well-converged diverse Pareto front. To address this problem, the above formulation was tested on two popular algorithms; real-coded NSGA-II (Non-dominated Sorting Genetic Algorithm) [13] and MOPSO (multi-objective particle swarm optimization) [14].

MOPSO: MOPSO is an extension of the basic particle swarm optimization [15] which is a behaviourally inspired algorithm based on insect swarm (idealized as particles) behaviour. Here potential solutions (particles) move rather than evolve through the search space unlike other evolutionary algorithms. PSO consists of several candidate solutions called particles each of which has a position and velocity, and experiences linear spring-like attractions towards two attractors:

1. Historical best position attained by that particle (P_{best}).
2. Historical best of all the particles (G_{best}).

The position and velocity of each individual is updated according to following equations:

$$v_t = w.v_t + c_1 r_1.(P_{\text{best}} - \phi_t) + c_2 r_2 (G_{\text{best}} - \phi_t) \tag{9}$$

$$\phi_{t+1} = v_t + \phi_t \tag{10}$$

where Φ denotes the current position of the particle, v denotes the velocity of the particle, and c_1, c_2 denotes personal and global acceleration coefficients, respectively.

Later PSO was extended to multi-objective optimization problems, i.e. multi-objective particle swarm optimization (MOPSO). In MOPSO, the PSO particles are evaluated for the individual objective functions based on Eqs. 9 and 10, and values for P_{best} and G_{best} are evaluated for minimizing each objective function; the non-dominated solutions are obtained from these P_{best} values [14].

NSGA-II: Elitist Non-dominated Sorting Genetic Algorithm (NSGA-II) [13] is a popularly used evolutionary algorithm (EA) for multi-objective optimization. Several prominent features like explicit diversity and elite preservation help in ensuring convergence and diversity in the generated population for every iteration. In NSGA-II, offspring populations (size N) are created using parent population (size N) by using genetic operators. The created offspring population is combined with parent population, to form combined population of size $2N$, and then a non-dominated sorting is carried out to classify the entire population into several non-dominated fronts. The new population (size N) is then obtained from the members of combined population belonging to different non-dominated levels.

One test case was run using both NSGA-II and MOPSO for same number of population, and iterations and the Pareto front for the above objectives were obtained and are shown in Figs. 4 and 5 (Table 2).

It is visibly observed that diversity of the Pareto front obtained by MOPSO was relatively less compared to the real-coded NSGA-II from Figs. 4 and 5. The NSGA-II

Fig. 4 Pareto front obtained using MOPSO

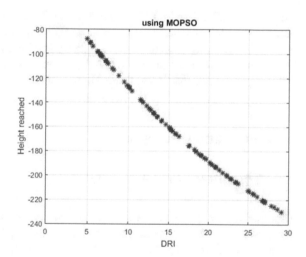

Table 2 Parameters for multi-objective optimization

Parameter	NSGA-II	MOPSO
No. of iterations	30	30
No. of population	100	100
Selection method	Roulette wheel	–
Crossover probability	0.9	–
Inertia weight	–	0.5
Inertia damping coefficient	–	0.99
Personal acceleration coefficient	–	1
Global acceleration coefficient	–	2

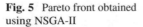

Fig. 5 Pareto front obtained
using NSGA-II

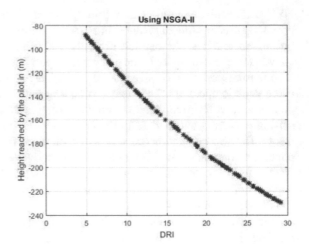

was 60.11% faster than MOPSO to give the results. Thus, the real-coded NSGA-II is observed to perform better for the given problem as it gave a good diversity in lesser time and hence was chosen for further analysis of the problem for different cases. Due to the stochastic nature of NSGA-II, ten runs were carried out for each of the following cases and the Pareto fronts shown are average of the 10 runs.

4   Results and Discussion

4.1   The Effect of Blank Angle

The blank angle (θ_p) shows us if the aircraft is moving upwards or downwards. The positive blank angle means the aircraft is moving upwards and vice versa. For the purpose of the study, three $\theta_p = \{-45°, 0°, 45°\}$ values were chosen and the Pareto-optimal fronts were obtained as shown in Fig. 6.

A low rate of acceleration as shown in Fig. 9 was used, and the velocity of plane (V_p) was taken as 50 m/s (180 km/h).

These Pareto-optimal fronts (Fig. 6) show that attitude of the aircraft at the instant of ejection plays an important role, as it can be seen that for the same injury level height reached is more when the aircraft is moving upwards. This can be attributed to the fact that high rate of ascend of the aircraft negates the height reached during the ejection process. Thus, ejection conditions were the nose of the aircraft is pointed upwards has a better chance of escape.

Fig. 6 Pareto-optimal fronts for different values of θ_p

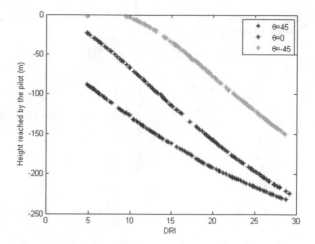

4.2 The Variation with Aircraft Speed

Even though aircraft speed may not be a controllable parameter at the time ejection, knowing how the Pareto-optimal fronts are for various flight speeds may help in determining the optimal parameters as the speed of the aircraft varies. The aircraft speed $V_p = \{25, 50, 75\}$ m/s was chosen for the study.

The blank angle (θ_p) was set to 0° and −45° meaning a horizontal flight and flight moving downwards and rate of loading was kept constant.

From the Pareto fronts as shown in Figs. 7 and 8, it can be found that for horizontal flights the flight speed makes minimal difference for a successful ejection as all the Pareto fronts lay close to each other. While when there are negative blank angles involved, the chances of successful ejection reduce as the flight speed increases. This can be attributed to the fact that the high rate of ascend is negated by the fast descend of the flight. It can be seen that for higher flight speeds even at higher impulses such as 12 g for 180 km/h flight and 14 g for 270 km/h, there is no significant height reached by the pilot thus these cases will fall out of the feasible region and has to be avoided.

4.3 The Effect of Rate of Loading

The rate of acceleration of the impulse also plays a major role in injury, but there are significant increases in fractures from lower to higher loading rates as studied experimentally by Stemper et al. [16]. The forcing functions used are as shown in Figs. 9 and 10. The rate of loading in Fig. 9 is increased twice from the rate of loading in Fig. 10.

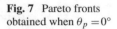

Fig. 7 Pareto fronts
obtained when $\theta_p = 0°$

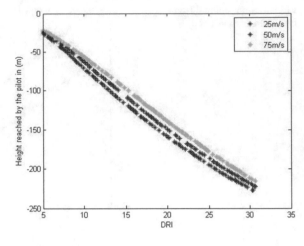

Fig. 8 Pareto fronts
obtained when $\theta_p = -45°$

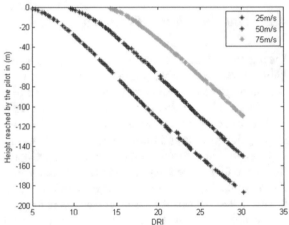

The blank angle (θ_p) was set to $0°$ meaning a horizontal flight, and the velocity of plane (V_p) was taken as 50 m/s.

Figures 11 and 12 show the Pareto front for different loading conditions. The rate of forcing doesn't show any difference in the obtained Pareto-optimal fronts in both Figs. 11 and 12 ideally the chances of injury increase with increasing loading rates, but this effect is not visible due to the approximation of the human body to single degree of freedom linear system. On investigating the DRI model separately, it was found that when the loading of the form Figs. 9 and 10, for a peak acceleration of 25 g, the DRI value increased marginally by 4.97% depicting the limitation of DRI model.

Fig. 9 Lower rate of loading

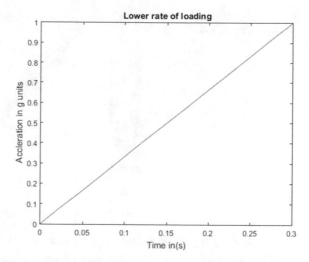

Fig. 10 Higher rate of loading

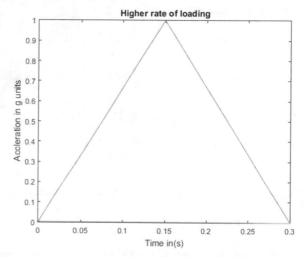

5 Conclusion and Future Work

It was found that positive blank angles gave higher chances of successful ejection; the difference in heights reached by the pilot for maximum forcing condition increased by 53% when the aircraft was pitched up from $-45°$ to $45°$. This result suggests that it is advisable for the pilot to go for a final boost in the top direction to keep the aircraft in positive blank angle before ejection to increase his chances of a successful ejection.

The flight velocity plays a crucial role when the flight is in an angled position rather than horizontal flight. It was noted that as the flight speed increased from 25 m/s (90 km/h) to 75 m/s (270 km/h), there was a 40% reduction in height reached

Fig. 11 Pareto fronts
obtained when $\theta_p = 0°$

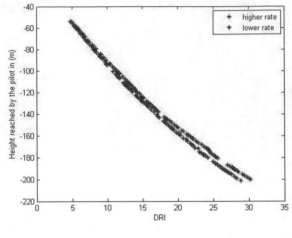

Fig. 12 Pareto fronts
obtained when $\theta_p = 45°$

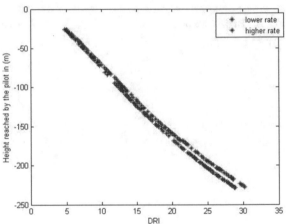

by the pilot for the maximum forces applied. Also, there was no noticeable height reached by the pilot below a peak impulse of 14 g due to the inertial effects.

The rate of loading showed minimal change in the obtained Pareto fronts, but the lower rate of loading was found to be better in that minimal change which is of practical sense. Though the DRI model has been an accepted parameter in literatures to predict the injuries in seat ejection scenario, its inability to capture the increase in risk of injury due to variation in rate of loading is a serious concern.

In lower altitude seat ejections to improve the chances of safe escape, the pilot has to be ejected in the best possible scenario. The delay in decision to eject is the usual mistake which leads to fatalities, but even under such critical circumstances pilot trying to give a final boost trying to keep the aircraft in nose up direction can improve his chances safe ejection.

In future, the authors plan to present an overall control law for a successful ejection process by considering various factors such as pitch, roll and yaw effects of the pilot and the applied force to avoid rollover and spin of the pilot.

References

1. Specker, L., Plaga, J., Santi, V.: Ejection seat capabilities to meet agile aircraft requirements. Air Force Research Laboratory TR015-09, pp. 121–129 (2001)
2. Newman, D.G.: Survival outcomes in low-level ejections from high performance aircraft. Aviat. Space Environ. Med. **84**(10), 1061–1065 (2013)
3. Lewis, M.E.: Spinal injuries caused by the acceleration of ejection. J. R. Army Med. Corps **148**(1), 22–26 (2002)
4. Latham, S.: A study in body ballistics: seat ejection. Proc. R. Soc. Lond. Ser. B Biol. Sci. **147**(926), 121–139 (1957)
5. Stech, E.L., Payne, P.R.:. Dynamic models of the human body. Frost Engineering Development Corporation, Englewood, CO (1969)
6. Ruff, S.: Brief acceleration: less than one second. In: German Aviation Medicine, World War II, vol. I. Department of the Air Force, pp. 584–598 (1950)
7. Eiband, A.M.: Human tolerance to rapidly applied accelerations: a summary of the literature. NASAMEMO-5-19-59E. National Aeronautics and Space Administration, Cleveland, OH (1959)
8. Brinkley, J.W., Schaffer, J.T.: Dynamic simulation techniques for the design of escape systems: current applications and future air force requirements. AMRL-TR-71-291971. Aerospace Medical Research Laboratory, Wright-Patterson Air Force Base, OH (1971)
9. Ejection Injury Criteria—USAF, 8 Nov 2016
10. Deb, K., Agarwal, R.B.: Simulated binary crossover for continuous search space. Complex Syst. **9**, 115–148 (1995)
11. Deb, K.: Multi-objective Optimization using Evolutionary Algorithms. Wiley, New York (2001)
12. Payne, P.R.: Dynamics of human restraint systems. In: Impact Acceleration Stress: A Symposium Washington, DC, 27–29 Nov 1962
13. Robinson, G., Jovanoski, Z.: Fighter pilot ejection study as an educational tool. Teaching Mathematics and Its Applications (2010): hrq011
14. Parsopoulos, K.E., Vrahatis, M.N.: Particle swarm optimization method in multiobjective problems. In: Proceedings of the 2002 ACM Symposium on Applied Computing. ACM (2002)
15. Kennedy, J., Eberhart, R: Particle swarm optimization. In: Proceedings of IEEE International Conference on Neural Networks IV, vol. 1000 (1995)
16. Stemper, B.D., et al.: Rate-dependent fracture characteristics of lumbar vertebral bodies. J. Mech. Behav. Biomed. Mater. **41**, 271–279 (2015)

Transmission Congestion Relief with Integration of Photovoltaic Power Using Lion Optimization Algorithm

Sadhan Gope, Subhojit Dawn, Rituparna Mitra, Arup Kr. Goswami and Prashant Kr. Tiwari

Abstract Transmission congestion is a vital problem in a deregulated power system. This paper proposes a novel transmission congestion management approach considering photovoltaic (PV) power using lion optimization algorithm (LOA). The main contributions of this paper have twofolds. Initially, the values of bus sensitivity factor (BSF) and generator sensitivity factor (GSF) are, respectively, used to select the optimal bus to integrate PV power and to select the participating generators for congestion management. Finally, LOA is used to determine the active power rescheduling amount and congestion cost. Test results on modified 39 bus New England system indicate that the LOA approach could provide a less active power rescheduling amount and congestion cost with integration of PV power compared to particle swarm optimization (PSO) and ant lion optimizer (ALO) algorithm.

Keywords Photovoltaic power · Bus sensitivity factor · Generator sensitivity factor · Lion optimization algorithm

S. Gope
Electrical Engineering Department, Mizoram University, Aizawl, Mizoram, India
e-mail: sadhan.nit@gmail.com

S. Dawn (✉)
Electrical Engineering Department, Siliguri Institute of Technology, Siliguri, West Bengal, India
e-mail: subhojit.dawn@gmail.com

R. Mitra · A. Kr. Goswami · P. Kr. Tiwari
Electrical Engineering Department, National Institute of Technology Silchar, Silchar, Assam, India
e-mail: rituparnamitra1990@gmail.com

A. Kr. Goswami
e-mail: gosarup@gmail.com

P. Kr. Tiwari
e-mail: prashant081.in@gmail.com

© Springer Nature Singapore Pte Ltd. 2019
J. C. Bansal et al. (eds.), *Soft Computing for Problem Solving*,
Advances in Intelligent Systems and Computing 816,
https://doi.org/10.1007/978-981-13-1592-3_25

1 Introduction

Violation of thermal limits or voltage limits or stability limits results in congestion in power system. Independent system operator (ISO) is responsible for relieving that transmission congestion by taking proper action. Most of the congestion management works concentrate on generator rescheduling technique and incorporation of FACTS devices in existing system network. To mitigate transmission congestion, lots of research works have been done in recent past years [1–6]. To alleviate line congestion, generators with strongest and non-uniform sensitivity indexes in the most sensitive zones are identified for rescheduling their generation [1]. Linear sensitivity factor-based approaches for congestion management are discussed in [2]. Paper [3] proposed a novel congestion management method based on the voltage stability margin sensitivities. Using this sensitivity, the system operator improves the voltage level and maintains the security level. To mitigate congestion, an optimal power flow (OPF) model is proposed in two-sided auction market structure with power demand elasticity. To determine the optimal solution for managing congestion, ISO runs an OPF taking into account the network constraints [4]. Optimal transmission dispatch in a competitive environment dominated by bilateral and multilateral transactions considering willingness to pay premium for minimum curtailment strategy is proposed in [5, 6].

Generator sensitivity factor is used for optimum selection of participating generators for congestion mitigation by rescheduling their output, and PSO algorithm is used for minimizing the deviations of rescheduled values of generator power outputs [7]. Thukaram et al. [8] have extensively studied the congestion management problem based on active power rescheduling amount considering relative electrical distance (RED) concept. He has proposed RED technique to minimize the system losses and to maintain the better voltage profile. Generators having the same RED but different price bids must reschedule their outputs in such a way that total cost of rescheduling becomes minimal. To alleviate transmission network congestion and to minimize rescheduling cost in a pool electricity market, generators active power rescheduling approach using firefly algorithm are employed considering load bus voltage and line loading [9]. Based on the generator sensitivity, generators are partially re-dispatched to eliminate the congestion in a pool electricity market and cost of congestion is distributed to the consumers connected to that congested parts [10]. If the generator rescheduling is not sufficient to remove the congestion, then load shedding approach is used as a final option to mitigate that congestion [11].

To relieve congestion, series FACTS devices are used as effective tools for congestion management studies. Optimally location and sizing of series FACTS devices are determined by the locational marginal prices (LMPs) on the congested branches [12]. In Ref. [13], authors presented an approach for short-term congestion management considering FACTS devices, demand response (DR) programs and generation re-dispatch subjected to minimization of congestion costs. Transmission congestion is managed by optimal placing the FACTS devices such as UPFC, TCSC and SSSC in deregulated power system [14–16]. Paper [17] deals with active and reactive power-based congestion management with FACTS devices in deregulated electricity market. Genetic algorithm and particle swarm optimization algorithm-based congestion management considering FACTS devices under deregulated power system is presented in [18–20].

Nowadays, renewable energy sources (RESs) power generation, especially solar and wind power generation, increases very fast because their eco-friendly behaviour. To manage congestion and minimize rescheduling cost of RES and conventional generators, optimal utilization of RES under day-ahead electricity market environment is presented based on the apparent power congestion index [21]. In [22], authors have used wind farm in the power system for mitigating transmission congestion. Congestion management is solved by placing one or more distributed generators (DGs) on congested lines. The main concern is given for determining the exact location of congested line and the placement of optimal size of DG considering reduce line losses and minimum congestion cost [23–25]. Due to the unpredictable nature of PV power, ISO is giving more concentration regarding the integration of PV power. Several research works have already been carried out to integrate the PV power in power system [26–28].

In the view of the above literature survey, this paper proposes a congestion management approach considering PV power with the help of LOA algorithm. The proposed congestion management technique is solved with the help of bus sensitivity factor (BSF) and generator sensitivity factor (GSF). To mitigate congestion, selective sensitive generators are rescheduled using LOA algorithm. The proposed congestion management method has been tested on modified 39 bus New England system. The generators' rescheduling amount and congestion cost, as obtained by LOA algorithm, are compared to those reported in the literature like PSO [7] and also with ALO algorithm.

2 Modelling of Photovoltaic Power Generation

Photovoltaic (PV) power depends on the radiation and rating of solar PV panel. The output power of a PV panel can be calculated as:

$$P_{pv}(t) = P_{pv}^r f_l^{pv} \frac{R_{pv}^i(t)}{R_{pv}^i} \tag{1}$$

where P_{pv}^r and $f_{l,}^{pv}$ respectively, denote the rating and loss factor of PV panel. $R_{pv}^i(t)$ and R_{pv}^i, respectively, denote the hourly incident radiation and standard incident radiation at the surface of PV panel. Battery is mainly used to store the PV power. Battery-stored energy can be measured with proper estimating the state of charge (SOC), which is expressed as follows:

$$\frac{B_T^{soc}(t)}{B_T^{soc}(t-1)} = \int_{T-1}^{T} \frac{P_{pv}(t)\eta_{bt}}{V_{bus}} dt \tag{2}$$

$$\eta_{bt} = \sqrt{\eta_{bt}^c \eta_{bt}^d} \tag{3}$$

where B_{soc}^T is the state of charge of battery, V_{bus} is the bus voltage and η_{bt} is the overall efficiency of the battery. η_{bt}^c and η_{bt}^d, respectively, represent the charging and discharging efficiency of a battery. Maximum value of $B_{T max}^{soc}$ is calculated by the aggregated capacity of battery bank ($B_{bank}^T(Ah)$).

$$B_{bank}^T(Ah) = \frac{N_{bt}}{N_{bt}^s} B_n(Ah) \tag{4}$$

$$N_{bt}^s = \frac{V_{bus}}{V_{bt}} \tag{5}$$

$$P_{max}^{bt} = \frac{N_{bt} V_{bt} I_{max}^{bt}}{1000} \tag{6}$$

where N_{bt} and N_{bt}^s, respectively, represent the total number of battery and number of battery connected in series. $B_n(Ah)$ and V_{bt}, respectively, denote capacity and voltage of a single battery. P_{max}^{bt} and I_{max}^{bt}, respectively, denote the maximum power and current of a battery. To feed the power in grid, suitable inverter is used to convert DC power into AC power. The inverter output power is calculated by using following equation:

$$P_{inv}^o(t) = P_{bt}^m(t)/\eta_{inv} \tag{7}$$

$$P_{bt}^m(t) = n P_{max}^{bt} \tag{8}$$

where $P_{bt}^m(t)$ is the total battery power inject in grid. n is the number of series-connected battery in inverter circuit and η_{inv} denotes the inverter efficiency.

3   Power Flow Through Congested Transmission Line

The active power flow through a congested line between bus-i and bus-j can be written as [7, 22]

$$P_{ij} = |V_i||V_j||Y_{ij}|\cos(\theta_{ij} - \delta_i + \delta_j) - V_i^2 Y_{ij} \cos\theta_{ij} \tag{9}$$

where V_i and δ_i, respectively, denote the voltage magnitude and angle at bus-i. Y_{ij} and θ_{ij}, respectively, denote the magnitude and angle of ijth element of Y_{Bus} matrix. From Eq. (9), BSF [7, 22] and GSF [7, 22] can be derived, which is denoted by the following mathematical equations:

$$BSF_{bus} = \frac{\Delta P_{ij}}{\Delta P_n} \qquad (10)$$

$$GSF_{gen} = \frac{\Delta P_{ij}}{\Delta P_{gk}} \qquad (11)$$

4 Problem Formulation

The main objective of this paper is to minimize the rescheduling cost by minimizing the active power rescheduling amount of thermal generators. Mathematically objective can be expressed by the following equation [7, 22]:

$$\text{Minimize} \sum_{k=1}^{N_g} C_{cost} \times \Delta P_{gk} \qquad (12)$$

ΔP_{gk} is the real power adjustment of the generator and C_{cost} are the incremental or decremented price submitted by generators. The solution of the above equation can be obtained subject to fulfil the following equality and inequality constraints:

Equality constraints:

$$P_{gk} - P_{dk} = \sum_{j=1}^{n} |V_k||V_j||Y_{kj}| \cos(\delta_k - \delta_j - \theta_{kj}) \quad k = 1, 2, 3 \ldots, N_g \qquad (13)$$

$$Q_{gk} - Q_{dk} = \sum_{j=1}^{n} |V_k||V_j||Y_{kj}| \sin(\delta_k - \delta_j - \theta_{kj}) \quad k = 1, 2, 3 \ldots, N_g \qquad (14)$$

$$P_{gk} = P_{gk} + P_{grid}^c \pm \sum_{k=1}^{N_g} \Delta P_{gk} \quad k = 1, 2, 3 \ldots, N_g \qquad (15)$$

$$P_{dmp}(t) = P_{inv}^o(t) - P_{grid}^c(t) \qquad (16)$$

Inequality constraint:

$$\sum_{k=1}^{N_g} ((GSF_{gk}) \times \Delta P_{gk}) + S_{kj}^0 \leq S_{kj}^{max} \qquad (17)$$

$$P_{gk}^{min} \leq P_{gk} + \Delta P_{gk} \leq P_{gk}^{max} \quad k = 1, 2, \ldots, N_g \qquad (18)$$

$$Q_{gk}^{min} \leq Q_{gk} \leq Q_{gk}^{max} \quad k = 1, 2, \ldots, N_g \qquad (19)$$

$$P_{gk} - P_{gk}^{min} = \Delta P_{gk}^{min} \leq \Delta P_{gk} \leq \Delta P_{gk}^{max} = P_{gk}^{max} - P_{gk} \quad k = 1, 2, \ldots, N_g \qquad (20)$$

$$V_j^{\min} \leq V_j \leq V_j^{\max} \quad j = 1, 2, 3 \ldots, N_b \tag{21}$$

where $P_{gk}^{\min}$ and $P_{gk}^{\max}$, respectively, denote the minimum and maximum real power limits of generators. $Q_{gk}^{\min}$ and $Q_{gk}^{\max}$, respectively, denote the minimum and maximum reactive power limits of generators. $\Delta P_{gk}^{\min}$ and $\Delta P_{gk}^{\max}$, respectively, denote the minimum and maximum limit of the change in generator active power output. P_{dk} and Q_{dk} are the active and reactive power demand at kth bus. P_{dmp} and P_{grid}^c are the active power consumed by dumped load and grid, respectively. S_k^0 denotes the power flow in the transmission line k; $S_k^{\max}$ denotes the maximum power flow limit of kth transmission line connected between bus-i and bus-j. N_g, N_b and N_L are the number of generators, number of buses and number of transmission lines.

5   Lion Optimization Algorithm (LOA)

Nature-inspired metaheuristic lion optimization algorithm (LOA) is proposed by Maziar Yazdani in 2016 [29]. The LOA algorithm is inspired by the special lifestyle and cooperation characteristics of lion. Lions typically hunt together with other members of their pride. LOA has a specific strategy to encircle the prey and catch it from the randomly generated population over the solution space. Several lionesses are work together and encircle the prey from different points and catch the victim with a fast attack [29].

By moving towards the selected safe area, LOA saves the best solutions and improves obtained so far. The mating operator and defence operator, respectively, share information between genders, while new cubs inherit character from both genders and retain powerful male lions as solutions in LOA. Migration operator enhances the diversity of the target pride by its position in the previous pride. The details description of LOA algorithm is given in [29]. Implementation flowchart of LOA algorithm for the proposed congestion management approach is shown in Fig. 1.

6   Result and Discussion

The proposed congestion management work is validated with modified 39 bus New England test system, and the test system data is obtained from [22]. In this test system, N-1 contingency analysis is carried out to make the system congested. To do this, line outage of 14–34 has been done, and as a result, 15–16 line is congested. The power flow through congested line pre- and post-rescheduling is shown in Table 1. The maximum power flow of the congested line is 500 MVA. From Table 1, it is observed that congested line power flow reduces below its maximum limit after rescheduling the thermal generators with integration of PV power.

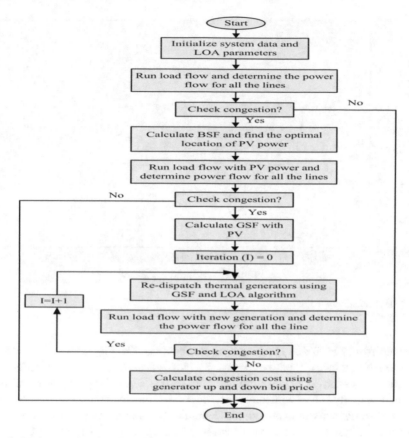

Fig. 1 Implementation flowchart of LOA algorithm

Table 1 Power flow through congested line pre- and post-rescheduling

Power flow (MVA)	Pre-rescheduling without PV	Pre-rescheduling with PV	Post-rescheduling with 30 MW PV		
		30 MW	PSO	ALO	LOA
L(15-16)	628	604	498	497	496

BSF takes very important role in this study because based on the BSF value, PV is optimally installed in the most sensitive buses. Figure 2 shows the BSF value of the system. From Fig. 2, it is observed that bus number 14 has strong negative GSF value and bus number 34 has strong positive GSF value compared to other buses of the system. So, optimal location of PV is decided in bus number 14.

The generators are selected based on the GSF value, and selected generators are rescheduled to mitigate congestion. Values of GSFs are calculated with and without PV power with respect to the congested line 15–16 and shown in Fig. 2. It is

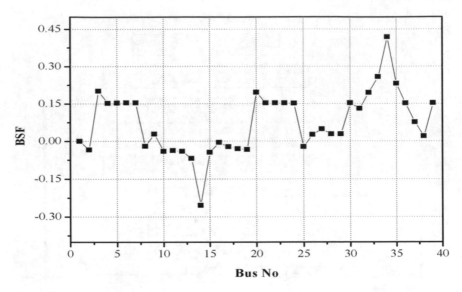

Fig. 2 Bus sensitivity factor

observed from Fig. 3 that the generator numbers G_4, G_5, G_6, G_7, G_9 have almost same sensitivity index in both the cases. So these generators are not rescheduled to mitigate congestion. Only generator numbers G_2, G_3, G_8 and G_{10} have non-uniform flow of sensitivity values with integration of PV. So these generators are rescheduled to minimize transmission congestion. Generator number G_1 is also rescheduled to minimize the system losses. For rescheduling these generators, ALO and LOA algorithms are used and results obtained from these algorithms are compared with earlier congestion management results PSO [7, 8].

Table 2 shows the active power rescheduling amount with and without PV power. To verify the impact of PV in congestion management, 30 MW PV power is considered here. Total rescheduling amount with PV power by using LOA algorithm is obtained 365.14 MW, which is very less compared to without PV rescheduling amount 530.87 MW. In the proposed method, less numbers of participating generator are required for congestion management as compared with result reported in [7, 8].

Table 3 shows the generators setting and congestion cost with and without PV power. For calculating the congestion cost, PV power investment cost and running cost is not considered here. From Table 3, it is observed that generators setting and congestion cost achieved by applying LOA algorithm is the least one as compared to PSO and ALO algorithms.

The comparative convergence profile of PSO, ALO and LOA algorithms for the congestion cost with PV power is presented in Fig. 4. From Fig. 4, it is observed that the convergence profile of congestion cost ($/h) for the LOA algorithm is promising one.

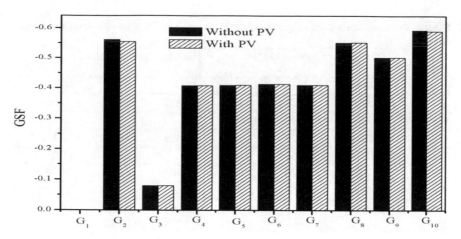

Fig. 3 Generator sensitivity factor with and without PV power

Table 2 Active power rescheduling amount of generator with and without PV power

Gen no.	Active power rescheduling without PV				Active power rescheduling with PV		
	Result [8]	PSO [7]	ALO	LOA	PSO	ALO	LOA
G_1	−99.59	−143.00	−147.40	−146.20	−154.62	−143.31	−141.22
G_2	98.75	72.40	63.70	69.30	NR	NR	NR
G_3	−159.64	−127.90	−119.80	−119.25	−46.40	−55.62	−56.36
G_4	12.34	NR	NR	NR	NR	NR	NR
G_5	24.69	NR	NR	NR	NR	NR	NR
G_6	24.69	NR	NR	NR	NR	NR	NR
G_7	12.34	NR	NR	NR	NR	NR	NR
G_8	24.69	71.90	72.60	72.20	NR	NR	NR
G_9	12.34	48.70	56.20	52.42	27.00	32.12	32.05
G_{10}	49.38	77.70	74.80	71.50	144.02	136.67	135.51
Total (MW)	518.45	541.60	534.6	530.87	372.04	367.8	365.14

NR not rescheduled

Table 4 shows the minimum bus voltage and total system loss before and after rescheduling the thermal generators. From Table 4, it is seen that total system loss is decreased to 56.15 MW and minimum bus voltage increases to 0.9478 p.u. after congestion management with PV power using LOA algorithm.

Table 3 Generator setting and congestion cost with and without PV power for mitigating congestion

Control variable	Generator setting without PV			Generator setting with PV		
	PSO	ALO	LOA	PSO	ALO	LOA
G_1	5.3487	5.3047	5.3167	5.2325	5.3456	5.3665
G_2	10.724	10.6370	10.693	NR	NR	NR
G_3	5.2210	5.3020	5.3075	6.0360	5.9438	5.9364
G_8	6.1190	6.1260	6.1220	NR	NR	NR
G_9	8.7870	8.8620	8.8242	8.5700	8.6212	8.6205
G_{10}	3.2770	3.2480	3.2480	3.9402	3.8667	3.8551
Resch. cost ($/h)	92.817	88.495	86.261	65.842	63.384	62.324

Fig. 4 Comparative convergence characteristics with PV power

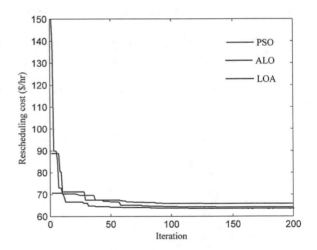

Table 4 System active power loss and minimum bus voltage

Algorithm name	Before rescheduling without wind farm		After rescheduling without PV power		After rescheduling with 30 MW PV power	
	P_{loss} (MW)	V_{min} (p.u)	P_{loss} (MW)	V_{min} (p.u)	P_{loss} (MW)	V_{min} (p.u)
PSO	59.34	0.935	57.31 [7]	0.9450 [7]	57.24	0.9518
ALO			56.84	0.9412	56.65	0.9524
LOA			56.75	0.9408	56.15	0.9478

7 Conclusion

This paper has presented PV power-based congestion management approach with the help of LOA algorithm. The main objective of this paper is to minimize the congestion cost by mitigating transmission congestion without violating system security constraints. Modified 39 bus New England system is used to analyse the proposed congestion management approach. Results obtained by proposed approach are better in terms of active power rescheduling amount and congestion cost as compared to the other reported results of congestion management.

References

1. Kumar, A., Srivastava, S.C., Singh, S.N.: A zonal congestion management approach using real and reactive power rescheduling. IEEE Trans. Power Syst. **19**(1), 554–562 (2004)
2. Fattahi, A., Ehsan, M.: Sensitivity based re-dispatching method for congestion management in a pool market. Int. J. Emerg. Electr. Power Syst. **3**(2), 1077–1085 (2005)
3. Esmaili, M., Shayanfar, H.A., Amjady, N.: Congestion management considering voltage security of power systems. Energy Convers. Manage. **50**, 2562–2569 (2009)
4. Marannino, P., Vailati, R., Zanellini, F., Bompard, E., Gross, G.: OPF tools for optimal pricing and congestion management in a two sided auction market structure. In: IEEE Porto Power Tech Conference, pp. 1–7 (2001)
5. Fang, R.S., David, A.K.: Optimal dispatch under transmission contracts. IEEE Trans. Power Syst. **14**(2), 732–737 (1999)
6. Fang, R.S., David, A.K.: Transmission congestion management in an electricity market. IEEE Trans. Power Syst. **14**(3), 877–883 (1999)
7. Dutta, S., Singh, S.P.: Optimal rescheduling of generators for congestion management based on particle swarm optimization. IEEE Trans. Power Syst. **23**(4), 1560–1568 (2008)
8. Yesuratnam, G., Thukaram, D.: Congestion management in open access based on relative electrical distances using voltage stability criteria. Electr. Power Syst. Res. **77**, 1608–1618 (2007)
9. Verma, S., Mukherjee, V.: Firefly algorithm for congestion management in deregulated environment. Int. J. Eng. Sci. Technol. **19**, 1254–1265 (2016)
10. Fattahi, A., Ehsan, M.: Sensitivity based re-dispatching method for congestion management in a pool model. Int. J. Emerg. Electr. Power Syst. **3**(2), 1–20 (2005)
11. Talukdar, B.K., Sinha, A.K., Mukhopadhyay, S., Bose, A.: A computationally simple method for cost-efficient generation rescheduling and load shedding for congestion management. Int. J. Electr. Power Energy Syst. **27**(5), 379–388 (2005)
12. Esmaili, M., Shayanfar, H.A., Moslemi, R.: Locating series FACTS devices for multi-objective congestion management improving voltage and transient stability. Eur. J. Oper. Res. **236**, 763–773 (2014)
13. Mohd Zin, A.A., Moradi, M., Khairuddin, A.: Short-term congestion management by strategic application of FACTS devices and demand response programs under contingency conditions. IETE J. Res. **63**(1), 109–123 (2017)
14. Reddy, K.R.S., Padhy, N.P., Patel, R.N., Congestion management in deregulated power system using FACTS devices. In: IEEE Power India Conference, pp. 1–8 (2006)
15. Retnamony, R., Jacob Raglend, I.: Congestion management is to enhance the transient stability in a deregulated power system using FACTS devices. In: International Conference on Control, Instrumentation, Communication and Computational Technologies, pp. 744–752 (2015)
16. Singh, J.G., Singh, S.N., Srivastava, S.C.: Congestion management by using FACTS controller in power system. In: IEEE Region 10 Humanitarian Technology Conference (R10-HTC), pp. 1–7 (2016)
17. Phichaisawat, S., Song, Y.H., Wang, X.L., Wang, X.F.: Combined active and reactive congestion management with FACTS devices. Electr. Power Compon. Syst. **30**, 1195–1205 (2002)
18. Singh, D., Verma, K.S.: GA-based congestion management in deregulated power system using FACTS devices. In: International Conference & Utility Exhibition on Power and Energy Systems: Issues and Prospects for Asia (ICUE), pp. 1–6 (2011)

19. Hosseinipoor, N., Nabavi, S.M.H.: Optimal locating and sizing of TCSC using genetic algorithm for congestion management in deregualted power markets. In: 9th International Conference on Environment and Electrical Engineering, pp. 136–139 (2010)
20. Hashemzadeh, H., Hosseini, S.H.: Locating series FACTS devices using line outage sensitivity factors and particle swarm optimization for congestion management. In: IEEE Power & Energy Society General Meeting, pp. 1–6 (2009)
21. Nesamalar, J.J.D., Venkatesh, P., Charles Raja, S.: Optimal utilization of renewable energy sources for congestion management. IFAC-Papers, pp. 264–269 (2015)
22. Deb, S., Gope, S., Goswami, A.K.: Congestion management considering wind energy sources using evolutionary algorithm. Electr. Power Compon. Syst. **43**, 723–732 (2015)
23. Singh, A.K., Parida, S.K.: Congestion management with distributed generation and its impact on electricity market. Electr. Power Energy Syst. **48**, 39–47 (2013)
24. Singh, K., Yadav, V.K., Padhy, N.P., Sharma, J.: Congestion management considering optimal placement of distributed generator in deregulated power system networks. Electr. Power Compon. Syst. **42**(1), 13–22 (2014)
25. Kashyap, M., Kansal, S.: Hybrid approach for congestion management using optimal placement of distributed generator. Int. J. Ambient Energy (2017). http://dx.doi.org/10.1080/01430750.2016.1269676
26. Mills, D.: Advances in solar thermal electricity technology. Sol. Energy **76**, 19–31 (2004)
27. Kabalci, E.: Design and analysis of a hybrid renewable energy plant with solar and wind power. Energy Convers. Manage. **72**, 51–59 (2013)
28. Mohammadi, M., Hosseinian, S.H., Gharehpetian, G.B.: Optimization of hybrid solar energy sources/wind turbine systems integrated to utility grids as microgrid (MG) under pool/bilateral/hybrid electricity market using PSO. Sol. Energy **86**, 112–125 (2012)
29. Yazdani, M., Jolai, F.: Lion Optimization Algorithm (LOA): a nature inspired meta-heuristic algorithm. J. Comput. Des. Eng. **3**, 24–36 (2016)

A Novel CPU Scheduling Algorithm Based on Ant Lion Optimizer

Shail Kumar Dinkar and Kusum Deep

Abstract In a multiprogramming environment, operating system plays a vital role to schedule the various user processes or tasks in different queues in efficient manner so that the system performance enhances in terms of increased throughput and reduced process waiting time. Processes carry varying time slices to be serviced by the processor. This variation of time slice authorizes the scheduler to schedule the processes so that it can provide an appropriate response time. Early response agreed by the processor after submitting a process in a queue ensures the less waiting time which suggests enhanced multiprogramming environment keeping more number of processes to get chance of early execution. This paper proposes a new CPU scheduling policy based on novel nature-inspired optimization technique, namely ant lion optimizer (ALO). This algorithm schedules the processes in such a way that the average waiting time is minimized. The proposed approach is compared with the widely used three CPU scheduling policies: first come first serve (FCFS), shortest computation time first (SCTF), and round robin (RR).

Keywords CPU scheduling · Waiting time · Optimization · Ant lion optimizer

1 Introduction

The allotment of resources is taken care by scheduling task while satisfying certain performance requirements [1]. The task or process scheduling has always been received a large amount of attention by research community. In general, scheduling is the activity to choose next request to be processed by the processor (or server). Servicing these requests by the processor is based on kind of scheduling policies [2].

S. K. Dinkar (✉) · K. Deep
Department of Mathematics, Indian Institute of Technology Roorkee, Roorkee,
Uttarakhand, India
e-mail: shailkdinkar@gmail.com

K. Deep
e-mail: kusumfma@iitr.ac.in

© Springer Nature Singapore Pte Ltd. 2019
J. C. Bansal et al. (eds.), *Soft Computing for Problem Solving*,
Advances in Intelligent Systems and Computing 816,
https://doi.org/10.1007/978-981-13-1592-3_26

Scheduling is performed by schedulers. The goal of scheduler is to allocate tasks to the available processors [3]. A task or process is said to arrive when it is submitted by the user. Then, it is admitted into the ready queue when scheduler considers it for execution. The admitted process either waits in the list of pending queue or gets executed. This scheduler schedules the processes based on various CPU scheduling policies.

Few policies execute the process until it gets executed completely. These policies are termed as non-preemptive while others do not ensure the complete execution of the process in one goes. Such policies are preemptive scheduling policies. In that situation, process has to enter into waiting queue and waits for their turn again for execution. The performance metrics which are influenced due to CPU scheduling can be defined as average waiting time of process, average turnaround time, maximum throughput, CPU utilization, and response time [4]. It is desirable to minimize waiting time, turnaround time, and response time while maximizing CPU utilization and throughput. There are number of CPU scheduling algorithms such as first come first serve (FCFS) [5] and shortest computation time first (SCTF) fall under non-preemptive scheduling. The round robin (RR), priority scheduling, multilevel feedback queue, etc., come under preemptive scheduling.

Most of the scheduling algorithms follow deterministic approach but recently probabilistic approaches are extensively used for scheduling. A well-established heuristic algorithm GA is being used for real-time task scheduling problems as it releases the practitioners from knowing how to guess a solution and the practitioners need to know about assessing the solution [6]. In this paper, ant lion optimizer (ALO) [7] is employed as a heuristic algorithm for scheduling task so that the average waiting time of tasks gets minimized.

The paper is organized as follows: Sect. 2 provides some conceptual preliminaries about the established scheduling algorithms. Section 3 defines the problem formulation as an optimization problem. Section 4 describes the method of solution and proposed algorithm. Section 5 depicts the computation study. Section 6 describes the analysis of the proposed algorithm in terms of degree of multiprogramming and data distribution using boxplot. Section 7 concludes the paper.

2 Background Materials

Resource allocation is the most significant task of operating system. This task is accomplished using diverse scheduling policies supported by various systems. Some prominent scheduling policies are first come first serve (FCFS), shortest computation time first (SCTF), and round-robin (RR) scheduling algorithms. All these scheduling policies follow deterministic approach to determine the various performance metrics. These strategies are discussed one by one as follow:

First come first serve (FCFS) is a non-preemptive scheduling algorithm which states that the process who requests first will get processor (CPU) first. This algorithm is implemented as first in first out (FIFO) queue [4]. If a process with larger burst time requests processor earlier than the process having less CPU burst time, then this algorithm is responsible for penalizing a process for large waiting time, thus reducing the CPU utilization.

Shortest computation time first (SCTF) is another non-preemptive CPU scheduling algorithm. Conceptually, it reduces the process waiting time as the scheduler selects the process from ready queue having smallest CPU burst time. This ensures the minimum average waiting time and also overcomes the drawback of FCFS where a process with small CPU burst time has to wait behind the processes having larger CPU burst time. But this algorithm requires the advance knowledge of the next CPU burst time [8].

The length of the next CPU burst is not known but approximated value can be predicted. It is generally predicted as an exponential average [4]. Let t_n be the length of nth CPU burst and let τ_{n+1} be the predicted value of next CPU burst. Then define

$$\tau_{n+1} = \alpha t_n + (1 - \alpha)\tau_n$$
$$\text{For all } 0 \leq \alpha \leq 1$$

Generally, t_n contains the most recent values and τ_n comprises the past history. In general, α is kept as $\frac{1}{2}$ to normalize the recent information and past information equally weighted.

Round robin (RR) [4] algorithm is designed for time-sharing system and similar to FCFS algorithm except that it follows preemption strategy to enable the system to switch among the processes. A time quantum q (small unit of time) is defined. The ready queue is now treated as the circular queue. The CPU scheduler goes around the ready queue and allocates the CPU to each process equal to one-time quantum q. The ready queue is implemented as FIFO and new processes are added to the tail of ready queue. The CPU scheduler picks the first process from the ready queue, sets timer to interrupt after one-time quantum and allocate to the CPU.

The above-discussed scheduling strategies are used to perform comparison with the proposed ALO algorithm-based CPU scheduling strategy.

3 Problem Formulation

The goal is to enhance the performance of CPU. One of the performance metrics which is affected by the CPU scheduling algorithm is average waiting time. The multiprogramming system frequently switches from one process to another while serving the process in various queues. More process waiting time means less CPU utilization. Thus, it is desirable to minimize the waiting time to increase the CPU utilization. Hence, it becomes an optimization problem to minimize the average waiting time of a process.

In this paper, the problem is modeled on the basis of arriving sequence of processes on first come first serve (FCFS) basis. Let $[P_1, P_2, \ldots, P_n]$ be the sequence of n processes arriving in same sequence and let $[b_1, b_2, \ldots, b_n]$ be the time required for each process to be executed by the processor (burst time). Assuming all processes arrive at time t_1 in the given sequence with δt difference of time with respect to each other which is assumed to be negligible. Then, the waiting time can be calculated as:

$$\text{Waiting time } w_i = t_1 + (b_1 - t_1) + (b_2 - t_1) + \cdots + (b_n - t_1) \tag{1}$$

where

$w_1 = t_1 = $ waiting time of process P_1,
$w_2 = (b_1 - t_1) = $ waiting time of process $P_2, \ldots,$
$w_n = (b_n - t_1) = $ waiting time of process P_n

Then, average waiting time can be defined as:

$$\text{Average waiting time} = \sum_{i=1}^{n} \frac{w_i}{n} \tag{2}$$

where w_i is waiting time of ith process, and n is the number of processes.

It is termed as objective function and can be defined as optimization problem to be minimized:

$$\text{Min } f = \sum_{i=1}^{n} \frac{w_i}{n} \tag{3}$$

where w_i is waiting time of ith process, and n is the number of processes.

4 Method of Solution: Ant Lion Optimizer

In this paper, a new nature-inspired optimization algorithm called Ant Lion optimizer (ALO) proposed by Mirjalili [7] is used for minimizing the average waiting time involved in processor for process execution. Conceptually, this nature-inspired optimization algorithm is based on hunting ants by antlions. The action of hunting is performed in an exceptional way in which antlion digs a cone-shaped pit in sand, sits at the bottom of the pit, and waits for an ant to be dropped into it so that it can consume it.

The functional aspect of this algorithm relies on two random walks performed by ants around the selected antlion by roulette wheel selection method and around the elite (best) antlion in current generation. Antlions build traps to catch the ants. Once an ant is caught by the antlion, trap is again rebuilt. This behavior of randomly picking antlion and random walk of ant around antlion ensures the exploration by visiting the search region thoroughly. Random walk is implemented using uniformly distributed random numbers. The performance of ALO is improved by modifying random walk in [9, 10].

Let N be the number of ants. Let the initial random population of ants is $K_{ant} = \left(K_{A,1}, K_{A,2}, \ldots K_{A,n}, \ldots, K_{A,N} \right)^{\mathrm{T}}$ in the D-dimensional search space and $K_{A,n} = \left(K_{A,n}^1, \ldots K_{A,n}^d, \ldots K_{A,n}^D \right)$ is the position of the nth ant where $K_{A,n}^d$ is the position of dth variable of the nth ant. The objective function values of all ants in the D-dimensional search space may be defined in a matrix $M_{ant} = \left(M_{A,1}, M_{A,2} \ldots M_{A,n}, \ldots M_{A,N} \right)^{\mathrm{T}}$ where $M_{A,n} = f\left(K_{A,n}^1, \ldots K_{A,n}^d, \ldots K_{A,n}^D \right)$ is the objective function value of nth ant.

Similarly, the randomly initialized ant lion population can be defined as $K_{antlion} = \left(K_{AL,1}, K_{AL,2}, \ldots K_{AL,n}, \ldots, K_{AL,N} \right)^{\mathrm{T}}$, $K_{AL,n} = \left(K_{AL,n}^1, \ldots K_{AL,n}^d, \ldots K_{AL,n}^D \right)$ is the position of the nth ant lion where $K_{AL,n}^d$ is the position of dth variable of the nth ant lion and fitness values of ant lions can be defined as a matrix $M_{antlion} = \left(M_{AL,1}, M_{AL,2} \ldots M_{AL,n}, \ldots M_{AL,N} \right)^{\mathrm{T}}$ where $M_{AL,n} = f\left(K_{AL,n}^1, \ldots K_{AL,n}^d, \ldots K_{AL,n}^D \right)$ is the objective function value of nth ant lion.

The random walk can be defined as

$$Rw\left(K_{A,n}^d \right) = [\text{cumsum}(2r(i_1) - 1), \ \text{cumsum}(2r(i_2) - 1) \ldots \text{cumsum}(2r(i_{max}) - 1)] \quad (4)$$

In this study $K_{A,n}^d(i)$ can be defined as the dth variable of nth ant at ith iteration.

where cumsum describes cumulative sum of uniformly generated random numbers and $r(i)$ is defined as follows

$$r(i) = \begin{cases} 1 & \text{if rand} > 0.5 \\ 0 & \text{if rand} \leq 0.5 \end{cases} \quad (5)$$

where rand generates uniformly distributed random number between 0 and 1.

The normalization of random walk using min-max normalization is described as per the following equation:

$$K_{A,n}^d(i) = \frac{\left(K_{A,n}^d(i) - \min Rw\left(K_{A,n}^d \right) \right)\left(U^d(i) - L^d(i) \right)}{\max Rw\left(K_{A,n}^d \right) - \min Rw\left(K_{A,n}^d \right)} + L^d(t) \quad (6)$$

where $\min Rw\left(K_{A,n}^d \right)$ and $\max Rw\left(K_{A,n}^d \right)$ define the minimum and maximum of random walk for dth variable of nth ant; $U^d(i)$ and $L^d(i)$ are the upper and lower bounds of dth variable at ith iteration

The random walk is performed around two antlions: (a) elite antlion and (b) antlion selected using roulette wheel selection method. The new population is then generated by taking the average of these two random walks. It can be defined as follows:

$$K_{A,n}^d(i) = \frac{Rw_A(i) + Rw_E(i)}{2} \quad (7)$$

where $Rw_A(i)$ a random walk around is selected antlion K_{sel}, and $Rw_E(i)$ is random walk around K_{elite} elite antlion.

In the same way, exploitation is ensured by adaptive shrinking of antlion's traps. This phenomenon is defined as the decrease in the lower and upper bounds adaptively and mathematically defined as:

$$L^d(i) = \frac{L^d(i)}{I} \tag{8}$$

$$U^d(i) = \frac{U^d(i)}{I} \tag{9}$$

where I is a ratio defined as $I = 10^w \frac{i_{current}}{i_{max}}$ where $i_{current}$ is the current iteration and i_{max} is the maximum number of iteration, w is a constant based on the value of current iteration ($w = 2$ when $i_{current} > 0.1\, i_{max}$, $w = 3$ when $i_{current} > 0.5 i_{max}$, $w = 4$ when $i_{current} > 0.75\, i_{max}$, $w = 5$ when $i_{current} > 0.9 i_{max}$, and $w = 6\, i_{current} > 0.95 i_{max}$). Here, the value of w is used to control the exploitation process.

When the fitness value of ant is better (less) than fitness value antlion, antlion catches the ant and then updates its position to the ant's position as defined below:

$$K_{AL,j}(i) = K_{A,i}(i) \text{ if } f\big(K_{A,i}(i)\big) < f\big(K_{AL,j}(i)\big) \tag{10}$$

where $K_{AL,j}(i)$ is the position of jth antlion at ith iteration, and $K_{A,i}(i)$ is the ith ant's position at ith iteration. This process is obtained by concatenating all M_{ant} and M and sorts them from smallest to largest. Then, first N rows are updated as $M_{antlion}$ and the corresponding position of $K_{antlion}$.

4.1 Proposed Algorithm

Population (antlions) is initialized randomly. The length of the population (antlions) represents number of processes entered in ready queue. In algorithm, the initialized values of antlions are termed as the CPU burst time of the respective process (antlion). The fitness of each antlion is calculated using objective function as in Eq. (3) which represents the corresponding waiting time of process. One process P_E with minimum waiting time (elite) and another process P_A using roulette wheel method are selected.

Adaptive shrinking of search region is performed by adaptively decreasing lower and upper bound of the region as shown in Eqs. (8) and (9). Then, random walks $Rw_E(i)$ and $R_{wA}(i)$ are determined around these processes and new position of ants $K_{A,n}^d(i)$ is calculated using Eq. (7). The fitness of ants is determined using objective function using Eq. (3) which is designated as the new waiting time of the processes. The determined waiting time (fitness) of these ants is compared with the waiting time of elite antlion P_E and then updated using Eq. (10). The elitism property is preserved by carrying elite to the next generation.

As the number of maximum iteration reached, the total waiting time is determined by sum up the waiting time of each antlion (process) and average waiting time is calculated as follows:

$$\text{Average waiting time} = \sum_{i=1}^{N} \frac{f\left(K_{AL,j}(i)\right)}{N}$$

The pseudo-code of proposed algorithm is defined in Table 1.

5 Computational Study

Ten different data sets are considered in this problem each of which contains randomly initiated population of size 6 antlions. It is to be noted that each antlion is termed as process entered into the ready queue for execution; that is, the length of the task queue is 6. This random initial value of each antlion is termed as the CPU burst time of the process. All the experiments have been performed on MATLAB 7.10.0(R2010a) on Intel(R) Core(TM) i5-7200 CPU with clock speed @ 2.50 GHz–2.71 Ghz with 8 GB RAM. The existing CPU scheduling algorithms FCFS, SCTF, and RR are implemented on same data set to conclude fair comparison.

Table 2 shows a snapshot of 10 data sets each of which contains 6 tasks or processes used for numerical simulation. The CPU burst time is randomly generated for each process of each data set. First column shows the CPU burst time and process sequence. Second column depicts the average waiting time for four algorithms. The proposed ALO-based CPU scheduling algorithm is compared with the three widely used CPU scheduling algorithms: first come first serve (FCFS), round robin (RR), and shortest computation time first (SCTF). It is to be noted that the time quantum for RR algorithm is chosen 2 in this implementation.

It is clearly evident from Table 2 that the proposed algorithm outperforms all three algorithms completely in terms of average waiting time. Figure 1 demonstrates comparison of the average waiting time determined using proposed algorithm, FCFS, RR, and SCTF. It shows that proposed algorithm outperforms other existing CPU scheduling policies.

Figure 2 shows the bar diagram depicting comparison between proposed algorithms and other scheduling algorithms. Figure shows that the proposed algorithm displays highly significant improvement of average waiting time as compared to other algorithms.

Table 1 Pseudo-code of proposed algorithm using ALO

Let waiting time=0, i_{max} = 2;
1. Initialize the random population of ants K_{ant} = $(K_{A,1}, K_{A,2}, \ldots K_{A,n}, \ldots K_{A,N})^T$ and ant lions $K_{antlion}$ = $(K_{AL,1}, K_{AL,2}, \ldots K_{AL,n}, \ldots K_{AL,N})^T$ of size N.
2. Find out the fitness(waiting time) value of all ants and ant lions using eq. (3)
3. Determine the best(minimum fitness) of ant lions as the elite K_{elite}
4. Initialize $i_{current}$ = 2
5. **while ($i_{current} \leq i_{max}$)**

    for every ant($i = 1 \ldots N$)

        Find an ant lion K_{sel} using Roulette wheel
        Update lower bound L and upper bound U using equations Eqs (8) and (9)
        Create a random walk $Rw_A(i)$ around K_{sel} and $Rw_E(i)$ around K_{elite}
        Normalize random walk using Equation (4) and (6)
        Update the position of ant using (7)

    end for
    Calculate the fitness(waiting time) of all ants using eq.(3)
    Replace an ant lion with its corresponding ant if it becomes fitter using Eq.(10)
    Update K_{elite} if an ant lion becomes fitter(less) than the elite
    Increment iteration i.e. $i_{current}$=$i_{current}$+1
    waiting time=waiting time+ fitness of K_{elite}
  end while
    average waiting time=waiting time/N
6. Return elite

Table 2 Comparison of average waiting time for each task sequence

CPU burst time $[P_1, P_2, P_3, P_4, P_5, P_6]$	Average waiting time			
	ALO	FCFS	RR	SCTF
[9.5512, 3.0802, 6.4616, 5.3738, 9.0217, 7.8589]	7.7006	16.53848	24.66787	13.50342
[9.575, 7.3366, 9.5849, 6.3834, 8.5667, 4.9854]	12.4634	21.21827	32.50507	16.5298
[8.8764, 3.8611, 3.4591, 7.0885, 1.6405, 2.7693]	4.1013	14.33687	13.08488	7.411283
[5.6455, 3.0272, 2.653, 2.9467, 4.8446, 9.7352]	6.6324	9.838883	13.43782	8.244683
[4.9237, 8.3912, 6.9824, 6.8099, 8.5959, 9.5448]	12.3093	16.89103	29.63205	16.36393
[8.7748, 9.1557, 2.8013, 8.7346, 8.1636, 4.0438]	7.0033	19.08892	26.96297	13.48608
[1.4523, 1.8884, 1.4088, 2.8677, 2.9839, 1.7231]	3.522	4.6268	1.9077	4.111167
[3.1432, 1.3494, 9.3879, 5.5561, 3.7296, 9.248]	3.3735	10.68652	12.39412	8.478133
[1.3063, 5.0949, 2.1575, 3.5324, 9.9184, 3.3008]	3.2658	8.394467	9.381917	6.203933
[3.228, 4.9222, 9.798, 8.7396, 3.7056, 3.393]	8.07	14.40127	16.88263	9.902133

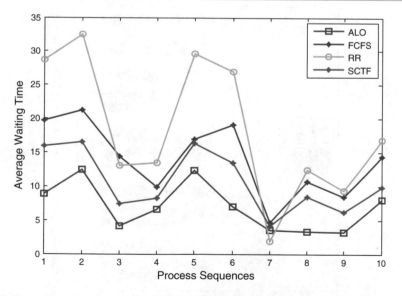

Fig. 1 Comparison of proposed method with existing scheduling algorithms

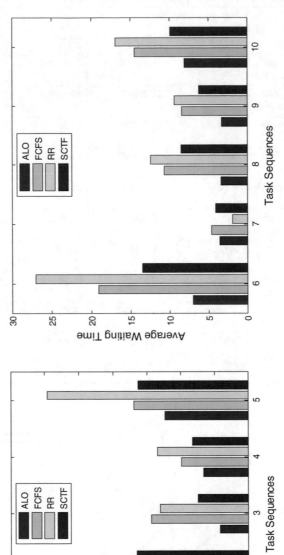

Fig. 2 Comparison of proposed method and other scheduling algorithms

6 Analysis of Proposed Algorithm

To analysis the behavior of proposed algorithm, the degree of multiprogramming is varied by taking varying length of task queue. The length of the task queue is chosen as 10, 20, 30,..., 100 on a fixed interval of 10 as shown in column 1. Column 2 shows the average waiting time determined by ant lion optimizer (ALO), column 3, 4, and 5 describes the average waiting time of FCFS, RR, and SCTF.

In Table 3, an average waiting time determined by the proposed algorithm and the three widely used scheduling algorithms FCFS, RR and SCTF is being shown which promises the significant improvement of proposed one.

Table 3 represents the effect of degree of multiprogramming. Row 1 where task length is 10, the waiting time determined by proposed algorithm is far better as compared to other algorithms. While increasing the length of task queue, it is clearly evident that the waiting time determined by ALO still decreases up to task length 30 while it goes up rapidly for other scheduling algorithms. When task length is in the range of 40–100, waiting time becomes constant and provides same results for ALO but it keeps increasing in the case of other scheduling algorithm. This analysis shows that the performance of proposed algorithm keeps going on better or fixed at same value while increasing the degree of multiprogramming. It also demonstrates the effectiveness of probabilistic approach over the deterministic approach.

Figure 3 represents the bar diagram depicting effect of increasing the length of task by enhancing the degree of multiprogramming in terms of average waiting time and comparing the proposed algorithm with other scheduling algorithms. Figure clearly shows that the proposed algorithm performs constantly better while increasing the length of task queue as compared to other algorithms.

The average, maximum, minimum, and standard deviation of average waiting time determined by ALO are also analyzed after taking 30 independent runs as shown in Table 4. It depicts that the minimum average waiting time is exceptionally decent. The standard deviation also illustrates that all the values are very close to the mean.

Table 3 Effect of degree of multiprogramming on average waiting time

Length of task queue	ALO	FCFS	RR	SCTF
10	2.9163	29.06708	37.7617	20.29046
20	2.8215	60.93202	89.19845	48.96218
30	2.7219	90.01367	110.1978	57.36765
40	2.7169	115.7006	148.2436	79.0185
50	2.7169	142.8851	185.0365	99.15694
60	2.7169	169.3457	224.4427	121.3109
70	2.7169	197.7345	270.0026	146.7467
80	2.716	226.4297	317.0744	173.1074
90	2.7169	256.1369	361.5264	198.7451
100	2.7169	285.2662	388.5582	212.3331

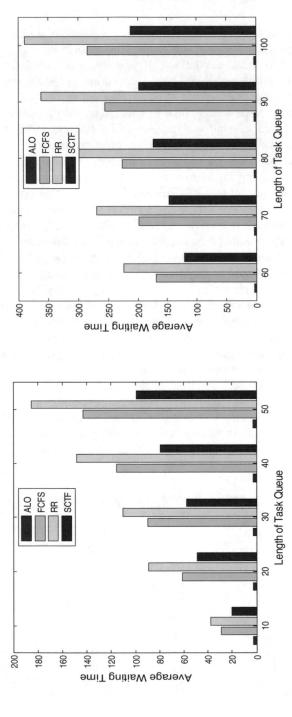

Fig. 3 Comparison of effect of degree of multiprogramming of proposed method and other scheduling algorithms

Table 4 Comparison of average waiting time for each task sequence

Length of task queue	ALO			
	Average	Max	Min	Std. Dev
10	2.64047	2.8956	2.5	0.127591
20	2.676403	2.9029	2.5	0.144764
30	2.541443	2.6993	2.5	0.068962
40	2.520923	2.6966	2.5	0.046916
50	2.512193	2.699	2.5	0.038991
60	2.51205	2.7169	2.5	0.042827
70	2.513107	2.7169	2.5	0.043016
80	2.511563	2.6876	2.5	0.037747
90	2.51756	2.7087	2.5	0.044944
100	2.51307	2.695	2.5	0.041815

6.1 Data Distribution Analysis Using Boxplot

Boxplot is a picturing technique employed for the determination of the distribution of sample data of average waiting time after employing ALO algorithm for each independent run. It is used to analyze the empirical distribution of data proficiently [11, 12]. This tool is used to exhibit two measures in the spread data: (1) Range, the vertical distance between smallest value and the largest value on boxplot diagram. (2) Interquartile range (IQR) which shows the width of the box. It can be clearly detected from Fig. 4 that the interquartile range and median of the proposed ALO algorithm is very low which ensures the proficiency of the proposed algorithm. The positive signs (+) as shown in boxplot diagram above median signify that the data distribution is positive skewed and it is highly significant.

7 Conclusion

Scheduling the tasks or processes in a way such that the waiting time of processes is minimum in a queue is very vital for researchers. Scheduling problems are generally NP-hard problems so there is always a possibility to enhance the optimal solutions for these algorithms. In this paper, a novel scheduling algorithm based on nature-inspired optimization algorithm called ant lion optimizer (ALO) is proposed for minimizing waiting time of the processes or tasks in ready queue. The proposed one is then compared with the widely used FCFS, SCTF, and RR scheduling algorithm and proven to be better in terms of minimizing waiting time. The proposed algorithm can be implemented for determining other performance metrics such as turnaround time, context switch. This algorithm can also be extended further for multiprocessor scheduling.

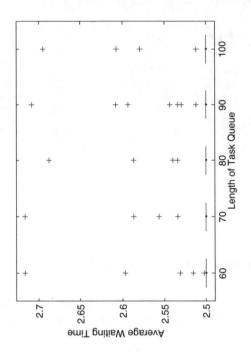

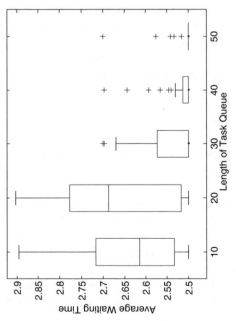

Fig. 4 Data distribution for task queues of different length

References

1. Dahal, K., Hossain, A., Varghese, B., Abraham, A., Xhafa, F., Daradoumis, A.: Scheduling in multiprocessor system using genetic algorithms. In: 7th Computer in Formation Systems and Industrial Management Applications, 2008. CISIM'08, pp. 281–286. IEEE (2008) (June)
2. Dhamdhare, D.M.: Operating System: A Concept Based Approach. McGraw Hill Higher Education (2009)
3. Omara, F.A., Arafa, M.M.: Genetic algorithms for task scheduling problem. J. Parallel Distrib. Comput. **70**(1), 13–22 (2010)
4. Silberschatzs, A., Galvin, P.B., Gagne, G.: Operating System Concepts. Addison-Wesley, Reading, MA (2009)
5. Maktum, T.A., Dhumal, R.A., Ragha, L.: A genetic approach for processor scheduling. In: Recent Advances and Innovations in Engineering (ICRAIE), pp. 1–4. IEEE (2014) (May)
6. Nossal, R., Galla, T.M.: Solving NP-complete problems in real-time system design by multi-chromosome genetic algorithms. In: Proceedings of the SIGPLAN 1997 Workshop on Languages, Compilers, and Tools for Real-Time Systems, pp. 6876. ACM SIGPLAN (1997) (June)
7. Mirjalili, S.: The ant lion optimizer. Adv. Eng. Softw. **83**, 80–98 (2015)
8. Cottet, F., Delacroix, J., Kaiser, C., Mammeri, Z.: Scheduling in Real-Time Systems, pp. 1–64. Wiley, England (2002)
9. Dinkar, S. K., Deep, K.: Opposition based Laplacian Antlion Optimizer. J. Comput. Sci, **23**, 71–90 (2017)
10. Dinkar, S. K., Deep, K.: Arab. J. Sci. Eng. https://doi.org/10.1007/s13369-018-3370-4 (2018)
11. Tukey, J.W.: Exploratory Data Analysis, vol. 2. (1977)
12. Williamson, D.F., Parker, R.A., Kendrick, J.S.: The box plot: a simple visual method to interpret data. Ann. Intern. Med. **110**(11), 916–921 (1989)

Design of Near-Optimal Trajectories for the Biped Robot Using MCIWO Algorithm

Ravi Kumar Mandava and Pandu R. Vundavilli

Abstract The present research paper concentrates on the development of optimal trajectories for the foot, hip, and wrist of the biped robot while walking on a flat surface. Cubic polynomial equations are used for generating the foot and wrist trajectories in sagittal plane and hip trajectory in frontal plane. The coefficients of the polynomial trajectories are used as seeds of the invasive weed optimization algorithm. It is important to note that the determination of the boundary conditions of the polynomial equation is a difficult task. To overcome this problem, a new variation of the invasive weed optimization (IWO) algorithm, i.e., modified chaotic invasive weed optimization (MCIWO) algorithm, has been used. Further, the dynamic balance margin (DBM) values obtained in x- and y-directions are compared with the values obtained using the standard IWO algorithm.

Keywords Optimal trajectories · Flat surface · Dynamic balance margin

1 Introduction

Human beings can perform dynamically balanced walk on various terrains such as flat, stair (ascending and descending), and slope (ascending and descending) surfaces. The present research work explains the generation of the dynamically balanced gait of the biped robot on flat terrain after considering optimal trajectories for foot, hip, and wrist of the biped robot. Jeon et al. [1, 2] proposed optimal trajectories for the biped robot while moving on a staircase (ascending and descending) using genetic algorithm (GA) and computed torque controller (CTC). The fourth-order polynomial coefficients were considered as the chromosomes in the problem. Alongside, Lee and Lee [3] had developed an optimal walking trajectory for a biped robot using multi-

R. K. Mandava (✉) · P. R. Vundavilli
School of Mechanical Sciences, IIT Bhubaneswar, Bhubaneswar 751013, Odisha, India
e-mail: rm19@iitbbs.ac.in

P. R. Vundavilli
e-mail: pandu@iitbbs.ac.in

© Springer Nature Singapore Pte Ltd. 2019
J. C. Bansal et al. (eds.), *Soft Computing for Problem Solving*,
Advances in Intelligent Systems and Computing 816,
https://doi.org/10.1007/978-981-13-1592-3_27

objective evolutionary algorithms. Moreover, Shrivastava [4] discussed the trajectory generation algorithm for 8-DOF biped robot using genetic algorithm (GA) with deformation of the sole at the foot. It was observed that the energy consumed by the biped robot was seen to be minimum when they used the soft sole with correction for deformation. Further, Yeon and Park [5] developed a walking trajectory generation method based on redundancy analysis. It was observed that a virtual spring-damper system had been used to achieve a balanced cyclic configuration of the robot. They also used a kinematic singularity method for the generation of more stable gaits. In addition to the above approaches, Ravi and Pandu [6] studied the influence of hip trajectory on the balance of the biped robot while moving on the flat surface. The PSO algorithm was used to optimize the parameters of the straight line and cubic polynomial trajectories of the biped robot. In [7], Jong and Moosung proposed an optimal gait trajectory for the 6-DOF biped robot using GA. The fourth-order polynomial has been used as the basic function for the locomotion gait, and the coefficients of the polynomial functions were considered as the design variables of the optimization algorithm. Further, Qiu and Fei [8] designed an optimal trajectory for the biped robot walking on an uneven terrain using neural network (NN) and fuzzy logic controller (FLC). They used PSO algorithm to train the weights of the NN and rules of FLC. Recently, Khusainov et al. [9] developed a kinematic and dynamic trajectory planning method after using an optimization approach. The kinematic optimization approach provided an unstable solution, which could be balanced by an upper body moment. Alongside, some of the researchers [10–13] used invasive weed optimization algorithm to optimize the process parameters of various engineering problems. Based on the above literature and with the best of the author's knowledge, no work is reported on the implementation of cubic polynomial trajectories for foot, hip, and wrist in both sagittal and frontal planes. It has also been observed that the optimization of these trajectories with the help of evolutionary algorithms is not yet realized. Therefore, in the present work, the coefficients used in the boundary conditions of the trajectories are being optimized with the help of a new stochastic optimization algorithm, i.e., MCIWO. Finally, the performances of the developed algorithm have been compared with the standard IWO algorithm in terms of dynamic balance in computer simulations.

2 Mathematical Model of the Biped Robot

In the present manuscript, an attempt is made to study the influence of coefficients used in the boundary conditions of foot, hip, and wrist trajectories of the biped robot (refer to Fig. 1). The foot and wrist in sagittal plane and hip joint in frontal plane are allowed to follow cubic polynomial trajectories. The MCIWO algorithm has been used to optimize the coefficients used in the boundary conditions (i.e., foot, hip, and wrist) of the biped robot.

The cubic polynomial equations and boundary conditions used for the foot, hip, and wrist trajectories of the biped robot are given below.

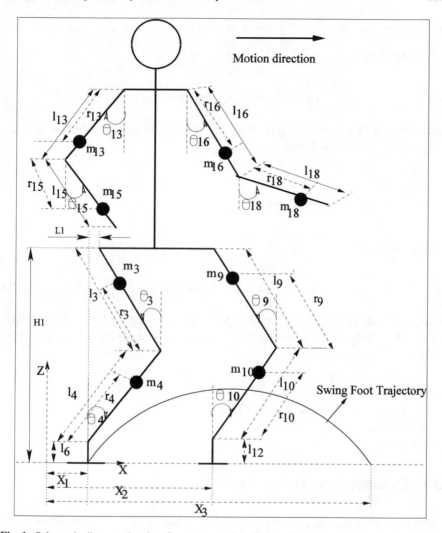

Fig. 1 Schematic diagram showing the structure of the biped robot

Foot Trajectory

$$z_f = a_{f0} + a_{f1}x_f + a_{f2}x_f^2 + a_{f3}x_f^3 \tag{1}$$

where z_f is the height of the swing foot at a distance x_f from the starting point of the foot and a_{f0}, a_{f1}, a_{f2}, and a_{f3} are the coefficients of the cubic polynomial equation.

- at $x_f = x_1$, $z_f = 0$;
- at $x_f = x_1 + (x_2 - x_1)/2$, $z_f = f_s/b_1$;
- at $x_f = x_2 + (x_3 - x_2)/2$, $z_f = f_s/b_1$;

- at $x_f = x_3$, $z_f = 0$;

Hip Trajectory

$$y_h = a_{h0} + a_{h1}x_h + a_{h2}x_h^2 + a_{h3}x_h^3 \tag{2}$$

where y_h is the width of the hip at a distance x_h from the starting point and a_{h0}, a_{h1}, a_{h2}, and a_{h3} are the coefficients.

- at $x_h = x_1$, $y_h = 0$;
- at $x_h = x_1 + (x_2 - x_1)/2$, $y_h = f_w/b_2$,
- at $x_h = x_2 + (x_3 - x_2)/2$, $y_h = f_w/b_2$,
- at $x_h = x_3$, $y_h = 0$;

Wrist Trajectory

$$z_w = a_{w0} + a_{w1}x_w + a_{w2}x_w^2 + a_{w3}x_w^3 \tag{3}$$

where z_w is the distance between the wrist to shoulder and x_w is the distance from the fixed reference point. Moreover, a_{w0}, a_{w1}, a_{w2}, and a_{w3} are the coefficients of the polynomial equation.

- at $x_w = x_{11}$, $z_w = h$;
- at $x_w = (x_{11} + x_{12})/2$, $z_w = h+f_w/b_3$;
- at $x_w = ((x_{13} - x_{12})/2)+x_{12}$, $z_w = h+f_w/b_3$;
- at $x_w = x_{13}$, $z_w = h$.

2.1 Dynamic Balance Margin

After evaluating the coefficients, the position of masses of the various joints of the biped robot is to be determined. These displacements of the biped robot are useful in determining the included angles made by various links of the biped robot. Further, the displacements and included angles obtained are used to calculate the velocity, acceleration, angular velocity, and angular acceleration of the joints. This information is helpful in determining the position of ZMP in x- and y-directions of the robot, and the equations that are used to serve the said purpose are given below.

$$x_{\text{ZMP}} = \frac{\sum_{i=1}^{n}\left(I_i\,\dot{\omega}_i - m_i\,\ddot{x}_i\,z_i + m_i x_i\left(g - \ddot{z}_i\right)\right)}{\sum_{i=1}^{n}\left(m_i(\ddot{z}_i - g)\right)} \tag{4}$$

$$y_{\text{ZMP}} = \frac{\sum_{i=1}^{n}\left(I_i\,\dot{\omega}_i - m_i\,\ddot{y}_i\,z_i + m_i y_i\left(g - \ddot{z}_i\right)\right)}{\sum_{i=1}^{n}\left(m_i(\ddot{z}_i - g)\right)} \tag{5}$$

where $\dot{\omega}_i$, I_i, g, and m_i denote the angular acceleration (rad/s^2), mass moment of inertia (kg-m^2) of the ith link, acceleration due to gravity (m/s^2), and mass (kg) of the link i, respectively. $\ddot{z}_i$ and $\ddot{x}_i$ indicate the acceleration of the ith link moving in z- and x-directions (m/s^2), respectively, and (x_i, y_i, z_i) signifies the coordinates of the ith lumped mass. Once the ZMP has been calculated, the dynamic balance margin in x- and y-directions is calculated by using the following equations.

$$x_{\text{DBM}} = \left(\frac{f_s}{2} - |x_{\text{ZMP}}| \right) \tag{6}$$

$$y_{\text{DBM}} = \left(\frac{f_w}{2} - |y_{\text{ZMP}}| \right) \tag{7}$$

where f_s and f_w represent the length and width of the foot.

3 Formulation of the Optimization Problem

To generate the optimal trajectories for the foot, hip, and wrist joints of the biped robot that will help in achieving the dynamically balanced gaits, the problem can be posed as an optimization problem as explained below:

$$\text{Maximize}(f) = \frac{1}{n} \sum_{i=1}^{n} X_{\text{DBM}_i} + Y_{\text{DBM}_i} \tag{4}$$

Subject to:
$$6 \leq b_1 \leq 14$$
$$4 \leq b_2 \leq 12$$
$$8 \leq b_3 \leq 16$$

where the variables b_1, b_2, and b_3 are having their usual meaning.

3.1 Modified Chaotic Invasive Weed Optimization Algorithm

In the present research, a new modified chaotic invasive weed optimization algorithm is used to optimize the coefficients used in the boundary conditions of the cubic polynomial trajectories of the biped robot. Invasive weed optimization [14] algorithm developed based on colonizing behavior of weeds is used to solve the said problem. The algorithm mimics the natural behavior of the weeds and spreads throughout the suitable search space for growth and reproduction. This algorithm has several advantages when adopted for searching solutions for systems of nonlinear equations.

It does not require a "good" initial point to perform the search, and the search space can be bounded by lower and upper values for each decision variable. Another added advantage of IWO algorithm is its reproduction stage. In this stage, some individuals with lower fitness value will carry useful information during the evolution process. Based on the above process, the system can reach optimal point quickly and easily. The following step-by-step procedure has been used to implement this algorithm.

Initialization: In a colony, initially, the weeds, i.e., coefficients, of the boundary conditions are dispersed randomly in D-dimensional search space.

$$x_{(i,j)} = x_{(l,j)} + (x_{(u,j)} - x_{(l,j)}) \times U[0, 1] \tag{5}$$

Reproduction: In this section, each weed is allowed to produce new seeds based on fitness. The higher the fitness of the weed, they will produce more seeds, and lesser fit weed will produce less seeds.

$$S = \text{Floor}\left[S_{\min} + \left(\frac{(f - f_{\min})}{(f_{\max} - f_{\min})} \right) \times S_{\max} \right] \tag{6}$$

where $f_{\min}$ and $f_{\max}$ indicate the minimum and maximum fitness in the colony, respectively, and $S_{\min}$ and $S_{\max}$ represent the minimum and maximum seeds produced by the plant, respectively.

Spatial Dispersal: The generated seeds are distributed randomly in a D-dimensional search space with mean equal to zero.

$$\sigma_{\text{Gen}} = \frac{(\text{Gen}_{\max} - \text{Gen})^n}{(\text{Gen}_{\max})^n} \times (\sigma_{\text{initial}} - \sigma_{\text{final}}) + \sigma_{\text{final}} \tag{7}$$

To improve the performance of the IWO algorithm, in this section the authors incorporated two new variables, namely chaotic [15] and cosine [16] variables. The chaotic variable is developed by Chebyshev map, and the expression is given below. The advantage of the chaotic variable is to minimize the chances to trap the solution into local optimum and to increase the spread of the new seeds in a dispersion area.

$$X_{k+1} = \cos(k \cos^{-1}(X_k)) \tag{8}$$

In addition to the chaotic variable, a new cosine variable has also been added in the standard deviation term. The added advantage of the cosine variable is to increase the search space by utilizing the minimum resources in the cropping field. The modified standard deviation (σ_{Gen}) term used in this study is given below:

$$\sigma_{\text{Gen}} = \frac{(\text{Gen}_{\max} - \text{Gen})^n}{(\text{Gen}_{\max})^n} \times |\cos(\text{Gen})| \times (\sigma_{\text{initial}} - \sigma_{\text{final}}) + \sigma_{\text{final}} \tag{9}$$

Competitive exclusion: Once the reproduction is over, the newly produced seeds grow to a flowering weed in a colony along with the parent weeds. Based on the fitness value in the colony, the ranks are assigned to all the flowering weed plants. Once it reaches the maximum population, the less fit weeds are eliminated from the colony and the better fitness weeds will be carried to the next generation. The above procedure is repeated until the termination condition is reached.

4 Results and Discussions

A parametric study has been conducted to obtain the optimal parameters of the MCIWO algorithm that produce the optimal trajectories for the foot, hip, and wrist of the biped robot. Figure 2 shows the variation of fitness with respect to the number of generations of MCIWO algorithm. The optimal values of $\sigma_{initial}$, σ_{final}, exponent, maximum number of seeds (S_{max}), minimum number of seeds (S_{min}), initial population (n_{pop0}), final population (n_{pop}), and number of generations are seen to be equal to 2%, 0.00001, 3, 4, 0, 10, 15, and 100, respectively. It can be observed that the modified chaotic invasive weed optimization algorithm is converged with a better value of DBM than compared with the IWO algorithm. The optimal values of the coefficients used in the boundary conditions, such as b_1, b_2, and b_3, are seen to be equal to 14.00, 5.11, and 8.00, respectively.

A parametric study is conducted to determine the influence of change in the values of the coefficients, such as b_1, b_2, and b_3 on the X- and Y-DBM values of the biped robot (refer to Fig. 3a, b). From Fig. 3a, it can be observed that the value of X-DBM increases with the increase in the value of b_1 and seen to be decreased with the increase in the value of b_2 and b_3 when the values of other coefficients are kept at their mid value. It may be due to the reason that the increase in the value of b_1 decreases the height of the swing foot trajectory. This results in the increase in the distance between the swing foot and the hip joint that results in producing dynamically more balanced gait. The other coefficients b_2 and b_3 are related to the hip and wrist trajectories, and

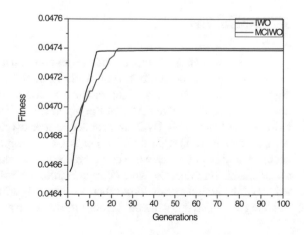

Fig. 2 Variation of cost function with generations

(a)   **(b)**

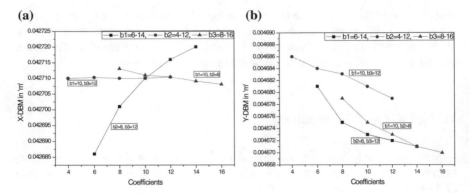

Fig. 3 Parametric study showing the variation of DBM with respect to change in the value of coefficients. **a** X-DBM, **b** Y-DBM

their influence on the DBM in x-direction is negligible. Similarly, Fig. 3b shows a decreasing trend in the value of Y-DBM with the increase in the values of b_1, b_2, and b_3. It may be due to the reason that the increase in the value of the coefficients pushes the hip joint away from the central plane in the lateral direction. This results in pushing the ZMP toward the edge of the foot support polygon and reduction in the value of DBM in y-direction.

Figure 4a, b, c shows the representation of near-optimal foot, hip, and wrist trajectories, respectively. Further, a comparative study (refer Fig. 5) has also been conducted for both the X- and Y-DBM after using IWO and MCIWO algorithms. From Fig. 5a, b, it can be observed that the MCIWO algorithm has provided more dynamically balanced gaits than the standard IWO algorithm. It may be due to the reason that in the case of MCIWO, the search space might have been increased after adding the chaotic and cosine variables.

5   Conclusions

In the present paper, an attempt is made to generate a near-optimal foot, hip, and wrist trajectories for the biped robot while walking on the flat surface by using MCIWO algorithm. The coefficients of the cubic polynomial trajectories are optimized to obtain the smooth walking gaits of the biped robot. The results of computer simulation show that the MCIWO has produced more dynamically balanced gaits when compared to the IWO algorithm. Further, MCIWO algorithm has also provided better accuracy and convergence when compared to the standard IWO algorithm. It may be due to the chaotic variable, which is responsible for its lower sensitivity to the initial values of the boundary conditions than the IWO algorithm. In future, the authors are planning to extend this work on various terrains such as stair, slop and ditch surfaces.

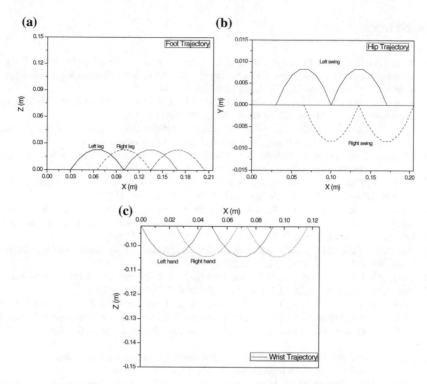

Fig. 4 Schematic diagram showing the variation of different trajectories, **a** swing foot, **b** hip, and **c** wrist

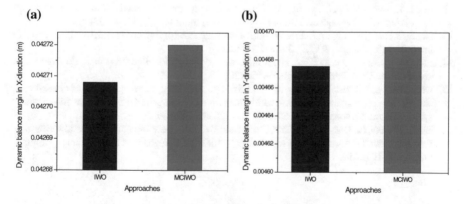

Fig. 5 Comparison of the dynamic balance margin. **a** X-DBM, **b** Y-DBM

References

1. Jeon, K.S., Kwon, O., Park, J.H.: Optimal trajectory generation for a biped robot walking a staircase based on genetic algorithms. In: IEEE/RSJ International Conference on Intelligent Robots and Systems, Sendai, Japan, Sept 28–Oct 2, 2004
2. Kwon, O., Jeon, K.S., Park, J.H.: Optimal trajectory generation for biped robots walking up- and –down stairs. J. Mech. Sci. Technol. **20**(5), 612–620 (2006)
3. Lee, J.Y., Lee, J.J.: Optimal walking trajectory generation for a biped robot using multi-objective evolutionary algorithms. In: 5th Asian Control Conference (2004)
4. Shrivastava, M., Dutta, A., Saxena, A.: Trajectory generation using GA for an 8 DOF biped robot with deformation at the sole of the foot. J. Intell. Robot Syst. **49**, 67–84 (2007)
5. Yeon, J.S., Park, J.H.: Variable walking trajectory generation method for biped robots based on redundancy analysis. J. Mech. Sci. Technol. **28**(11), 4397–4405 (2014)
6. Mandava, R.K., Vundavilli, P.R.: Study on influence of hip trajectory on the balance of a biped robot. In: Emerging Trends in Electrical, Communications and Information Technologies. Lecture Notes in Electrical Engineering, vol. 394, pp. 265–272 (2017)
7. Park, J.H., Choi, M.: Generation of an optimal gait trajectory for biped robots using a genetic algorithm. JSME Int. J. **47**(2), 715–721 (2004)
8. Zhong, Q., Chen, F.: Trajectory planning for biped robot walking on uneven terrain—taking stepping as an example. CAAI Trans. Intell. Technol. **1**, 197–209 (2016)
9. Khusainov, R., Klimchik, A., Magid, E.: Comparison of kinematic and dynamic leg trajectory optimization techniques for biped robot locomotion. International Conference on Information Technologies in Business and Industry, IOP Conference Series. J. Phys.: Conf. Ser. **803**, 1–6 (2017)
10. Karimkashi, Shaya, Kishk, Ahmed A.: Invasive weed optimization and its features in electromagnetics. IEEE Trans. Antennas Propag. **58**(4), 1269–1277 (2010)
11. Nagib, M.M., Othman, M.M., Naiem, A.A., Hegazy, Y.G.: Invasive weed optimization algorithm for solving economic load dispatch. In: IEEE International Conference on Environment and Electrical Engineering (EEEIC) (2016)
12. Zhou, Y., Luo, Q., Chen, H., He, A., Wu, J.: A discrete invasive weed optimization algorithm for solving traveling salesman problem. Neurocomputing **151**(3), 1227–1236 (2015)
13. Naidu, Y.R., Ojha, A.K.: A hybrid version of invasive weed optimization with quadratic approximation. Soft. Comput. **19**(12), 3581–3598 (2015)
14. Mehrabian, A.R., Lucas, C.: A novel numerical optimization algorithm inspired from weed colonization. Ecol. Inform. **1**(4), 355–366 (2006)
15. Ghasemi, M., Ghavidel, S., Aghaei, J., Gitizadeh, M., Falah, H.: Application of chaos-based chaotic invasive weed optimization techniques for environmental OPF problems in the power system. Chaos Solitons Fractals **69**, 271–284 (2014). Elsevier
16. Basak, A., Pal, S., Das, S., Abraham, A.: A modified invasive weed optimization algorithm for time-modulated linear antenna array synthesis. In: IEEE Congress on Evolutionary Computation, pp. 1–10 (2010)

Python-Based Fuzzy Classifier
for Cashew Kernels

Snehal Singh Tomar and V. G. Narendra

Abstract Fuzzy logic is a well-known branch of mathematics which provides a quantitative framework to discuss uncertain events and hence make logical estimations for uncertain outcomes. In this work, the key objective is to explore and illustrate the tools and techniques required to perform fuzzy operations and hence realize a basic fuzzy classifier in Python and assert its applicability over other conventional fuzzy logic tools such as the fuzzy logic toolbox in MATLAB. The above-mentioned classifier took real-world data of physical parameters such as length, width and thickness of white wholes cashew kernels which had highly overlapping data ranges as input and classified them into suitable categories. The observed computation time for successful (crisp) classification of the kernels into WW-320, WW-240, WW-210 and WW-180 categories using the said classifier was 0.43, 0.43, 0.42 and 0.46 s, respectively, whereas the fuzzy logic toolbox in MATLAB took minimum 0.58 s only to obtain a fuzzy output on the same computing system.

Keywords Fuzzy logic · Python · Classification systems

1 Introduction

Fuzzy logic is a robust tool that can be applied to a wide range of data to give decently accurate results. In works like [1–4] its applicability to act as a standalone system for varied applications as well as that to significantly improve the performance and robustness of conventional algorithms like PI, PD, PID, adaptive-PID (pertaining to

S. S. Tomar (✉)
Department of Electronics and Communication Engineering, Manipal Institute of Technology, Manipal Academy of Higher Education (MAHE), Manipal 576104, Karnataka, India
e-mail: snehalstomar@gmail.com

V. G. Narendra
Department of Computer Science and Engineering, Manipal Institute of Technology, Manipal Academy of Higher Education (MAHE), Manipal 576104, Karnataka, India
e-mail: narendra.vg@manipal.edu

© Springer Nature Singapore Pte Ltd. 2019
J. C. Bansal et al. (eds.), *Soft Computing for Problem Solving*,
Advances in Intelligent Systems and Computing 816,
https://doi.org/10.1007/978-981-13-1592-3_28

control systems) etc. by fuzzification of certain parameters followed by their defuzzification has been argued and proved. Also in works like [5–7], the exact theory and mathematical steps involved in application of fuzzy logic in real-life problems have been described clearly. Also in works like [8, 9], the advantages of a fuzzy classifier have been argued and proved. However, appropriate literature regarding the application of these principles to develop a software with cross-platform compatibility that incorporates fuzzy logic based classification system is difficult to find. The primary objective of this work is to fathom this research gap and create a simple fuzzy classifier in Python while giving a detailed description of the mathematical steps to be taken and their programming analogues. This work also asserts the edge, a Python-based fuzzy classifier has over other conventional methods.

2 The Problem

As discussed in [10], Table 1 represents the basic characteristics of certain varieties of white wholes cashew kernels.

The goal here is to successfully classify the kernels as per the given specifications.

3 The Algorithm

The overlapping nature of the various input data ranges makes fuzzy logic a natural choice for solution to the above problem. In works like [11], various techniques of fuzzy classification have been explored. This sections aims to present a broad structure of the problem-solving process adopted to code and create the Python-based fuzzy classifier developed in this work. Figure 1 illustrates the steps of the fuzzy classification algorithm.

Fuzzy Set Definition: This step involves definition of fuzzy sets depending upon the problem and the input–output variables pertaining to it. A range (universe) for the fuzzy sets to exist within is also designated. The universe is decided based upon the approximate range of values in which the input–output values of the membership

Table 1 Average minimum and maximum values of length, width and thickness of cashew kernels

Grades	Samples size	Length, L_p, cm		Width, W_p, cm		Thickness, T_p, cm	
		Min.	Max.	Min.	Max.	Min.	Max.
WW-180	500	2.5	3.27	1.49	2.22	0.95	1.59
WW-210	500	2.24	2.95	1.42	1.79	0.94	1.56
WW-240	500	2.05	2.89	1.36	1.71	0.89	1.45
WW-320	500	1.91	2.72	1.21	1.62	0.91	1.43

Fig. 1 Fuzzy classification algorithm

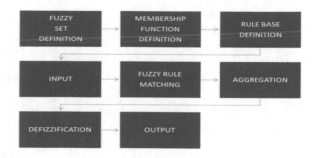

functions are expected to lie. It is also specified whether the fuzzy set belongs to input–output category.

Membership Function Definition: This step involves definition of the various membership functions that are a part of the fuzzy sets, their range and the type of mathematical function that defines them. In this work, trapezoidal membership functions have been used, wherever the membership functions were highly overlapping in nature and triangular membership functions have been used where they had relatively distinct boundaries.

Rule Base Definition: In this step, fuzzy rules which are basically the linguistic relationships between the membership functions of one or more input fuzzy sets and their corresponding membership functions in the output fuzzy set are defined.

Input: Input values (length, width and thickness) from the user are matched (assigned degrees of membership) with the various membership functions in the input fuzzy sets, depending on the region of the universe in which their crisp values lie.

Fuzzy Rule Matching: Depending upon the degree of membership assigned to the various input membership functions in the previous step, they are matched to appropriate rules in the rule base and conclusions are drawn with proportional weightage.

Aggregation: In this step, a controller is defined in the code which basically acts as a one-stop collection of the entire rule base. It gives a single fuzzy output by combining all the conclusions obtained from the previous steps on the basis of their weightage. In Python, this happens in the same way as it happens in a traditional Mamdani [5] controller by default.

Defuzzification and Output: This procedure is done to defuzzify the fuzzy output obtained in the previous step so as to return a crisp value either as output to the user or to a further block of code. There are various methods [12] to defuzzify a fuzzy input, some of them being:

- Centroid Method: This method returns the centroid of the area under the plot of the membership function to which the fuzzy parameter belongs as the defuzzified output.
- Bisector Method: This method returns the bisector value, which divides the area under the plot of the membership function under consideration into two subplots of equal area as the defuzzified output.

Fig. 2 Defuzzification
methods at a glance

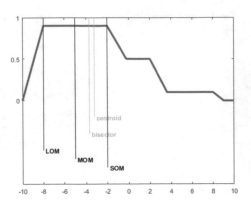

- Middle, Smallest and Largest of Maximum methods: These methods return the
  magnitude-wise middle, smallest and largest value (respectively) of the region in
  which the membership function under consideration attains maximum membership value.

The centroid method was found to be the most suitable defuzzification method
(shown in Fig. 2) for the classifier's output results since it ensured that all parameters
were given equal weightage.

The defuzzified output obtained at this stage was the final output of the classifier
and the category to which the specified kernel belonged to was decided on the basis
of this value.

4 Execution

The solution in Python is as follows:

(a) The solution presented in this work, required creation of four fuzzy sets, namely
    length (refer Fig. 3), width (refer Fig. 4), thickness (refer Fig. 5) and output (refer
    Fig. 6 which depicts the output fuzzy set when the sample belonged to WW-
    180 category). Length, width and thickness act as input fuzzy sets and take-in
    the actual input values, whereas output acts as the output set and returns crisp
    values between 1 and 5. Categorization of the cashew kernels is done on the
    basis of the range in which this output value lies. The input sets were designed
    to have four trapezoidal membership functions; WW-320, WW-240, WW-210
    and WW-180 (refer Fig. 3 for length; Fig. 4 for width; Fig. 5 for thickness)
    with suitable universes (in accordance with the range of their values in Table 1).
    The output fuzzy set had four symmetric triangular membership functions in
    the universe [1, 7] (refer Fig. 6).

Fig. 3 Input fuzzy
set—length

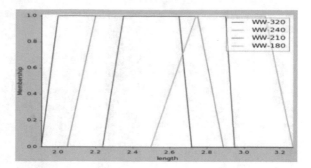

Fig. 4 Input fuzzy
set—width

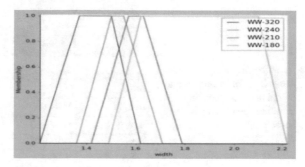

Fig. 5 Input fuzzy
set—thickness

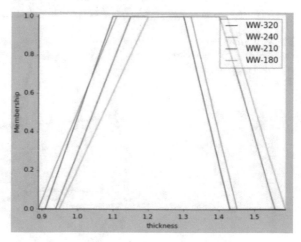

(b) These membership functions were then matched to the user input and were
    subjected to the control rule base mentioned in Table 2 and depicted graphically
    by Fig. 7, to obtain fuzzy outputs which were later defuzzified.
(c) Results Obtained:
    Upon execution of the code, a cashew kernel was successfully classified as
    belonging to the WW-180 category in **0.43 s** (including user's input time) on an

Fig. 6 Output fuzzy
set—output

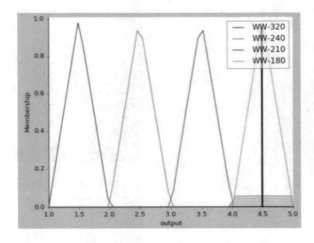

Table 2 Control rule base to
obtain fuzzy outputs for
classification of cashew
kernels

Input			Output
Length	Width	Thickness	
WW-320	WW-320	WW-320	WW-320
WW-240	WW-240	WW-240	WW-240
WW-210	WW-210	WW-210	WW-210
WW-180	WW-180	WW-180	WW-180

Fig. 7 A graphical
representation of the control
rule base mentioned in
Table 2

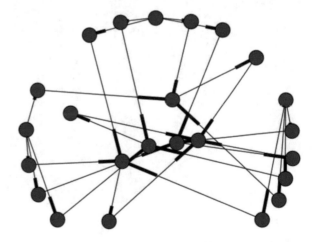

Intel Core–i5 6200U CPU @ 2.30 GHz machine (shown in Fig. 8), when the
following input was given:

- Length = 3 cm
- Width = 2 cm
- Thickness = 1.58

```
→ python_practice time python cashew_trap_classification.py
length(cm) =
3
width(cm) =
2
thickness(cm) =
1.58
Sample belongs to WW-180 category
python cashew_trap_classification.py  0.43s user 0.32s system 15% cpu 4.764 total
```

Fig. 8 Run-time view of the code in terminal

Table 3 Computation time and output for successful classification of cashew kernels

Input			Output	Classification	Computation time (s)
Length	Width	Thickness			
2	1.43	1	WW-320	Successful	0.43
2.4	1.48	0.98	WW-240	Successful	0.43
2.8	1.6	1.36	WW-210	Successful	0.42
3	2	1.58	WW-180	Successful	0.46

Fig. 9 Fuzzy set and membership function definition

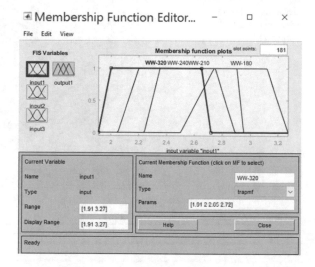

The details of a few more executions in Python are listed in Table 3.

The solution in **MATLAB** is as follows:

All the steps mentioned in section (a) and (b) were performed in MATLAB using the fuzzy logic toolbox (refer Figs. 9 and 10) and it was observed that it took **0.58 s** on an Intel Core–i5 6200U CPU @ 2.30 GHz machine to generate just the control surface (Fig. 11) showing the relation between output, length and width only.

In this solution, the designated values are:

- Input 1: Length
- Input 2: Width

Fig. 10 Rule base definition

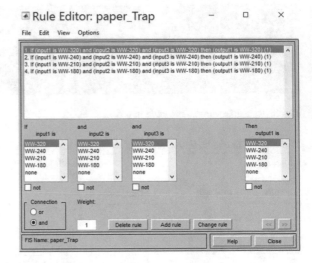

Fig. 11 Surface obtained (depicting relationship between Input 1, 2 and Output 1)

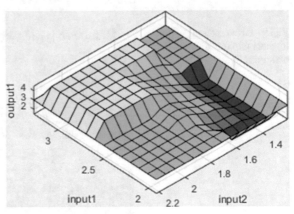

Table 4 Computation time for surface generation

Input variables	Computation time for obtaining surface (s)
1 and 2	0.58
1 and 3	1.09
2 and 3	1.22

- Input 3: Thickness
- Output 1: Unlike the Python code, the output is a fuzzy value in [1, 7] which needs to be grouped into crisp sub-ranges for further division into categories based on the sub-range to which it belongs.

Details of all the **MATLAB** executions are listed in Table 4.

5 Discussion

Thus, the proposed classifier is successfully realized in Python using basic programming techniques and predefined modules. Implementation of this classifier in Python is highly advantageous over doing it in a GUI environment like the fuzzy logic toolbox in MATLAB. Few parameters which showcase the same are:

a. **Applicability**: This implementation finds real-life application unlike a MATLAB implementation which is just for the purpose of mathematical simulation.
b. **Cross-Platform support**: Python codes are executable on most computing systems regardless of it being a Windows or a Linux-based environment. Moreover, various high-performance microcontrollers are also capable of interpreting Python code. Thus, platform does not remain a barrier for the realized classifier.
c. **Results Obtained**: Python is capable of giving a **crisp, defuzzified** value as output and can provide it as an input to a further block of code, neither of which is possible in the MATLAB-based implementation. Also attempts to incorporate the classifier designed in fuzzy logic toolbox for further use using a Simulink block fail, as the block can take only one input. Hence, a fuzzy logic classifier in MATLAB has limited utility.
d. **Faster Computation**: Upon execution of the controller in Python as well as MATLAB on an Intel Core–i5 6200U CPU @ 2.30 GHz machine, the following results were obtained:

  (i) The MATLAB (fuzzy logic toolbox)-based classifier took minimum **0.58 s** just to obtain the control surface when the range (universe) of the fuzzy sets was predefined.
  (ii) The classifier in Python took maximum **0.46 s** (including time taken by user for input) to assign dynamic values to the input membership functions and return a crisp, defuzzified output value (refer Table 3).

6 Conclusion

Thus the process of implementing the discussed fuzzy classifier in Python, a critical analysis of the results obtained and the advantages of such an implementation have been depicted successfully. The classifier realized in this work can also be realized in real life using industry grade sensors and microcontrollers which can lead to a fully engineered product. Further, the realized classifier can be made more adaptive by utilizing the principles mentioned in works like [13, 14]. To conclude, it can be said conveniently that, in a world where entropy, complexity and uncertainty continue to rise; fuzzy logic is an immensely powerful tool and its applications through Python remain boundless.

References

1. Mitra, P., Dey, C., Mudi, R.K.: Dynamic set-point weighted fuzzy PID controllers. In: International Symposium on Computational and Business Intelligence (ISCBI), 24–26 Aug 2013. https://doi.org/10.1109/iscbi.2013.29
2. Lee, C.C.: Fuzzy logic in control systems: fuzzy logic controller: part 1. IEEE Trans. Syst. Man Cybern. **20**(2), 404–418 (1990)
3. Li, H.H., Gupta, M.M. (eds.): Fuzzy Logic and Intelligent Systems. Springer Publications (1995)
4. Mendel, J.M.: Uncertain Rule-Based Fuzzy Logic Systems: Introduction and New Directions. Springer Publications (2017)
5. Mamdani, E.H.: Application of fuzzy logic to approximate reasoning using linguistic synthesis. IEEE Trans. Comput. **26**(12), 1182–1191 (1977)
6. Ross, T.J.: Fuzzy Logic with Engineering Applications, 3rd edn. Wiley, New York (2010)
7. Jang, J.-S.R.: Neuro-Fuzzy and Soft Computing: A Computational Approach to Learning and Machine Intelligence. Prentice Hall Inc, Upper Saddle River (1997)
8. Ghosh, B.: Using fuzzy classification for chronic disease management. Indian J. Econ. Bus. (2012). ISSN: 09725784
9. Tarzee, A., Masud, A., et al.: A semantic image classifier based on hierarchical fuzzy association rule mining. Multimedia Tools Appl. **69**, 921–949 (2014) (Springer Science+Business Media)
10. Narendra, V.G., Hareesh, K.S.: Computer vision system to estimate cashew kernel (white wholes) grade colour and geometric parameters. Agric. Eng. (EJPAU) **17**(4), #5 (2014). ISSN 1505-0297. http://www.ejpau.media.pl/volume17/issue4/art-05.html
11. Mozaffari, A., et al.: An evolvable self-organising neuro fuzzy multilayered classifier with group method data handling. Appl. Intell. (Boston). ISSN: 0924669X
12. http://in.mathworks.com/help/fuzzy/examples/defuzzification-methods.html
13. Ganesh Kumar, P., Devraj, D.: Improved genetic algorithm for optimal design of fuzzy classifier. J. Comput. Appl. Technol. (2009). ISSN: 09528091
14. Crockett, K.A., O' Shea, J., et al.: Genetic tuning of fuzzy inference within fuzzy classifier systems. Expert Syst. **23**(2) (2006) (Oxford). ISSN: 02664720

Linking Brainstem Cholinergic Input to Thalamocortical Circuitry

Madhuleena Dasgupta, Basabdatta Sen Bhattacharya and Atulya Nagar

Abstract Building computational models of brain parts at realistic scale is critical for prediction and treatment of neurological and psychiatric disorders and trauma. The contribution of this paper is the use of the Runge–Kutta–Fehlberg method to incorporate a kinetic model of synaptic transmission mediated by the neurotransmitter Acetylcholine (ACh) into a neural mass model of the thalamocortical circuitry, thereby imitating brainstem inputs to the thalamus. The result is a model of the thalamocortical oscillations as observed in electroencephalogram (EEG), firstly by introducing ACh into the existing model and secondly by varying synaptic parameters of the cholinergic pathway. Results show that embedding cholinergic input to the existing thalamocortical model leads to a change in postsynaptic voltages in the Lateral Geniculate Nucleus, as compared to the postsynaptic voltages without cholinergic input. In addition, it is observed that varying the cholinergic input to the Thalamic Reticular Nucleus (TRN) cell population around basal values deals with bifurcation in the model behaviour, implicating the crucial role of brainstem inputs to the TRN. This may underpin neural correlates of visuomotor deficiencies, which is a potential biomarker for early detection of Alzheimer's disease (AD).

Keywords Neural mass models · Kinetic modelling · Thalamic reticular nucleus Alzheimer's disease

M. Dasgupta (✉) · A. Nagar
Department of Mathematics and Computer Science,
Liverpool Hope University, Liverpool, UK
e-mail: dasgupm@hope.ac.uk; 15010524@hope.ac.uk

A. Nagar
e-mail: nagara@hope.ac.uk

B. S. Bhattacharya
School of Computer Science, University of Manchester, Manchester, UK
e-mail: basab@manchester.ac.uk

© Springer Nature Singapore Pte Ltd. 2019
J. C. Bansal et al. (eds.), *Soft Computing for Problem Solving*,
Advances in Intelligent Systems and Computing 816,
https://doi.org/10.1007/978-981-13-1592-3_29

1 Introduction

Neural mass modelling (NMM) represents collective behaviour of a mesoscopic [18] mass of 10^4–10^7 neurons so densely packed that they are assumed a single entity. Although single neuron modelling enables investigating functional attributes at cellular and molecular level [22], population-level modelling, such as the neural mass [8], allows a more global view. Moreover, it is believed that understanding population behaviour of neurons is more appropriate in studying diseases affected by cognition [7] such as Alzheimer's disease (AD). Mathematical models defining the population activity of thalamic and cortical neurons and their interconnectivity are considered appropriate abstractions, gaining increasing popularity towards modelling brain oscillations observed via electroencephalogram (EEG). However, due to computational constraints, such models are simulated at simplistic scales. We have therefore worked towards building biologically realistic neural mass modelling, simulating EEG signals by introducing Acetylcholine (ACh) in the existing model and analysing its role. The aim is to find reasons for the early symptoms of AD.

This study observes the time series behaviour of the model output after being affected by ACh and varying synaptic parameters of the cholinergic pathway, highlighting the influence of the brainstem to thalamus cholinergic pathway on the thalamic oscillatory behaviour. This is a key to understanding visuomotor deficiencies in AD. The remaining of this paper is organized as follows: Sect. 2 reviews the prior literature, and in Sect. 3, we represent the neural mass model with embedded kinetic models for cholinergic synapses. The results are represented in Sect. 4, followed by a brief discussion in Sect. 5. Lastly, Sect. 6 gives ongoing and future work.

2 Related Work

Prior work [25] suggests that optometric observations report multifarious impairments in the eye in Dementia, including anomalies in the retinal ganglion cells, whereas experimental study shows normal response to stimuli in early stages of AD [1], implying that the retinal cells may not be affected in these early stages. Experiments and evidence [19, 28] further show impaired visual motor integration in early stages, i.e. ability to integrate motor tasks with visual feedback. An example is crossing the road, which is seemingly impaired, supported by reports of motor dysfunction and structural damage to sensory-motor pathways [11, 23]. Moreover, latency in cortical response corresponding to an event is increased in AD compared to young healthy adults.

In this above context, research suggests that a copy of each thalamic sensory input is also sent to the brainstem, a vital organ integrating sensory and motor pathways [14]. Although there are few model-based studies relating motor activity to brain rhythms observed in EEG [5, 32], prior work suggests a lack of model-based studies on the role of thalamocortical circuitry in linking motor centres in the brainstem to motor and perceptual centres in the cortex.

The Lateral Geniculate Nucleus (LGN) is known to be one of the most densely innervated thalamic nuclei by brainstem cholinergic inputs (synaptic transmission mediated by the ACh neurotransmitter) [17] which may implicate a vital role of the thalamus in visuomotor deficiencies, as seen in AD. Literature surveys, such as [26], find that neurotransmission imbalances are the possible cause of AD and propose the "Cholinergic hypothesis of Alzheimer's Disease", which states that reduced synthesis of Acetylcholine may have led to the disease. Hence, the main drug-based treatment for symptomatic relief in AD is associated with the cholinergic hypothesis [10]. Therefore, the cholinergic pathway from the brainstem to the thalamus is identified as a vital area of research [16, 30].

In a model-based study on event-related synchronization and desynchronization of alpha rhythms in AD [5], it is postulated that a reduced cholinergic input to the TRN cell population may have led to sensory-motor-related symptoms observed in AD, but the main drawback was the use of an "alpha function" to represent synaptic transmission associated with all neurotransmitter types.

An alternate method of modelling synaptic transmission function in neural mass models is the replacement of the "alpha function" with a kinetic framework [4, 6], which is reported as more biologically plausible and at the same time computationally efficient [3]. In this work, the neural mass computational model of the LGN in [3] is enhanced by introducing a kinetic framework of cholinergic synapse based on the model in [15].

3 Proposed Method

Figure 1 shows the schematic model used in this work, whose synaptic layout is dependent on the intra-thalamic connectivity given in [3]. This is, in turn, presented in [24], based on experimental data obtained from the LGN of mammals and rodents and is as reported in [29]. The mean activity of the retinal spiking neurons when the brain is in an awake state with eyes closed is regarded as the "model input".

The excitatory (glutamatergic) synapses from retinal spiking neurons to the TCR and IN cell populations in LGN are mediated by ionotropic AMPA neuroreceptors. The TCR cell population makes excitatory synapses on the TRN cell population, also mediated by ionotropic AMPA neuroreceptors. The TRN cell population, in turn, sends inhibitory (GABA-ergic) feedback to the TCR cell population, mediated by both ionotropic $GABA_A$ and metabotropic $GABA_B$ neuroreceptors.

Inhibitory (GABA-ergic) feedforward synapses are made by the IN cell population on the TCR cell population, mediated by ionotropic $GABA_A$ neuroreceptors. The feedback connections in both TRN and IN cell populations on their respective self-populations are mediated by the ionotropic $GABA_A$ neuroreceptors [3].

In accordance with the work in [20], depending on the receptor type on the adjoining cell, ACh can be excitatory or inhibitory. Hence, TCR cells being excitatory in nature, the brainstem cell population makes excitatory cholinergic synapses on the

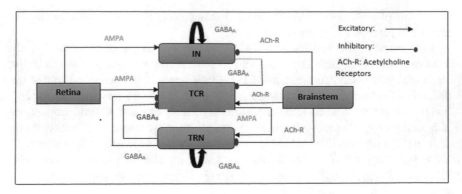

Fig. 1 Schematic of the NMM of thalamic LGN with three cell populations, i.e. IN, TCR and TRN and brainstem cholinergic inputs to LGN

Table 1 Selected values for AMPA- and GABA-based neurotransmission

Parameters	Values
α (AMPA, GABA$_A$)	1000 $(\text{mM})^{-1}$ $(\text{s})^{-1}$
β (AMPA)	50 $(\text{s})^{-1}$
β (GABA$_A$)	40 $(\text{s})^{-1}$
g_{max} (AMPA) [retina to TCR]	300 μS/cm^2
g_{max} (AMPA) [retina to IN]; [TCR to TRN]	100 μS/cm^2
g_{max} (GABA$_A$)	100 μS/cm^2
E_{rev} (AMPA)	0 mV
E_{rev} (GABA$_A$) [TRN/IN to TCR]	-85 mV
E_{rev} (GABA$_A$) [TRN(IN) to TRN(IN)]	-75 mV

TCR cell population and inhibitory cholinergic synapses on the IN and TRN cell populations, since they are inhibitory in nature.

The mathematical representation of the model is calculated in Sect. 3.1, model parameterization along with the "base values", i.e. the reference values of the model parameters used in this work which are sorted down in tabular forms in Tables 1, 2, 3 and 4 is in Sect. 3.2, and lastly Sect. 3.3 describes the simulation methodologies.

3.1 Mathematical Framework

The schematic model in Fig. 1 can be expressed mathematically by first-order differential Eqs. (1)–(11). The expressions with respect to AMPA and GABA neurotransmission are implemented as described in [3]. The equations corresponding

Table 2 Synaptic connectivity parameter values	Synaptic connectivity parameters	Base values
	C_{nte}	35
	C_{tni}	11.58
	C_{tre}	7.1
	C_{nsi}	20
	C_{ire}	47.4
	C_{isi}	23.6
	C_{tii}	15.45

Table 3 Selected values for cholinergic receptor-based neurotransmission	Parameters	Values
	α (ACh-R)	$3.75\ (\text{mM})^{-1}\ (\text{s})^{-1}$
	β (ACh-R)	$0.26\ (\text{s})^{-1}$
	$g_{\max}$ (ACh-R)	$1\ \mu\text{S/cm}^2$
	E_{rev} (ACh-R) [BRF to IN]	$-70\,\text{mV}$
	E_{rev} (ACh-R) [BRF to TCR]	$-65\,\text{mV}$
	E_{rev} (ACh-R) [BRF to TRN]	$-72\,\text{mV}$

Table 4 Cholinergic synaptic connectivity parameter values	Synaptic connectivity parameters	Base values
	C_{tbe}	31
	C_{nbi}	25
	C_{ibi}	14.5

to the inclusion of Acetylcholine in the existing retino-thalamic model and cholinergic receptor-based neurotransmission are introduced in this paper and are explained explicitly with the help of the following equations basing on the concepts in [3, 15] (the GABA$_B$ pathway is ignored, as model output does not seem to be dependent on this pathway [3]; however, future study will analyse its impact).

The neurotransmitter concentration in the synaptic cleft released by the brainstem is given by:

$$[T]_{\text{brf}} = \frac{T_{\max}}{1 + e^{-\frac{V_{\text{brf}} - V_{\text{thr}}}{\sigma}}} \tag{1}$$

where V_{thr} is the threshold voltage when the neurotransmitter concentration crosses 50% of its maximum value $T_{\max}$. V_{brf} is the voltage in the brainstem, and σ is the steepness parameter.

The dynamics of the synapses, mediated by cholinergic neuroreceptors, is given by:

$$\frac{dr^{ACh}}{dt} = \alpha[T]_{brf}(1 - r(t)) - \beta r(t) \tag{2}$$

where r is the proportion of the open ion channels on postsynaptic population caused by binding of Acetylcholine neurotransmitters with cholinergic neuroreceptors. α and β are the forward and reverse rates of chemical reactions, respectively.

The postsynaptic current in IN due to brainstem is given by:

$$I_{PSC\_IN}^{brf} = C_{ibi} g_{max}^{ACh-R}(1 - T_{brf})r(t)\left[V_{PSP\_IN} - E_{rev}^{ACh-R}\right] \tag{3}$$

Similarly, the postsynaptic currents in TCR and TRN due to the brainstem are given by Eqs. (4) and (5), respectively.

$$I_{PSC\_TCR}^{brf} = C_{tbe} g_{max}^{ACh-R}(1 + T_{brf})r(t)\left[V_{PSP\_TCR} - E_{rev}^{ACh-R}\right] \tag{4}$$

$$I_{PSC\_TRN}^{brf} = C_{nbi} g_{max}^{ACh-R}(1 - T_{brf})r(t)\left[V_{PSP\_TRN} - E_{rev}^{ACh-R}\right] \tag{5}$$

where $I_{PSC\_IN}^{brf}$, $I_{PSC\_TCR}^{brf}$ and $I_{PSC\_TRN}^{brf}$ represent the postsynaptic currents in IN, TCR and TRN, respectively, due to brainstem. V_{PSP_IN}, V_{PSP_TCR} and V_{PSP_TRN} are the postsynaptic potentials in IN, TCR and TRN, respectively. C_{ibi} is the normalized figure representing the percentage of the synaptic contacts made on the postsynaptic cell population, IN (denoted by i) by the presynaptic cell population, brainstem (denoted by b), and i is the sign of the synapse which in this case is inhibitory. Similarly, C_{tbe} is the normalized figure representing the percentage of the synaptic contacts made on the postsynaptic cell population, TCR (denoted by t) by the presynaptic cell population, brainstem, and e represents the sign of the synapse, which in this case is excitatory, and C_{nbi} is the normalized figure representing the percentage of the synaptic contacts made on the postsynaptic cell population, TRN (denoted by n) by the presynaptic cell population, brainstem, and i represents the sign of the synapse which in this case is inhibitory in nature. g_{max} represents the maximum conductance, and E_{rev} represents reverse potential whose values depend on the mediating synapse (in this case cholinergic neuroreceptors).

Therefore, the total postsynaptic current in IN is given in (6) by the summation of currents obtained with respect to retina and brainstem.

$$I_{PSC\_IN}^{total} = I_{PSC\_IN}^{ret} + I_{PSC\_IN}^{brf} \tag{6}$$

Similarly, the total postsynaptic current in TCR and TRN is given in Eqs. (7) and (8), respectively, which sum currents obtained with respect to retina, brainstem, IN and TRN in case of TCR and current summation obtained with respect to TCR and brainstem in case of TRN.

$$I_{\text{PSC\_TCR}}^{\text{total}} = I_{\text{PSC\_TCR}}^{\text{ret}} + I_{\text{PSC\_TCR}}^{\text{brf}} + I_{\text{PSC\_TCR}}^{\text{in}} + I_{\text{PSC\_TCR}}^{\text{trn}} \tag{7}$$

$$I_{\text{PSC\_TRN}}^{\text{total}} = I_{\text{PSC\_TRN}}^{\text{tcr}} + I_{\text{PSC\_TRN}}^{\text{brf}} \tag{8}$$

Therefore, total postsynaptic voltage in IN, TCR and TRN are given below in Eqs. (9), (10) and (11), respectively.

$$k_m \frac{dV_{\text{PSP\_IN}}}{dt} = I_{\text{PSC\_IN}}^{\text{total}} - I_{\text{Leak\_IN}} \tag{9}$$

$$k_m \frac{dV_{\text{PSP\_TCR}}}{dt} = I_{\text{PSC\_TCR}}^{\text{total}} - I_{\text{Leak\_TCR}} \tag{10}$$

$$k_m \frac{dV_{\text{PSP\_TRN}}}{dt} = I_{\text{PSC\_TRN}}^{\text{total}} - I_{\text{Leak\_TRN}} \tag{11}$$

where k_m is the ensemble membrane capacitance of the postsynaptic cell populations, IN, TCR and TRN and $I_{\text{Leak\_IN}}$, $I_{\text{Leak\_TCR}}$ and $I_{\text{Leak\_TRN}}$ are the ensemble membrane leak currents of the postsynaptic cell populations, IN, TCR and TRN, respectively.

3.2 Model Parameterization

A computer-generated random noise having low variance is simulated as model input. The resting state membrane potential for the excitatory retinal cells, V_{ret}, is set as -65 mV, as mentioned in [3]. The parameter $[T]_{\text{max}}$ is taken as 1 mM as in [3, 9]. The base values for threshold voltage V_{thr} is taken as -32 mV, and steepness parameter σ is 3.8 mV as in [3]. k_m, representing the ensemble membrane capacitance, is set at 1 μF/cm^2 as in [3].

The resting membrane potential for brainstem V_{brf} is chosen as -50 mV by trial and error so that the concentration function $[T]_{\text{brf}}$ covers all the regions of the sigmoid. The resting state membrane potential values for IN, TCR and TRN, which are -85, -65 and -75 mV, respectively, as in [3], are initialized into the variables $V_{\text{PSP\_IN}}$, $V_{\text{PSP\_TCR}}$ and $V_{\text{PSP\_TRN}}$, respectively.

Tables 1 and 2 give data corresponding to AMPA- and GABA-based neurotransmission, as given in [3, 12, 29, 31]. Tables 3 and 4 give data corresponding to cholinergic receptor-based neurotransmission, as mentioned in [4, 12, 21, 27, 29].

The leak conductance g_{leak} for TCR, TRN and IN is taken as 10 μS/cm^2. The leak potential E_{leak} for TCR is taken as -55 mV and for TRN and IN as -72.5 mV, as given in [3, 12, 31].

The nomenclature for the synaptic connectivity used in this work is consistent with the pattern represented in [3]. The parameter values are in accordance with the ranges in those studies.

Nomenclature for synaptic connectivity parameters: t represents TCR, n represents TRN, i represents IN, b represents brainstem, and r represents retina. Excitatory and inhibitory synapses are denoted by e and i, respectively. s is used to explain synaptic contacts by a cell population on itself. Moreover, it is worth mentioning that some parameter values are different from references and are scaled down to a range that is computable in MATLAB and consistent with kinetic model parameters.

3.3 Simulation Method

The fourth and fifth order Runge–Kutta–Fehlberg method (RKF45) is used for solving the ordinary differential equations in MATLAB. The total duration is 40 s with 1 ms resolution. The output voltage with respect to time is obtained by taking an average over 20 simulations and parameter variation with respect to base values (given in the above tables) is carried out to study the model behaviour.

4 Results

A comparative study on the effects of incorporating cholinergic input to the LGN circuitry and variation of the parameter in cholinergic synaptic pathways are studied. The observed effects and consequences are given in this section. These results may correspond to an altered brain condition.

4.1 Comparison of the Postsynaptic Voltages

The postsynaptic voltages of all three neural populations in LGN circuitry with cholinergic input and without cholinergic input are determined and compared.

It is observed that cholinergic input in the existing model leads to a significant change in the postsynaptic potentials in all three neural populations, leading us to the conclusion that ACh plays an essential role in the thalamocortical oscillations observed in EEG.

4.2 Varying the Synaptic Connectivity Parameters

Table 5 gives the results of varying the synaptic connection between brainstem and TRN around the basal value, and the corresponding effect on the model behaviour is shown.

It is observed that a reduced feeding of the cholinergic input from brainstem to TRN cell population, i.e. progressive decrease of C_{nbi} from its base value, shows a progressive decrease in postsynaptic potential in the TRN cell population. Moreover, when brainstem input is disconnected from the TRN cell population by making

Table 5 Variation of synaptic connectivity parameter to TRN

Connectivity parameter	Base value	Below base value	Effect on membrane potential	Above base value	Effect on membrane potential
C_{nbi}	25	20, 15, 10	Decrease	30, 50, 100	Increase

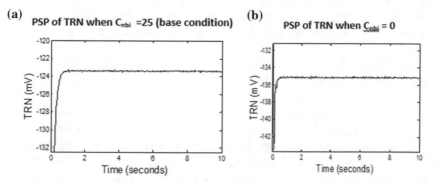

Fig. 2 **a** Postsynaptic potential of TRN cell population at base condition. **b** Postsynaptic potential of TRN cell population when brainstem input is disconnected

$C_{nbi} = 0$, which simulates the blocking of ACh into the TRN cell population, it leads to distinct bifurcation. In this condition, the output shows hyperpolarization in postsynaptic potential with respect to the base condition. This is evident in Fig. 2a, b.

Note that the synaptic connection between the brainstem and IN and TCR cell populations is also being varied below and above the reference values. In the case of the IN cell population, when there is a reduction in the cholinergic input from brainstem to the IN cell population, there is a slight decrease in postsynaptic potential in the IN cell population output.

When the brainstem input is disconnected from the IN cell population by making $C_{ibi} = 0$, this leads to slight hyperpolarization in postsynaptic potential with respect to the base condition. On the other hand, variation as well as disconnection of the cholinergic input to TCR cell population did not lead to any significant change in the excitatory TCR population output.

5 Discussion

It is observed firstly that the brainstem input to the LGN circuitry, which deals with supply of Acetylcholine to the LGN circuitry, leads to a significant change in the postsynaptic voltages in all three neural population outputs, as compared to the condition when there was no cholinergic input to the model. Therefore, it can be concluded that ACh plays a vital role in the thalamocortical oscillations observed in EEG.

Secondly, varying the cholinergic connectivity parameter to IN, TRN and TCR led, on one hand, to a distinct bifurcation in the case of TRN cell population output and on the other, to a slight change in the output of the IN cell population and almost no significant change in the TCR population output.

It is observed that progressive decrease of C_{nbi} from its base value shows a progressive decrease in postsynaptic potential in the TRN cell population. The TRN cell population shows distinct bifurcation when the brainstem input is disconnected from the TRN cell population by making $C_{nbi} = 0$, thereby simulating the blocking of ACh.

The bifurcation (hyperpolarization) in postsynaptic potential of TRN implies the crucial role of brainstem inputs to the TRN. It may also be associated with reduced cognition in the early stages of AD which can be explained explicitly as follows.

In [19, 28], it is suggested that experimental studies and evidence show impaired visual motor integration in these early stages. It is known that TRN, the brain structure divided into a number of sections, each concerned and associated with performing different functions like seeing, touching, hearing and movement, and connected with the corresponding cortical area, is associated with the regulation of sensory processing and formation of rhythmic activity in the thalamocortical system [2, 13].

Hence, bifurcation in the TRN voltages in LGN circuitry (which is considered as a relay centre for visual pathway) [24] during reduced feeding of ACh or when ACh is removed, i.e. brainstem input to TRN, when removed, might be a cause for "visual motor integration" delaying or impairments in AD.

6   Conclusion and Ongoing Research

This study found the following:

- ACh plays a vital role in the thalamocortical oscillations observed in EEG since the brainstem input to the LGN circuitry leads to a significant change in the postsynaptic voltages in all three neural population outputs.
- The bifurcation (hyperpolarization) in postsynaptic potential of TRN implies the crucial role of brainstem inputs to the TRN which is associated with reduced cognition in the early stages of AD.

Therefore, the present work concentrated mainly on the influence of the brainstem to the thalamus cholinergic pathway on thalamic oscillatory behaviour. This is a key to understanding the visuomotor deficiencies or impairment in AD. This model-based study concerned the postsynaptic effect of the brainstem input into the LGN circuitry.

However, the presynaptic effects should also be considered. This includes anomalies in the release of ACh, as it may also contribute to the change in model behaviour. It might also be the control parameter for model bifurcation and lead to the conclusion that these are the main attributes which cause the model to behave differently.

Acknowledgements We would like to convey our sincere thanks and gratitude to Dr Neil Buckley of Department of Mathematics and Computer Science, Liverpool Hope University, UK, for his active help during the preparation of this manuscript.

References

1. Armstrong, R.A., Syed, A.B.: Alzheimer's disease and the eye. Opthalmology and Physiol. Opt. **16**(1), S2–S8 (1996)
2. Beierlein, M.: Synaptic mechanisms underlying cholinergic control of thalamic reticular nucleus neurons. J. Physiol. **592**(19), 4137–4145 (2014)
3. Bhattacharya, B.S., Bond, T.P., Hare, L.O., Turner, D., Durrant, S.J.: Causal role of thalamic interneurons in brain state transitions: a study using a neural mass model implementing synaptic kinetics. Front. Comput. Neurosci. **10**, 115 (2016)
4. Bhattacharya, B.S., Coyle, D., Maguire, L.P., Stewart, J.: Kinetic modelling of synaptic functions in the alpha rhythm neural mass model. In: Villa, A.E.P., Duch, W., Erdi, P., Masulli, F., Palm, G. (eds.) Lecture Notes in Computer Science, Volume 7552 Part I, pp. 645–652. Springer, Berlin (2012)
5. Bhattacharya, B.S., Coyle, D., Maguire, L.P.: Assessing alpha band event related synchronisation/desynchronization with a mutually coupled thalamo-cortical circuitry model. J. Univ. Comput. Sci. **18**(13), 1888–1904 (2012)
6. Bhattacharya, B.S.: Implementing the cellular mechanisms of the synaptic transmission in a neural mass model of the thalamocortical circuitry. Front. Comput. Neurosci. **7**, 81 (2013)
7. Bressler, S.L., Menon, V.: Large-scale brain networks in cognition: emerging methods and principles. Trends Cogn. Sci. **14**, 287–290 (2010)
8. David, O., Friston, K.J.: A neural mass model for MEG/EEG: coupling and neuronal dynamics. NeuroImage **20**, 1743–1755 (2003)
9. Destexhe, A., Mainen, Z., Sejnowski, T.: An efficient method for computing synaptic conductances based on a kinetic model of receptor binding. Neural Comput. **6**, 14–18 (1994)
10. Francis, P.T., Palmer, A.M., Snape, M., Wilcock, G.K.: The cholinergic hypothesis of Alzheimer's Disease: a review of progress. J. Neurol. Neurosurg. Psychiatry **66**, 137–147 (1999)
11. Goldman, W.P., Baty, J.D., Buckles, V.D., Sahrmann, S., Morris, J.C.: Motor dysfunction in mildly demented AD individuals without extrapyramidal signs. Neurology **53**(5), 956–962 (1999)
12. Golomb, D., Wang, X.-J., Rinzel, J.: Propagation of spindle waves in a thalamic slice model. J. Neurophysiol. **75**(2), 750–769 (1996)
13. Guillery, R.W., Feig, S.L., Lozsádi, D.A.: Paying attention to the thalamic reticular nucleus. TINS **21**(1), 28–32 (1998)
14. Guillery, R.W.: Branching thalamic afferents link action and perception. J. Neurophysiol. **90**(2), 539–548 (2003)
15. Hasselmo, M.E., Schnell, E.: Laminar selectivity of the cholinergic suppression of synaptic transmission in rat hippocampal region CA 1: computational modelling and brain slice physiology. J. Neurosci. **14**(6), 3898–3914 (1994)
16. Kar, S., Slowinkowski, P.M., Westaway, D., Mount, H.T.J.: Interactions between β-amyloid and central cholinergic neurons: implications for Alzheimer's disease. J. Psychiatry Neurosci. **29**, 427–441 (2004)
17. Kasa, P., Rakonczay, Z., Gulya, K.: The cholinergic system in Alzheimer's disease. Prog. Neurobiol. **52**, 511–535 (1997)
18. Liljenström, H.: Mesoscopic brain dynamics. Scholarpedia **7**(9), 4601 (2012)
19. Macpherson, K., Weston, P.S.J., Nicholas, J.M., Donnachie, E.R., Ariti, C., Crutch, S.J., Fox, N.C., Henley, S.M.D.: Visuomotor integration in presymptomatic familial Alzheimer's Disease. Poster Presentations P2-458, Cross Mark, p. 815 (2017)

20. McDowall, J.: Acetylcholine Receptors. Interpro, RCSB Protein Data Bank (PDB) (2005)
21. Miftakhov, R.N., Christensen, J.: A physiochemical basis of synaptic transmission in the myenteric plexus. In: Poznanski, R.R. (ed.) Biophysical Neural Networks: Foundations of Integrative Neuroscience, pp. 147–176. Mary Anne Liebert, New York (2007)
22. Neymotin, S.A., Lee, H., Park, E., Fenton, A.A., Lytton, W.W.: Emergence of physiological oscillation frequencies in a computer model of neocortex. Front. Comput. Neurosci. **5**, 19 (2011)
23. O'Leary, T.P., Robertson, A., Chipman, P.H., Rafuse, V.F., Brown, R.E.: Motor function deficits in the 12 month-old female 5xFAD mouse model of Alzheimer's disease. Behav. Brain Res. (2017) (Elsevier)
24. Sherman, S.M.: Thalamus. Scholarpedia **1**(9), 1583 (2006)
25. Solomons, H.: Vision and dementia. Clinical, pp. 25–28 (2005)
26. Steriade, M., McCarley, R.W.: Brain Control of Wakefulness and Sleep, 2nd edn. Springer, Berlin (2005)
27. Suffczynski, P., Kalitzin, S., Silva, F.L.D.: Dynamics of non-convulsive epileptic phenomena modelled by a bistable neuronal network. Neuroscience **126**, 467–484 (2004)
28. Tippett, W.J., Sergio, L.E.: Visuomotor integration is impaired in early stage Alzheimer's Disease. Brain Res. **1102**, 92–102 (2006)
29. Van Horn, S.C., Erisir, A., Sherman, S.M.: Relative distribution of synapses in the A-laminae of the lateral geniculate nucleus of the cat. J. Comp. Neurol. **416**, 509–520 (2000)
30. Venkatraman, A., Edlow, B.L., Immordino-Yang, M.H.: The brainstem in emotion: a review. Front. Neuroanat. **11**, 15 (2017)
31. Wang, X.-J., Golomb, D., Rinzel, J.: Emergent spindle oscillations and intermittent burst firing in a thalamic model: specific neuronal mechanisms. Proc. Natl. Acad. Sci. U.S.A. **92**, 5577–5581 (1995)
32. Zavaglia, M., Astolfi, L., Babiloni, F., Ursino, M.: A model of rhythm generation and functional connectivity during a simple motor task: preliminary validation with real scalp EEG data. Int. J. Bioelectromagnetism **10**(1), 68–75 (2008)

Genetic Algorithm-Based Oversampling Technique to Learn from Imbalanced Data

Puneeth Srinivas Mohan Saladi and Tirtharaj Dash

Abstract Availability of data from many different applications such as surveillance systems, security appliances, finances has been continuously expanding. Many machine learning (ML) and data mining models have shown promising power in learning from the available data. However, the problem of learning an ML classifier from imbalanced data is still a challenging problem. This problem is often regarded as the imbalanced learning problem. In this problem, there is more amount of information known from the majority classes than the minority classes. In such a learning environment, the classifier during training over-fits to the former classes and under-fits to the minority classes. Distance-based strategy, for example, SMOTE, has been quite useful to oversample the minority classes that essentially uses nearest neighbor samples from the available samples. In this paper, we propose a notion of employing genetic algorithm (GA) that would essentially learn the probability distribution from the available data to generate the minority class samples for binary classification problems. We validate and test our proposed oversampling strategy by training three different kinds of classifiers. The comparative analysis with SMOTE-based oversampling and the proposed GA-based oversampling shows promising results for a selected ten very popular imbalanced datasets.

Keywords Genetic algorithm · SMOTE · Machine learning · Imbalanced classification · Minority class · Majority class

P. S. M. Saladi · T. Dash (✉)
Data Science Research Group, Department of Computer Science,
Birla Institute of Technology and Science Pilani, K.K. Birla Goa Campus,
Zuarinagar 403726, Goa, India
e-mail: tirtharaj@goa.bits-pilani.ac.in

P. S. M. Saladi
e-mail: f20140075@goa.bits-pilani.ac.in

© Springer Nature Singapore Pte Ltd. 2019
J. C. Bansal et al. (eds.), *Soft Computing for Problem Solving*,
Advances in Intelligent Systems and Computing 816,
https://doi.org/10.1007/978-981-13-1592-3_30

1 Introduction

There was a time when data scarcity was a bottleneck for research in the field of machine learning and data mining. Presently, data is abundantly available which may come from many different real-world applications of which few well-known areas are social networking, sensor networks, home monitoring systems, finances, bioinformatics, fraudulent banking transactions, malware analysis, intrusion detection. However, it still creates new challenges to discover knowledge embedded among these huge chunks of data and then further creates situations in which advancing learning from raw data toward decision making becomes critical [1]. Although the new knowledge discovery tools have significantly succeeded in the field, the problem of learning from the data with highly skewed distribution remains still a critical problem. This problem is popularly regarded as the imbalanced learning problem [2]. For binary classification, imbalanced learning problem can be stated as a problem with more number of samples of one class, called 'majority' class, and very limited number of samples from the other class, called 'minority' class. In a similar sense, for the multiclass (number of classes is more than 2) classification problems [3], there are a higher number of samples for some classes (majority classes) and lower number of samples for the other classes (minority classes).

There are two possible and well-accepted re-balancing approaches in the imbalanced learning problem, (a) under-sampling (down-sampling) of the majority class samples and (b) oversampling of the minority class samples. These re-balancing approaches are carried out before training the intended classifier to reduce the possibility of over- or under-fitting. The former approach is quite highly applied in the research [4]. The latter approach is difficult, and the performance of the intended classifier would highly be based on the quality of the oversampling. If the oversampling engine generates well-qualified samples which possibly follows closely the underlying probability distribution, $P(\cdot)$, then it would contribute in a better manner during the training of the classifier that attempts to learn this distribution $P(\cdot)$. However, finding an optimal mixture of the minority and majority class samples to learn an optimal classifier is still an unresolved problem in ML. Moreover, there is always a nonlinear relationship between available features and the class labels. In such a scenario, brute-force combination of the samples is NP-hard which keeps the problem open-ended and only approximation heuristics may exist such as evolutionary algorithms [5]. Moreover, the available literature suggests that if the complex real-world problems are formulated adequately, evolutionary metaheuristics are indeed capable of generating the best approximate solution for complex real-world problems [6–8]. Nevertheless, researchers have attempted to improve the quality of oversampling of the presently available method such as synthetic minority oversampling technique (SMOTE) [9] using various interesting approaches [10–12].

Our present work is different from other oversampling methods in the following sense. This work does not focus toward the improvement of SMOTE for any problem or a set of problems. Rather, it attempts to provide a procedure to oversample minority class of various problems based on genetic algorithm (GA). Hence, it proposes a notion of employing the genetic algorithm (GA) heuristic [13] that adapts itself continuously to closely follow the underlying distribution of the originally available data samples. Our GA-based oversampling engine learns classifiers for the highly imbalanced binary classification problems.

The rest of the paper is organized as follows. Section 2 briefly presents our proposed GA-based oversampling engine and the materials for testing this experiment. Three different classifiers trained with the generated oversampled data have been evaluated in Sect. 3, and the results are compared with the same classifiers trained with SMOTE [9]-based oversampling strategy. The inferences and conclusions have been drawn in Sect. 4 followed by a set of references.

2 Materials and Methods

2.1 Materials

To effectively evaluate our proposed GA-based oversampling hypothesis-based implemented engine, we used ten imbalanced benchmark datasets obtained from the KEEL dataset repository [14]. Some abstract properties of these datasets have been summarized in Table 1. The table also lists the number of samples in each dataset, the number of majority class (negative) samples, and the number of majority class samples (positive).

Each dataset was divided into two major independent groups (a) training-and-validating set ($\mathcal{D}_{learn}$) and (b) testing set ($\mathcal{D}_{test}$). The testing set has been considered to test the performance of generalization for the trained classifiers. The training-and-validation set was used to learn the classifiers. For each dataset, the testing set $\mathcal{D}_{test}$

Table 1 Characteristics of datasets (the imbalanced ratio (n) can be read as 1 : n against minority and majority classes, respectively)

Dataset	#Samples	Negative	Positive	Imbalance ratio
ecoli-0-1-4-7_vs_2-3-5-6	336	307	29	10.59
ecoli-0-1-4-7_vs_5-6	332	307	25	12.28
ecoli2	336	284	52	5.46
winequality-red-8_vs_6	656	638	18	35.44
winequality-red-8_vs_6-7	855	837	18	46.5
winequality-white-3-9_vs_5	1482	1457	25	58.28
yeast-1-2-8-9_vs_7	947	917	30	30.57
yeast-1-4-5-8_vs_7	693	663	30	22.1
yeast4	1484	1433	51	28.10
yeast-0-3-5-9_vs_7-8	506	456	50	9.12

contains a randomly selected 20% from both the class samples. The dataset $\mathcal{D}_{\text{learn}}$ was oversampled by (a) SMOTE [9] and (b) the proposed GA-based oversampling engine. The comparative performance of the implemented classifiers on these over-sampled datasets (SMOTE-based dataset and GA-based dataset) would provide a clear demonstration of our proposed evolutionary technique.

2.2 Development of the GA-based Oversampling Engine

A chromosome of the GA[1] can be understood as a sample that belongs to $\mathcal{D}_{\text{learn}}$ in this problem. There was no specific encoding (rather can be called real-coded GA) applied to avoid any unprecedented information loss. Further, keeping the samples as available in original dataset allows us to follow the distribution throughout without any additional bias of representation and encoding and makes the approach straight-forward. This incorporates an inherent assumption that the samples in the dataset $\mathcal{D}_{\text{learn}}$ have been generated by a probability distribution $P(\cdot)$ and the samples that would be seen by the trained classification model would also follow the same dis-tribution. Since the oversampled datasets are generated using a non-distance-based approach, it can be hypothesized that these samples would also follow the same distribution leading to superior learning of the intended classification models on the oversampled data.

Let us represent two chromosomes (minority samples of a dataset) as x_i and x_j. We apply the Eqs. (1) and (2) to generate two new chromosomes using these two chromosomes.

$$x_1^{\text{new}} = \alpha x_i + (1 - \alpha)x_j \tag{1}$$

$$x_2^{\text{new}} = (1 - \alpha)x_i + \alpha x_j \tag{2}$$

where α can be considered as a constant of distribution learning and is allowed in the range $(0, 1)$. This value is chosen at random in the specified range during the generation of the new samples from the available two samples. It is believed that a suitable value for α should be less than 1. The α also decides how much information regarding the distribution should be propagated to the new set of samples from the parent, thereby making our approach different from the standard distance- or neighbor-based oversampling heuristics.

The new minority samples are generated until there is an equal number of the pos-itive and negative samples. This balanced dataset, denoted as $\mathcal{D}_{\text{osLearn}}$, is considered as a candidate solution in the GA population. The population size is set to 10 in which each member of the population is an oversampled set (a set of chromosomes). The GA-based oversampling engine is iterated until the best set of oversampled data is obtained. In each iteration, a classifier is learned from the generated candidate solu-

---

[1]We intentionally avoid common terminologies in GA such as population, chromosome, crossover, mutation without any loss of generality in understanding this methodology.

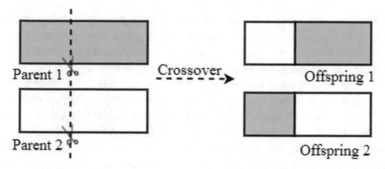

Fig. 1 Crossover strategy showing the creation of the offspring from the parents (Parents are the candidate solutions, and the crossover point shown as scissor with dotted line is chosen at random.)

tion and validated. It is trivial to say that $|\mathcal{D}_{osLearn}| \gg |\mathcal{D}_{learn}|$. In each iteration, 5CV is performed on the presently obtained candidate solutions ($\mathcal{D}_{osLearn}$), and the average cross-validation accuracy is taken to be the fitness estimate for each candidate solution. To support the theory of natural selection and survival of the fittest [13], 40% of the candidate solutions that have achieved worse fitness estimate are discarded. To maintain the population size, existing solutions are selected with a probability proportional to their fitness estimate for crossover. The crossover point is selected at random, and new offspring is produced. The crossover strategy has been demonstrated in the Fig. 1. One iteration of GA is complete once the crossover is carried out, and the original population size is obtained. This process is repeated for 15 such iterations. The GA-based oversampling engine has been presented in Algorithm 1.

Algorithm 1 GA-based oversampling engine

1: **procedure** GA_OVERSAMPLER($\mathcal{D}_{learn}$)
2:　　Create 10 candidate solutions from the given positive instances $\mathcal{D}^+_{learn}$
3:　　**for** $iter \leftarrow 1$ to 15 **do**
4:　　　　Calculate the fitness estimate of each candidate solution
5:　　　　Remove 40% of the candidate solutions with low fitness estimate
6:　　　　Select parents for crossover with the probability $\propto$ fitness estimate
7:　　　　Crossover twice so that the population is maintained
8:　　Return $\mathcal{D}_{osLearn}$ from the available candidates with best fitness value

---

3　Experimental Evaluation

3.1　Implementation and Performance Metrics

So while evaluating each candidate solution, we performed 5CV and took the average of accuracies obtained in 5CV to be the fitness function. Algorithm 1 is followed to generate the best-oversampled set that is equivalent to the probability distribution of the original data. The number of the evolving generations, i.e., the number of iterations of GA-based oversampling engine, has been set to a maximum 15. The outcome of the engine is the best-oversampled set called $\mathcal{D}_{\text{osLearn}}$ that is used to learn the final and optimal classifier. The trained and validated classifier is then tested against the independent test $\mathcal{D}_{\text{test}}$ for testing its generalization performance. Comparisons have been made in this work with the SMOTE-based oversampling method for the same classifiers. We used three different classifiers, namely k-nearest neighbor (kNN), random forest, and AdaBoost. In this work, for preliminary testing of our hypothesis, we tried to avoid iterative learners such as neural networks. However, we believe that the GA-based oversampling may still be suitable to learn a multilayer perceptron (MLP), and there is no limitation on the use of classifiers.

3.2 Performance Metrics

Many different performance metrics have been developed for the evaluation of the adequacy of classification models owing to imbalance training which are beyond accuracy measurements. Some useful metrics can be found in the recent literature [15]. These measures are briefly explained as follows. The principal evaluation, a binary classification model, is a confusion matrix that has been shown in Table 2 which lists a sensible trade-off between actual and predicted class samples. All other performance measures can be obtained using these four values true positive (TP), true negative (TN), false positive (FP), and false negative (FN).

Accuracy can be written as

$$\text{Accuracy} = \frac{\text{TN} + \text{TP}}{\text{TN} + \text{TP} + \text{FN} + \text{FP}} \tag{3}$$

Precision ($=\frac{\text{TP}}{\text{TP}+\text{FP}}$), recall ($=\frac{\text{TP}}{\text{TP}+\text{FN}}$), F-measure, and Cohen's Kappa [16] are very useful and crucial measures to quantify the performance of classifiers in case of imbalanced learning. These metrics are presented in the following equations.

$$F\text{-measure} = \frac{(1 + \beta^2) \times \text{Recall} \times \text{Precision}}{\beta^2 \times \text{Recall} + \text{Precision}} \tag{4}$$

Table 2 Confusion matrix for binary classification problem

	Pred (+)	Pred (−)
Actual (+)	TP	FN
Actual (−)	FP	TN

where β dictates the importance of precision versus recall. We will be using balanced F-measure with $\beta = 1$ also called as $F1$ score.

$$\text{Kappa} = \frac{P_o - P_c}{1 - P_c} \tag{5}$$

where P_o is essentially the accuracy and P_c is obtained as

$$P_c = \frac{(TP + FP) \times (TP + FN) \times (FN + TN) \times (FP + TN)}{(TP + TN + FP + FN)^2} \tag{6}$$

3.3 Results

All the simulations have been carried out using Python 2.7.12 (Scikit-learn [17]) in a personal laptop with 4 GB main memory and quad-core processor architecture with an equal speed of 2.60 GHz per processor. As mentioned earlier, the value of the parameter α has been randomly selected between (0, 1). However, based on interest, this can be fixed initially as discussed earlier. The number of possible neighbors in kNN approach is set to 5 ($k = 5$). For random forest classifier, the number of trees is set to 10. The maximum number of estimators at which boosting is terminated for AdaBoost is fixed at 30.

3.3.1 Performance Summary

Tables 3, 4, and 5 record the *accuracy*, $F1$, and Kappa scores, respectively, for all the ten datasets used in this work. It should be noted that the recorded values are only from the testing phase of the model where the generalization of the trained classifier is evaluated with the available balanced independent test dataset ($\mathcal{D}_{\text{test}}$). From the $F1$ scores and Kappa values in the tables, we observe that in almost all of the cases our GA-based oversampling approach-based techniques performs at least as good as SMOTE, and in some cases, it performs better than SMOTE-based oversampling method.

3.3.2 Performance Visualization

We show a plot of the evolution of the GA-based oversampling during the learning phase of the random forest classifier while using ecoli-0-1-4-7_vs_5-6 dataset over 15 iterations in Fig. 2. In this figure, 'Min' represents the minimum fitness value (the average CV accuracy) of a particular generation. Similarly, 'Avg' represents the average CV accuracy of the generation, and 'Max' represents the maximum average CV accuracy. It should be noted that there are ten candidate solutions in a

Table 3 Accuracy scores obtained for various models on all the datasets

Dataset	SMOTE			GA-OverSampling		
	kNN	Random forest	AdaBoost	kNN	Random forest	AdaBoost
ecoli-0-1-4-7_vs_2-3-5-6	0.455	0.439	0.606	0.53	0.439	0.758
ecoli-0-1-4-7_vs_5-6	0.394	0.667	0.5	0.439	0.652	0.576
ecoli2	0.636	0.621	0.788	0.636	0.712	0.773
winequality-red-8_vs_6	0.846	0.969	0.923	0.815	0.969	0.946
winequality-red-8_vs_6-7	0.819	0.977	0.942	0.813	0.965	0.965
winequality-white-3-9_vs_5	0.777	0.983	0.953	0.767	0.973	0.959
yeast-1-2-8-9_vs_7	0.73	0.889	0.794	0.82	0.947	0.836
yeast-1-4-5-8_vs_7	0.719	0.906	0.849	0.755	0.928	0.871
yeast4	0.899	0.953	0.939	0.932	0.959	0.953
yeast-0-3-5-9_vs_7-8	0.637	0.833	0.775	0.706	0.882	0.863

Table 4 F1 scores obtained for various models on all the datasets

Dataset	SMOTE			GA-OverSampling		
	kNN	Random forest	AdaBoost	kNN	Random forest	AdaBoost
ecoli-0-1-4-7_vs_2-3-5-6	0.217	0.213	0.278	0.162	0.213	0.385
ecoli-0-1-4-7_vs_5-6	0.13	0.214	0.195	0.178	0.207	0.222
ecoli2	0.429	0.39	0.563	0.429	0.424	0.516
winequality-red-8_vs_6	0.0	0.0	0.0	0.0	0.0	0.0
winequality-red-8_vs_6-7	0.0	0.0	0.0	0.0	0.0	0.0
winequality-white-3-9_vs_5	0.057	0.286	0.222	0.08	0.0	0.25
yeast-1-2-8-9_vs_7	0.136	0.087	0.133	0.227	0.167	0.162
yeast-1-4-5-8_vs_7	0.133	0.235	0.16	0.15	0.167	0.25
yeast4	0.318	0.3	0.4	0.412	0.25	0.3
yeast-0-3-5-9_vs_7-8	0.178	0.0	0.207	0.25	0.4	0.364

generation. The recorded learning performance shows that the generation is indeed evolving with time to produce better individuals (majority samples) in each of the candidate solutions in each generation.

Table 5 Kappa values obtained for various models on all the datasets

Dataset	SMOTE			GA-OverSampling		
	kNN	Random Forest	AdaBoost	kNN	Random Forest	AdaBoost
ecoli-0-1-4-7_vs_2-3-5-6	0.095	0.089	0.169	0.036	0.089	0.299
ecoli-0-1-4-7_vs_5-6	−0.005	0.103	0.072	0.05	0.093	0.106
ecoli2	0.257	0.209	0.447	0.257	0.27	0.391
winequality-red-8_vs_6	−0.041	−0.012	−0.033	−0.042	−0.012	−0.027
winequality-red-8_vs_6-7	−0.033	−0.009	−0.025	−0.033	−0.018	−0.018
winequality-white-3-9_vs_5	0.027	0.279	0.203	0.05	−0.013	0.232
yeast-1-2-8-9_vs_7	0.083	0.042	0.083	0.182	0.139	0.115
yeast-1-4-5-8_vs_7	0.063	0.19	0.101	0.083	0.129	0.198
yeast4	0.281	0.276	0.372	0.382	0.231	0.276
yeast-0-3-5-9_vs_7-8	0.03	−0.088	0.09	0.121	0.335	0.287

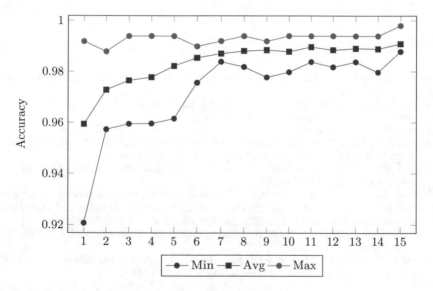

Fig. 2 Evolution of candidate solution with time (generation)

4 Conclusion

In this paper, we proposed a self-adaptive GA-based oversampling approach to deal with imbalanced learning problem, specifically in the area of with binary classification. We used the oversampled dataset obtained by our GA-based oversampling engine to cross-validate three different classifiers such as kNN, random forest, and AdaBoost. These classifiers were tested for their generalization performance against an independent test dataset. The obtained results have been compared with SMOTE-based oversampling strategy leading to the learning for the mentioned classifiers. The recorded comparative performance reports that the proposed GA-based oversampling is at par and sometimes better than the SMOTE oversampling for the evaluated ten imbalanced datasets. Further, it would be interesting to look at imbalanced classification problem commonly arising in the field of life science [18, 19]. As a challenging future work, one could extend the present work to develop GA-based oversampling strategy for highly imbalanced real-world binary and multiclass problems often arising in the field of medical diagnosis, robotics, security, and so forth [20–23].

References

1. He, H., Garcia, E.A.: Learning from imbalanced data. IEEE Trans. Knowl. Data Eng. **21**(9), 1263–1284 (2009)
2. Barandela, R., Sánchez, J.S., García, V., Rangel, E.: Strategies for learning in class imbalance problems. Pattern Recogn. **36**(3), 849–851 (2003)
3. Aly, M.: Survey on multiclass classification methods. Neural Netw. **19**, 1–9 (2005)
4. Liu, X.Y., Wu, J., Zhou, Z.H.: Exploratory undersampling for class-imbalance learning. IEEE Trans. Syst. Man Cybern. Part B (Cybern.) **39**(2), 539–550 (2009)
5. Li, J., Fong, S., Wong, R.K., Chu, V.W.: Adaptive multi-objective swarm fusion for imbalanced data classification. Inf. Fusion (2017)
6. Dash, T., Nayak, T., Swain, R.R.: Controlling wall following robot navigation based on gravitational search and feed forward neural network. In: Proceedings of the 2nd International Conference on Perception and Machine Intelligence, pp. 196–200. ACM (2015)
7. Boussaïd, I., Lepagnot, J., Siarry, P.: A survey on optimization metaheuristics. Inf. Sci. **237**, 82–117 (2013)
8. Dash, T., Sahu, P.K.: Gradient gravitational search: an efficient metaheuristic algorithm for global optimization. J. Comput. Chem. **36**(14), 1060–1068 (2015)
9. Chawla, N.V., Bowyer, K.W., Hall, L.O., Kegelmeyer, W.P.: Smote: synthetic minority oversampling technique. J. Artif. Intell. Res. **16**, 321–357 (2002)
10. Li, J., Fong, S., Zhuang, Y.: Optimizing smote by metaheuristics with neural network and decision tree. In: 2015 3rd International Symposium on Computational and Business Intelligence (ISCBI), pp. 26–32. IEEE (2015)
11. Jiang, K., Lu, J., Xia, K.: A novel algorithm for imbalance data classification based on genetic algorithm improved smote. Arab. J. Sci. Eng. **41**(8), 3255–3266 (2016)
12. Zorić, B., Bajer, D., Martinović, G.: Employing different optimisation approaches for smote parameter tuning. In: International Conference on Smart Systems and Technologies (SST), pp. 191–196. IEEE (2016)
13. Goldberg, D.E., Holland, J.H.: Genetic algorithms and machine learning. Mach. Learn. **3**(2), 95–99 (1988)

14. Alcalá-Fdez, J., Fernández, A., Luengo, J., Derrac, J., García, S., Sanchez, L., Herrera, F.: Keel data-mining software tool: data set repository, integration of algorithms and experimental analysis framework. J. Multiple-Valued Logic Soft Comput. **17** (2011)
15. Dinesh, S., Dash, T.: Reliable evaluation of neural network for multiclass classification of real-world data. arXiv preprint arXiv:1612.00671 (2016)
16. Cohen, J.: A coefficient of agreement for nominal scales. Educ. Psychol. Measur. **20**(1), 37–46 (1960)
17. Pedregosa, F., Varoquaux, G., Gramfort, A., Michel, V., Thirion, B., Grisel, O., Blondel, M., Prettenhofer, P., Weiss, R., Dubourg, V., et al.: Scikit-learn: machine learning in python. J. Mach. Learn. Res. **12**(Oct), 2825–2830 (2011)
18. Pai, P.P., Dash, T., Mondal, S.: Sequence-based discrimination of protein-RNA interacting residues using a probabilistic approach. J. Theor. Biol. **418**, 77–83 (2017)
19. Wang, S., Yao, X.: Multiclass imbalance problems: analysis and potential solutions. IEEE Trans. Syst. Man Cybern. Part B (Cybern.) **42**(4), 1119–1130 (2012)
20. Wan, X., Liu, J., Cheung, W.K., Tong, T.: Learning to improve medical decision making from imbalanced data without a priori cost. BMC Med. Inform. Decis. Mak. **14**(1), 111 (2014)
21. Nayak, T., Dash, T., Rao, D.C., Sahu, P.K.: Evolutionary neural networks versus adaptive resonance theory net for breast cancer diagnosis. In: Proceedings of the International Conference on Informatics and Analytics, p. 97. ACM (2016)
22. Dash, T.: Automatic navigation of wall following mobile robot using adaptive resonance theory of type-1. Biologically Inspired Cogn. Archit. **12**, 1–8 (2015)
23. Dash, T.: A study on intrusion detection using neural networks trained with evolutionary algorithms. Soft Comput. **21**(10), 2687–2700 (2017)

Using NSGA-II to Solve Interactive Fuzzy Multi-objective Reliability Optimization of Complex System

Hemant Kumar and Shiv Prasad Yadav

Abstract In the real-world problem, reliability enhancement is one of the primary concerns in the system design. A conflicting situation often occurs when the cost of the system is reduced and its reliability is improved simultaneously. Practically, design data included in the system are not found specific. Various types of uncertainty such as vagueness, qualitative statements, expert's information character etc. are found in the multi-objective optimization of reliability problems. Multiple solutions (Pareto-optimal solutions) are obtained in multi-objective optimization problem (MOP) where a decision-maker (DM) plays a crucial role in decision-making process. In view of such things, a fuzzy multi-objective reliability optimization model is developed interactively. Numerical examples of complex systems are given for the illustrations. To solve the problems, an efficient multi-objective evolutionary algorithm (MOEA), namely NSGA-II is employed. Finally, we get the solutions according to the preference of the DM.

Keywords Multi-objective optimization problem (MOP) · System reliability
Membership function · Pareto-optimal front · NSGA-II

1 Introduction

Multi-objective reliability optimization problem in the system design is to enhance the system reliability with reduction of other resource consumptions like cost, weight, and volume simultaneously. The main conflicting objectives of the system design are considered as the system reliability and system cost. A DM always demands to diminish the cost and enhance the reliability of the system simultaneously. During the designing phase of the system, unreliable and incomplete input information are

H. Kumar (✉) · S. P. Yadav
Department of Mathematics, I.I.T. Roorkee, Roorkee 247667, Uttarakhand, India
e-mail: hemantkumar2654@gmail.com

S. P. Yadav
e-mail: spyorfma@gmail.com

© Springer Nature Singapore Pte Ltd. 2019
J. C. Bansal et al. (eds.), *Soft Computing for Problem Solving*,
Advances in Intelligent Systems and Computing 816,
https://doi.org/10.1007/978-981-13-1592-3_31

needed to tackle. This may be due to unfamiliar with environmental factors or the system itself. Various types of uncertainty such as vagueness, uncertainty in judgments, expert's information character etc. are found in the decision-making of reliability problems. So, the models of reliability design system should have more flexible and adaptable to the human decision-making process. Fuzzy set theory [1] is able to tackle such types of situations. Weighted sum method [2] is a well-known classical optimization method to solve MOP, where all the objectives are transformed into a single objective by supplying a user-supplied weight to each objective. If we require multiple solutions, then the problem is solved multiple times with a different combination of weights. Due to this reason, we need to choose the correct vector in each run of the program. The other drawback is the inability to get the Pareto-optimal solutions, especially for non-convex and non-differentiable functions. In principle, an MOP gives multiple solutions (Pareto-optimal solutions) to the DM. A DM cannot claim to one of these solutions to be better than another in the absence of any further information [3]. So, an ideal technique is to explore the Pareto-optimal front. In the past decades, MOEAs have been performed to be powerful techniques for finding the non-dominated solutions (Pareto-optimal front) in a single simulation run. Such types of solution approaches can be seen in Konak et al. [4] and Zhou et al. [5]. NSGA-II [3] is one of such MOEAs. It performs better than two other elitist MOEAs such as PAES [6] and SPEA [7] and gets better convergence and spread of solutions near the true Pareto-optimal front [3]. Therefore, NSGA-II is a well-known technique, growing applications due to its parameter-less sharing approach, elitism and low computational requirements [3]. The advantages of NSGA-II have been successfully shown by Salazar et al. [8], Kishore et al. [9], Safari [10] and Sharifi et al. [11]. In this paper, interactive fuzzy multi-objective reliability optimization problems of complex system are solved by NSGA-II. Various Pareto-optimal fronts have been shown in keeping the views of DM's interest. The remaining sections of this paper are categorized as follows. Section 2 contains some basic definitions. Section 3 presents a brief description of NSGA-II. Section 4 presents the proposed approach with illustrations. Section 5 contains the results and discussion and Sect. 6 gives the conclusion.

2 Some Basic Definitions

Definition 1 (*Multi-objective optimization problem (MOP)*) [2] In general, an MOP minimizes $F(X) = [f_1(X), f_2(X), \ldots, f_k(X)]^{\mathrm{T}}$ subject to $g_i(X) \leq 0$, $i = 1, 2, \ldots, m$, $X \in \Omega$. A decision variable vector $X = [x_1, x_2, \ldots, x_n]^{\mathrm{T}}$ is called a solution from the feasible region $\Omega \subseteq R^n$ (Euclidean n-space); the objective functions $f_p(X)$, $p = 1, 2, \ldots, k$, where $f_p: \Omega \to R$ and the constrained functions $g_i(X)$, where $g_i: \Omega \to R$; $F(X)$ is called a multi-objective vector or criterion vector; $x_j^{(L)}$ and $x_j^{(U)}$ are the lower and upper bounds of the decision variable x_j respectively. If all f_p's and g_i's are linear then the problem is called a multi-objective linear programming problem (MOLPP), otherwise it is called a multi-objective nonlinear programming problem (MONLPP). It is evident that the notion of single-objective optimization problem (SOP) is not relevant to MOP. Due to this reason, the solutions of an MOP are studied in terms of Pareto optimality.

Definition 2 *Pareto dominance*: A solution [12] vector X^1 dominates another solution vector X^2 denoted as $X^1 \succ X^2$ iff

- $f_p(X^1) \leq f_p(X^2) \; \forall p = 1, 2, \ldots, k$,
- $f_q(X^1) < f_q(X^2)$ for at least one $q \in \{1, 2, \ldots, k\}, \; p \neq q$.

If there does not exist such solutions which dominate X^1, then X^1 is called a non-dominated solution.

Definition 3 (*Pareto-optimal set*) A set of non-dominated solutions [12] $P = \{X^* \in \Omega | \neg \exists X \; s.t. \; X \succ X^*\}$ is said to be a Pareto-optimal set.

Definition 4 (*Pareto-optimal solution*) A point $X^* \in \Omega$ is called a Pareto-optimal solution [12] if there does not exist another point $X \in \Omega$ such that $f_p(X) \leq f_p(X^*) \; \forall p = 1, 2, \ldots, k$ and $f_q(X) < f_q(X^*)$ for at least one $q \in \{1, 2, \ldots, k\}$, $p \neq q$.

Definition 5 (*Pareto-optimal front*) The set of vectors in the objective space that are images of elements of a Pareto-optimal set [12] under F, i.e. $PF = \{F(X^*) | \neg \exists X \; s.t. \; X \succ X^*\}$.

3 NSGA-II

Non-dominated sorting genetic algorithm (NSGA) was initially suggested by Srinivas and Deb [13]. It uses Goldberg's domination criterion [14] to assign ranks for the solutions and utilization of fitness sharing for maintaining the diversity in the solution set. It has some difficulty regarding the computational complexity, non-elitist approach and highly dependent on the parameters of fitness sharing. Deb et al. [3] extended this algorithm in form of NSGA-II by giving some new features like fast non-dominated sorting, crowding distance, and comparison operator.

NSGA-II assigns a rank for solutions employing non-dominated sorting procedure and emphasizes good solutions throughout this algorithm. The overall complexity governed by this process is $O(kN^2)$, where k and N denote the no. of objectives and population size, respectively [3].

For maintaining the diversity in the solution set, NSGA-II calculates the crowding distance of each solution. It is basically defined as those solutions that contain the same rank. A partial order comparison operator is applied to determine a better solution between two solutions. According to this operator, if both the solutions belong to the same rank, then preference is given to the solution that contains a higher crowding distance value. A higher crowding distance value gives the lesser crowded region and vice versa [3] (Fig. 1).

(a)   **(b)**

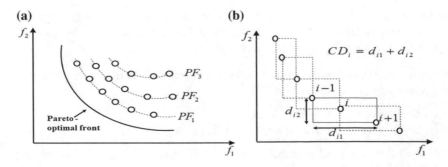

(c)

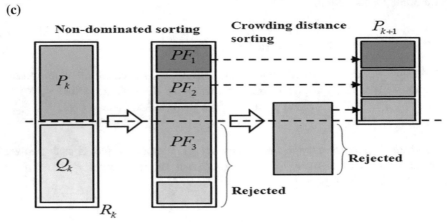

Fig. 1 a Sorting procedure of a population. **b** Crowding distance estimation of a solution. **c** Evaluation cycle of the NSGA-II algorithm

The pseudo-code of NSGA-II algorithm is given as follows:

Step 1.   Initializing randomly a parent population P_0 of size N. Setting $k = 0$.

Step 2.   Assigning fitness (rank) according to non-domination level and crowded-comparison operator.

Step 3.   **while** $k <$ number of maximum generation **do**

   (i)   Creating an offspring population Q_k of size N applying reproduction, crossover and mutation.

   (ii) Combining via $R_k = P_k \cup Q_k$.

   (iii) Sorting on R_k and classifying them into non-dominated fronts (Pareto-front) PF_i, $i = 1, 2, \ldots$, etc.

   (iv) Setting a new population $P_{k+1} = \emptyset$ and $i = 1$.
         while the parent population size $|P_{k+1}| + |PF_i| < N$ **do**
            (i)   Calculating the crowding distance of PF_i.
            (ii) Adding the ith non-dominated front PF_i to the parent population P_{k+1}.

  (iii) $i = i + 1$.
         end while
  (v) Sorting the PF_i using the crowding distance-based comparison operator.
  (vi) Filling the parent population P_{k+1} with the first $N - |P_{k+1}|$ solutions of PF_i.
  (vii) Generating the offspring population Q_{k+1}.
  (viii) Setting $k = k + 1$.
         end while

Step 4. Collecting the non-dominated solutions in the vector P.

4 Proposed Methodology

4.1 Mathematical Model of the Problem

In reliability system design, a complex system possesses certain components connected to one another, not in exactly serial or parallel [15]. Here, we have two complex systems as shown in Fig. 2.

Mathematical model

$$\left.\begin{array}{l} \text{Maximize } R_s \\ \text{Minimize } C_s \end{array}\right\} \text{ or, Minimize } (-R_s, C_s) \tag{1}$$

subject to $R_{i,\min} \le R_i \le R_{i,\max} \le 1$, $i = 1, 2, \ldots, n$, and $R_{s,\min} \le R_s \le R_{s,\max} \le 1$, where R_s and C_s are system reliability and system cost, $R_{i,\min}$ and $R_{l,\max}$ are the lower and upper bounds on the reliability of the ith component, $R_{s,\min}$ and $R_{s,\max}$ are the lower and upper bounds on the system reliability, respectively.

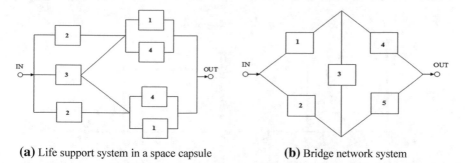

(a) Life support system in a space capsule   **(b)** Bridge network system

Fig. 2 System configuration of complex systems

4.2   Fuzzy MOP Model

$$
\mu_{\widetilde{R}_s} = \begin{cases} 0, & R_s \leq R_s^l, \\ h_1(R_s), & R_s^l \leq R_s \leq R_s^u, \\ 1, & R_s \geq R_s^u, \end{cases} \tag{2}
$$

$$
\mu_{\widetilde{C}_s} = \begin{cases} 1, & C_s \leq C_s^l, \\ h_2(C_s), & C_s^l \leq C_s \leq C_s^u, \\ 0, & C_s \geq C_s^u, \end{cases} \tag{3}
$$

where R_s^l and R_s^u are the lower and upper limits on the system reliability, respectively. These values are decided according to the given situation. $h_1(R_s)$ is a monotonically increasing function of R_s. Similarly, C_s^l and C_s^u are determined by the DM according to the given condition. $h_2(C_s)$ is a monotonically decreasing function of C_s. In Fig. 3, the shape of $\mu_{\widetilde{R}_s}$ and $\mu_{\widetilde{C}_s}$ is shown. The fuzzy MOP model is given as follows.

$$
\left. \begin{array}{l} \text{Maximize } \left(\mu_{\widetilde{R}_s}, \mu_{\widetilde{C}_s} \right) \\ \text{subject to } R_{i,\min} \leq R_i \leq R_{i,\max} \leq 1, \ i = 1, 2, \ldots, n \end{array} \right\} \tag{4}
$$

Theorem *The Pareto-optimal solutions of the MOP (4) are also Pareto-optimal solutions of the MOP (1).*

Proof Let $\boldsymbol{R}^*$ be a Pareto-optimal solution vector of (4). Then by definition of Pareto-optimal solution, we get $\Leftrightarrow \nexists \boldsymbol{R} \in \Omega$ (feasible region) such that $-\mu_{\widetilde{R}_s}(\boldsymbol{R}) \leq -\mu_{\widetilde{R}_s}(\boldsymbol{R}^*)$ and $-\mu_{\widetilde{C}_s}(\boldsymbol{R}) < -\mu_{\widetilde{C}_s}(\boldsymbol{R}^*) \Leftrightarrow \nexists \boldsymbol{R} \in \Omega$ such that $-h_1[R_s(\boldsymbol{R})] \leq -h_1[R_s(\boldsymbol{R}^*)]$ and $-h_2[C_s(\boldsymbol{R})] < -h_2[C_s(\boldsymbol{R}^*)] \Leftrightarrow \nexists \boldsymbol{R} \in \Omega$ such that $h_1[R_s(\boldsymbol{R})] \geq h_1[R_s(\boldsymbol{R}^*)]$ and $h_2[C_s(\boldsymbol{R})] > h_2[C_s(\boldsymbol{R}^*)] \Leftrightarrow \nexists \boldsymbol{R} \in \Omega$ such that $R_s(\boldsymbol{R}) \geq R_s(\boldsymbol{R}^*)$ and $C_s(\boldsymbol{R}) < C_s(\boldsymbol{R}^*)$ (since h_1 is monotonically increasing and h_2 is monotonically decreasing function). $\Leftrightarrow \nexists \boldsymbol{R} \in \Omega$ such that $-R_s(\boldsymbol{R}) \leq -R_s(\boldsymbol{R}^*)$

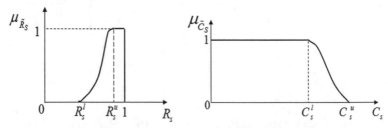

(a) Monotonically increasing function   **(b)** Monotonically decreasing function

Fig. 3   Membership function for **a** system reliability and **b** system cost

and $C_s(R) < C_s(R^*) \Leftrightarrow R^* \in \Omega$ is a Pareto-optimal solution of the MOP given by (1). Similarly, we can prove by taking the second objective C_s first. $\qquad\qquad\square$

4.3 Constructing the Membership Functions Through Interaction of the DM

Case (i): If the DM is not biased towards the objectives
Linear membership function (neutral behaviour)

$$
\mu_{\tilde{R}_s} = \begin{cases} 0, & R_s \leq R_s^l, \\ \left(\frac{R_s - R_s^l}{R_s^u - R_s^l}\right), & R_s^l \leq R_s \leq R_s^u, \\ 1, & R_s \geq R_s^u, \end{cases} \qquad \mu_{\tilde{C}_s} = \begin{cases} 1, & C_s \leq C_s^l, \\ \left(\frac{C_s^u - C_s}{C_s^u - C_s^l}\right), & C_s^l \leq C_s \leq C_s^u, \\ 0, & C_s \geq C_s^u, \end{cases}
$$

Case (ii): If the DM is biased towards minimizing the system cost
Biased to minimizing the system cost (convex–concave shape)

$$
\mu_{\tilde{R}_s} = \begin{cases} 0, & R_s \leq R_s^l \\ \left(\frac{R_s - R_s^l}{R_s^u - R_s^l}\right)^{n_1}, & R_s^l \leq R_s \leq R_s^u, n_1 > 1 \\ 1, & R_s \geq R_s^u, \end{cases} \qquad \mu_{\tilde{C}_s} = \begin{cases} 1, & C_s \leq C_s^l, \\ \left(\frac{C_s^u - C_s}{C_s^u - C_s^l}\right)^{n_2}, & C_s^l \leq C_s \leq C_s^u, n_2 < 1 \\ 0, & C_s \geq C_s^u, \end{cases}
$$

Case (iii): If the DM is biased towards maximizing the system reliability where system reliability approaches to 1 as far as possible.
Biased to maximizing the system reliability (concave–convex shape)

$$
\mu_{\tilde{R}_s} = \begin{cases} 0, & R_s \leq R_s^l, \\ \left(\frac{R_s - R_s^l}{R_s^u - R_s^l}\right)^{n_1}, & R_s^l \leq R_s \leq R_s^u, n_1 < 1 \\ 1, & R_s \geq R_s^u, \end{cases} \qquad \mu_{\tilde{C}_s} = \begin{cases} 1, & C_s \leq C_s^l, \\ \left(\frac{C_s^u - C_s}{C_s^u - C_s^l}\right)^{n_2}, & C_s^l \leq C_s \leq C_s^u, n_2 > 1 \\ 0, & C_s \geq C_s^u, \end{cases}
$$

Case (iv): If the DM wishes to get compromise solutions towards the objectives (sigmoidal shape)

$$
\mu_{\tilde{R}_s} = \begin{cases} 0, & R_s \leq R_s^l, \\ \frac{1}{1 + e^{-\lambda_1(R_s - R_{s,\text{avg}})}}, & R_s^l \leq R_s \leq R_s^u, \lambda_1 \in \mathbb{R}^+, \\ 1, & R_s \geq R_s^u, \end{cases}
$$

$$\mu_{\tilde{C}_s} = \begin{cases} 1, & C_s \leq C_s^l, \\ \dfrac{1}{1+e^{\lambda_2(C_s-C_{s,\text{avg}})}}, & C_s^l \leq C_s \leq C_s^u, \lambda_2 \in \mathbb{R}^+ \\ 0, & C_s \geq C_s^u, \end{cases}$$

4.4 Numerical Examples

Example 1 (Life-support system in a space capsule) [15] "Maximize system reliability as close as possible to 1 with approximate system cost of 641.8 (cost units)"

Here $R_s^l = 0.9$, $R_s^u = 0.99$, $C_s^l = 641$, $C_s^u = 700$, $R_{j,\min} = 0.5$, $R_{j,\max} = 0.99$.

For Case (ii): $n_1 = 5, n_2 = 1/5$, Case (iii): $n_1 = 1/5, n_2 = 5$, Case (iv): $\lambda_1 = 120$, $\lambda_2 = 0.25$.

Example 2 (Bridge network system) [15] "Maximize system reliability as close as possible to 1 with approximate system cost of 5.01993 (cost units)"

Here $R_s^l = 0.99$, $R_s^u = 0.9905$, $C_s^l = 5.0$, $C_s^u = 5.02047$, $R_{j,\min} = 0$, $R_{j,\max} = 1$.

For Case (ii): $n_1 = 9$, $n_2 = 1/9$, Case (iii): $n_1 = 1/9$, $n_2 = 9$, Case (iv): $\lambda_1 = 7000$, $\lambda_2 = 2000$ (Figs. 4, 5, 6 and 7).

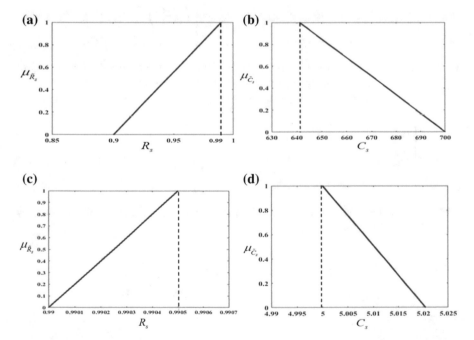

Fig. 4 Membership functions for R_s and C_s on the basis of case (i): Example 1 (**a–b**); Example 2 (**c–d**)

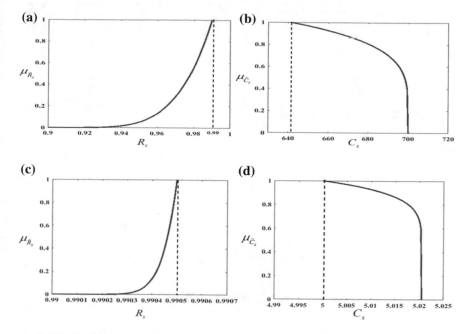

Fig. 5 Membership functions for R_s and C_s on the basis of case (ii): Example 1 (**a–b**); Example 2 (**c–d**)

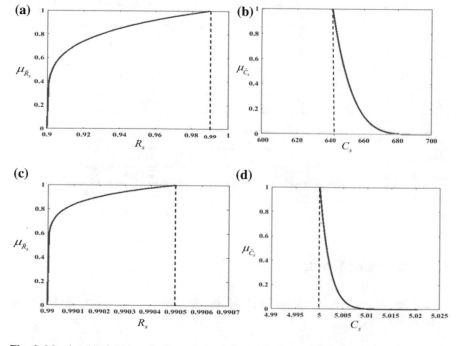

Fig. 6 Membership functions for R_s and C_s on the basis of case (iii): Example 1 (**a–b**); Example 2 (**c–d**)

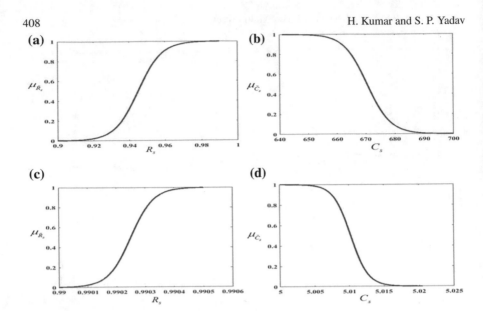

Fig. 7 Membership functions for R_s and C_s on the basis of case (iv): Example 1 (**a–b**); Example 2 (**c–d**)

Table 1 NSGA-II parameter settings

Parameters	Life-support system in a space capsule	Bridge network system
No. of variables (V)	4	5
Population size (N)	40	50
Maximum no. of generations (t_{max})	80	100
Crossover rate (p_c)	0.8	0.9
Mutation rate (p_m)	0.2	0.1
Crossover index (η_c)	10	10
Mutation index (η_m)	50	100
Random seed	0.9876	0.1234

4.5 Parameter Settings

On the basis of rigorous experimentation, the parameter settings of NSGA-II are given in Table 1.

5 Results and Discussion

After applying the proposed approach, we get the Pareto-optimal fronts in different conditions raised by the DM. NSGA-II algorithm, a second generation MOEA, is employed effectively in each case shown in Figs. 8, 9, 10 and 11. We compare numerical examples of two complex systems with different number of variables. In Fig. 8, we have solutions equally distributed on the Pareto-optimal front that are obtained by DM's neutral behaviour towards the objectives. In Fig. 9, DM wants the biased solution towards the system cost and so finds the clustered solutions on the lower side of the Pareto-optimal front. In Fig. 10, DM wants the biased solution towards system reliability and finds the clustered solutions upper side of the Pareto optimal front. In Fig. 11, DM wants the compromise solution and finds the solutions in the intermediate region of the Pareto-optimal front.

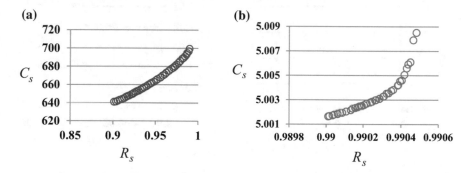

Fig. 8 Pareto-optimal front obtained on the basis of case (i): **a** Example 1; **b** Example 2

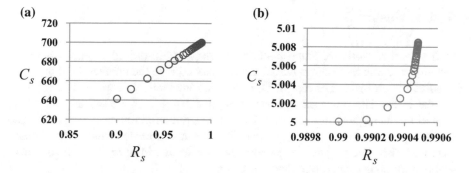

Fig. 9 Pareto-optimal front obtained on the basis of case (ii): **a** Example 1; **b** Example 2

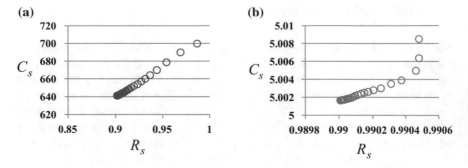

Fig. 10 Pareto-optimal front obtained on the basis of case (iii): **a** Example 1; **b** Example 2

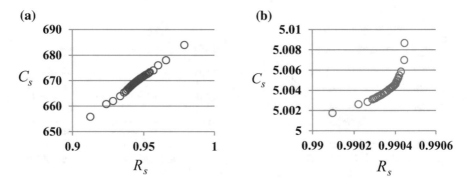

Fig. 11 Pareto-optimal front obtained on the basis of case (iv): **a** Example 1; **b** Example 2

6   Conclusion

In this piece of work, we propose interactive fuzzy multi-objective reliability optimization of complex systems. NSGA-II is employed with the interaction of the DM
to solve the problem. The role of DM has been interactively exploited in achieving the goals. The proposed approach has advantages over the existing methods
such as producing multiple solutions in a single simulation run, not using any kind
of aggregation operators, getting information of non-dominated solutions and their
characteristics. The proposed approach can be effective in fuzzy decision-making
problem for other engineering design system.

Acknowledgement   The authors acknowledge the MHRD, Government of India, for the financial
grant.

Appendix

Life-support system in a space capsule: [15]

The system reliability R_s and system cost C_s of system configuration given in Fig. 2a are expressed as follows.

$$R_s = 1 - R_3[(1 - R_1)(1 - R_4)]^2 - (1 - R_3)[1 - R_2\{1 - (1 - R_1)(1 - R_4)\}]^2$$

$$C_s = 2 \sum_{i=1}^{4} K_i R_i^{\alpha_i},$$

where vectors of coefficients K_i and α_i are $K = \{100, 100, 200, 150\}$ and $\alpha = \{0.6, 0.6, 0.6, 0.6\}$, respectively.

Bridge network system: [15]

The algebraic expression of R_s and C_s of system configuration given in Fig. 2b is given as follows.

$$R_s = R_1 R_4 + R_2 R_5 + R_2 R_3 R_4 + R_1 R_3 R_5 + 2 R_1 R_2 R_3 R_4 R_5 - R_1 R_2 R_4 R_5$$
$$- R_1 R_2 R_3 R_4 - R_1 R_3 R_4 R_5 - R_2 R_3 R_4 R_5 - R_1 R_2 R_3 R_5$$

$$C_s = \sum_{i=1}^{5} a_i \exp\left[\frac{b_i}{1 - R_i}\right],$$

where $a_i = 1$ and $b_i = 0.0003 \,\forall i, i = 1, 2, \ldots, 5$.

References

1. Zadeh, L.A.: Fuzzy sets. Inf. Control **8**(3), 338–353 (1965)
2. Deb, K.: Multi-objective Optimization Using Evolutionary Algorithms. Wiley, New York (2001)
3. Deb, K., Pratap, A., Agarwal, S., Meyarivan, T.: A fast and elitist multiobjective genetic algorithm: NSGA-II. IEEE Trans. Evol. Comput. **6**(2), 182–197 (2002)
4. Konak, A., Coit, D.W., Smith, A.E.: Multi-objective optimization using genetic algorithms: a tutorial. Reliab. Eng. Syst. Saf. **91**, 992–1007 (2006)
5. Zhou, A., Qu, B.Y., Li, H., Zhao, S.Z., Suganthan, P.N., Zhang, Q.Z.: Multiobjective evolutionary algorithms: a survey of the state of the art. Swarm Evol. Comput. **1**, 32–49 (2011)
6. Knowles, J., Corne D.: The Pareto archived evolution strategy: a new baseline algorithm for multiobjective optimization. In: Proceedings of the 1999 Congress on Evolutionary Computation, pp. 98–105. IEEE Press, Piscataway, NJ (1999). https://doi.org/10.1109/cec.1999.78191 3
7. Zitzler, E., Thiele, L.: An evolutionary algorithm for multi-objective optimization: The strength Pareto approach. Technical report 43, Zurich, Switzerland: Computer Engineering and Networks Laboratory (TIK), Swiss Federal Institute of Technology (ETH) (1998)
8. Salazar, D., Rocco, C.M., Galvan, B.J.: Optimization of constrained multiple objective reliability problems using evolutionary algorithms. Reliab. Eng. Syst. Saf. **91**, 1057–1070 (2006)

9. Kishore, A., Yadav, S.P., Kumar, S.: A multi-objective genetic algorithm for reliability optimization problem. Int. J. Performability Eng. **5**(3), 227–234 (2009)
10. Safari, J.: Multi-objective reliability optimization of series-parallel systems with a choice of redundancy strategies. Reliab. Eng. Syst. Saf. **180**, 10–20 (2012)
11. Sharifi, M., Guilani, P.P., Shahriari, M.: Using NSGA-II algorithm for a three-objective redundancy allocation problem with k-out-of-n sub-systems. J. Optim. Ind. Eng. **19**, 87–95 (2016)
12. Garg, H., Sharma, S.P.: Multi-objective reliability-redundancy allocation problem using particle swarm optimization. Comput. Ind. Eng. **64**, 247–255 (2013)
13. Srinivas, N., Deb, K.: Multi-objective optimization using non-dominated sorting in genetic algorithms. Evol. Comput. **2**(3), 221–248 (1994)
14. Goldberg, D.E.: Genetic Algorithms for Search, Optimization, and Machine Learning. Addison-Wesley, Reading, MA (1989)
15. Ravi, V., Reddy, P.J., Zimmermann, H.J.: Fuzzy global optimization of complex system reliability. IEEE Trans. Fuzzy Syst. **8**(3), 241–248 (2000)

Fuzzy Time Series Forecasting Model Using Particle Swarm Optimization and Neural Network

Mahua Bose and Kalyani Mali

Abstract In recent years, there are several ongoing efforts to develop models for forecasting fuzzy time series using classical or artificial intelligence (AI) techniques in different application areas. A major challenge lying with the fuzzy time series forecasting model is efficient partitioning of data. It has significant effect on forecasting accuracy. Proposed work overcomes the difficulty of searching appropriate interval length for partitioning the data. In this study, a hybrid model using particle swarm optimization and backpropagation neural network (BPNN) is applied for forecasting fuzzy time series. Particle swarm optimization (PSO) searches for optimal partitioning of data, and weights of neural network are adjusted using gradient descent technique. The neural network takes fuzzy membership values as input, and every particle represents a set of boundaries between two adjacent intervals. This hybrid procedure is iterated until stopping condition is reached. The experiment is carried out on standard datasets, and results are compared with related models including neuro-fuzzy models applied on the same dataset. Proposed idea shows best performance in terms of accuracy in prediction.

Keywords Backpropagation · Interval · Neural network · Position · Particle
Triangular membership

1 Introduction

Forecasting is essential for policy planning and decision-making. It also helps in disaster management activities. During the last twenty years, many forecasting techniques have been developed using fuzzy time series. These techniques have been widely applied in the area of finance, weather, etc. Designing forecasting model

M. Bose (✉) · K. Mali
Department of Computer Science & Engineering, University of Kalyani, Kalyani, India
e-mail: e_cithi@yahoo.com

K. Mali
e-mail: kalyanimali1992@gmail.com

© Springer Nature Singapore Pte Ltd. 2019
J. C. Bansal et al. (eds.), *Soft Computing for Problem Solving*,
Advances in Intelligent Systems and Computing 816,
https://doi.org/10.1007/978-981-13-1592-3_32

413

using fuzzy time series has two main advantages: (1) In addition to numeric data, it can process linguistic terms like little, medium, big. (2) It does not require large dataset for prediction.

First significant effort in this research area was made by Song and Chissom [28–30]. Drawback of his model is its high computational overheads due to complex matrix operations for the calculation of min–max composition. Chen [10] proposed a simplified model including only simple arithmetic operations. This model is considered as most important milestone in this particular field of research. Later many researchers contributed for the development and advancement of research in this direction. Fuzzy time series model designing focuses on two main issues: (1) creation of intervals and (2) formulation of fuzzy logical relationships between current and past data.

Partitioning of data is the key factor related to the performance of forecast. A lot of research work addressed this problem. Data partitioning techniques can be broadly classified into two categories: (1) equal-sized intervals partitioning technique and (2) unequal-sized intervals partitioning technique. Variable length partitioning methods are developed using mathematical models and soft computing techniques such as clustering algorithms or evolutionary techniques. Models using evolutionary techniques namely PSO [6, 12, 14, 17, 24], ant colony optimization [9], genetic algorithm [7, 13] have been implemented successfully.

Formulation of fuzzy logical relationships is also a challenging task. It can also be done using mathematical models and soft computing techniques. Hybrid models using artificial neural network (ANN) and fuzzy time series have shown superior performances as they are efficient in handling nonlinear systems. BPNN-based models [1, 11, 15, 16, 25, 26, 31–33], single multiplicative neuron (SMNM-ANN) model [2], and self-organizing feature map-based models [27] have been designed in the past. A hybrid model based on autoregressive moving average (ARIMA) and neural networks [37] has been proposed. A genetic algorithm-based recurrent neuro-fuzzy model [4] has been presented also.

In the past, efforts have been made to combine PSO and ANN for updating link weights of neural networks hybrid PSO-BPNN algorithm [38], integrated PSO-radial basis function (RBF) [22], hybrid approach using adaptive neuro-fuzzy inference system (ANFIS) with PSO [5, 6, 21] have been implemented successfully. All of these previous works applied PSO for training of all parameters of ANFIS structure. This study intends to propose variable size interval creation technique based on PSO, and prediction is done using a neuro-fuzzy model. Novelty of the proposed research is that it overcomes the difficulty of searching appropriate interval length for partitioning the universe of discourse. PSO algorithm is employed to choose best partition of data to generate the fuzzy membership values. Every particle vector represents a set of positions of boundaries between two adjacent intervals. BPNN is used to compute fitness of the PSO algorithm. The proposed model performs optimal partition creation and training at a time. Weights of neural network are adjusted using gradient descent technique. Proposed hybrid technique shows better performance than existing neuro-fuzzy models.

This paper describes the following sections: Basic idea of FTS model is given in Sect. 2. Description of proposed methodology is given in Sect. 3. In Sect. 4, comparative study of the FTS algorithms is shown. Conclusion is presented in Sect. 5.

2 Fuzzy Time Series

A fuzzy set [35, 36] X on the universe of discourse U, $U = \{u_1, u_2, \ldots, u_k\}$, is defined as follows:

$$X = f_X(u_1)/u_1 + f_X(u_2)/u_2 + \cdots + f_X(u_k)/u_k$$

where $f_X(u_i)$ represents membership degree of u_i to the fuzzy set X, and $1 \leq i \leq k$ and $0 \leq f_X(u_i) \leq 1$.

Let us consider, a subset $Z(t)$ $(t = \ldots, 0, 1, 2, \ldots)$, of real time R_t is the universe of discourse on which fuzzy sets $f_k(t)$ $(k = 1, 2, \ldots)$ are defined. $T(t)$ is called a fuzzy time series defined on $Z(t)$ if $T(t)$ is a collection of $f_1(t), f_2(t), \ldots$ [10, 19, 28–30].

If $T(t)$ is dependent on $T(t - 1)$ only, then the first-order relationship is denoted by $T(t - 1) \rightarrow T(t)$.

In all fuzzy time series forecasting models, there are four common steps to be performed [10]: (1) partitioning the dataset into intervals, (2) fuzzification of the dataset, (3) defining fuzzy logical relationships (FLR) and fuzzy logical relationship groups (FLRG), and (4) forecasting.

3 Proposed Methodology Using PSO and Neuro-Fuzzy Method

In this study, two major tasks are performed: (1) PSO [18] is used for proper partitioning of data. PSO is applied to optimize the interval length by determining proper position of boundary, and it has main role in accuracy of results. It has great impact on proper fuzzification of data. (2) Forecast is obtained using the neuro-fuzzy procedure.

First of all, data domain D is to be defined. Lower bound and upper bound of the data domain are LB and UB, respectively. X_1 and X_2 are two positive integers.

$$D = \left[(\text{Data}_{\text{min}} - X_1) - (\text{Data}_{\text{max}} + X_2) \right]$$
$$\text{LB} = \text{Data}_{\text{min}} - X_1$$
$$\text{UB} = \text{Data}_{\text{max}} + X_2$$

Let the number of intervals be n. Position of boundary between two adjacent intervals is to be optimized. A particle vector consists of $n - 1$ (boundary) elements excluding LB and UB, i.e., $b_1, b_2, \ldots, b_{n-1}$, are position of boundaries between two adjacent intervals. Intervals are defined as follows:

$$I_1 = [\text{LB}, \ b_1], \ I_2 = [b_1, \ b_2], \ldots, I_n = \left[b_{n-1}, \ \text{UB}\right].$$

Elements in each particle vector are sorted in ascending order within the range of the input space. Position of the ith particle is represented as $b_i = (b_{i1}, b_{i2}, \ldots, b_{id})$ and the velocity of the ith particle is given by: $V_i = (v_{i1}, v_{i2}, \ldots, v_{id})$. In this case, $d = n - 1$.

Initially, particle's local best-known position is its initial position.

$$\text{pbest}_i = (\text{pbest}_{i1}, \ \text{pbest}_{i2}, \ldots, \text{pbest}_{id}).$$

Set local best-known position with minimum MSE as global best position: gbest_k ($k = 1, 2, \ldots, d$).

Based on the n intervals, fuzzy sets are to be defined (Fig. 1). Fuzzy membership of all data values are calculated using triangular membership function. For each particle, a 2D membership matrix is created. In case of m data values and n fuzzy sets, fuzzy membership vector is an $m \times n$ matrix. Each row of fuzzy membership matrix is a pattern.

For each particle, membership vector is passed to Neuro_fuzzy procedure. In this step, a backpropagation neural network [23] with one hidden layer in feedforward topology has been used for prediction. Unipolar Sigmoidal or squashed S-function has been used as nonlinear function. Number of inputs and outputs are equal to the number for fuzzy set defined. The inputs and the outputs of BPNN consist of the membership degrees. Two consecutive values are taken as one input–output (target) pair. For example, at first, the neural network takes fuzzy membership values at time '$t - 1$' as input. Fuzzy membership values at time 't' are desired output. Then, membership values of next two consecutive time periods are considered as input–output pair and this way all patterns are presented to network and weights are updated using gradient descent method.

Output needs to be defuzzified to obtain predicted value. The fuzzy set having the highest membership value is considered for defuzzification. Predicted value is the middle point of the said fuzzy set. Let us consider, an input value whose maximum output membership value occurs in fuzzy set defined by Triangle (x: b_1; b_2; b_3). Parameters b_1, b_2, b_3 are estimated by PSO. In this case, crisp output will be b_2.

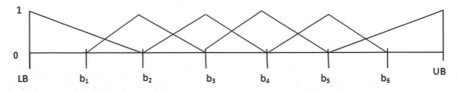

Fig. 1 Assigning triangular membership values

The Neuro_fuzzy procedure returns fitness value of the corresponding particle. For N observation, fitness value is calculated using Eq. 1.

$$\text{Fitness} = \sum_{i=1}^{N} (\text{Desired\_Value} - \text{Predicted\_Value})^2 / N \tag{1}$$

The entire procedure can be described by the following algorithm:

Steps:

1. For the dataset, determine lower and upper bounds.
2. Generate m particles with dimension $n-1$, where n is the number of intervals.
3. Initialize position and velocity for each particle p_i ($i = 1, 2, \ldots m$).
4. Initialize local best position of each particle.
5. Set iteration $= 1$.
6. For each particle, perform the steps 7 to 10.
7. Calculate fuzzy membership of data values using triangular membership function and initialize the inertia weight.
8. Call Neuro_fuzzy (fuzzy membership vector).
   //It returns the fitness value of particle i (global error of BPNN).
9. Set $f(p_i) = \text{fitness}(i)$.
10. Update the personal best position of the particle.
11. Calculate global best $f(g)$.
12. For each particle p_i, update its position and velocity.
13. Increment number of iterations by one.
14. If iteration $<$ MAX Limit, go to step 6 else Display Optimum positions.
15. Calculate final output using optimal interval length.
16. Stop.

Neuro_fuzzy (fuzzy membership vector)

1. For each training pattern (set of fuzzy membership values), calculate network output.
2. Compare it with desired output.
3. Calculate error and backpropagate the updated weights.
4. If maximum iteration reached stop and return else go to step 1.
5. Return global error.

4　Performance Evaluation

4.1　Datasets

Experiments are carried out using three standard datasets. The present study considers in sample data only.

(a)   Enrollment dataset [10] from 1971 to 1992 (yearly data).
(b)   Daily stock price of State Bank of India [25] (6/5/2012–7/31/2012) available at
      (http://in.finance.yahoo.com/).
(c)   Daily temperature (for the month of June) data [26] of Taipei, Taiwan.

4.2   Results and Discussion

Performance of the model is evaluated using different hidden layer nodes for each
dataset. Number of inputs and outputs are equal to the number for fuzzy set defined.
Numbers of hidden layer nodes are kept nearly equal to half of the input nodes.
Number of iteration in neuro-fuzzy procedure is 1500. At the beginning of the PSO
algorithm, $\omega = 1.4$. Random positive numbers are: rand1 $= 0.7$ and rand2 $= 1$; acceleration coefficients are: C1 $= 2$ and C2 $= 2$; maximum generation $= 50$; number of
particles $= 5$. In order to examine the improvements made by the proposed method
in forecasting accuracy, performance of the model is compared with autoregression
model (of order one) and existing models using the same datasets. Error estimation
is done using root mean square error (RMSE) and mean absolute percentage error
(MAPE). Results are analyzed using Figs. 2, 3, 4, 5, 6 and 7.

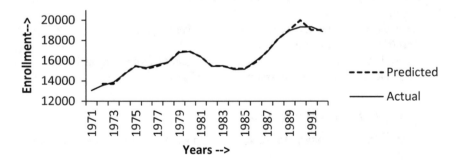

Fig. 2 Enrollment graph

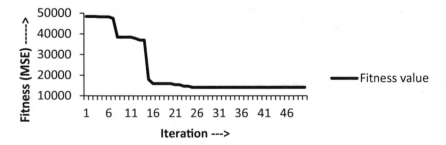

Fig. 3 Fitness value (enrollment data)

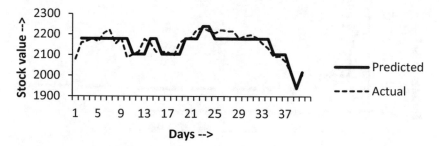

Fig. 4 Actual versus predicted stock values

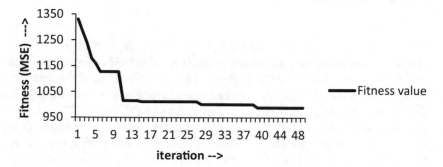

Fig. 5 Fitness value (SBI data)

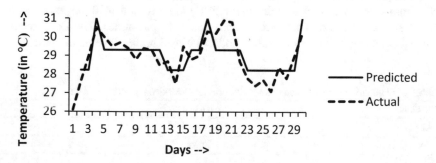

Fig. 6 Temperature graph

$$RMSE = \sqrt{\sum_{i=1}^{n} ((Actual - predicted) * (Actual - predicted)) \Big/ n} \qquad (2)$$

$$MAPE = \left(\sum_{i=1}^{n} |Actual - predicted|/Actual\right) * 100/n \qquad (3)$$

where n is the number of observations.

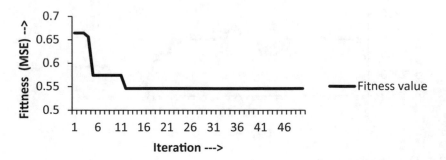

Fig. 7 Fitness value (temperature data)

The algorithm has been run on an Intel (R) Core (TM) i3-4005U CPU.

For enrollment dataset, $Data_{min} = 13055$ and $Data_{max} = 19337$. Here, $X_1 = 55$ and $X_2 = 663$. So, $D = [13000–20000]$. There are 13 intermediate boundaries/partitions between two adjacent intervals. Here, 13 fuzzy sets have been created on the universe of discourse U. Maximum and minimum velocities (for PSO) are $+100$ and -100, respectively. Minimum RMSE value is obtained using 7 hidden layer nodes. Using 6, 7, and 8 hidden layer nodes RMSE are 213.92, 118.92, and 180.51, respectively. MAPE values are 0.78, 0.53, and 0.62 for 6, 7, and 8 hidden layer nodes, respectively (Table 1).

In case of SBI dataset, $D = [1940–2240]$, $Data_{min} = 1941.2$ and $Data_{max} = 2232.2$. Maximum and minimum velocities (for PSO) are $+10$ and -10, respectively. Six fuzzy sets are created. There are 6 input and 6 output nodes. The experiment is carried out using 3 and 4 hidden layer nodes. Using 3 hidden layer nodes RMSE and

Table 1 Comparison of proposed model with previous works on enrollment data

Models	RMSE
AR(1)	542.46
Chen [10]	638.5
Chen [11]	383
Aladag et al. [1]	279.42
Aliev et al. [3]	194
Egrioglu et al. [15]	181.24
Bose and Mali [8]	141.66
Lu et al. [20]	120.6
Proposed	118.92

Table 2 Comparison of result of proposed model with previous works on SBI data

Models	AR(1)	Chen [10]	Yu [34]	Singh [25]	Proposed
RMSE	36.63	54.89	82.32	32.26	31.45

Table 3 Comparison using temperature data (June)

Models	AR(1)	Aliev et al. [3]	Singh and Borah [26]	Proposed
RMSE	0.82	0.9	1.22	0.74

MAPE values are 33.06 and 1.17, respectively. Using 4 hidden layer nodes RMSE and MAPE values are 31.45 and 1.06, respectively (Table 2).

For temperature dataset, $D = [26–31]$. Here, $Data_{min} = 26.1$ and $Data_{max} = 30.9$. Four fuzzy sets are created. There are 4 input and 4 output nodes. Maximum and minimum velocities (for PSO) are $+1$ and -1, respectively. The experiment is carried out using 2 hidden layer nodes. RMSE and MAPE values are 0.74 and 2.17, respectively. Performance of the model using temperature data is compared (Table 3) with existing (univariate) models including neural network-based models.

Drawback of the model is its efficiency. At each iteration, the proposed model needs to train one BPNN. A multilayer perceptron with n inputs, h hidden units, and m outputs has $h (n + 1)$ weights in the first layer and $m (h + 1)$ weights in the second layer. For each epoch, both the space complexity and time complexity are $O (h \cdot (m + n))$ [4]. So, the amount of calculation is dependent on the size of network. For the large dataset, training speed decreases. For large dataset, it will lead to an unacceptable amount of time complexity. In this case, retaining the basic structure of the algorithm and modular network architecture can be developed.

5 Conclusion and Future Work

In this paper, we present a hybrid method using PSO and BPNN for fuzzy time series forecasting. In the past, particle swarm optimization-based neural network models are developed, where weights of neural network are adjusted/updated by employing PSO. But in this model, PSO determines the proper interval length of universe of discourse and weights of neural network are adjusted using gradient descent technique. The hybrid algorithm creates proper intervals of data that increases forecast accuracy. The study reveals that proposed technique performs better than other existing models applying same dataset. This is a first-order model, i.e., only past one year's value is taken into consideration to compute predicted value. Result can further be improved by efficient selection of particle size, number of intervals, and number of hidden layer nodes. The algorithm has been implemented using small dataset with in sample observations. But it can be extended for out-of-sample dataset.

References

1. Aladag, C.H., Basaran, M.A., Egrioglu, E.: A high order fuzzy time series forecasting model based on adaptive expectation and artificial neural networks. Math. Comput. Simul. **81**(4), 875–882 (2010)

2. Aladag, C.H.: Using multiplicative neuron model to establish fuzzy logic relationships. Expert Syst. Appl. **40**(3), 850–853 (2013)
3. Aliev, R., Fazlollahi, B., Aliev, R., Guirimov, B.: Fuzzy time series prediction method based on fuzzy recurrent neural network. In: King, I., et al. (eds.) ICONIP, Part II. LNCS, vol. 4233, pp. 860–869. Springer, Berlin (2006)
4. Alpaydın, E.: Introduction to Machine Learning, 2nd edn. The MIT Press, Cambridge, Massachusetts (2010)
5. Basser, H., Karami, H., Shamshirband, S., Akib, S., Amirmojahedi, M., Ahmad, R., Jahangirzadeh, A., Javidnia, H.: Hybrid ANFIS-PSO approach for predicting optimum parameters of a protective spur dike. Appl. Soft Comput. **30**, 642–649 (2015)
6. Bas, E., Egrioglu, E., Aladag, C.H., Yolcu, U.: Fuzzy-time-series network used to forecast linear and nonlinear time series. Appl. Intell. **43**, 343–355 (2015)
7. Bas, E., Uslu, V.R., Yolcu, U., Egrioglu, E.: A modified genetic algorithm for forecasting fuzzy time series. Appl. Intell. **41**(2), 453–463 (2014)
8. Bose, M., Mali, K.: High order time series forecasting using fuzzy discretization. Int. J. Fuzzy Syst. Appl. **5**(4) (2016). https://doi.org/10.4018/ijfsa.2016100107
9. Cai, Q., Zhang, D., Zheng, W., Leung, S.C.H.: A new fuzzy time series forecasting model combined with ant colony optimization and auto-regression. Knowl.-Based Syst. **74**, 61–68 (2015)
10. Chen, S.M.: Forecasting enrollments based on fuzzy time series. Fuzzy Sets Syst. **81**, 311–319 (1996)
11. Chen, M.Y.: A high-order fuzzy time series forecasting model for inter net stock trading. Future Gener. Comput. Syst. **37**, 461–467 (2014)
12. Chen, S.M., Kao, P.Y.: TAIEX forecasting based on fuzzy time series, particle swarm optimization techniques and support vector machines. Inf. Sci. **247**, 62–71 (2013)
13. Cheng, C.H., Chen, T.L., Wei, L.Y.: A hybrid model based on rough sets theory and genetic algorithms for stock price forecasting. Inf. Sci. **180**(9), 1610–1629 (2010)
14. Cheng, S.-H., Chen, S.-M., Jian, W.-S.: Fuzzy time series forecasting based on fuzzy logical relationships and similarity measures. Inf. Sci. **327**, 272–287 (2016)
15. Egrioglu, E., Aladag, C.H., Yolcu, U.: Fuzzy time series forecasting with a novel hybrid approach combining fuzzy c-means and neural networks. Expert Syst. Appl. **40**, 854–857 (2013)
16. Huarng, K., Yu, H.-K.: The application of neural networks to forecast fuzzy time series. Phys. A **363**(2), 481–491 (2006)
17. Hsu, L.-Y., Horng, S.-J., Kao, T.-W., Chen, Y.-H., Run, R.-S., Chen, R.-J., Lai, J.-L., Kuo, I.-H.: Temperature prediction and TAIFEX forecasting based on fuzzy relationships and MTPSO techniques. Expert Syst. Appl. **37**, 2756–2770 (2010)
18. Kennedy, J., Eberhart, R.: Particle swarm optimization. In: Proceedings of IEEE International Conference on Neural Networks, IV, Piscataway, NJ, pp. 1942–1948 (1995)
19. Lee, L.W., Wang, L.H., Chen, S.M., Leu, Y.H.: Handling fore casting problems based on two-factors high-order fuzzy time series. IEEE Trans. Fuzzy Syst. **14**(3), 468–477 (2006)
20. Lu, W., Chen, X., Pedrycz, W., Liu, X., Yang, J.: Using interval information granules to improve forecasting in fuzzy time series. Int. J. Approximate Reasoning **57**, 1–18 (2015)
21. Oliveira, M.V., Schirru, R.: Applying particle swarm optimization algorithm for tuning a neuro-fuzzy inference system for sensor monitoring. Prog. Nucl. Energy **51**, 177–183 (2009)
22. Qasem, S.N, Shamsuddin, S.M.: Hybrid learning enhancement of RBF network based on particle swarm optimization. In: Yu, W., He, H., Zhang, N. (eds.) Part III. LNCS, vol. 5553, pp. 19–29 Springer, Berlin (2009)
23. Rumelhart, D.E., Mcclelland, J.L. (eds.): Parallel Distributed Prosessing, vol. 1. MIT press, Cambridge, MA (1986)
24. Singh, P., Borah, B.: Forecasting stock index price based on M-factors fuzzy time series and particle swarm optimization. Int. J. Approximate Reasoning **55**, 812–833 (2014)
25. Singh, P.: High-order fuzzy-neuro-entropy integration-based expert system for time series forecasting. Neural Comput. Appl. (2016a). https://doi.org/10.1007/s00521-016-2261-4

26. Singh, P., Borah, B.: High-order fuzzy-neuro expert system for daily temperature Forecasting. Knowl.-Based Syst. **46**, 12–21 (2013)
27. Singh, P.: Rainfall and financial forecasting using fuzzy time series and neural networks based model. Int. J. Mach. Learn. Cybern. (2016b). https://doi.org/10.1007/s13042-016-0548-5
28. Song, Q., Chissom, B.S.: Fuzzy time series and its models. Fuzzy Sets Syst. **54**, 269–277 (1993)
29. Song, Q., Chissom, B.S.: Forecasting enrollments with fuzzy time series—part I. Fuzzy Sets Syst. **54**, 1–9 (1993)
30. Song, Q., Chissom, B.S.: Forecasting enrollments with fuzzy time series—part II. Fuzzy Sets Syst. **64**, 1–8 (1994)
31. Sun, B., Guo, H., Karimi, H.R., Ge, Y., Xiong, S.: Prediction of stock index futures prices based on fuzzy sets and multivariate fuzzy time series. Neurocomputing **151**, 1528–1536 (2015)
32. Yu, H.-K., Huarng, K.: A bivariate fuzzy time series model to forecast TAIEX. Expert Syst. Appl. **34**, 2945–2952 (2008)
33. Yu, T.H.-K., Huarng, K.: A neural network-based fuzzy time series model to improve forecasting. Expert Syst. Appl. **37**, 3366–3372 (2010)
34. Yu, H.-K.: Weighted fuzzy time series models for TAIEX fore casting. Phys. A **349**(3–4), 609–624 (2005)
35. Zadeh, L.A.: Fuzzy set. Inf. Control **8**, 338–353 (1965)
36. Zadeh, L.A.: The concept of a linguistic variable and its application to approximate reasoning—part I. Inf. Sci. **8**, 199–249 (1975)
37. Zhang, G.P.: Time series forecasting using a hybrid ARIMA and neural network model. Neurocomputing **50**, 159–175 (2003)
38. Zhang, J.-R., Zhang, J., Lok, T.-M., Lyu, M.R.: A hybrid particle swarm optimization–backpropagation algorithm for feed forward neural network training. Appl. Math. Comput. **185**, 1026–1037 (2007)

A Genetic-Based Bayesian Framework for Stateless Group Key Management in Mobile Ad Hoc Networks

V. S. Janani and M. S. K. Manikandan

Abstract This paper addresses the issue in managing a group key among dynamic group of nodes in mobile ad hoc networks (MANETs), where the participants frequently miss the group key update, commonly known as rekeying. In this paper, we propose a broadcast stateless and distributed group key management (GKM) framework: genetic-based Bayesian networks group key agreement (GBKA) scheme, for supporting dynamic rekeying mechanism in MANET. The proposed framework includes two main functionalities: First to initialize the construction of group key and second is to optimize the keys to maintain the group communication with less overhead, whenever membership changes. The genetic algorithm selects individuals with high genetic characteristics that reflect the node's ability to adapt and survive the MANET environments. The rekeying scheme operates on genetic computation over finite area with Bayesian network modelling. Whenever the group membership changes, the public messages are updated accordingly with an unchanged private key. Our proposed scheme is very efficient with hash and Lagrange interpolation polynomial implementation. The simulation results show that the proposed group key management scheme achieved higher performance and security compared with other existing key management schemes.

Keywords Group key management · Genetic · Bayesian network
Stateless rekeying · MANET

1 Introduction

In many potential applications of mobile ad hoc networks (MANETs) such as battlefield, rescue operations and other civilian commercial applications, providing a secure group communication is significant. The group key management (GKM) is a well-established cryptographic technique to authorize and to maintain group key

V. S. Janani (✉) · M. S. K. Manikandan
Department of ECE, Thiagarajar College of Engineering, Madurai 15, India
e-mail: jananivs@tce.edu

© Springer Nature Singapore Pte Ltd. 2019
J. C. Bansal et al. (eds.), *Soft Computing for Problem Solving*,
Advances in Intelligent Systems and Computing 816,
https://doi.org/10.1007/978-981-13-1592-3_33

425

in a group communication, through secured channels. While considering a highly dynamic network like MANET, the group membership in the communication changes frequently with the addition and deletion of nodes in the group. This introduces group rekeying issue, wherein the group keys should be updated and reconstructed in a secure and timely manner. An important security requirement that should be maintained while performing the rekey operation is access control in terms of forward and backward secrecy. This access control mechanism prevents: (i) the new members from obtaining the past group key (backward secrecy) and (ii) the revoked or evicted members from accessing the present or future group keys (forward secrecy). Therefore, providing an efficient and scalable immediate group keying procedure for dynamic groups is a tedious problem, especially in dynamic group communication with minimal size of rekey messages.

Genetic algorithms (GA) are iterative optimization techniques that represent a heuristic tool of adaptive search mechanisms which have been studied in recent years. This technique selects the best-fitted individuals (i.e. nodes) to survive any mischievous network conditions. The works with a pool of binary strings called chromosomes, in the iterative process. The suitable nodes are selected by evolution of fitness function among the MANET nodes. To develop an eventually reliable population of individuals GA uses three processes, namely fitter individual selection, crossovers and mutations. The intrinsically parallel feature of the algorithm makes it well suitable to scan the large MANET environment to manage multi-optimization issues. The GA works as follows: with a pool of individuals that are encoded as a gene in binary code segments, a population is chosen randomly for evaluation. Based on the fitness value, the individuals with highest fitness are selected to generate new offspring. Mutations are performed with these offspring to produce a new population of nodes. Bayesian network (BN) model for genetic-based GKM structure analyses the set of gene derived from the evolution to investigate the chromosome dependency in a population. The hidden dependencies make constraints to the chromosomes which lead to the superior features. This problem is handled by accelerating the genetic process in a MANET population.

This paper introduces stateless GKM schemes in order to resolve the inherent key update issues for ad hoc networks. For secure communication, each cluster member is accomplished to produce a transient subgroup. Genetic algorithms are evolutionary methodology that works on MANET population to choose the best gene using fitness rate. In our approach, a BN is used by the nodes to analyse data obtained in the genetic process. The proposed schemes have performance advantages when compared with the existing approaches. Moreover, the schemes possess several distinguishable features as: (i) stateless rekey mechanism and (ii) practical solution for out-of-sync problem. Moreover, the proposed scheme in the paper ensures access control with forward and backward secrecy. The proposed schemes integrate GKA protocol with efficient constant round in order to reduce the time for rekeying in a hierarchical access control.

2 Related Works

The key management is a vital part of any secure group communication. Most cryptographic-based security systems rely on some underlying robust and efficient key management system. This section of the paper discusses some of the earlier proposed GKM schemes for secure communication in mobile ad hoc networks. The most well-known distributed group key establishment protocols is the key agreement schemes where all the communication members determine session keys in a collaboratively fashion [1]. The last decade has witnessed an exciting growth of research on group key agreement and its dynamic rekeying [2–5]. However, these existing protocols establishes only a single group key for group communication and failed to show its efficiency in rekeying mechanisms whenever group membership changes. Thus, these schemes are inappropriate in large and dynamic network like MANET. The group key should be updated and redistributed to all the nodes in a secure and timely manner, for every membership changes. This issue has been widely studied in wireless networks, and several schemes have been proposed [6–11]. These stateful protocols result in the out-of-sync issue if the node members fail to obtain the rekey update message in time. On security aspect, these schemes do not guarantee a perfect forward and backward secrecy. These drawbacks make stateful rekeying protocols practically inapt for large and highly networks where the nodes are characterized by dynamic mobility. A stateless protocol was introduced as a powerful rekeying mechanism to overcome the update loss issue of the stateful scheme [12–23]. These protocols increased the overhead in the communication of key messages. Furthermore, these schemes are designed with centralized CA to manage the entire GKM operations. Therefore, an efficient group rekey mechanism should be designed to overcome all the mentioned drawbacks with distributed and self-organized GKM.

3 Motivation of the Research Work

With respect to the discussions in the literature survey, the merits and demerits of all the existing mechanisms are compared in order to choose the best-suited mechanisms to establish group key and its dynamic rekey mechanism in MANET. The group rekeying remains as an attractive research topic and presenting an advantageous stateless protocol that manages MANET features is a challenging issue. Such an optimal solution is proposed in this paper for providing security in MANET by overwhelming the following drawbacks in the existing mechanisms.

- The conventional GKA schemes are vulnerable to out-of-sync problem.
- Most of the existing solutions are expensive in computation and communication.
- The traditional GKM schemes provide lower access control.
- The size of a rekey message should be minimized for a larger and highly dynamic network.

It is therefore clear that the drawbacks of the long-established GKM techniques should be minimized in order to make the PKI framework viable for secured multicast communication. On this pursuit, the proposed research work concentrated in developing a distributed stateless group key agreement schemes for self-organized MANET environment, which quantifies nodes behaviour in the form of trust in hexagonal non-overlapping clusters. In this paper, we propose a broadcast stateless and distributed group key management (GKM) framework: genetic-based Bayesian networks group key agreement (GBKA) scheme to provide public key infrastructural security in MANET framework. This stateless GKM schemes work on genetic algorithm, Boolean operation and Bayesian networks, to manage group key. In GBKA scheme, genetic algorithm is used to search appropriate key generating parameters (or code segments) and efficient stateless rekeying from the code segments. The GBKA applies BNs to analyse the characteristics of relationships among the selected node population. The proposed scheme shows efficient rekey mechanisms whoever misses the key update messages frequently for a large and dynamic group of members. Moreover, the distributed MANET setting allows the nodes to establish group key by themselves in a self-organized and co-operative fashion. This system design avoids the single point of failure issue and network bottleneck. The proposed scheme shows better performance efficiency compared to the existing GKM schemes stands ahead forward in rekeying operations.

4   Proposed Methodology: GBKA Scheme

The GBKA model is developed over a hierarchical cluster architecture where the nodes are grouped into non-overlapping hexagonal structure with spatial reuse benefits. It includes two types of nodes including CH and member nodes monitored by a distributed certificate authority (CA). The CA sends the information monitored in the network to the more powerful and larger memory size CH nodes. The CH combines the received data from the CA and distributes the aggregated results to member nodes within each cluster. The CH in the cluster-based MANET generates the key pair and transmits among the other header nodes securely. The member nodes generate key generating parameters (KGP), which are used for session group key (SGK) generation in rekeying operation. The parameter is further divided into code segments that are embedded into cluster members and CHs before deployment. The CH further assembles the segments to rebuild the KGP at definite interval, which would be harder for an attacker to know the segments. The member node delivers m number of code segments to the CH from which the CH chooses a permutation of certain code segments and broadcasts to the members. Each cluster member generates a new key with the permutation of code segments. Thus, the member nodes establish a common session key with a CH in each cluster. The GA analyses the appropriate key generating code segments. The operation of GBKA is divided into two phases: key agreement with an appropriate code segments set and efficient stateless rekeying. For key agreement, a population of N individuals with m code segment is maintained. Initially, the segment is encoded in a chromosome (here CM) as a gene.

The CH chooses m segment to assign an initial chromosome to the segment pool, which is further increased exponentially with all member contribution. GBKA then selects a relevant chromosome for a simple crossover mutation to generate new offsprings. The code segments of the offsprings are experienced to repetitive evolution throughout the network functionality. The best among such offsprings are considered as the key generating parameters to perform rekeying in the next phase. To measure the quality of gene from initial population, GBKA selects segments with high rate of fitness. The rate of fitness is defined as the actual rate, which an individual concludes up being tested in contributing to the subsequent generation. In the stateless rekeying phase, the selected code segments are encoded randomly and distribute to the CHs and the CA.

A. Group key management

The prime goal of the proposed key management system is to generate keys to the nodes to manage keys, to encrypt/decrypt the messages and thus preventing the illegal usage of handling certified keys. That is, it facilitates to provide highest security to networks, which will manage several attacks. The code segments are encoded as chromosomes to make it suitable for applying the genetic mechanism. The group members generate the code segments which are then used for constructing the key generating parameters. The pool of code segments (considered as gene) accumulates all the operations and operands that are operated to form the segments. With these genes, different permutations and combinations of keys are generated. For example, with k number of genes, $k!$ combinations of keys are generated. In order to measure the uncertainty of an outcome and to compute the distribution function of key values, the predictability of the code segment pool is measured. To construct an efficient quality of gene from the initial population, the GA chooses parent code segment of high rate of fitness. The fitness function thus conveys the feedback of the GBKA scheme in managing the appropriate segment searching to improve the quality of chromosomes. This fitness is evaluated as in (1):

$$F_R(\mathrm{Ch}_m) = \sum_{j=1}^{t} \mathbb{P}_j \log_2\left(1/\mathbb{P}_j\right) \tag{1}$$

where j: randomly selected k keys from the pool of m chromosomes. $\mathbb{P}$: probability function of k chromosomes, with $\sum_{j=1}^{t} \mathbb{P}_j = 1$; $j = 1, 2, 3 \ldots k$. If the EF value is large, the distribution key is objective or more likely to be uniform. Fitness function chooses the strongest node, whose chromosomes are considered in the evolutionary time segment. The rate of fitness represents the ability of each node to compete the dynamic network condition. The nodes with the optimal fitness rate are sought. To apply GA, the population of possible genes is initialized and refreshed during chromosome evolution. The common point mutation and crossover mutation is combined in the proposed GBKA scheme. From the code segment pool, a crossover point is chosen randomly by the simple crossover function with the parent chromosomes (Ch). The offspring is generated with two parent genes and kth gene as the point of crossover, as follows in (2) and (3):

$$\text{Offspring}_1^{j+1} = (\text{Ch}_{a1}^{j+1}, \ldots, \text{Ch}_{ak}^{j+1}, \text{Ch}_{b(k+1)}^{j+1}, \ldots, \text{Ch}_{bM}^{j+1}) \tag{2}$$

$$\text{Offspring}_2^{j+1} = (\text{Ch}_{b1}^{j+1}, \ldots, \text{Ch}_{bk}^{j+1}, \text{Ch}_{a(k+1)}^{j+1}, \ldots, \text{Ch}_{aM}^{j+1}) \tag{3}$$

where ath chromosome and bth chromosome are the parent chromosomes randomly selected from the jth generation.

$$\text{Ch}_a^j = (\text{Ch}_{a1}, \ldots \text{Ch}_{ak}, \ldots, \text{Ch}_{aM}) \tag{4}$$

$$\text{Ch}_b^j = (\text{Ch}_{b1}', \ldots \text{Ch}_{bk}', \ldots, \text{Ch}_{bM}') \tag{5}$$

During each successive generation of offspring, a portion of the initial population is chosen to breed a new generation. The fitness process selects individual solutions with fitter genes are selected. The existing selection methodologies perform a random selection operation which is an expensive procedure. To control the population, the relative distance for each solution pair is measured in the solution pool, which is given as in (6)

$$\text{Distance}(s_1) = \min_{s' \in \mathbb{P}} d(s_1, s') \tag{6}$$

The population s_1 is either added or discarded from the population pool with respect to the diversity factor ω. That is, the population is added if Distance $(s_1) \geq \omega$, and otherwise, it is discarded from the pool. During key generation and communication period, the proposed genetic-based system will use various random key generators for minimizing the vulnerabilities of KM methods.

B. Dynamic rekeying

This phase includes the member joining eviction and multiple secrets key update (rekey). Usually, each secure group management is associated with the group controllers, i.e. CH, who manage the cluster membership. The major challenge in group management is the keying data distribution in a secure manner, whenever a node leaves or joins the group. If there is any modification on membership in the group, then the process say, the group rekeying will create novel group keys in a secure manner. Each node has a set of supplementary keys for decrypting the rekeying messages. The nodes in a secure group management are signified as $\{N_1, N_2, N_3 \ldots N_l\}$, where $l = 2i$, 'l' mentions the maximum number of nodes in the secure group communication and 'i' is the binary string length that represents the code segments of each node. The code segments can be written as binary string (7),

$$\text{Code segments}(N) = \{A_1, A_2, A_3 \ldots A_i\} \tag{7}$$

Table 1 Code segments and values

Sl. no.	Segment	Value
1	00/01	1
2	10/11	2
3	100/101/110/111	3
4	Else	4

Here, the variables 'A_i', '$i = 1, 2, 3 \dots$ m' are either 0 or 1. Whenever the node joins a session, a secret key for the SGK is specified by the CH to the node. To identify the relationship among the best-fittest chromosome is precisely done with BNs. To put the code segment into usable form, some computations are required. Firstly, the code segment containing the genetic information for the chromosome is segmented into 12 on the basis of bit size. Further, the code segments are assigned with real numbers which are considered as the nominal values as given in Table 1.

The Bayesian network uncovers the survival rule that is embedded in the segment. Accordingly, only selected combinations of gene will permit the individual to survive. In MANET, the priority of level of survival of every gene varies with the mobility factor. Therefore, even the generations with weaker genes may sometimes survive and stay for a longer lifespan to reproduce new offspring. The BN indicates mainly on a factor that makes an individual to stay longer, defend and reproduce as mobility with efficient energy level. Each individual follows their own mobility characteristic to select behaviours: attack, defend or mutation. Randomness determines which chromosome will be chosen to run at the start time primarily, but it is genetic fitness which decides who will survive in the environment. The code segment merely reveals the rules; nevertheless, BN gives the relationship between the factors and how these relations influence the way the whole group key management works.

5 Performance Analysis

The simulation analysis of the proposed schemes has been performed and evaluated in the QualNet 4.5 environment along with the visual studio 2013 (IDE), VC++(programming language), NSC_XE-NETSIMCAP (SDK). The comparison among different methods is run in this simulation environment of 40 nodes. This analysis follows a random walk mobility (RWM) model with the 'T_{halt}' value of '0'. At various time intervals, each node varies its rate of mobility. In this simulation environment, all mobile nodes allocate 4 Mbps as its bandwidth or the channel capacity. The proposed scheme is compared with the existing ID-based multiple secrets key management (IMKM) scheme [19], GKA [20] and GKM [21], in terms of key updates (rekeying).

The proposed schemes, whenever a node adds or revokes the cluster, the rekeying operation is carried out by employing a one-way lightweighted key generation method. In the key update process, the nodes in the path are directly involved from the newly joined or revoked node to the cluster head. Herein, the header node does not broadcast the updated message as well as initialize the operation. As shown in Fig. 1, this process can minimize the rekeying completion time in the proposed scheme. At various node speeds, the average completion time is implemented for 40 nodes in which the GBKA scheme reveals the lowest time than others. The overhead of key management is also counted with respect to the average number of messages sent. As shown in Fig. 2, this comparison analysis comprises all the key requests and responses in the existing IMKM scheme with the proposed methodologies. From Fig. 2, it is clearly revealed that the overhead is similar at all probable node mobility. It makes both the existing IMKM scheme and the proposed methodologies robust towards dynamic mobility. Herein, the existing IMKM method necessitates higher overhead than the proposed methodology. Among these, the proposed scheme possesses the lowest overhead, at different node mobility.

The proposed schemes are compared with other schemes with respect to the computation and communication costs, as depicted in Table 2. From the analysis observations, it can be identified that the proposed scheme has better performance while compared to the other two existing protocols. Therefore, the proposed schemes demonstrate an absolute improvement in the above-mentioned factors. This can minimize the computation as well as the communication costs. The cost efficiency of the proposed schemes is calculated in terms of communication and computation cost in rekey mechanism. The communication cost of the group rekeying process comprises the cost of distributing node revocation information and the cost of secure group key

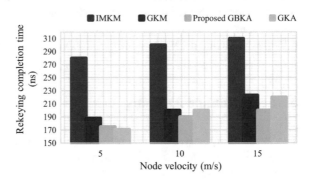

Fig. 1 Rekeying completion time

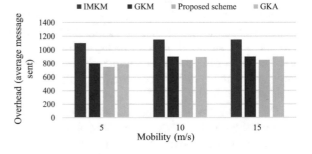

Fig. 2 Overhead versus mobility

Table 2 Comparison of key management schemes

Factors	Schemes			
	GKA [20]	GKM [21]	IMKM [19]	GBKA
Communication rounds	2	2	1	1
Pairing computations	$2x$	$2x$	x	x
Memory	$x - 1$	$2x - 2$	$(2x/g) - 2$	$(2x/g) - 2$
Power consumption (mW)	1207	1114	860	272
Communication complexity	Two broadcast	One broadcast	One broadcast	One broadcast

packets distribution. On computing all the computational and communication operations, we obtain the overall cost in rekey scheme as $O(M)$ with M number of group members in a cluster. As shown in Table 2, our scheme has better performance than the other protocols. Even though GBKA needs a little more memory, our scheme has absolute advantages in communication rounds, pairing, power consumption and communication complexity.

6 Conclusions

In MANET, secure key management and rekeying is quiet problematic than classical networks due to the number of nodes and the lack of infrastructure. This research addresses a secure and efficient group key management scheme that combines the beneficial feature of genetic algorithm and Bayesian network (BN) in MANETs. The nodes with high attack/defence and survival properties are selected to perform rekey mechanism, in order to adapt the high mobility and varying link characteristics of MANET. From the experimental results, it is revealed that the proposed scheme enhances the security as well as the performance factors (i.e. power consumption, overhead, rekeying, revocation rate and time) than the other schemes. Moreover, the proposed schemes attain cost efficiency of $O(M)$ in rekeying with M number of group members in a cluster. The proposed methodology thus shows better performance in MANET, when compared to the existing techniques.

References

1. Boyd, C.: On key agreement and conference key agreement. In: Proceedings of Second Australasian Conference on Information Security and Privacy (ACISP '97). LNCS, vol. 1270, pp. 294–302 (1997)
2. Tiloca, M., Dini, G.: GREP: a group rekeying protocol based on member join history. In: IEEE Symposium on Computers and Communication (ISCC). https://doi.org/10.1109/iscc.2016.75 43761
3. Ermis, O., Bahtiyar, S., Anarım, E., Çaglayan, M.U.: A secure and efficient group key agreement approach for mobile ad hoc networks. Ad Hoc Netw. **67**, 24–39 (2017)
4. Jarecki, S., Kim, J., Tsudik, G.: Flexible robust group key agreement. IEEE Trans. Parallel Distrib. Syst. **22**(5), 879–886 (2011)
5. Harn, L., Hsu, C.-F., Li, B.: Centralized group key establishment protocol without a mutually trusted third party. Mobile Netw. Appl. (2016) (Springer)
6. Park, Y., Je, D., Park, M., Seo, S.: Efficient rekeying framework for secure multicast with diverse-subscription period mobile users. IEEE Trans. Mobile Comput. **13**(4) (2014)
7. Jing, L., Liu, M., Wang, C., Yao, S.: Group rekeying in the exclusive subset-cover framework. Theor. Comput. Sci. **678**, 63–77 (2017)
8. Hur, J., Lee, Y.: A reliable group key management scheme for broadcast encryption. J. Commun. Netw. **18**(2), 246–260 (2016)
9. Sherman, T., McGrew, D.A.: Key establishment in large dynamic groups using one-way function trees. IEEE Trans. Softw. Eng. **29**(5), 444–458 (2003)
10. Liu, J., Liu, M., Wang, C., Yao, S.: Group rekeying in the exclusive subset-cover framework. Theor. Comput. Sci. **678**(C), 63–77 (2017)
11. Je, D.H., Lee, J., Park, Y., Seo, S.: Computation-and-storage-efficient key tree management protocol for secure multicast communications. Comput. Commun. **33**(2), 136–148 (2010)
12. Yanji, P., Kim, J., Tariq, U., Hong, M.: Polynomial-based key management for secure intra-group and inter-group communication. Comput. Math Appl. **65**(9), 1300–1309 (2013)
13. Chen, Y.-R., Tzeng, W.-G.: Efficient and provably-secure group key management scheme using key derivation. In: Proceedings of the IEEE Conference on Trust, Security and Privacy in Computing and Communications TrustCom, pp. 295–302 (2012)
14. Sun, Y., Chen, M., Bacchus, A., Lin, X.: Towards collusion-attack-resilient group key management using one-way function tree. Comput. Netw. **104**(20), 16–26 (2016)
15. Park, Y., Je, D., Park, M., Seo, S.: Efficient rekeying framework for secure multicast with diverse-subscription-period mobile users. IEEE Trans. Mob. Comput. **13**(4), 783–796 (2014)
16. Daghighi, B., Kiah, M.L.M., Shamshirband, S., Iqbal, S., Asghari, P.: Key management paradigm for mobile secure group communications: issues, solutions, and challenges. Comput. Commun. **72**, 1–16 (2015)
17. Mapoka, T.T., Shepherd, S.J., Abd-Alhameed, R.A.: A new multiple service key management scheme for secure wireless mobile multicast. IEEE Trans. Mob. Comput. **14**(8), 1545–1559 (2015)
18. Chen, Y.-R., Tzeng, W.-G.: Group key management with efficient rekey mechanism: a semi-stateful approach for out-of-Synchronized members. Comput. Commun. **98**(15), 31–42 (2017)
19. Li, L.-C., Liu, R.-S.: Securing cluster-based ad hoc networks with distributed authorities. IEEE Trans. Wireless Commun. **9**(10), 3072–3081 (2010)
20. Lin, C.-H., Lin, H.-H., Chang, J.-C.: Multiparty key agreement for secure teleconferencing. In: IEEE International Conference on Systems, Man and Cybernetics, pp. 3702–3707 (2006)
21. Wu, B., Wu, J., Dong, Y.: An efficient group key management scheme for mobile ad hoc networks. Int. J. Secure. Netw. **4**(1–2), 125–134 (2009)
22. Ramesh, C.P.: Viability analysis of Two Ray Ground and Nakagami model for vehicular ad-hoc networks. Int. J. Appl. Evol. Comput. **8**(2), 44–57 (2017)
23. Poonia, R.C., Sharma, V.P., Goyal, P.: Routing protocol in MANET: a survey. Int. J. Modern Comput. Sci. (IJMCS) **2**(3), 28–31 (2014) (June). ISSN: 2320-7868 (Online)

GEP Algorithm for Oil Spill Detection and Differentiation from Lookalikes in RISAT SAR Images

Ashoka Vanjare, C. S. Arvind, S. N. Omkar, Jandhyala Kishore
and Vijaya Kumar

Abstract Earth is covered with three fourth of water and one fourth of land. Ninety percent of world cargo transportation happens via ships that sail across great waters. Increase in sea traffic at the ports, natural disasters, technical, human errors may lead to oil spilling on oceanic surface. These spills will cause a lot of damage to marine ecosystem. Estimating the damage is one of the challenging tasks that can be addressed using remote sensing technology. In this paper, detection and differentiating look-alike image features of four different oceanic regions are studied using gene expression programming (GEP) algorithms on RISAT-1 SAR satellite images. GEP algorithm clearly differentiates lookalike image feature pixel from oil spill image feature pixel with classification accuracy on four different oil spill datasets is more than 98%. Proving GEP can be used for two class oil spill detection and classification problem.

Keywords RISAT-1 · Gene expression programming algorithm · Look-alike image pixels · Oil spill image pixels

A. Vanjare (✉) · S. N. Omkar
Department of Aerospace Engineering, Indian Institute of Science, Bengaluru, India
e-mail: ashokavajare@gmail.com

S. N. Omkar
e-mail: omkar@aero.iisc.ernet.in

C. S. Arvind
Cognitive Computing and Data Science Research Lab, Global Technology Office, Cognizant Technology Solutions Private Limited, Bengaluru, India
e-mail: csarvind2000@gmail.com

J. Kishore · V. Kumar
Indian Space Research Organization, Bengaluru, India
e-mail: jkk@isac.gov.in

V. Kumar
e-mail: Vijayakumarlj2012@gmail.com

© Springer Nature Singapore Pte Ltd. 2019
J. C. Bansal et al. (eds.), *Soft Computing for Problem Solving*,
Advances in Intelligent Systems and Computing 816,
https://doi.org/10.1007/978-981-13-1592-3_34

1 Introduction

Oil spill is one of the major environmental threats causing severe damage to oceanic environment. Radar Imaging Satellite (RISAT-1) from ISRO can be used for marine oil spill monitoring, because it has capabilities to work all-day and all-weather. Understanding oceanic RISAT-1 imagery is a complex task for researcher, because look-alike and oil spill image pixels are very similar and traditional classification algorithms perform very poorly. Frate et al. proposed a method to detect oil spills using neural network on synthetic aperture Radar (SAR) images first detects dark spots over the sea and compute there features based on physical-geometrical characteristics and train the network. The trained neural network classifier is used for classification of oil spill image pixels [1]. Bertacca et al. proposed new novel technique where fractionally integrated autoregressive-moving average model is used to distinguish low-wind and oil slick areas in high-resolution sea SAR image [2]. Brekke et al. proposed a new feature extraction method aimed to improve the classification performance of oil spill image pixel detection from 89 to 97% [3]. Mercer et al. proposed a method to detect oil slick using support vector machine in wavelet decomposition of a SAR image which increases the detection without considering signal stationary nor the strong backscatters (such as ships) [4]. Solbergetal. developed a framework to detection oil spill image pixels using adaptive threshold method to extract dark oil image pixels which will distinguish in classifying oil spills and look-alike marine phenomena which resemble oil spills [5]. Zhanget et al. extracted textural features using co-occurrence matrix method and combined with support vector machine (SVM) for classification of oil spill image pixels using Synthetic Aperture Radar (SAR) image [6].

In this paper, GEP model is been used to classify lookalike image pixels and oil spill image pixels from oceanic image pixels as two class problems. GEP model is applied for different datasets and results from GEP produce expression tree to identify the desired image features. GEP model helps in developing a mathematical model or mathematical expressions for distinguishing two classes. Oil spill detection problem involves understanding features of oil spills, lookalike with reference to the oceanic image pixels. Hence, GEP is used to evolve a mathematical model to illustrate its capability for oil spill image pixels' detection.

2 Automatic Oil Spill Region Detection

In this research work, gene expression programming model is used for automatic detection and differentiation of oil spill image pixels using RISAT-1 Hybrid polarity SAR images. GEP method will provide optimal threshold values which will differentiate oil spill regions from look-alike regions and classify oil spill accurately. This paper is divided into four sections-RISAT-1 hybrid SAR image dataset, Study Area, Methodology, Results and Conclusion.

2.1 Overview on RISAT-1 Hybrid Satellite Image

Different space agencies around the world have their own SAR imaging products to collect earth surface features. Indian Space Research Organization (ISRO) on April 26, 2012, launched a satellite mission with SAR sensor as payload called RISAT-1 satellite. RISAT-1 is a radar remote sensing satellite [7] which is having the capability to collect earth surface features. RISAT-1 uses scan SAR strip and modes methods to provide images with different resolutions like coarse, fine, and high spatial radar satellite images. These images are useful in remote sensing applications like urban region mapping, forest resource assessment, and water resource management. In this paper, RISAT-1 fine resolution image of 3 m (Azimuth) × 2 m (Range) resolution and swath 25 km is used for study of oil spill regions.

2.2 Study Area

In this paper, we are trying to detect and classify two different categories of oceanic oil spill challenges. We have taken four different oil spill regions' satellite images across the world for our study.

In category 1, detection and classification of lookalike image features (low wind area, rain cells, and ship wakes) which resembles oil spill image pixels and oceanic pixels are studied under dataset 1 and dataset 2 using RISAT-1.

- Dataset 1: Lookalike and oceanic image pixels of region of Gulf of Mexico.
- Dataset 2: Lookalike and oceanic image pixels of Mumbai region.

In category 2, detection and classification of normal oil spill regions without lookalike signatures are studied under dataset 3 and dataset 4 of RISAT-1 satellite images.

- Dataset 3: Oil spill region and oceanic image pixels of Norway region [8]
- Dataset 4: Oil spill region and oceanic image pixels of Mumbai region

Four different oil spill regions across the world used in the present research are shown in Fig. 1. Table 1 describes different oil spill regions with its swath area and total image pixels.

2.2.1 Problem Formulation

The problem that is been addressed is described in Table 2 using RISAT-1 satellite images for detection and classification of lookalike, oceanic, and oil spill image pixels accurately for different oil spill datasets.

Polarimetric combination gives insight about the features of radar image. In Fig. 2, we can see image containing oceanic image pixels, oil spill image pixels, and looka-

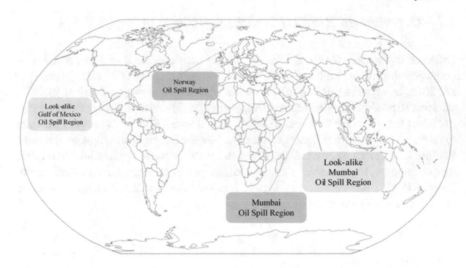

Fig. 1 Study area of four oil spill regions across the world used in this present research

Table 1 Different oceanic regions of RISAT-1 satellite images with lookalike and oil spill image datasets

Category		Region	Total image pixels	Swath area (3*2) SQ meters
Category 1	Dataset 1	Look alike gulf of Mexico	25,611,201	153,667,206
	Dataset 2	Look alike Mumbai	24,000,000	144,000,000
Category 2	Dataset 3	Oil spill Norway	15,003,000	90,018,000
	Dataset 4	Oil spill Mumbai	24,000,000	144,000,000

Table 2 Different classification problems addressed using RISAT-1 datasets

Datasets	Regions	Class 1	Class 2
Dataset 1	Look alike Gulf of Mexico	Look alike image pixels	Oceanic image pixels
Dataset 2	Look alike Mumbai	Look alike image pixels	Oceanic image pixels
Dataset 3	Oil spill Norway	Oil spill image pixels	Oceanic image pixels
Dataset 4	Oil spill Mumbai	Oil spill image pixels	Oceanic image pixels

like image pixels. The problem is been formulated such that we are separate oceanic image pixels with oil spill and look alike image pixels.

Ground truth preparation is done using geo-referenced RISAT-1 satellite images overlaid on maps from space agencies. To validate the overlay, we have once again overlaid RISAT image on Google maps. Initially, all the coordinates are marked and

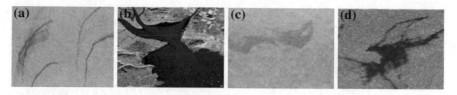

Fig. 2 Different oil spill RISAT-1 satellite images **a** Dataset 1—Look-alike Gulf of Mexico oil spill RISAT-1 image **b** Dataset 2—Look-alike Mumbai Oil spill RISAT-1 Image **c** Dataset 3—Oil Spill region of Norway regions RISAT-1 satellite image **d** Dataset 4—Oil Spill region of Mumbai regions RISAT-1 satellite Image

Table 3 Details of four subset data used of training and testing GEP algorithms to detect oil spill from the given datasets images

Category	Datasets	Region	Training pixels	Testing pixels	Total pixels
Category 1	Subset—dataset 1	Look alike Gulf of Mexico	483	80,587	81,070
	Subset—dataset 2	Look alike Mumbai	398	102,202	102,600
Category 2	Subset—dataset 3	Oil spill Norway	446	70,954	71,400
	Subset—dataset 4	Oil spill Mumbai	431	43,499	43,930

ground truth data is prepared which is considered as training pixels. These training sets are used for classifying the entire regions. Subset region is considered from a given full swath for our experiment. Table 3 shows the details of four different regions subset data used by GEP algorithm.

2.3 Methodology

In this section, methodology used to detect oil spills for four different datasets is been discussed. Block diagram in Fig. 3 shows different stages of processing. This comprises subregion data extraction from four different oil spill regions, followed by image filtering and enhancement to remove noise from the satellite images and detection and classification of oil spill using gene expression programming algorithm.

2.3.1 Image Filtering and Enhancement

RISAT-1 is a hybrid polarimetric radar satellite sensor. Composite RGB image is formed by combining delta (degrees), orientation (degrees), and ellipticity (degrees) of polorimetric RISAT-1 satellite image [9]. Median filter is applied to remove any

Fig. 3 Different stage of processing to detect oil spills using RISAT-1 satellite image

noise on composite RGB images on all of the four different datasets. For lookalike Mumbai dataset, contrast enhancement is done to improve the features of the image.

2.3.2 Gene Expression Programming (GEP)

The gene expression programming is a genotype and phenotype system, where linear chromosomes functioning as genotype and expression trees as phenotype. In gene expression programming, individuals are encoded as linear strings of fixed length chromosomes which are expressed later as nonlinear entities of different size and shape expression trees. Chromosomes are easily manipulated as they are compact and linear in nature. Selection process acts on chromosomes to reproduce with modification according to fitness function. Gene expression programming is simple in encoding any modification made in the chromosome results in syntactically correct expression trees. Every GEP chromosome is composed of genes of same length, and each gene is further composed of a head and a tail. The head symbols are selected randomly from function set F and terminal set T, whereas the tail contains just terminal. Sub-expression trees are general symbolic regression problems [10]. Multi-subunit expression trees are generated by linking sub-expression trees. These expression trees are created using addition or multiplication operator. The gene expression programming algorithm starts with random generation of chromosomes of initial population. The chromosomes are expressed as expression trees and fitness of each individual are assessed. Individual chromosomes are then selected according to fitness to reproduce with genetic operations. The new individuals are in their turn subjected to similar evolution process as mentioned above. This iterative process is repeated for a specified number of generations or until a solution has been obtained [11].

3 Experiments and Results

Classification results of oil spill regions of four different RISAT-1 images dataset are discussed in this section. Datasets were prepared and experimentation is carried using MATLAB 2015 with windows 7 operating system using 32 GB RAM. In this paper, two different categories of oil spills and lookalike RISAT satellite image datasets are experimented using gene expression programming. In category-1: Look-alike and oceanic image regions are classified while in category-2: oil spill and oceanic image pixels are classified. Receiver operating characteristics (ROC) measure is used as fitness function throughout the experiment for determining the classification accuracy.

3.1 Gene Expression Programming Algorithm

Four different oil spill regions' samples are trained with ground truth samples using gene expression models and tested with testing image pixels.

Dataset 1

Look alike Gulf of Mexico dataset, trained using GEP model. After training it is producing gene expression tree is represented in Eqs. (1–2) which is used for classifying look alike and oceanic image pixels.

$$C = (4 * R) - 8.5 \tag{1}$$

$$D = 4.5 - (15.8 * R) \tag{2}$$

where R represents RISAT-1 satellite image.

Here, GEP algorithm is trained with 483 records for training and the rest 80,587 records for testing in the dataset-1. The trained samples are trained using GEP model in order to identify and distinguish between two classes—lookalike and oceanic image pixels.

In Fig. 4, we can see the small training set is fed into GEP model, where the model gets trained and produces expression trees which are shown in Fig. 6. Using these expressions, a threshold value can use for separating lookalike image pixels with oceanic image pixels. Here in Fig. 5, we can see the gene expression tree which is obtained after evaluating the training samples using GEP mathematical model. This expression is used for identifying and classifying the other datasets. Here, the same methodology is used for identifying different features in different datasets like lookalike—Mumbai region, oil spill—Norway region, and oil spill—Mumbai region.

After applying the GEP model rules with mathematical expressions are generated, this is used for image classification. When these expressions are simplified, Eqs. 1 and 2 are obtained; this is applied for the testing image for identifying and classifying the datasets. Results obtained after training and testing are shown in Fig. 6.

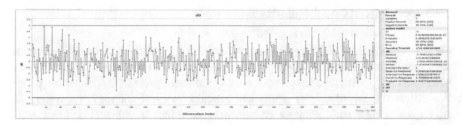

Fig. 4 483 image pixels are used as training set for dataset-1 with GEP model parameters

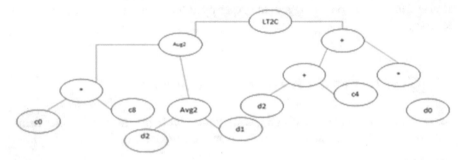

Fig. 5 Expression tree obtained for Lookalike Gulf of Mexico region (dataset-1) using GEP model after applying training set

	Best of Run - Training	Best of Run - Validation
Fitness	999.658971099319	998.36403999359
Accuracy	99.79% (482)	99.16% (79914)
Max Fitness	1000	1000

Fig. 6 Results obtained using GEP model by both training and testing for dataset 1 lookalike Gulf of Mexico dataset

Results are getting better if the training samples collected are better if noise is removed. Same methodology is carried out for all the other datasets.

Dataset 2

Look alike Mumbai dataset 2 gene expression tree is represented in Eqs. (3–4)

$$C = (8 * R) - 16.5 \tag{3}$$
$$D = 8.5 - (18.8 * R) \tag{4}$$

where R represents RISAT-1 satellite image.

Dataset 3

Oil Spill Norway region dataset 3 gene expression tree is represented in Eqs. (5–6)

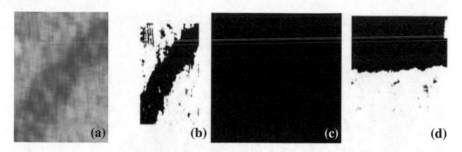

Fig. 7 **a** Look alike and oceanic image pixels of Gulf of Mexico region. **b** Extracted GEP algorithm output. **c** Look alike and oceanic image pixels of Lookalike of Mumbai region. **d** Extracted using GEP expression algorithm output

$$C = (9 * R) - 15.5 \tag{5}$$

$$D = 9.5 - (16.8 * R) \tag{6}$$

where R represents RISAT-1 satellite image.

Dataset 4

Oil Spill Mumbai region dataset 4 gene expression tree is represented in Eqs. (7–8)

$$C = (10 * R) - 12.5 \tag{7}$$

$$D = 12.5 - (19.8 * R) \tag{8}$$

where R represents RISAT-1 satellite image.

For classifying, category 1 and satellite image datasets we have used same steps. Optimal threshold values 0.5 and 0.75 are obtained from gene expression tree as mentioned in the above Eqs. (1–4) which differentiate oceanic image pixels and lookalike image pixel values. Generating optimal threshold values helps in differentiate image pixels and automate the procedure using GEP model. Threshold value is obtained when training samples are fed to GEP model and evaluated using fitness functions. These functions are used for differentiating two classes. After obtaining threshold value, same training model is applied and validated for complete image. In category 2 satellite image, datasets are classified as oil spill and oceanic image pixels. If the probability of C and D for that image pixel value >0.6, Optimal threshold value 0.6 is obtained from gene expression tree as mentioned are giving in the above Eqs. (5–8).

Figure 7 shows lookalike image pixels which are identified and extracted accurately using GEP algorithm for dataset 1 and dataset 2.

Figure 8 shows oil spill regions which are identified and extracted accurately using GEP algorithm for dataset 3 and dataset 4. Oil spill and lookalike features are very similar to extract, and GEP gives expressions that gives optimal threshold that separates desired pixels from oceanic image pixels.

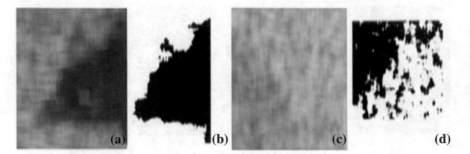

Fig. 8 a Oil Spill and oceanic image pixels of Norway region. **b** Extracted using GEP algorithm output. **c** Oil Spill and oceanic image pixels of Mumbai region. **d** Extracted GEP algorithm output

Table 4 Classification table for different dataset of lookalike and normal oil spill dataset using GEP model

Category	Dataset	Total dataset points	Training dataset	Testing dataset	Classification		Accuracy (%)
					True positive	False positive	
Category 1	Look alike Gulf of Mexico	81,070	483	80,587	79,914	673	99.16
	Look alike Mumbai	102,600	398	102,202	101,633	569	99.14
Category 2	Oil spill Norway	71,400	446	70,954	69,956	998	98.59
	Oil spill Mumbai	43,930	431	43,499	43,166	333	99.23

3.2 Results

In results section, results obtained from four different oceanic image datasets using gene expression programming is evaluated Receiver Operating Characteristics (ROC) measure.

According to Table 4, Category (1) look-alike oil spill dataset accuracy is 99.15%. Category (2) Oil Spill dataset accuracy is nearly 98.91%.

Figure 9 shows results obtained for four different oceanic regions on application of GEP algorithm. In our experimentation, we found GEP model predicts accurately which are nearly 98%. So, we can use this model for differentiating lookalike image pixels from oceanic image pixels in category 1. While in category 2, oil spill image pixels are differentiated from oceanic image pixels. Proving GEP has effectively detected and differentiates look-alike oil spill signature features and correctly classified oil and non-oil regions. But there are few false classifications caused by very

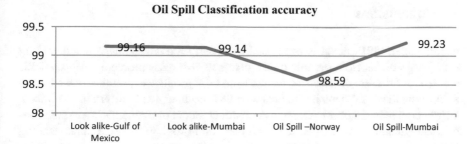

Oil Spill Classification accuracy

Fig. 9 Graph of all the results obtained for four different oceanic regions

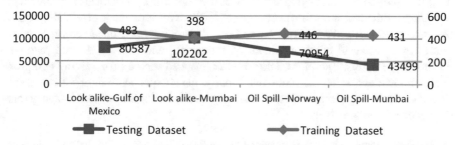

Fig. 10 Blue color training dataset legend shows number of training samples used to train GEP algorithm. Red color testing dataset legend shows number of testing samples used to check classification accuracy of GEP algorithm for four different oceanic oil spill regions

near look-alike features and non-optimal threshold value obtained might be due to slight overfitting of the model. The expression tree helps in better classification as it evolves. The expression trees are simplified to equations. These equations can be used a methods for threshold for different image classes.

Figure 10 shows training samples used which is very low in comparison with testing samples. GEP performs better in case of small training samples. In Fig. 2, we can see different oceanic image regions. Here, the challenge is differentiating lookalike, oil spill image, and oceanic image features. By differentiating, these pixels feature correctly. GEP has performed complex data analysis tasks better. Our experiment hence proved that GEP can classify and differentiate lookalike features effectively for complex radar sensor satellite images. This method will help in automation of image classification for a given complex radar sensor satellite data. Hence, we present novel methodology for image classification for complex data using less training samples.

4 Conclusions

In this paper, GEP model is been used to classify lookalike image pixels and oil spill image pixels from oceanic image pixels as two class problem. GEP model is applied for different datasets and results from GEP produce expression tree which in identifying the desired image features with 98% accuracy with uncertain conditions. GEP model helps in developing a mathematical model or mathematical expressions for distinguishing two classes. Oil spill detection problem involves understanding features like lookalike and oil spill image pixels values which are very similar to oceanic image pixels. So, complex mathematical model is explored and confusion matrix is plotted for calculating accuracy of the algorithm. Classification problem is explained with illustrating the functioning of GEP for oil spill image pixels' detection. Advanced evolutionary algorithm like GEP uses concept of linear chromosomes with fitness value to efficiently identify, recognize, and classify desired features in the image pixels. Therefore, GEP offers advanced method for solving complex technological and scientific problems. Finally, gene expression algorithms present mathematical computer models to detect and classify lookalike, oil spill image pixels from oceanic image pixels.

Acknowledgements This work is supported by the Space Technology Cell, Indian Institute of Science, Bangalore, and Indian Space Research Organization (ISRO). We also acknowledge the RISAT mission scientists and associated ISRO personnel for the production of the remote sensing data which is used in this paper.

References

1. Del Frate, F., Salvatori, L., Sistemi, I.: Oil spill detection by means of neural networks algorithms: a sensitivity analysis **00**(C), 1370–1373 (2004)
2. Bertacca, M., Member, S., Berizzi, F., Mese, E.D.: A FARIMA-based technique for oil slick and low-wind areas discrimination in sea SAR imagery **43**(11), 2484–2493 (2005)
3. Brekke, C., Solberg, A.H.S.: Feature extraction for oil spill detection based on SAR images, 75–84 (2005)
4. Mercer, G., Girard-ardhuin, F., Brest-iroise, T., Quentin, M., Plouzane, F.: Oil slick detection by SAR imagery using support vector machines, 90–95 (2005)
5. Solberg, A.H.S.: Automatic detection of oil spills in envisat, radarsat and ERS SAR images **00**(C), 2747–2749 (2003)
6. Zhang, F., Shao, Y., Tian, W., Wang, S., Box, P.O. (2008) Oil spill identification based on textural information of SAR image, 1308–1311
7. Handbook of Risat-1 Satellite, ISRO, Bangalore
8. www.nofo.no, Norway Oil spill experiment website
9. Kumar, L.J.V., Kishore, J.K., Rao, P.K. (2013) Unsupervised classification based on decomposition of RISAT-1 images for oil spill detection
10. Shao, S., Liu, X., Zhou, M., Zhan, J., Liu, X., Chu, Y., Chen, H.: A GPU-based implementation of an enhanced GEP algorithm. In: Proceedings of the fourteenth international conference on Genetic and evolutionary computation conference
11. Ferreira, C.: Gene Expression Programming: Mathematical Modeling by an Artificial Intelligence. Springer-Verlag. ISBN 3-540-32796-7 (2006)

Estimation of Interfacial Heat Transfer Coefficient for Horizontal Directional Solidification of Sn-5 wt%Pb Alloy Using Genetic Algorithm as Inverse Method

P. S. Vishweshwara, N. Gnanasekaran and M. Arun

Abstract In the present work, a one-dimensional transient solidification heat transfer problem is solved to determine the unknown interfacial heat transfer coefficient (IHTC) at the mold–metal interface using genetic algorithm (GA), an evolutionary and widely known algorithm, as an inverse method. The forward model is numerically solved to obtain the exact temperatures by incorporating the appropriate correlation for the IHTC that varies with time. In order to mimic experiments, the exact temperatures are then perturbed with the standard deviations of 0.01, 0.02, and 0.03. In the inverse estimation, genetic algorithm is used to minimize the objective function, thereby reducing the error between the measured and the simulated temperatures. The study on the performance parameters of the algorithm is also discussed in detail.

Keywords Inverse heat transfer · Genetic algorithm · Interfacial heat transfer coefficient · Solidification

1 Introduction

Casting is one of the oldest processes which is used to obtain the components required for various engineering applications. The studies of microstructure, mechanical properties, and heat transfer in casting and solidification have helped the foundry industry to improve the quality of the products. Among various studies, the study pertinent to stages of solidification in metal casting has a direct influence on the microstructure of the casting product. The mechanical properties of the materials are related to the evolution of microstructure of casting which depends on type of solidification and the stability of the metal–mold interface. During the solidification, the metal shrinks as it loses its superheat to the mold. Due to the relative motion between metal and expansion of the mold, an air gap is formed between the mold and the solidified

P. S. Vishweshwara · N. Gnanasekaran · M. Arun (✉)
Department of Mechanical Engineering, National Institute of Technology Karnataka, Surathkal, Mangalore 575025, Karnataka, India
e-mail: m.arun1978@gmail.com

J. C. Bansal et al. (eds.), *Soft Computing for Problem Solving*,
Advances in Intelligent Systems and Computing 816,
https://doi.org/10.1007/978-981-13-1592-3_35

metal which creates resistance to the heat flow. The heat flux liberated to the mold is called interfacial heat flux and the corresponding heat transfer coefficient is termed as interfacial heat transfer coefficient (IHTC). It is very onerous to study the interfacial heat transfer aspects in solidification because there is moving boundaries; hence, the location of thermocouple at these boundaries will distort the thermal field at the interface. The thermophysical properties are dependent on the phase of the casting that are used to solve the governing heat transfer equations; therefore, it requires more knowledge about the metallurgical as well as heat transfer aspects.

Two methods are commonly used to find the values of IHTC during solidification. One method is directly measuring the air gap thickness using the displacement meters and by knowing the thermal conductivity of air. The other methods are use of Beck's nonlinear estimation method, control volume approach, finite difference method, finite element method, etc., as inverse methods. The factors like cast geometry, chill thickness, chill material, mold coating, superheat, type of casting method, and the direction of solidification affect the values of IHTC. Ho and Pehkle [1] provided the insight of the effect of the IHTC on various metal–mold systems for the upward and downward castings. In upward casting, the contact between the mold and metal was found to be good and further decrease in the IHTC values was due to the formation of the oxides on the metal. In downward direction solidification, the air gap evolves and stays for longer time, and hence, there will be more resistance to heat transfer at the mold–metal interface. Nishida et al. [2] observed the phenomenon of the air gap evolution at the interface for the flat and the cylindrical castings. Displacement meters were used to measure the mold and casting movements. In flat casting, the movement of the mold was inwards in first few seconds after pouring and later away from casting; hence, higher IHTC values of the range 2900–4500 W/m^2K was found compared to cylindrical casting. Rajaraman and Velraj [3] compared two different methods, Beck's algorithm and control volume approach, to obtain the IHTC values for aluminum and sand mold. A deviation of 57% in the values of the IHTC was found between the two methods. Designing a good quality mold is a primary concern in foundries as the thermophysical property of the mold and the geometry of the casting has a direct effect on controlling the solidification rate. The studies on IHTC on dependence of chill thickness and chill materials show that higher the thermal conductivity of the material more is the capacity to receive more heat from the casting in turn bearing higher values of IHTC [4–6]. The coating thickness and coating material will cause the resistance to the heat flow. Hence, a higher thermal conductivity coating material with thinner coating is effective as noticed by [7–9]. Santos et al. [10] performed experiments on Al-Cu and Sn-Pb alloys for horizontal and vertical directional solidification. They proposed correlation based on IHTC and also studied the dependence of IHTC on superheat where the increase in superheat causes higher IHTC values as there will be more latent heat present in the melt. Jose Silva et al. [11] conducted theoretical and experimental studies of the IHTC during the solidification of Sn-Pb alloys with water-cooled molds. Their study on the alloy compositions with different percentages of Pb for Sn-Pb alloy and direction of solidification showed that the increase in Pb solute content decreases the IHTC values as the increase of percentage weight of the alloy decreases the mushy zone.

With the increase in the mushy zone length, the interdendritic fluid can feed the solidification contraction which causes a continuation of the presence of liquid at the interface. They proposed IHTC values ranging from $3500\ t^{-0.33}$ to $6000\ t^{-0.33}$. The IHTC values differ for different casting process like investment casting, pressure die casting, and blade casting. For high-pressure die casting, the metal will be in firm contact with the die for longer time; hence, a good contact is ensured between the mold and the metal [12–14].

With the recent development in the use of simulation software like Anycast, Procast, Magmasoft, several studies show the reduction in experimental cost [15, 16]. Wong and Pao [17] used GA for the estimation of heat transfer coefficient in gravity die casting. It is found from several researches that the use of evolutionary algorithms can be productively used to solve inverse problems. But only few works have been found where the evolutionary algorithms are utilized in the field of solidification for analyzing IHTC. Dousti et al. [18] determined the IHTC values for Al-5% Si alloy against steel mold using PSO algorithm. Two-dimensional approach was considered and a power law form of equation for IHTC variation with time is assumed which is used for solving the direct problem. A similar work to determine IHTC values for Al-4.5% Cu alloy using genetic algorithm can be found for squeeze casting process [19]. Similarly, a study on cost reduction in experiments was analyzed using GA by Vasileiou et al. [20] where they performed several simulations for varying casting geometry using ProCAST by assuming the IHTC correlation with temperature as stepwise and exponential as well as stepwise function of time. It was observed from their work that the estimation of IHTC was possible for variable region in casting geometry using GA as inverse which was found to be fast. Yu and Lou [21] estimated IHTC values for continuous casting of steel billets by using differential evolution algorithm and the results were validated with the industrial data.

In the present work, IHTC for horizontal solidification of Sn-5 wt%Pb alloy against steel mold is numerically investigated to estimate the unknown IHTC. The exact temperatures are then added with noise in order to mimic the actual experiments. GA is used as inverse method to retrieve the unknown interfacial heat transfer coefficient by minimizing the error between simulated and measured temperatures.

2 Direct Problem

Direct problem is solved based on one-dimensional unsteady heat conduction equation for solidification process to obtain the exact temperatures for the assumed value of IHTC. Figure 1 shows the schematic view of the problem. The domain is considered in such a way that the heat transfer from the cast to the mold is unidirectional. The fluid flow effects during solidification are neglected and heat transfer through the air gap is assumed only through by conduction. Carbon steel mold of length 110 mm with 60 mm width with water-cooled system is considered for horizontal solidification of Sn-5 wt%Pb alloy casting. The mold walls are insulated and no heat is transferred through the walls. Three sensors for temperature measurement are

Fig. 1 Schematic view of
experimental setup

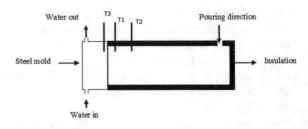

Fig. 2 Discretisation of
mold and metal system

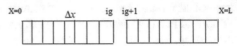

located in the computational domain. Two sensors T_1 and T_2 are located inside the
mold cavity at a distance of 6 and 12 mm, respectively, from the mold–metal interface. One more sensor T_3 is located inside the mold at a distance of 2 mm from the
mold–metal interface. The discretisation of solidification is as shown in the Fig. 2.

2.1   Modeling of Interfacial Heat Transfer Coefficient

The interfacial heat transfer coefficient 'h_i' (IHTC) between the metal and mold
surface is given by,

$$h_i = \frac{q}{T_C - T_M} \tag{1}$$

where q is the average heat flux across the metal–mold interface in W/m^2, and T_C
and T_M are the casting and mold surface temperature, respectively. The heat flux
at the casting and mold interface is calculated using the temperature gradient at the
surface and subsurface nodes which is given by,

$$q = -k\frac{dT}{dx} = -k\frac{T_i^p - T_{i-1}^p}{\Delta x} \tag{2}$$

where k is the thermal conductivity of the material (W/mK).

2.1.1   Governing Equations for the Heat Transfer in the Mold

The unsteady state one-dimensional conduction heat transfer is given by

$$\frac{\partial^2 T(x,T)}{\partial x^2} = \frac{1}{\alpha}\frac{\partial T(x,T)}{\partial t} \tag{3}$$

where T is the temperature in °C, t is the time in seconds, and α is thermal diffusivity

$$\alpha = \frac{k}{\rho C} \tag{4}$$

where k is thermal conductivity, ρ is the density, and C is the specific heat capacity.

2.1.2 Governing Equation for the Heat Flow in the Casting

The governing equation for heat transfer in the casting contains heat source term added to the left-hand side of Eq. (3). The heat source term includes latent heat, fraction rate terms which should be solved by Schiel's equation. The final form of the equation is given by Eq. (5).

$$k\frac{\partial^2 T(x, T)}{\partial x^2} = \rho C'\frac{\partial T(x, T)}{\partial t} \tag{5}$$

where

$$C' = C - l\frac{\partial f_s}{\partial T} \tag{6}$$

2.1.3 Boundary Conditions

$$\text{At}\, x = 0,\ \frac{\mathrm{d}T_C}{\mathrm{d}x} = 0$$

$$\text{At}\, x = i_g,\ \text{(casting surface)};\, -k_c\frac{T_c}{\mathrm{d}X} = h_i(T_C - T_M)$$

$$\text{At}\, x = i_{g+1},\ \text{(mold surface)};\, k_M\frac{T_M}{\mathrm{d}X} = h_i(T_C - T_M)$$

$$\text{At}\, x = L,\ -k_M\frac{T_M}{\mathrm{d}X} = h_a(T_L - T_\alpha)\ \text{and}\ T_\alpha = 27\,°\text{C}.$$

In the present study, the Sn-5 wt%Pb alloy is chosen as the cast metal and the mold material is made of steel. The thermophysical properties of the materials are mentioned in Table 1 [11]. The initial temperature of the melt is taken as 0.1 superheat of its liquidus temperature. The mold is initially at an atmospheric temperature 27 °C. The IHTC (h_i) is time dependent; hence in this study, it is assumed to vary as power law form with time by considering a and b as the unknown parameters as shown in Eq. (7).

Table 1 Therrmophysical properties of the Sn-5 wt%Pb and steel materials [11]

Properties	k_s	k_l	σ_s	σ_l	C_s	C_l	T_f	T_s	T_l	k_p
Alloy	65.6	32.8	7475	7181	217	253	232	183	226	0.0656
Steel	46		7860		527					

$h_a = 5.7t^{0.15}$ (SI units: T—°C, σ—kg/m^3, C—J/KgK, K—W/mK), l-liquidus, s-solidus, f-fusion

$$h_i = at^{-b} \tag{7}$$

Here, in the present work, in order to simulate the experiments, IHTC values are presumed based on literature data which is as shown in Eq. (8) [11].

$$h_i = 6000\, t^{-0.33} \tag{8}$$

When the pre-chosen values of IHTC are given as input to solve the direct problem, the simulated temperatures are obtained. Generally, experimental temperatures are prone to errors. In order to mimic the measured data, Gaussian white noise of 0.01 T_{max}, 0.02 T_{max}, and 0.03 T_{max} is added as shown in Eq. (9) to the simulated temperatures, respectively.

$$Y_{iM} = T_{exact}(t_i, \text{Sensor}_m) + \varepsilon\sigma \tag{9}$$

where M is the number of sensors, σ is the standard deviation of the temperature measurements, and ε is the random numbers varying between -2.576 and 2.576 for normally distributed errors with zero mean and 99% confidence bounds [19].

The maximum temperature is considered which is the initial temperatures of the molten metal and mold. The noisy data is now considered as measure data Y_{iM}, which in turn used to solve the inverse problem.

3 Inverse Problem

In the present work, the objective function is expressed similar to least square method. The solution of the parameter estimation problem is obtained by minimizing Eq. (10). GA is used as the inverse method to minimize this objective function. The objective or fitness function is expressed as,

$$S(h_i) = \sum_{m=1}^{M} \sum_{i=1}^{N} [Y_{im} - T_{im}(h_i)]^2 \tag{10}$$

where h_i is the unknown parameter. Y_{im} is the ith observation from the mth measurement; M and N are the number of measurements and observations, respectively. T_{im}

(h_i) is the simulated temperature obtained from the forward model. The unknown parameter is estimated by minimizing the sum of the squared difference between the simulated and measured temperatures [22].

4 Genetic Algorithm

Genetic algorithm was developed by Goldberg in 1989 [23]. It is a powerful search works on the principle of evolution of species where in the best individual in the species survives and will be predominant, hence becoming a successor in continuing its generation. The above concept can effectively be used in solving various engineering problems. To solve such problems, an objective function has to be specified where the purpose might be maximizing or minimizing it. The basic steps of GA are population initialization, evaluation of the fitness function, selection, crossover, mutation, and generation of offsprings. Population initialization is the creation of set of individuals which is called chromosomes. Each chromosome is made of genes. The fitness function values are evaluated for each chromosome and the ranking of the chromosomes is carried out based on the fitness values. The chromosome with the best fitness value will be kept for the next generations and worst chromosome is rejected. The selected set of chromosomes will be allowed to produce a new offsprings by the process called crossover where the chromosomes are randomly swapped at random

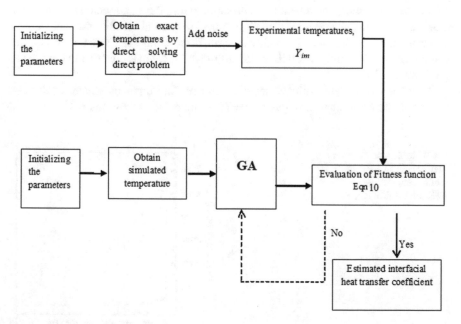

Fig. 3 Overview of the present work

crossover sites to generate new offsprings. Even two point or multipoint crossover can be used. This new set of chromosomes has a great tendency to produce good performance in retrieving the solution. Mutation process is the modification in the genes, where few genes of particular chromosome which has a capability to enhance the liability to vary the population. From the new population, every location of bit in the chromosome will go through random change in with identical probability. Thus, the newborn population is used to find the fitness function and the process is continued until the stopping criterion is reached. The overview of the present work is shown in Fig. 3.

5 Results

The inverse one-dimensional solidification heat transfer problem is solved using GA as inverse method. The inverse estimation is accomplished using in-house codes developed using MATLAB and executed in computer with configuration of 12 GB RAM, INTEL i5 Core, 1.70 GHz.

The direct model is solved by using Eq. (8) for the known boundary conditions to obtain the exact transient temperatures. Figure 4 shows the assumed transient mold–metal interfacial heat transfer coefficient for horizontal solidification. The exact transient temperatures are as shown in Fig. 5. It can be seen that the temperature inside the casting decreases as time progresses. When the metal comes in contact with the mold, the metal loses its superheat to the mold which increases the mold temperature. A thin solidified skin is first formed at the interface which acts as a resistance for the incoming molten metal. Temperature at 12 mm shows higher value than that of 6 mm as it will be still in mushy form and contains latent heat to give away.

To solve the above-mentioned inverse problem, the input parameter of the GA is initially set as number of population = 15, number of generations = 30, mutation

Fig. 4 Mold–metal interfacial heat transfer coefficients versus time for horizontal solidification

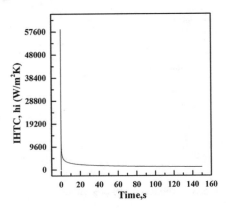

Fig. 5 Temperature curves of the sensors T_1, T_2, and T_3 respectively

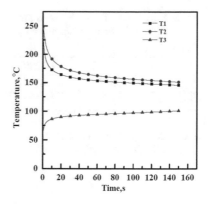

Table 2 Retrieval of a and b values for exact temperature values using GA with population = 15, generations = 30, and mutation rate = 0.1

Runs	a	b	Time, s	% Error a	% Error b
1	5637.6	0.312	23,258	6.04	5.45
2	6513.1	0.349	22,746	8.55	5.76
3	6366.4	0.345	22,586	6.11	4.55
Average	6172.36	0.33		6.9	5.25

rate = 0.1, crossover probability = 0.9. According to Eq. 7, a and b are the unknown parameters to be estimated. The values of a and b are assumed in the range of [1000 10000] and [0 0.6], respectively. The range of a and b was selected from the correlation data available based on the work of Silva et al. [11]. In each iteration, the population is randomly generated within the specified range. Each chromosome in the set of population represents a and b values. After obtaining the simulated temperatures, the error is minimized using the least square method and the fitness function is calculated using Eq. (10). The inverse estimation is first conducted for exact temperatures. Later, the analysis is performed for measured temperature with added noise of $\sigma = 0.01\ T_{max}$, $\sigma = 0.02\ T_{max}$, and $\sigma = 0.03\ T_{max}$, respectively. With the same assumed input parameters, three cases were run and the retrieved values are noted in Table 2. It was found that for every run, GA estimates near to the assumed values of a and b ($a = 6000$ and $b = 0.33$). The average percentage error in the estimation of unknown parameters a and b was found to be 6.9 and 5.25%, respectively. The performance of the GA is affected by the input parameters like population size and mutation rate. Two runs were carried out by changing the mutation rate to 0.06. It can be observed from the Table 3 that GA estimates best results within 2% error. Figure 6 shows the corresponding minimum, mean, and maximum values of the fitness function, and Fig. 7 shows the estimation of average values a and b, respectively.

Table 3 Retrieval of a and b values for exact temperature values using GA with population = 15, generations = 30, and mutation rate = 0.06

Runs	a	b	Time, s	% Error a	% Error b
1	5999.2	0.329	21,407	0.013	0.303
2	5895.10	0.324	27,404	1.75	1.82
Average	5947.15	0.327		0.88	1.06

Fig. 6 Maximum, mean and mean of the fitness function

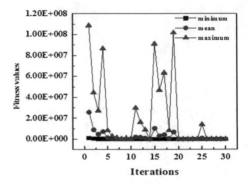

Fig. 7 Convergence of average values of a and b from different runs for 0.01 T_{max} noise data

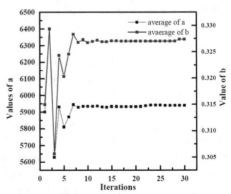

Table 4 Retrieval of a and b values using noisy temperature values using GA with population = 15, generations = 30, and mutation rate = 0.06 ($\sigma = 0.01\ T_{max}$)

Runs	a	b	Time, s	% Error a	% Error b
1	6016.60	0.33	21,236.00	0.28	0.00
2	5792.04	0.32	20,913.00	3.47	3.03
3	6009.70	0.33	23,015.64	0.16	0.00
Average	5939.45	0.33		1.30	1.01

Now, the estimation process is carried out for three different noisy temperature data. The population size = 15, iterations = 30, and mutation rate = 0.1 are assumed. Table 4 shows the retrieval of a and b values for the actual values of 6000 and 0.33.

Table 5 Comparison of estimation with increase in population for noisy measurement with $\sigma = (0.01\ T_{max})$

Population size	a	b	Time, s
15	5939.45	0.33	21,236
20	6004.32	0.33	34,968

Table 6 Comparison of retrieved values for three different noisy data

Sl No	Noise	a	b	Time, s	% Error a	% Error b
1	$0.01\ (T_{max})$	5939.45	0.33	21,236	1.01	0
2	$0.02\ (T_{max})$	6098.38	0.33	23,849	1.64	0
3	$0.03\ (T_{max})$	5867.89	0.323	22,648	2.2	2.12

Fig. 8 Comparison of temperature distribution

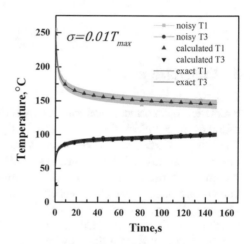

Increase in the population size increases the accuracy of the solution as there will be a provision for large number of chromosome to predict the result within the specified range. Though the accurate results are obtained, it requires more computational time as seen in Table 5. Similarly, different test runs were conducted for noisy measurement of $\sigma = (0.02\ T_{max})$ and $\sigma = (0.03\ T_{max})$, respectively, and the values of the estimates are noted in Table 6. The aim of the present study is to prove the ability of GA to retrieve the unknown parameters for different noise levels and such investigations are shown in Figs. 8 and 9. Thus, GA is capable of handling noisy data and the results of the estimation of a and b were found satisfactory with an error less than 3%. Similar result was found in [19] where GA has been adopted for the solidification of Al-4.5 wt%Cu.

Fig. 9 Comparison of
temperature error at T1

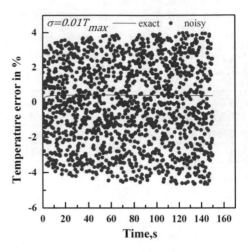

6 Conclusion

Estimation of interfacial heat transfer coefficient for horizontal directional solidification of Sn-5 wt%Pb alloy was successfully carried out using Genetic Algorithm. The values of IHTC were taken from the literature and incorporated in the forward model in order to obtain the cooling curves. The estimation was initially attempted for the simulated measurements to check the robustness of the proposed methodology. Later, noise-added simulated measurements were used in the estimation process. It was found that even for the noisy data, GA was able to retrieve the unknown parameters thus ensuring the proposed method can effectively be used for solving the inverse heat transfer problem for the actual measurement data.

References

1. Ho, K., Pehlke, R.D.: Metal-mold interfacial heat transfer. Metall. Mater. Trans. B **16**(3), 585–594 (1985)
2. Nishida, Y., Droste, W., Engler, S.: The air-gap formation process at the casting-mold interface and the heat transfer mechanism through the gap. Metall. Mater. Trans. B **17**(4), 833–844 (1986)
3. Rajaraman, R., Velraj, R.: Comparison of interfacial heat transfer coefficient estimated by two different techniques during solidification of cylindrical aluminum alloy casting. Heat Mass Transf. **44**(9), 1025–1034 (2008)
4. Meneghini, A., Tomesani, L.: Chill material and size effects on HTC evolution in sand casting of aluminum alloys. J. Mater. Process. Technol. **162**, 534–539 (2005)
5. Prabhu, K.N., Kumar, S.T., Venkataraman, N.: Heat transfer at the metal/substrate interface during solidification of Pb-Sn solder alloys. J. Mater. Eng. Perform. **11**(3), 265–273 (2002)
6. Gafur, M.A., Haque, M.N., Prabhu, K.N.: Effect of chill thickness and superheat on casting/chill interfacial heat transfer during solidification of commercially pure aluminium. J. Mater. Process. Technol. **133**(3), 257–265 (2003)

7. Hallam, C.P., Griffiths, W.D.: A model of the interfacial heat-transfer coefficient for the aluminum gravity die-casting process. Metall. mater. Trans. B **35**(4), 721–733 (2004)
8. Hamasaiid, A., Dargusch, M.S., Davidson, C.J., Tovar, S., Loulou, T., Rezai-Aria, F., Dour, G.: Effect of mold coating materials and thickness on heat transfer in permanent mold casting of aluminum alloys. Metall. Mater. Trans. A **38**(6), 1303–1316 (2007)
9. Prabhu, K.N., Chowdary, B., Venkataraman, N.: Casting/mold thermal contact heat transfer during solidification of Al-Cu-Si alloy (LM 21) plates in thick and thin molds. J. Mater. Eng. Perform. **14**(5), 604–609 (2005)
10. Santos, C.A., Siqueira, C.A., Garcia, A., Quaresma, J.M., Spim, J.A.: Metal–mold heat transfer coefficients during horizontal and vertical unsteady-state solidification of Al–Cu and Sn–Pb alloys. Inverse Problems Sci Eng. **12**(3), 279–296 (2004)
11. Silva, J.N., Moutinho, D.J., Moreira, A.L., Ferreira, I.L., Rocha, O.L.: Determination of heat transfer coefficients at metal–mold interface during horizontal unsteady-state directional solidification of Sn–Pb alloys. Mater. Chem. Phys. **130**(1), 179–185 (2011)
12. Konrad, C.H., Brunner, M., Kyrgyzbaev, K., Völkl, R., Glatzel, U.: Determination of heat transfer coefficient and ceramic mold material parameters for alloy IN738LC investment castings. J. Mater. Process. Technol. **211**(2), 181–186 (2011)
13. Dong, Y., Bu, K., Dou, Y., Zhang, D.: Determination of interfacial heat-transfer coefficient during investment-casting process of single-crystal blades. J. Mater. Process. Technol. **211**(12), 2123–2131 (2011)
14. Sun, Z., Hu, H., Niu, X.: Determination of heat transfer coefficients by extrapolation and numerical inverse methods in squeeze casting of magnesium alloy AM60. J. Mater. Process. Technol. **211**(8), 1432–1440 (2011)
15. Zhang, L., Li, L., Ju, H., Zhu, B.: Inverse identification of interfacial heat transfer coefficient between the casting and metal mold using neural network. Energy Convers. Manag. **51**(10), 1898–1904 (2010)
16. Palumbo, G., Piglionico, V., Piccininni, A., Guglielmi, P., Sorgente, D., Tricarico, L.: Determination of interfacial heat transfer coefficients in a sand mould casting process using an optimised inverse analysis. Appl. Therm. Eng. **78**, 682–694 (2015)
17. Wong, M.D., Pao, W.K.: A genetic algorithm for optimizing gravity die casting's heat transfer coefficients. Expert Syst. Appl. **38**(6), 7076–7080 (2011)
18. Dousti, P., Ranjbar, A.A., Famouri, M., Ghaderi, A.: An inverse problem in estimation of interfacial heat transfer coefficient during two-dimensional solidification of Al 5% Wt-Si based on PSO. Int. J. Numer. Meth. Heat Fluid Flow **22**(4), 473–490 (2012)
19. Ranjbar, A., Ghaderi, A., Dousti, P., Famouri, M.: A transient two-dimentional inverse estimation of the metal-mold heat transfer coefficient during squeeze casting of AL-4.5 wt% CU. Int. J. Eng.-Trans. A: Basics **23**(3&4), 273 (2011)
20. Vasileiou, A.N., Vosniakos, G.C., Pantelis, D.I.: Determination of local heat transfer coefficients in precision castings by genetic optimisation aided by numerical simulation. Proc. Inst. Mechanical Eng. Part C: J. Mechanical Eng. Sci. **229**(4), 735–750 (2015)
21. Yu, Y., Luo, X.: Identification of heat transfer coefficients of steel billet in continuous casting by weight least square and improved difference evolution method. Appl. Therm. Eng. **114**, 36–43 (2017)
22. Ozisik, M.N.: Inverse Heat Transfer: Fundamentals and Applications. CRC Press (2000)
23. Goldberg, D.E., Holland, J.H.: Genetic algorithms and machine learning. Mach. Learn. **3**(2), 95–99 (1988)

Conventional and AI Models for Operational Guidance and Control of Sponge Iron Rotary Kilns at TATA Sponge

Chaitanya Shah, Puneet Choudhary, Brahma Deo, Parimal Malakar,
Susil Kumar Sahoo, Gyanrajan Pothal and Partho Chattopadhyay

Abstract Prediction models for temperature, pressure, and quality control in rotary sponge iron kilns are developed from operational data. The conventional and AI-based methods which are used to develop the models include extreme learning machine (ELM), artificial neural net (ANN), and multiple linear regression (MLR). The performance of the developed models is tested on shop floor in actual operation and compared. Extensive plant data is used to develop and validate the models on day-to-day basis of operation so as to take care of the dynamically changing situation inside the kiln, giving first preference to quality control and then to accretion control. Accretion control increases the life of lining and thus also the available time for production. Automatic pressure control greatly helps in chaos control inside the kiln. Dynamically changing Lyapunov exponent acts a guide line for automatic pressure control.

Keywords Rotary kiln · Sponge iron · Operational control · AI models · Chaos control

1 Introduction

TATA Sponge employs coal fired rotary kilns to produce sponge iron. Iron ore, coal, and dolomite are charged from one end of rotary kiln (schematic diagram in Fig. 1), and the coal for combustion is fired from the other end. Rotary kiln is slightly inclined so that material flows in forward direction. Partial combustion of coal is assisted by the air blown through primary air blower and root blower from firing end, and, in addition, several secondary air blowers are suitably placed along the entire length of the kiln. The rotary kiln is a counter current reactor (opposite movement of charge

C. Shah · P. Choudhary · B. Deo (✉)
Indian Institute of Technology Bhubaneswar, Bhubaneswar, Odisha, India
e-mail: bdeo@iitbbs.ac.in

P. Malakar · S. K. Sahoo · G. Pothal · P. Chattopadhyay
TATA Sponge Iron Limited, Joda, Odisha, India

© Springer Nature Singapore Pte Ltd. 2019
J. C. Bansal et al. (eds.), *Soft Computing for Problem Solving*,
Advances in Intelligent Systems and Computing 816,
https://doi.org/10.1007/978-981-13-1592-3_36

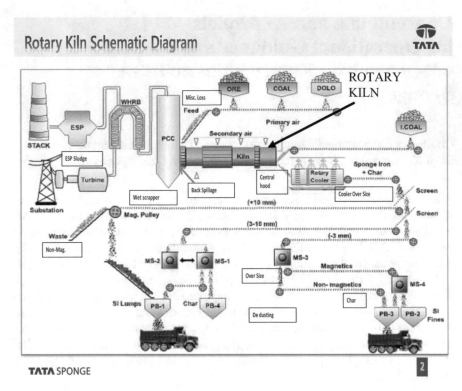

Fig. 1 Schematic diagram of plant and the rotary kiln (see arrow mark) for production of sponge iron

and air). PID controllers are used to maintain desired pressure difference between inlet and outlet gas pressure inside the kiln. The reducing gases produced inside the kiln need to be maintained within a desired window of temperature, pressure, and composition range to enable optimal reduction of iron oxide to iron. The degree of reduction of iron oxide to iron is one of the key quality parameters for deciding the selling price of sponge iron in the market. Several research papers on sponge iron production in rotary kilns have been published describing both heat and mass transfer aspects as well as expert systems based on fuzzy logic and ANN.

The earlier work [1] on mathematical modeling of heat and mass transfer in rotary kiln helped to predict the concentration and temperature profiles inside the kiln. A series of research publications following the same line of approach [2–5] finally culminated in [6] in which some suggestions were made regarding optimum steady state operability of kiln. Process control of kiln cannot be achieved by heat and mass transfer-based models alone. The main reason is that the fluctuations inside the kiln are chaotic, whereas the earlier researchers assumed convective flow conditions. Recent approaches are based on fuzzy logic [7, 8] and ANN [9, 10] to take care of nonlinear phenomenon inside the kiln. The aspect of constant presence of chaos inside the kiln has still eluded these researchers.

Chaos, by definition, is bounded, and control schemes can be devised for obtaining the results within a desired window in the presence of chaos as well. In the present work, it is shown that control of chaotic variation of pressure variations inside the kiln is the first requirement of accretion and quality control, in addition to kiln rotation speed, feed rate of ore and coal, and the quality of raw material (ash content of coal and percentage of fines in ore and coal). The heat and mass transfer models are still needed because they help to decide the basic mass and thermal requirements of kiln, but the control of rotary kiln so as to minimize accretions and maintain quality is much more involved because of the nonlinear dynamical nature of kiln reflected in chaotic variation of pressure inside the kiln.

The models developed for the kilns at TATA sponge, along with chaos control procedures, have been embedded in the decision support system of kiln control. The conventional models are developed by using multiple linear regression (MLR), and the AI models are developed by using artificial neural net (ANN) and extreme learning machine (ELM) [11]. Selection of appropriate model to use is made on the basis of its efficacy. A separate routine is implemented for dynamic pressure control. In the present work, the aim is to compare the predictive abilities of a combination of AI and conventional models and pressure control in the real-life plant environment with an aim of accretion and quality control.

2 Plant Data

A large amount of reliable data is required for developing the conventional and AI models. Plant data is collected (average of 8 h, depending upon nature of data and equation to be developed, but over 30 days) of inlet pressure, outlet pressure, primary air blow rate and secondary air flow rate (in eight secondary air blowers placed along the length, (80+ meters, of kiln), ore and coal charging rate from feed end, coal firing rate from firing end, and temperature measured by ten thermocouples (again, placed along the length of kiln).

Typical values of measured pressure (PR, mm of water column), primary air blown (PAB, m^3 per hour), ore feed rate (ORE, tons per hour), kiln rotation (RPM, rotation per hour) feed carbon in injection coal (FCIC, tons/hour), feed carbon in feed foal (FCFC, tons per hour), air blow through secondary air blower (SAB 1–8, m^3 per hour) are given in Table 1.

Corresponding to the values in Table 1, the typical values of measured temperatures at 8 h intervals and at different thermocouple locations (TC3–TC11) are given in Table 2.

The campaign life of kiln may vary starting from 60 to 300 days, depending upon how it is operated. Operational control should be such that the product quality and production rate are maintained, subject to control of the growth of accretion inside the kiln. Accretions form due to sintering (or fusion) of charge materials at high temperature (800–1200 °C). The growth rate of accretion is predicted by a separate thermal model. The decision to change operating parameters is guided by growth

Table 1 Typical values of control parameters of rotary kiln

PR.	PAB	ORE	RPM	FCIC	FCFC
7.8	11,315	33.5	376.9	4.8	4.8
7.5	11,479.7	35.4	388.2	5.4	4.6
8.3	10,307.5	32.0	390.3	4.6	4.7
7.0	9395.2	30.5	347.0	4.3	4.7

(continued)

Table 1 (continued)

SAB1	SAB2	SAB3	SAB4	SAB5	SAB6	SAB7	SAB8
6734.0	7280.0	6760.0	3744.0	4680.0	7670.0	7748.0	7540.0
6838.0	7774.0	6708.0	3822.0	4810.0	7540.0	7670.0	7618.0
6292.0	6578.0	5512.0	4238.0	3952.0	6240.0	6864.0	7176.0
5590.0	5850.0	5460.0	4238.0	4186.0	5902.0	7150.0	7540.0

Table 2 Typical values of temperatures, TC3–TC11, along the length of kiln

TC3	TC4	TC5	TC6	TC7	TC8	TC9	TC10	TC11
989.6	1029.3	940.4	819.7	947.2	997.1	1024.6	517.9	866.9
949.5	1010.5	1045.1	1088.5	1088.2	1027.3	1067.9	1115.3	972.3
1003.4	1014.5	1063.6	1035.5	1036.2	984.6	1032.2	833.3	907.6
964.9	1020.5	964.7	822.2	911.3	932.2	1019.7	536.7	868.5

rate of accretions at different locations inside the kiln. A typical growth profile of accretion at formed at TC7 and TC 10 (change in accretion thickness with number of days of operation) is shown in Fig. 2; it can be seen that the accretion at TC7 has formed and broken because of the control strategy adopted through the models of present work; accretion at TC10 is also seen to increase and decrease in size. The effort is to restrict (or reduce) the growth of accretion while maintaining quality.

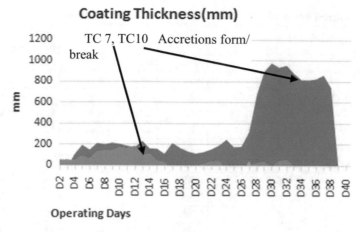

Fig. 2 Progress of formation of accretion at two different locations, TC7 and TC10

3 Results of Models Developed in Present Work

The results of different models for prediction of temperature (TC3–TC11), quality, and pressure obtained and tested on actual plant data using ANN, multiple linear regression (MLR), and extreme learning machine (ELM) are compared in Table 3. SEE means the standard error of estimate of parameter from independent variables, and Adj R^2 is adjusted correlation coefficient. The number of neurons finally selected for one middle layer, both in ANN and ELM, is given in the last two columns of Table 1, respectively. It can be seen that ELM out performs both ANN and MLR, both in terms of SEE, Adj R^2, except for the case of pressure where MLR gives the best results. An hourly analysis of pressure has revealed that fluctuations in pressure are actually chaotic with Lyapunov exponent greater than +2. For the present, MLR, by virtue of its better averaging effect through linearization, pressure prediction by MLR is better that by ELM and ELM under chaotic conditions. The chaotic pressure variations and its control will be discussed in a separate work.

The usual practice in plant is to employ MLR for determining different prediction equations to account for short-term changes; it is always convenient to assume a linear behavior over short-term periods. Typical results for TC6 temperature by using MLR are given in Table 3, and the results of quality prediction using MLR are given in Table 3. The parameters which are not contributing to prediction of a particular variable are dropped from regression equation on the basis of their relative importance. The typical equation for prediction of TC6 by MLR, for example, contains only four parameters, namely air flow through secondary air blowers (SAB1), (SAB2), (SAB3), and primary air blower (PAB).

$$Tc6 = 0.018 * SAB1 + 0.008 * SAB2 + 0.008 * SAB3 - 0.002 * PAB \quad (1)$$

Table 3 Prediction results

	SEE (±)			Adj R2			Number of neurons in	
	ANN	MLR	ELM	ANN	MLR	ELM	ANN	ELM
TC3	11.06	14.16	6.62	0.66	0.42	0.88	11	85
TC4	9.71	10.23	8.06	0.29	0.19	0.43	10	55
TC5	21.92	40.26	12.39	0.82	0.37	0.94	11	85
TC6	35.33	40.15	9.16	0.85	0.81	0.99	12	85
TC7	21.06	24.58	8.15	0.78	0.70	0.97	11	85
TC8	41.16	70.95	27.25	0.82	0.45	0.92	13	85
TC9	23.53	28.93	13.37	0.51	0.25	0.84	10	80
TC10	83.96	58.64	12.51	0.85	0.94	0.90	11	80
TC11	19.52	15.27	7.97	0.80	0.87	0.97	12	80
Quality	1.81	2.01	1.43	0.26	0.09	0.55	11	90
Pressure	0.56	0.33	0.62	0.59	0.72	0.55	13	85

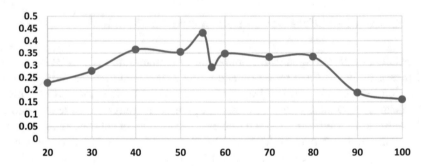

Fig. 3 Selection of optimal number of neurons in ELM

When artificial neural net (ANN) and (extreme learning machine (ELM) are used for prediction, then all operational parameters are incorporated in the prediction set to account for their possible nonlinear contribution to the system and obtain better results of prediction.

A number of neurons in the middle layer have to be found out by trial and error, both in case of ELM and ANN. For example, for the worst prediction case of TC4 in Table 3, the effect of number of neurons in ELM is shown in Fig. 3; it can be seen that 55 neurons are optimal for ELM in the case of TC4, whereas in other cases, 80–90 neurons are required for ELM.

It is observed during the shop floor application of models that the accuracy of prediction of all models decreases when accretion breaks down due to physical movement of charge. An advance indication of this can be obtained from chaos analysis (dynamic nonlinear chaotic changes through Lyapunov exponent) of pressure. In order to maintain the Lyapunov exponent in a desired window, it is found that the pressure difference between inlet and outlet should be kept in the window of

4–6 mm. However, even then when the accretions break down, the pressure inside the kiln can change unpredictably and that is a very special feature and nature of coal fired rotary kilns. Nature always keeps the last card to play in it its own hand to respond to non-equilibrium conditions, similar to the phenomenon of formation and breaking of glaciers, except that the glaciers form when temperatures go down and break when the temperatures go up, whereas the accretions inside the kiln form and grow when temperatures go up. In contrast to glaciers, accretions are always of a wave form on a cylindrical base due to constant and counter current movement of charge versus reducing gases inside the kiln.

If pressure is carefully measured and controlled, then it is observed on the shop floor that accretions grow at a slower rate in a wave form (increasing in height gradually, see Fig. 2) and then eventually breaking down when the peak height reaches a particular level, depending upon the region of temperature. Accretions break when they are unable to resist the stresses (due to impact) of the rolling charge which is continuously striking at their base and on their side surfaces due to rotation of kiln. Besides pressure, therefore, the rotation rate of kiln is another important control parameter to be adjusted suitably, both for maintaining quality and also for restricting the accretion growth rate.

The feed rate of ore and coal is the next important parameters. The fines and ash content of coal enhance the accretion formation and must be regulated within a certain window for the process to be steady (for prediction purposes as well). All these features together make the control of rotary kiln a challenging task for the operator to produce and for the manufacturer to optimize the cost. Cost of coals with lower ash content and lower fines is progressively prohibitive to use in kiln.

The next step in improvement of control of kilns at TATA Sponge would be automatic regulation of air blown through the secondary air blowers. Stable kiln operations directly determine stable power generation too because the waste gases from the kiln are used for steam generation which is then used for power generation. It is observed at TATA Sponge that after implementation of automatic pressure regulation the power generation has also stabilized proportionately, in spite of the fact that the total volume of secondary air blown has been reduced by at least 10% compared to previous practices in which heat and mass transfer models alone were the guiding criteria.

Models developed and implemented in the present work have helped in accretion control and quality control. Models were made by taking 4, 8, 16, 24 h, and one month periods. Best results were obtained when the average response time of the kiln was taken as 8 h and the models are tuned on day-to-day basis accordingly. Pressure control is, however, done automatically (continuously, on-line) through PID controller. These features lying on two extremes of control (few seconds for pressure control to 8 h duration for kiln) render the rotary kiln operation unique in its own way.

4 Conclusions

Application of ELM, ANN, and MLR multiple-based models along with dynamic pressure control is demonstrated for quality and accretion control in rotary kiln on shop floor. Quality is maintained, and accretions are minimized. It is recommended to use ELM, in preference to ANN and MLR when data size is adequate, else MLR is to be used over short time intervals (10 days plus). For extrapolation into a new region over a short time period, MLR is to be preferred. It is observed that the accretions grow and eventually break in a wave form. Pressure fluctuation is chaotic with an average Lyapunov exponent of approximately +2. For a given raw material input quality (ash in coal + its fine size fraction, and reducibility and Fe content of ore), the important control parameters are pressure, rotation speed of kiln, feed rate of ore and coal, and ratio of feed coal to injection coal, in that order. As a result of pressure regulation and better process control, the volume of secondary air requirement has been reduced by 10%. As an added benefit, the power generation has also been stabilized proportionately. Since the growth rate of accretions has been reduced, it is expected that the campaign life of kiln will increase. The trials are continuing for further improvements.

Acknowledgements The authors (SC, PC, BD) from IIT Bhubaneswar are very thankful to the management of TATA Sponge Iron Limited (TSIL) for giving an opportunity to work with TSIL engineers and learn from them the practical aspects of kiln operation. Perhaps it will serve as an encouraging example to showcase how an industry takes a step forward to promote joint industry-academia efforts.

References

1. Venkateswaran, V.: Mathematical model of SL/RN direct reduction process, M.Sc. Thesis. Univ. British Coumbia (1976)
2. Runkana, V.: Model-based optimization of industrial gas-solid reactors. KONA Powder Part. J. **32**, 115–130 (2015)
3. Srivastava, M.P., Bandopadhyaya, A., Prasad, K.K., Chaudhury, B.R..: Operating aspects of sponge pilot plant-some aspects. Trans. Indian Inst. of Metals **41**(2), 177–186 (1988)
4. Srivastava, M.P., Bandopadhyaya, A., Prasad, K.K., Chaudhury, B.R.: Operating aspects of sponge pilot plant. Trans. Indian Inst. Met. **41**(2), 177–186 (1988)
5. Baikadi, A., Venkataraman, R., Subranmanian, S.: Operaability analysis of direct reduction of iron ore by coal in an industrial rotary kiln. Sci. Direct IFAC Papers Line **49**(1), 468–473 (2016)
6. Mharakurwa, E.T, Nyakoe, G.N., Ikua, B.N.: Accretion control in sponge iron production kiln using fuzzy logic. Innovative system Design and Engineering, ISSN 2222–1727 (paper) ISSN 2222–2871 (online) **5**(7), 42–51 (2014)
7. Garikayi, T., Nyanga, L., Mushiri, T., Mhlanga, S., Kuipa, P.K.: Designing an intelligent Fuzzy logic system for accretion prevention in sponge iron SL/RN Rotary Kiln Based 100TPD DRI process. In: SAIIE proceedings, pp. 523(1)–523(14), SAIIE, Stellenbosch (2013)
8. Poonia, A.K., Soni, B. Khanam, S.: Optimization of operating parameters for sponge iron production process using neural network. International J. Chem. Tech. Res. CODEN (USA): IJCRGG ISSN: 0974-4290, 9(2), 20-34 (2016).

9. Baikadi, A., Runkana, V., Subramanian, S.: Operability analysis of direct reduction of iron ore by coal in an industrial rotary kiln. Science Direct, IFAC Papers online **49**(1), 468–473 (2016)
10. Poonia, A.K., Shabina, K.: Simulation of Rotary Kiln Used in Sponge Iron Process Using ANN. IACSIT Int. j. Engg. And Tech. **6**(2), 95–98 (2014)
11. Guang-Bin, H., Qin-Yu, Z., Chee-Kheong, S.: Extreme Learning Machine: Theory and applications. Neurocomputing **70**, 489–501 (2006)

Reconstruction of the State Space Figure of Indian Ocean Dipole

Swarnali Majumder, T. M. Balakrishnan Nair and N. Kiran Kumar

Abstract State space reconstruction is an important index for describing nonlinear time series. However, reconstruction of state space figure is difficult if the data is noisy. Hence, noise reduction is an important step for reconstructing state space figure. In this study, we propose a method which can reconstruct state space picture from a noisy time series. This method is used for reconstructing state space figure from the data of Indian Ocean Dipole. Dimension of the reconstructed attractor is measured by computing correlation dimension. The dynamics of Indian Ocean Dipole is not well understood. The reconstruction of state space figure indicates that there is chaos in Indian Ocean Dipole. Positive Lyapunov exponent reconfirms that the dynamics of Indian Ocean Dipole is chaotic.

Keywords Singular value decomposition · Fast Fourier transform
Indian Ocean Dipole

1 Introduction

Indian Ocean Dipole (IOD) is an interesting phenomenon in which the western Indian Ocean becomes alternately warmer and then colder than the eastern part of the Indian Ocean. It is a challenging job to forecast IOD event [1]. A positive IOD is the phenomena when eastern equatorial Indian Ocean is cooler than normal SST and the western tropical Indian Ocean is warmer than normal SST. A negative IOD is characterized by warmer than normal SST in the eastern equatorial Indian Ocean and

S. Majumder (✉) · T. M. Balakrishnan Nair · N. Kiran Kumar
Indian National Centre for Ocean Information Services, Pragathi Nagar,
Nizampet, Hyderabad 500 090, India
e-mail: swarnali.majumder48@gmail.com

T. M. Balakrishnan Nair
e-mail: bala@incois.gov.in

N. Kiran Kumar
e-mail: kirankumar@incois.gov.in

© Springer Nature Singapore Pte Ltd. 2019
J. C. Bansal et al. (eds.), *Soft Computing for Problem Solving*,
Advances in Intelligent Systems and Computing 816,
https://doi.org/10.1007/978-981-13-1592-3_37

cooler than normal SST in the western tropical Indian Ocean. During positive IOD, equatorial surface winds, which in a normal condition blows toward east, weakens and blows toward west. Rainfall increases over the western Indian Ocean, while over the Indonesia and Australia it decreases, resulting in severe drought. During negative IOD, westerly winds intensify along the equator. Eastern Indian Ocean flooded whereas dried condition is observed in the western Indian Ocean [2]. Since several oceanic and atmospheric parameters are intertwined in the mechanisms of IOD, intuitively it appears that it is a complex system. However, the dynamics of IOD is not understood yet. Researchers have differing opinions regarding the nature of the dynamics of the Indian Ocean SST, because of its complexity compared to that of the tropical Pacific. Some study has shown that SST variability in the Indian Ocean can be compared to a spatial first-order autoregressive process [3]. Penland 1996 has argued that Indo-Pacific SST anomalies can be represented as a stable linear process driven by spatially coherent stochastic forcing [4]. However, several researchers believe that IOD, the dominant mode of climate variability in the Indian Ocean, is independent of major Pacific phenomena and grown by complex ocean–atmosphere interaction in the Indian Ocean [5, 6]. Till now no study has been made to explore the chaotic properties of Indian Ocean SST. In this study, we explore the chaotic characteristics of IOD by analyzing its time series.

State space reconstruction is a major index for describing nonlinear time series. According to this concept, a state space is constructed that is topologically equivalent to the original dynamical system. Since the dynamics of the system is unknown, we cannot reconstruct the original attractor from the observed time series.

Instead, we look for an embedding space where we can reconstruct an attractor from the time series that preserves the invariant characteristics of the original unknown attractors. State space figure was initially reconstructed from the time series of fluid flow by creating time-delay vectors [7]. Afterward, it has been mathematically proved that if any data resides in an M-dimensional manifold, then from a given time series, the state space figure can be reconstructed by creating $(2M + 1)$ time-delay vectors $(x(t), x(t + T), \ldots, x(t + T + 2M))$, where x is a time series $(x(1), x(2), \ldots, x(t), x(t + 1), \ldots, x(t + T), \ldots, x(t + T + 2M), \ldots)$, T is the time lag and M is the embedding dimension [8]. State space picture can be obtained by plotting $x(t)$ and $x(t + T + 2M)$ along x- and y-axis, respectively. This is known as time-delay method. The trajectories of the state space figure describe the evolution of the system from some initial state (assumed to be known) and hence represent the history of the system. State space reconstruction plays an important role in forecasting a signal. By using this method, many techniques of system characterization and identification are made possible. However, reconstruction of state space figure is difficult if the data is noisy. There are several factors which complicate the reconstruction problem for real-world data: observational noise, dynamic noise, and estimation errors [9]. Hence, noise reduction is an important step for reconstructing state space figure. Recently, several researches have emerged on reducing noise from a time series [10, 11]. In this study, we propose a noise reduction method based on singular value decomposition and Fast Fourier transform for reconstructing state space figure. This method is applied to the time series of IOD to reconstruct its state space figure. We compute correlation dimension of IOD time series to measure the dimension of its attractor. Finally, we compute the Lyapunov exponent of IOD time series and show that its Lyapunov exponent is positive.

2 Data

First, our proposed method of reconstructing state space figure is implemented to a numerically simulated data, obtained from Van der Pol equation. Van der Pol oscillator is a two-dimensional dynamical system. Its equation is of the following form:

$$\frac{dx_0}{dt} = x_1$$

$$\frac{dx_1}{dt} = cx_1 - cx_1x_0^2 - kx_0$$

In this case, c is chosen as 2, k is chosen as 1, initial conditions are $x_0 = 1.82$ and $x_1 = -0.17$, and the resulting trajectory is shown in Fig. 2. Random noise is added to this numerically simulated data (say, $x1$) to make it noisy (say, x).

Dipole mode index (DMI) is the measure of the intensity of the IOD. DMI is represented by anomalous SST gradient between the western equatorial Indian Ocean (50 °E–70 °E and 10 °S–10 °N) and the southeastern equatorial Indian Ocean (90 °E–110 °E and 10 °S–0 °N). Both weekly and monthly DMI data are downloaded from Japan Agency for Marine-Earth Science and Technology [12, 13]. Weekly DMI is considered from November 1981 to September 2016. Weekly DMI dataset is derived from NOAA OI SST Ver.2 [12, 13]. The optimum interpolation (OI) sea surface temperature (SST) analysis is produced weekly on a one-degree grid. The analysis uses in situ and satellite SST's plus SST's simulated by sea-ice cover. Figure 1 shows the time series of weekly DMI. Monthly DMI is considered from January 1958 to December 2010. SST data of Hadley Centre is used to compute monthly DMI.

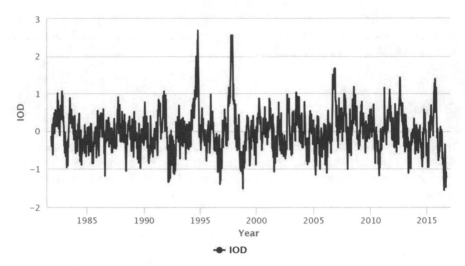

Fig. 1 Time series of dipole mode index

3 Methodology

3.1 Singular Value Decomposition

Singular value decomposition (SVD) is a novel technique for reducing noise from data [14–16]. The first step of this method is to create a Hankel matrix, say X, from a given time series. Let $x = (x_0, x_1, ..., x_{N-1})$ be a time series of length N, and K be an integer, which will be called as the "embedding dimension." We set $L = N - K + 1$ and define the L-lagged vectors $Xj = (x_{j-1}, x_j, x_{j+1}, ..., x_{j+L-2})^T$, $j = 1, 2, ..., K$, and the trajectory matrix $X = [X1 ... XK]$. Each Xj ($j = 1, 2, ..., K$) constitutes the columns of X. The trajectory matrix X is a Hankel matrix, which means that all the elements along the diagonal $i + j = $ constant are equal.

	$x(0)$	$X(1)$	...	$x(K-1)$
	$x(1)$	$x(2)$	...	$x(K)$
$X=$	...	...	...	...
	$x(L-1)$	$x(L)$	...	$x(K+L-2)$

Next SVD is performed on X, which decomposes X into three matrices U, W, and V, which satisfies the relation $X = UWV^T$. Here U is an L by K-dimensional orthonormal matrix, W is a K-dimensional diagonal matrix, and V is a K by K-dimensional orthogonal matrix. U is the matrix of the eigenvectors of XX^T and V is the matrix of the eigenvectors of X^TX. Diagonal elements of W are the singular values of X. The column vectors of U matrix, say, $(U1, U2,...,UK)$ are called "empirical orthogonal functions" or simply EOFs. The idea behind SVD is that most of the variance (power) is contained within the first few components. Diagonal elements of W matrix, say, $\{w_1, w_2,....,w_K\}$ are arranged in a decreasing order, i.e., $w_1 \geq w_2 \geq w_3 \cdots \geq w_K$. It is assumed that the diagonal elements with lower magnitude are associated with the noise of the signal. Hence, we omit the negligible singular values and consider the first few significant values, say, $\{w_1, w_2, w_3,...,w_p\}$, where p is less than K. Now we assign 0 to the $(p+1)$th to Kth diagonal elements and let us call this new matrix as W'. The signal is reconstructed by the relationship $X' = UW' V^T$. This reconstructed signal X' is supposed to be smooth and free from noise. We consider the filtered data after SVD and use this noise-free data for further filtering.

3.2   FFT-Based Filter

A method has been proposed for reconstructing state space figure, which is based on Fast Fourier transform [17]. In this method, we compute the Fourier Transform of the time series X. Then, we find the first two dominating peaks in the frequency–amplitude diagram. Let $b1$ and $b2$ be the bin numbers corresponding to the dominating peaks and $\omega1$ and $\omega2$ be the corresponding angular frequencies. Let us consider a 100 by 2 matrix A. An nth element in the first column of A is of the form $e^{-i\omega 1nh/100}$, an nth element in the second column of A is of the form $e^{-i\omega 2nh/100}$, n is the row number, h is the sampling interval, $\omega1$ and $\omega2$ are defined above. Here the row length of A is chosen arbitrarily. We construct a Hankel matrix H from X, of m number of rows, where m is greater than 100.

Let us consider the relationship $K = HA$. K has same number of rows as H and 2 columns. The above relationship converts each of the Takens vectors of H into a two-dimensional vector in the complex domain. The significance of this transformation is that it filters in the most prominent frequencies of the signal. This filter maps each time-delay vector into a point in the two-dimensional complex domain (C^2). Basically, it embeds the data series X in a two-dimensional complex domain $(C0, C1)$, which can be treated as a four-dimensional real-time series (u, v, y, z), where u and v are the real and imaginary parts of $C0$, respectively, y and z are the real and imaginary parts of $C1$, respectively. If we plot any one of these four variables against the other one, we get a state space picture.

	$x'(1)$	$x'(2)$	...	$x'(100)$
	$x'(2)$	$x'(3)$	...	$x'(101)$
$H =$	...	...	...	...
	$x'(m)$	$x'(m+1)$	...	$x'(m+10)$

$$
A = \begin{array}{|c|c|}
\hline
1 & 1 \\
\hline
e^{\frac{-i\omega 1 h}{100}} & e^{\frac{-i\omega 2 h}{100}} \\
\vdots & \vdots \\
e^{\frac{-i 99\omega 1 h}{100}} & e^{\frac{-i 99\omega 2 h}{100}} \\
\hline
\end{array}
$$

3.3 Modified Approach Based on SVD and FFT-Based Filtering

In this study, the above-mentioned FFT-based filter is modified by incorporating SVD to it. FFT-based filter is applied to the filtered data X', which is obtained by SVD in Sect. 3.1. This complex transformation transforms X' into a complex domain $(C0, C1)$. This modification should produce a better state space figure. Basically, two distinct filters are applied in this technique. The idea behind SVD is that the frequency–amplitude diagram of FFT remains same after SVD. Dominant peaks remain in the same frequency for X' as they are for X. So the effectiveness of the complex filter is not affected because of SVD.

3.4 Correlation Dimension and Lyapunov Exponent

Dimension of a system is defined as the power of the radius of the hypersphere (ε) with which the volume of the system within the hypersphere changes. Correlation dimension [18] is a measurable parameter similar to the dimension of a system; the only difference is that here we count data points of the system as we increase the volume of the hypersphere. First, we count the number of points within a distance ε of each other:

$$
N(\varepsilon) = \sum_{i,j} H\big(\varepsilon - |x_i - x_j|\big)
$$

In this expression, x_i is the point on an attractor in an n-dimensional phase space, H is the Heaviside theta function: $H(y) = 0$ if $y < 0$ and $H(y) = 1$ if $y \geq 0$. This number decreases exponentially as ε decreases. Hence, $N(\varepsilon) \sim \varepsilon^{d_c}$. If so, then the ratio $d_c = \lim_{\varepsilon \to 0} \frac{\log N(\varepsilon)}{\log(\varepsilon)}$ should exist. This limit defines the correlation dimension.

Lyapunov exponent describes the rate at which close trajectories diverge or converge in the state space diagram of a time series. So it is a direct measure of chaos of a system, and it is a measure of predictability also. Let us assume that l_0 to be the initial separation between two points in a state space diagram. Now, if $l(t) \sim l_0 e^{\lambda t}$, then λ is the Lyapunov exponent. In this study, Lyapunov exponents are obtained by using Rosenstein's technique [19]. The first step of this method is the reconstruction of the attractor from a single time series. After reconstructing the attractor, nearest neighbor is computed of each point on the trajectory. The nearest neighbor, X_j', is located by finding the point that minimizes the distance to the particular reference point, X_j. This is expressed as $d_j(O) = \min \text{norm}(X_j - X_j')$, (min is over X_j'), where $d_j(O)$ is the initial distance from the jth point to its nearest neighbor, and "norm" denotes the Euclidean norm. An additional constraint is that the nearest neighbors have a temporal separation greater than the mean period of the time series, i.e., $|j-j'| > $ mean period. This allows one to consider each pair of neighbors as nearby initial conditions for different trajectories. The largest Lyapunov exponent is then estimated as the mean rate of separation of the nearest neighbors. Since sensitive dependence on initial condition is a significant property of chaotic dynamics, presence of positive Lyapunov exponent is an indication of chaos in a system. One limitation of this technique is that Lyapunov exponent cannot distinguish chaos from noise. So noise reduction is an important step before computing Lyapunov exponent [20, 21]. Noise might influence the correlation dimension also. Hence, in this study, both correlation dimension and Lyapunov exponent are calculated for the noise-free filtered signal X', obtained in Sect. 3.1.

4 Results and Discussions

Figure 2 shows that state space figure can be obtained from noise-free data of Van der Pol by time-delay embedding method (by plotting $X1_j$ against $X1_{j+8}$). Figure 3 illustrates the limitation of time-delay embedding method. Figure 3 shows that state space reconstruction is not possible by the same method, when noise is added to the Van der Pol data. FFT based filter is applied to the noisy Van der Pol data. Reconstructed state space figure is shown in Fig. 4. This method reveals an indication of the underlying dynamics, but the reconstructed figure is still noisy. So any further analysis of this attractor is not possible. This leads us to apply our modified approach to the noisy Van der Pol data. The reconstructed state space figure is shown in Fig. 5. This reconstructed figure is free from noise, and it captures the dynamics of Van der Pol equation.

FFT-based filter and our modified method are applied to weekly and monthly DMI data. Figure 6 shows the reconstructed state space figures of weekly DMI by FFT-based filter and its modification. FFT-based filtering shows a clumsy picture, whereas our approach produces a clear attractor. Time lag is empirically chosen as one for weekly DMI data. If we take time lag higher than one, state space figure becomes obscure. Figure 7 shows that state space reconstruction is possible with monthly DMI data also. However, this figure is not as informative as it is for weekly data, as the length of the monthly data set is much shorter than the weekly data.

One limitation of the time-delay method is that it is always not possible to find out the embedding dimension M of a real-world system. M is chosen empirically for the practical signals of our real-world phenomena. If M is taken very high, it might cause ambiguity [22]. This problem can be avoided in FFT based filtering and its modified approach. The concept of singular spectrum analysis is applied for reconstructing

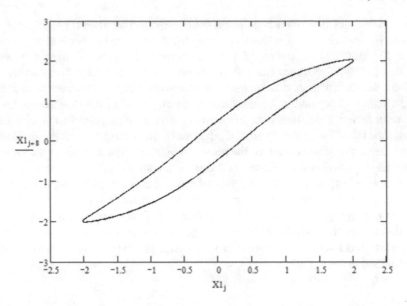

Fig. 2 Reconstructed state space figure of Van der Pol Oscillator by time-delay method when noise is not added to the data

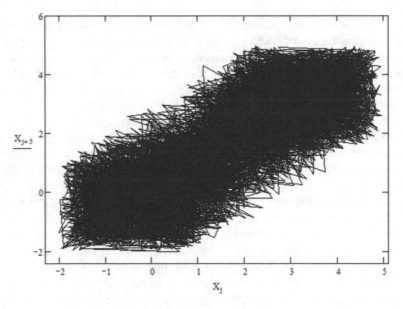

Fig. 3 State space figure of noisy Van der Pol data cannot be reconstructed by time-delay method

attractor from a noisy data [23]. The first two columns of U matrix (say, $U1$ and $U2$) were plotted for reconstructing the state space picture. State space picture can also be reconstructed by plotting $U1$ (t) versus $U1$ $(t+d)$, where t is any time step and d is

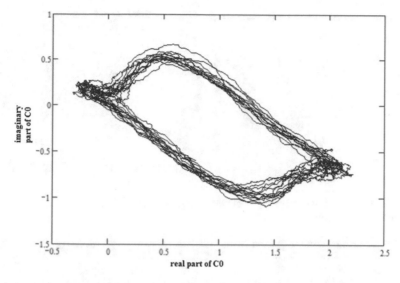

Fig. 4 Reconstructed state space figure from noisy Van der Pol data by FFT-based filtering

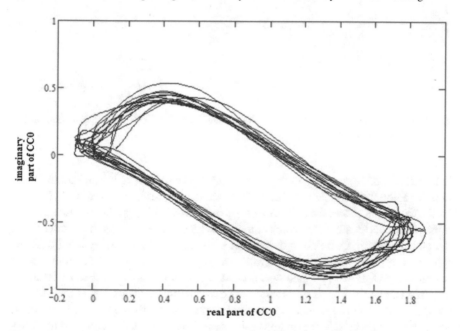

Fig. 5 Reconstructed state space figure from noisy Van der Pol data by the modified technique

the time lag. d is known as the embedding dimension. There are several drawbacks in this method [24]. One more limitation of this method is that it is not always possible to determine d for a system [22]. Interestingly, our noise reduction method reveals

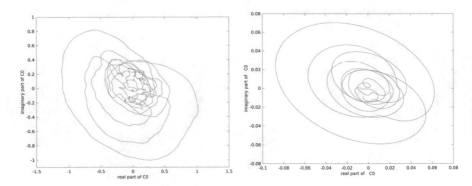

Fig. 6 Reconstructed state space figure of weekly IOD signal by FFT-based filtering (left) and its modified technique (right)

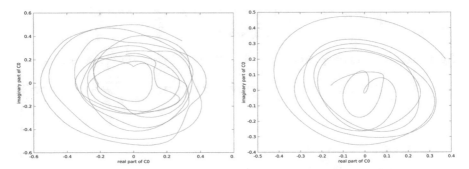

Fig. 7 Reconstructed state space figure of monthly IOD signal by FFT-based filtering (left) and its modified technique (right)

the evolution of a trajectory in the case of IOD. The dynamics of the Indian Ocean Dipole region is debatable and till now it is not really possible to predict IOD events [1, 2]. Our result shows that the state space picture of DMI signal is not like random noise. It reveals an attractor, which is indication that there is chaos in the underlying dynamical system. Correlation dimension is found as 3.4 for weekly DMI data and 4 for monthly DMI data. Finally, Lyapunov exponent is computed for both weekly and monthly DMI data. Lyapunov exponent is 0.3 for weekly IOD data and 0.34 for monthly DMI data. Positive Lyapunov exponent indicates chaos, but sometimes the computational technique of Lyapunov exponent cannot distinguish between noise and chaotic dynamics. Since, in this case filtered data is used, one can reconfirm that the dynamics of IOD is chaotic because of positive Lyapunov exponent of DMI.

5 Conclusions

The key finding of this work is revealing the existence of chaos in the dynamics of IOD. A filtering technique is proposed in this study, which can reconstruct state space picture from a noisy time series. Our proposed noise reduction technique is applied to the time series of IOD, and it can reconstruct its state space figure. The dimension of this attractor is measured by computing correlation dimension. The reconstruction of state space figure indicates that there is chaos in the dynamics of Indian Ocean Dipole. Positive Lyapunov exponent reconfirms the fact that IOD is chaotic. Since this work reveals the underlying dynamics of IOD, it might help forecasters to develop an appropriate method for forecasting IOD.

Acknowledgements One of the authors acknowledge Department of Science & Technology, Government of India, for financial support vide reference no. SR/WOS-A/EA3/2016 under Women Scientist Scheme to carry out this work. We thank Director, INCOIS for supporting this work. This is ESSO-INCOIS contribution No. 325.

References

1. Shin, L., Hendon, H.H., Alves, O., Luo, J.J., Balmaseda, M., Anderson, D.: How Predictable is Indian Ocean Dipole? Mon. Weather Rev. **140**(12), 3867–3884 (2012)
2. http://www.bom.gov.au/climate/iod/
3. Dommenget, D.: Evaluating EOF modes against a stochastic null hypothesis. Clim. Dyn. **28**, 517–531 (2007)
4. Penland, C.: A stochastic model of IndoPacific sea surface temperature anomalies. Phys. D **98**, 534–558 (1996)
5. Saji, N.H., Goswami, B.N., Vinaychandran, P.N., Yamagata, T.: A dipole mode in the tropical Indian Ocean. Nature **401**, 360–363 (1999)
6. Yamagata, T., Behera, S.K., Luo, J-J., Masson, S., Jury, M., Rao, S.A.: Coupled ocean- atmosphere variability in the tropical Indian Ocean. In: Wang, C., Xie, S.P., Carton, J.A. (eds.) Earth climate: the Ocean–Atmosphere Interaction. Geophs. Monogr. vol. 147, pp. 189–212. AGU, Washington (2004)
7. Packard, N.H., Crutchfield, J.P., Farmer, J.D., Shaw, R.S.: Geometry from a time series. Phys. Rev. Lett. **45**, 712–716 (1980)
8. Takens, F.: Detecting strange attractors in turbulence. In: Rand, D.A., Young, L.S. (eds.) Lecture Notes in Mathematics, vol. 898, pp. 366–381. Springer, Berlin (1981)
9. Casdagli, M., Eubank, S., Farmer, J.D., Gibson, J.: State space reconstruction in the presence of noise. Phys. D **51**, 52–98 (1991)
10. Hu, B., Li, Q., Smith, A.: Noise reduction of hyperspectral data using singular spectral analysis. Int. J. Remote Sens. **30**, 2277–2296 (2009)
11. Tan, J.P.L.: Simple noise-reduction method based on nonlinear forecasting. Phys. Rev. E **95**, 032218 (2017)
12. http://www.jamstec.go.jp/frsgc/research/d1/iod/iod/dipole_mode_index.html
13. http://www.metoffice.gov.uk/hadobs/hadisst/
14. Golyandina, N., Nekrutkin, V., Zhigljavsky, A.: Analysis of time series structure: ssa and related techniques. Chapman & Hall/CRC, USA (2001)
15. Vautard, R., Yiou, P., Ghil, M.: Singular-spectrum analysis: a toolkit for short, noisy chaotic signals. Phys. D **58**, 95–126 (1992)

16. Yiou, P., Sornette, D., Ghil, M.: Data Adaptive wavelets and multi-scale singular-spectrum analysis. Phys. D **142**, 254–290 (2000)
17. Majumder, S.: Application of Topology in Inverse Problem: Getting Equations from data. Ph.D. Thesis. Manipal University, India (2013)
18. Grassberger, P., Procaccia, I.: Characterization of strange attractors. Phys. Rev. Lett. **50**, 346–349 (1983)
19. Rosenstein, M.T., Collins, J.J., De Luca, C.J.: A practical method for calculating largest Lyapunov exponents from small data sets. Phys. D **65**(1–2), 117–134 (1993)
20. Dammig, M., Mitschke, F.: Estimation of Lyapunov exponents from time series: the stochastic case. Phys. Lett. A **178**, 385–394 (1993)
21. Eckmann, J.-P., Ruelle, D.: Fundamental limitations for estimating dimensions and lyapunov exponents in dynamical systems. Phys. D **56**, 185–187 (1992)
22. Vaidya, P.G., Majumder, S.: Embedding in higher dimension causes ambiguity for the problem of determining equation from data. Eur. Phys. J. Spec. Top. **165**, 15–24 (2008)
23. Broomhead, D.S., King, G.P.: Extracting qualitative dynamics from experimental data. Phys. D **20**, 217–236 (1986)
24. Palus, M., Dvorak, I.: Singular-value decomposition in attractor reconstruction: pitfalls and precautions. Phys. D **55**, 221–234 (1992)

VMSSS: A Proposed Model for Cloud Forensic in Cloud Computing Using VM Snapshot Server

Shaik Sharmila and Ch. Aparna

Abstract Cloud computing besides being used in industries it is also used in academics; existing cloud computing architectures do not support cloud forensic investigations and are also not forensic ready to a remarkable extent, and also the present tools which are being used in the cloud forensics do not support the elastic nature of cloud. We explore and expose several issues related to cloud forensics in cloud computing by keeping an eye on the concepts of cloud computing which are being developed and are utilized along with latest technologies and also the investments which are being made on cloud computing. Latest developments in technologies have created certain challenges which are emerging and have exposed that cloud has the potential to handle most computing technologies which are being transformative, one such challenging concepts which have been increasing its wait is cloud forensics. In this paper we have traced out the concepts which revolves around cloud forensics and here we propose a model which includes a VM SnapShot Server which continuously stores the snapshots of the cloud service provider and certain servers involved in computing moment by moment so that this would be useful for any digital crime related to cloud, as this plays a key role in identifying the correct cause of the mischief task which resulted in the loss or damage of the original data; this is also helpful during the cases where either the cloud service provider or the suspect gives an incorrect information during the investigation carried out in digital crime; this model also has certain advantages over the present existing models. When certain new activities such as uploading a malware in the cloud, downloading more files then the permissible number, more access from a location, cracking the saved passwords, launching and deleting malicious files, creating corrupted files on the sensitive data stored in the cloud such kind of things can also be traced out easily.

S. Sharmila
Dept. of CSE, Acharya Nagarjuna University, Guntur, AP, India
e-mail: sharmilamca2011@gmail.com

Ch. Aparna (✉)
Dept. of CSE, RVR & JC Engineering College, Guntur, AP, India
e-mail: chaparala_aparna@yahoo.com

© Springer Nature Singapore Pte Ltd. 2019
J. C. Bansal et al. (eds.), *Soft Computing for Problem Solving*,
Advances in Intelligent Systems and Computing 816,
https://doi.org/10.1007/978-981-13-1592-3_38

Keywords Cloud computing (CC) · Cloud forensic (CF) · Cloud service provider (CSP) · Digital forensic readiness (DFR) · Forensic virtual machine (FVM) Forensic toolkit (FTK) · Virtual machine introspection (VMI) · Main memory (MM) · Virtual machine monitor (VMM)

1 Introduction

Cloud computing is standing as one among many solutions for the upcoming technologies which has laid the path related to which the services and information are created, managed, accessed, and delivered. CC has grown to a mark more than $121 billion in the year 2015 [1]. The existing methods for digital forensic cannot be accepted in order to perform CF investigations due to lack of forensic readiness. There exist various challenges related to cloud forensic which are being discussed in [2]. Having knowledge of various deployment models in cloud services is very important in order to perform CF [3]. Though it may be the case of digital or traditional forensics process the following methods are carried out which compromises identification, collection, analysis/examination, and presentation/reporting [4]. Digital investigators are facing challenges in various dimensions of the cloud which may fall either in areas related to legal, technical, and also organizational cloud forensic dimensions [5]. Defensive techniques should be developed which are dynamic in nature due to which recovery action can be taken before any serious damage occurs. Forensic virtual machines described by [6] give basic idea about them. Malicious components can be reused as described by [7, 8] for writing better malware. Many malwares set more than one registry key on windows which are different from the regular values [9, 10]. Using virtual machine introspection, a FVM acts as small virtual machines (VMs) using which other VMs are monitored. Each FVM is dedicated to identifying only one malicious symptom. When one particular FVM detects any symptom in a virtual machine it will immediately inform other FVMs in order to detect further symptoms, if more symptoms are being traced out then more will be the possibility of spiteful behavior. This report is then forwarded to a central snapshot server which collects the information, this acts as a communication channel to an autonomous system for a dynamic defense, in order to produce FVMs for Xen, here FVMs are considered as small because para-virtualization for which Mini OS are used [11]. These FVMs can also be used to detect symptoms in Zeus [9], Gauss [12], and Spyeye [13]. Clear copies of forensic images are being collected during a Forensic investigation. The entire investigation estimation is carried out with the help of these digital evidences by maintaining the integrity of the original hardware [14]. The shifting from physical nature to virtual nature is considered as the main differentiation between traditional and cloud computing.

VMI characteristics As VMI is applied on many domains so they are having more characteristics, though there are variations in the properties of VMI depending on different domains, VMI tools are having important properties in all the domains, they exhibit the following properties.

A. *Impact of performance is minimum*: Sharing the resources between the available guests is the main goal of virtualization. Resources of real hardware and hypervisors are less affected by the introspection techniques; existing system observes a little burden when introspection techniques are being implemented on these systems.
B. *Modification to hypervisor is minimum*: A negligible modification of the hypervisor code is carried out when introspection techniques are implemented independently these changes are carried out during a minor revision of code.
C. *Transparency in operations*: All VMI techniques exhibit transparency in nature to the hypervisor.
D. *Guest OS not modified*: Guest OS is supported by the hypervisors in the real world, even a small change in the code of OS is going to create a trouble, as the OS is being supported by the hypervisor so even when there occurs a small change in the code due to introspection it can be considered as negligible or none.
E. *No side effects*: When introspection tools are being implemented they do not cause any side effects to the systems, if so then systems components show unwanted behavior.
F. *Hypervisor independence*: VMI techniques are independent on the hypervisor architectures, and this is applicable to any kind of architecture.
G. *Security of monitoring components*: VMI modules which are located in the hypervisor are secured from the external attacks. Special protections should be provided when a VMI module is present in the guest VM.

2 Literature Survey

The two main factors which are guiding and developing the research around cloud computing are security and safety. Virtualization concept plays a key role in cloud to a far extent. The virtual machines which are being created in the cloud can deliver, manage, and migrate different technologies remotely, even a single VM can sometimes handle an array of physical systems, and this virtual concept of cloud has seized the physical hardware in cloud environment to a large extent. On a virtual machine code can be analyzed and inspected using virtual machine introspection; in the research field of computer security, VMI has attained a significant attention. From the past years, VMI has been applied in the field of intrusion detection, analysis of malwares for the complete monitoring of the cloud platforms. In major VMI application techniques, the VM which is being introspected is different from the VM which is observing the results of introspections. Both deployed and monitoring code exhibits a different nature. while monitoring code which exists on guest VM, it initiates certain limitations; in concern to cloud security with respect to VMI, there evolves certain applications which are pathbreaking. With the successful implementation of VMI, the evidences clearly show results of the intrusion detection in certain systems [15, 16, 17]. Virtualization framework is being applied with different technologies by the virtualization of hardware processors. An I/O virtual environment for computing is developed with the help of virtualization techniques. VM monitor is the special layer of software with which virtual environment is possible. VMMs can be classified with respect to their logical position into two categories.

Type I Hypervisor
Type II Hypervisor

On the available hardware, Type I hypervisor is made to run directly without the need of OS thereby providing high efficiency, such as Microsoft Hyper [18], VM Ware ESX, Xen [19], Type II hypervisor have interface with OS in order to communicate with the hardware QEMU [20], VMware workstation [21], KVM [22]. These things can also be carried out on high-performance computing also [25]. The below table shows the comparison of various VMI techniques (Table 1).

3   Requirements for Achieving DFR in the Cloud Environment

Beside large usage of cloud in business, scientific, and academics as well as in other areas, with respect to open and distributed nature of cloud there arise lots of challenges many of them are security concern. DFR is an approach of formulation regarding moving a step toward finding the potential incidents of security in the cloud. Digital forensic consists of components which are proactive and reactive in nature. Data forensic tools are being used for the collection of digital evidences. Steps required for the forensic readiness planning

- Without disturbing the business process evidences must be gathered legally.
- Evidences are gathered by targeting the crimes.
- Investigation should be carried out without disturbing business.
- The collected evidence should produce a positive impact.

Consider a scenario which gives a clear cut idea about the concept: A Web site owner John who runs his business successfully by running an online e-commerce site by servers kept in cloud, which are being maintained by CSP. Marry want to create a disturbance in the e-commerce site of John, so she takes rent of certain servers in the cloud and with the help of generation of multiple requests being generated by these servers are sent to the commerce site, due to the heavy created traffic the commerce site servers goes down for which John has to pay a lot in the form of facing loss of original requests and data, when the investigation teams enter the scenario they trace out about the heavy created traffic, and issues a subpoena to the particular CSP in order to provide the log information of the servers. Now either Marry may manage CSP to provide false report or even though CSP provides a correct result Marry is

Table 1 Comparison of VMI Techniques

Category	Technique	Location of code			Virtual machine memory transparency	Guest support	Virtual machine memory alteration	Advantages
		Guest virtual machine	Secure virtual machine	Virtual machine memory				
Memory introspection	Using Xen libraries	No	Yes	Guests	No	Guests	Required	Safety of VMI code
Process introspection	Using hooks	Yes	Yes	AU Types	Yes	AU Types	Required	Reverse remote control possible
	Using CFG	Y Yes		AU Types	Yes	AU Types	Required	Novel approach for code malfunction detection
IO Introspection		No	No	AU Types	No	AU Types	Required	Driver and IO access inspection
Other techniques	Code injection	Yes	No	AU Types		AU Types	Required	Secure and less prone to attacks
	Function call injection	Yes	Yes	AU Types	No	AU Types	Required	Novel approach
	Page flag inspection	Yes	No	PV guests	No	PV guests	No	Detects packed and encrypted malwares.
	Process out getting	Yes	Yes	AU Types	Yes	AU Types	Required	A novel approach
System call introspection	Using VI support	No	Yes	AU Types	No	AU Types	Required	Process support makes introspection less complicated
	By hardware rooting	N	Y	AU Types	No	AU Types	Required	Protection from DKSM attacks

successful in escaping by deleting the log information, there comes the importance of cloud forensic, which is then used to bring back the original deleted data. In this case, preserving the integrity of data is very important. The concept of cloud forensic process is carried out as, first the identification of the theft data is found and then it is labeled, collected, and its integrity is preserved, manipulated data is identified and the same is preserved in the form of snapshots, many such snapshots are collected from different CSP's associated with the same cloud where the disturbance of data has occurred, then the collected data is organized in an order to identify how the task of data theft, data modification has been carried out and at last with the help of available and suitable forensic toolkits (FTK) the lost or original data which has been modified is traced out successfully.

4   Classification of VMI Techniques

VMI techniques are mainly classified into three types as shown in the below classifications as (Fig. 1).

A. *Memory Introspection*: It deals with live memory analysis; all important data structures are stored in the main memory during the running of OS. MM consists of process control block, loadable kernel modules, page tables, registry entries, kernel data structures, code segments, and data segments which are related to the pages. Main memory can be accessed by using various VMI techniques.

B. *I/O Introspection*: It deals with utilities of hardware communication and device drivers [23]. For the machines which are running windows which encounters any kind of malware activities it is helpful for the generation of reports, it is also helpful for the detection of direct kernel object manipulation (DKOM) and call hooking along with kernel patching. Information related to exported symbols, layouts, and data structures are being extracted from Windows OS. With the help of the techniques described by [24], reconstruction of information, data structures are extracted from Windows OS.

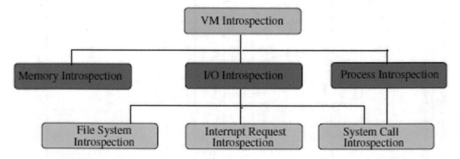

Fig. 1 Classification of virtual machine introspection

C. *Process Introspection*: It is used for analysis of code, useful for the behavioral analysis of malware, this kind of introspection helps in debugging any process at any point of time anywhere during the execution of the entire execution cycle, for any module invocation or code snippet are also detected.

5   Usage of Snapshot to Manage Virtual Machines

A snapshot is a short-term solution with respect to testing software, main usage of these snapshots is to reduce contamination and to acquire correct evidence legally, in order to carry out this, the collection of digital evidences should be taken in an appropriate way, the corrupted digital data may be either volatile or non-volatile, in order to carry out the collection of data from a non-volatile device firstly non-volatile device is disconnected and then it is attached to a forensic analysis machine this can be carried out either by using forensic tools such as dd, FTK Imager, Encase, etc. At

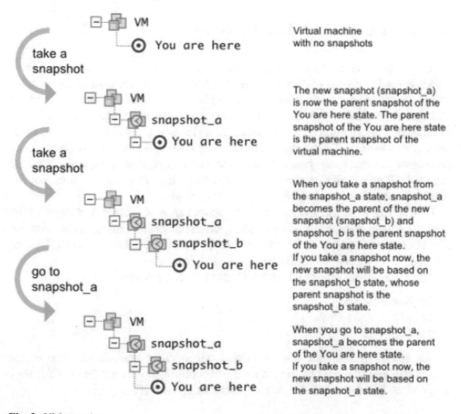

Fig. 2   VM snapshot

the time of taking a snapshot, the data and the state of the virtual machine are stored. When multiple snapshots are taken then restorations positions are considered in a linear way, they are operated on individual machines. When a snapshot is created its corresponding delta disk file is created in the database and required changes are carried out in that delta disk, snapshot contains the below information (Fig. 2).

- Power state: the virtual machine can be suspended, powered ON, and powered OFF.
- Disk states: all the states of the virtual machines in the virtual disks.
- VM Settings: information about various disks which are either added or changed.
- Memory State: represents virtual machines memory.

    Limitations of snapshot: they exhibit the following limitations.

- VM performance is affected.
- They do not exhibit robust nature for backup and recovery of data.
- Suspended VM machine snapshots are not working properly.
- VM Snapshots with bus configuration are not supported.

6   Our Proposed Model Architecture

In the proposed model we have traced out the importance of VM Snapshot Server which keeps the record of data for every second in the cloud, if some unexpected things are carried out by the intruder or if the intruder deletes the data or due to the volatile nature of data if the data is being lost in such extreme cases also depending on the collected information from VM snapshot server we get back unchanged original data. With a proper FTK use, one can easily determine the cause for the loss of data and reasons related to the cause. In our proposed model, there is an advantage that even if CSP or Intruder provides false information then also one can trace out the correct information with help of VM snapshot server (Fig. 3).

    Here to carry out the above-mentioned model an algorithm is necessary which is in the development process. Depending on the algorithm and collected data on checking proper conditions if they are true then with the help of proper choice of FTK one can successfully conclude.

    There are various factors which affect cloud forensic readiness, they are

a. Technical factors
b. Organizational factors
c. Legal factors

    Forensics investigations can be carried out in cloud environment due to the dynamic nature exhibited by the cloud; if a compromise is done with the CSP, then the produced evidence will not be considered as genuine (Fig. 4).

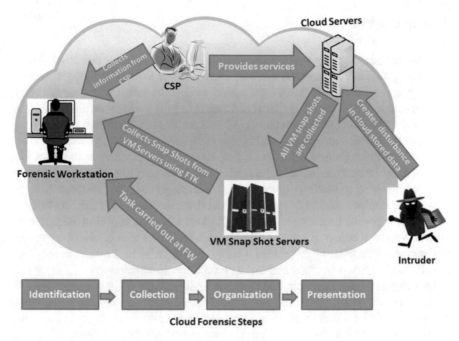

Fig. 3 Our proposed model for cloud forensic

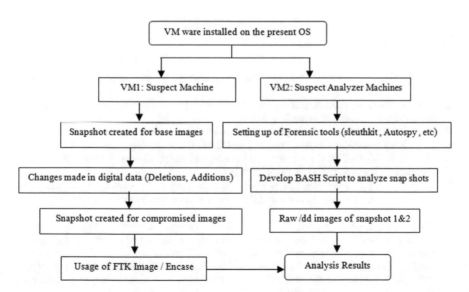

Fig. 4 Proposed model flow of control for cloud forensic investigation

7 Conclusion

Though forensic tools have been introduced into the market in the late 1990s, which are lagging with major architectural changes and also by Digital Forensic Readiness. In our proposed work which we have stated with an initiation of the problem, we are working to take forward our proposed model in future where we are planning to discuss the algorithm related to the model and outcomes of results by overcoming the limitations which are standing as hurdle at the present moment, present tools, and architectural models on which cloud forensics are being carried out does not give expected results because lot has to be upgraded with respect to architectural point of view, with respect to platform point of view and many more, this idea can be taken forward by implementation of further methods by overcoming the existing hurdles. In order to support a forensic ready model in cloud with consideration of high scope with respect to computing power which is available in the area of cloud computing environment with these opportunities one can come up with better ways to minimize the effect of intruders in cloud and make cloud forensic methods to be friendly with cloud computing environments, and these methods not only stand as supportive with respect to cloud architectures and also should not cross the jurisdiction limit, where the result of entire cloud forensic is declared to a right destination. Snapshots play an important role as a source for digital evidence which is being collected from the cloud environment during the evidence collection in cloud crime.

References

1. http://cloud.cioreview.com/news/cloud-market-to-reach-121-billion-by-2015-nid-866-cid-17. html
2. Dykstra, Sherman, A.T.: Acquiring forensic evidence from infrastructure-as-a-service cloud computing: exploring and evaluating tools
3. Ruan, K., Carthy, J.: Cloud computing reference architecture and its forensic implications: a preliminary analysis. In: Digital Forensics and Cyber Crime. Springer, pp. 1–21 (2013)
4. Martini, B., Choo, K.K.R.: An integrated conceptual digital forensic framework for cloud computing. Digital Invest. 9(2), 71–80 (2012)
5. Ruan, K., Carthy, J., Kechadi, T., Crosbie, M.: Cloud forensics. In: Advances in Digital Forensics VII, vol. 361, no. IFIP Advances in Information and Communication Technology pp. 35–46 (2011)
6. Harrison, K., Bordbar, B., Ali, S.T.T., Dalton, C.I., Norman, A.: A framework for detecting malware in cloud by identifying symptoms. In: Enterprise Distributed Object Computing Conference (EDOC), IEEE 16th International, pp. 164–172 (2012)
7. Wu, Z., Gianvecchio, S., Xie, M., Wang, H.: Mimimorphism: a new approach to binary code obfuscation. In: Proceedings of the 17th ACM conference on Computer and communications security. ACM, pp. 536–546 (2010)
8. Baltazar, J., Costoya, J., Flores, R.: The real face of koobface: the largest web 2.0 botnet explained. In: Trend Micro Research, vol. 5, no. 9, p. 10 (2009)
9. Falliere, N., Chien, E.: Zeus: king of the bots. In: Symantec Security Response, Technical Report 2009, available at: www.symantec.com/content/en/us/enterprise/media/securityrespon se/whitepapers/zeuskingofbots.pdf

10. Porras, P., Saidi, H., Yegneswaran, V.: Conficker c analysis. SRI International, Technical Report 2009, Accessed 13 Mar 2013
11. Citrix Systems, Xen Hypervisor, xen.org/products/xenhyp.html
12. Kaspersky: Gauss: abnormal distribution. Kaspersky Lab Global Research and Analysis Team, Technical Report, 2012, available at: www.securelist.com/en/downloads/vlpdfs/kaspersky-la b-gauss.pdf
13. Symantec,"Infostealer.Banker.C," www.symantec.com/securityresponse/writeup.jsp?docid = 2010-020216-0135-99
14. Nelson, B.: Guide to computer forensics and investigations, 4, illustrated 3ed. Cengage Learning **2010**, 42–48 (2010)
15. Arbone, M. et al.: Research in attacks, intrusions, and defenses. Lecture Notes in Computer Science. Springer, Berlin Heidelberg, vol. 7462, pp. 22–41
16. Butt, S. et al., (2012) Proceedings of the ACM Conference on Computer and Communications Security. Raleigh, North Carolina. ACM, Raleigh, New York, NY, pp. 253–264
17. Harrison, C. et al.: IEEE 11th International Conference on. IEEE, Liverpool, pp. 163–169 (2012)
18. Microsoft: Microsoft hyper -v homepage. http://www.microsoft.com/en-us/server-cloud/hype r-v-server/default.aspx (2012)
19. Xen: Xen homepage. http://www.xen.org/. Accessed date 15 March 2013 (2012)
20. Qemu: Qemu homepage. http://wiki.qemu.org/Main_Page (2012). Accessed date 15 Mar 2013
21. Ware, V.M.: Vmware workstation overview. http://www.vmware.com/products/workstation/o verview.html (2012). Accessed date 15 March 2013
22. VM: Linux kvm homepage. http://www.linux-kvm.org/page/Main_Page (2012). Accessed date 15 Mar 2013
23. Neugschwandtner, M., et al.: Detection of Intrusions and Malware, and Vulnerability Assessment, volume 6201 of Lecture Notes in Computer Science. Springer, Berlin Heidelberg pp. 41–60. ISBN 978-3-642-14214-7 (2010)
24. Jiang, X., et al.: Proceedings of the 14th ACM conference on Computer and communications security, CCS'07. ACM, New York, NY, USA. pp. 128–138. ISBN 978-1-59593-703-2. http:// doi.acm (2007)
25. Mohiddin, S.K. et al.: Proceedings of 2016 3rd International Conference on Computing for Sustainable Global Development (INDIACom), A practical approach to overcome glitches in achieving high performance computing

Maximization of Social Welfare by Enhancement of Demand-Side Bidding in a Deregulated Power Market

Subhojit Dawn and Sadhan Gope

Abstract This paper presents a productive, coherent, and efficient approach to maximize the social welfare and minimize the system losses of an electrical system by incorporating power pool model in a fully deregulated power environment. Generation-side bidding and demand-side bidding both are considered in this work with the help of three evolutionary algorithms like particle swarm optimization (PSO) algorithm, artificial bee colony (ABC) algorithm, and BAT algorithm (BA) to check the potential and effectiveness of the presented approach. Investigation of the presented work clearly reveals that the increment in the demand-side bidding reduces the system losses and improves the voltage profile. Modified IEEE 14 bus and modified IEEE 30 bus test systems are considered for analyzing and validating the presented approach.

Keywords Deregulated power market · Social welfare · Particle swarm optimization (PSO) · Artificial bee colony (ABC) algorithm · BAT algorithm (BA) · Demand-side bidding

1 Introduction

Deregulation in energy market introduces competition among energy supplier and energy buyers. In deregulated market, the price of the energy generation is not fixed by any governing body, but after the completion of the bidding process independent system operator (ISO) sets the prices with reviewing both generation companies (GENCOs) and distribution companies (DISCOs) bids. In this market, customer can

S. Dawn (✉)
Electrical Engineering Department, Siliguri Institute of Technology, Siliguri,
West Bengal, India
e-mail: subhojit.dawn@gmail.com

S. Gope
Electrical Engineering Department, Mizoram University, Aizawl, Mizoram, India
e-mail: sadhan.nit@gmail.com

© Springer Nature Singapore Pte Ltd. 2019
J. C. Bansal et al. (eds.), *Soft Computing for Problem Solving*,
Advances in Intelligent Systems and Computing 816,
https://doi.org/10.1007/978-981-13-1592-3_39

choose their power suppliers. But the marketing of this service is regulated by ISO. The electricity market is mainly divided into three parts corresponding to the power contract basis: pool market, bilateral contract, and multilateral contract. In the power market, both single-side bidding (generation-side bidding) and double-side bidding (both generation- and demand-side biddings) are presented for selling and buying the power.

Reference [1] presents an approach to optimize the system reactive power by using ABC algorithm. In that paper, authors compared the result from ABC with some other evolutionary algorithm like genetic algorithm (GA), differential evolution (DE), PSO, interior point method (IPM), and quantum-inspired evolutionary algorithm (QEA). In [2], authors have proposed a methodology for maximum power point tracking (MPPT) of PV solar systems using ABC algorithm under the partially shaded conditions. Attia [3] presents an approach for optimal placement of static shunt capacitors by ABC algorithms in a radial distribution network. Reference [4] proposed a method for optimal placement and sizing of DSTATCOM and distributed generator using PSO algorithm in a radial distributed electrical system. Paper [5] proposed a method to determine the optimal size of a battery energy storage system, by minimizing the total energy storage cost of a micro-grid system using PSO algorithm. Singh et al. [6] present an approach to solve the optimal reactive power dispatch problem of an electrical system using the PSO algorithm. Paper [7] presents an approach to minimize the real power loss by using BAT algorithm in the presence of UPFC. Reference [8] proposed a technique for optimal design of a power system stabilizer with the help of BAT algorithm. In [9], authors represent the optimal MPPT control design of a PV solar system to maximize the output power by using the BAT algorithm.

From the literature, it is reveals that use of evolutionary algorithm in the power system operation is not new. But, in best of author's knowledge no one has focused on the effect of evolutionary algorithm in deregulated power market. This paper depicts an efficient optimization approach for maximizing the social welfare and minimizing the system loss by maximizing the amount of demand-side bidding of an electrical system. The optimal power flow problem has been solved by three evolutionary processes: PSO, ABC, and BAT algorithms. In this work, we give our focus mainly on the double-side bidding only. The programming part of this work has been done in MATLAB.

2 Mathematical Formulation

A. *Pool Market*

In this model, both GENCOs and DISCOs participate in the power auction. GENCOs have submitted their price bid for the maximum capacity of power which they want to deliver to the power pool, and in the other hand DISCOs have also submitted their price bid for the maximum capacity of power which they want to take from the pool. The ISO and power exchange (PX) operate this power pool to implement the price-based dispatch and deliver a forum for large option, and they settled the market price at locational marginal price (LMP) or market clearing price (MCP) in both the single auction and double auction markets.

B. *Social Welfare*

Social welfare is defined as 'the difference between the benefits of consumers and total generation cost.' For an electrical system, if total generation cost is $C_i(PG_i)$ and consumer benefit is $B_j(PD_j)$, then the social welfare is stated as [10]

$$
SW = \sum_{j=1}^{N_D} B_j(P_{Dj}) - \sum_{i=1}^{N_G} C_i(P_{Gi}) \tag{1}
$$

C. *Particle Swarm Optimization*

PSO is originally attributed to Dr. Kennedy and Dr. Eberhart and was first intended for simulating the social behavior, as a representation of the movement of organisms of a fish schooling or bird flocking [5]. PSO is a population-based meta-heuristic optimization techniques, as it makes a few or not a single assumptions about the problem being optimized and can search very large spaces of candidate solution.

Pseudocode of PSO algorithm:

1. Begin.
2. Initialize particles and calculate the fitness values for each particle.
3. If the current fitness value is better than 'best fitness value (pBest)', then set current fitness value as new pBest.
   If not, then keep previous pBest.
4. Choose the best particle's pBest value and assign this to global best (gBest).
5. Calculate the velocity for each particle and update particle position using each particles velocity values.
6. End.

D. *Artificial Bee Colony*

Artificial bee colony algorithm is defined by Dervis Karaboga in 2005, motivated by the intelligent behavior of honey bees. It uses only common control parameters like maximum cycle number and colony size. ABC is an optimization tool which provides a population-based search approach in which individuals called foods positions are modified by the artificial bees with time and the main aim of the bees is to discover the places of food sources with high nectar amount and finally the one with the highest nectar [2].

Pseudocode of ABC algorithm:

1. Begin.
2. Initial Population ().
3. While remain iteration do.
4. Select sites for the local search.
5. Employ bees for the particular chosen sites and to evaluate fitness.
6. Select the bee with the best fitness and assign the remaining bees to looking for randomly.
7. Examine the fitness of remaining bees and update optimum ().
8. End while.
9. Return Best Solution.
10. End.

E. *BAT Algorithm*

Like PSO, firefly, harmony search algorithms, BAT algorithm is also a meta-heuristic algorithm. BAT algorithm is first developed by Xin-She Yang in 2010. This algorithm operates or works based on the echolocation behavior of microbats with varying pulse rates of loudness and emission [8].

Pseudocode of BAT algorithm:

1. Begin.
2. Take objective function $f(x)$, where $x = (x_1,..., x_d)^T$.
3. Set the initial bat population x_i ($i = 1, 2... n$) and velocity v_i.
4. Define pulse frequency f_i at position xi and initialize pulse rates r_i and the loudness A_i.
5. while ($t <$ maximum number of iterations)
   Generate new solutions by adjusting the frequency and updating locations and velocities
   if (r and $> r_i$)
   Choose a solution among the all best solutions and generate a local solution around the chosen best solution
   end if
   Generate a new solution by flying randomly
   if (rand $< A_i$ & $f(x_i) < f(x*)$)
   Accept the new solutions and increase the value of r_i & reduce the value of A_i
   end if
   Rank the bats and find the current best $x*$
   end while
6. Post-process results and visualization.

3 Problem Formulation

A. *Objective Function*

Let an electrical system be N_G number of generators, N_b number of buses, N_{TL} number of transmission lines, and N_D number of total loads. The main objective function of this work is to maximize the social welfare. Here, we considered a function 'F' which is to be minimized. This function can also be called as 'negative social welfare.'

$$\text{Minimize } F = \sum_{i=1}^{N_G} C_i(P_{Gi}) - \sum_{j=1}^{N_D} B_j(P_{Dj}) \tag{2}$$

From Eqs. (1) and (2), we can see that if the function F is minimized, then social welfare of the system (Eq. 1) is automatically maximized, which is beneficial to the consumers. And the second objective of this work is to minimize the system losses [shown in Eq. (4)], which is very much advantageous for the economic operation of an electrical system.

B. *Constraints*

There are mainly two types of constraints used for solving the optimal power flow problem—equality and inequality constraints.

Equality Constraints

$$\sum_{i=1}^{N_G} P_{Gi} - P_{\text{loss}} - P_D = 0 \tag{3}$$

$$P_{\text{loss}} = \sum_{J=1}^{N_{TL}} G_J \left[|V_i|^2 + |V_j|^2 - 2|V_i||V_j|\text{Cos}(\delta_i - \delta_j) \right] \tag{4}$$

$$P_i - \sum_{k=1}^{N_b} [|V_i V_k Y_{ik}|\text{Cos}(\theta_{ik} - \delta_i + \delta_k)] = 0 \tag{5}$$

$$Q_i + \sum_{k=1}^{N_b} [|V_i V_k Y_{ik}|\text{Sin}(\theta_{ik} - \delta_i + \delta_k)] = 0 \tag{6}$$

Inequality Constraints

$$V_i^{\min} \leq V_i \leq V_i^{\max} \quad i = 1, 2, 3 \ldots N_b \tag{7}$$

$$\phi_i^{\min} \leq \phi_i \leq \phi_i^{\max} \quad i = 1, 2, 3 \ldots N_b \tag{8}$$

$$TL_l \leq TL_l^{\max} \quad l = 1, 2, 3 \dots N_{\text{TL}} \tag{9}$$

$$P_{Gi}^{\min} \leq P_{Gi} \leq P_{Gi}^{\max} \quad i = 1, 2, 3 \dots N_b \tag{10}$$

$$Q_{Gi}^{\min} \leq Q_{Gi} \leq Q_{Gi}^{\max} \quad i = 1, 2, 3 \dots N_b \tag{11}$$

Here, P_{Gi}, Q_{Gi} are real and reactive power generation at ith generation unit. P_i, Q_i are real and reactive power injected into the system at bus number 'i'. P_{Gi}^{min}, $P_{Gi}^{\max}$ are lower and upper limit of real power of bus 'i'. Q_{Gi}^{min}, $Q_{Gi}^{\max}$ are lower and upper limit of reactive power of bus 'i'. Y_{ik}, θ_{ik} are magnitude and angle of element of ith row and kth column of bus admittance matrix. P_{loss}, P_D are transmission loss and power demand. G_J is conductance of line, connected between buses 'i' and 'j'$|V_i|,|V_j|$, V_k are voltage magnitude of bus 'i', bus 'j', and bus 'k'. δ_i, δ_j, δ_k are voltage angles of bus 'i', bus 'j', and bus 'k', respectively. $V_i^{\min}$, $V_i^{\max}$ are lower and upper voltage limits of bus 'i'. $\phi_i^{\min}$, $\phi_i^{\max}$ are lower and upper limits of phase angle of voltage at bus 'i'. TL$_l$, $TL_l^{\max}$ are actual and maximum line flow limits of line 'l'. L is the total load after demand-side bidding.

C. Presented Approach

This paper presents an efficient optimization approach to maximize the social welfare and minimize the loss by using the evolutionary algorithm by maximizing the amount of demand-side bidding in an electrical system. The steps associated with the presented approach are as follows

[1] Take all system data including bus data, line data, generation data, and cost coefficient data of modified IEEE 14 bus and modified IEEE 30 bus test systems.
[2] Solve optimal power flow problem using evolutionary algorithms (PSO, ABC, and BAT) by rescheduling of generators, and study the values of total generation cost and system losses.
[3] Calculate the social welfare after completing the double-side bidding (at first demand-side bidding done for only single load bus).
[4] The amount of demand-side bidding is increased and calculates social welfare for every case. (Starting from single demand-side bus bidding to double demand-side bus bidding to triple demand bus bidding has been done).
[5] Steps 3 and 4 has been done for the entire evolutionary algorithm (PSO, ABC and BAT), and compare the results.

4 Results and Discussions

In this work, all optimization techniques have been applied to the modified IEEE 14 bus and modified IEEE 30 bus test systems for checking the effectiveness of the proposed approach. In modified IEEE 14 bus system, 5 generators, 14 buses, 20 transmission lines, and 10 loads are present. The generators are placed at the bus no. 1, 2, 3, 6, and 8. The system data is acquired from [11]. Bus 1 and 100 MVA have been taken as reference throughout this paper. On the other hand, system data of modified IEEE 30 bus test system has been taken from [12]. This system consists of 30 buses, 6 generators, 19 loads, and 41 transmission lines. This system has also been used in this work to examine the potential of the proposed approach. Generators are placed at bus no. 1, 2, 5, 8, 11, and 13 in the system. There are three cases considered in this work

Case I. OPF problem is solved using PSO, ABC, and BAT algorithms in the presence of single demand-side bus bidding.

Case II. Run OPF problem using PSO, ABC, and BAT algorithms with double demand-side bus bidding.

Case III. Solve OPF problem with triple demand-side bus bidding using PSO, ABC, and BAT algorithms.

In Case I, OPF problem has been solved using PSO, ABC, and BAT algorithms by rescheduling of all generators in fully deregulated power environment. At first, demand-side bidding has been done in bus no. 9 for modified IEEE 14 bus system and bus no. 21 for modified IEEE 30 bus system. We can see from Tables 1 and 2 that objective function value is minimum in the BAT algorithm and maximum in the PSO algorithm for both systems. After applying optimization techniques, we also get minimum active power loss in BAT algorithm and maximum loss in PSO algorithm.

For Case II, PSO, ABC, and BAT algorithms have been used for solving the optimal power flow problem by considering the double demand-side bus bidding in deregulated power environment. Bus no. 5 and 9 and bus no. 15 and 21 have been chosen for demand-side bidding for modified IEEE 14 bus and modified IEEE 30 bus systems, respectively.

In this case, we see that the nature of the result is not changed but the value of objective function and system losses is minimized more for every optimization algorithms corresponding to the previous case study. Because the value of objective function is decreased, so that the social welfare is automatically increased for every case.

In Case III, OPF problem has been solved using PSO, ABC, and BAT algorithms like previous two cases. The amount of the demand-side bidding has also been increased in this case from previous two cases. Bus no. 5, 9, and 11 of modified IEEE 14 bus system and bus no. 15, 17, and 21 of modified IEEE 30 bus system have been considered for demand-side bidding.

If we compare all three cases, then we can conclude that the maximum demand-side bidding gives the maximum social welfare and minimum system loss for all algorithms, and also BAT algorithm gives the most beneficial result among the all three algorithms for both test systems. Figures 1a and 1b shows the comparison of social welfare for IEEE 14 bus and IEEE 30 bus systems. BAT algorithm gives the maximum social welfare, and PSO algorithm gives the minimum social welfare for both the cases. The voltage profile of the system has also been improved after incrementing the amount of demand-side bidding. Figures 2 and 3 show the comparative study of voltage profile of IEEE 14 bus and IEEE 30 bus systems, respectively.

Table 1 Result for modified IEEE 14 bus system

Parameters	PSO algorithm			ABC algorithm			BAT algorithm		
	Single demand-side bus bidding (bus no. 9)	Double demand-side bus bidding (bus no. 5 and 9)	Triple demand bus bidding (bus no. 5, 9 & 11)	Single demand-side bus bidding (bus no. 9)	Double demand-side bus bidding (bus no. 5 and 9)	Triple demand-side bus bidding (bus no. 5, 9, and 11)	Single demand-side bus bidding (bus no. 9)	Double demand-side bus bidding (bus no. 5 and 9)	Triple demand-side bus bidding (bus no. 5, 9, and 11)
PG_1 (MW)	53.3216	52.4502	52.3786	53.1401	52.0569	51.4259	52.4767	49.9435	51.3365
PG_2 (MW)	69.1693	68.4929	68.1359	69.6041	67.6929	67.4284	71.6521	71.536	67.2133
PG_3 (MW)	29.4647	25.9015	24.5426	27.2598	26.5702	26.6048	25.2494	26.7971	26.3088
PG_6 (MW)	56.7557	53.3593	51.9367	56.8557	53.9093	51.4834	58.1799	52.6828	52.528
PG_8 (MW)	23.761724	24.5965	24.3709	25.5244	24.4369	24.2906	24.6421	23.5031	23.621
L (MW)	−0.01	−0.0334	−0.03	−0.01	−0.02	−0.0302	−0.01	−0.02	−0.03
Obj. func. ($/hr.)	748.706	715.719	701.495	747.6311	715.3008	701.1501	746.7633	714.8415	700.123
Loss (MW)	2.9634	2.8668	2.9258	2.8741	2.7462	2.8029	2.6902	2.5425	2.5776

Table 2 Result for modified IEEE 30 bus system

Parameters	PSO algorithm			ABC algorithm			BAT algorithm		
	Single demand-side bus bidding (bus no. 21)	Double demand-side bus bidding (bus no. 15 and 21)	Triple demand-side bus bidding (bus no. 15,17, and 21)	Single demand-side bus bidding (bus no. 21)	Double demand-side bus bidding (bus no. 15 and 21)	Triple demand-side bus bidding (bus no. 15,17, and 21)	Single demand-side bus bidding (bus no. 21)	Double demand-side bus bidding (bus no. 15 and 21)	Triple demand-side bus bidding (bus no. 15, 17, and 21)
PG_1 (MW)	167.9954	162.1275	157.2269	168.7601	163.8875	159.3974	173.0196	170.5784	159.6833
PG_2 (MW)	48.3256	44.9199	45.6932	45.8586	45.5199	44.4819	47.7325	47.3073	44.7976
PG_5 (MW)	23.5869	22.6471	20.405	21.0773	20.1471	20.154	21.8	15.9361	19.605
PG_8 (MW)	12.2654	14.387	10.0156	16.3809	15.084	10	9.98	10.02	10.02
PG_{11} (MW)	10.1069	10.15	10.0369	10.2093	10	10	9.8256	10.034	10.014
PG_{13} (MW)	12.0565	12	13.069	12	12	12.2742	11.91	12	12.06
L (MW)	−0.01	−0.2584	−0.0438	−0.01	−0.6992	−0.03	−0.01	−0.02	−0.03
Obj. func. ($/hr.)	739.3901	712.08358	679.48737	738.4954	711.3092	678.4478	737.6119	709.6104	677.8343
Losses (MW)	8.4236	8.2662	7.6987	8.3862	8.2393	7.5275	8.3505	8.1615	7.4659

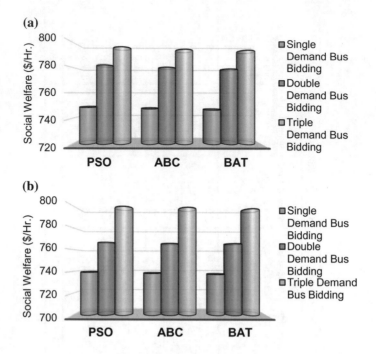

Fig. 1 Comparative study of social welfare: **a** IEEE 14 bus system; **b** IEEE 30 bus system

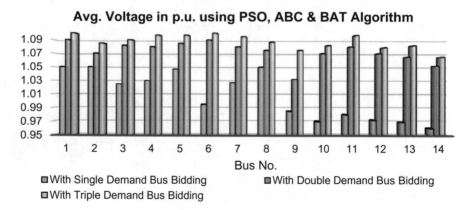

Fig. 2 Comparative study of voltage profile for IEEE 14 bus system

5 Conclusion

This paper presents an optimization strategy for maximizing the social welfare of an electrical system in a fully deregulated power market. PSO, ABC, and BAT algorithms have been incorporated to optimize the objective function and system loss

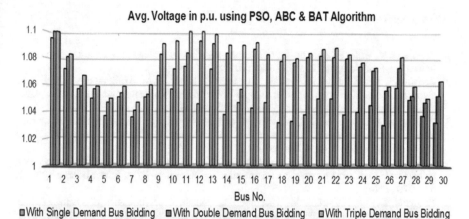

Fig. 3 Comparative study of voltage profile for IEEE 30 bus system

by maximizing the amount of demand-side bidding. The test results obtained using BAT algorithm were compared with PSO and ABC algorithms, and performance of BAT algorithm was found superior compared to other two implemented algorithms in the proposed study. It is also noted from the investigation that the voltage profile and social welfare are increased by adopting the proposed bidding strategy in a deregulated electricity market. This proposed approach is a generalized approach which can be applied to any small and large integrated deregulated power system.

Appendix

Table 3 Parameters used in PSO, ABC, and BAT algorithms

PSO algorithm			ABC algorithm			BAT algorithm		
Sl.	Parameter	Value	Sl.	Parameter	Value	Sl.	Parameter	Value
1	Population size	50	1	No. of scout bees	100	1	Population size	30
2	Acceleration constant	2	2	No. of bee colony size	20	2	Loudness	0.3
3	Initial inertia weight	1.2	3	No. of food source	10	3	Pulse rate	0.5
4	Final inertia weight	0.4	4	No. of iterations	300	4	No. of iteration	300
5	No. of iteration	300						

References

1. Ayan, Kursat, Kilic, Ulas: Artificial bee colony algorithm solution for optimal reactive power flow. Appl. Soft Comput. **12**, 1477–1482 (2012)
2. Benyoucef, A.S., Chouder, A. Kara, K., Silvestre, S., Sahed, O.A.: Artificial bee colony based algorithm for maximum power point tracking (MPPT) for PV systems operating under partial shaded conditions. Appl. Soft Computing, **32**, 38–48 (2015)
3. El-Fergany Attia, A.: Involvement of cost savings and voltage stability indices in optimal capacitor allocation in radial distribution networks using artificial bee colony algorithm. Electr. Power and Energy Syst. **62**, 608–616 (2014)
4. Devi, S., Geethanjali, M.: Optimal location and sizing determination of distributed generation and DSTATCOM using particle swarm optimization algorithm. Elect. Power Energy Syst. **62**, 562–570 (2014)
5. Kerdphol, T., Fuji, K., Mitani, Y., Watanabe, M., Qudaih, Y.: Optimization of a battery energy storage system using particle swarm optimization for stand-alone micro-grids. Electr. Power and Energy Syst. **81**, 32–39 (2016)
6. Singh, R.P., Mukherjee, V., Ghoshal, S.P.: Optimal reactive power dispatch by particle swarm optimization with an aging leader and challengers. Appl. Soft Comput. **29**, 298–309 (2015)
7. Venkateswara Rao, B., Nagesh Kumar, G.V.: Optimal power flow by BAT search algorithm for generation reallocation with unified power flow controller. Electr. Power Energy Syst. **68**, 81–88 (2015)
8. Ali, E.S.: Optimization of power system stabilizers using BAT search algorithm. Electr. Power Energy Syst. **61**, 683–690 (2014)
9. Oshaba, A.S., Ali, E.S., Abd Elazim, S.M.: MPPT control design of PV system supplied SRM using BAT search algorithm. Sustainable Energy, Grids Networks **2**, 51–60 (2015)
10. Dawn, S., Tiwari, P.K.: Improvement of economic profit by optimal allocation of TCSC & UPFC with wind power generators in double auction competitive power Market. Electr. Power Energy Syst. **80**, 190–201 (2016)
11. Dawn, S., Tiwari, P.K.: Improvement of social welfare and profit by optimal allocation of TCSC with wind power generator in double auction competitive power market. In: 2014 Fourth International Conference on Advances in Computing and Communications (ICACC), pp. 362–365, https://doi.org/10.1109/icacc.2014.95
12. Zimerman, R.D., Murillo-Sanchez, C.E., Gam, D.: Matpower-A MATLAB Power System Simulation Package, Version 3, available at://www.pserc.cornell.edu/matpower

An Orthogonal Symbiotic Organisms Search Algorithm to Determine Approximate Solution of Systems of Ordinary Differential Equations

Arnapurna Panda and Sabyasachi Pani

Abstract Determining exact solution of systems of ordinary differential equations (ODEs) is a challenging task in many real-life problems of science and engineering. In this paper, an attempt is made to determine approximate solutions for such complicated ODEs. The Fourier series expansion is used as an approximator. The coefficients of Fourier series expansion are determined by nature-inspired algorithms. The Symbiotic Organism Search (SOS) is an evolutionary algorithm proposed by Cheng and Prayogo in 2014. It is inspired by natural phenomenon of organisms interaction in an ecosystem for their survival. Recently, Panda and Pani in 2017 reported an Orthogonal SOS (OSOS) algorithm by incorporating orthogonal array strategies in SOS, which enhances the exploration capability of original algorithm. Here, the OSOS algorithm is used to compute the coefficients of Fourier series. Simulation studies on two real-life examples using systems of ODEs reported superior performance of the proposed OSOS learning over the same model trained by three recently reported nature-inspired algorithms OCBO, OPSO, and WCA in terms of close response matching and minimal generalized distance achieved.

Keywords Symbiotic organisms search · Orthogonal array · Systems of differential equation

1 Introduction

A number of coupled ODEs together form a system of ordinary differential equations. They find extensive applications in science and engineering problems, e.g., behavior of voltage and current in any electrical/electronic circuit [1]; the behavior of irregular

A. Panda (✉) · S. Pani
School of Basic Sciences, Indian Institute of Technology Bhubaneswar,
Bhubaneswar 752050, Odisha, India
e-mail: arnapurna.math@gmail.com

S. Pani
e-mail: spani@iitbbs.ac.in

© Springer Nature Singapore Pte Ltd. 2019
J. C. Bansal et al. (eds.), *Soft Computing for Problem Solving*,
Advances in Intelligent Systems and Computing 816,
https://doi.org/10.1007/978-981-13-1592-3_40

heartbeat and corresponding drug behavior (lidocaine) with respect to time [2]; the behavior of variation of temperature in attic, main floor, and basement of a room [2]. Determining exact solution of system of ODEs at times is challenging as the complexity involves with the process increasing with (i) increase in number of DOEs and (ii) increase in order of ODEs. Under such scenarios, approximate solution of the problem is also helpful.

The Fourier series expansion has been a benchmark approximator to all kind of functions. Recently in [3], a formulation is given to find out the approximate solution of regular ODEs using Fourier series. Here, the author determined the Fourier coefficient with particle swarm optimization (PSO). Later on, the coefficients of similar formulation are computed by cuckoo search and water cycle algorithm (WCA) [1].

Several literature [4, 5] reported use of orthogonal array (OA) improves the exploration capabilities of nature-inspired algorithms. Recently, Panda and Pani reported an Orthogonal Colliding Bodies Optimization (OCBO) algorithm by hybridizing OA with CBO [6]. They applied the OCBO for identification of parameters of Hammerstein plant [7] and to determine approximate solutions of regular nonlinear ODEs [8]. In 2017, Panda and Pani also proposed another algorithm Orthogonal Symbiotic Organism Search (OSOS) [9] which combines the good features of OA with SOS [10]. Authors have used OSOS along with Augmented Lagrange Multiplier to solve various engineering constrained optimization problems [9].

In this manuscript, the Fourier series is used as an approximator to determine solutions of system of ODEs. The Fourier coefficients are computed by the recently reported hybrid orthogonal algorithm OSOS and OCBO. Simulation studies are reported for two problems: response of voltage and current with time in an electrical circuit [1] and nature of irregular heartbeat and corresponding drug behavior (lidocaine) with respect to time [2]. The results obtained by the proposed algorithms are compared with the same Fourier approximator trained by Orthogonal PSO [4, 5] and water cycle algorithm [1].

The paper is outlined as follows. Section 2 describes the proposed formulation of determining approximate solution for system of ODEs with OSOS algorithm. Section 3 discusses two real-life problems based on system of ODEs, simulation environments, and comparative algorithms used for analysis. The obtained results are highlighted in Sect. 4. Concluding remarks are given in Sect. 5.

2 Proposed Approach to Determine Approximate Solution for Systems of ODEs

2.1 Fourier Series as an Approximator

Consider a system of ODEs comprises of two linear/nonlinear ODEs

$$\begin{cases} F_1(t, x, z, x', z', \ldots, x^{(n)}, z^{(n)}) = 0 \\ F_2(t, x, z, x', z', \ldots, x^{(n)}, z^{(n)}) = 0 \end{cases} \tag{1}$$

The characteristics of Eq. (1) are dependent on linearity/nonlinearity behavior of F_1 and F_2. Eq. (1) is subject to the below boundary conditions:

$$\begin{cases} x(t_0) = x_0 \\ z(t_0) = z_0 \end{cases}, \begin{cases} x'(t_0) = x_0' \\ z'(t_0) = z_0' \end{cases}, \ldots, \begin{cases} x(t_n) = x_n \\ z(t_n) = z_n \end{cases}, \begin{cases} x'(t_n) = x_n' \\ z'(t_n) = z_n' \end{cases} \tag{2}$$

or the following initial conditions:

$$\begin{cases} x(t_0) = x_0 \\ z(t_0) = z_0 \end{cases}, \begin{cases} x'(t_0) = x_0' \\ z'(t_0) = z_0' \end{cases}, \ldots, \begin{cases} x^{(n-1)}(t_0) = x_0^{(n-1)} \\ z^{(n-1)}(t_0) = z_0^{(n-1)} \end{cases} \tag{3}$$

Here, x and z are functions of variable t. The approximation for exact solutions $x(t)$ and $y(t)$ is given by Fourier series as follows:

$$\begin{cases} x(t) \approx X_{apx}(t) = a_0 + \sum_{m=1}^{T} \left[a_m \cos\left(\frac{m\pi(t-t_0)}{N}\right) + b_m \sin\left(\frac{m\pi(t-t_0)}{N}\right) \right] \\ z(t) \approx Z_{apx}(t) = c_0 + \sum_{m=1}^{T} \left[c_m \cos\left(\frac{m\pi(t-t_0)}{N}\right) + d_m \sin\left(\frac{m\pi(t-t_0)}{N}\right) \right] \end{cases} \tag{4}$$

The N is length of the interval solution. The t_0 and t_n are the start point and end point of the interval, respectively.

The aim is to determine the Fourier coefficients in Eq. (4), i.e., $(a_0, a_m, b_m, c_0, c_m,$ and $d_m)$ with optimization algorithm such that it has minimal error compared to the exact solution.

2.2 Orthogonal SOS to Compute the Fourier Series Coefficients

The recently proposed Orthogonal SOS (OSOS) algorithm by Panda and Pani [9] is used here to compute the Fourier series coefficients as defined in Eq. (4). The stepwise procedure is outlined below:

Step 1: Initialization of Ecosystem: Initialize an ecosystem having 'n' number of organisms, and each organism has dimension 'D' (the number of parameters to optimize is determined by number of Fourier coefficients in Eq. (4), i.e., $(a_0, a_m, b_m, c_0, c_m,$ and $d_m)$. The ecosystem is represented by

$$\vec{E} = \begin{bmatrix} E_1 \\ E_2 \\ \vdots \\ E_n \end{bmatrix} = \begin{bmatrix} e_{1,1} & e_{1,2} & .. & e_{1,D} \\ e_{2,1} & e_{2,2} & .. & e_{2,D} \\ \vdots & \vdots & \vdots\vdots & \vdots \\ e_{n,1} & e_{n,2} & .. & e_{n,D} \end{bmatrix} \tag{5}$$

Set a predefined maximum number of iteration for optimal solution.

Step 2: Fitness Evaluation: Each row of $\overrightarrow{E}$ represents a possible solution. For n number of organisms in the ecosystem, for each organism compute the fitness value by using the error function between the true and approximate solutions for set of ODEs given by

$$\begin{cases} \epsilon_1 = x(t) - X_{apx}(t) \\ \epsilon_2 = z(t) - Z_{apx}(t) \end{cases} \tag{6}$$

$$\epsilon_{total} = |\epsilon_1| + |\epsilon_2| \tag{7}$$

The objective is to minimize ϵ_{total}. Determine the best fitness value, i.e., the row of $\overrightarrow{E}$ for which ϵ_{total} is minimum, termed it as E_{best}.

Step 3: Mutualism Phase: In this phase, two organisms in an ecosystem participate and both get benefited, e.g., interaction between a cow and grass. Cow graze the grass to fill its stomach as well as grass, also get the benefit to grow again with time. Mathematically, two organisms E_i and E_j with $\{E_i, E_j\} \in \overrightarrow{E}$ essentially $i \neq j$ perform the mutualism phase.

$$E_{i-new} = E_i + rand(0, 1) * (E_{best} - MV * BF_1) \tag{8}$$

$$E_{j-new} = E_j + rand(0, 1) * (E_{best} - MV * BF_2) \tag{9}$$

where

$$MV = \frac{E_i + E_j}{2} \tag{10}$$

The MV is termed as 'mutual vector' which corresponds to the average characteristics of organism E_i and E_j. The $rand(0, 1)$ signify a random number in $[0, 1]$. The BF_1 and BF_2 are benefit factors randomly taken as either 1 or 2 representing partial or full advantage received by an organism during the process of mutualism. After mutualism, the output is given by

$$\overrightarrow{ME} = \begin{bmatrix} E_{i-new} \\ E_{j-new} \end{bmatrix}_{2n \times D} ; \quad \forall i, j \in [1, n] \tag{11}$$

Step 4: Commensalism Phase: In this phase, out of the two participating organisms one gets benefited. The other one is not affected by the process, e.g., interaction between shark and remora fish. The remora fish follows behind the huge shark and takes the leftover food by shark, thus getting the benefit. The shark is not being affected by this interaction.

Mathematically two different organisms E_i and E_j belong to ecosystem involve in Commensalism phase is given by

$$E_{i-new} = E_i + rand(-1, 1) * (E_{best} - E_j) \tag{12}$$

The $rand(-1, 1)$ is a random number in range $[-1, 1]$. Here, organism E_i is benefited by E_j. The difference $(E_{best} - E_j)$ is the benefit received by organism E_i. The output after commensalism is given by

$$\overrightarrow{CE} = \left[E_{i-new}\right]_{n \times D}; \quad \forall i \in [1, n] \tag{13}$$

Step 5: Parasitism Phase: In this phase of interaction between two organisms, one survives and the other one gets eliminated from the ecosystem, e.g., interaction between bacteria and an animal. In this case, if the bacteria proliferate, the animal becomes sick and dies. If the immune system of the animal is strong, then bacteria get eliminated.

Considering the formulation suppose two different organisms E_i and E_j that belong to ecosystem undergo parasitism phase as follows:

1. Let E_i be considered as a bacteria which proliferates to create a 'parasite vector (PV).' This is carried out by cloning parent organism E_i to multiple times and bringing random change at selected dimension values.
2. Let E_j be the animal.
3. The fitness value of 'PV' and E_j is evaluated and compared. If fitness of PV is better than that of $(E_j$, then E_j is eliminated. The 'PV' overtakes the position of E_j.
4. Otherwise, if fitness of E_j is better, then the 'PV' gets eliminated.

After Parasitism, the output is given by

$$\overrightarrow{PE} = [A_i]_{n \times D}; \quad \forall i \in [1, n] \tag{14}$$

where

$$A_i = \begin{cases} E_j & \text{if Fitness}(E_j) \prec \text{Fitness}(PV) \\ PV & \text{Otherwise} \end{cases}; \quad \forall j \in [1, n] \text{ and } j \neq i \tag{15}$$

Step 6: Selection of Best Organisms: The fitness of original chromosomes before and after interaction is compared. Among $[\overrightarrow{E}, \overrightarrow{ME}, \overrightarrow{CE}, \overrightarrow{PE}]_{5n \times D}$, only $\overrightarrow{SE}_{n \times D}$ best-fit organisms are selected.

Step 7: Creation of Orthogonal Positions: Create orthogonal array positions for the selected best-fit organisms $\overrightarrow{SE}_{n \times D}$ using the procedure defined by Panda and Pani in Sect. 2.4 of [9]. The output is

$$\overrightarrow{OE}_{Q \times D} \text{ where } Q = (2 \times n)^{\lceil \log_D(D+1) \rceil} \tag{16}$$

Step 8: Upgradation of Organisms Position to Orthogonal Position: The fitness of all orthogonal positions $\overrightarrow{OE}$ is computed, and among the $(Q \times D)$ only best-fit $(n \times D)$ are replaced as parent organisms for next-generation $\overrightarrow{E}$.

Step 9: Stopping Criteria: The steps 1–8 are repeated for fixed number of iteration set in step 1 till a desired low fitness value is achieved.

3 Systems Simulation

3.1 Simulation Environment and Comparative Algorithms

In order to evaluate the performance of proposed approach, simulation studies are carried out on two practical problems in MATLAB2011 platform in Dell Latitude E5430 laptop with processor Intel 3540M, 3GHz, RAM size 8GB, Windows 10 (64-bit) environment. The proposed OSOS-based learning to determine Fourier coefficients is compared with the coefficients determined by Orthogonal CBO (OCBO) [8], Orthogonal PSO (OPSO) [4, 5], and water cycle algorithm (WCA) [1]. In all the four algorithms, the initial population size taken for both examples is 200. The maximum number of iteration each algorithm is allowed to run is 200. The rest parameter settings are kept as it is as reported in the corresponding journal papers. All the four algorithms are set for twenty independent runs, and best, average, and worst values obtained are reported.

3.2 Practical Problems Based on Systems of ODEs

– **Example 1: Response of voltage and current with time in an electrical circuit**

In electrical circuits, the voltage (V) and current (I) decay with time (t) [1]. The simultaneous behavior of voltage and current is represented by a system of ODEs given below:

$$\frac{dV}{dt} = 2I - V \tag{17}$$

$$\frac{dI}{dt} = -I - V$$

where $I(0) = 2$, $V(0) = 2$ are the initial conditions. The exact solutions for the above system of ODEs are given by

$$V(t) = 2\sqrt{2}e^{-t}\sin(\sqrt{2}t) + 2e^{-t}\cos(\sqrt{2}t)$$
$$I(t) = 2e^{-t}\cos(\sqrt{2}t) - \sqrt{2}e^{-t}\sin(\sqrt{2}t) \tag{18}$$

Let us consider the time interval for solution is between 0 and 1.5 s. The approximate solutions for the voltage and current are approximated as follows

$$V_{apx}(t) = a_0 + a_1 \cos\left(\frac{\pi t}{1.5}\right) + b_1 \sin\left(\frac{\pi t}{1.5}\right) + a_2 \cos\left(\frac{2\pi t}{1.5}\right)$$
$$+ b_2 \sin\left(\frac{2\pi t}{1.5}\right) + a_3 \cos\left(\frac{3\pi t}{1.5}\right) + b_3 \sin\left(\frac{3\pi t}{1.5}\right) \tag{19}$$

$$I_{apx}(t) = a_0 + a_1 \cos\left(\frac{\pi t}{1.5}\right) + b_1 \sin\left(\frac{\pi t}{1.5}\right) + a_2 \cos\left(\frac{2\pi t}{1.5}\right)$$
$$+ b_2 \sin\left(\frac{2\pi t}{1.5}\right) + a_3 \cos\left(\frac{3\pi t}{1.5}\right) + b_3 \sin\left(\frac{3\pi t}{1.5}\right) \tag{20}$$

- **Example 2: Nature of irregular heartbeat and corresponding drug behavior (lidocaine) over time**

Let $x(t)$ and $z(t)$ represent the amount of lidocaine present in the bloodstream and body tissue, respectively. Their response represented by system of ODEs [2] is given by

$$\frac{dx}{dt} = -0.09x + 0.038z$$

$$\frac{dz}{dt} = 0.066x - 0.038z \tag{21}$$

At the initial situation, there is no drug in the blood system given by $x(0) = 0$ and the injection dosage of the drug $y(0) = y_0$. The exact solution of these two ODEs is

$$x(t) = -0.3367y_0 e^{-0.1204t} + 0.3367y_0 e^{-0.0076t} \tag{22}$$
$$z(t) = 0.2696y_0 e^{-0.1204t} + 0.7304y_0 e^{-0.0076t}$$

The approximate solutions for the $x(t)$ and $z(t)$ are as follows

$$x_{apx}(t) = a_0 + a_1 \cos\left(\frac{\pi t}{2}\right) + b_1 \sin\left(\frac{\pi t}{2}\right) + a_2 \cos\left(\frac{2\pi t}{2}\right)$$
$$+ b_2 \sin\left(\frac{2\pi t}{2}\right) + a_3 \cos\left(\frac{3\pi t}{2}\right) + b_3 \sin\left(\frac{3\pi t}{2}\right)$$
$$+ a_4 \cos\left(\frac{4\pi t}{2}\right) + b_4 \sin\left(\frac{4\pi t}{2}\right) + a_5 \cos\left(\frac{5\pi t}{2}\right)$$
$$+ b_5 \sin\left(\frac{5\pi t}{2}\right) \tag{23}$$

$$z_{apx}(t) = a_0 + a_1 \cos\left(\frac{\pi t}{2}\right) + b_1 \sin\left(\frac{\pi t}{2}\right) + a_2 \cos\left(\frac{2\pi t}{2}\right)$$
$$+ b_2 \sin\left(\frac{2\pi t}{2}\right) + a_3 \cos\left(\frac{3\pi t}{2}\right) + b_3 \sin\left(\frac{3\pi t}{2}\right)$$
$$+ a_4 \cos\left(\frac{4\pi t}{2}\right) + b_4 \sin\left(\frac{4\pi t}{2}\right) + a_5 \cos\left(\frac{5\pi t}{2}\right)$$
$$+ b_5 \sin\left(\frac{5\pi t}{2}\right) \tag{24}$$

3.3 Performance Evaluation

The following four criteria are used to evaluate the performance of algorithms:

- **Response Matching**: The response matching indicates closeness between exact analytical solution and obtained approximate solution [8].
- **Generalized Distance**: The Euclidean distance between points lying on the true and approximate solutions is termed as generalized distance [1]:

$$GD = \sqrt{\sum_{s=1}^{S} (e_s - E_s)^2} \tag{25}$$

where e_s are points of exact solution, E_s are points obtained by approximate solution, and S is total number of points in both the curves. For best results, the value of GD should be minimum.
- **Estimated Fourier Coefficients**: The values of coefficients obtained by the proposed and comparative algorithms reflect an effective value of the approximation to the ODE.
- **Computational Time**: The average runtime over fixed number of fitness evaluation reflects the computational complexity associated with the algorithm.

4 Results and Discussion

The approximated values of Fourier series coefficients obtained with proposed OSOS algorithm and that obtained by OCBO, OPSO, and WCA for both the examples are presented in Table 1. It is observed that the obtained coefficient values for each example by different algorithms are close to each other. This signifies the closeness in system modeling.

The response matching of the true exact analytical response and achieved approximate response by the algorithms for voltage and current variation in Ex.1 is shown in Fig. 1a and b, respectively. Similarly, the response matching curves for drug values in bloodstream $x(t)$ and in body tissue $z(t)$ are shown in Fig. 2a and b, respectively. In all the four figures, it is observed that the approximate solution by the proposed OSOS training (red one) is close to the exact solution (black one). The response of comparative learning models by OCBO, OPSO, and WCA also does not deviate much.

The best, average, worst, and standard deviation values of generalized distance achieved over 20 runs for both the examples are reported in Table 2. In both the examples, the proposed OSOS model gives lower GD values compared to other three algorithms which are highlighted in bold letters.

Similarly for computational complexity, the best, average, worst, and standard deviation values of runtime taken over 20 runs for both the examples are reported in Table 3. It is observed that the WCA is computationally efficient represented by bold letters. It is noted that by addition of OA to the algorithms enhance its accuracy by compromising with the computational burden. Among the orthogonal algorithms, OCBO has lower computation followed by OPSO and OSOS.

Table 1 Comparative results of obtained Fourier coefficients after optimization with proposed OSOS and comparative OCBO, OPSO, and WCA

Ex.	Coefficients	OSOS	OCBO	OPSO	WCA
Ex.1	Voltage $V(t)$				
	a_0	0.6619	0.6620	0.6623	0.6618
	a_1	0.8186	0.8185	0.8187	0.8179
	b_1	1.3268	1.3270	1.3277	1.3265
	a_2	0.4821	0.4820	0.4822	0.4821
	b_2	8.5914E−02	8.5915E−02	8.5921E−02	8.5909E−02
	a_3	3.7431E−02	3.7430E−02	3.7433E−02	3.7436E−02
	b_3	−0.1134	−0.1135	−0.1133	−0.1137
	Current $I(t)$				
	a_0	0.8384	0.8385	0.8392	0.8381
	a_1	1.3626	1.3625	1.3633	1.3621
	b_1	−1.0560	−1.0562	−1.0567	−1.0557
	a_2	−9.8533E−02	−9.8535E−02	−9.8542E−02	−9.8533E−02
	b_2	−0.4600	−0.4600	−0.4611	−0.4597
	a_3	−0.1024	−0.1025	−0.1027	−0.1021
	b_3	1.1919E−02	1.1920E−02	1.1931E−02	1.1915E−02
Ex.2	Drug presence in bloodstream $x(t)$				
	a_0	0.0295	0.0293	0.0294	0.0294
	a_1	−0.0247	−0.0247	−0.0246	−0.0247
	b_1	0.0099	0.0101	0.0100	0.0101
	a_2	0.0066	0.0065	0.0066	0.0065
	b_2	−0.0056	−0.0056	−0.0057	−0.0057
	a_3	−0.0109	−0.0109	−0.0108	−0.0108
	b_3	−0.0046	−0.0043	−0.0046	−0.0043
	a_4	−0.0025	−0.0022	−0.0024	−0.0023
	b_4	0.0069	0.0069	0.0067	0.0067
	a_5	0.0021	0.0021	0.0020	0.0020
	b_5	0.0008	0.0007	0.0008	0.0008
	Drug presence in body tissue $z(t)$				
	a_0	0.7962	0.8011	0.7956	0.7688
	a_1	0.0343	0.0379	0.0354	0.1605
	b_1	0.2901	0.2845	0.2870	0.3472
	a_2	0.1821	0.1856	0.1847	0.2407
	b_2	−0.0099	−0.0146	−0.0168	−0.1874
	a_3	−0.0064	−0.0094	−0.0086	−0.1486
	b_3	−0.0752	−0.0843	−0.0784	−0.1245
	a_4	−0.0147	−0.0225	−0.0183	−0.0442
	b_4	0.0064	0.0072	0.0075	0.0772
	a_5	0.0042	0.0039	0.0051	0.0222
	b_5	−0.0017	0.0015	0.0023	0.0085

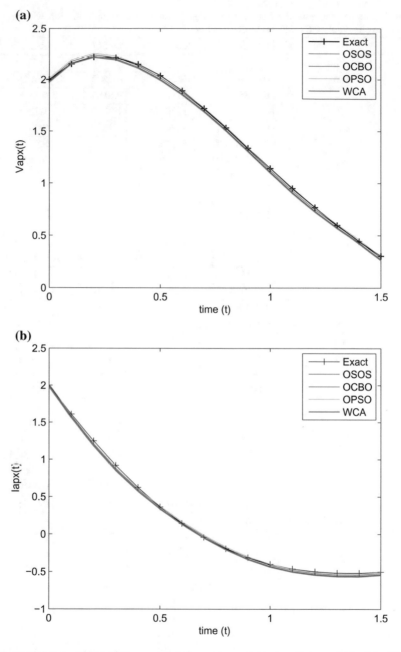

Fig. 1 Response matching of the exact solution and approximate solutions obtained by OSOS, OCBO, OPSO, and WCA for Ex-1: **a** approximation voltage and **b** approximation current

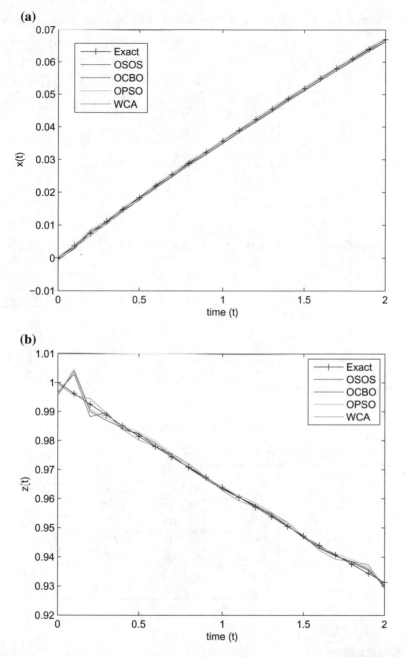

Fig. 2 Response matching of the exact solution and approximate solutions obtained by OSOS, OCBO, OPSO, and WCA for Ex-2: **a** approximation of lidocaine in bloodstream $x(t)$ and **b** approximation of lidocaine in body tissue $z(t)$

Table 2 Comparative results of generalized distance obtained for 20 independent runs with OSOS, OCBO, OPSO, and WCA

Ex.	GD value	OSOS	OCBO	OPSO	WCA
Ex.1	Voltage $V(t)$				
	Best	**3.90E−04**	4.08E−04	4.18E−04	4.25E−04
	Average	**7.42E−03**	7.62E−03	8.12E−03	9.52E−03
	Worst	**1.55E−02**	1.59E−02	1.98E−02	2.17E−02
	SD	**2.05E−03**	2.38E−03	3.12E−03	5.34E−03
	Current $I(t)$				
	Best	**2.53E−03**	2.61E−03	3.24E−03	3.75E−03
	Average	**8.82E−03**	9.25E−03	1.14E−02	1.23E−02
	Worst	**1.26E−02**	1.42E−02	1.94E−02	2.73E−02
	SD	**3.30E−03**	3.45E−03	4.12E−03	6.48E−03
Ex.2	Drug presence in bloodstream $x(t)$				
	Best	**6.19E−04**	6.22E−04	7.36E−04	8.17E−04
	Average	**7.36E−04**	7.42E−04	8.55E−04	9.47E−04
	Worst	**2.58E−03**	2.70E−03	3.15E−03	3.59E−03
	SD	**1.06E−04**	1.19E−04	1.72E−04	2.18E−04
	Drug presence in body tissue $z(t)$				
	Best	**7.68E−01**	8.07E−01	9.13E−01	9.84E−01
	Average	**9.32E−01**	9.86E−01	1.2303	1.7205
	Worst	**2.43E−01**	2.40E−01	2.6804	2.8506
	SD	**1.57E−01**	1.74E−01	1.94E−01	2.10E−01

Table 3 Comparative results of runtime obtained for 20 independent runs with OSOS, OPSO, WCA, and HS

Ex.	Comp. time	OSOS	OCBO	OPSO	WCA
Ex.1	Best	20.4320	15.7344	18.5017	**11.3008**
	Average	20.8125	15.7912	18.5869	**11.3120**
	Worst	20.9312	15.8413	18.6210	**11.3224**
	SD	0.1023	0.0714	0.0902	0.0821
Ex.2	Best	19.8201	14.9115	17.7526	**10.9025**
	Average	19.9432	14.9743	17.7920	**10.9504**
	Worst	20.1290	15.0520	17.8621	**10.9805**
	SD	0.0832	0.0591	0.0635	**0.0412**

5 Conclusion

In this paper, a new approximation method is proposed to solve system of ODEs. Fourier series expansion is used as the approximation function whose coefficients are optimized by meta-heuristic algorithm OSOS. Simulation studies are demonstrated on two real-life problems: variation of voltage and current in an electrical circuit and response of a drug behavior over time in bloodstream and body tissue. Result analysis conveys the OSOS-based training has superior accuracy over that achieved by OCBO, OPSO, and WCA. It is observed that addition of OA to evolutionary algorithms enhances their accuracy but simultaneously increases computational burden. Thus, computational complexity of WCA is low followed by OCBO, OPSO, and OSOS. Thus, the proposed OSOS-based training is applicable where the system requires greater accuracy with compromise over runtime of couple of seconds.

References

1. Sadollah, A., Eskandar, H., Yoo, D.G., Kim, J.H.: Approximate solving of nonlinear ordinary differential equations using least square weight function and metaheuristic algorithms. Eng. Appl. Artif. Intell. **40**, 117–132 (2015)
2. Gustafson, G.B.: Chapter 11: Systems of Differential Equations. Available : http://www.math.utah.edu/~gustafso/2250systems-de.pdf, Mathematics Department University of Utah, Salt Lake City. (2017)
3. Babaei, M.: A general approach to approximate solutions of nonlinear differential equations using particle swarm optimization. Appl. Soft Comput. **13**, 3354–3365 (2013)
4. Ho, S.Y., Lin, H.S., Liauh, W.H., Ho, S.J.: OPSO: Orthogonal Particle Swarm Optimization and Its Application to Task Assignment Problems. IEEE Transactions on Systems, Man Cybernetics, part-A: Systems and Humans, vol. 38, no. 2, pp. 288–298.(2008)
5. Zhan, Z.H., Zhang, J., Li, Y., Shi, Y.H.: Orthogonal learning particle swarm optimization. IEEE Trans. on Evolutionary Comput. **15**(6), 832–847 (2011)
6. Kaveh, A., Mahdavi, V.: Colliding bodies optimization: a novel meta-heuristic method. Comput. Struct. **139**, 1827 (2014)
7. Panda, A., Pani, S.: A WNN Model trained with Orthogonal Colliding Bodies Optimization for Accurate Identification of Hammerstein Plant. IEEE World Congress on Evolutionary Computation, Vancouver Convention Center, Canada, pp. 1100–1106 (2016)
8. Panda A., Pani S.: Determining approximate solutions of nonlinear ordinary differential equations using Orthogonal Colliding Bodies Optimization, Accepted Manuscript Neural Processing Letters, Springer (2017). https://doi.org/10.1007/s11063-017-9711-6
9. Panda A., Pani S.: An Orthogonal Parallel Symbiotic Organism search algorithm embodied with Augmented Lagrange Multiplier for solving constrained optimization problems. Soft Comput. 1–19 (2017) https://doi.org/10.1007/s00500-017-2693-5
10. Cheng M.Y., Prayogo, D.: Symbiotic organisms search: a new metaheuristic optimization algorithm. Comput. Struct. **139**, 98112 (2014)

Salp Swarm Algorithm (SSA) for Training Feed-Forward Neural Networks

Divya Bairathi and Dinesh Gopalani

Abstract   Artificial neural networks (ANNs) have shown efficient results in statistics and computer science applications. Feed-forward neural network (FNN) is the most popular and simplest neural network architecture, capable of solving nonlinearity. In this paper, feed-forward neural networks' weight and bias figuring using a newly proposed metaheuristic Salp Swarm Algorithm (SSA) are proposed. SSA is a swarm-based metaheuristic inspired by the navigating and foraging behaviour of salp swarm. The performance is evaluated for some of the benchmarked datasets and compared with some well-known metaheuristics.

Keywords   Optimization · Metaheuristics · Salp swarm algorithm
Feed-forward neural networks—training · Classification · Regression

1   Introduction

An artificial neural network (ANN) is a mathematical model based on computational intelligence [1]. The model imitates biological neural networks, which consist of net of interconnected biological neurons. In a similar way, an artificial neural network is an interconnected group of artificial neurons and capable of processing information using intelligence-based approaches of computation. ANNs are successfully used to perform classification, pattern recognition and many other machine learning jobs, having complex relationships between inputs and outputs.

Learning is an important process, which makes ANN work efficiently. Several ANN learning algorithms have been proposed and developed over the last few decades. These algorithms can be bifurcated into two major categories: deterministic approaches and stochastic approaches. The most popular deterministic approach is

D. Bairathi (✉) · D. Gopalani
MNIT Jaipur, Jaipur, India
e-mail: divyabairathijain@yahoo.co.in

D. Gopalani
e-mail: dgopalani.cse@mnit.ac.in

© Springer Nature Singapore Pte Ltd. 2019
J. C. Bansal et al. (eds.), *Soft Computing for Problem Solving*,
Advances in Intelligent Systems and Computing 816,
https://doi.org/10.1007/978-981-13-1592-3_41

521

backpropagation (BP) [2], which uses gradient-based methods. Although deterministic approaches like BP are fast, they have weakness of getting trapped in non-desirable solutions, i.e. non-optimum or local optimum solutions. Apart from this, most of these algorithms cannot explore non-differentiable and non-continuous multi-modal surfaces. Due to single solution deterministic approach, solution provided by these algorithms is highly dependent on initial solution. Therefore, stochastic optimization techniques, popularly called metaheuristics [3], are required for training an ANN. Some well-known metaheuristics which are applied for neural network training are genetic algorithm (GA) [4, 5], evolution strategy (ES) [6], particle swarm optimization (PSO) [7, 8], differential evolution (DE) [9, 10], ant colony optimization (ACO) [11, 12], artificial bee colony (ABC) [13], Grey wolf optimizer (GWO) [14], etc.

These stochastic techniques use randomness, which make the approaches less dependent on initial solution and less probable to local minima entrapment. Stochastic techniques have shown good acceptance for neural networks' training because of powerful optimization capacity, ability to explore non-differentiable and non-continuous multi-modal surfaces and local optimum avoidance. The performance of metaheuristics can be further improved by employing multiple solutions in parallel. Swarm intelligence-based optimization techniques are an example of such kind of metaheuristics. These metaheuristics are based on mathematical modal of social behaviour of animals, birds and insects.

Feed-forward neural network (FNN) is most applied and popular ANN structure. It contains layered architecture of neurons with each neuron linked to all neurons in previous and next layer. In this paper, a swarm intelligence-based metaheuristic Salp Swarm Algorithm (SSA) [15] is applied for feed-forward neural network (FNN) training. The performance is evaluated for some well-known datasets.

2 Salp Swarm Algorithm

Salp is a tunicate, having barrel-shaped transparent body. Salp life cycle consists of two phases: solitary phase and aggregate phase. During aggregate phase, tens to hundreds of individual salps form a chain by attaching with each other. They feed, swim and grow in these chains. Salp Swarm Algorithm mimics this chain formation behaviour of salps. Mathematical model of these salp chains given by Mirjalili et al. [15] consists of two groups of salps: leader and followers. Leader salp is the salp at front of the chain and lead the swarm towards the food location. The rest of salps are considered as followers, which follow each other and leader during the aggregate phase. The positions of leader and followers are given by Eqs. 1 and 2, respectively.

$$x_j^1 = \begin{cases} f_j + c_1 \left(\left(ub_j - lb_j \right) c_2 + lb_j \right) & c_3 \geq 0 \\ f_j - c_1 \left(\left(ub_j - lb_j \right) c_2 + lb_j \right) & c_3 < 0 \end{cases} \tag{1}$$

$$x_j^i = \left(x_j^i + x_j^{i-1} \right) / 2 \tag{2}$$

Here, x_j^1 shows the spatial dimension (positioning) of the first salp (leader) (in the jth dimension) and x_j^i shows position of ith salp (follower) (in the jth dimension). Position of the food source in the jth dimension is represented by f_j. ub_j and lb_j indicate the upper bound and lower bound on positions in jth dimension, respectively. c_2 and c_3 are uniformly generated random numbers in range [0,]. Value of c_1 is calculated as follows.

$$c_1 = 2e^{-(4t/T)^2} \tag{3}$$

where t and T depict current iteration and maximum number of iterations respectively. The steps of SSA are as follows:

1. Initialize a randomly generated salp population constrained by the upper and lower bounds
2. Calculate the objective function value for each salp
3. Select the best salp as f
4. Update parameter c_1
5. for each salp
   if(i==1)

   Modify the positioning of the leading salp using Eq. 1

   Else

   Modify the positioning of the follower salp using Eq. 2

6. If the stopping criterion is not satisfied, go to step 2
7. Return the positioning of f as the best estimated optimum solution.

Mirjalili et al. [15] depicted that the SSA algorithm provides good exploration and exploitation [16–18], which are key factors to avoid local optima and find better solutions, respectively. In exploration, whole search space is looked over to have the idea about the areas where solution can be found. In exploitation, these areas are traversed thoroughly in order to find optimum solution. Effective and balanced exploration–exploitation properties of SSA optimization algorithm are favourable to apply this algorithm for neural network training.

3   SSA-Based FNN Trainer

Feed-forward neural network (FNN) also known as multilayer perceptron (MLP) is the simplest neural network architecture, capable of solving nonlinear problems, where data is not linearly separable. Feed-forward neural network is multi-layered architecture of ANNs, which is bounded to have only forward directing connections. No backward connection is allowed for FNNs; hence, the information travels unidirectionally in forward direction, from input layer to output layer through the intermediary layers called hidden layers as shown in Fig. 1. Output of jth neuron in

Fig. 1 Feed-forward neural
network topology

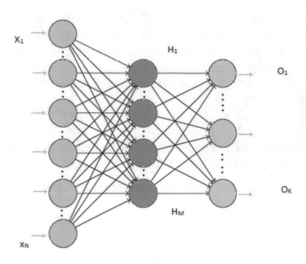

Input Layer   Hidden Layer   Output Layer

hidden layer, output of kth neuron in output layer and mean square error (MSE) are
given by following equations.

$$O_j = \Theta_j[\sum_{i=1}^{N} w_{ij}x_i + b_j] \tag{4}$$

$$O_k = \Theta_k[\sum_{j=1}^{M} w_{jk}O_j + b_k] \tag{5}$$

$$MSE = \sum_{d=1}^{D} [\sum_{k=1}^{K} [\frac{1}{2}[O_{kd} - Y_{kd}]^2]]/DK \tag{6}$$

Here,

x_i ith input
w_{ij} Weight of edge connecting ith input of input layer to jth neuron of hidden
layer
w_{jk} Weight of edge linking jth neuron of hidden layer to kth neuron of output
layer
b_j Bias of jth neuron in hidden layer
b_k Bias of kth neuron in output layer
Θ_k Transfer function of kth neuron of output layer
Θ_j Transfer function of jth neuron of hidden layer
O_{kd} Output of FNN
Y_{kd} Expected output
N Number of inputs in input layer

M Number of neurons in hidden layer
K Number of neurons in output layer
D Total number of training example

Transfer function also called activation function incorporates nonlinearity at each neuron. It converts the linear output ($\sum(wx) + b$) of neurons into nonlinear output, which is taken as final outputs of the neurons. Due to these nonlinear transfer functions, FNN learns nonlinear relationships. Sigmoid function is mostly used nonlinear transfer function given by Eq. 7.

$$f(S) = \frac{1}{1 + e^{-S}} \qquad (7)$$

For performing efficiently, FNNs are trained first using a training set corresponding to the problem. During the training process, some learning algorithm is employed, which continuously changes the weight and bias values until some stopping criteria are satisfied. Training of FNNs aims to assign weights to each connection between two neurons, such that mean square error (MSE) is minimum. Hence, MSE acts as objective function for this optimization problem. Weight vector w and bias vector b combine to make the solution vectors. In our work, SSA is taken as training algorithm. SSA initializes and evolves the solution vector to find best-approximated optimum (minimum MSE). Training the FNN using SSA can be explained in following order:

1. Initialize predefine number of randomly generated solution vector (consist of weight vector and bias vector) constrained by the upper and lower bounds
2. Compute the objective function value for each solution vector. For this purpose, the solution vectors produced are applied on FNN and MSE is calculated for each solution vector as shown in Fig. 2. These calculated MSEs corresponding to the solutions are returned to SSA as objective function value.
3. Select the best solution vector as f
4. Update parameter c_1

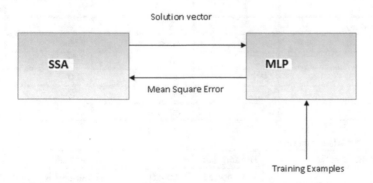

Fig. 2 FNN training using SSA

5. for each solution vector
   if (i==1)

   Modify the positioning of the solution vector (leader salp) using Eq. 1

   Else

   Modify the positioning of the solution vector (follower salp) using Eq. 2

6. If the stopping criterion is not satisfied, go to step 2
7. Return the positioning of f as the best estimated optimum solution vector

4 Results and Experiments

After the learning (by training process), the performance of FNN is examined with the help of test set. Test set is the set of samples corresponding to the problem, different to those used during the training phase. The solution vector obtained by the learning process is applied and evaluated on test samples. For analysis purpose, training and test datasets are selected from Machine Learning Repository [19] of UCI (University of Califomia, Irvine): XOR, balloon, iris, wine, breast cancer and heart are benchmarked. These datasets are used as classification datasets. A part of these, function-approximation datasets: cosine, sigmoid and sine are taken from [20] and used for regression analysis. Description of these datasets is represented in Tables 1 and 2.

FNN structure used for experiment is given in Table 3. FNN topology is represented in (i-j-k) form, where i represents number of inputs in input layer. Value of i taken is equal to number of attributes for corresponding dataset. Here, j and k show

Table 1 Datasets for classification

Classification datasets	Attribute count	Training samples	Test samples	Class count
XOR (3-bits)	3	8	8	2
Balloon	4	16	16	2
Iris	4	150	150	3
Breast cancer	9	599	100	2
Wine	13	100	78	3
Heart	22	80	187	2

Table 2 Datasets for regression

Regression datasets	Training samples	Test samples
Cosine	51 samples in [0:0.04:2]	101 samples in [0:0.02:2]
Sine	126 samples in [−2:0.1:2]	252 samples in [−2:0.05:2]
Sigmoid	61 samples in [3:0.1:3]	121 samples in [3:0.05:3]

Table 3 Structure information of FNNs

Datasets	Number of connections (#w)	Number of neurons (#b)	Size of solution vector (#w + #b)	FNN structure
XOR (3-bits)	28	8	36	3-7-1
Balloon	45	10	55	4-9-1
Iris	63	12	75	4-9-3
Breast cancer	190	20	210	9-19-1
Wine	432	30	462	13-27-3
Heart	1035	46	1081	22-45-1
Cosine	30	16	46	1-15-1
Sine	30	16	46	1-15-1
Sigmoid	30	16	46	1-15-1

number of neurons in hidden layer and output layer, respectively. Value of j taken is proportional to i, whereas value of k depends on class count for the dataset. Size of weight vector (number of connections), bias vector (number of neurons) and solution vector is also represented in the table. Sigmoid function is adopted as activation function.

The SSA algorithm is compared with PSO, DE, GA, ACO and ES. For implementation, population size used is 50 for XOR and Balloon datasets and 200 for the rest. Maximum iteration count is 250. Weights and biases values are kept in range $[-10, 10]$. For normalization of inputs data, following min-max normalization is used in input layer, which assigns values of x from interval [e, f] to [g, h].

$$X = \frac{(x - e)(h - g)}{(f - e)} + g \tag{8}$$

MATLAB R2010b is used to implement the proposed SSA trainer and other trainers. Parameter settings used for PSO, GA, ACO and ES are taken from [14] and for SSA are taken from [15]. Experimental design or other parameter tuning methods can be used for parameter setting of metaheuristic-based FNN trainers. The other parameters for algorithms are as follows:

SSA settings—c_2 and c_3 are calculated by uniform random numbers' generator in range [0,1].

PSO settings—Inertia value (ω), cognitive (local) constant value (C1) and social constant value (C2) for PSO are 0.03, 1 and 1, respectively.

DE settings—Mutation factor is calculated by uniform random numbers generator in range [0.2,0.8] and crossover rate taken is 0.2.

GA settings—Real-coded GA with roulette wheel selection, uniform mutation with probability 0.01, single point crossover with probability 0.8 are employed.

ACO settings—ACO with Visibility sensitivity (β) value = 5, Initial pheromone (τ) = 1e−06, Pheromone constant (q) value=1, Pheromone update constant (Q) value =

20, Local pheromone decay rate $(p_t) = 0.5$, Global pheromone decay rate $(p_g) = 0.9$ and Pheromone sensitivity (α) value = 1 is used for experiment.

ES settings—For ES, values of λ and σ taken are 10 and 1, respectively.

Each algorithm runs 20 times. Best solution (with lowest estimated MSE for training dataset) found by algorithms in each run is recorded. Average (AVG. MSE) and standard deviation (Std. deviation) of MSE corresponding to these solutions are shown as statistical results. In addition, accuracy (classification rate)/test error for classification test dataset/regression test dataset is also compared. Accuracy and test error are calculated as follows.

$$Accuracy = \frac{Number\ of\ correctly\ classified\ test\ samples}{Total\ number\ of\ test\ samples} \tag{9}$$

$$Test\ error = \sum_{d=1}^{D} |O_d - Y_d| \tag{10}$$

Here

O_d Output of FNN
Y_d Expected output
D Total number of test sample

Results for classification datasets and regression dataset are shown in Tables 4, 5, 6, 7, 8, 9 and Tables 10, 11, 12, respectively. Bar graph for classification datasets is also depicted in Fig. 3. Results show that for XOR, iris, heart, sigmoid, cosine and sine datasets, SSA achieves best performance amongst all algorithms. For cancer dataset

Table 4 XOR dataset results

Training algorithm	AVG. MSE	Std. deviation	Accuracy (%)
SSA	0.000120	0.000383	100
PSO	0.083955	0.036038	37.45
DE	0.104932	0.009863	65.50
GA	0.000203	0.000421	100.0
ACO	0.179348	0.026108	62.75
ES	0.120098	0.012119	63.50

Table 5 Balloon dataset results

Training algorithm	AVG. MSE	Std. deviation	Accuracy (%)
SSA	6.39e−05	9.96e−05	100.0
PSO	0.000631	0.000812	100.0
DE	6.72e−08	2.0599e−07	100.0
GA	8.56e−23	6.34e−21	100.0
ACO	0.005384	0.006980	100.0
ES	0.023487	0.169467	100.0

Table 6 Iris dataset results

Training algorithm	AVG. MSE	Std. deviation	Accuracy (%)
SSA	0.054104	0.077826	90.67
PSO	0.230261	0.060362	37.00
DE	0.105690	0.032736	89.33
GA	0.091783	0.139356	88.67
ACO	0.397632	0.052787	33.00
ES	0.309887	0.049874	47.00

Table 7 Breast cancer dataset results

Training algorithm	AVG. MSE	Std. deviation	Accuracy (%)
SSA	0.006319	0.004932	94.5
PSO	0.035232	0.003124	11.50
DE	0.028437	0.004679	19.00
GA	0.003145	0.001761	97.00
ACO	0.017547	0.002852	41.50
ES	0.047452	0.007682	06.00

Table 8 Wine dataset results

Training algorithm	AVG. MSE	Std. deviation	Accuracy (%)
SSA	0.18382	0.099353	64.5
PSO	0.287361	0.254932	51.50
DE	0.143965	0.190458	64.00
GA	0.094616	0.073940	67.00
ACO	0.375390	0.289771	41.50
ES	0.484330	0.223767	45.00

Table 9 Heart dataset results

Training algorithm	AVG. MSE	Std. deviation	Accuracy (%)
SSA	0.115944	0.0073682	73.25
PSO	0.123649	0.007645	38.75
DE	0.140986	0.0134254	32.00
GA	0.100327	0.008936	58.00
ACO	0.285967	0.038739	26.25
ES	0.197692	0.018743	29.25

Table 10 Cosine dataset results

Training algorithm	AVG. MSE	Std. deviation	Test error
SSA	0.010144	0.006786	0.7001
PSO	0.063257	0.020233	2.0113
DE	0.025768	0.014729	0.9020
GA	0.011024	0.006476	0.7176
ACO	0.051590	0.010812	2.4521
ES	0.085466	0.026264	3.1457

Table 11 Sine dataset results

Training algorithm	AVG. MSE	Std. deviation	Test error
SSA	0.331572	0.050013	105.21
PSO	0.495253	0.067816	121.84
DE	0.484638	0.053947	106.98
GA	0.419470	0.057239	109.25
ACO	0.498820	0.063480	113.89
ES	0.776581	0.073548	144.24

Table 12 Sigmoid dataset results

Training algorithm	AVG. MSE	Std. deviation	Test error
SSA	0.000944	0.000842	0.40395
PSO	0.0230012	0.010236	3.40653
DE	0.003998	0.006127	1.62569
GA	0.000981	0.000901	0.43490
ACO	0.024013	0.009786	4.00240
ES	0.081907	0.017942	9.01055

and wine dataset, SSA is providing second best results after GA. It can be observed that test error for sine dataset is very high than of other datasets. Sine dataset is very complex dataset in caparison of cosine and sigmoid dataset. Due to this complexity, sine dataset shows high test error.

For evaluating the overall performance of SSA-based trainer against other training algorithms and for assuring the significance of results, Friedman statistical test [21, 22] is conducted. Trainer method and classification/regression dataset are taken as two impact factors. Average rank (the lower is better) of each algorithm according to the Friedman test is shown in Table 13. Significant differences amongst the six techniques can be observed from Table 13. SSA has attained highest rank amongst the six FNN trainer, which again represents merits of SSA in training FNNs.

FNN training is considered as complex multi-modal problem and the search space for this kind of problems contains multiple local optima. Optimization technique with

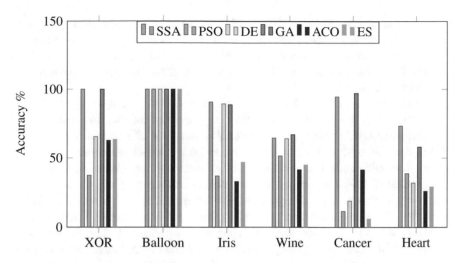

Fig. 3 Results for clssification dataset

Table 13 Average rank of algorithms according to Friedman test

Training algorithm	Average ranking
OSCA	1.5556
PSO	4.3889
GA	2.1111
ACO	4.8333
ES	5.0556
DE	3.0555

efficient and powerful exploration is required to find the promising basins, which may contain the optimum solution(s). Also, a mature balance between exploratory and exploitative components is required; that is, the exploitative component should be powerful enough to examine the promising basins thoroughly. Results depict that SSA-based trainer outperforms other algorithms for classification- and regression-based problems. It proves highly efficient exploration and local optimum avoidance quality of SSA algorithm. This originates from the fact of high exploration during initial iterations and moving food location. High value of c_1 causes exploration of larger search area. Simultaneously moving food location causes searching of new areas. Food location represents the best solution found so far. When a new food location is found, the leader and follower gradually move towards this food location. During this gradual move, generally some other better solution(s) is (are) found. In this case, the best solution found is treated as food location for next iteration and the salp chain move towards this location. In this way, the salp chain explores new areas. Local optimum avoidance is provided by mutual interaction of the artificial salps, which make the salps able to provide informations about search space structure to each other and avoid local optima. SSA is also able to attain balance between exploration and exploitation components. The final iterations in SSA algorithm are exploitation intensive. The low value of c_1 highly promotes exploitation around food location in the final steps. The interaction between the salps tends to converge the swarm towards the global optimum.

Performance of GA- and DE-based trainers is also noticeable. GA and DE are evolutionary-based algorithms and inherit high exploration due to crossover and mutation operators. These operators cause abrupt motion of agents in search space. Hence, these algorithms have high exploratory power, which make them suitable to employ for multi-modal problem like neural network training. On the basis of critical examination of ES-based trainer, reason account of its poor performance is worth discussing. Although ES is an evolutionary algorithm, selection of individuals is less random in comparison with other evolutionary algorithms. Hence, local optimum avoidance is less in this algorithm, which degrades its performance for FNN training. Similarly, PSO also shows poor performance. Reason for this poor performance is stagnation. Stagnation is the state when solutions got entrapped at local optima and no further progress is possible. Due to stagnation, PSO is unable to provide good results.

One more important observation, which should be considered, can be found in result Table 9 for heart dataset. Here, MSE average for GA is minimum but still it is not providing best classification rate. This observation could be explained by the over-fitting phenomenon of machine learning. In over-fitting, the learning algorithm over-train the system according to training data and incorporate the noise and error of training data instead of the underlying relationship. In this case, the modal is not general enough for application purpose, hence gives higher error for test data. Many methods are reported in literature to deal with over-fitting problem in ANNs. Any of the best suited can be opted as solution. One of them is to select multiple solutions with minimum MSE from the solutions generated (not difficult, as these are population-based methods and generate multiple solution) and compute the error for test data. Choose the solution vector with minimum error.

Results represent that SSA possesses powerful exploratory component and is also capable to provide balanced between exploratory and exploitation phases. Hence, SSA can be consider as a useful and efficient optimization technique for feed-forward neural network (FNN) training.

5 Conclusion

In this paper, salp swarm algorithm (SSA) is employed for training feed-forward neural networks (FNNs). The proposed approach is applied to six standard classification datasets and three regression datasets. The results are also compared to other metaheuristic trainers: PSO, DE, GA, ACO and ES. Results represent SSA as an effective trainer for neural network training. Results also show the good and balanced exploration–exploitation properties of SSA optimization algorithm, which make the algorithm favourable to apply for neural network training.

For future work, SSA algorithm can be applied for training of other kind of neural networks like recurrent network, convolutional neural network.

References

1. Hertz J.: Introduction to the theory of neural computation. Basic Books 1 (1991)
2. Rumelhart, D.E., Williams, R.J., Hinton, G.E.: Learning internal representations by error propagation. Parallel Distributed Process.: Explorations Microstruct. Cognition 1, 318–362 (1986)
3. Glover, F.W., Kochenberger, G.A. (eds.) Handbook of metaheuristics, vol. 57. Springer Science & Business Media (2006)
4. Gao, Q., Lei, K.Q.Y., He, Z.: An Improved Genetic Algorithm and Its Application in Artificial Neural Network, Information, Communications and Signal Processing, 2005. In: Fifth International Conference on, December 06–09, pp. 357–360 (2005)
5. Tsai, J.T., Chou, J.H., Liu, T.K.: Tuning the structure and parameters of a neural network by using hybrid taguchi-genetic algorithm. IEEE Trans. Neural Networks 17(1) (2006)
6. Pavlidis, N.G., Tasoulis, D.K., Plagianakos, V.P., Nikiforidis, G., Vrahatis, M.N.: Spiking Neural Network Training Using Evolutionary Algorithms, Neural Networks, 2005. In: IJCNN 05. Proceedings 2005 IEEE International Joint Conference, vol. 4, pp. 2190–2194 (2005)
7. Mendes, R., Cortez, P., Rocha, M., Neves, J.: Particle swarm for feedforward neural network training. In: Proceedings of the International Joint Conference on Neural Networks, vol. 2, pp. 1895–1899 (2002)
8. Meissner, M., Schmuker, M., Schneider, G.: Optimized Particle Swarm Optimization (OPSO) and its application to artificial neural network training. BMC Bioin- Formatics 7, 125 (2006)
9. Fan, H., Lampinen, J.: A trigonometric mutation operation to differential evolution. J. Global Optim. 27, 105–129 (2003)
10. Slowik, A., Bialko, M.: Training of artificial neural networks using differential evolution algorithm. Human System Interactions, pp. 60–65 (2008)
11. Blum, C., Socha, K.: Training feed-forward neural networks with ant colony optimization: an application to pattern classification. In: 5th international conference on, Hybrid Intelligent Systems, 2005. HIS05, p. 6 (2005)
12. Socha, K., Blum, C.: An ant colony optimization algorithm for continuous optimization: application to feed-forward neural network training. Neural Comput. Appl. 16, 235–247 (2007)
13. Ozturk, C., Karaboga, D.: Hybrid Artificial Bee Colony algorithm for neural network training. In: 2011 IEEE Congress on, Evolutionary Computation (CEC), pp. 84–88 (2011)
14. Mirjalili, S.: How effective is the Grey Wolf optimizer in training multi-layer perceptrons. Appl. Intell. 43(1), 150–161 (2015)
15. Mirjalili, S., Gandomi, A.H., Mirjalili, S.Z., Saremi, S., Faris, H., Mirjalili, S.M.: Salp swarm algorithm: a bio-inspired optimizer for engineering design problems. Adv. Eng. Software (2017)
16. Olorunda, O., & Engelbrecht, A.P.: Measuring exploration/exploitation in particle swarms using swarm diversity. In: Evolutionary Computation, 2008. CEC 2008. (IEEE World Congress on Computational Intelligence). IEEE Congress on (pp. 1128–1134). IEEE (2008, June)
17. Alba, E., Dorronsoro, B.: The exploration/exploitation tradeoff in dynamic cellular genetic algorithms. IEEE Trans. Evolutionary Comput. 9(2), 126–142 (2005)
18. Crepinsek, M., Liu, S.H., Mernik, M.: Exploration and exploitation in evolutionary algorithms: a survey. ACM Comput. Surveys (CSUR) 45(3), 35 (2013)
19. Blake, C., Merz, C.J.: UCI Repository of machine learning databases (1998)
20. Mirjalili, S., Mirjalili, S.M., Lewis, A.: Let a biogeography-based optimizer train your multilayer perceptron. Inf. Sci. 269, 188–209 (2014)

21. Demar, J.: Statistical comparisons of classifiers over multiple data sets. J. Machine Learning Res. 7(Jan), 1–30 (2006)
22. Garca, S., Fernndez, A., Luengo, J., Herrera, F.: Advanced nonparametric tests for multiple comparisons in the design of experiments in computational intelligence and data mining: experimental analysis of power. Inf. Sci. **180**(10), 2044–2064 (2010)

Optimizing Integrated Production–Inventory Model for Time-Dependent Deteriorating Items Using Analytical and Genetic Algorithm Approach

Poonam Mishra and Isha Talati

Abstract Proposed model is a sincere effort to compare analytical method and evolutionary algorithm for cost minimization of an integrated production–inventory model. Since, many decades researchers were using analytical methods for optimization of inventory and supply chain models. But from last decade, researchers have started paying attention towards other search algorithms that are inspired by nature, and genetic algorithm is one among them. We have proposed an integrated model comprising of manufacturer and retailer. Proposed inventory system assumes time-dependent deterioration governed by two parameters—Weibull distribution. Aim of this paper is to minimize joint as well as individual inventory cost of the system. Cost has been optimized (minimized) using analytical as well as by GA. Algorithms for both analytical method and GA have been proposed. Observations obtained shows that GA gives global minimum, whereas analytical method stucks with local minimum. We have used MATLAB 13a for running genetic algorithm and MAPLE 18 for analytical solution. Sensitivity of crucial inventory parameters has been studied. Results of this paper give an insight into researchers and managers that are involved in the inventory management of fashion goods, electronic gadgets and other FMCG, etc. …

Keywords Integrated inventory · Weibull distribution · Genetic algorithm
Lot-size dependent ordering cost · Salvage value

1 Introduction

Globalization, limited resources and bottleneck competition have made business a serious affair. Business houses are consistently thriving for survival and growth. Inventory and supply chain management play an important role in overall business growth and revenue generation. This scenario demands proper coordination and

P. Mishra (✉) · I. Talati
Pandit Deendayal Petroleum University, Raisan Gandhinagar 382007, India
e-mail: poonam.mishra@sot.pdpu.ac.in

© Springer Nature Singapore Pte Ltd. 2019   535
J. C. Bansal et al. (eds.), *Soft Computing for Problem Solving*,
Advances in Intelligent Systems and Computing 816,
https://doi.org/10.1007/978-981-13-1592-3_42

communication among supply chain players which can be achieved through the analysis of integrated inventory models. First time, Goyal [9] formulated a model for single supplier and single customers. Further, Banerjee [2] found a joint optimal ordering policy with an appropriate price adjustment so that it can be beneficial for both the purchaser and vendor. Goyal and Gunasekaran [8] generated model for deteriorating items with respect to production and marketing aspects. Chung and Crdenas-Barrn [4] extended the same for stock-dependent demand and two-level trade credit. Further, Chung et al. [3] extended it for items that exponentially deteriorate. Shah et al. [26] optimized price and order quantity for deteriorating items using promotional tools trade credit- and price-dependent demand. Sarkar et al. [23] formulated model for deteriorating items with different set-up cost. Shah [27] formulated model for price- and trade credit-dependent demand and two-level trade credit. Mishra and Talati [13] formulated model with optimizing backorder and order quantity for advertisement and stock-dependent demand. Shah et al. [25] generated model for time-dependent deteriorating items with controllable deterioration using preservation technology.

In classical EOQ model, it is implicitly assumed that items deplete only through demand. But in real life, items also deplete due to deterioration. In earlier literature, Ghare and Schrader [5] formulated model for items that deteriorate exponentially. For the first time, Philip and Covert [18] assumed the deterioration governed by Weibull distribution, and further, Philip [19] generalized that model. Since then, many researchers studied effect of deterioration in optimal inventory policies. Manna and Chaudhuri [11] generated model for shortages with fix demand and production rate. Shah and Mishra [24] formulated model for retailer with salvage value. Bakker et al. [1] gave an up to date review of inventory models with deteriorating items from year 2001 to 2012. Sarkar and Saren [22] used partial trade credit for exponentially deteriorating items. Janssen et al. [10] gave a review on inventory models with deteriorating items from year 2012 to 2015. Mishra and Talati [14] maximized profit with price- and trade credit-dependent demand for items that deteriorate with respect to time. Mohan [15] formulated model for items that deteriorate with respect to time and used quadratic demand and variable holding cost.

Genetic algorithm (GA) is heuristic search algorithm which is inspired by process of evolution and natural selection. It is based on Darwins theory of evolution "survival of the fittest". Many of the times, it is found that the use of traditional gradient method in multi-objective, nonlinear complex optimization problems stuck with local optimum. But un-conventional heuristic methods like genetic algorithm gives global optimum for such optimization problem. Goldberg [6] proposed algorithm of GA. Then, researchers like Murata et al. [16], Goren et al. [7], Radhakrishnan et al. [21], Radhakrishnan et al. [20], Narmadha et al. [17], Woarawichai et al. [28], Mishra and Talati [12] used genetic algorithm for optimization.

This paper considers integrated production–inventory model for items that deteriorate with respect to time and follow Weibull distribution. Proposed supply chain model assumes single manufacturer and single retailer for its study. Ordering cost is lot-size dependent, and salvage value is associated with deteriorating units. Aim of this paper is to optimize (minimize) total as well as individual costs of the inventory. Inventory total cost has been minimized by both traditional gradient technique and genetic algorithm. Analysis of model also studies two scenarios: (i) independent decision of manufacturer and retailer (ii) joint decision. Model is validated by numerical example. Sensitivity is carried to understand the effect of different parameters on the joint total cost.

2 Notations and Assumptions

2.1 Notations

Inventory parameters for manufacturer

C_m	Production cost/unit($)
h_m	Holding cost/unit/annum($)
D	Demand/unit time
P	Production rate
b_1	Deteriorating cost/unit($)
γ	Salvage cost/unit($)
A_m	Set-up costs($)
TC_m	Total cost for manufacturer($)

Inventory parameters for retailer

C_0	Fix ordering costs($)
Q_r	Retailer's order quantity per order
T	Cycle time(years)(decision variable)
h_r	Holding cost/unit/annum
$\theta(t)$	Time-dependent deterioration of inventory follows Weibull distribution
TC_r	Total cost for retailer($)
TC	Joint total cost($)

2.2 Assumptions

1. In this model, we consider single manufacturer and single retailer for single item.
2. Demand is deterministic and constant.
3. Replenishment rate is infinite.
4. Lead time is zero, and no shortages are allowed.
5. The production rate say $P = \lambda D, \lambda > 1$
6. Ordering cost is lot-size dependent.

7. The on-hand inventory deteriorates with respect to time and follows Weibull distribution;
   $\theta(t) = \alpha\beta t^{\beta-1}$, where $0 < \alpha < 1$ is shape parameter and $\beta \geq 1$ is scale parameter.
8. Deteriorated items can neither be repaired nor be replaced during the cycle time.

3   Mathematical Model

3.1   Manufacturer's Total Cost

Here we assumed production dominates demand and deterioration during cycle time T. So the governing differential equation for on-hand inventory of manufacturer is defined as

$$\frac{dQ_m}{dt} + \theta(t)Q_m = P - D; \qquad 0 \leq t \leq T \tag{1}$$

Solving Eq. (1) using boundary conditions $Q_m(0) = 0$ and $Q_m(T) = Q_m$ we get

$$Q_m(t) = (P - D)(t - \frac{\alpha\beta t^{\beta+1}}{\beta+1}) \tag{2}$$

So total quantity produce by manufacturer per cycle is

$$Q_m = (P - D)(T - \frac{\alpha\beta T^{\beta+1}}{\beta+1}) \tag{3}$$

Basic costs:

- Set-up Cost

$$SC_m = A_m \tag{4}$$

- Inventory holding cost per unit is given by

$$HC_m = h_m \int_0^T Q_m(t)\,dt$$

So

$$HC_m = h_m P[\frac{T^2}{2} - \frac{\alpha\beta T^{\beta+2}}{(\beta+1)(\beta+2)}] \tag{5}$$

Now number of deteriorating units during cycle time T

$$DE(T) = Total quantity - \int_0^T D\,dt = P - DT \tag{6}$$

- The deteriorating cost is given by

$$DC_m = b_1(P - DT) \tag{7}$$

- Salvage value is given by

$$SV_m = \gamma b_1(P - DT) \tag{8}$$

The total cost of manufacturer is

$$TC_m(T) = SC_m + HC_m + DC_m - SV_m \tag{9}$$

3.2 Retailer's Total Cost

Retailer's on-hand inventory depletes with demand- and time-dependent deterioration. So the differential equation describse the inventory level at any time t which is given by

$$\frac{dQ_r}{dt} + \theta(t)Q_r = -D; \quad 0 \le t \le T \tag{10}$$

Using boundary condition $Q_r(T) = 0$ and $Q_r(0) = Q_r$, we obtain solution of Eq. (10)

$$Q_r(t) = D[T - t + \frac{\alpha T}{\beta + 1}(T^\beta - (1 + \beta)t^\beta) + \frac{\alpha \beta t^\beta}{\beta + 1}] \tag{11}$$

There for total quantity purchase by retailer per cycle which is

$$Q_r = D(T + \frac{\alpha T^{\beta+1}}{\beta + 1}) \tag{12}$$

Basic costs associated with retailer total cost are

- Here ordering cost is lot-size dependent, so it is defined as

$$OC_r = C_0 Q_r^\eta \tag{13}$$

- The holding cost per time unit is given below

$$HC_r = h_r \int_0^T Q_r(t)\, dt$$

$$HC_r = h_r D[\frac{T^2}{2} - \frac{\alpha \beta T^{\beta+2}}{(\beta + 1)(\beta + 2)}] \tag{14}$$

- The deteriorating cost per time unit is

$$DC_r = b_1(Q_r - DT) \tag{15}$$

- Salvage value per time unit is

$$SV_r = \gamma b_1(Q_r - DT) \tag{16}$$

The total cost for retailer is given by

$$TC_r(T, \eta) = OC_r + HC_r + DC_r - SV_r \tag{17}$$

3.3 Joint Cost for System

Total cost for system is

$$TC(T, \eta) = TC_m(T) + TC_r(T, \eta) \tag{18}$$

4 Computational Algorithms

4.1 Algorithm for Analytical Approach

1. Set all parameters values in mathematical model.
2. Find optimal T^* from $\frac{\partial TC_m}{\partial T} = 0$
3. Using T^*, find optimize lot size and total cost for manufacturer.
4. Compute optimal T^* and η^* from $\frac{\partial TC_r}{\partial T} = 0$ and $\frac{\partial TC_r}{\partial \eta} = 0$ simultaneously.
5. Using T^* and η^*, find optimize lot size and total cost for retailer.
6. Obtain optimal T^* and η^* from $\frac{\partial TC}{\partial T} = 0$ and $\frac{\partial TC}{\partial \eta} = 0$ simultaneously.
7. Using T^* and η^*, find optimize lot size and total cost for system.

4.2 Genetic Algorithm Approach

A genetic algorithm (or GA) is a search technique inspired by evolution theory used to compute true or approximate solutions to optimization problems. It is a parallel computation process; the GA explores the solution space in many directions and from many points. Complex multi-objective problems with nonlinear behaviour can be optimized with GAs. GA can be applied for constraint problems too. Genetic algorithms are a particular class of evolutionary algorithms that use techniques motivated by Darwins evolution theory and thus follow principles of biology such as inheritance, mutation, selection and crossover. Basic idea of GA is demonstrated in Fig. 1 that includes initialization, selection, crossover, mutation and termination.

Initialization: In the beginning, randomly generated many solutions form an initial population. This is also known as search space.

Selection: In each successive generation, a proportion of the present population is selected to produce a new generation. Selection of individuals is on the basis of fitness score achieved by them. Many of the times, less fit solutions are also included to prevent premature convergence to poor solutions. Roulette wheel selection and tournament are widely used methods.

Reproduction: Chromosomes of selected individuals recombine with each other to produce new generation by the process of crossover. Mutation also takes place in some cases which offer change (alteration) within the chromosome of an individual. This process generates new parents and then new children in each iteration. Process continues till a feasible solution set of appropriate size is generated.

Crossover:

Parent

| Parent 1 | 1 | 1 | 0 | 0 | 1 | 0 | 1 | 0 | 0 | 1 |
| Parent 2 | 0 | 1 | 0 | 1 | 0 | 1 | 0 | 1 | 0 | 1 |

Two possible offspring

| Child 1 | 1 | 1 | 0 | 0 | 1 | 1 | 0 | 1 | 0 | 1 |
| Child 2 | 0 | 1 | 0 | 1 | 0 | 0 | 1 | 0 | 0 | 1 |

Mutation: To ensure that individuals do not remain same in next generation, there is provision of some alterations in the chromosomes itself called mutation which is as shown below.

| Before mutation | 1 | 1 | 0 | 0 | 1 | 1 | 0 | 1 | 0 | 1 |
| After mutation | 1 | 1 | 0 | 0 | 0 | 1 | 0 | 1 | 0 | 1 |

It can be flipped also

| Before mutation | 1 | 1 | 0 | 0 | 1 | 1 | 0 | 1 | 0 | 1 |
| After mutation | 0 | 0 | 1 | 1 | 0 | 0 | 1 | 0 | 1 | 0 |

Termination: Above-discussed process gets repeated until a solution with desired tolerance is not achieved. A flow chart to illustrate working of genetic algorithm is shown in Fig. 1.

Total cost function and individual cost function are considered to be fitness function and minimized using below-mentioned algorithm. MATLAB 13a is used for running iterations.

1. Set numerical values for different parameters except for decision variables T and η in the fitness function.
2. Start with an initial population of 20 chromosomes.
3. Rank the chromosomes on the basis of their fitness score.
4. Chromosomes with good fitness score will get entry in mating pool.

Fig. 1 Genetic algorithm
flow for the proposed model

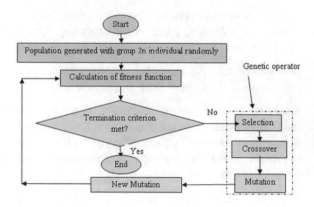

5. Perform stochastic uniform crossover for reproduction. Crossover fraction is considered 0.8, and 2-Elites are considered at each generation.
6. Again rank members of new generation on the basis of their fitness function and select members which can create next generation.
7. Perform step 3 and step 4 till absolute difference between two successive members is negligible, i.e $|x_{i+1} - x_i| < \varepsilon (\varepsilon = 10^{-5})$.

5 Numerical Examples

Example Consider one integrated inventory system with $h_m = 2$; $h_r = 2$; $\alpha = 0.1$; $\beta = 1.5$; $D = 5000$; $C_0 = 5$; $A_m = 1000$; $\gamma = 0.1$; $\lambda = 6$ Optimize using analytical method, then we get some computational results that are shown in Table 1

Here in individual decision, convexity of total cost is

$$
\begin{vmatrix} \frac{\partial^2 TC_r}{\partial \eta^2} & \frac{\partial^2 TC_r}{\partial T\eta} \\ \frac{\partial^2 TC_r}{\partial T\eta} & \frac{\partial^2 TC_r}{\partial T^2} \end{vmatrix} = 6.39902857 * 10^9 > 0 \text{ and } \frac{\partial^2 TC_r}{\partial T^2} = 29{,}742.39585 > 0
$$

Table 1 Computational results obtained by using gradient- based analytical approach

	Independent decision	Joint decision
Optimal cycle time (Year)	0.6055090338	0.7542325995
Optimal η	0.7553682148	0.8420638952
Optimal lot size	10,712	22,230
Total cost for manufacturer ($)	18,705.66874	6298.883576
Total cost for retailer ($)	10,826.59853	5626.866559
Total cost for system ($)	29,532.26727	11,925.750135

And for joint decision, convexity of total cost is

$$\begin{vmatrix} \frac{\partial^2 TC}{\partial \eta^2} & \frac{\partial^2 TC}{\partial T \eta} \\ \frac{\partial^2 TC}{\partial T \eta} & \frac{\partial^2 TC}{\partial T^2} \end{vmatrix} = 1.570856646 * 10^9 > 0 \text{ and } \frac{\partial^2 TC}{\partial T^2} = 19{,}849.20902 > 0$$

Above example is also solved using genetic algorithm, and results of the same have been shown in Table 2. Here for individual decision, genetic algorithm took 30 iterations to reach this solution. Best fitness plot of manufacturer and retailer by GA is shown in Figs. 2 and 3, respectively. For joint decision, genetic algorithm took 20 iterations to reach this solution. Best fitness plot of system by GA is shown in Fig. 4.

Sensitivity analysis for the above example is carried out to find behaviour of different inventory and supply chain parameter related to total cost in joint decision by varying inventory parameters as −10, −5, 5 and 10%. The results are shown in Fig. 5.

Table 2 Computational results obtained by using genetic algorithm

	Independent decision	Joint decision
Iteration	30	20
Optimal cycle time (Year)	0.408	0.709
Optimal η	0.8	0.5
Optimal lot size	12,504	17,302
Total cost for manufacturer ($)	18,705.7	6500.02
Total cost for retailer ($)	7291.12	4058.04
Total cost for system ($)	25,996.82	10,558.06

Fig. 2 Best fitness solution for manufacturer total cost in independent scenario

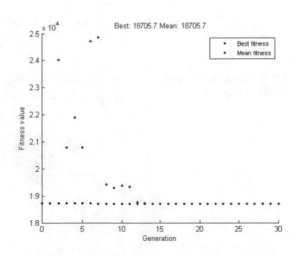

Fig. 3 Best fitness solution for retailer total cost in independent scenario

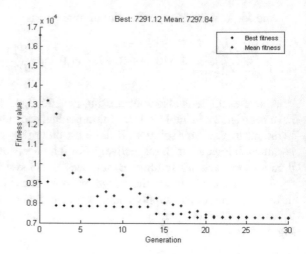

Fig. 4 Best fitness solution for system total cost in joint scenario

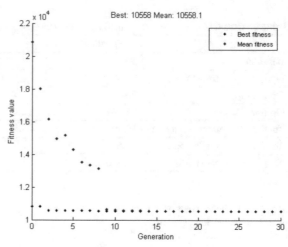

Fig. 5 Sensitivity analyses for different parameters

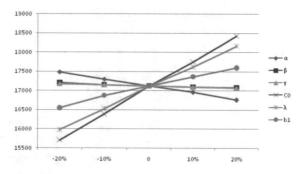

Observations

- Results from Tables 1 and 2 clearly suggest that GA minimizes both individual and total joint inventory cost.
- Increment in shape parameter, salvage cost per unit, scale parameter increase reduction in total cost.
- As deteriorating cost per unit, production constant and fix part of ordering cost increases, total cost also increases.

6 Conclusion

An integrated production–inventory model for time-dependent deteriorating items is studied for both manufacturer and retailers viewpoint. Individual as well as joint cost is analysed to attain long-term sustainable supply chain. Analysis shows that decision taken in collaboration reduces the cost compared to decisions in isolation. Models are optimized with analytical as well as with genetic algorithm. Observations clearly show that genetic algorithm minimizes the total cost in joint as well as independent decision as compared to analytical method. Convexity is explained mathematically for analytical technique. Results clearly show that analytical method fails to provide global minima which are attained by evolutionary algorithm genetic algorithm. Proposed model can be further extended to minimize the total cost of supply chain network with multi-objective fitness function using genetic algorithm.

References

1. Bakker, M., Reizebos, J., Teunter, R.H.: Review of inventory system with deterioration since 2001. Eur. J. of Ope. Res. **221**(2), 275–284 (2012)
2. Banerjee, A.: A joint economic-lot-size model for purchaser and vendor. Dec. Sci. **17**(3), 292–311 (1986)
3. Chung, K.J., Crdenas-Barrn, L.E., Ting, P.S.: An inventory model with non-instantaneous receipt and exponentially deteriorating items for an integrated three layer supply chain system under two levels of trade credit. Int. J. Prod. Econ. **155**, 310317 (2014)
4. Chung, K.J., Crdenas-Barrn, L.E.: The simplified solution procedure for deteriorating items under stock-dependent demand and two-level trade-credit in the supply chain management. Appl. Math. Model. **37**(7), 46534660 (2013)
5. Ghare, P.M., Schrader, G.P.: A model for an exponentially decaying inventory. J. of Indu. Eng. **14**(5), 238–243 (1963)
6. Goldberg, D.E.: Genetic Algorithms in Search, Optimization and Machine Learning. Addison-Wesley, New York (1989)
7. Goren, G.H., Tunali, S., Jans, R.: A review of applications of genetic algorithms in lot sizing. J. of Inte. Manu. Springer, Netherlands (2008)
8. Goyal, S.K., Gunasekaran, A.: An integrated production inventorymarketing model for deteriorating items. Com. and Ind. Eng. **28**(4), 755–762 (1995)
9. Goyal, S.K.: An integrated inventory model for a single supplier-single customer problem. Int. J. of Pro. Res. **15**(1), 107–11 (1976)

10. Janssen, L., Claus, T., Sauer, J.: Literature review of deteriorating inventory models by key topics from 2012–2015. Art. in Inter. J. of Pro. Eco, **182**, 86–112 (2016)
11. Manna, S.K., Chaudhuri, K.S.: An economic order quantity model for deteriorating items with time-dependent deterioration rate, demand rate, unit production cost and shortages. Inter. J. of Pro. Res. **32**(8), 1003–1009 (2001)
12. Mishra, P.,Talati, I.: A Genetic algorithm approach for an inventory model when ordering cost is lot size dependent. Int. J. of Latest Tech. in Eng., Man. and App. Sci. **4**(2), 92–97 (2015)
13. Mishra, P., Talati, I.: An integrated and coordinated supply chain model with backorder for advertisement and stock dependent demand. Glo. J. Pur. and App. Math. **13**, 4035–4054 (2017)
14. Mishra, P.,Talati, I.: Optimal supply chain policies for two - echelon players with credit time and price sensitive demand when inventory is subjected to time dependent deterioration. AMSE J. Ser.-Mode.-D, **37**(1), 77–96 (2017)
15. Mohan, R.: Quadratic demand. Variable holding cost with Time dependent deterioration without shortages and salvage value **13**(2), 59–66 (2017)
16. Murata, T., Ishibuchi, H., Tanaka, H.: Multi objective algorithm and its applications to Flow shop Scheduling. Com. Indu. Eng., **30**(4), 954 968 (1996)
17. Narmadha, S., Selladurai, V, Sathish, G.: Multiproduct Inventory Optimization using Uniform crossover Genetic algorithm. IJCSIS **7**(1), 170–179 (2010)
18. Philip, G.C., Covert, R.P.: An EOQ model for items with weibull distribution deterioration. All E Transzctions **5**(4), 323–326 (1973)
19. Philip, G.C.: A generalized EOQ model for items with weibull distribution. All E Transzctions **6**(2), 159–162 (1974)
20. Radhakrishnan, P, Prasad, V.M., Jeyanthi, N.: Design of Genetic algorithm based supply chain inventory optimization with lead time. Int. J. of Com. Sci. and Net. Sec. **10**(3), 1820–1826 (2010)
21. Radhakrishnan, P, Prasad, V. M., Jeyanthi, N.: Inventory optimization in supply chain management using genetic algorithm. Int. J. of Com. Sci. and Net. Sec. **9**(1), 33–40 (2009)
22. Sarkar, B., Saren, S.: Partial trade-credit policy of retailer with exponentially deteriorating items. Int. J. Appl. Comput. Math. **1**, 343368 (2015)
23. Sarkar, B., Sett, B.K., Roy, G., Goswami, A.: Flexible setup cost and deterioration of products in a supply chain model. Int. J. Appl. Comput. Math, **2**(1), 25–40 (2015)
24. Shah, N.H., Mishra, P.: Inventory management of time dependent deteriorating items with salvage value. App. Math. Sci. **2**(16), 793–798 (2008)
25. Shah, N.H., Chaudhari, U., Jani, M.Y.: Optimal policies for time-varying deteriorating item with preservation technology under selling price and trade credit dependent quadratic demand in a supply chain. Int. J. Appl. Comput. Math. **3**(2), 363–379 (2017)
26. Shah, N.H., Shah, D.B., Patel, D.G.: Optimal pricing and ordering policies for inventory system with two-level trade credits under price-sensitive trended demand. Int. J. Appl. Comput. Math. **1**(1), 101–110 (2015)
27. Shah, N.H.: Manufacturer-retailer inventory model for deteriorating items with price-sensitive credit-linked demand under two-level trade credit financing and profit sharing contract. Cogent. Eng. **2**(1), (2015)
28. Woarawichai, C., Kuruvit, K., Vashirawongpinyo, P.: Applying genetic algorithms for inventory lot-sizing problem with supplier selection under storage capacity. Constraints **9**(1), 18–23 (2012)

Modified Three-Layered Artificial Neural Network-Based Improved Control of Multilevel Inverters for Active Filtering

Soumyadeep Ray, Nitin Gupta and R. A. Gupta

Abstract Multilevel inverter (MLI) especially cascaded H-bridge-type MLI provides transformer-less solution in case of medium-voltage and high-power distribution scenario for mitigating current-related power quality (PQ) problems. Performance of MLI-based shunt filtering unit is dependent on control and extraction technique used. Therefore, a lot of control algorithms are already proposed for conventional two-level inverter-based shunt active power filter (SAPF). However, most control algorithms and their capabilities are yet to be explored for MLI-based SAPF. In view of this concern, this paper aims at presenting an advanced control technique for cascaded H-bridge multilevel inverter-based shunt active power filter. This control scheme composed of three parts, i.e., sensing of source voltage, load and source current, DC-link voltage regulation, and reference current generation technique. DC-link voltage regulation and its characteristics are important in terms of settling time and peak overshoot during load changing condition. Normally, PI compensators are used for DC-link voltage regulation during transient and steady-state conditions. However, soft computing techniques give better response in case of DC voltage regulation. Therefore, modified three-layered artificial neural network (M3L-ANN) is used in this paper for regulating DC-link voltages. The proposed control technique is simulated in MATLAB/Simulink with three-phase and five-level CHB-MLI-based SAPF. Extensive simulation analysis is presented and verified. It shows better performance in terms of settling time and peak overshoot during load changing condition. Simultaneously, source current also becomes well below 5% as per IEEE-519 standard.

Keywords Multilevel inverter · Cascaded H-bridge inverter · Shunt active power filter · Artificial neural network

S. Ray · N. Gupta (✉) · R. A. Gupta
Department of Electrical Engineering, MNIT Jaipur, Jaipur, India
e-mail: nitingupta.ee@mnit.ac.in

S. Ray
e-mail: write2prithu@gmail.com

R. A. Gupta
e-mail: ragupta.ee@mnit.ac.in

© Springer Nature Singapore Pte Ltd. 2019
J. C. Bansal et al. (eds.), *Soft Computing for Problem Solving*,
Advances in Intelligent Systems and Computing 816,
https://doi.org/10.1007/978-981-13-1592-3_43

547

1 Introduction

Power quality (PQ) problems and its solution become a habitual choice among researcher and engineer community as large-scale grid integration are taking place in existing distribution sector. Conventional and non-conventional energy, both are integrated into large volumes in order to maintain sustainable development. Therefore, solving PQ problems of this existing and future grid scenario is a major concern for electrical researcher community [1]. At the same instant, industrial loading based on power electronics converters are increasing in industries as well as in distribution sector. Process control and heavy industries are using medium-voltage and high-power industrial drive system comprising of power electronics converters. This nonlinear behavior of loads is main sources of current harmonics in the present distribution sector. Power factor degradation, generation of voltage harmonics due to propagation of current harmonics through PCC, neutral current burden are other major disadvantages due to usage of nonlinear and unbalanced loading. These PQ problems are responsible for problems like malfunctioning of transformer, heating of wire, insulation failure, and breakdown of protection equipment [2, 3]. Therefore, researchers found different possible ways of solution to counter these above-mentioned problems. Combination of inductor and capacitors is called passive filters which are capable of solving these problems. However, these passive filters are having disadvantages like bigger size, space, fixed compensation. Active power filters are designed and rigorously researched over the last decade. Series, shunt, and hybrid active filters are different types of available active filters explored by researchers. Among these, shunt active filters are capable of current harmonics compensation and power factor correction in different loading conditions [3, 4].

Medium-voltage distribution sector (e.g., 11 kV line) is generally used for industrial sector motor drive applications. This sector is now growing as large-scale nonconventional energy sources are integrating to conventional grid. High-power electric motor drives (both AC and DC) are used in industries which are driven by medium-voltage distribution system. Therefore, current harmonics are produced by these loads. SAPF system can take care of this problem. Normally, two-level inverters are main component of SAPF system. However, these inverters are having limitations like low voltage and power handling capability, high di/dt and dv/dt ratings of switches, use of transformer, while using in medium-voltage grid. Therefore, multilevel inverters are primary choice among all possible solutions as it is having distinct advantages like low di/dt and dv/dt rating, low voltage and current stress, transformerless interconnection. So, MLI-based SAPF is becoming a very effective solution for medium-voltage and high-power distribution system. Mainly, three major types of multilevel inverters are there in industrial applications. Among these three available options, cascaded H-bridge multilevel inverter proves its effectiveness in case of reactive power compensation application as this structure is modular in nature and it does not require any source for supplying active component [5, 6].

A qualitative amount of research is already done in the area of different control algorithm and reference current generation techniques in case of SAPF [7–10]. Artificial intelligence-based different control algorithms are also proposed by various researchers in recent past. A fuzzy logic controller-based system is used for single-phase multilevel inverter system by Tolbert et al. [11]. An optimization technique like generic algorithm is used for determination of conduction angle of IGBTs in CHB-MLI [12]. A combined approach including fuzzy logic and generic algorithm is used for harmonic content elimination which is designed for fuel cell-based system [13]. Fuzzy logic-based control for diode-clamped MLI is compared with conventional PI compensator-based system [14]. Optimization techniques like bacterial foraging optimization and particle swarm optimization are used to determine the optimal parameters of PI compensator in case of filtering applications. However, most of the system dynamics have been checked with two-level inverters in case of filtering applications. ANN is very much useful for solving DC voltage regulation problem as it is having very simple structure and minimal output error is produced among all soft computing techniques. A three-layer feed-forward ANN is presented for the determination of voltage variation in case of phasor measurement unit by Khare et al. [15]. However, its performance is yet to be tested in case of active filtering and multilevel inverter-based applications for mitigating various PQ problems.

Therefore, a modified three-layered ANN-based approach is presented in this paper for generation of current reference in case of MLI-based filtering applications. Generally, ANN can be applicable in a large number of systems including power electronics converter-based system. No further system or mathematical modeling is required for the implementation of ANN. A comprehensive training is required with the help of input data and best-chosen output data. Training and validation will be repeated until the system error minimizes to an acceptable level. Finally, trained system is fed with newer input parameters for getting optimal outputs from this. The complete proposed MLI-based filtering system is designed and implemented in MATLAB.

Rest of the paper is presented under the subsequent sections. Five-level CHB-MLI-based SAPF is shown in Sect. 2, whereas modified three-level ANN technique along with advanced control algorithm is depicted and explained in Sect. 3. The proposed system architecture is designed in Simulink platform of MATLAB software. Results and comparative analysis with conventional controller are shown in Sect. 4. Finally, the proposed work is concluded with significant remarks in Sect. 5.

2 System Configuration

Single-phase inverter produces two numbers of levels at the output voltage terminal. Higher levels in output voltage can be formed by cascaded combination of single-phase H-bridge cells. Output voltage becomes almost of sine shape as number of H-bridges increases. Simultaneously, total harmonic distortion of output side voltage becomes smaller. Different number of voltages can be produced by using different combination of H-bridge modules and PWM techniques. In this paper, five-level

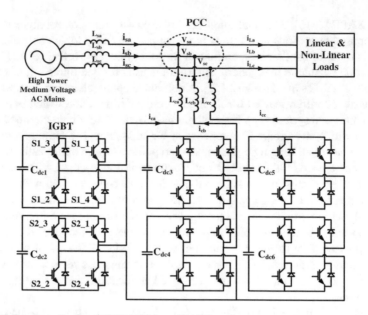

Fig. 1 Three-phase five-level CHB-MLI-based SAPF

three-phase CHB-MLI is designed as a SAPF module. Separate active DC sources are main requirement for five-level H-bridge inverter. However, in case of active filtering application, DC sources are not required as only reactive power needs to be compensated by this system. Therefore, DC-link capacitors can be used as a source. Separate active DC sources requirement is not there in MLI-based SAPF unit. Two numbers of H-bridges are used in cascaded manner to form five-level voltage output at the output terminal of the inverter output. This combination is used for three different phases. One H-bridge comprises four number of power electronic switches. Each leg in H-bridge contains two IGBTs. These two IGBTs are triggered in such a manner that each can be complementary to each other. Three different voltage levels can be developed by applying different combinations of switching. Five-levels of voltage are generated with the help of phase-shifted pulse width modulation technique. Figure 1 depicts five-level CHB-MLI-based three-phase SAPF unit for medium-voltage distribution sector. 11 kV grid voltage is used, and transformer-less interconnection is utilized between SAPF unit and line where L_{Sa}, L_{Sb}, and L_{Sc} are source impedance of three respective phases; i_{Sa}, i_{Sb}, and i_{Sc} are different source currents; i_{Ca}, i_{Cb}, and i_{Cc} are compensating current for three phases; and i_{La}, i_{Lb}, and i_{Lc} are load currents for phase a, b, and c, respectively. Four power electronic switches are denoted by S1_1, S1_2, S1_3, and S1_4 of upper H-bridge inverter and S2_1, S2_2, S2_3 and S2_4 are IGBTs of lower bridge. Load current, source voltage, and DC-link voltages are being sensed by using current and voltage sensors, respectively.

3 Modified Three-Level Artificial Neural Network-Based Advanced Control Technique

Synchronous reference frame (SRF) theory is a well-known technique for obtaining current reference. Three-phase load currents are sensed with the help of current sensors. The sensed load currents are first transformed to alpha–beta component (rotating frame) with the help of Clark's transformation. This is depicted as per Eq. 1.

$$
\begin{bmatrix} i_{L\alpha} \\ i_{L\beta} \\ i_{L0} \end{bmatrix} = \sqrt{2/3} \begin{bmatrix} 1 & -1/2 & -1/2 \\ 0 & \sqrt{3}/2 & -\sqrt{3}/2 \\ 1/\sqrt{2} & 1/\sqrt{2} & 1/\sqrt{2} \end{bmatrix} \begin{bmatrix} i_{La} \\ i_{Lb} \\ i_{Lc} \end{bmatrix} \tag{1}
$$

Rotating frame components of load currents are further transformed to d-$q0$ or synchronous reference frame with the help of park's transformation which can be shown from Eq. 2.

$$
\begin{bmatrix} i_{Ld} \\ i_{Lq} \\ i_{L0} \end{bmatrix} = \begin{bmatrix} \cos \omega t & \sin \omega t & 0 \\ -\sin \omega t & \cos \omega t & 0 \\ 0 & 0 & 1 \end{bmatrix} \begin{bmatrix} i_{L\alpha} \\ i_{L\beta} \\ i_{L0} \end{bmatrix} \tag{2}
$$

Both oscillatory and fundamental component of load current signals are present in this d-q-0 system. Low-pass filter of cutoff frequency 25 Hz is used for extracting DC quantity from these d-axis and q-axis signals. As system is balanced one, zero component is not considered for further calculation. For generation of reference component of source current, d-axis and q-axis components are calculated separately. This control of d- and q-axis component is composed of DC voltage control loop and PCC voltage control loop. Main aim of SAPF unit is to compensate reactive power. However, some portion of active power is required for compensating switching losses of inverter unit. DC-link voltage of capacitor is sensed by voltage sensor, and it is compared with reference value of DC-link voltage. Error of voltage signal is traditionally passed through PI compensator, and finally, this component is added with i_{Ld}. The reference 'd'-axis current is expressed by Eq. 3.

$$
i_d = i_{d-DC} + i_{\text{loss}} \tag{3}
$$

However, in order to get faster and smoother response in terms of settling time and peak overshoot of DC-link voltages, ANN-based compensator is used to extract the loss component from the proposed system. This use of ANN avoids the complicated method of PI compensator parameter tuning. PI parameters are also sensible with respect to load variation; i.e., it requires rigorous tuning. ANN can be adjusted itself according to the variation of load parameters. The loss calculation method which is extracted with the help of ANN is shown in Fig. 2.

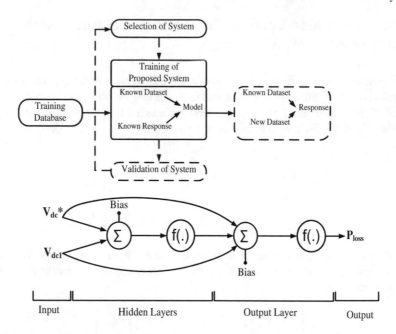

Fig. 2 Modified three-layered ANN for DC voltage regulation and loss calculation

The ANN method requires three steps, i.e., learning process, training, and test-ing of pre-consumed data for completion of the total process. ANN can be applied to this proposed system after completion of design. The proposed system is simu-lated with PI compensator-based control algorithm. Best data are chosen for further implementation of ANN controller. System data stored in workspace is used for ANN implementation as guidance. Efficiency and adaptability of the proposed ANN structure are dependent on in-depth training and testing. A diverse range of learn-ing methods and structures is used by researchers for designing of ANN structure. However, three-level modified ANN is used in this paper. The implementation of the proposed ANN approach includes choice of quantity of layers, neuron quantity in every layer, relationship among different neurons, and transfer function selection for each layer. Two feed-forward networks are chosen for implementation of the proposed ANN as this kind of network can model complex functions easily and in an effective way. This modified ANN network is dedicated for loss component prediction of DC-link voltages. Three layers are present in the proposed ANN struc-ture, i.e., input layer, hidden, and output layer. Input layer of the proposed structure consists of two inputs, i.e., sensed DC-link voltage and reference DC-link voltage; and output layer is having one number of neuron, i.e., loss component of current. The proposed ANN structure is modified from fundamental three-layered ANN. Inputs of the ANN structure is not only fed to hidden layer, but it is fed to out-put layer also at the same instant. Tangent sigmoid function and linear function are used as an activation function in the proposed ANN for hidden and output layer,

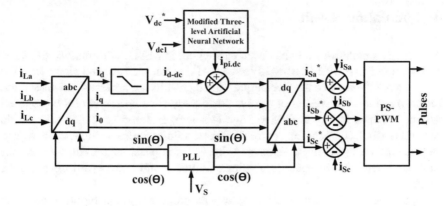

Fig. 3 Advanced control used for CHB-MLI-based SAPF

respectively. Gradient descent algorithm is used for upgradation of weight and bias which is required for training process. The training can be declared completed and stopped when least means square error criteria of sensed and actual DC-link voltage is suitably small enough. Backpropagation NN [15] is used by using feed-forward phase after completion of the training process.

Generated reference source current should be in phase with source voltage. Therefore, PLL plays a vital role for deciding this factor. Phase angle calculation is done by PLL itself.

Finally, alpha–beta to a-b-c component transformation is done with the help of Eq. 4 and reference source current is converted back from d-q-0 frame to alpha–beta frame with the help of inverse park transformation as per Eq. 5.

$$
\begin{bmatrix} i_a \\ i_b \\ i_c \end{bmatrix} = \sqrt{2/3} \begin{bmatrix} 2/3 & 0 & \sqrt{2}/3 \\ -1/3 & 1/\sqrt{3} & \sqrt{2}/3 \\ -1/3 & -1/\sqrt{3} & -\sqrt{2}/3 \end{bmatrix} \begin{bmatrix} i_\alpha \\ i_\beta \\ i_0 \end{bmatrix} \tag{4}
$$

$$
\begin{bmatrix} i_\alpha \\ i_\beta \\ i_0 \end{bmatrix} = \frac{1}{[\sin^2(\omega t) + \cos^2(\omega t)]} \begin{bmatrix} \cos \omega t & -\sin \omega t & 0 \\ \sin \omega t & \cos \omega t & 0 \\ 0 & 0 & 1 \end{bmatrix} \begin{bmatrix} i_d \\ i_q \\ i_0 \end{bmatrix} \tag{5}
$$

This reference current component is compared with sensed source current signal, and error signal is compared with triangular signal with the help of phase-shifted pulse width modulation technique. Generated gate pulses are used for triggering IGBTs of CHB-MLI. The flow diagram of total control theory along with the operation of three-phase PLL is depicted in Fig. 3.

4 Simulation Result

A thorough simulative study has been carried out in MATLAB/Simulink software. The proposed five-level CHB-MLI-based SAPF system is designed along with the modified ANN-based control technique. Finally results are presented in this section for proving the effectiveness of the system. Steady-state operation of the proposed system is shown in Fig. 4. Grid voltage (Vs-abc), grid current (Is-abc), load side current (IL-abc), compensating current of phase a (Ic-a), inverter output voltage (Vt), and DC-link voltage (Vdc) are shown in Fig. 4. Three-phase source current becomes sinusoidal while load current is nonlinear in nature after successful operation of SAPF.

Figures 5 and 6 show transient behavior of the proposed system. Figure 5 shows the system behavior along with Vdc with step decrement of load while Fig. 6 depicts the system performance with step increment of load. Grid-side current is non-sinusoidal and full of harmonic component while nonlinear loads are connected in the distribution side and harmonic profile is shown in Fig. 7a. After successful operation of SAPF unit, grid-side current becomes sinusoidal and its harmonic profile is shown in Fig. 7b. Table 1 shows comparative analysis among conventional control technique,

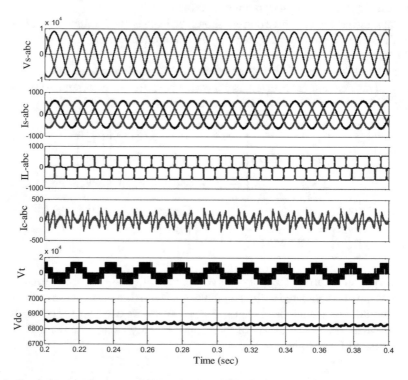

Fig. 4 Steady-state performance of CHB-MLI-based SAPF with M3L-ANN

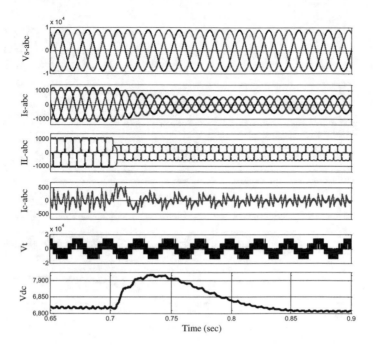

Fig. 5 Transient performance of CHB-MLI-based SAPF during load decrement with M3L-ANN

Table 1 Comparative analysis of the proposed controller with conventional controller

Parameters	Conventional controller	ANN-based controller	Modified ANN-based controller
Peak overshoot during decrement of loading (Volt)	158	126	95
Settling time during decrement of loading (sec)	0.26	0.18	0.15
Peak undershoot during increment of loading (Volt)	163	134	105
Settling time during increment of loading (sec)	0.25	0.2	0.15
THD of source current (%)	3.37	2.76	2.13

ANN-based control technique, and modified ANN-based control technique to show the effectiveness of the system under variable loading condition.

It can be seen from the above-mentioned table that M3L-ANN-based controller provides better response in terms of settling time and peak overshoot/undershoot

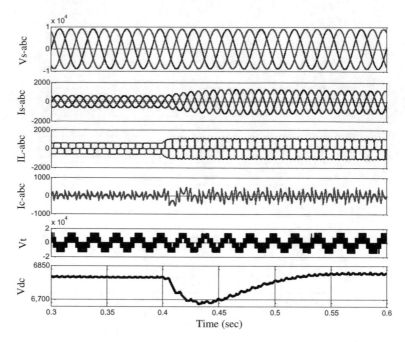

Fig. 6 Transient performance of CHB-MLI-based SAPF during load increment with M3L-ANN

during load changing condition. As operating line voltage is 11 kV, 95 V overshoot is minimal. System becomes stable again after 6–8 cycles. From the above analysis, it can be clearly stated that system is having better transient stability as well as steady-state performance.

5   Conclusion

In this paper, a modified ANN-based control algorithm is proposed for three-phase five-level CHB-MLI-based SAPF system which is capable of mitigating current harmonics and reactive power in distribution sector. The system is simulated in MAT-LAB, and its results and calculative analysis prove the effectiveness of the proposed system. Steady-state response of the proposed system found satisfactory, and source current THD becomes within IEEE recommended standard. Transient response of the CHB-MLI-based SAPF system is also better than conventional control algorithm. So it can be concluded that the proposed M3L-ANN-based control algorithm gives better and faster dynamic response than conventional PI controller and ANN-based control system.

(a)

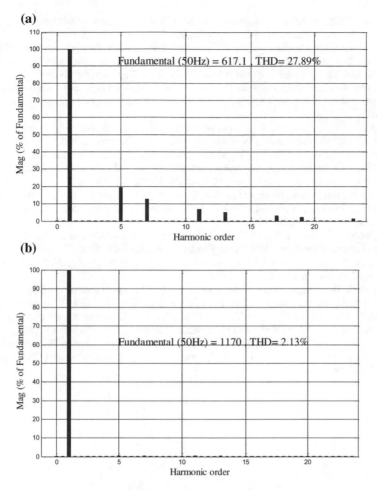

(b)

Fig. 7 **a** Harmonic profile of source current before compensation, **b** harmonic profile of source current after compensation with M3L-ANN-based controller

References

1. Hong, Y.Y., Chen, Y.Y.: Placement of power quality monitors using enhanced genetic algorithm and wavelet transform. IET Gener. Transm. Distrib. **5**(4), 461–466 (2011)
2. Dalai, S., Chatterjee, B., Dey, D., Chakravorti, S., Bhattacharya, K.: Rough-set-based feature selection and classification for power quality sensing device employing correlation techniques. IEEE Sens. J. **13**(2), 563–573 (2013)
3. Teke, A., Saribulut, L., Tumay, M.: A novel reference signal generation method for power-quality improvement of unified power-quality conditioner. IEEE Trans. Power Delivery **26**(4), 2205–2214 (2011)
4. Mahela, O.P., Shaik, A.G.: Topological aspects of power quality improvement techniques: a comprehensive overview. Renew. Sustain. Energy Rev. **58**, 1129–1142 (2016)

5. Ray, S., Gupta, N., Gupta, R.A.: A comprehensive review on cascaded h-bridge inverter based large-scale photovoltaic. IETE Technical Review. **34**(5), 463–477 (2017)
6. Ray, S., Sreedhar, M., Dasgupta, A.: ZVCS based high frequency link grid connected SVPWM applied three phase three level diode clamped inverter for photovoltaic applications. In: 2014 Power and Energy Systems: Towards Sustainable Energy, pp. 1–6 (Feb. 2014)
7. Rammos, G.A., Costa-Castello, R.: Power factor correction and harmonic compensation using second-order odd-harmonic repetative control. IET Control Theory Appl. **6**(11), 1633–1644 (2012)
8. Singh, B., Arya, S.R., Jain, C.: Simple peak detection control algorithm of distribution static compensator for power quality improvement. IET Power Electronics. **7**(7), 1736–1746 (2014)
9. Arya, S.R., Singh, B., Chandra, A., Al-Haddad, K.: Power factor correction and zero voltage regulation in distribution system using DSTATCOM. In: IEEE International Conference on Power Electronics, Drives and Energy Systems (PEDES), pp. 1–6 (2012)
10. Kumar, S., Verma, A.K., Hussain, I., Singh, B., Jain, C.: Better control for a solar energy system: using improved enhanced phase-locked loop-based control under variable solar intensity. IEEE Ind. Appl. Mag. **23**(2), 24–36 (2017)
11. Ozpineci, B., Tolbert, L.M., Chaisson, J.N.: Harmonic optimization of multilevel converters using Generic algorithms. IEEE Power Electron. Lett. **3**, 92–95 (2005)
12. Altin, N., Sefa, A.B.: dSPACE based adaptive neuro- fuzzy controller of grid interactive inverter. Energy Convers. Manage. **56**, 130–139 (2012)
13. Jurado, F., Valverde, M.: Generic fuzzy control applied to inverter of solid oxide fuel cell for power quality improvement. Electric Power Syst. Res. **76**, 93–105 (2005)
14. Ravi, A., Manoharan, P.S., Vijay Anand, J.: Modeling and simulation of three phase multilevel inverter for grid connected photovoltaic systems. Sol. Energy **85**, 2811–2818 (2011)
15. Khare, G., Singh, S.N., Mohapatra A., Sunitha, R.: Prediction of missing PMU measurement using artificial neural network. In: 2016 National Power Systems Conference (NPSC), pp. 1–6, IEEE (2016)

Probabilistic Histogram-Based Band Selection and Its Effect on Classification of Hyperspectral Images

Ram Narayan Patro, Subhashree Subudhi, Pradyut Kumar Biswal and Harish Kumar Sahoo

Abstract Hyperspectral images are a series of images, which are captured for a specific region over a range of wavelengths. This makes the classification process computationally more expensive. For reducing the computational complexity, instead of considering all bands, it is essential to select the most informative bands. In this paper, a probabilistic histogram-based band selection approach is proposed. Here, adjacent band fusion with a class-specific deviation is computed followed by extraction of fused band intra- and inter-class histogram properties, to rank the bands with ensemble probability. In both the steps, median measure is used to half the total dimension. So finally, one-fourth of the optimal bands are obtained. Both spectral and spatial features of the reduced bands are considered for classification using KNN with different distance measures. Performance measures like accuracy and execution time are compared. Even by considering only 5% of optimal bands, the proposed approach maintains reference accuracy with reduced computational complexity.

1 Introduction

Hyperspectral images have a large number of spectral and spatial information, where spectral information possesses strong correlation between various bands. A lot of researchers proposed algorithms where classification is carried out without reduction [1, 12, 19]. Reduction in dimension is a necessary step before classification in order to

R. N. Patro · S. Subudhi · P. K. Biswal (✉)
Department of Electronics and Telecommunication, IIIT Bhubaneswar, Bhubaneswar, India
e-mail: pradyut@iiit-bh.ac.in

R. N. Patro
e-mail: ram_patro@rediffmail.com

S. Subudhi
e-mail: ssubudhi84@gmail.com

H. K. Sahoo
Department of Electronics and Telecommunication, VSSUT, Burla, India
e-mail: harish_sahoo@yahoo.co.in

© Springer Nature Singapore Pte Ltd. 2019
J. C. Bansal et al. (eds.), *Soft Computing for Problem Solving*,
Advances in Intelligent Systems and Computing 816,
https://doi.org/10.1007/978-981-13-1592-3_44

reduce overfitting and computational complexity. Dimension reduction can be carried out by various linear and nonlinear dimension reduction methods or by selecting an optimal set of dimension which strongly represents class-specific information. In case of linear or nonlinear models, the spectral information is transformed to lower-dimensional space. Reduction in dimension can be carried out with the help of spatial, spectral, or both information [3, 4]. But in the case of band selection, the original spectral information is retained without affecting their physical properties. Again the selection of bands can be supervised or unsupervised. A framework for the hyperspectral reduction as a preprocessing step is presented in Fig. 1.

Various unsupervised band reduction approaches such as principal component analysis (PCA) [9]; orthogonal subspace and projection-based reduced similarity indexed reduction [2]; linear prediction-based reduction [15]; and affinity propagation and wavelet shrinkage-based reduction in band are presented in [7]. Similarly, a range of supervised band selection methods such as mutual information and rejection bandwidth-based selection [13] and paired selection [16] make use of the spectral signature in order to predict the optimal bands. In [6, 11], visual and spectral shape-similarity-based unsupervised band selection is proposed. Hybrid with spatial-based morphological profile information is used to select optimal bands in [8]. A series of heuristic approaches are also proposed by many researchers, where a third classifier or model is used to find the objective, such as accuracy and the time consumed. Search-based algorithms like genetic algorithm (GA), particle swarm optimization (PSO), and GA with support vector machine (GASVM) are presented in [10, 14, 17, 18].

Depending on the availability of ground truth, classifiers can be broadly divided into three types: supervised, unsupervised, and semi-supervised. The supervised algorithms consider the ground truth information for classification. But if ground truth is not available, unsupervised classification is preferred, while semi-supervised classifier considers both labeled and unlabeled data for classification.

Embedding spectral with spatial information can improve the classification results. So in the current approach, histogram-based band selection and hybrid classification

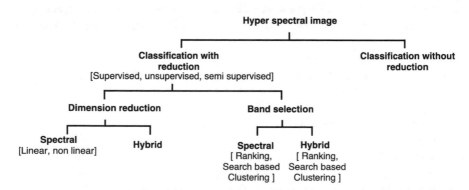

Fig. 1 Architectural flow diagram for hyperspectral imagery

are carried out with KNN for different distance measures. Here, more emphasis is given to band selection rather than classifier design. So, this paper focuses on attaining the reference accuracy even with reduced number of bands.

Description of the band selection approaches is given in Sect. 2. In Sect. 3, the detailed experimental setup and result are discussed. Finally, Sect. 4 provides the conclusion.

2 Methodology

In the current literature, n and m denote the number of classes and bands, respectively. Ω_n^m represents a subspace (region) of band m with class n in the ground truth. Optimal selection of band is carried out in a supervised manner in two subsequent steps. These steps are described briefly in the following subsections.

2.1 Adjacent Band Fusion

For a range of wavelength, images are captured with some sampling rate. These sampled images may have similarities over adjacent bands. So the reduction in band can be done in spectral domain by finding the spectral similarity. To calculate the similarity measure, absolute deviation d^b of the class-specific region between adjacent bands is evaluated using (Eq. 1). Median of the deviation $\tilde{d}^b$ is used as a threshold to select the initial 50% of the optimal bands B^{opt^1} (Eq. 2).

$$d^b = \sum_{c=1}^{n} |\Omega_c^b - \Omega_c^{b+1}| \tag{1}$$

where $b = 1, 2, \ldots (m-1)$.

$$\left[B^{opt^1} = b \right]_{d^b \geq \tilde{d}^b} \tag{2}$$

where $opt^1 = 1, 2, \ldots m/2$ and $\tilde{d}^b$ represent the *median* of d^b.

The information about the class-specific region is obtained from ground truth, so the process becomes supervised. If the classes are unknown for a region where the image is captured, it is quite difficult to select the optimal bands. But with a prior information of class labels, optimal band selection can be achieved.

2.2 Probabilistic Histogram-Based Ranking

After obtaining the optimal bands B^{opt^1} from the previous approach, for each band, intersection of intraclass discrete histogram and interclass-specific standard deviation of histogram is calculated (Eqs. 3 and 4). The histogram co-relation of the classes within a band is summed, and $P_{\cap}^b$ is updated for every band (Eq. 3). Later probability of intersection $\widehat{P_{\cap}^b}$ for each band is calculated by normalizing it over the bands (Eq. 4). Finally, optimal bands are obtained where intraclass histogram intersection is maximized and interclass histogram standard deviation is minimized.

$$P_{\cap}^b = \sum_{i=1}^n \sum_{j=1}^n \left(1 - \left(h_{\Omega_i^b} \cap h_{\Omega_j^b}\right)\right) \tag{3}$$

$$\widehat{P_{\cap}^b} = P_{\cap}^b / max(P_{\cap}^b) \tag{4}$$

where $b = B^{opt^1}$ is the initial set of optimal bands $(opt^1 = 1, 2, \ldots m/2)$; $h_{\Omega_i^b}$ is the subspace histogram of band b for class i; and $\widehat{P_{\cap}^b}$ is the probability of intersection over the band b.

Interclass-specific standard deviation of histogram will reveal the spread of spectral information. So to get the information of bands where concentric histogram is found, standard deviation for each class is considered. So, the overall standard deviation for classes is inversely proportional to the information contained in the band. Sum of standard deviation of class-specific histogram P_σ^b for every band is obtained using (Eq. 5), which is converted to probability $\widehat{P_\sigma^b}$ by normalizing it over the band (Eq. 6).

$$P_\sigma^b = \sum_{i=1}^n \sigma_{h_{\Omega_i^b}} \tag{5}$$

$$\widehat{P_\sigma^b} = P_\sigma^b / max(P_\sigma^b) \tag{6}$$

where $\sigma_{h_{\Omega_i^b}}$ is the standard deviation of subspace histogram for band b and class i; $\widehat{P_\sigma^b}$ is the probability of standard deviation over band b.

As intraclass histogram processing and interclass histogram processing are two different events, the intersection probability P^b of $P_{\cap}^b$ and P_σ^b can be obtained by their multiplication (Eq. 7).

$$P^b = \widehat{P_{\cap}^b} \times \widehat{P_\sigma^b} \tag{7}$$

In order to select second optimal set of bands, median of P^b (i.e., $\widetilde{P^b}$) is calculated as a threshold. Bands below or equal to the threshold are selected as the final set of optimal bands which reduces the size of B^{opt^1} to fifty percent, and in turn seventy-five percent of the original bands are reduced. So the size of final optimal bands B^{opt} is equal to one-fourth of the size of original bands, i.e., $m/4$.

$$\left[B^{opt} = b\right]_{P^b \leq \widetilde{P^b}} \tag{8}$$

where B^{opt} is the final set of optimal bands ($opt = 1, 2, \ldots m/4$) and $\widetilde{P^b}$ is the median of P^b.

$$B_{sort}^{opt} = sort(B^{opt}) \tag{9}$$

A final set of selected B^{opt} (having $m/4$ bands) is used as a reduced set. B^{opt} are ranked according to their associated probability in ascending order (Eq. 9). Sorted bands B_{sort}^{opt} are used to evaluate performance measures with the reference to the original data having m bands.

2.3 Spatial Inclusion

Along with spectral information, spatial information is included to form a hybrid classification model [5]. Spatial information is simply the spatial locations (row, column) of the selected labeled samples. The final feature vector can be represented as a concatenation of obtained spectral signature and spatial locations by a scaling factor μ in Eq. 10.

$$F = [S_b, \mu \times S_r, \mu \times S_c] \tag{10}$$

where F represents a final feature vector, row vector S_b represents the spectral signature of an original or reduced bands, and S_r and S_c represent spatial row and column indices, respectively. Scaling parameter μ decides importance given to the spatial features.

Band-specific spectral features are normalized before classification for both training and testing data in order to form an equilibrium between the spectral and spatial features.

3 Experimental Setup

3.1 Dataset Used

For validation of algorithmic results, Indian pines dataset is used. It is a hypercube with 220 bands ranging from wavelength 0.4 to 2.5 μm. The ground truth contains

Fig. 2 For Indian pines dataset: false gray representation of band 23 (left) and false color representation of ground truth (right)

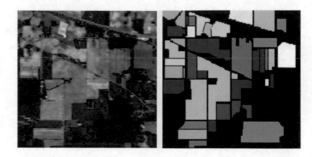

sixteen different classes. Figure 2 contains the false color composite image and ground truth image of the dataset. For validation purpose, the training and testing data are kept fixed. 70% of total sample was considered for training and rest 30% for testing.

3.2 KNN Classifier

In order to verify the classification accuracy over a number of bands, the training data, testing data and the hyperparameters are kept fixed for KNN classifier. k is selected as 1, scaling factor for spatial feature μ is considered as 0.035, and five different distance measures are evaluated. Seventy percent of all individual classes are selected as training, and 30% data are used for testing (Table 1). Performance measures like accuracy and execution time are evaluated for different distance measures as mentioned in Table 2 over selected (sorted optimal) bands. Original data (Indian pines with 220 bands) are also evaluated with the same hyperparameters and kept as a reference for performance comparison over reduced bands.

For distance measures in Table 2, x_s and y_t are vectors. r_s and r_t are the coordinate-wise rank vectors of x_s and y_t; $\tilde{r}_s = \tilde{r}_t = (n + 1)/2$, and n is the number of dimensions.

3.3 Selection of Optimal Bands

Information about the absolute deviation d^b of adjacent bands (Indian pines) across the original band clearly depicts the similarity measure between successive frames (Fig. 3). Absolute deviation d^b of adjacent bands is plotted along with its median $\tilde{d}^b$; selected and rejected bands are colored with red and blue, respectively, across the original band $m = 220$.

Initial optimal bands B^{opt^1} are processed for probabilistic calculation, which are depicted in Fig. 4. This is mainly the rejection probability of bands; i.e., lower the

Table 1 Indian pines dataset

Class	Name	Total sample	Training sample	Testing sample
1	Alfalfa	46	32	14
2	Corn-notill	1428	1000	428
3	Corn-mintill	830	581	249
4	Corn	237	166	71
5	Grass-pasture	483	338	145
6	Grass-trees	730	511	219
7	Grass-pasture-mowed	28	20	8
8	Hay-windrowed	478	335	143
9	Oats	20	14	6
10	Soybean-notill	972	680	292
11	Soybean-mintill	2455	1719	737
12	Soybean-clean	593	415	178
13	Wheat	205	144	62
14	Woods	1265	886	380
15	Buildings-grass-trees-drives	386	270	116
16	Stone-Steel-Towers	93	65	28

Table 2 Distance measures

Distance	Equation
Euclidean	$d_{st}^2 = (x_s - y_t)(x_s - y_t)'$
Correlation	$d_{st} = 1 - (x_s - \tilde{x}_s)(y_t - \tilde{y}_t)'/((x_s - \tilde{x}_s)(x_s - \tilde{x}_s)')^{1/2}((y_t - \tilde{y}_t)(y_t - \tilde{y}_t)')^{1/2}$
Spearman	$d_{st} = 1 - (r_s - \tilde{r}_s)(r_t - \tilde{r}_t)'/((r_s - \tilde{r}_s)(r_s - \tilde{r}_s)')^{1/2}((r_t - \tilde{r}_t)(r_t - \tilde{r}_t)')^{1/2}$
Cosine	$d_{st} = 1 - x_s y_t'/((x_s x_s')(y_t y_t'))^{1/2}$
Cityblock	$d_{st} = \sum_{j=1}^{y_i} |x_{sj} - y_{tj}|$

probability, higher the acceptance and vice versa. Bands having probability p^b below or equal to median probability $\tilde{p}^b$ are selected as the final optimal bands B^{opt}. Then, according to the probability of B^{opt}, bands are sorted and ranked in ascending order to obtain B^{opt}_{sort}. For Indian pines dataset, the optimal selected bands B^{opt}_{sort} are: [19 20 21 29 25 42 39 41 27 18 23 26 43 31 48 52 38 14 22 44 45 17 34 16 50 33 30 12 47 46 53 35 10 13 37 9 54 70 71 11 55 72 8 56 64 40 73 32 63 7 74 36 62 6 57].

Fig. 3 Absolute deviation of adjacent bands

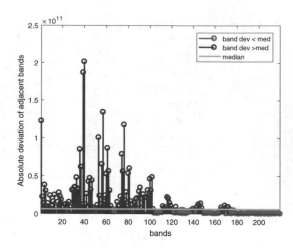

Fig. 4 Rejection probability of bands with their median as a threshold

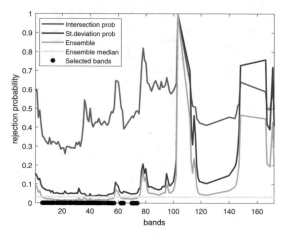

3.3.1 Performance of Selected Bands on Classifier

Original Indian pines dataset is classified with fixed hyperparameters, and its accuracy and processing time are kept as a reference to validate the band selection approach. The final set of optimal bands is selected sequentially in increasing order of percentage (normalized), combined with their respective spatial features and classified using KNN classifier which is compared with the reference measures. Figures 5 and 6 shows the results of KNN classifier over the variation in selection of number of optimal bands with five different distance measures and comparison with the reference measure.

As the accuracy is almost closer toward the reference accuracy with very minimal (5–10) bands (i.e., 3% of original band) (Fig. 5; Table 3), our algorithm can be a good approach for supervised band selection. Similarly, in Fig. 6 and Table 3, there is a larger deviation in the reference time and elapsed time, reducing around ninety-five

Fig. 5 Accuracy variation
over optimal bands

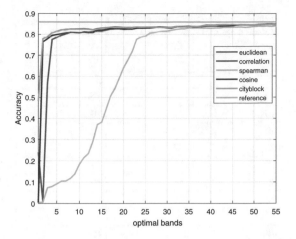

Fig. 6 Elapsed time for the
increase in bands

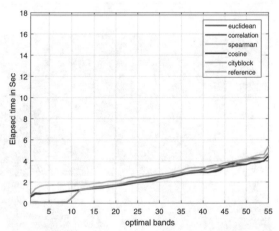

percent of the reference time. Figure 7 shows the classification map for the optimal 17
bands using different distance measures. By using Euclidean and cityblock, improved
classification results are achieved, but in case of Spearman distance, poor result is
obtained.

4   Conclusion

In this paper, band selection is performed in two stages. In the first stage, the most
informative adjacent bands with less similarity are selected and then the class-specific
histogram information based on the ranking of bands is carried out. In each approach,
median is used as a threshold to reduce half of the band. So finally, one-fourth of the
total optimal bands are selected and sorted. The sorted optimal bands are incorporated

Table 3 Accuracy and time comparison

Selected optimal band (%)	No. of band selected	Selected original band (%)	Highest accuracy	Lowest time
10	6	2.5	0.8252	0.0898
20	11	5	0.826	1.2573
30	17	7.5	0.8354	1.5839
40	22	10	0.8351	1.9017
50	28	12.5	0.8341	2.367
60	33	15	0.8392	2.7982
70	39	17.5	0.8467	3.0046
80	44	20	0.8464	3.5633
90	50	22.5	0.8488	3.7882
100	55	25	0.8498	4.4385
400	**220**	**100**	**0.8595**	**17.7704**

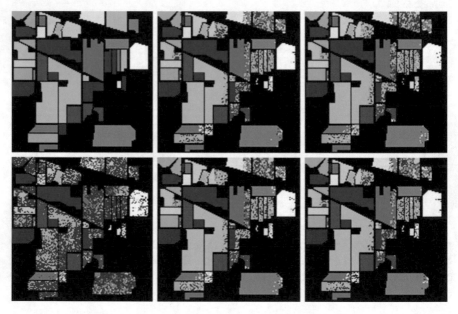

Fig. 7 False color representation of ground truth (upper left) and algorithmic results for distance measures like Euclidean (upper imiddle), correlation (upper right), Spearman (lower left), cosine (lower middle), cityblock (lower left) for reduced 17 bands

with their spatial information, which is used to evaluate the performance measure in terms of accuracy and time complexity using KNN classifier. Five different distance measures are used to validate the results. With less than 3% of the original band also, the desired accuracy is achieved (above 80%). The cityblock distance measure is found as a more effective distance measure. However, other datasets can also be cross-checked to find the global significance of proposed approach, along with other classifiers. Various heuristic approaches can also be applied with different objective functions.

References

1. Dobigeon, N., Tourneret, J.Y., Chang, C.I.: Semi-supervised linear spectral unmixing using a hierarchical Bayesian model for hyperspectral imagery. IEEE Trans. Sig. Process. **56**(7), 2684–2695 (2008)
2. Du, Q., Yang, H.: Unsupervised band selection for hyperspectral image analysis. Int. Geosci. Remote Sens. Symp. (IGARSS) **5**(4), 282–285 (2007)
3. Feng, F., Li, W., Du, Q., Zhang, B.: Dimensionality reduction of hyperspectral image with graph-based discriminant analysis considering spectral similarity. Remote Sens. **9**(4), 323 (2017)
4. Franchi, G., Angulo, J.: Morphological principal component analysis for hyperspectral image analysis. ISPRS Int. J. Geo-Inf. **5**(6), 83 (2016)
5. Huang, K., Li, S., Kang, X., Fang, L.: Spectral-spatial hyperspectral image classification based on knn. Sens. Imaging **17**(1), 1 (2016)
6. Ifarraguerri, A., Prairie, M.W.: Visual method for spectral band selection. IEEE Geosci. Remote Sens. Lett. **1**(2), 101–106 (2004)
7. Jia, S., Ji, Z., Qian, Y., Shen, L.: Unsupervised band selection for hyperspectral imagery classification without manual band removal. IEEE J. Sel. Top. Appl. Earth Observ. Remote Sens. **5**(2), 531–543 (2012)
8. Kun, T., Erzhu, L., Qian, D., Peijun, D.: Hyperspectral image classification using band selection and morphological profiles. IEEE J. Sel. Top. Appl. Earth Observ. Remote Sens. **7**(1), 40–48 (2014)
9. Laparra, V., Malo, J., Camps-Valls, G.: Dimensionality reduction via regression in hyperspectral imagery. IEEE J. Sel. Top. Sig. Process. **9**(6), 1026–1036 (2015)
10. Li, S., Wu, H., Wan, D., Zhu, J.: An effective feature selection method for hyperspectral image classification based on genetic algorithm and support vector machine. Knowl.-Based Syst. **24**(1), 40–48 (2011)
11. Li, S., Qiu, J., Yang, X., Liu, H., Wan, D., Zhu, Y.: A novel approach to hyperspectral band selection based on spectral shape similarity analysis and fast branch and bound search. Eng. Appl. Artif. Intell. **27**, 241–250 (2014)
12. Plaza, A., Benediktsson, J.A., Boardman, J.W., Brazile, J., Bruzzone, L., Camps-Valls, G., Chanussot, J., Fauvel, M., Gamba, P., Gualtieri, A., Marconcini, M., Tilton, J.C., Trianni, G.: Recent advances in techniques for hyperspectral image processing. Remote Sens. Environ. **113**(SUPPL. 1), S110–S122 (2009)
13. Sarhrouni. E., Hammouch, A., Aboutajdine, D.: Band selection and classification of hyperspectral images by minimizing normalized mutual information. In: Second International Conference on the Innovative Computing Technology (INTECH 2012), pp. 184–189 (2012)
14. Serpico, S.B., Bruzzone, L.: A new search algorithm for feature selection in hyperspectral remote sensing images. IEEE Trans. Geosci. Remote Sens. **39**(7), 1360–1367 (2001)
15. Yang, H., Du, Q., Chen, G.: Unsupervised hyperspectral band selection using graphics processing units. IEEE J. Sel. Top. Appl. Earth Observ. Remote Sens. **4**(3), 660–668 (2011)
16. Yang, H., Du, Q., Su, H., Sheng, Y.: An efficient method for supervised hyperspectral band selection. IEEE Geosci. Remote Sens. Lett. **8**(1), 138–142 (2011)
17. Yang, H., Du, Q., Chen, G.: Particle swarm optimization-based hyperspectral dimensionality reduction for urban land cover classification. IEEE J. Sel. Top. Appl. Earth Observ. Remote Sens. **5**(2), 544–554 (2012)

18. Zhang, X., Sun, Q., Li, J.: Optimal band selection for high dimensional remote sensing data using genetic algorithm. In: Second International Conference on Earth Observation for Global Changes, vol. 7471, pp. 74711R–74711R–7 (2009)
19. Zortea, M., Plaza, A.: Spatial preprocessing for endmember extraction. IEEE Trans. Geosci. Remote Sens. **47**(8), 2679–2693 (2009)

Enhanced Particle Swarm Optimization Technique for Interleaved Inverter Tied Shunt Active Power Filter

Vijayakumar Gali, Nitin Gupta and R. A. Gupta

Abstract This paper proposes an advanced search-based enhanced particle swarm optimization technique (EPSO) for interleaved inverter tied shunt active power filter (SAPF) to tune the integral and proportional (PI) controller gain values. The present power system is connected with inevitable nonlinear loads which cause harmonic pollution in the system. The unpredictable loses of interleaved inverter tied SAPF like switching loses, inductor power losses have to be supplied by DC-link capacitor under steady-state operation and load real power during transient condition of the load. This is effectively controlled by the PI controller to maintain the reference value under any circumstances. Conventionally, PI gain values had been tuned by the linearized model of SAPF. However, this technique gives inadequate results under voltages and transient condition of the loads. The soft computing techniques play a tremendous role in optimization of the gain parameters of PI controller values. The EPSO reduces the search process to obtain the best position which reduces the convergence time, memory and improves the convergence speed compared to conventional PSO. The complete system is modeled using MATLAB/Simulink software to show the comparative analysis of analytical PI, conventional PSO, and EPSO performance under the steady-state and transient condition of nonlinear loads. The simulation results have been validated by developing the hardware prototype model in the laboratory using dSPACE1104 controller. Complete results have been presented.

Keywords Artificial intelligent techniques · Conventional PSO
Interleaved inverter tied SAPF · Proposed enhanced PSO · Power quality

V. Gali · N. Gupta (✉) · R. A. Gupta
Department of Electrical Engineering,
Malaviya National Institute of Technology Jaipur, Jaipur, India
e-mail: nitingupta.ee@mnit.ac.in

V. Gali
e-mail: vijaykumar209@gmail.com

R. A. Gupta
e-mail: ragupta.ee@mnit.ac.in

© Springer Nature Singapore Pte Ltd. 2019
J. C. Bansal et al. (eds.), *Soft Computing for Problem Solving*,
Advances in Intelligent Systems and Computing 816,
https://doi.org/10.1007/978-981-13-1592-3_45

1 Introduction

Intelligent controllers are being used in different applications like automobiles, electrical power system, robotics which gives optimized solutions for various problems of multi-objective functions. The economic growth of country depends on the policies they make to mitigate the power quality (PQ) problems due to industrial and commercial appliances. Continued proliferation of power electric-based equipment creates harmonic pollution such as current harmonics, reactive power burden and voltage swell, voltage sag in the electrical distribution and transmission system, respectively. This harmonic pollution creates huge problems or malfunctioning of various equipment like create humming noise in transformers, damage the winding of electrical machines and increase the rating of machine, failure of insulation, create interferences in the telecommunication system, malfunctioning of digital meters. There are many researchers across the globe finding solutions to solve these PQ problems [1, 2]. The shunt active power filters (SAPFs) are being developed in this paper to mitigate current harmonics, reactive power to improve the power factor of the system.

The shunt APF works by incorporating with the reference current generation techniques. There are many reference current generation techniques which are available in the literature such as p-q theory and also named as instantaneous reactive power theory (IRPT) [3], unit template technique, power balance theory, $Icos\Phi$ control algorithm, synchronous reference frame (SRF) theory, instantaneous symmetrical component theory (ISCT), enhanced phase-locked loop (EPLL)-based control algorithm [4]. These control techniques are highly influenced by the supply voltage and load conditions. Practically, the supply voltage and loads are disturbed due to large usage of nonlinear loads. The DC-link voltage stabilization under these supply voltage and load conditions is very a big challenge for the researchers. Proportional and integral (PI) controllers are well-established controllers in commercial and industrial applications. The interleaved inverter tied SAPF makes use of PI controller to stabilize DC-link voltage, overcome the losses occurring due to switching action of SAPF switches, inductor losses, etc. [5]. The PI controller gain values can be obtained by conventional linearization model of shunt APF. However, these gain values may not work under any variation in the operating conditions. The optimization of PI controller gain values under different supply voltage and load conditions is more prominent scope for the researchers.

The artificial intelligent techniques play an immense role to solve multi-objective function due to nonlinearity of the system. There are many researchers across the globe who presented many evolutionary algorithm-based approaches/intelligent control techniques [6]: Artificial neural network (ANN), bacteria foraging (BF), BF with swarming (BFS), ant colony optimization, particle swarm optimization (PSO), fuzzy logic controllers, etc. [7, 8]. The complexity of program, optimum number of iterations, adopting the behavior of the system under large variations are the major considerations while designing controllers. The above-mentioned evolutionary theories work effectively under undesirable operations of the system. However, these intelligent techniques uses more complex calculations which reduce the speed of the processor and hence slower the execution of the program. In contrary to overcome these challenges, author proposed enhanced particle swarm optimization (EPSO) in this paper [9, 10]. The EPSO increases the convergence speed by considering only global best positions obtained by each particle. This leads to reduce the calculation burden on the processor and greater diversity among the solutions.

This paper is organized in the following sections. The interleaved inverter tied SAPF architecture is depicted in Sect. 2. Section 3 depicts the generalized p-q theory for reference current generation. Section 4 describes soft computation of DC-link voltage stabilization using conventional PSO and EPSO. The performance of interleaved inverter tied SAPF is represented in Sect. 5 and followed by the conclusions.

2 Interleaved Inverter Tied SAPF Architecture

The construction of interleaved inverter-based SAPF is different from the conventional voltage source inverter-based SAPF. The block diagram of the interleaved inverter tied SAPF is shown in Fig. 1.

This converter has six legs where each of the legs consists of one power electronic switch and power diode which has fast recovery capability. This type of construction avoids shoot-through state which is a common problem in the conventional VSI topology. The two legs of the interleaved inverter combine with the interleaved inductors (L_{a1}, L_{a2}) as one phase. The three phases are formed with the six legs of the interleaved inverter. The interleaved inductors (L_{a1}, L_{a2}, L_{b1}, L_{b2}, L_{c1}, L_{c2}) act as interfacing inductors which help to remove ripples in the compensating current and connect as medium between interleaved inverter tied SAPF and grid. The DC-

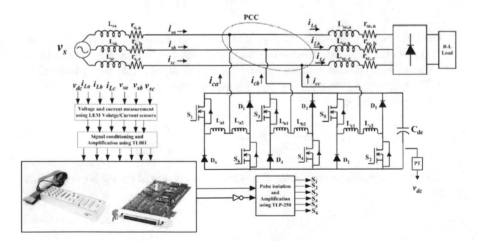

Fig. 1 Block diagram of interleaved inverter tied SAPF

link electrolytic capacitor is connected at the DC side of the interleaved inverter which supplies real power during the transient condition of the load and required compensating currents for harmonic mitigation.

3 Reference Current Generation

There are various reference current generation techniques available in the literature. However, generalized p-q theory is often easy to implement [3]. In generalized p-q theory, two types of approaches are present: direct and indirect current control techniques. In this paper, indirect current control method is adopted to mitigate current harmonics and reactive power.

3.1 Indirect Current Control Technique

According to indirect current control method, first, the three-phase supply voltages and load currents are to be converted from a-b-c to α-β coordinates by using Clerk's transformation. The orthogonal coordinates of supply voltages are as follows:

$$
\begin{bmatrix} v_{s\alpha} \\ v_{s\beta} \end{bmatrix} = \sqrt{\frac{2}{3}} \begin{bmatrix} 1 & \frac{-1}{2} & \frac{-1}{2} \\ 0 & \frac{\sqrt{3}}{2} & \frac{-\sqrt{3}}{2} \end{bmatrix} \begin{bmatrix} v_{sa} \\ v_{sb} \\ v_{sc} \end{bmatrix}
\tag{1}
$$

$$
\begin{bmatrix} i_{L\alpha} \\ i_{L\beta} \end{bmatrix} = \sqrt{\frac{2}{3}} \begin{bmatrix} 1 & \frac{-1}{2} & \frac{-1}{2} \\ 0 & \frac{\sqrt{3}}{2} & \frac{-\sqrt{3}}{2} \end{bmatrix} \begin{bmatrix} i_{la} \\ i_{lb} \\ i_{lc} \end{bmatrix}
\tag{2}
$$

The orthogonal coordinates of supply voltages $v_{s\alpha}$, $v_{s\beta}$ and the load currents are $i_{L\alpha}$, $i_{L\beta}$, respectively. The reference source currents can be obtained from (1) and (2). According to this technique, the source should supply only real power of the load. Therefore, the reference source currents in respective to α-β coordinates can be expressed as follows:

$$
i_{s\alpha}^* = \frac{v_{L\alpha}}{v_{L\alpha}^2 + v_{L\beta}^2} \overline{p} \quad \text{and} \quad i_{s\beta}^* = \frac{v_{L\beta}}{v_{L\alpha}^2 + v_{L\beta}^2} \overline{p}
\tag{3}
$$

These α-β coordinates of source currents can be converted into a-b-c coordinates as follows:

$$\begin{bmatrix} i_{sa}^* \\ i_{sb}^* \\ i_{sc}^* \end{bmatrix} = \sqrt{\frac{2}{3}} \begin{bmatrix} 1 & 0 \\ \frac{-1}{2} & \frac{\sqrt{3}}{2} \\ \frac{1}{2} & \frac{-\sqrt{3}}{2} \end{bmatrix} \begin{bmatrix} i_{s\alpha}^* \\ i_{s\beta}^* \end{bmatrix} \tag{4}$$

The error will be generated by comparing the actual source current and reference source current which further processed through hysteresis controller. This hysteresis controller generates switching signals for the interleaved inverter tied SAPF.

4 DC-Link Voltage Stabilization

The compensation efficiency highly depends on the DC-link voltage stabilization [5]. The DC-link capacitor experiences the overshoot and undershoot under the transient condition of the load. This PI controller minimizes the switching losses occurring while switching of interleaved inverter tied SAPF, inductor losses, etc., and real power required by the load during the transient condition. The conventional linearized model of SAPF PI tuning gives inadequate results under the transient condition of the load. The controller has to be fast to respond if any undesirable operation happens in the system. Moreover, conventionally, system modelling is extremely nonlinear due to involving large use of inequality constraints in the non-stationary system. Thus, enhanced operation of interleaved inverter tied SAPF is archived by using various optimization techniques. The gain parameters of PI controller can be tuned under any state of load. However, the loads in the electrical distribution system are not constant, but vary frequently. Therefore, there is enormous difference between real power supplied by the source and load demand. The DC-link capacitor has to retain quickly its reference value to balance the real power supplied by the source and load demand. Thus, the DC-link capacitor acts as an energy source to the shunt APF to balance the load demand and supply. The following equation shows the average energy stored in the DC-link capacitor

$$E_{dc} = C_{dc} \frac{(V_{dc,avg}^*)^2}{2} \tag{5}$$

where $V_{dc,avg}^*$ and C_{dc} are the average DC-link voltage and capacitor value, respectively. The energy loss in the DC-link capacitor is as follows:

$$\Delta E_{dc} = E_{dc}^* - E_{dc} = C_{dc} \frac{(V_{dc}^*)^2 - (V_{dc,avg}^*)^2}{2} \tag{6}$$

where ΔE_{dc} is the energy loss, V_{dc}^* is the DC-link capacitor voltage. The PI controller works effectively to maintain DC-link voltage constant; therefore, these undesirable losses can be minimized. However, conventional PI controller gives inadequate results under the transient condition of load. To overcome these problems of PI tuning, various optimization techniques have been used to tune the PI controller gain values. The output of PI controller can be obtained as follows:

$$
\left| \begin{array}{l}
u(t) = K_p.e(t) + K_i \int\limits_0^t e(t).\mathrm{d}t \\[4mm]
= K_p.[w(t) - c(t)] + K_i \int\limits_0^t [w(t) - c(t)].\mathrm{d}t
\end{array} \right|
\tag{7}
$$

Here, $w(t)$, $c(t)$, and $e(t)$ are the PI controller desired output, actual output, and generated error, respectively.

The conventional PSO-based PI and enhanced PSO-based PI controllers are discussed in this paper.

4.1   Conventional PSO-Based PI Controller

The artificial intelligent techniques play an immense role to solve multi-objective function due to nonlinearity of the system. The PSO was developed by Kennedy and Eberhart in 1995 [8]. This was inspired by observing social behavior of the bird flock or fish schooling. PSO employs based on population of particles in the search space. The swarm having N number of particles that fly n-dimensional domain in the search space to explore optimized solutions. Every particle in the search space explores its own best position and best swarm overall experience that is the swarm intelligence. The interleaved inverter tied SAPF undergoes undesirable operation under the transient condition of the load which increases the nonlinearity in the system. The PSO has been proven to be very efficient in solving nonlinearity problems, non-differentiability, multiple objective function, and multi-dimensional problem. It is comprehensively used because of its simple structure, ease of implementation, less computational burden, and well-defined mechanism to explore both local and global maxima. The following equations are used for updating the positions and velocities of the particles in accordance with the social only and cognition only components.

$$
u_{k+1}^i = \chi u_k^i + \lambda_1.c_1(q_{Lbest}^i - x_k^i) + \lambda_2.c_2(q_{gbest}^i - x_k^i)
\tag{8}
$$

$$
x_{k+1}^i = x_k^i(1 - \lambda_2) + \lambda_2 q_{gbest} + \eta\beta
\tag{9}
$$

where q_{Lbest}^i and q_{Gbest}^i are local best and global best obtained by the particles, respectively. k and i are the number of iterations and particle number, respectively. x_k^i and u_k^i are the present position and velocity of ith particle at kth iteration, respectively. x_{k+1}^i and u_{k+1}^i are the position and velocity of the ith particle at $(k+i)$th iteration. χ, λ_1 and λ_2 are the coefficient of inertia, cognitive and social constraints, respectively. The random numbers c_1 and c_2 are in the interval $[-1, 1]$. The speed of the particles is decided by the λ_1 and λ_2 where the balance is provided by χ between local and global in the search space. The search process will be terminated if the pre-defined maximum number of iterations completed or additional best solution is not obtained. The number of iterations and increased number of premature convergence of particle's best position will increase the complexity of program and degrade the convergence speed. The EPSO is proposed in this paper to overcome the problems of conventional PSO and to improve the performance of interleaved SAPF.

4.2 Proposed Enhanced Particle Swarm Optimization

The EPSO works to eliminate the premature convergence and reduce the computational complexity by avoiding each particle's best position obtained during the search process [9, 10]. This improves the convergence speed of the particle to obtain the best solution.

The modified EPSO equations for the ith particle's position after $(k + 1)$th iterations can be written as follows:

$$x_{k+1}^i = x_k^i(1 - \lambda_2) + \lambda_2 q_{g\text{best}} + \eta\beta \tag{10}$$

where η is the random number which decreases after successive iterations and can be defined as follows:

$$\eta = \eta_0 e^{-\alpha t} \quad (0 < \alpha < 1), (0.5 < \alpha_0 < 1) \tag{11}$$

The process of EPSO is as follows:

Step 1 Initialize the position and velocities of each particle.
Step 2 Find the fitness of each particle.
Step 3 Find the best position obtained by each particle.
         Update the particles with the best position obtained, re-initialize the position of each particle, and analyze the lowest value obtained whether its new position is suitable; if it is yes, update its position, otherwise assign other position randomly.
Step 4 Compare each particles global best fitness value with the $q_{g\text{best}}$; if the present value is greater, update its fitness value.
Step 6 Check whether Eqs. (6) and (7) satisfy, quit the iteration otherwise and return to step 3.

5 Simulation Results and Discussion

The simulation performance of the proposed interleaved inverter tied SAPF is modeled using MATLAB/Simulink environment. A three-phase full wave diode bridge rectifier with resistive and inductive elements (R-L) is connected as nonlinear load. The nonlinear load injects current harmonics and draws reactive power from the system. The performance of three-phase interleaved inverter tied SAPF is shown in Fig. 2. It is observed that the interleaved inverter is working as a shunt APF without shoot-through effect and compensates current harmonics in the source current by bringing the total harmonic distortion from 24.18 to 2.59% as shown in Fig. 3a, b, respectively. The steady-state DC-link stabilization has been observed for conventional PI, PSO-based PI, and EPSO-based PI. The simulation convergence characteristics of the PI, conventional PSO, and EPSO are shown in Fig. 4. It is contemplated that the EPSO converges faster due to elimination of finding local maxima where the conventional PSO searches for local maxima. Therefore, EPSO gives fast harmonic compensation action.

The practical electrical power system is subjected to different load transients which vary the output power demand. The DC-link capacitor has to be regulated to supply the load demand under this transient condition of the load for efficient performance of the interleaved inverter tied SAPF.

A comparison study of conventional PI, PSO-based PI, and EPSO-based PI have been carried out to test the system performance under the transient condition of the load and DC-link voltage stabilization and is shown in Fig. 5a, b, respectively.

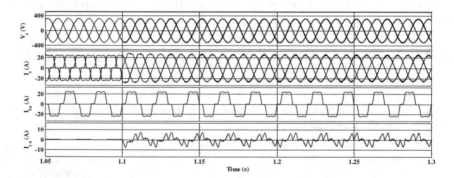

Fig. 2 Performance of interleaved inverter tied SAPF under steady-state of nonlinear load condition

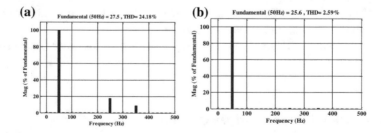

Fig. 3 FFT analysis of source current **a** before compensation and **b** after compensation

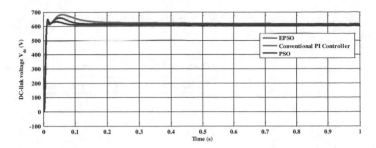

Fig. 4 DC-link voltage stabilization

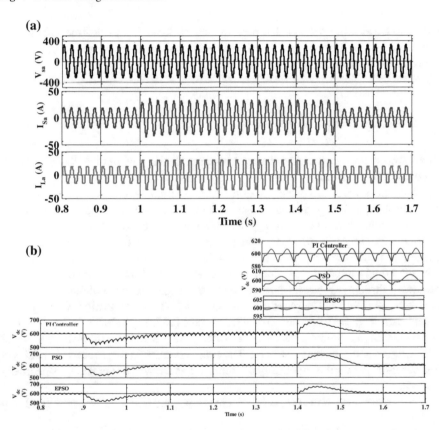

Fig. 5 **a** Performance parameters under transient condition of the load **b** conventional PI, conventional PSO, and EPSO-based PI controller DC-link voltage stabilization

It is contemplated that the step load is increasing and decreasing at 1 and 1.5 s, respectively. The EPSO-based PI controller sets back the DC-link voltage at its reference value within extremely short period of time where the PSO and conventional method of PI controller tuning take longer time. It is evident that the proposed

Table 1 Comparison of conventional PI, PSO-based PI, and EPSO-based PI controller

Controller type	System characteristics					
	Rise time (ms)		Overshoot (%)		Settling time (ms)	
	Simulation	Hardware	Simulation	Hardware	Simulation	Hardware
Conventional PI	0.05	0.1	30	40	0.3	0.5
PSO-PI	0.03	0.05	20	25	0.2	0.4
EPSO-PI	0.01	0.03	10	5	0.1	0.2

EPSO-based PI tuning gives better performance with quick settling time and reduced overshoot and undershoot under the steady-state and transient condition of the load. The complete comparison is given in Table 1.

6 Hardware Implementation

The hardware prototype model of three-phase interleaved inverter tied SAPF is developed to test the performance under the steady-state and transient condition of the load. The laboratory hardware prototype of three-phase interleaved inverter tied SAPF is shown in Fig. 5. The performance parameters of three-phase supply voltages (v_{sa}, v_{sb}, v_{sc}), source currents (i_s, i_{sb}, i_{sc}) before compensation are shown in Fig. 7a, b, respectively. The generalized p-q theory is used for calculating the reference source currents as mentioned in Sect. 3. The calculated reference source current is compared with the actual source current which produces error signal. This error signal is further processed through hysteresis current controller for generating switching signals for the interleaved SAPF. The phase-a source current after compensation (CH1), phase-a load current (CH4), and compensating current (CH3) are shown in Fig. 7c. It is contemplated that the interleaved inverter tied SAPF working successfully as shunt APF without shoot-through problem. The source current THD reduced from 25.8 to 3.95% which shows the superiority of controller as shown in Fig. 8a, b, respectively (Fig. 6).

The performance of the proposed system is analyzed under the transient condition of the load.

The EPSO-based PI controller sets back the DC-link voltage at its reference value within extremely short period of time where the PSO and conventional method of PI controller tuning take longer time as shown in Fig. 9a–c, respectively. The rise time, % overshoot, and settling time are 0.1 ms, 40%, 0.5 ms using PI, 0.05 ms, 25%, 0.4 ms using conventional PSO, and 0.03 ms, 5%, 0.2 ms using EPSO. It is evident that the proposed EPSO-based PI tuning gives better performance with quick settling time and reduced overshoot and undershoot under the steady-state and transient condition of the load. The complete comparison is shown in Table 1.

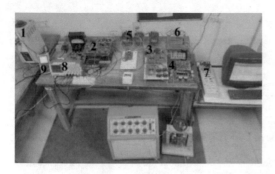

1. Three-phase Supply
2. Voltage/ Current sensor circuit
3. Voltage source inverter
4. GATE driver circuit
5. Interfacing inductors
6. Diode bridge rectifier
7. dSPACE1104 Controller
8. DSO
9. Power Quality Analyzer

Fig. 6 Laboratory hardware prototype model of interleaved inverter tied SAPF

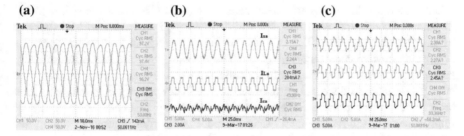

Fig. 7 **a** Three-phase source voltages **b** three-phase source currents before compensation **c** phase-a source current after compensation (CH1), phase-a load current (CH4) and compensating current (CH3)

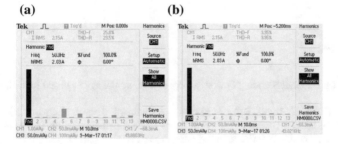

Fig. 8 Phase-a source current THD **a** before compensation and **b** after compensation

7 Conclusion

Soft commutating techniques are applied to solve the multi-objective function of DC-link voltage stabilization under the steady-state and transient state of the non-linear loads. A fair comparison of the conventional PI, conventional PSO and EPSO

(a)   (b)   (c)

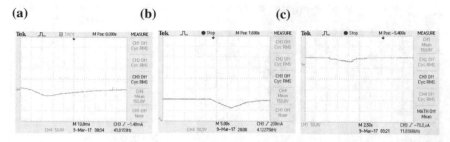

Fig. 9 DC-link voltage stabilization under transient condition of the load by using **a** conventional PI controller **b** PSO-based PI controller **c** EPSO-based PI controller

are tested under the transient condition of the load. The electrical power system is interconnected with various nonlinear loads which inject harmonics into the system. The unpredictable losses of inductor and switching power losses will be supplied by the DC-link capacitor. The DC-link capacitor undergoes inadequate operation under the transient condition of the load. The PI controller has been used for stabilizing the DC-link voltage under these conditions of the load. However, this gives poor results under sudden changing of loads. The PI controller gain parameters have been tuned with different conventional PSO and EPSO approaches. The performance improvement of the proposed EPSO is compared with conventional PI tuning and conventional PSO approaches using MATLAB/Simulink environment. A prototype model has been developed in the laboratory to validate the simulation results. It can be concluded from both simulation and hardware results that the proposed EPSO converges fast to find the best optimum PI gain parameters under the steady-state and transient condition of the load with less overshoot and undershoot.

References

1. Bollen, M. H.: Understanding Power Quality Problems: Voltage Sags and Interruptions. Wiley (1999)
2. Gali, V., Gupta, N., Gupta, R.A.: Application of shunt active power filters in medical diagnosis and critical lab equipment. In: Proceedings of IEEE International Conference on Power and Embedded Drive Control (ICPEDC), Chennai, pp. 290–295 (2017)
3. Akagi, H., Watanabe, E.H., Aredes, M.: Instantaneous Power Theory and Applications to Power Conditioning. IEEE Press, Piscataway (2007)
4. Singh, B., Chandra, A., Al-Haddad, K.: Power Quality: Problems and Mitigation Techniques. Wiley (2014)
5. Wang, Y., Xie, Y.X.: Adaptive DC-link voltage control for shunt active power filter. Int. J. Power Electron. **14**(4), 764–777 (2014)
6. Gupta, N., Singh, S. P., Dubey, S. P.: Neural network based shunt active filter for harmonic and reactive power compensation under non-ideal mains voltage. In: Proceedings of IEEE International Conference on Industrial Electronics and Applications (ICIEA), Taiwan, pp. 370–375 (2010)

7. Mishra, S., Bhende, C.N.: Bacterial foraging technique-based optimized active power filter for load compensation. IEEE Trans. Power Del. **22**(1), 457–465 (2007)
8. Patnaik, S.S., Panda, A.K.: Real-time performance analysis and comparison of various control schemes for particle swarm optimization-based shunt active power filters. Int. J. Electr. Power Energy Sys. **52**, 185–197 (2013)
9. Lu, Y., Liang, M., Ye, Z., Cao, L.: Improved particle swarm optimization algorithm and its application in text feature selection. Int. J. Appl. Soft Comput. **35**, 629–636 (2015)
10. de Souza, J.S., Molina, Y.P., de Araujo, C.S., de Farias, W.P., de Araujo, I.S.: Modified particle swarm optimization algorithm for sizing photovoltaic system. IEEE Latin Am. Trans. **15**(2), 283–289 (2017)

A New Improved Hybrid Algorithm for Multi-objective Capacitor Allocation in Radial Distribution Networks

S. Mandal, K. K. Mandal, B. Tudu and N. Chakraborty

Abstract Capacitor allocation is one of the complex and challenging problems in modern power system operation. Several benefits including reduction of system loss and improvement of bus voltage profile can be achieved by optimal capacitor allocation. It is a complex nonlinear constrained optimization problem, and many modern meta-heuristic techniques have been applied to solve the problem. However, one of the major difficulties in modern population-based meta-heuristic techniques is the selection of control parameters which are normally problem-dependent. As reported in the literature, a wrong parameter selection may result in premature convergence and even lead to stagnation. This paper introduces a new improved algorithm using chaos enhanced differential evolution (DE) for optimal capacitor allocation to avoid premature convergence. Logistic map chaotic sequence is used for self-adjustment of control parameters of DE. A multi-objective framework for the capacitor allocation problem is presented considering active power loss, cumulative voltage deviation (CVD) and system cost with a set of equality and inequality constraints. A 69-bus radial distribution is selected for demonstration and validation purpose.

1 Introduction

Distribution system is an important part of power system and normally operates at low voltage. Capacitors are widely employed for reduction of system loss, improvement of bus voltage profile and power factor. However, capacitors of optimal sizes are to be placed at optimal locations in order to achieve maximum societal benefits in terms of reduced cost, loss reduction and voltage profile improvement.

Optimal capacitor allocation is a complex combinatorial optimization problem. Numerous methods including traditional numerical methods and modern heuristic

S. Mandal
Department of Electrical Engineering, Jadavpur University, Kolkata 700032, India

K. K. Mandal (✉) · B. Tudu · N. Chakraborty
Department of Power Engineering, Jadavpur University, Kolkata 700098, India
e-mail: kkm567@yahoo.co.in

© Springer Nature Singapore Pte Ltd. 2019
J. C. Bansal et al. (eds.), *Soft Computing for Problem Solving*,
Advances in Intelligent Systems and Computing 816,
https://doi.org/10.1007/978-981-13-1592-3_46

techniques have been applied to solve the problem. Many traditional analytic methods have been reported as earlier works in the literature [1]. In recent years, many algorithms and techniques have been proposed to address the problem of optimal capacitor allocation including mixed integer nonlinear programming [2], genetic algorithm [3], artificial bee colony algorithm [4], bacteria foraging optimization [5], teaching–learning-based optimization [6], cuckoo search algorithm [7], ant colony search algorithm [8], particle swarm optimization [9] and firefly algorithm [10]. Recently, a new and comprehensive objective function formulation for capacitor placement in distribution networks was proposed by Karimi et al. [11] and a particle swarm optimization algorithm was used for demonstration purpose. More recently, Abril [12] proposed a new search algorithm of inclusion and interchange of variables for obtaining the sizes, the placements and the control schemes of the capacitors banks in distribution systems. A new problem formulation was presented, and the algorithm was tested on 69- and 85-bus system for validation purpose.

Differential evolution (DE) is one of the efficient members of evolutionary algorithms. It was first proposed by Storn and Price in the year 1995 [13]. One of the major drawbacks for almost all modern heuristic techniques is the setting of control parameters on which the success of the particular method is largely dependent. Further, these parameters are in general problem-dependent. A new self-adaptation strategy of the control parameters of DE is proposed in this work using chaotic sequence. For the present work, logistic map chaotic sequence is utilized and it is termed as logistic map differential evolution (LMDE). The proposed algorithm is used for optimal capacitor allocation in distribution networks.

2 Problem Formulation

The primary aim of the present work is to allocate capacitors of optimal size and place them at optimal locations so that system loss is minimized with enhancement of bus voltage profile. A multi-objective problem formulation is presented considering both loss and cumulative voltage deviation (CVD). Finally, yearly savings are calculated. This section describes the problem, and finally, objective function is formulated.

2.1 Calculation of Power Loss

Power loss in a radial distribution network system can be calculated by summing up the losses of all the lines of the system. Figure 1 shows the single-line diagram of a radial distribution network with n number of buses.

From Fig. 1, one can write the following set of equations for determination of real power and reactive power loss and bus voltage.

$$P_{i+1} = P_i - P_{Li+1} - R_{i,i+1}\left[\left(P_i^2 + Q_i^2\right)/|V_i|^2\right] \tag{1}$$

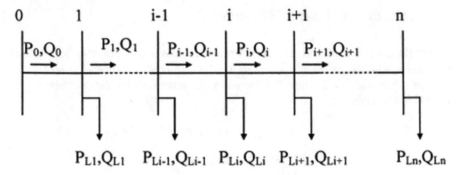

Fig. 1 Single-line diagram of a simple radial distribution system

$$Q_{i+1} = Q_i - Q_{Li+1} - X_{i,i+1} \cdot \left[(P_i^2 + Q_i^2)/|V_i|^2 \right] \tag{2}$$

$$|V_{i+1}|^2 = |V_i|^2 - 2(R_{i,i+1} \cdot P_i + X_{i,i+1} \cdot Q_i) + (R_{i,i+1}^2 + X_{i,i+1}^2) \frac{(P_i^2 + Q_i^2)}{|V_i|^2} \tag{3}$$

where P_i is the real power flowing out of bus i, Q_i is reactive powers flowing out of bus i, P_{Li} represents active load at ith bus, Q_{Li} represents reactive load at ith bus, $R_{i,i+1}$ indicates the resistance and reactance of the line section between buses i and $i+1$, $X_{i,i+1}$ indicates the resistance and reactance of the line section between buses i and $i+1$, and V_i is the voltage of ith bus.

The active power loss of the line section between buses i and $i+1$ can be found as

$$P_{\text{Loss}}(i, i+1) = R_{i,i+1} \cdot \frac{P_i^2 + Q_i^2}{|V_i|^2} \tag{4}$$

Therefore, total active and reactive power loss for the system can be determined as

$$P_{T,\text{Loss}} = \sum_{i=0}^{n} P_{\text{Loss}}(i, i+1) \tag{5}$$

$$Q_{T,\text{Loss}} = \sum_{i=0}^{n} Q_{\text{Loss}}(i, i+1) \tag{6}$$

where $P_{T,\text{Loss}}$ indicates total real power loss of the system under consideration and $Q_{T,\text{Loss}}$ indicates total reactive power loss of the system under consideration.

2.2 Cumulative Voltage Deviation

One of the objectives of optimal capacitor allocation is improvement of bus voltage profile for enhancing system performance. Thus, voltage deviation is to be kept as smaller as possible. CVD can be considered as a measure for bus voltage enhancement [7] and is expressed as

$$
\text{CVD} = \begin{cases} 0 & \text{for} \quad 0.95 \leq V_i \leq 1.05 \\ \sum_{i=1}^{n} |1 - V_i| & \text{otherwise} \end{cases} \tag{7}
$$

2.3 Objective Function Formulation

The primary objective of optimal capacitor allocation is to minimize system loss and enhancement of bus voltage profile subject to several equality and inequality constraints. Thus, it is expressed as

$$
\text{Minimize } f = \text{Minimize}\left(w_1 \times \frac{P_{L,WC}}{P_{L,WOC}} + w_2 \times \frac{\text{CVD}_{CWC}}{\text{CVD}_{WOC}} \right) \tag{8}
$$

where $P_{L,WC}$ indicates the real power loss of the system with capacitor placed at optimal location, $P_{L,WOC}$ represents the real power loss of the original system without capacitor, and w_1 and w_2 are the two weight factors. For loss minimization $w_1 = 1$ and $w_2 = 0$, whereas $w_1 = 0$ and $w_2 = 1$ result in minimization of CVD. When both loss and voltage deviation are considered, both are set to 1, i.e. $w_1 = 1$ and $w_2 = 1$. The above function is to be minimized subject to the several constraints.

2.4 Constraints

The above objective function is to be minimized subject to the several equality and inequality constraints.

(i) Power balance constraints:

At every instant, both active and real power constraints must be satisfied and can be represented as

$$
P_{\text{Slack}} = \sum_{i=1}^{n} P_{Di} + \sum_{j=1}^{n_L} P_{Lj} \tag{9}
$$

$$Q_{\text{Slack}} + \sum_{i=1}^{n_B} Q_{Ci} = \sum_{i-1}^{n} Q_{Di} + \sum_{j=1}^{n_L} Q_{Li} \qquad (10)$$

where P_{Slack} and Q_{Slack} are the active and reactive power of the slack bus respectively, P_{Di} and Q_{Di} are the active reactive power demand of ith bus respectively, P_{Lj} and Q_{Lj} represent active and reactive power loss at jth line respectively, n, n_B and n_L indicate the total bus number, number of buses where capacitors are placed and number of lines, respectively.

(ii) Bus voltage constraints:

For maintaining service quality, bus voltages must be maintained within specified limits and it is expressed as

$$V_i^{\min} \leq V_i \leq V_i^{\max}$$
$$i = 1, 2, \ldots, n \qquad (11)$$

where $V_i^{\min}$ and $V_i^{\max}$ are the minimum and maximum bus voltage limits.

(iii) Limits on capacitor size:

For a practical system, size of capacitor to place at the buses must be limited by lower and upper limits and it is represented as

$$Q_{Ci}^{\min} \leq Q_{Ci} \leq Q_{Ci}^{\max}$$
$$i = 1, 2, \ldots, n_B \qquad (12)$$

where $Q_{Ci}^{\min}$ and $Q_{Ci}^{\max}$ are the lower limit and upper limit of capacitor to be placed.

(iv) Limits on maximum VAr compensation:

For practical point of view, maximum reactive power injection by using capacitor must be less than reactive power demand and is expressed as

$$\sum_{i=1}^{n_B} Q_{Ci} \leq \sum_{j=1}^{n_L} Q_{Dj} \qquad (13)$$

3 Power Loss Index

Power loss index (PLI) is used to identify the potential candidates for placing the capacitors. Capacitors are to be placed at relatively weaker buses for achieving maximum economic benefits. Real power loss is calculated for all buses except the slack bus for the uncompensated system. Thus, PLI for the kth bus can be expressed as

$$\mathrm{PLI}(k) = \frac{\mathrm{LR}(k) - \mathrm{LR}^{\min}}{\mathrm{LR}^{\max} - \mathrm{LR}^{\min}} \tag{14}$$

where $\mathrm{LR}(k)$ is the loss reduction (LR) for the kth bus, and $\mathrm{LR}^{\min}$ and $\mathrm{LR}^{\max}$ are the minimum and maximum loss reduction, respectively.

The buses with higher loss are characterized with lower bus voltage. Therefore, these buses are the potential candidates for placing capacitors. The buses are arranged in decreasing order depending on PLI, and the list of candidates is prepared.

4 Calculation of Yearly Savings

Once the optimal size and locations of capacitors are determined by the proposed algorithm, the yearly savings can be calculated considering equivalent monetary benefits due to reduction in active power loss, capacitor purchase, cost of installation and maintenance cost. This can be expressed as

$$C_e \times \left(P_{L,\,WOC} - P_{L,\,WC}\right) \times T - C_c \times \sum_{j=1}^{n_B} Q_{Cj} - (Q_{ci} + Q_{Co}) \tag{15}$$

where C_e is average rate of energy cost, C_c is capacitor purchase cost, C_{ci} is capacitor installation cost, C_{co} is capacitor operating cost, $P_{L,\,WC}$ is total active power loss after placement of capacitor, $P_{L,\,WOC}$ is total active power loss of the original system, i.e. before the placement of capacitor, T is total time interval, Q_{Cj} is the amount of reactive power of installed by capacitor bank at bus j, and n_B is number of buses where capacitors are placed.

5 Differential Evolution (DE)

DE is one of the efficient and powerful global optimization techniques and was proposed by Storn and Price in 1995 [13, 14]. It is easy to implement and has only four steps: initialization, mutation, crossover and selection.

5.1 Initialization

Let there be d-dimensional decision variables in the problem search space. It begins randomly the initialization process of N_P individuals with d-dimensional decision variables. Thus, if $x_{i,j}^{(o)}$ represents the jth component of the ith decision vector then it can be expressed as

$$x_{i,j}^{(0)} = x_{\min,j} + \mathrm{rand}_{i,j}()\left(x_{\max,j} - x_{\min,j}\right)$$

$$i = 1, 2, \ldots N_P \quad j = 1, 2, \ldots d \tag{16}$$

where $\mathrm{rand}_{i,j}()$ is a random number within 0 and 1 including both the upper and lower values. If $P^{(G)}$ denotes the population that evolves after Tth iteration, then it is expressed as

$$P^{(T)} = \left[X_i^{(T)}, \ldots, X_{N_P}^{(T)}\right] \tag{17}$$

$$X_i^{(T)} = \left[X_{1,i}^{(T)}, \ldots X_{d,i}^{(T)}\right]^T$$

$$i = 1, \ldots N_P \tag{18}$$

5.2 Mutation Operation

Mutation operation is the method which is utilized to create mutant or donor vectors (V_i) with the help of vector difference. Several variants of DE are available in the literature depending on vector difference of parent or target vectors (X_i). For this work, $DE/\mathrm{rand}/1$ is used. The mutation operation can be described as follows:

$$V_i^{(T)} = X_k^{(T)} + F_M\left(X_l^{(T)} - X_m^{(T)}\right) \tag{19}$$

where X_k, X_l, X_m are chosen randomly from the set of parent vectors (X_i) and $k \neq l \neq m \neq i$. The mutation factor F_M is one of the important control parameters of DE and is used for scaling the vector difference. It is normally selected by the user within the range [0, 2].

5.3 Crossover Operation

Crossover operation is used for mixing the components of mutation or donor vectors with parent or target vectors, and thus, trial vectors (U_i) are created. For the present work, binomial method is used which can be expressed as

$$U_{j,i}^{(T)} = \begin{cases} V_{j,i}^{(T)}, & \text{if } \mathrm{rand}_{i,j}() \leq C_R \text{ or } j = N \\ X_{j,i}^{(T)}, & \text{otherwise} \end{cases} \tag{20}$$

where $\mathrm{rand}_{i,j}()$ is a uniform random number within [0, 1] and N is any randomly chosen natural number from [1, 2, \ldots d]. This ensures that at least one component of trial vector U_i is selected from mutant vector V_i. The crossover factor is another important predefined control parameter of DE. Diversity of the population is controlled by C_R.

5.4   Selection Operation

Selection operation is the last step of DE which determines whether parent vector X_i or trail vector U_i is to be selected for the next iteration. If f is the objective function, then it is described as

$$X_i^{(T+1)} = \begin{cases} U_i^{(T)}, \text{ if } f\left(U_i^{(T)}\right) \leq f\left(X_i^{(T)}\right) \\ \\ X_i^{(T)}, \text{ otherwise} \end{cases} \tag{21}$$

6   Logistic Map Differential Evolution (LMDE)

Various search strategies such as simplex crossover search strategy, orthogonal search strategy and chaotic search strategy have been used to enhance the performance of many meta-heuristic techniques like DE, particle swarm optimization [15]. For the present work, chaotic search strategy is used. Actually, chaos is typical classic nonlinear dynamical system which is characterized with ergodicity and randomicity and very sensitive to initial conditions. Because of its ergodicity and randomicity, a chaotic system is capable in generating long-time sequence randomly which can traverse to every state if a long time duration is allowed. This feature of chaotic sequence is utilized for self-adaptation of two important controlling parameters, mutation factor and crossover ratio, of DE to enhance its performance.

In this paper, logistic map chaotic sequence [16] is used which can be represented as

$$y(k) = ay(k-1)[1 - y(k-1)] \tag{22}$$

where k is the sample and a is the control parameter, $0 \leq a \leq 4$. The above expression is deterministic displaying chaotic dynamics when $a = 4$ and $y(0) \notin \{0, 0.25, 0.5, 0.75, 1\}$. In this case, $y(k)$ is distributed in the range of $(0,1)$ provided the initial $y(0) \in (0, 1)$.

In this work, mutation factor and crossover ratio are controlled dynamically using (22) as follows:

$$F_M(T) = a \cdot F_M(T-1) \cdot [1 - F_M(T-1)] \tag{23}$$

$$C_R(T) = a \cdot C_R(T-1) \cdot [1 - C_R(T-1)] \tag{24}$$

where T is the current iteration number.

7 Simulation Results

The effectiveness of the proposed chaos enhanced DE-based algorithm is tested on 69-bus radial distribution system [17]. The proposed algorithm is tested using MATLAB on 3.0 GHz, 8.0 GB RAM PC.

Initial values of mutation factor and crossover ratio are chosen as 0.65 and 0.85, respectively. Maximum iteration number is set at 100 by after several run. The value of the capacitor banks is chosen in between the limits 0 and 1500 kVAr with a step of 50. Table 1 shows the rate of energy cost, purchase and maintenance cost of capacitors.

Using PLI as described in Sect. 3, the buses are arranged in descending order as $\{61, 64, 59, 65, 21, 62, 18, 17, 16, 24, 27, 26, 22, 20, 63\ldots\}$. For simplicity and direct comparison, only five higher potential buses are considered to be the candidates for capacitor allocation.

For the present work, both w_1 and w_2 are taken as 1 for a direct comparison. In other words, both the loss and cumulative voltage are given equal importance. Optimal size and location of capacitors (location, size) obtained by the proposed algorithm based on LMDE are shown in Table 2. Table 2 also compares the same with that obtained by other methods like DE-PS [17] and PSO [18].

Table 1 Rate of energy, purchase, operation and maintenance cost of capacitors

Serial no.	Item	Rate
1	Average energy cost	$0.06 kWh
2	Purchase cost	$3.00 kVAr
3	Installation cost	$1000/location
4	Operating cost	$300/year/location
5	Hours per year	8760

Table 2 Optimal size and locations of capacitor obtained by proposed LMDE

Method	Proposed method	DE-PS [17]	PSO [18]
Location and size	(21, 200)	(21, 700)	(59, 1015)
	(59, 250)	(59, 150)	(61, 240)
	(61, 250)	(61, 950)	(65, 365)
	(62, 600)	(64, 200)	
	(65, 300)	(65, 50)	
Total kVAr placed	1600	1650	1620

Table 3 Optimal results obtained by the proposed method

Parameters of comparison	Uncompensated system	Proposed method (LMDE)	DE-PS [17]	PSO [18]
$V_{min}(p.u.)$	0.9092	0.9310	0.9310	0.9340
$V_{max}(p.u.)$	0.9999	0.9999	1.000	1.000
CVD	0.7313	0.5063	0.500	0.500
P_{loss} (kW)	224.92	146.31	146.13	156.14
Reduction in P_{loss} (%)	–	34.95	35.02	30.57
Q_{loss} (kVAr)	102.13	68.20	68.06	71.95
Reduction in Q_{loss} (%)	–	33.39	33.35	29.54
Total compensation (kVAr)	–	1600	1650	1620
Net savings/year		$30,022.00	$29,951.50	$29,301.00
CPU time (s)		35.12	48.46	NA

Minimum bus voltage, active and reactive power losses and CVD for both the uncompensated and compensated system are shown in Table 3. Table 3 also compares the results with other modern heuristic methods. It is observed from Table 3 that active power loss is 146.31 kW with a reduction of 34.95% in comparison with uncompensated system. The computation time is found to be 35.12 s. Net yearly savings are found to be $30,022.00.

Variation of bus voltage is shown in Fig. 2. Figure 2 also compares the bus voltage of the uncompensated system with that of compensated system. Bus voltage is seen to be improved substantially with respect to uncompensated system.

The convergence characteristic for minimum loss is shown in Fig. 3. Figure 4 shows the convergence characteristic for the objective function. Figure 3 also compares the minimum loss obtained by the proposed LMDE and classical DE where the loss is found to be 157.31 kW. It is clearly observed that the proposed algorithm is capable in enhancing the performance of classical DE.

8 Conclusion

A new improved hybrid algorithm using logistic map chaotic sequence and DE is proposed in this paper for optimal capacitor allocation in distribution networks. The problem is formulated as multi-objective one considering loss, CVD and cost. PLI is used to find the potential candidates for allocation of capacitor. It applied on 69-bus distribution network to test its effectiveness. It is observed that proposed algorithm based on LMDE is really capable in producing good-quality solutions. Simulation

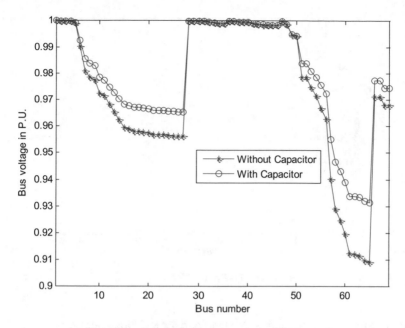

Fig. 2 Comparison of bus voltages of 69-bus radial distribution system

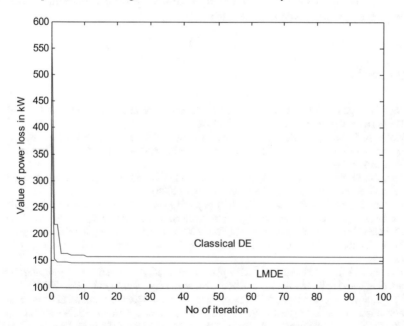

Fig. 3 Convergence characteristic for minimum loss

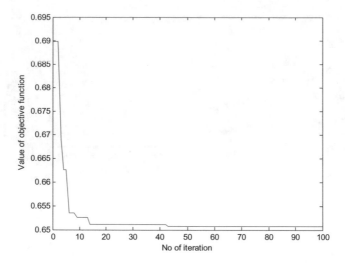

Fig. 4 Convergence characteristic for objective function

results and a comparative study with other heuristic techniques are presented. It is also found that that proposed LMDE-based algorithm is capable in avoiding premature convergence in comparison with classical DE.

References

1. Ng, H.N., Salama, M.M.A., Chikhani, A.Y.: Classification of capacitor allocation techniques. IEEE Trans. Power Del. **15**(1), 387–392 (2000)
2. Nojavan, S., Jalali, M., Zare, K.: Optimal allocation of capacitors in radial/mesh distribution systems using mixed integer nonlinear programming approach. Int. J. Electr. Power Energy Syst. **107**, 119–124 (2014)
3. Das, D.: Optimal placement of capacitors in radial distribution system using a fuzzy-GA method. Electr. Power Energy Syst. **30**, 361–367 (2008)
4. El-Fergany, A.A., Abdelaziz, A.Y.: Multi-objective capacitor allocations in distribution networks using artificial bee colony algorithm. J. Electr. Eng. Technol. **9**(2), 441–451 (2014)
5. Mohamed Imran, A., Kowsalya, M.: Optimal size and siting of multiple distributed generators in distribution system using bacteria foraging optimization. Swarm Evol. Comput., pp. 58–65 (2014)
6. Sultana, S., Roy, P.K.: Optimal capacitor placement in radial distribution systems using teaching learning based optimization. Int. J. Electr. Power Energy Syst. **54**, 387–398 (2014)
7. El-Fergany, A.A., Abdelaziz, A.Y.: Cuckoo search-based algorithm for optimal shunt capacitors allocations in distribution networks. Electr. Power Compon. Syst. **41**, 1567–1581 (2013)
8. El-Ela, A.A.A., El-Sehiemy, R.A., Kinawy, A.M., Mouwafi, M.T.: Optimal capacitor placement in distribution systems for power loss reduction and voltage profile improvement. IET Gener. Transm. Distrib. **10**(5), 1209–1221 (2016)
9. Lee, C.S., Ayala, H.V.H., Coelho, L.S.: Capacitor placement of distribution systems using particle swarm optimization approaches. Int. J. Electr. Power Energy Syst. **64**, 839–851 (2015)

10. Das, P., Banerjee, S.: Optimal sizing and placement of capacitor in a radial distribution system using loss sensitivity factor and firefly algorithm. Int. J. Eng. Comput. Sci. **3**(4), 5346–5352 (2014)
11. Karimi, H., Dash, R.: Comprehensive framework for capacitor placement in distribution networks from the perspective of distribution system management in a restructured environment. Int. J. Electr. Power Energy Syst. **82**, 11–18 (2016)
12. Abril, I.P.: Algorithm of inclusion and interchange of variables for capacitors placement. Electr. Power Syst. Res. **148**, 117–126 (2017)
13. Storn, R., Price, K.: Differential evolution: a simple and efficient adaptive scheme for global optimization over continuous spaces. Technical Report TR-95-012, Berkeley, USA: International Computer Science Institute (1995)
14. Storn, R., Price, K.: Differential evolution—a simple and efficient heuristic for global optimization over continuous spaces. J. Global Optim. **11**(4), 341–359 (1997)
15. Jia, D., Zhenga, G., Khan, M.K.: An effective memetic differential evolution algorithm based on chaotic local search. Inf. Sci. **181**, 3175–3187 (2011)
16. Gandomia, A.H., Yang, X.-S., Talatahari, S., Alavi, A.H.: Firefly algorithm with chaos. Commun. Nonlinear Sci. Numer. Simulat. **18**, 89–98 (2013)
17. El-Fergany, A.: Optimal capacitor allocations using integrated evolutionary algorithms. IET Gener. Transm. Distrib. **7**(6), 593–601 (2013)
18. Prakash, K., Sydulu, M.: Particle swarm optimization based capacitor placement on radial distribution systems. IEEE PES General Meeting, Tampa, FL, pp. 1–5 (2007)

A Hybrid Algorithm Based on Particle Swarm and Spotted Hyena Optimizer for Global Optimization

Gaurav Dhiman and Amandeep Kaur

Abstract In this paper, a novel hybrid metaheuristic optimization algorithm which is based on Particle Swarm Optimization (PSO) and recently developed Spotted Hyena Optimizer (SHO) named as Hybrid Particle Swarm and Spotted Hyena Optimizer (HPSSHO) is presented. The main concept of this algorithm is to improve the hunting strategy of Spotted Hyena Optimizer using particle swarm algorithm. The proposed algorithm is compared with four metaheuristic algorithms (i.e., SHO, PSO, DE, and GA) and benchmarked it on thirteen well-known benchmark test functions which include unimodal and multimodal. The convergence analysis of the proposed as well as other metaheuristics has also been analyzed and compared. The algorithm is tested on 25-bar real-life constraint engineering design problem to demonstrate its applicability. The experimental results reveal that the proposed algorithm performs better than other metaheuristic algorithms.

Keywords Optimization techniques · Metaheuristics · SHO · Constrained optimization · Benchmark test functions · Engineering design problem

1 Introduction

Last few decades, the computational complexity of real-world problems has been increased which becomes a major issue in the field of optimization. To reduce these problems, there is a need to develop some novel metaheuristic technique which has been used for obtaining the optimal solutions in low computational efforts.

Basically, metaheuristics are broadly classified into three categories: evolutionary, physical, and swarm intelligence-based algorithms. The first technique is a

G. Dhiman (✉) · A. Kaur
Department of Computer Science and Engineering,
Thapar University, Patiala, Punjab, India
e-mail: Gdhiman0001@gmail.com; Gaurav.dhiman@thapar.edu

A. Kaur
e-mail: Kaur.amandeep@thapar.edu

© Springer Nature Singapore Pte Ltd. 2019   599
J. C. Bansal et al. (eds.), *Soft Computing for Problem Solving*,
Advances in Intelligent Systems and Computing 816,
https://doi.org/10.1007/978-981-13-1592-3_47

population-based metaheuristic which is inspired by the biological evolution: muta-tion, reproduction, selection, and recombination. The popular evolutionary tech-niques are evolution strategy (ES) [4], genetic algorithms (GAs) [6], biogeography-based optimizer (BBO) [45], and genetic programming (GP) [31].

The second category is physical-based algorithms. In these algorithms, the search agents can move around the search space according to the physics rules: gravita-tional force, inertia force, and electromagnetic force. The name of few algorithms are gravitational search algorithm (GSA) [40], simulated annealing (SA) [30], charged system search (CSS) [27], black hole (BH) [20] algorithm, small-world optimiza-tion algorithm (SWOA) [11], ray optimization (RO) algorithm [24], big-bang big-crunch (BBBC) [12], artificial chemical reaction optimization algorithm (ACROA) [1], curved space optimization (CSO) [34], central force optimization (CFO) [14], and galaxy-based search algorithm (GBSA) [43].

The last category is swarm intelligence-based algorithms which are based on the collective behaviors of social creatures. These collective behaviors are inspired by the interaction of swarm with each other. The well-known algorithm of SI technique is Particle Swarm Optimization (PSO). Another popular swarm intelligence-based technique is ant colony optimization [10], bat-inspired algorithm (BA) [49], monkey search [36], dolphin partner optimization (DPO) [44], wolf pack search algorithm [47], hunting search (HUS) [37], bee collecting pollen algorithm (BCPA) [33], firefly algorithm (FA) [48], and cuckoo search (CS) [50].

In particular, there are also other metaheuristic techniques which are inspired by human behaviors. The popular algorithms are harmony search (HS) [16], group search optimizer (GSO) [21, 22], tabu (taboo) search (TS) [13, 18, 19], fuzzy-based approach [7], imperialist competitive algorithm (ICA) [3], firework algorithm [46], mine blast algorithm (MBA) [41], interior search algorithm (ISA)[15], colliding bodies optimization (CBO) [25], soccer league competition (SLC) algorithm [35], social-based algorithm (SBA) [39], seeker optimization algorithm (SOA) [8], league championship algorithm (LCA) [23], and exchange market algorithm (EMA) [17].

However, exploration and exploitation are the two main components to design a novel metaheuristic algorithm [2, 38]. Exploration ensures that the algorithm should reach to different promising regions of the search space while exploitation ensures that the searching of optimal solutions is always within the given region [32]. Hence, there is a need to fine-tune these two components to achieve the near-best optimal solutions. This fact motivates me to develop a novel hybrid metaheuristic algorithm which is inspired by the swarm intelligence for solving real-life engineering design problems.

Particle Swarm Optimization (PSO) is a swarm intelligence-based algorithm which is developed by [29]. The fundamental concepts of this algorithm are the selec-tion of local and global fitness values, whereas Spotted Hyena Optimizer (SHO) is a recently developed swarm intelligence-based algorithm by [9] which is inspired by the social hierarchy and hunting behaviors of spotted hyenas. The swarm intelligence-based algorithms are easier to implement and include very few operators than evolutionary algorithms and require less computational efforts. The main objective of this paper is to develop a hybrid metaheuristic algorithm named as Hybrid Particle Swarm and Spotted Hyena Optimizer (HPSSHO). The performance of the HPSSHO algorithm is tested on thirteen benchmark test functions and one real constrained engineering design problem. The results show that the performance of HPSSHO is better than the other competitor algorithms.

The rest of this paper is structured as follows: Sect. 2 presents the concepts of Particle Swarm Optimization (PSO) algorithm. Sect. 3 presents the concepts of recently developed Spotted Hyena Optimizer (SHO) algorithm. Section 4 presents the concepts of the proposed HPSSHO algorithm. The experimental results and discussion are discussed in Sect. 5. In Sect. 6, the performance of the proposed HPSSHO is tested on 25-bar constrained engineering design problem. Finally, the conclusion and some future research works are given in Sect. 7.

2 Particle Swarm Optimization (PSO)

The PSO algorithm was developed by Eberhart and Kennedy [29]. It is a popular swarm intelligence technique, inspired by the social behaviors of animals: fish schooling, bird flocking, and so on. The PSO algorithm is applied to a set of particles, where each particle has assigned a randomized velocity. Further, each particle can track its own best fitness solution as well as the neighboring particles best solution which is known as local best solution. Then each particle is attracted for finding the global best solution and obtains global fitness value known as global best. During the iteration process, the velocity of each particle is changed by the following equations:

$$
\begin{aligned}
V_{id,k} = \alpha \times V_{id,k} + M_1 \times Rand() \times (PB_{id} - CP_{id,k}) \\
+ M_2 \times Rand() \times (PG_{best} - CP_{id,k})
\end{aligned}
\tag{1}
$$

Whereas, the position of new particle is described as follows:

$$
CP_{id,k} = CP_{id,k} + V_{id,k}
\tag{2}
$$

where i represents the ith particle, α is the inertia weight, d represents the dimension of the given search space, PB_{id} is the previous best position that experiences the local best fitness value, $CP_{id,k}$ is the current position of ith particle in kth iteration, PG_{best} is the best global fitness value, $Rand()$ is the random value in the range of [0, 1], M_1 and M_2 are the self-confidence coefficient and the social coefficient, respectively, and $V_{id,k}$ represents the velocity of the particle in a limited range $[-Vmax, Vmax]$ where $Vmax$ is a user-defined constant.

3 Spotted Hyena Optimizer (SHO)

In this section, the four main steps of SHO algorithm is described which are: searching, encircling, hunting, and attacking prey.

3.1 Encircling Prey

To encircle the prey, the best solution is considered as the target prey and the other search agents can update their positions about this optimal solution which is defined as:

$$D_h =\mid B \cdot P_p(x) - P(x) \mid \tag{3}$$

$$P(x+1) = P_p(x) - E \cdot D_h \tag{4}$$

where D_h is the distance between the prey and spotted hyena, x is the current iteration, P_p defines the position of prey, B and E are coefficients, P indicates the position of spotted hyena.

The variables B and E are calculated as follows:

$$B = 2 \cdot rd_1 \tag{5}$$

$$E = 2h \cdot rd_2 - h \tag{6}$$

$$h = 5 - (Iteration \times (5/Max_{Iteration})) \tag{7}$$
$$\text{where, } Iteration = 0, 1, 2, \ldots, Max_{Iteration}$$

3.2 Hunting

The next step is hunting to make a cluster of optimal solutions against the best search agent and save the best solutions to update the positions of other search agents. The following equations are described in this mechanism:

$$D_h =\mid B \cdot P_h - P_k \mid \tag{8}$$

$$P_k = P_h - E \cdot D_h \tag{9}$$

$$C_h = P_k + P_{k+1} + \cdots + P_{k+N} \tag{10}$$

where P_h defines the position of first best spotted hyena, P_k is the position of other spotted hyenas, and N defines the number of spotted hyenas which is calculated as follows:

$$N = count_{nos}(P_h, P_{h+1}, P_{h+2}, \ldots, (P_h + M)) \tag{11}$$

where M is a random variable in [0.5, 1], nos is the number of solutions, after addition with M, in a given search space, and C_h is a group of N number of optimal solutions.

3.3 Attacking Prey (Exploitation)

To describe this behavior, there is a need to decrease the value of h. Therefore, the variation in E is also decreased due to changing the value in h which can decrease from 5 to 0 during simulation runs. The mathematical formulation for attacking behavior is defined as follows:

$$P(x + 1) = \frac{C_h}{N} \tag{12}$$

where $P(x + 1)$ saves the best solution and updates the positions of other search agents.

3.4 Search for Prey (Exploration)

The searching mechanism describes the exploration capability of an algorithm. The SHO algorithm ensures thus capability using E with random values which are greater than 1 or less than -1. To show more randomized behavior of SHO algorithm, B is also responsible and avoids local optimum.

The pseudocode of the SHO algorithm is discussed in Algorithm 1.

4 Proposed Hybrid Particle Swarm and Spotted Hyena Optimizer (HPSSHO)

This paper presents an improved hunting strategy of Spotted Hyena Optimizer using particle swarm as shown in Fig. 1. Therefore to see this effect, the hunting mechanism of spotted hyena shown in Eq. (9) has been modified using particle swarm velocity update mechanism as given in Eq. (1). The following equation represents the updated hunting mechanism:

$$P_k = \alpha \times P_k + M_1 \times Rand() \times (P_h - D_h) + M_2 \times Rand() \times (P_h - D_h) \tag{13}$$

Algorithm : (1) Spotted Hyena Optimizer

Input: the population of spotted hyenas P_i ($i = 1, 2, \ldots, n$)
Output: the obtained best search agent
1: **procedure** SHO
2: Initialize the h, B, E, and N parameters
3: Calculate the fitness value of each search agent
4: P_h= the first best search agent
5: C_h= the group or cluster of all obtained optimal solutions
6:     **while** ($x < Max_{Iteration}$) **do**
7:         **for** each search agent **do**
8:             Update the position of current search agent by using Eq. (12)
9:         **end for**
10:        Update the parameters h, B, E, and N
11:        Check the search agent which goes beyond the search space and then amend it
12:        Calculate the fitness value of each search agent
13:        Update P_h if there is a better solution than previously obtained optimal solution
14:        Update the group C_h with respect to P_h
15:        $x=x+1$
16:     **end while**
17: return P_h
18: **end procedure**

---

where P_h is the position of first best obtained spotted hyena, P_k is the position of other spotted hyenas, D_h defines the distance between the prey and spotted hyena, α defines the inertia weight, M_1 and M_2 represent the self-confidence and social coefficients, respectively, and $Rand()$ is the random number in range [0, 1]. The pseudocode of the proposed HPSSHO algorithm is discussed in Algorithm 2.

5   Experimental Results and Discussion

This section describes the experimentation of the proposed HPSSHO algorithm to evaluate the performance on highly scalable environment and compared it with four well-known metaheuristic algorithms.

5.1   Functions Evaluation F1 − F7 (Exploitation)

The unimodal test functions $F1 - F7$ allow to assess the exploitation ability of the optimization algorithms. The characteristics formulation of unimodal benchmark test functions is given in Table 1. In Table 2, HPSSHO is very competitive as compared with other metaheuristics. Despite, HPSSHO is the most efficient algorithm for functions $F1$, $F2$, $F3$, $F5$, and $F7$ and provides best exploitation capability throughout the search spaces. Figure 2 shows the convergence analysis of the proposed as well

Fig. 1 Flowchart of the
proposed HPSSHO

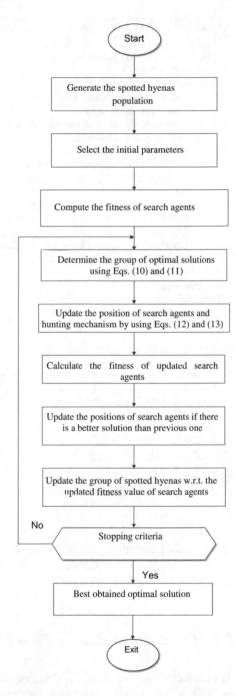

Algorithm : (2) Hybrid Particle Swarm and Spotted Hyena Optimizer

---
Input: the population of spotted hyenas P_i $(i = 1, 2, \ldots, n)$
Output: the obtained best search agent
1: **procedure** HPSSHO
2: Initialize the $h, B, E,$ and N parameters
3: Calculate the fitness value of each search agent
4: P_h= the first best search agent
5: C_h= the group or cluster of all obtained optimal solutions
6:    **while** $(x < Max_{Iteration})$ **do**
7:        **for** each search agent **do**
8:            Update the position of current search agent and hunting mechanism
                by using Eqs. (12) and (13), respectively
9:        **end for**
10:       Update the parameters $h, B, E,$ and N
11:       Check if any search agent goes beyond the search space and then amend it
12:       Calculate the fitness value of each search agent
13:       Update P_h if there is a better solution than previously obtained optimal solution
14:       Update the group C_h with respect to P_h
15:       $x=x+1$
16:    **end while**
17: return P_h
18: **end procedure**

Table 1 Unimodal benchmark test functions

Function	Dim	Range	f_{min}
$F_1(z) = \sum_{k=1}^{N} z_i^2$	30	$[-100, 100]$	0
$F_2(z) = \sum_{k=1}^{N} |z_k| + \prod_{k=1}^{N} |z_k|$	30	$[-10, 10]$	0
$F_3(z) = \sum_{k=1}^{N} (\sum_{j-1}^{k} z_j)^2$	30	$[-100, 100]$	0
$F_4(z) = max_k\{|z_k|, 1 \leq k \leq N\}$	30	$[-100, 100]$	0
$F_5(z) = \sum_{k=1}^{N-1}[100(z_{k+1} - z_k^2)^2 + (z_k - 1)^2]$	30	$[-30, 30]$	0
$F_6(z) = \sum_{k=1}^{N} (|z_k + 0.5|)^2$	30	$[-100, 100]$	0
$F_7(z) = \sum_{k=1}^{N} kz_k^4 + random[0, 1]$	30	$[-1.28, 1.28]$	0

as competitor algorithms on unimodal benchmark test functions. The convergence behaviors of the proposed HPSSHO algorithm are superior to other optimization algorithms during simulation runs.

5.2   Functions Evaluation F8 − F13 (Exploration)

The characteristics formulation of unimodal benchmark test functions is summarized in Table 3. Table 4 shows the results for functions $F8 - F13$ which can indicate that the proposed HPSSHO is the most efficient optimizer in two of the multimodal test problems (i.e., $F9$ and $F13$) and also very competitive in other test problems. These

Table 2 Results on unimodal benchmark test functions

F	HPSSHO		SHO		PSO		DE		GA	
	Avg	Stdv	Avg	Stdv	Avg	Stdv	Avg	Stdv	Avg	Stdv
F1	**0.00E+00**	**0.00E+00**	0.00E+00	0.00E+00	4.98E−09	1.40E−08	1.11E−04	5.10E−03	1.95E−12	2.01E−11
F2	**0.00E+00**	**0.00E+00**	0.00E+00	0.00E+00	7.29E−04	1.84E−03	2.60E−01	6.21E−01	6.53E−18	5.10E−17
F3	**0.00E+00**	**0.00E+00**	0.00E+00	0.00E+00	1.40E+01	7.13E+00	3.12E+02	1.06E+02	7.70E−10	7.36E−09
F4	4.58E−11	7.84E−09	**7.78E−12**	**8.96E−12**	6.00E−01	1.72E−01	3.10E+00	5.09E−01	9.17E+01	5.67E+01
F5	**4.11E+00**	**2.99E−02**	8.59E+00	5.53E−01	4.93E+01	3.89E+01	3.15E+01	2.41E+01	5.57E+02	4.16E+01
F6	9.70E−05	7.86E−02	2.46E−01	1.78E−01	**9.23E−09**	**1.78E−08**	2.38E−03	3.01E−02	3.15E−01	9.98E−02
F7	**9.05E−07**	**5.91E−06**	3.29E−05	2.43E−05	6.92E−02	2.87E−02	4.61E−01	2.70E+00	6.79E−04	3.29E−03

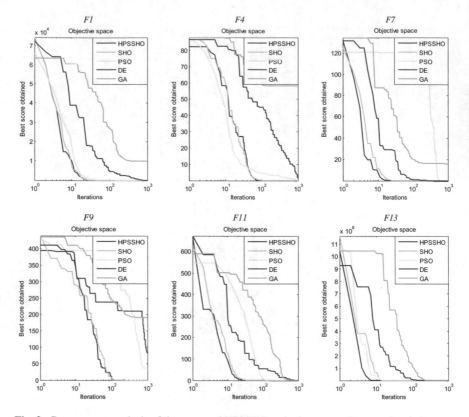

Fig. 2 Convergence analysis of the proposed HPSSHO and other competitor metaheuristics

results reveal that the HPSSHO algorithm has good worth regarding exploration. The convergence behaviors of multimodal benchmark test functions are shown in Fig. 2. After analyzing these behaviors, it has been concluded that the proposed HPSSHO algorithm performs better than other competitor algorithms and converges very efficiently during the whole simulation process using low computational costs.

6 25-Bar Truss Engineering Design Problem

The 25-bar truss is a popular optimization problem in the literature. As shown in Fig. 3, there are fixed 10 nodes and 25 bars cross-sectional members which are grouped into eight categories:

- Group 1: $A1$
- Group 2: $A2, A3, A4, A5$
- Group 3: $A6, A7, A8, A9$

Table 3 Multimodal benchmark test functions

Function	Dim	Range	f_{min}
$F_8(z) = \sum_{k=1}^{N} -z_k \sin(\sqrt{|z_k|})$	30	[-500, 500]	-418.982×5
$F_9(z) = \sum_{k=1}^{N}[z_k^2 - 10\cos(2\pi z_k) + 10]$	30	[-5.12, 5.12]	0
$F_{10}(z) = -20exp\left(-0.2\sqrt{\frac{1}{N}\sum_{k=1}^{N} z_k^2}\right) - exp\left(\frac{1}{N}\sum_{k=1}^{N}\cos(2\pi z_k)\right) + 20 + e$	30	[-32, 32]	0
$F_{11}(z) = \frac{1}{4000}\sum_{k=1}^{N} z_k^2 - \prod_{k=1}^{N}\cos\left(\frac{z_k}{\sqrt{k}}\right) + 1$	30	[-600, 600]	0
$F_{12}(z) = \frac{\pi}{N}\{10\sin(\pi x_1) + \sum_{k=1}^{N-1}(x_k - 1)^2[1 + 10\sin^2(\pi x_{k+1})] + (x_n - 1)^2\}$ $+ \sum_{k=1}^{N} u(z_k, 10, 100, 4)$ $x_k = 1 + \frac{z_k+1}{4}$ $u(z_k, a, k, m) = \begin{cases} k(z_k - a)^m & z_k > a \\ 0 & -a < z_k < a \\ k(-z_k - a)^m & z_k < -a \end{cases}$	30	[-50, 50]	0
$F_{13}(z) = 0.1\{\sin^2(3\pi z_1) + \sum_{k=1}^{N}(z_k - 1)^2[1 + \sin^2(3\pi z_k + 1)]$ $+ (z_n - 1)^2[1 + \sin^2(2\pi z_n)]\} + \sum_{k=1}^{N} u(z_k, 5, 100, 4)$	30	[-50, 50]	0

Table 4 Results on multimodal benchmark test functions

F	HPSSHO		SHO		PSO		DE		GA	
	Avg	Stdv	Avg	Stdv	Avg	Stdv	Avg	Stdv	Avg	Stdv
F8	−4.98E+03	4.99E+02	−1.16E+03	**2.72E+02**	**−6.01E+03**	1.30E+03	−2.70E+03	5.12E+02	−5.11E+03	4.37E+02
F9	**0.00E+00**	**0.00E+00**	0.00E+00	0.00E+00	4.72E+01	1.03E+01	3.31E+01	1.13E+01	1.23E−01	4.11E+01
F10	8.00E−07	6.22E−06	2.48E+00	1.41E+00	3.86E−02	2.11E−01	8.15E−08	1.20E−09	**5.31E−11**	**1.11E−10**
F11	7.55E−13	5.00E−11	**0.00E+00**	**0.00E+00**	5.50E−03	7.39E−03	8.11E+00	3.45E+00	3.31E−06	4.23E−05
F12	8.10E−04	7.59E−03	3.68E−02	1.15E−02	**1.05E−10**	**2.06E−10**	2.63E−01	3.04E−01	9.16E−08	4.88E−07
F13	**1.47E−04**	**6.45E−04**	9.29E−01	9.52E−02	4.03E−03	5.39E−03	3.33E−02	3.01E−02	6.39E−02	4.49E−02

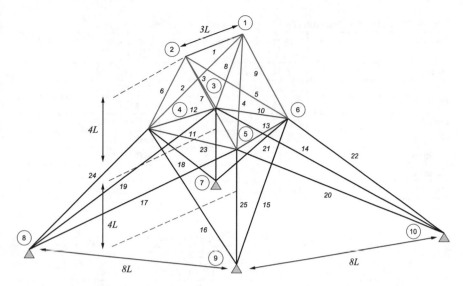

Fig. 3 Schematic view of 25-bar design problem

- Group 4: $A10, A11$
- Group 5: $A12, A13$
- Group 6: $A14, A15, A17$
- Group 7: $A18, A19, A20, A21$
- Group 8: $A22, A23, A24, A25$

The other variables are described as follows:

- $p = 0.0272 \, \text{N/cm}^3 \, (0.1 \, \text{lb/in.}^3)$
- $E = 68947 \, \text{MPa} \, (10000 \, \text{Ksi})$
- Displacement limitation = 0.35 in.
- Maximum displacement = 0.3504 in.
- Design variable set = $\{0.1, 0.2, 0.3, 0.4, 0.5, 0.6, 0.7, 0.8, 0.9, 1.0, 1.1, 1.2, 1.3,$

   $1.4, 1.5, 1.6, 1.7, 1.8, 1.9, 2.0, 2.1, 2.2, 2.3, 2.4, 2.6, 2.8, 3.0, 3.2, 3.4\}$

The member stress limitations are shown in Table 5. The leading conditions of 25-bar truss are listed in Table 6. However, Table 7 reveals that HPSSHO obtains best optimal solution which is superior to other metaheuristic optimization algorithms. The statistical results also reveal that HPSSHO outperforms other competitor metaheuristics.

Table 5 Member stress limitations for 25-bar truss design problem

Groups	Compressive stress limitations Ksi (MPa)	Tensile stress limitations Ksi (MPa)
Group 1	35.092 (241.96)	40.0 (275.80)
Group 2	11.590 (79.913)	40.0 (275.80)
Group 3	17.305 (119.31)	40.0 (275.80)
Group 4	35.092 (241.96)	40.0 (275.80)
Group 5	35.092 (241.96)	40.0 (275.80)
Group 6	6.759 (46.603)	40.0 (275.80)
Group 7	6.959 (47.982)	40.0 (275.80)
Group 8	11.082 (76.410)	40.0 (275.80)

Table 6 The two loading conditions for 25-bar truss design problem

Nodes	Case 1			Case 2		
	$P_xKips(kN)$	$P_yKips(kN)$	$P_zKips(kN)$	$P_xKips(kN)$	$P_yKips(kN)$	$P_zKips(kN)$
1	0.0	20.0 (89)	−5.0 (22.25)	1.0 (4.45)	10.0 (44.5)	−5.0 (22.25)
2	0.0	−20.0 (89)	−5.0 (22.25)	0.0	10.0 (44.5)	−5.0 (22.25)
3	0.0	0.0	0.0	0.5 (2.22)	0.0	0.0
6	0.0	0.0	0.0	0.5 (2.22)	0.0	0.0

Table 7 Statistical results of HPSSHO with literature for 25-bar truss design problem

Group	HPSSHO	SHO	ACO [5]	PSO [42]	CSS [28]	BBBC [26]
A1	0.01	0.01	0.01	0.01	0.01	0.01
A2-A5	1.985	1.842	2.042	2.052	2.003	1.993
A6-A9	3.000	3.000	3.001	3.001	3.007	3.056
A10-A11	0.01	0.01	0.01	0.01	0.01	0.01
A12-A13	0.01	0.01	0.01	0.01	0.01	0.01
A14-A17	0.653	0.659	0.684	0.684	0.687	0.665
A18-A21	1.625	1.626	1.625	**1.616**	1.655	1.642
A22-A25	2.669	2.663	2.672	2.673	**2.66**	2.679
Best	**542.97**	544.81	545.03	545.21	545.10	545.16
Average	**545.07**	546.11	545.74	546.84	545.58	545.66
Std. dev.	**0.368**	0.375	0.94	1.478	0.412	0.491

7 Conclusion

This paper presented a new hybrid swarm intelligence-based optimization algorithm called the Hybrid Particle Swarm and Spotted Hyena Optimizer (HPSSHO). In this paper, the HPSSHO algorithm is benchmarked on thirteen well-known test functions to analyze the exploration and exploitation capability of the proposed algorithm. Moreover, one real-life 25-bar truss engineering design problem is employed to further examine the effectiveness of the proposed HPSSHO algorithm. The results demonstrate that the proposed HPSSHO provides very competitive results as compared to other optimization algorithms such as SHO, PSO, DE, and GA. The proposed algorithm is used for solving the real-life optimization problems using low computational costs. However, further research is required to investigate the effectiveness of the proposed HPSSHO algorithm on large-scale optimization problems.

There are several research directions which can be recommended as future work. To extend this algorithm and hybridize it with various evolutionary as well as bio-inspired algorithms can be seen as a future work. Also, to apply this algorithm to solve various real-life optimization problems can also be seen as a future contribution.

References

1. Alatas, B.: Acroa: artificial chemical reaction optimization algorithm for global optimization. Expert Syst. Appl. **38**(10), 13170–13180 (2011)
2. Alba, E., Dorronsoro, B.: The exploration/exploitation tradeoff in dynamic cellular genetic algorithms. IEEE Trans. Evolutionary Comput. **9**(2), 126–142 (2005)
3. Atashpaz-Gargari, E., Lucas, C.: Imperialist competitive algorithm: An algorithm for optimization inspired by imperialistic competition. In: IEEE Congress on Evolutionary Computation pp. 4661–4667 (2007)
4. Beyer, H.-G., Schwefel, H.-P.: Evolution strategies - a comprehensive introduction. Nat. Comput. **1**(1), 3–52 (2002)
5. Bichon, C.V.C.B.J.: Design of space trusses using ant colony optimization. J. Struct. Eng. **130**(5), 741–751 (2004)
6. Bonabeau, E., Dorigo, M., Theraulaz, G.: Swarm intelligence: from natural to artificial systems. Oxford University Press, Inc. (1999)
7. Chandrawat, R.K., Kumar, R., Garg, B.P., Dhiman, G., Kumar, S.: An Analysis of Modeling and Optimization Production Cost Through Fuzzy Linear Programming Problem with Symmetric and Right Angle Triangular Fuzzy Number. pp. 197–211. Springer Singapore, Singapore (2017)
8. Dai, C. Zhu, Y., Chen, W.: Seeker optimization algorithm. In: International Conference on Computational Intelligence and Security, pp. 167–176 (2007)
9. Dhiman, G., Kumar, V.: Spotted hyena optimizer: a novel bio-inspired based metaheuristic technique for engineering applications. Advances in Engineering Software (2017)
10. Dorigo, M., Birattari, M., Stutzle, T.: Ant colony optimization - artificial ants as a computational intelligence technique. IEEE Comput. Intell. Mag. **1**, 28–39 (2006)
11. Du, H., Wu, X., Zhuang, J.: Small-world optimization algorithm for function optimization, pp. 264–273. Springer, Berlin Heidelberg (2006)
12. Erol, O.K., Eksin, I.: A new optimization method: big bang-big crunch. Adv. Eng. Software **37**(2), 106–111 (2006)
13. Fogel, D.B.: Artificial intelligence through simulated evolution. Wiley-IEEE Press, pp. 227–296 (1998)

14. Formato, R.A.: Central force optimization: a new deterministic gradient-like optimization meta-heuristic. Opsearch **46**(1), 25–51 (2009)
15. Gandomi, A.H.: Interior search algorithm (isa): a novel approach for global optimization. ISA Transactions **53**(4), 1168–1183 (2014)
16. Geem, Z.W., Kim, J.H., Loganathan, G.V.: A new heuristic optimization algorithm: harmony search. Simulation **76**(2), 60–68 (2001)
17. Ghorbani, N., Babaei, E.: Exchange market algorithm. Appl. soft comput. **19**, 177–187 (2014)
18. Glover, F.: Tabu search-part i. ORSA J. Comput. **1**(3), 190–206 (1989)
19. Glover, F.: Tabu search-part ii. ORSA J. Comput. **2**(1), 4–32 (1990)
20. Hatamlou, A.: Black hole: a new heuristic optimization approach for data clustering. Inf. Sci. **222**, 175–184 (2013)
21. He, S., Wu, Q.H., Saunders, J.R.: A novel group search optimizer inspired by animal behavioural ecology. In: IEEE International Conference on Evolutionary Computation, pp. 1272–1278 (2006)
22. He, S., Wu, Q.H., Saunders, J.R.: Group search optimizer: an optimization algorithm inspired by animal searching behavior. IEEE Trans. Evolutionary Comput. **13**(5), 973–990 (2009)
23. Kashan, A.H.: League championship algorithm: a new algorithm for numerical function optimization. In: International Conference of Soft Computing and Pattern Recognition, pp. 43–48 (Dec 2009)
24. Kaveh, A., Khayatazad, M.: A new meta-heuristic method: ray optimization. Comput. Struct. **112–113**, 283–294 (2012)
25. Kaveh, A., Mahdavi, V.: Colliding bodies optimization: a novel meta-heuristic method. Comput. Struct. **139**, 18–27 (2014)
26. Kaveh, A., Talatahari, S.: Size optimization of space trusses using big bang-big crunch algorithm. Comput. Struct. **87**(17–18), 1129–1140 (2009)
27. Kaveh, A., Talatahari, S.: A novel heuristic optimization method: charged system search. Acta Mechanica **213**(3), 267–289 (2010)
28. Kaveh, A., Talatahari, S.: Optimal design of skeletal structures via the charged system search algorithm. Struct. Multidisciplinary Optim. **41**(6), 893–911 (2010)
29. Kennedy, J., Eberhart, R.C.: Particle swarm optimization. In: Proceedings of IEEE International Conference on Neural Networks, pp. 1942–1948 (1995)
30. Kirkpatrick, S., Gelatt, C.D., Vecchi, M.P.: Optimization by simulated annealing. Science **220**(4598), 671–680 (1983)
31. Koza J.R.: Genetic programming: on the programming of computers by means of natural selection. MIT Press (1992)
32. Lozano, M., Garcia-Martinez, C.: Hybrid metaheuristics with evolutionary algorithms specializing in intensification and diversification: overview and progress report. Comput. Oper. Res. **37**(3), 481–497 (2010)
33. Lu, X., Zhou, Y.: A novel global convergence algorithm: bee collecting pollen algorithm. In: 4th International Conference on Intelligent Computing, Springer, pp. 518–525 (2008)
34. Moghaddam, F.F., Moghaddam, R.F., Cheriet, M.: Curved space optimization: a random search based on general relativity theory. Neural and Evolutionary Comput. (2012)
35. Moosavian, N., Roodsari, B.K.: Soccer league competition algorithm: a novel meta-heuristic algorithm for optimal design of water distribution networks. Swarm and Evolutionary Comput. **17**, 14–24 (2014)
36. Mucherino, A., Seref, O.: Monkey search: a novel metaheuristic search for global optimization. AIP Conference Proc. **953**(1) (2007)
37. Oftadeh, R., Mahjoob, M., Shariatpanahi, M.: A novel meta-heuristic optimization algorithm inspired by group hunting of animals: hunting search. Comput. Mathe. Appl. **60**(7), 2087–2098 (2010)
38. Olorunda, O., Engelbrecht, A.P.: Measuring exploration/exploitation in particle swarms using swarm diversity. IEEE Congress on evolutionary computation, pp. 1128–1134 (2008)
39. Ramezani, F., Lotfi, S.: Social-based algorithm. Appl. Soft Comput. **13**(5), 2837–2856 (2013)

40. Rashedi, E., Nezamabadi-pour, H., Saryazdi, S.: GSA: a gravitational search algorithm. Inf. Sci. **179**(13), 2232–2248 (2009)
41. Sadollah, A., Bahreininejad, A., Eskandar, H., Hamdi, M.: Mine blast algorithm: a new population based algorithm for solving constrained engineering optimization problems. Appl. Soft Comput. **13**(5), 2592–2612 (2013)
42. Schutte, J., Groenwold, A.: Sizing design of truss structures using particle swarms. Struct. Multidisciplinary Optim. **25**(4), 261–269 (2003)
43. Hosseini, S.H.: Principal components analysis by the galaxy-based search algorithm: a novel metaheuristic for continuous optimisation. Int. J. Comput. Sci. Eng. **6**, 132–140 (2011)
44. Shiqin, Y., Jianjun, J., Guangxing, Y.: A dolphin partner optimization. In: Proceedings of the WRI Global Congress on Intelligent Systems, pp. 124–128 (2009)
45. Simon, D.: Biogeography-based optimization. IEEE Trans. Evolutionary Comput. **12**(6), 702–713 (2008)
46. Tan, Y., Zhu, Y.: Fireworks Algorithm for Optimization, pp. 355–364. Springer, Berlin Heidelberg (2010)
47. Yang, C., Tu, X., Chen, J.: Algorithm of marriage in honey bees optimization based on the wolf pack search. In: International Conference on Intelligent Pervasive Computing, pp. 462–467 (2007)
48. Yang, X.-S.: Firefly algorithm, stochastic test functions and design optimisation. Int. J. Bio-Inspired Comput. **2**(2), 78–84 (2010)
49. Yang, X.-S.: A New Metaheuristic Bat-Inspired Algorithm, pp. 65–74. Springer, Berlin Heidelberg (2010)
50. Yang, X.S., Deb, S.: Cuckoo search via levy flights. In: World congress on nature biologically inspired computing, pp. 210–214 (2009)

PSO-Based Synthetic Minority Oversampling Technique for Classification of Reduced Hyperspectral Image

Subhashree Subudhi, Ram Narayan Patro and Pradyut Kumar Biswal

Abstract In recent years, hyperspectral image (HSI) classification has become a popular topic of the research. The common problem with HSI is imbalance between limited number of available samples and high dimensionality. To deal with this issue, several linear and nonlinear feature reduction approaches can be used. In HSI, another issue is the imbalance in number of labeled samples present in different classes. This paper presents a novel approach for dealing with imbalanced learning problem in HSI. In this proposed approach, first principal component analysis (PCA) algorithm is applied for feature reduction, followed by application of synthetic minority oversampling technique (SMOTE) on the reduced dimensional dataset. In order to estimate the percentage of oversampling required for each class, the particle swarm optimization technique (PSO) is used. After wisely oversampling the samples present in each class, the oversampled training data is fed into the k-nearest neighbor (KNN) classifier. The obtained results revealed that by properly oversampling the training samples per class, the classification accuracy is increased with reduced time complexity. The proposed approach was tested on the widely used Indian Pines dataset.

S. Subudhi · R. N. Patro · P. K. Biswal (✉)
Department of Electronics and Telecommunication, IIIT Bhubaneswar,
Bhubaneswar, India
e-mail: pradyut@iiit-bh.ac.in

S. Subudhi
e-mail: ssubudhi84@gmail.com

R. N. Patro
e-mail: ram_patro@rediffmail.com

© Springer Nature Singapore Pte Ltd. 2019
J. C. Bansal et al. (eds.), *Soft Computing for Problem Solving*,
Advances in Intelligent Systems and Computing 816,
https://doi.org/10.1007/978-981-13-1592-3_48

617

1 Introduction

Recently, there has been a lot of improvement in hyperspectral imaging technology. Nowadays, hyperspectral sensors are capable of recording information about a single scene over several hundreds of spectral bands. Discrimination of objects of interest is possible in HSI because of its higher spectral resolution. Due to aforementioned advantages, HSI is widely used for a variety of applications like mapping, monitoring, and target detection [10]. However, because of the imbalance in high dimensionality of data and limited number of training samples, HSI suffers from curse of dimensionality or Hughes effect [3, 6]. It is not advisable to directly use the complete set of bands for processing, as it results in higher computational complexity and poor generalization capability of the classifier. The training accuracy may be good but accuracy will fall on test dataset due to limited number of training samples and large number of spectral bands.

In order to deal with this problem, feature reduction (FR) methods must be adopted. In this process, high-dimensional data are transformed into lower dimensionality consisting of all key features. FR is one of the basic preprocessing steps prior to classification. Several FR approaches are available in the literature, which can be broadly categoried into linear and nonlinear FR techniques. Some of the widely used linear techniques are linear discriminant analysis (LDA) [1], principal component analysis(PCA) [8], and independent component analysis (ICA) [16]. Popular nonlinear transformation techniques are diffusion maps [7], local linear embedding (LLE) [11], Sammon mapping [12], kernel principal component analysis (KPCA) [13], isomap [15], autoencoder [17], maximum variance unfolding (MVU) [18], and local tangent space alignment) LTSA [19].

In addition to imbalance between number of training samples and dimension, there exist another issue of class imbalanced learning problem [5] in HSI. This problem is very common; however, this issue is often neglected. In HSI dataset, some classes are severely under-represented (called minority classes) as compared to other classes (majority class). This is one of the main cause for poor classification results in HSI. Synthetic minority oversampling technique (SMOTE) algorithms is a well-known sampling strategy to provide balanced class distribution; i.e., it oversamples the minority classes and undersamples majority classes [2]. In machine learning community, class imbalance problem is studied extensively, while in remote sensing community, only a few approaches are proposed to solve this issue [14].

Generally, heuristic approaches are used for selecting optimal set of bands. For example, in [4, 9], PSO algorithm has been implemented for the same. In this paper, we have proposed a new PSO-based SMOTE method to predict the amount of oversampling required per class in HS data. This resulted in improved classification accuracy. Here to reduce computational complexity, we applied SMOTE algorithm on the PCA-reduced HSI instead of applying it on the original image. The organization of the paper is as follows. The methodology involved in this paper is described in Sect. 2. In Sect. 3, the detailed experimental setup and result is discussed. Finally, Sect. 4 provides the conclusion.

2 Methodology

2.1 Feature Reduction

HSI contains several highly correlated contiguous spectral bands, resulting in lot of redundant information. Therefore, to transform data from high- dimensional space to lower-dimensional space, various feature extraction techniques can be adopted. After feature reduction, the generalization ability of classifier increases and its classification accuracy on unknown test dataset increases. In addition to that the computational complexity of classifier also reduces drastically.

In this paper, we have used the basic PCA algorithm for feature reduction; however, other FR methods may also be used. As PCA is a linear FR method, it transforms higher-dimensional data into lower-dimensional data by applying suitable linear transformation technique. Here the principal components are those components which have the maximum variance. Hence, only the principal components contain the majority of information. As a result, other components can be rejected. Detailed information about PCA algorithm can be found in [8].

2.2 SMOTE

Most often HSI datasets contain unbalanced number of training samples. For example, in Indian Pines dataset class 9 contains only 20 samples while class 11 contains 2455 samples (see Table 1). Because of such disparity, the misclassification rate is very high and the accuracy falls. In order to improve the classifier accuracy, the training samples should be balanced prior to classification. To oversample the training samples for each class, we have used the well-known SMOTE algorithm [2]. The parameter N in SMOTE represents the desired rate of oversampling required based on which synthetic samples are generated (see Table 1). In this algorithm, for generation of synthetic samples, euclidean distance is used; however, other distance measures like *spectral angle* may also be used.

2.3 PSO

Deciding the amount of oversampling required per class is a very crucial step. The classifier accuracy greatly depends on this. In this paper, PSO algorithm is used to decide the percent of required oversampling. Oversampling is applied on the 80% of training data, and 20% of training data are considered as validation data for finding the accuracy, which is taken as the objective function for PSO algorithm. The pseudocode for the algorithm is given below.

Table 1 Classification result for different classes of Indian Pines dataset

Class	Name	Total sample	Testing sample	SMOTE with PSO	Training sample			Accuracy		
					Original	Reduced	Over SI	Original	Reduced	Over SI
1	Alfalfa	46	14	156	32 × 200	32 × 29	50 × 29	93.4783	97.826	**100**
2	Corn-notill	1428	428	414	1000 × 200	1000 × 29	4140 × 29	82.2129	81.7226	**82.4229**
3	Corn-mintill	830	249	273	581 × 200	581 × 29	1586 × 29	78.6747	78.4337	**79.8795**
4	Corn	237	71	200	166 × 200	166 × 29	332 × 29	81.0127	83.1223	**83.5444**
5	Grass-pasture	483	145	317	338 × 200	338 × 29	1071 × 29	70.1863	69.9792	**70.1863**
6	Grass-trees	730	219	271	511 × 200	511 × 29	1385 × 29	93.9726	94.7945	**95.2054**
7	Grass-pasture-mowed	28	8	209	20 × 200	20 × 29	42 × 29	96.4285	96.4285	**96.4285**
8	Hay-windrowed	478	143	141	335 × 200	335 × 29	472 × 29	98.954	98.7447	**98.9541**
9	Oats	20	6	168	14 × 200	14 × 29	24 × 29	75	**75**	75
10	Soybean-notill	972	292	249	680 × 200	680 × 29	1693 × 29	91.2551	**91.3589**	90.7407
11	Soybean-mintill	2455	737	354	1719 × 200	1719 × 29	6085 × 29	80.5703	80.5702	80.6517
12	Soybean-clean	593	178	163	415 × 200	415 × 29	676 × 29	**74.8735**	**74.8735**	**74.8735**
13	Wheat	205	62	242	144 × 200	144 × 29	348 × 29	**97.0732**	96.5853	96.5853
14	Woods	1265	380	499	886 × 200	886 × 29	4421 × 29	95.4941	95.415	**95.8899**
15	Buildings-Grass-Trees-Drives	386	116	128	270 × 200	270 × 29	346 × 29	**78.2383**	77.7202	**78.2383**
16	Stone-Steel-Towers	93	28	281	65 × 200	65 × 29	183 × 29	97.8495	98.9247	**98.9247**
Average Accuracy (AA) ::								0.8499	0.8499	**0.8531**
Elapsed Time in seconds ::								14.8975	**2.4423**	10.8799

In the Algorithm 1, X is the sequence of oversampling; $f(X)$ is the error $(1 - accuracy)$ of the oversampled data (oversampling done by $SMOTE$ with each generated sequence of X) using KNN classifier over the validation set. w, r_p, r_g, c_1 and c_2 are the weight parameters in the range $[0, 1]$; p is the population, and d is the dimension (here d is the number of classes to find the oversampling rate N for $SMOTE$).

Algorithm 1 PSO for optimal sequence generation

1: **procedure** PSO	
Initialize : $w, c_1, c_2, Ub, Lb, p, d$	Size
Initialize : position $X \sim U[Ub, Lb]$	$[p \times d]$
Initialize : velocity $V \sim U([-\|Ub - Lb\|, \|Ub - Lb\|])$	$[p \times d]$
$P_{best} = min(f(X))$	$[1 \times 1]$
$P_{sol} = X(argmin(f(X)))$	$[1 \times d]$
$G_{best} = P_{best}$	$[1 \times 1]$
$G_{sol} = P_{sol}$	$[1 \times d]$
2: **while** termination criteria not met **do**	
3: $V = w \times V + c_1 \times r_p \times (P_{sol} - X) + c_2 \times r_g \times (G_{sol} - X)$	$[p \times d]$
4: $X = X + V$	$[p \times d]$
5: $X = X [Ub, Lb]$	$[p \times d]$
6: $P_{best} = min(f(X))$	$[1 \times 1]$
7: $P_{sol} = X(argmin(f(X)))$	$[1 \times d]$
8: **if** $G_{best} > P_{best}$ **then**	
9: $G_{best} = P_{best}$	$[1 \times 1]$
10: $G_{sol} = P_{sol}$	$[1 \times d]$
11: **end if**	
12: **end while**	
Final G_{sol} is the optimal sequence of oversampling rate N for $SMOTE$	
13: **end procedure**	

---

3 Experimental Setup

3.1 Dataset Used

Indian Pines dataset is used to validate the performance of the proposed algorithm. This dataset was captured by airborne visible/infrared imaging spectrometer (AVIRIS) sensor. It consists of 145×145 pixels and 220 spectral bands in the wavelength range of 0.4–2.5 µm. After removing the water absorption bands, the number of bands reduces to 200. Figure 1 contains the false color composite image and ground truth image of the dataset. The ground truth contains 16 different classes as shown in Table 1. For validation purpose, the training and testing data are kept fixed.

Fig. 1 Indian Pines dataset:
(Left) False gray
representation of band 23
and (Right) False color
representation of ground
truth

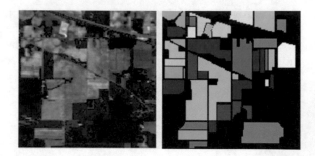

3.2 Classifier Performance Analysis

To demonstrate the effect of oversampling on classification accuracy of HSI, we performed the experiment in three different stages. Each of these stages are explained in detail below.

3.2.1 Stage-1

To verify the classification accuracy over number of bands, the training data, testing data, and hyperparameters were fixed for KNN classifier. Seventy percent of individual classes of original Indian Pines data was selected for training, and the thirty percent samples were used for testing. The obtained classifier accuracy was 84.99%, and the time taken for classification was 14.8975 s. These results were considered as a reference for further comparisons (see Figs. 2 and 3).

3.2.2 Stage-2

Due to imbalance between training samples and dimension of HSI, the computational complexity is very high. To overcome the drawbacks of high-dimensional HSI data, we perform dimensionality reduction on original dataset using PCA algorithm. Figure 2 shows the accuracy of KNN algorithm versus number of features. From Fig. 2, it can be observed that even by reducing the number of features to 20, we are able to achieve the same accuracy as in original image with 200 bands. Thus, computational complexity is highly reduced by using feature reduction. The classification accuracy can still be increased by oversampling the samples of each class. The oversampling process is explained in the following section.

3.2.3 Stage-3

We used SMOTE algorithm to oversample the training samples per class. However, it is not mandatory that only the minority samples are to be oversampled. It can be shown from Table 1 that class 6 containing 511 training samples were oversampled

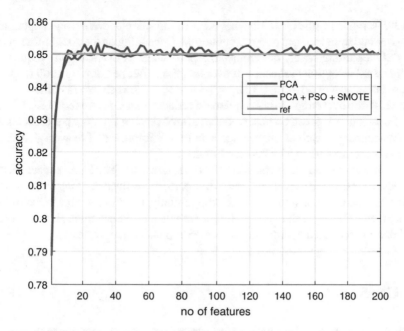

Fig. 2 Accuracy versus number of features plot for Indian Pines dataset

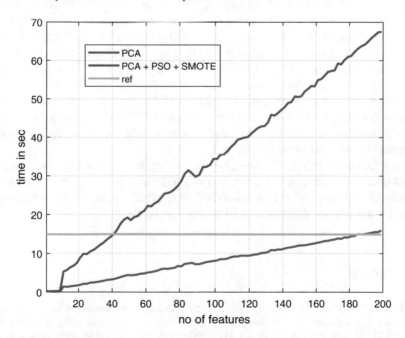

Fig. 3 Time versus number of features plot for Indian Pines dataset

to 1385 samples, while class 1 having 32 training samples were oversampled to 50 samples, and improved accuracy was obtained. Depending on the distribution of data, the PSO algorithm wisely determines the percentage of oversampling required for each class. The optimal sequence of oversampling after performing 1000 runs, with a population size of 10, $w = 0.7$, $c_1 = 0.3$ and $c_2 = 0.6$ is shown in Table 1.

At last, the oversampled data applied on reduced dataset were fed to KNN classifier. From Table 1, it can be observed that after oversampling, the per class classification accuracy as well as the average accuracy is improved. The average accuracy increased from 84.99 to 85.31% (see Fig. 2).

Figure 3 shows the time versus number of features plot. The PCA reduced features take less time than the reference value. After performing oversampling on the reduced features, time taken increases almost exponentially, but still the time taken to reach the desired accuracy with oversampling is less than reference time. Hence, it can be said that the proposed approach has lesser computational complexity.

4 Conclusion

Even though hyperspectral images have several advantages like higher spectral resolution, they suffer from many limitations. The major drawbacks of HSI are Hughes effect and unbalanced number of samples for different classes. In order to deal with the aforementioned problems, classification of HSI is performed in 3 stages. In stage one, the original dataset was classified directly using KNN classifier and its accuracy was taken as reference. After that feature reduction was carried out and classification was done using reduced dimensionality data. Finally, the reduced dimensional training samples were oversampled using SMOTE algorithm. The percentage of oversampling required for each class was decided by PSO algorithm. From the performed experiment, it can be concluded that by wisely oversampling the training samples, the classifier accuracy can be increased.

As a future extension, this work can be applied to other HSI datasets, and for generation of oversampling data, we may use other metaheuristic approaches along with different distance measures in SMOTE. For evaluating the performance of oversampling algorithms, different classifiers can also be used.

References

1. Bandos, T.V., Bruzzone, L., Camps-Valls, G.: Classification of hyperspectral images with regularized linear discriminant analysis. IEEE Trans. Geosci. Remote Sens. **47**(3), 862–873 (2009)
2. Chawla, N.V., Bowyer, K.W., Hall, L.O., Kegelmeyer, W.P.: Smote: synthetic minority oversampling technique. J. Artif. Intell. Res. **16**, 321–357 (2002)
3. David, L.: Hyperspectral image data analysis as a high dimensional signal processing problem. IEEE Sig. Process. Mag. **19**(1), 17–28 (2002)

4. Ghamisi, P., Couceiro, M.S., Benediktsson, J.A.: A novel feature selection approach based on FODPSO and SVM. IEEE Trans. Geosci. Remote Sens. **53**(5), 2935–2947 (2015)
5. He, H., Garcia, E.A.: Learning from imbalanced data. IEEE Trans. Knowl. Data Eng. **21**(9), 1263–1284 (2009)
6. Hughes, G.: On the mean accuracy of statistical pattern recognizers. IEEE Trans. Inf. Theory **14**(1), 55–63 (1968)
7. Lafon, S., Lee, A.B.: Diffusion maps and coarse-graining: a unified framework for dimensionality reduction, graph partitioning, and data set parameterization. IEEE Trans. Pattern Anal. Mach. Intell. **28**(9), 1393–1403 (2006)
8. Licciardi, G., Marpu, P.R., Chanussot, J., Benediktsson, J.A.: Linear versus nonlinear pca for the classification of hyperspectral data based on the extended morphological profiles. IEEE Geosci. Remote Sens. Lett. **9**(3), 447–451 (2012)
9. Paoli, A., Melgani, F., Pasolli, E.: Clustering of hyperspectral images based on multiobjective particle swarm optimization. IEEE Trans. Geosci. Remote Sens. **47**(12), 4175–4188 (2009)
10. Plaza, A., Benediktsson, J.A., Boardman, J.W., Brazile, J., Bruzzone, L., Camps-Valls, G., Chanussot, J., Fauvel, M., Gamba, P., Gualtieri, A., et al.: Recent advances in techniques for hyperspectral image processing. Remote Sens. Environ. **113**, S110–S122 (2009)
11. Roweis, S.T., Saul, L.K.: Nonlinear dimensionality reduction by locally linear embedding. Science **290**(5500), 2323–2326 (2000)
12. Sammon, J.W.: A nonlinear mapping for data structure analysis. IEEE Trans. Comput. **100**(5), 401–409 (1969)
13. Schölkopf, B., Smola, A., Müller, K.R.: Nonlinear component analysis as a kernel eigenvalue problem. Neural Comput. **10**(5), 1299–1319 (1998)
14. Sun, T., Jiao, L., Feng, J., Liu, F., Zhang, X.: Imbalanced hyperspectral image classification based on maximum margin. IEEE Geosci. Remote Sens. Lett. **12**(3), 522–526 (2015)
15. Tenenbaum, J.B., De, Silva V., Langford, J.C.: A global geometric framework for nonlinear dimensionality reduction. Science **290**(5500), 2319–2323 (2000)
16. Villa, A., Benediktsson, J.A., Chanussot, J., Jutten, C.: Hyperspectral image classification with independent component discriminant analysis. IEEE Trans. Geosci. Remote Sens. **49**(12), 4865–4876 (2011)
17. Wang, J., He, H., Prokhorov, D.V.: A folded neural network autoencoder for dimensionality reduction. Procedia Comput. Sci. **13**, 120–127 (2012)
18. Weinberger, K.Q., Sha, F., Saul, L.K.: Learning a kernel matrix for nonlinear dimensionality reduction. In: Proceedings of the Twenty-First International Conference on Machine Learning, p. 106. ACM, New York (2004)
19. Zhang, Z., Zha, H.: Principal manifolds and nonlinear dimensionality reduction via tangent space alignment. SIAM J. Sci. Comput. **26**(1), 313–338 (2004)

Hybridized Cuckoo–Bat Algorithm for Optimal Assembly Sequence Planning

Balamurali Gunji, B. B. V. L. Deepak, Amruta Rout, Golak Bihari Mohanta and B. B. Biswal

Abstract Assembly sequence planning (ASP) problem is one of the NP-hard combinatorial problems in manufacturing, where generating a feasible sequence from the set of finite possible solutions is a difficult process. As the ASP problem is the discrete optimization problem, it takes a major part of the time in the assembly process. Many researchers have implemented different algorithms to get optimal assembly sequences for the given assembly. Initially, mathematical models have been developed to solve ASP problems, which are very poor in performance. Later on, soft computing techniques have been developed to solve ASP problems, which are very effective in achieving the optimal assembly sequences. But these soft computing techniques consume more time during execution to get optimal assembly sequence. Sometimes these algorithms fall in local optima during execution. Keeping the above things in mind in this paper, a new algorithm namely hybrid cuckoo–bat algorithm (HCBA) is implemented to obtain the optimal assembly sequences. The proposed algorithm is compared with two different assemblies (gear assembly and wall rack assembly) with the algorithms like genetic algorithm (GA), ant colony optimization (ACO), grey wolf optimization (GWO), advanced immune system (AIS) and hybrid ant–wolf algorithm (HAWA). The results of the different algorithms are compared in terms of CPU time and fitness values with the proposed algorithm. The results show that the proposed algorithm performs better than the compared algorithms.

B. Gunji (✉) · B. B. V. L. Deepak · A. Rout · G. B. Mohanta · B. B. Biswal
Industrial Design Department, NIT Rourkela, Rourkela 769008, Orissa, India
e-mail: bmgunji@gmail.com

B. B. V. L. Deepak
e-mail: bbv@nitrkl.ac.in

A. Rout
e-mail: 516id6003@nitrkl.ac.in

G. B. Mohanta
e-mail: 516id1001@nitrkl.ac.in

B. B. Biswal
e-mail: bbbiswal@nitrkl.ac.in

© Springer Nature Singapore Pte Ltd. 2019
J. C. Bansal et al. (eds.), *Soft Computing for Problem Solving*,
Advances in Intelligent Systems and Computing 816,
https://doi.org/10.1007/978-981-13-1592-3_49

627

Keywords Assembly sequence planning problem · Objective constraints
Input constraints · Soft computing techniques

1 Introduction

Assembly is the process of joining the parts in an order one after the other to form
an assembly. To join the parts in an order, one requires assembly sequence, which
will give the information about the parts that are to be joined to form an assembly. As the assembly sequence planning (ASP) problem is the discrete optimization
problem, achieving the optimal assembly sequence is a difficult process. To achieve
the optimal assembly sequence, the generated sequence from any method has to
undergo two criteria. One is feasibility criteria; to satisfy this criterion, the assembly sequence has to check with the input constraints (liaison data, stability data).
If the assembly sequence satisfies the feasibility criterion, then it has to check for
the second criterion to increase the quality of the sequence by evaluating through
objective constraints. Many researchers used different methodologies/algorithms to
obtain the optimum assembly sequence. At the initial stage of developments in the
ASP problems, researchers are used mathematical models like liaison graph/liaison
tree to check the feasibility of the sequences, which is time-consuming process. Later
on, computer-aided methods have been developed to extract the input constraints
automatically [1–3]. In order to have good-quality sequence; initially, researchers
followed the mathematical algorithms like cut-set methods, AND/OR questions to
obtain the optimal assembly sequence but these are very tedious and time consuming
[4, 5]. Later on, researchers developed the computer-aided techniques to obtain the
optimal assembly sequences. Generally, these methods are classified into two types:
one is graph search algorithm and the second one is artificial intelligence (AI)-based
algorithms. Even though these methods are successful to achieve optimal assembly
sequences, but sometimes these methods fall in the local optima during execution.

To overcome this, researchers are attracted towards the hybrid algorithms. As the
hybrid algorithms are the combination of two or more algorithms desired features
top obtain the optimal assembly sequence [6–8].

In the current research, a new HCBA has been developed to obtain the optimal
assembly sequence. The developed algorithm is compared with the different well-
known algorithms like GA, ACO, AIS, GWO and HAGA.

2 Literature Review

Solving the ASP problem started in the late 1980s by Ayoub and Doty [9] . Later on,
De Mello and Sanderson developed AND/OR graph to obtain the optimal assembly sequence [5]. The researchers like Chakrabarty and Wolter [10] developed a
hierarchy of assembly structure to reduce the complexity of the problem. Later, the

researchers like Xiaoming and Pingan [11] developed object-oriented method to obtain the optimal assembly sequence.

The above-discussed methods consume lot of execution time as well as search space also. To avoid this problem, researchers are motivated towards the soft computing techniques. Till now many soft computing techniques are applied by different researchers to obtain the optimal assembly sequences. Out of those, GA is used by most of the researchers because of its simplicity in implementing. Initially, GA is implemented by Wong and Leu [12] to obtain the optimal assembly sequence. In this, he implemented adaptive GA by continuously varying the genetic operators. Later, the researchers like Boizneville et al. [13], Dini et al. [14], Hong and Cho [15] and Smith and Liu [16] developed the GA to solve optimal ASP problem. Out of those, Smith and Liu use multi-level genetic algorithm, in which the sequence obtained from the level-1 will be given as input to the level-2 by which infeasible solutions will reduce and quality of the solution will increase. Apart from GA, the next most used algorithm by the researchers is ACO algorithm. This algorithm is initially implemented by Failli and Dini [17] to solve ASP problem. Later on, it was developed by Wang et al. [18], McGovern and Gupta [19] and Wang et al. [20]. Out of them, Wang uses disassembly feasibility graph to obtain the optimal assembly sequence. Apart from these algorithms, many recently developed algorithms like advanced immune strategy proposed by Bahubalendruni [21], grey wolf optimization (GWO) algorithm proposed by Mirjalili et al. [22] and many more have been implemented to solve ASP problem.

The rest of the paper is arranged as follows: Sect. 3 deals with the proposed algorithm, Sect. 4 deals with the results and comparisons of the proposed algorithm and Sect. 5 deals with the conclusion of the research paper.

3 Proposed Algorithm

In this section, a new hybrid algorithm is proposed to obtain the optimum assembly sequence. In this, cuckoo search and bat algorithms are combined to form a hybrid algorithm, to achieve optimal assembly sequence. In the proposed algorithm, two fitness functions have been considered for two separate assemblies to evaluate the quality of the assembly sequence.

The detailed flow chart of the developed algorithm is shown in Fig. 1. The fitness functions for both the assemblies are as follows:

For the first assembly, three objective constraints are considered to evaluate the fitness of the sequence. In this, directional changes ($D.C$), gripper changes ($G.C$), and part movement ($P.M$) are considered for developing the fitness function.

$$\text{Fitness Function} \quad F.F = \sum_{i=1}^{n-1} \frac{w}{3} * (D.C_i) + \frac{w}{3} * (G.C_i) + \frac{w}{3} * (P.M_i) \quad (1)$$

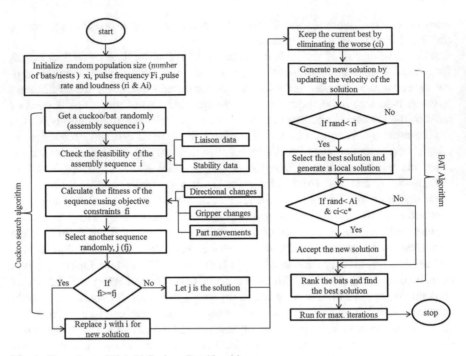

Fig. 1 Flow chart of Hybrid Cuckoo–Bat Algorithm

where w is the weight function of the different constraints, which depends on the industry requirement. n is the number of parts in the assembly

If $w = \left. \begin{cases} 1 \\ 3/2 \\ 3 \end{cases} \right\}$
Three objective constraints are having equal priority
Any two objective constraints are having equal priory and other is '0'
Only one objective constrains is having full priority and rest are '0's

The formulation of the fitness function by giving equal priority to the three objective constraints is as follows:

$$F.F = \sum_{i=1}^{n-1} 0.33 * (D.C_i) + 0.33 * (G.C_i) + 0.33 * (P.M_i) \tag{2}$$

To compare the proposed algorithm, the second assembly considered is wall rack. For the wall rack assembly to evaluate the quality of the sequence, the fitness function is formulated by considering the directional changes and gripper changes as objective constraints. The formulation of the equation is as follows:

$$F.F = \sum_{i=1}^{n-1} w * (D.C_i) + (1 - w) * (G.C_i) \tag{3}$$

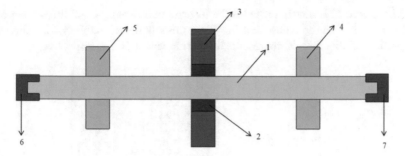

Fig. 2 Gear assembly

Let us consider both objective constraints are having the equal priority, then

$$w = 0.5 \Rightarrow F.F = \sum_{i=1}^{n-1} 0.5 * (D.C_i) + 0.5 * (G.C_i) \qquad (4)$$

4   Results and Comparisons

This section deals with the extracted results of the two industrial products, namely gear assembly and wall rack assembly, from the proposed algorithm. In this, the results are compared with the different algorithms like genetic algorithm (GA), ant colony optimization (ACO), grey wolf optimizer (GWO), hybrid ant–wolf algorithm (HAWA) and advanced immune-based strategy. In this, two assemblies are been considered to compare with all algorithms.

In this, two cases have been considered separately for two assemblies. In the first case, gear assembly shown in Fig. 2 is considered for evaluating the quality of the sequence by the proposed methodology. The results of the proposed algorithm are compared with the advanced immune algorithm [21].

Case-1: Gear Assembly

Contact Data: This data provides the information about the contact of the parts in the assembly. In the below matrix, '0' represents no contact between the parts and '1' represents the contact between the parts.

Liaison data ➡

	1	2	3	4	5	6	7
1	0	1	0	1	1	1	1
2	1	0	1	0	0	0	0
3	0	1	0	0	0	0	0
4	1	0	0	0	0	0	0
5	1	0	0	0	0	0	0
6	1	0	0	0	0	0	0
7	1	0	0	0	0	0	0

Stability Data: This matrix provides the information about the stability of the parts in the assembly. In the below matrix, '0' represents the no stability, '1' represents the partial stability and '2' represents the permanent stability, respectively.

Stability data ➡

	1	2	3	4	5	6	7
1	0	1	0	1	1	2	2
2	1	0	1	0	0	0	0
3	0	1	0	0	0	0	0
4	1	0	0	0	0	0	0
5	1	0	0	0	0	0	0
6	2	0	0	0	0	0	0
7	2	0	0	0	0	0	0

Geometrical feasibility matrices: This data provides the information about the feasibility direction of part for the assembly. These matrices are total six in six principle axes. In this, '0' represents feasible in that direction and '1' represents not feasible in that direction.

+x

	1	2	3	4	5	6	7
1	0	0	0	0	0	0	0
2	0	0	0	1	1	1	1
3	0	0	0	1	1	1	1
4	0	1	1	0	1	1	1
5	0	1	1	1	0	1	1
6	0	1	1	1	1	0	1
7	0	1	1	1	1	1	0

-x

	1	2	3	4	5	6	7
1	0	0	0	0	0	0	0
2	0	0	0	1	1	1	1
3	0	0	0	1	1	1	1
4	0	1	1	0	1	1	1
5	0	1	1	1	0	1	1
6	0	1	1	1	1	0	1
7	0	1	1	1	1	1	0

+y

	1	2	3	4	5	6	7
1	0	1	1	1	1	1	0
2	1	0	1	1	0	1	0
3	1	1	0	1	0	1	1
4	1	0	0	0	0	1	0
5	1	1	1	1	0	1	0
6	0	0	1	0	0	0	0
7	1	1	1	1	1	1	0

-y

	1	2	3	4	5	6	7
1	0	1	1	1	1	0	1
2	1	0	1	0	1	0	1
3	1	1	0	0	1	1	1
4	1	1	1	0	1	0	1
5	1	0	0	0	0	0	1
6	1	1	1	1	1	0	1
7	0	0	0	0	0	0	0

+z

	1	2	3	4	5	6	7
1	0	0	0	0	0	0	0
2	0	0	0	1	1	1	1
3	0	0	0	1	1	1	1
4	0	1	1	0	1	1	1
5	0	1	1	1	0	1	1
6	0	1	1	1	1	0	1
7	0	1	1	1	1	1	0

-z

	1	2	3	4	5	6	7
1	0	0	0	0	0	0	0
2	0	0	0	1	1	1	1
3	0	0	0	1	1	1	1
4	0	1	1	0	1	1	1
5	0	1	1	1	0	1	1
6	0	1	1	1	1	0	1
7	0	1	1	1	1	1	0

The optimal assembly sequences obtained for the gear assembly using developed algorithm are tabulated in Table 1. In this two optimal assembly, sequences are obtained with minimum number of directional changes, gripper changes and part movement. In the part movement, generally distance has been considered as the

Table 1 Optimal assembly sequences for the gear assembly

S. No.	Assembly sequence	No. of directional changes	No. of gripper changes	Part movement	Fitness value
1	3 2 1 4 5 7 6	2	4	1.946	2.622
2	3 2 1 5 4 6 7	2	4	1.946	2.622

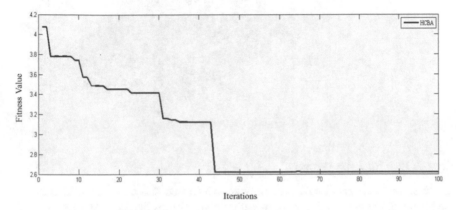

Fig. 3 Graph between number of iterations and fitness value

Table 2 Comparison of results

Type of assembly	No. of sequences	Fitness value	Execution time (s)
Gear assembly			
Advanced immune strategy [21]	2	2.622	0.64
HCBA	2	2.622	0.61

objective constraint. As the other two objective constraints are unitless, so the part movement distance is multiplied with a large constant (e^{10}) to convert it into unitless.

A graph shown in Fig. 3 is plotted between number of iterations and fitness value. In this, the minimum fitness value is obtained after 44 iterations, which is less compared to the advanced immune strategy algorithm.

The results shown in Table 2 of the developed algorithm are compared with the advanced immune strategy in terms of number of optimal assembly sequences, fitness value and execution time. Out of those, execution time to get the optimal assembly sequences is less compared to the advanced immune strategy.

Fig. 4 Wall rack assembly

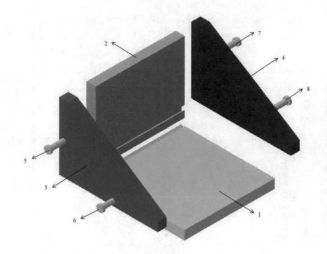

Case-2: Wall Rack Assembly

In the second case, wall rack assembly shown in Fig. 4 is considered to compare the results of the proposed algorithm with the algorithms like GA, ACO, GWO and HAWA. In this, two objective constraints shown in Eq. (4) is considered to evaluate the quality of the sequence.

Liaison data

	1	2	3	4	5	6	7	8
1	0	1	1	1	0	1	0	1
2	1	0	1	1	1	0	1	0
3	1	1	0	0	1	1	0	0
4	1	1	0	0	0	0	1	1
5	0	1	1	0	0	0	0	0
6	1	0	1	0	0	0	0	0
7	0	1	0	1	0	0	0	0
8	1	0	0	1	0	0	0	0

Stability data

	1	2	3	4	5	6	7	8
1	0	2	1	1	0	2	0	2
2	2	0	1	1	2	0	2	0
3	1	1	0	0	2	2	0	0
4	1	1	0	0	0	0	2	2
5	0	1	1	0	0	0	0	0
6	1	0	1	0	0	0	0	0
7	0	1	0	1	0	0	0	0
8	1	0	0	1	0	0	0	0

Geometrical feasibility matrices

x+

	1	2	3	4	5	6	7	8
1	0	0	0	1	1	0	1	1
2	1	0	0	1	0	1	1	1
3	1	1	0	1	0	0	1	1
4	0	0	1	0	1	1	1	1
5	1	1	1	1	0	1	1	1
6	1	1	1	1	1	0	1	1
7	1	0	1	0	1	1	0	1
8	0	1	1	0	1	1	1	0

x-

	1	2	3	4	5	6	7	8
1	0	1	1	0	1	1	1	0
2	0	0	1	0	1	1	0	1
3	0	0	0	1	1	1	1	1
4	1	1	1	0	1	1	0	0
5	1	0	0	1	0	1	1	1
6	0	1	0	1	1	0	1	1
7	1	1	1	1	1	1	0	1
8	1	1	1	1	1	1	1	0

y+

	1	2	3	4	5	6	7	8
1	0	0	1	1	1	0	1	0
2	0	0	1	1	0	0	0	1
3	1	1	0	1	0	0	1	1
4	1	1	1	0	1	1	0	0
5	1	0	0	1	0	1	1	1
6	0	1	0	1	1	0	1	1
7	1	0	1	0	1	1	0	1
8	0	1	1	0	1	1	1	0

y-

	1	2	3	4	5	6	7	8
1	0	0	1	1	1	0	1	0
2	0	0	1	1	0	1	0	1
3	1	1	0	1	0	0	1	1
4	1	1	1	0	1	1	0	0
5	1	0	0	1	0	1	1	1
6	0	0	0	1	1	0	1	1
7	1	0	1	0	1	1	0	1
8	0	1	1	0	1	1	1	0

z+

	1	2	3	4	5	6	7	8
1	0	0	1	1	0	0	0	0
2	0	0	1	1	0	1	1	1
3	1	1	0	1	0	0	1	1
4	1	1	1	0	1	1	1	1
5	1	1	1	1	0	1	1	1
6	0	1	1	1	1	0	1	1
7	1	0	1	0	1	1	0	1
8	0	1	1	0	1	1	1	0

z-

	1	2	3	4	5	6	7	8
1	0	0	1	1	1	0	1	0
2	0	0	1	1	1	1	0	1
3	1	1	0	1	1	1	1	1
4	1	1	1	0	1	1	0	0
5	0	0	0	0	1	0	1	1
6	0	1	0	1	1	0	1	1
7	0	1	1	1	1	1	0	1
8	0	1	1	1	1	1	1	0

The results of the proposed algorithm are shown in Table 3. In this, eight optimal assembly sequences with minimum number of directional changes and gripper changes are obtained.

A graph shown in Fig. 5 is plotted between number of iterations and fitness values. The algorithm is run for 300 iterations; fitness value is converged after 49 iterations only.

Table 3 Represents the assembly sequences for wall rack assembly

S. no.	Assembly sequence								No. of directional changes	No. of gripper changes	Fitness value
1	1	2	3	4	7	8	5	6	2	2	2
2	1	2	3	4	7	8	6	5	2	2	2
3	1	2	3	4	8	7	5	6	2	2	2
4	1	2	3	4	8	7	6	5	2	2	2
5	2	1	4	3	5	6	7	8	2	2	2
6	2	1	4	3	5	6	8	7	2	2	2
7	2	1	4	3	6	5	7	8	2	2	2
8	2	1	4	3	6	5	8	7	2	2	2

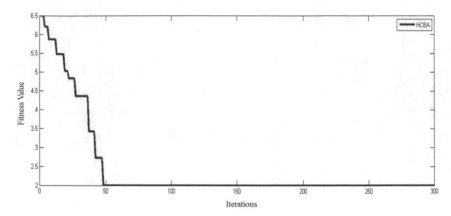

Fig. 5 Graph between number of iterations and fitness value

Table 4 Represents the assembly sequences for wall rack assembly

Indicator	GA [8]	ACO [8]	GWO [22]	HAWA [8]	HCBA
Rack assembly					
Fitness value	5.2727	5.4545	5.4090	5	2
Average CPU time (s)	4.7861	4.0984	5.1246	4.4258	2.9263

The results obtained from the developed algorithm (HCBA) are compared with the several well-known algorithms like GA, ACO, GWO and HAWA in terms of fitness value and CPU time, which is shown in Table 4.

5 Conclusions

In this paper, a new hybrid algorithm is developed by the combination of the cuckoo search and bat algorithms, respectively. Mainly in this, the searching nest in cuckoo search algorithm and updating the velocities and positions of the bats in the bat algorithm are combined to form the hybrid algorithm. The following conclusions are been observed.

1. The developed hybrid algorithm (HCBA) is able to obtain the optimal assembly sequences with less average CPU time compared to the other algorithms, compared in the above Sect. 4. Similarly, it generates more optimal sequences with less fitness value compared to the other algorithms, which is shown in the Sect. 4.
2. The algorithm is compared with two different assemblies to evaluate the quality of the solution in terms of fitness value and CPU execution time. The proposed algorithm is compared with the advanced immune-based strategy algorithm for

gear assembly. In this, the average CPU time for the developed algorithm is less compared to the advanced immune-based strategy algorithm.

3. The developed algorithm is also compared to GA, ACO, GWO and HAWA algorithms for wall rack assembly. In this, the average CPU time and fitness value of the proposed algorithm is less compared to the other algorithms.

As a future work, this algorithm can be implemented for the more number of part assemblies. Moreover, the algorithm can be extended to the flexible part assemblies and the parts which are to be assemble other than principle axes.

References

1. Bahubalendruni, M.V.A.R., Biswal, B.B.: Computer aid for automatic liaisons extraction from cad based robotic assembly. In: IEEE 8th International Conference on Intelligent Systems and Control (ISCO), Coimbatore, pp. 42–45 (2014)
2. Raju Bahubalendruni, M.V.A., Biswal, B.B.: An intelligent method to test feasibility predicate for robotic assembly sequence generation. In: Jain, L., Patnaik, S., Ichalkaranje, N. (eds.) Intelligent Computing, Communication and Devices. Advances in Intelligent Systems and Computing, vol. 308. Springer, New Delhi (2015)
3. Bala Murali, G., et al.: An Intelligent Strategy for Automated Assembly Sequence Planning While Considering DFA Concept (2017)
4. De Fazio, T.L., Whitney, D.E.: Simplified generation of all mechanical assembly sequences. IEEE J. Robot. Autom. 3(6), 640–658 (1987)
5. Homem de Mello, L.S., Sanderson, A.C.: AND/OR graph representation of assembly plans. IEEE Trans. Robot. Autom. 6(2), 188–199 (1990)
6. Bala Murali, G., Deepak, B.B.V.L., Raju Bahubalendruni, M.V.A., Biswal, B.B.: Optimal assembly sequence planning towards design for assembly using simulated annealing technique. In: Chakrabarti, A., Chakrabarti, D. (eds.) Research into Design for Communities, vol. 1. ICoRD 2017. Smart Innovation, Systems and Technologies, vol. 65. Springer, Singapore (2017)
7. Gunji, B., et al.: Hybridized genetic-immune based strategy to obtain optimal feasible assembly sequences. Int. J. Ind. Eng. Comput. 8(3), 333–346 (2017)
8. Ab Rashid, M.F.F.: A hybrid Ant-Wolf Algorithm to optimize assembly sequence planning problem. Assembly Autom., 37(2), 238–248 (2017)
9. Ayoub, R.G., Doty, K.L.: Representation for discrete assembly sequences in task planning. In: Proceedings -IEEE Computer Society's International Computer Software & Applications Conference, Orlando, pp. 746–753 (1989)
10. Chakrabarty, S., Wolter, J.: A structure-oriented approach to assembly sequence planning. IEEE Trans. Robot. Autom. 13(1), 14–29 (1997)
11. Xiaoming, Z., Pingan, D.: A model-based approach to assembly sequence planning. Int. J. Adv. Manuf. Technol. 39(9/10), 983–994 (2008)
12. Wong, H., Leu, M.C.: Adaptive genetic algorithm for optimal printed circuit board assembly planning. CIRP Ann. Manuf. Technol. 42(2), 17–20 (1993)
13. Bonneville, F., Perrard, C., Henrioud, J.M.: A genetic algorithm to generate and evaluate assembly plans. In: ETFA'95, Proceedings. 1995 INRIA/IEEE Symposium on Emerging Technologies and Factory Automation, 1995, vol. 2, pp. 231–239. IEEE (1995)
14. Dini, G., Failli, F., Lazzerini, B., Marcelloni, F.: Generation of optimized assembly sequences using genetic algorithms. CIRP Ann. Manuf. Technol. 48(1), 17–20 (1999)
15. Hong, D.S., Cho, H.S.: A genetic-algorithm-based approach to the generation of robotic assembly sequences. Control Eng. Pract. 7(2), 151–159 (1999)

16. Smith, S.S.F., Liu, Y.J.: The application of multi-level genetic algorithms in assembly planning. J. Ind. Technol. **17**(4), 1–4 (2001)
17. Failli, F., Dini, G.: Ant colony systems in assembly planning: a new approach to sequence detection and optimization. In: Proceedings of the 2nd CIRP International Seminar on Intelligent Computation in Manufacturing Engineering, pp. 227–232 (2000)
18. Wang, J.F., Liu, J.H., Li, S.Q., Zhong, Y.F.: Intelligent selective disassembly using the ant colony algorithm. AI EDAM: Artif. Intell. Eng. Des. Anal. Manufact. **17**(04), 325–333 (2003)
19. McGovern, S.M., Gupta, S.M.: Ant colony optimization for disassembly sequencing with multiple objectives. Int. J. Adv. Manufact. Technol. **30**(5-6), 481–496 (2006)
20. Wang, H., Rong, Y., Xiang, D.: Mechanical assembly planning using ant colony optimization. Comput. Aided Des. **47**, 59–71 (2014)
21. Raju Bahubalendruni, M.V.A., Deepak, B.B.V.L., Biswal, Bibhuti Bhusan: An advanced immune based strategy to obtain an optimal feasible assembly sequence. Assembly Autom. **36**(2), 127–137 (2016)
22. Mirjalili, S., Mirjalili, S.M., Lewis, A.: Grey Wolf optimizer. Adv. Eng. Softw. **69**, 46–61 (2014)

A Variable ε-DBSCAN Algorithm for Declustering Earthquake Catalogs

Rahul Kumar Vijay and Satyasai Jagannath Nanda

Abstract This paper introduces a two-stage clustering model to determine the seismic activities of a region in spatio-temporal domain. In the spatial domain for cluster analysis, a K-means algorithm based on "Haversine distance" is introduced. With this, a seismic region is classified into distinct zones which are correlated in space. In each zone, events' temporal activities are analyzed. This temporal domain analysis is carried out using a variable "ε" density-based clustering algorithm. In this algorithm, the neighborhood radius "ε" is varied to determine the core points. The variation of "ε" is a time-dependent function of "magnitude" of the event (empirical relation makes out higher magnitude leads to a larger value of "ε", i.e., number of days in time). The proposed model is applied to analyze the seismic activities of Himalaya and Sumatra–Andaman region for the time interval between 1965 and 2015 (51 Years). Simulation results reveal that the de-clustered catalogs obtained for both the regions follow linear trend which justifies the background events are homogeneous with respect to time. Corresponding clustered catalogs which reflect the presence of foreshock, mainshock, and aftershock events follow the behavior of true catalog with time.

1 Introduction

Cluster analysis is an efficient approach to find the groups of homogeneous elements in a large database which helps to extract the valuable information or hidden patterns among the elements. Most clustering algorithms are built to classify the ordinary datasets into groups which have non-spatial and non-temporal characteristics. These algorithms are not well suited for clustering the spatial temporal datasets in variety of applications such as geographic information systems, medical imaging, weather

R. K. Vijay (✉) · S. J. Nanda
Malaviya National Institute of Technology Jaipur, Jaipur, India
e-mail: vijay.rahul1986@gmail.com

S. J. Nanda
e-mail: nanda.satyasai@gmail.com

© Springer Nature Singapore Pte Ltd. 2019
J. C. Bansal et al. (eds.), *Soft Computing for Problem Solving*,
Advances in Intelligent Systems and Computing 816,
https://doi.org/10.1007/978-981-13-1592-3_50

forecasting, and seismic signal processing. In seismology, spatio-temporal cluster-ing algorithms are also designed to identify the fault patterns present in a specific geographic region. It helps to evaluate regional seismic hazards, determining after-shocks related to main quakes, rise of seismic activities prior to a large earthquake and developing prediction methodology about earthquakes [1]. The seismic cata-log of an earthquake-active region comprises a large database of events recorded in terms of time of occurrence, magnitude in Richter scale, location in form of latitude and longitude, and depth in Kms [2]. Zaliapin et al. [3] proposed a statistical-based clustering approach to show the existence of the clustered and non-clustered part of seismicity. Recently, Vijay and Nanda [4] developed an gray wolf optimizer-based de-clustering approach to analyze seismicity of California.

Density-based clustering algorithms have been more popularized due to their potential to detect clusters of arbitrary shape and ability to handle outliers present in a spatio-temporal domain. In this regard, a ST-DBSCAN algorithm [5] is proposed to discover clusters with non-spatial, spatial, and temporal attributes of the objects and applied to spatial temporal data of warehouse system. This algorithm assigns a density factor to each cluster to detect the noise points of clusters with different densities. A weight-based density estimation approach is reported by [6] considering the accumulated seismic mass for isolation of earthquake clusters in time and space. It also employs single linkage agglomerative hierarchical approach in second stage for spatial distribution of the seismic events. Nanda et al. [7] proposed a computa-tionally efficient DBSCAN considering a new merging criterion at the initial stage of clustering. It considers the correlation coefficient as similarity measure and applied effectively on historical earthquake catalog of Japan.

This paper presents a new density-based algorithm called variable ε-DBSCAN by modified the original DBSCAN [8]. In DBSCAN, the density associated with a point or an object is determined by counting the number of points (MinPts) in a region of specified radius (ε) around the point. Points with a density above a specified thresh-old are considered as core points (clusters). The paper extended existing DBSCAN algorithm in two important directions. First, the proposed method is able to cluster the non-spatial, spatial, and temporal attributes present in the data. Second, existing density-based algorithms have constant threshold value (MinPts, ε) for the entire algorithm. The proposed algorithm have time-dependent variable ε neighborhood distance parameter (different for each point) by keeping the other threshold param-eters (MinPts) constant. The proposed algorithm is applied to cluster analysis of the seismic catalog of Himalaya and Sumatra–Andaman region. The algorithm detects aftershocks (triggered events) related to mainshock and evaluates the regional seis-mic hazard in the catalog. The significant achievement of the algorithm is obtained the outliers which represent the background seismicity (indepedent, non-triggered events) present in the catalog.

The rest of the paper is organized as follows. Section 2 describes the catalog of the Himalaya and Sumatra–Andaman region. The proposed methodology are discussed in Sect. 3 which is build on two sub-stages. The spatial domain clustering based on haversine distance function is mentioned in Sect. 3.1. The temporal clustering based on proposed variable "ε"-based density algorithm is discussed in Sect. 3.2. The clustering results obtained from the proposed model is discussed in Sect. 4 for both the region. The comparison with benchmark declustering method is discussed in Sect. 5. Finally, the conclusions are drawn and summarized in Sect. 6.

2 Seismic Catalog of Himalaya and Sumatra–Andaman Region

The seismic catalog of Himalaya and Sumatra–Andaman region considered for analysis is obtained from Advanced National Seismic System (ANSS). The ANSS earthquake catalog is accessed through the Northern California Earthquake Data Center [9]. The parameter setting to download the catalogs for last 51 years is presented in Table 1. The Himalayan earthquake catalog consists of 17,770 earthquake events during the period 1965–2015 and few of them having magnitude greater than 7 (in Richter scale). They are well separated in time and having great impact on the earth and identified as a mainshocks shown in Fig. 1a. The frequency of events in 0.3-year time interval is shown in Fig. 1b. Sumatra–Andaman catalog has total 19,721 earthquake events among which 17 great earthquakes are considered as the mainshocks shown in Fig. 1c in Sumatra–Andaman catalog. The frequency of events in 0.3-year time interval is shown in Fig. 1b. This catalog has very prone to seismic activities which leads to the destructive tsunamis.

Table 1 Parameter setting to download the true catalogs of Himalaya and Sumatra–Andaman from ANSS

Parameters	Himalaya	Sumatra–Andaman
Start time	1965/01/01,00:00:00	1965/01/01,00:00:00
End time	2015/12/31,00:00:00	2015/12/31,00:00:00
Min latitude	20	−10
Max latitude	40	20
Min longitude	70	90
Max longitude	100	110
Min magnitude	1.5	2
Max magnitude	10	10
Min depth	0	0
Max depth	1000	1000
Event type	E	E

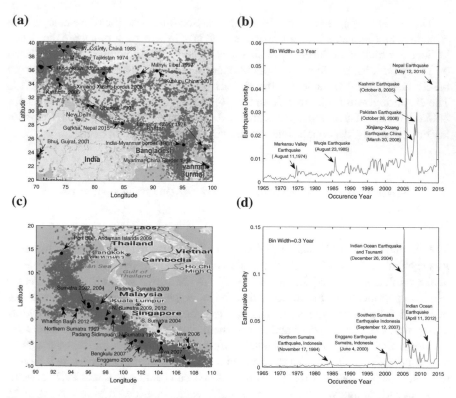

Fig. 1 Distribution of seismic events with period from 1965 to 2015: **a** epicenter plot with identified mainshocks, **b** frequency count with bin width 0.3-year for Himalayan catalog: **c** epicenter plot with identified mainshocks, **d** frequency count with same bin width in Sumatra–Andaman catalog

3 Proposed Methodology

The method is based on hybrid approach to analyze the seismic catalog in spatiotemporal domain and it is built on two stages. In the first stage, the model finds the clusters (seismic zones) by considering the distance criterion between cluster centers (mainshocks) and events in 2D (coordinates). After that model obtained, the finite number of significant clusters along with outliers present in each seismic zones. In this regard, the model uses the modified DBSCAN algorithm which works in time domain with varied ε (neighbor radius) concept. Finally, the model figure out the space-time correlated clusters or foreshock–aftershock pattern around the mainshock with background seismicity (outliers) in entire region. It is helpful to de-cluster the catalog which is excluded from foreshock and aftershocks (only having independent background events (Outliers)). This de-cluster catalog is further used in hazard analysis and earthquake prediction models because of having randomness characteristics.

Fig. 2 Proposed two-stage clustering model for spatio-temporal domain analysis of seismic catalog

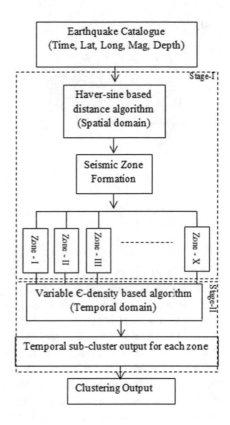

The irregular-shaped temporal clusters tell the presence of earthquake fault network in a particular zone. The proposed clustering model to classify the event either a background or foreshock–aftershock is shown in Fig. 2 and it is described in detail into two different section as follows.

3.1 Stage: I Spatial Domain Clustering

The spatial domain clustering is applied to classify the events into distinct clusters and zones. This paper used selective mainshocks (cluster prototypes) to determine the number of spatial clusters and the distance between mainshocks and events decide the grouping of the data. This distance-based approach takes care of the spatial orientation of the mainshock epicenters. This approach defines a distance metric by calculating the similarity measure between each event and cluster prototypes in the spatial domain (Coordinates). Because of the irregularity in the surface of the earth, geographical coordinate (Latitude (ϕ), Longitude (λ)) points and distances among them plays an important role in the formation of seismic zones. The distance calcu-

(a)

(b)

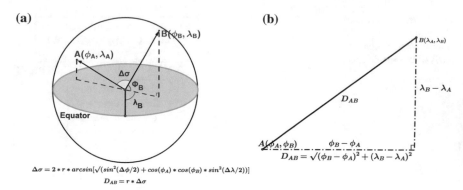

$$\Delta\sigma = 2 * r * arcsin[\sqrt{(sin^2(\Delta\phi/2) + cos(\phi_A) * cos(\phi_B) * sin^2(\Delta\lambda/2))}]$$

$$D_{AB} = r * \Delta\sigma$$

Fig. 3 Distance calculation based on: **a** laws of haversines (spherical trigonometry), **b** laws of Cosines (plane trigonometry)

lation is basically based on some level of abstraction (Earth surfaces), and it does not provide an exact distance. Flat surface approximations are good for short distances, and this approximation becomes inaccurate if the separation between the points are greater and a point becomes closer to a geographic pole. So, Pythagoras theorem (Law of Cosines) works as a distance function only on 2D Euclidean plane (see Fig. 3b) because at small scales the surface of a sphere looks very much like a plane. By assuming the Earth surface nearly spherical, the haversine formula [10] (Law of Haversines) gives "as-the-crow-flies" distance. It is the shortest distance along a great circle on a sphere from their longitudes and latitudes. It is easy to implement and a nice balance of accuracy over complexity in spherical trigonometry. That is why the paper uses haversine distance function (see Fig. 3a) to obtain the spatial clusters. This distance function is also computed by transforming the geodetic (latitude-longitude) coordinates into Cartesian coordinates system (X, Y, Z) as follows:

$$X = r * \cos(\phi) * \cos(\lambda), \ Y = r * \cos(\phi) * \sin(\lambda), \ Z = r * \sin(\phi) \qquad (1)$$

where r is the approximate radius of the earth (6371 km). ϕ and λ represent the longitude and latitude in radians. Here, Z-coordinate does not correspond to the altitude, it is the altitude at the North Pole, but at the equator it is in the north-south direction. Now, the Euclidean distance D_{AB} in $\mathbb{R}^3$ between the two point A and B can be calculated as follows:

$$D_{AB} = \sqrt{[(X_A - X_B)^2 + (Y_A - Y_B)^2 + (Z_A - Z_B)^2]} \qquad (2)$$

The procedure to obtain the spatial clusters and seismic zones based on the described distance function is summarized as follows:

Step 1. The earthquake data obtained from NCEDC is represented mathematically as follows:

$$E_{N \times D} = \begin{bmatrix} e_{11} & e_{12} & e_{13} & \dots & e_{1D} \\ e_{21} & e_{22} & e_{23} & \dots & e_{2D} \\ \vdots & \vdots & \vdots & \ddots & \vdots \\ x_{d1} & e_{N2} & e_{N3} & \dots & e_{ND} \end{bmatrix} \tag{3}$$

where N is the total number of earthquake events, and D is the dimension/features in the formulated catalog. The identified mainshocks are treated as cluster prototypes/centroids for analysis of the given catalog. The number of cluster centers are same as the number of identified mainshocks.

$$\langle M \rangle = \{m_1, m_2, m_3 \dots m_{Count}\} \tag{4}$$

Step 2. Now, calculate the X, Y, and Z points which corresponds to latitude–longitude using (1), (2) and (3), respectively, and then find the haversine distance metric between each event and cluster prototypes.

$$D_{hav}(e_i, m_j) = \sqrt{\sum_{d=1}^{3}(e_{i,d} - m_{j,d})^2} \quad \forall i = 1, 2, 3 \dots N \ \& \ j = 1, 2, 3, \dots Count \tag{5}$$

Step 3. Assign a label C_i to each event to nearest cluster prototype for which distance is minimum.

$$C_i = arg \ \min_i \ D_{hav}(e_i, m_k) \quad k \in 1, 2, 3, \dots Count \tag{6}$$

Finally, the entire region is classified into distinct spatial clusters with unique cluster Ids. Fig. 4a, c represents the spatial clusters for Himalaya and Sumatra–Andaman region, respectively. In this process, some of the clusters are very near to each other because their corresponding mainshock is tightly closed. Basically, these mainshocks occurred on the same fault but their inter-occurrence time is large. To discover the seismic zone, these corresponding clusters are merged with another cluster which occurs on the same fault. In the Himalaya region, all the clusters which belong to the Hindukush region (Cluster-1, 4, 8, 13, 17) and Cluster-2 & 5 are combined (see Fig. 4b). Similarly, C5 & C11 are grouped together and cluster 6, 8, and 12 also to make the seismic zone for Sumatra–Andaman region (see Fig. 4d).

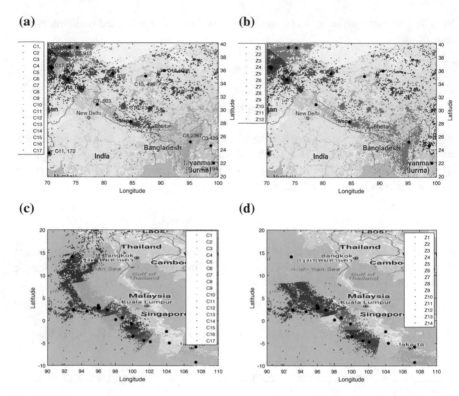

Fig. 4 Spatial domain clustering output: **a** (C1–C17) spatial clusters identified with corresponding mainshocks in Himalaya, **b** obtained (Z1–Z12) distinct seismic zones represented in different color for Himalaya, **c** (C1–C17) spatial cluster identified with corresponding mainshocks, and **d** obtained (Z1–Z14) distinct seismic zones in Sumatra–Andaman region

3.2 Stage: II Variable ε–Density-Based Algorithm in Time Domain

Density-based clustering algorithms are very popularized due to their potential to detect clusters of arbitrary shape and ability to handle outliers present in a dataset. The original DBSCAN [8] uses the minimum number of points and ε neighborhood radius as inputs to find the high-density regions in multidimensional space. The appropriate parameter selection of these two in the algorithm decides the optimal number of clusters and outliers in a given dataset. Here, the paper deals with the seismic catalog (data) which has the non-spatial, spatial, and temporal attributes (magnitude in Richter scale, coordinates in terms of latitude–longitude and occurrence time). The data present in a seismic catalog has the strong magnitude-based correlation in space and time due to the presence of aftershocks related to mainshocks.

In this stage, the modified version of the original well-known DBSCAN clustering algorithm [8] is presented to find the density-based temporal clusters in each zone effectively. The earthquake catalogs inherently show the space-time clustering and these events are highly correlated in both the domain. The proposed density-based algorithm is considered the magnitude-based temporal aspect of an event where the spatial criterion is already satisfied in stage-I. In the proposed algorithm, "MinPts" selection criterion is same but neighborhood radius ε is made variable instead of keeping constant as in original DBSCAN algorithm. A weight (magnitude in earth-quake catalog) is given to each event and ε is selected based on these weights. Hence, the ε of each event is tuned in accordance with an empirical relation given as follows:

$$T = 10^{(p*M)} \; days \tag{7}$$

where T is the time window for each event's magnitude. The values of constant p depend on the seismological region under investigation [11, 12]. This relation signifies the temporal extent of the each event and generation of triggered events under this temporal bound. The size of the window depends on the event's magnitude as a power law. This magnitude-dependent temporal window is used as a neighborhood radius and applied to find core points in the algorithm. This variable temporal window heavily influences the shape of the resulting clusters and the outcome.

4   Result and Discussions

The modified variable ε density-based algorithm determines the significant number of clusters in each seismic zone for both the catalogs. These clusters have strong correlation in space-time-energy domain and belong to a specific fault network in a region. Figure 5a, c represents the clustering output obtained from the model for the Himalaya and Sumatra–Andaman catalog, respectively. The highly dense region around the mainshock (in black circles) and a small dense region around the high magnitude event (in red stars) is shown in Fig. 5a, c. They represent the density variation w.r.t. magnitude of the events. The cluster's density indicates the change in length of time window that depends on event's magnitude as a power law. There are total 159 for Himalaya and 244 earthquake events for Sumatra–Andaman hav-ing a magnitude greater than 6.0 in Richter scale (shown in red stars in Fig. 5a, c). They also satisfy the spatio-temporal criterion but their temporal window is short as compared to mainshock and have less dense region around them. These events describe the presence of triggered phenomenon which led to the space-time cluster-ing. It is observed that zone 2 and 12 are most hazardous and have higher seismic occurrences among all in Himalaya catalog. All the earthquakes occurred in zone 12 belongs to the Hindu kush region and have deep hypocenter. Seismic activities present in Sumatra–Andaman triggers tsunamis in Indian Ocean. Geographically, it is one of the longest faults present which covers the entire Sumatra and Andaman islands.

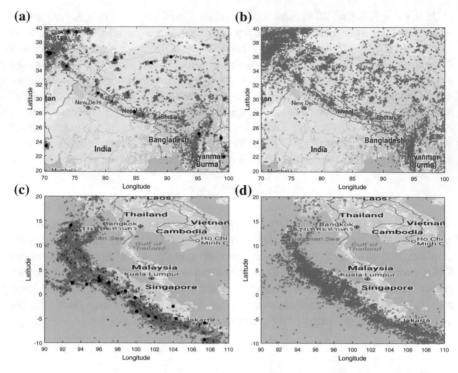

Fig. 5 Clustering output obtained from the proposed model: **a** epicenter plot of clustered events treated as foreshock–aftershocks (clustered catalog), **b** epicenter plot of outliers treated as background event (de-clustered catalog), **c** clustered events and **d** de-clustered events for Sumatra–Andaman

The outliers obtained from the proposed algorithm for each zone are shown in Fig. 5b, d for Himalaya and Sumatra–Andaman catalog, respectively. It reflects the non-clustered part of the seismicity. These events are neither associated with any fault nor triggered by any event. These events are treated as backgrounds which do not satisfy the algorithm criterion. They are uniform throughout the region (see Fig. 5b, d).

The proposed model also compares the cumulative number of events with time obtained from the original, clustered, and de-clustered catalog for both the region. Fig. 6a, b are both reveal that the clustered events resemble the same characteristics as true catalog whereas de-clustered catalog (Background seismicity) has the linear cumulative rate with time.

The potential advantages of the proposed model are that it not only finds the clusters for the identified mainshocks, but it also locates the several small individual seismic clusters that exist in their vicinity. A large number of clusters which occupy a particular zone are obtained due to the inclusion of time and magnitude as an extra feature in the algorithm. These clusters are automatically combined if they are occurred in a very short duration. The triggered nature of the earthquake events

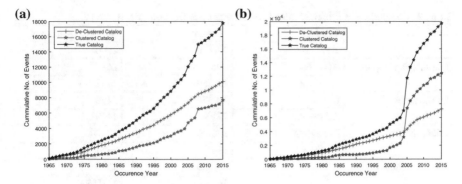

Fig. 6 Cumulative number of events with occurrence year for true, clustered, de-clustered catalog for: **a** himalaya and **b** Sumatra–Andaman catalog

is incorporated in the proposed algorithm to make it suitable for finding the highly dense regions as well as significant clusters. The time zone of each individual event is correlated with event magnitude according to the Eq. (7). In most cases, this empirical relation has great impact on in the investigation of seismic zones. But in few cases, triggering effect of events also depends on the fault mechanism, soil condition, and terrain of the environment.

5 Comparison with Benchmark De-clustering Algorithm

This paper also compares the results with the three other benchmark declustering method proposed by Gardner and Knopoff [13], Uhrhammer [14], and Reasenberg [15] model. The number of background events identified by Reasenberg declustering method is very high, whereas few background events obtained by the Gardner & Knopoff method as mentioned in Table 2. Although the Gardner and Knopoff satisfied the linearity criterion for background seismicity, it gives the very fewer background events for the catalog. Whereas Reseanberg declustering method has excess amount of background events with a nonlinear cumulative rate. The proposed method almost follows the linear trend in time with moderate background seismicity as shown in Table 2 for Himalaya and Sumatra–Andaman. The Sumatra–Andaman catalog has burst seismic activity in 2004–2005 due to the occurrence of 26 Dec 2004 Indian Ocean Earthquake. The overall background seismicity for the catalog is linear in time.

Table 2 Comparison of proposed model with Benchmark Declustering Method

Method	Catalogs	Himalaya	Sumatra–Andaman
	Total events	17,770	19,721
Gardner & Knopoff	BG events	4988	2689
	AF events	12,782	17,032
Uhrhammer	BG events	10,377	5159
	AF events	7393	14,562
Reasenberg	BG events	15,830	15,370
	AF events	1940	4351
Proposed model	BG events	**10,089**	**7266**
	AF events	**7681**	**12,455**

6 Conclusion

In this paper, a new clustering model is proposed to group seismic events based on their spatial and temporal domain characteristics. The haversine distance between the event's coordinate with respect to the mainshock clusters the events in space and obtained the distinct seismic zones. A variable ε-density-based algorithm is proposed which finds the density variant and arbitrary-shaped temporal clusters and non-clustered events (outliers/backgrounds) present in each seismic zone very accurately. The obtained clusters (comprise foreshock–aftershock events) also describe the underlying faults structures beneath the earth surfaces. The de-cluster catalog (non-clustered events) has a significant role to get the unbiased estimation of seismicity and shows the random behavior in nature. The linear trend observed in de-clustered catalog shows uniform background seismicity and may be characterized with Poisson distribution.

References

1. Nanda, S.J., et al.: A tri-stage cluster identification model for accurate analysis of seismic catalogs. Nonlinear Process. Geophys. **20**(1), 143–162 (2013)
2. Vijay, R., Nanda, S.J.: A tetra-stage cluster identification model to analyze the seismic activities of Japan. Himalaya and Taiwan, IET Signal Processing (2017)
3. Zaliapin, I., et al.: Clustering analysis of seismicity and aftershock identification. Phys. Rev. Lett. **101**(1) (2008)
4. Vijay, R., Nanda, S.J.: Declustering of an earthquake catalog based on ergodicity using parallel grey wolf optimization. In: 2017 IEEE Congress on Evolutionary Computation (CEC), pp. 1667–1674 (2017)
5. Birant, D., Kut, A.: ST-DBSCAN: an algorithm for clustering spatial-temporal data. Data Knowl. Eng. **60**(1), 208–221 (2007)
6. Georgoulas, G., et al.: Seismic-mass density-based algorithm for spatio-temporal clustering. Expert Syst. Appl. **40**(10), 4183–4189 (2013)

7. Nanda, S.J., Panda, G.: Design of computationally efficient density-based clustering algorithms. Data Knowl. Eng. **95**, 23–38 (2015)
8. Ester, M., Kriegel, H., Sander, J., Xu, X.: A density-based algorithm for discovering clusters in large spatial databases with noise. In: Kdd, vol. 96, pp. 226–231 (1996)
9. NCEDC. Northern California Earthquake Data Center.: UC Berkeley Seismological Laboratory. Dataset (2017)
10. Sheba, S., Ramadoss, B., Balasundaram, S.R.: Geo distance-based event detection in social media. Int. J. Comput. Intell. Stud. **4**(1), 87–101 (2015)
11. Drakatos, G., Latoussakis, J.: A catalog of aftershock sequences in Greece (1971–1997): their spatial and temporal characteristics. J. Seismol. **5**(2), 137–145 (2001)
12. Zubkov, S.I.: The appearance times of earthquake precursors. Izv. Akad. Nauk SSSR Fiz. Zemli (Solid Earth) **5**, 87–91 (1987)
13. Gardner, J.K., Knopoff, L.: Is the sequence of earthquakes in southern California, with aftershocks removed, Poissonian. Bull. Seismol. Soc. Am **64**(5), 1363–1367 (1974)
14. Uhrhammer, R.: Characteristics of northern and southern California seismicity. Earthq. Notes **57**(1), 21 (1986)
15. Reasenberg, P.: Second-order moment of central California seismicity, 1969–1982. J. Geophys. Res. Solid Earth (1978–2012) **90**(B7), 5479–5495 (1985)

A Digital Image Processing Tool for Size and Number Density Distribution of Precipitates in Creep-Exposed Material

Minati Kumari Sahu, Chandan Dutta, Arpita Ghosh and S. Palit Sagar

Abstract Present paper deals with developing an advanced digital image processing tool for determination of size and number density of precipitates present in creep-exposed P92 steel. The image processing toolbox in MATLAB is usually employed for noise detection and removal, edge detection, cropping, histogram of the region of interest as well as size wise distribution of the desired objects from the micrograph. This tool helps in fast and accurate acquisition of information. In this investigation, creep testing has been carried out on two P92 steel specimens at a temperature of 650 °C and stress of 120 MPa till rupture. Scanning electron microscopy (SEM) was used to capture images of the specimen in as-tempered condition and in the near-rupture gage region after creep rupture. The developed software has been used to analyze the micrographs for quantification of precipitates in terms of area fraction, size, and number density. The use of edge detection technique in the developed software helps in avoiding human intervention during image thresholding, thereby increasing accuracy in precipitate sizing. Findings from this investigation have been used to evaluate the role of precipitation morphology for specimen failure due to creep especially in precipitate strengthened steel like P92 steel.

Keywords Image processing toolbox · Graphical user interface · MATLAB · P92 steel · Precipitates

1 Introduction

Generally, images contain different types of objects and structures which convey information about the material [1, 2]. Detection followed by subsequent counting estimates the number of objects in an image. Counting arises in many real-time appli-

M. K. Sahu (✉) · C. Dutta · A. Ghosh · S. P. Sagar
CSIR-National Metallurgical Laboratory, Jamshedpur 831007, India
e-mail: minati.sahu@gmail.com

M. K. Sahu · C. Dutta · A. Ghosh · S. P. Sagar
NDE & MM Group, Jamshedpur, India

© Springer Nature Singapore Pte Ltd. 2019
J. C. Bansal et al. (eds.), *Soft Computing for Problem Solving*,
Advances in Intelligent Systems and Computing 816,
https://doi.org/10.1007/978-981-13-1592-3_51

cations such as counting grains in agriculture industry, counting cells in microscopic images, counting of number diamonds in industry. Existing methods for counting involves a large amount of hardware which also adds to the cost or manual counting which is time consuming and may give erroneous results. The rapid development of computer vision and pattern recognition technology has been used in quality assessment, detecting and counting object details using image processing tool [3, 4]. Automatic counting of objects is a subject that has shown its application with objects as varied as cells [3], RBCs [4], fish [5], eggs [6], etc. Because automatic counting is objective, reliable, and reproducible, comparison of different objects in a specimen, it gives more accurate results with automatic programs than with manual counting.

Components operating in extreme environment are prone to several damage mechanisms like creep, fatigue, corrosion, and thermal aging. Damage induced in the material is mainly due to microstructural changes in terms of grains and precipitation morphology leading to voids and crack formation, followed by material failure. So, microstructural property evaluation is needed to know the material status. All the microstructural properties have been evaluated previously either through manual scaling or with some software which can mark the boundary irregularly in which the targeted object detection is obscured or eliminated by the program as they are not clearly expressed or may be overlapped. Using color processing, lighting intensity also affects original image [7–10]. In the existing software for microstructural analysis, calibration of scale provided in the microstructure is calibrated manually. As the microstructure is the major strength of any components, their measurement needs a higher accuracy level of access. To address this issue, we have proposed a MATLAB graphical user interface (GUI)-based method capable of quantifying all the microstructural features like grain size, precipitate size, and number density, voids, crack sizing from the micrograph of a specimen. We have applied the software to quantify microstructure of creep-ruptured P92 steel specimen, where the material has undergone high-temperature deformation at 650 °C and stress of 120 MPa. This method is based on morphological operations used in digital image processing in order to make its implementation simple obtaining result accuracy. Development of a method to characterize the distribution of precipitate particles and study the impact of these particles on change of material properties by taking their microstructure is the prime objective of the present research. Effectiveness of this developed tool was analyzed by using the micrographs of creep-damaged P92 steel.

2　Experimental

Steel plates of P92 steel specimen were collected in as-tempered condition. Samples were prepared for creep-testing machine. Creep testing was carried out on two P92 steel specimens at a temperature of 650 °C and stress of 120 MPa till rupture. This test condition was chosen according to the operating condition in power plants where this material is used. After creep rupture, failed specimens were cut near the rupture region for microstructural analysis. Micrographs of specimen were taken in as-received condition and near the rupture region of gage length of creep failed specimen using scanning electron microscopy (SEM). A MATLAB-based GUI was developed (in-house) for microstructural analysis using the image processing toolbox. The SEM images were then analyzed using the proposed software. Brightness and contrast thresholds were adjusted automatically for better separation of the precipitates from the surrounding grain structure. The SEM scale bar on the image was used to calibrate the number of pixels per micrometer to accurately measure particle size. Depending on the size of the object, the software then measured the number of particles (count), average particle size (average size), and area fraction of particles (% area) for each image. Each sample was examined to determine the effectiveness of this technique in analyzing the correlation between precipitates and creep damage parameters in P92 steel.

The fundamental steps of image processing and analysis in the proposed algorithm showing steps are presented in Fig. 1. Brightness and contrast thresholds were adjusted automatically for better separation of the precipitates from the surrounding matrix grain structure. The scale bar on the image was used to calibrate the number of pixels per micrometer for accurate particle size measurement. The software then used to measure the number of precipitated particles (count), their actual size, and area fraction (% area) for each image. But, the output result has the capability to output statistics for each individual particle, providing an exhaustive study of precipitate size and distribution for future research. Averaging of data was carried out considering 10 micrographs for each specimen.

3 Results and Discussion

The microstructure of P92 steel specimen in as-tempered condition and after creep is shown in Fig. 2. From the microstructure, it is revealed that there is a change in size of precipitates (in terms of growth and coarsening) as moving from as-tempered condition toward creep rupture. So, evaluation of precipitation morphology is the major determining parameter for creep life evaluation of this precipitate-strengthened material. Using the developed software number density and size distribution of precipitates (with both minimum and maximum size), area fraction was determined. Image thresholding was done in each image using edge detection technique for background noise reduction. Figure 3 shows the steps followed for precipitate counting using the developed software.

After loading the micrograph, pixel size was calibrated using the provided micrograph scale length. In the next step, background noise reduction was done by edge detection technique which reveals only the precipitates present in the matrix. The image processed through the software was able to identify the pixels of precipitated particles (white) against the surrounding matrix (black). The precipitates in each image were evaluated using the "count precipitate" function. Size-wise distribution was carried out on these particles depending on the requirement. The report generation module was used for saving data for future reference.

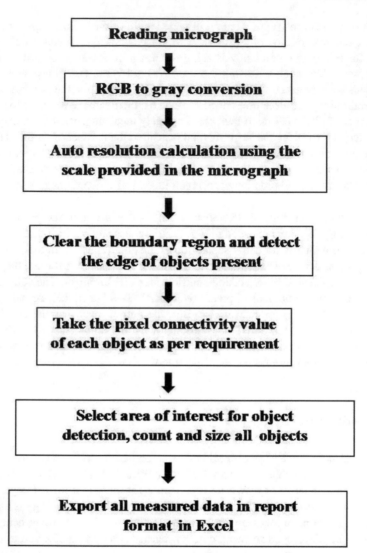

Fig. 1 Proposed algorithm for operation of the GUI

It was observed that the number of precipitates ranged from approximately 200 to 800 moving from as-received specimen toward crept specimen. The average area of each precipitate was found between 0.3 and 0.9 μm^2. Area fractions which are calculated taking the percentage of ratio of total area of precipitate to the total area of the image ranged approximately from 1 to 8%. The precipitates were divided into four different categories, i.e., fine, finer, coarse, and coarser based on size ranges. Figure 4 shows the area fraction of all the four category of precipitates according to their size range in as-received and crept specimen. It is observed that area occupied

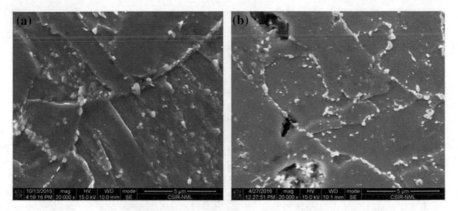

Fig. 2 SEM micrographs of P92 specimen in **a** as-tempered condition, **b** near-rupture region across the gage length after creep failure

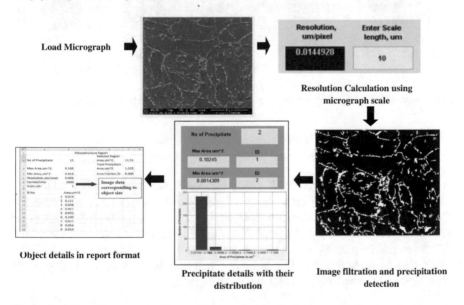

Fig. 3 Steps for object counting from a micrograph using the indigenously developed software in MATLAB interface

by coarser precipitates in crept specimen was maximum as compared to coarse, finer, and fine category. This distribution study was possible to perform only after developing the indigenously developed software. This study helped to quantify the precipitate effect on creep-ruptured specimen.

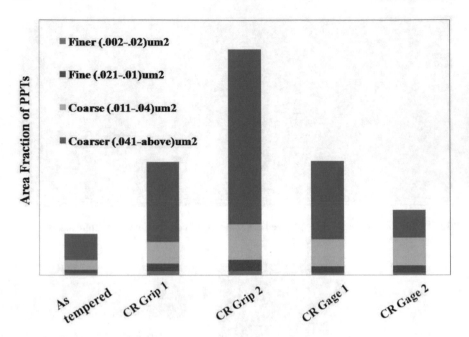

Fig. 4 Area fraction of precipitates in as-received and creep-ruptured (CR) specimen

4 Summary

In this study, digital image processing based on MATLAB is effectively used for distribution and counting of different objects of any micrograph with enhanced accuracy. The developed program is fast, low-cost, and user-friendly software which can be helpful for counting of object details. Through this study, an analytical technique was developed to quantify precipitate population and distribution by measuring their count, size, and area fraction. This technique will be helpful for the analyzing effect of precipitation morphology in P92 steel during creep.

Acknowledgements The authors are thankful to the Director, CSIR-National Metallurgical Laboratory, Jamshedpur, India, for his permission to publish the work. The first author is grateful to Council of Scientific and Industrial Research for granting financial support.

References

1. Lee, S.G., Mao, Y., Gokhale, A.M., Harris, J., Horstemeyer, M.F.: Application of digital image processing for automatic detection and characterization of cracked constituent particles/inclusions in wrought aluminum alloys. Mater. Charact. **60**(9), 964–970 (2009)
2. Deb, D., Hariharan, S., Rao, U.M., Ryu, C.-H.: Automatic detection and analysis of discontinuity geometry of rock mass from digital images. Comput. Geosci. **34**(2), 115–126 (2008)

3. Guo, X., Yu, F.: A method of automatic cell counting based on microscopic image. In: 5th International Conference on Intelligent Human-Machine Systems and Cybernetics, vol. 1, pp. 293–296 (2013)
4. Venkatalaksmi, B., Thilagavathi, K.: Automatic red blood cell counting using hough transform. In: Proceedings of 2013 IEEE Conference on Information and Communication Technology, pp. 267–271 (2013)
5. Fabic, J.N., Turla, I.E., Capacillo, J.A., David, L.T., Naval, P.C. Jr.: Fish population estimation and species classification from underwater video sequences using blob counting and shape analysis. In: 2013 International Underwater Technology Symposium (UT), pp. 1–6 (2013)
6. Mello, C.A.B., dos Santos, W.P., Rodrigues, M.A.B., Candeias, A.L.B., Gusmao, C.M.G.: Image segmentation of ovitraps for automatic counting of Aedes Aegypti Eggs. In: 30th Annual International IEEE EMBS Conference. Vancouver, British Columbia, Canada, pp. 3103–3106 (2008)
7. Reddy, D.V.K.: Sorting of objects based on colour by pick and place robotic arm and with conveyor belt arrangement. Int. J. Mech. Eng. Robot. Res. India 3(1) (2014)
8. Shen, Z., Luo, J.C., Zhou, C., Huang, G., Ma, W., Ming, D.: System design and implementation of digital-image processing using computational grids. Comput. Geosci. 31(5), 619–630 (2005)
9. Moolman, D.W., Aldrich, C., Van Deventer, J.S.J., Stange, W.W.: Digital image processing as a tool for online monitoring of froth in flotation plants. Miner. Eng. 7(9), 1149–1164 (1994)
10. Tremaine, A.: Characterization of internal defects in open die forgings. FIERF Grant Project for Undergraduate Research 17 Nov 2005

On (i, j) Generalized Fuzzy γ-Closed Set in Fuzzy Bitopological Spaces

Birojit Das and Baby Bhattacharya

Abstract In this present treatise, we introduce the notion of (i, j) generalized fuzzy γ-closed $((i, j)\ gf\gamma$-closed) set and $(i, j)\ \gamma$-generalized fuzzy closed $((i, j)\ \gamma\text{-}gf$ closed) set in a fuzzy bitopological space. We show that $(i, j)\ \gamma$-generalized fuzzy closed set and (i, j) generalized fuzzy γ-closed set are totally independent of each other. Different properties related to those sets are studied. Also, (i, j) generalized fuzzy γ-continuous functions and (i, j) generalized fuzzy γ-irresolute functions based on $(i, j)\ gf\gamma$-closed set are introduced and interrelationship among them are established. Finally, we characterize these functions in various directions for different purposes.

1 Introduction and Preliminaries

The notion of generalized closed set was first introduced by N. Levine in the year 1970 [6], and then after, N. Palaniappan introduced a weaker form of this set, namely regular generalized closed set [7] in a topological space. After that, many generalizations have been made in this direction in different environments. In 1997, G. Balasubramaniam and P. Sundaram first originated the idea of generalized closed set in fuzzy topological space (in short, fts) and named it generalized fuzzy closed set [1] and after that J. H. Park and J. K. Park further studied regular generalized closed set in fuzzy topological structure [8]. In this paper, we are going to initiate the idea of generalized fuzzy closed set in a fuzzy bitopological space.

Here, we state the definition of regular generalized fuzzy closed set given by Park and Park [8] which is as follows:

A fuzzy set λ in a fts X is called regular generalized fuzzy closed (in short, rgf-closed) if $cl(\lambda) \leq \mu$, whenever $\lambda \leq \mu$ and μ is a fuzzy regular open set.

B. Das · B. Bhattacharya (✉)
Department of Mathematics, National Institute of Technology, Agartala, India
e-mail: babybhatt75@gmail.com

B. Das
e-mail: dasbirojit@gmail.com

© Springer Nature Singapore Pte Ltd. 2019
J. C. Bansal et al. (eds.), *Soft Computing for Problem Solving*,
Advances in Intelligent Systems and Computing 816,
https://doi.org/10.1007/978-981-13-1592-3_52

In 2011, S. Bhattacharya introduced the notion of generalized regular closed set [3] in topological space by defining it in a different approach.

Very recently, B. Bhattacharya and J. Chakraborty extended the above work in fuzzy topological space [2] as follows.

A fuzzy set λ in a fts (X, τ) is called generalized regular fuzzy closed (in short, grf-closed) set if $R\text{-}cl(\lambda) \leq \mu$, whenever $\lambda \leq \mu$ and μ is a fuzzy open set. Here, $R\text{-}cl(\lambda)$ is the intersection of all regular fuzzy closed set containing λ.

We are going to define (i, j) generalized fuzzy γ-closed set in a fuzzy bitopolgial space following the above definition. More specifically, we will consider (i, j) γ-closure operator and (i, j) fuzzy open set to define it.

Moreover, T. Fukutake, in 1986, introduced the concept of generalized closed set in bitopological space [4]. Surprisingly, it is observed that no further research work has been done by any researcher in this direction in a fuzzy bitopological space.

All these above mentioned research work motivated us to examine the behaviour of generalized closed set in the light of γ-open set [10] in fuzzy bitopological space. In this paper, we define the concept of (i, j) generalized fuzzy γ-closed (in short (i, j) $gf\gamma$-closed) set, (i, j) γ-generalized fuzzy closed (in short, (i, j) γ-gf closed) set, following the definition of Park et al. and Bhattacharya et al. respectively. In this work, our main objective is to study the characterization of (i, j) $gf\gamma$-closed set and (i, j) γ-gf closed set and their interrelationship in a fuzzy bitopological space. Throughout this paper, we denote a fuzzy bitopological space (FBTS) by $(X, \tau_i, \tau_j), i, j = 1, 2, i \neq j$ and more rare simply denote it as X.

Before going to the main contribution, we present some existing related definitions available in the literature as ready references for the work.

Definition 1.1 [5] Let X be a non-empty set and τ_i, τ_j be two fuzzy topologies defined on X. Then, (X, τ_i, τ_j) is called a FBTS. Any fuzzy set λ in X is said to be (i, j) fuzzy open if $\lambda \in \tau_i \vee \tau_j$. The complement of a (i, j) fuzzy open set is said to be a (i, j) fuzzy closed set.

Here, (i, j) $cl(\mu)$ denotes the intersection of all super (i, j) fuzzy closed set of μ.

Definition 1.2 [9] Let (X, τ) be a fuzzy topological space. A fuzzy set λ in X is called a generalized fuzzy closed (in short, gf closed) set if $cl(\lambda) \leq \mu$, whenever $\lambda \leq \mu$ and μ is a fuzzy open set in X.

Definition 1.3 [10] A fuzzy subset λ of a FBTS (X, τ_i, τ_j) for $i, j = 1, 2$ and $i \neq j$ is called (i, j) fuzzy γ-open if $\lambda \wedge \mu$ is (i, j) fuzzy pre-open for every (i, j) fuzzy pre-open set μ in X. A fuzzy subset η of X is called (i, j) fuzzy γ-closed set if its complement, $1_X - \eta$ in X is a (i, j) fuzzy γ-open set.

We use (i, j) $cl_\gamma(\lambda)$ to denote the intersection of all (i, j) fuzzy γ-closed sets containing λ.

Definition 1.4 [1] A map $f : (X, \tau) \to (Y, \sigma)$ from a fuzzy topological space into another fuzzy topological space is called a generalized fuzzy continuous (in short, gf continuous) function if the inverse image of every fuzzy closed set in Y is a gf closed set in X.

2 (i, j) Generalized Fuzzy γ-Closed Sets and (i, j) Generalized Fuzzy γ-Open Sets

Definition 2.1 A fuzzy subset μ in a FBTS (X, τ_i, τ_j) for $i, j = 1, 2$ and $i \neq j$ is called a (i, j) generalized fuzzy closed (for short (i, j) gf closed) set if (i, j) $cl(\mu) \leq \eta$, whenever $\mu \leq \eta$ and $\eta \in (i, j)$ $FO(X)$, where (i, j) $FO(X)$ is the family of all (i, j) fuzzy open sets.

Definition 2.2 Let (X, τ_i, τ_j) for $i, j = 1, 2$ and $i \neq j$ be a FBTS and λ be an arbitrary fuzzy set in X. Then, λ is said to be (i, j) generalized fuzzy γ-closed (in short, (i, j) $gf\gamma$-closed) set if for any (i, j) fuzzy open set $\mu \in X$, $\lambda \leq \mu$ implies (i, j) $cl_\gamma(\lambda) \leq \mu$. The family of all (i, j) $gf\gamma$-closed sets is denoted by (i, j) $gf\gamma$-$C(X)$.

Definition 2.3 Let (X, τ_i, τ_j) for $i, j = 1, 2$ and $i \neq j$ be a FBTS and $\lambda \in X$. Then, λ is said to be (i, j) γ-generalized fuzzy closed (in short, (i, j) γ-gf closed) set if for any (i, j) fuzzy γ-open set μ in X, (i, j) $cl(\lambda) \leq \mu$, whenever $\lambda \leq \mu$. The family of all (i, j) γ-gf closed sets is denoted by (i, j) γ-gf $C(X)$.

Definition 2.4 In a FBTS (X, τ_i, τ_j) with $i, j = 1, 2, i \neq j$, let λ be any fuzzy set in X. Then, λ is said to be (i, j) γ-generalized fuzzy γ-closed (in short, (i, j) γgf-γ-closed) set if for any (i, j) fuzzy γ-open set μ in X, $\lambda \leq \mu$ implies (i, j) $cl_\gamma(\lambda) \leq \mu$ where (i, j) $cl_\gamma(\lambda)$ is the (i, j) γ-closure of λ in X. The family of all (i, j) $\gamma gf\gamma$-closed sets is denoted by (i, j) $\gamma gf\gamma$-$C(X)$.

Theorem 2.1 *If λ_1 and λ_2 are any two (i, j) $gf\gamma$-closed sets in a FBTS X, then $\lambda_1 \vee \lambda_2$ is also a (i, j) $gf\gamma$-closed set in X.*

Proof Let us consider two (i, j) $gf\gamma$-closed sets λ_1 and λ_2 and $\lambda_1 \vee \lambda_2 \leq \mu$, where μ is a (i, j) fuzzy open set. Then, $\lambda_1, \lambda_2 \leq \mu$, and then, (i, j) $cl_\gamma(\lambda_1) \leq \mu$, $(i, j) cl_\gamma(\lambda_2) \leq \mu$. Therefore, $(i, j) cl_\gamma(\lambda_1 \vee \lambda_2) = (i, j) cl_\gamma(\lambda_1) \vee (i, j) cl_\gamma(\lambda_2) \leq \mu$. Hence, $\lambda_1 \vee \lambda_2$ is a (i, j) $gf\gamma$-closed set in X.

Theorem 2.2 *If λ and η are (i, j) γ-gf closed sets, then $\lambda \vee \eta$ is (i, j) γ-gf closed.*

Proof Let $\lambda \vee \eta \leq \mu$ and μ be any (i, j) fuzzy γ-open set. Then, $\lambda \leq \mu$ and $\eta \leq \mu$, and thus, (i, j) $cl(\lambda)$, (i, j) $cl(\eta) \leq \mu$. Hence, (i, j) $cl(\lambda \vee \eta) \leq \mu$, since (i, j) $cl(\lambda) \vee (i, j)$ $cl(\eta) = (i, j)$ $cl(\lambda \vee \eta)$. Therefore, $\lambda \vee \eta$ is a (i, j) γ-gf closed set.

Remark 2.1 Intersection of any two (i, j) $gf\gamma$-closed sets may not be a (i, j) $gf\gamma$-closed set.

Example 2.1 Let us consider a FBTS (X, τ_i, τ_j) with $i, j = 1, 2, i \neq j$ such that $X = \{x, y\}$, $\tau_i = \{0_X, 1_X, \{(x, 0.2), (y, 0)\}, \{(x, 0.2), (y, 0.3)\}\}$ and $\tau_2 = \{0_X, 1_X, \{(x, 0), (y, 0.3)\}\}$. Then, we have (i, j) $FO(X) = \{0_X, 1_X, \{(x, 0.2), (y, 0.3)\}, \{(x, 0), (y, 0.3)\}, \{(x, 0.2), (y, 0)\}\}$ and so (i, j) $FC(X) = \{0_X, 1_X, \{(x, 0.8),$

$(y, 0.7)\}, \{(x, 1), (y, 0.7)\}, \{(x, 0.8), (y, 1)\}\}$. Then, (i, j) $F\gamma$-$O(X) = \{0_X, 1_X, \{(x, \alpha), (y, \beta)\} : \alpha \geq 0.8, \beta \geq 0.7\}$ and $(i, j) F\gamma$-$C(X) = \{0_X, 1_X, \{\{(x, \alpha), (y, \beta)\} : \alpha \leq 0.2, \beta \leq 0.3\}$. Now, we consider two fuzzy sets $\mu_1 = \{(x, 0.8), (y, 0.3)\}$ and $\mu_2 = \{(x, 0.2), (y, 0.5)\}$. Obviously, μ_1 and μ_2 are both (i, j) $gf\gamma$-closed sets. Again, $\mu_1 \wedge \mu_2 = \{(x, 0.2), (y, 0.3)\} \leq \{(x, 0.2), (y, 0.3)\}$, which is an (i, j) fuzzy open set, but $(i, j) cl_\gamma(\mu_1 \wedge \mu_2) = \{(x, 0.8), (y, 0.7)\} \nleq \{(x, 0.2), (y, 0.3)\}$. Thus, $\mu_1 \wedge \mu_2$ is not a (i, j) $gf\gamma$-closed set in X.

Remark 2.2 The intersection of two (i, j) γ-gf closed sets is not a (i, j) γ-gf closed set and it is demonstrated in the following example.

Example 2.2 We consider a FBTS (X, τ_i, τ_j) for $i, j = 1, 2$ and $i \neq j$ with $X = \{x, y\}$ and $\tau_i = \{0_X, 1_X, \{(x, 0.2), (y, 0)\}\}$ and $\tau_j = \{0_X, 1_X, \{(x, 0), (y, 0.3)\}\}$. Then, we have (i, j) $FO(X) = \{0_X, 1_X, \{(x, 0.2), (y, 0)\}, \{(x, 0), (y, 0.3)\}, \{(x, 0.2), (y, 0.3)\}\}$. Therefore, $(i, j) FC(X) = \{0_X, 1_X, \{(x, 0.8), (y, 1)\}, \{(x, 1), (y, 0.7)\}, \{(a, 0.8), (b, 0.7)\}\}$. Now we get, (i, j) $F\gamma$-$O(X) = \{0_X, 1_X, \{\{(x, \alpha), (y, \beta)\} : 0 \leq \alpha \leq 0.2, 0 \leq \beta \leq 0.3\}, \{\{(x, \alpha), (y, \beta)\} : \alpha > 0.8, \beta > 0.7\}\}$. Let $\mu = \{(x, 0.8), (y, 0.1)\}$ and $\eta = \{(x, 0.1), (y, 0.5)\}$. Here, μ and η are (i, j) γ-gf closed sets, but their intersection $\mu \wedge \eta = \{(x, 0.1), (y, 0.1)\}$ is not a (i, j) γ-gf closed set because for the (i, j) fuzzy γ-open set $\{(x, 0.1), (y, 0.1)\}$ we have $\mu \wedge \eta < \{(x, 0.1), (y, 0.1)\}$, whereas $(i, j) cl(\mu \wedge \eta) \nleq \{(x, 0.1), (y, 0.1)\}$.

Theorem 2.3 Every (i, j) fuzzy γ-closed set in a FBTS X is a (i, j) $gf\gamma$-closed set.

Proof Let us suppose that λ is an (i, j) fuzzy γ-closed set in a FBTS X. Let $\lambda \leq \mu$, where μ is a (i, j) fuzzy γ-open set. Then, $(i, j) cl_\gamma(\lambda) = \lambda \leq \mu$. Consequently, λ is a (i, j) $gf\gamma$-closed set in X and hence the result.

Remark 2.3 An (i, j) $gf\gamma$-closed set in a FBTS may not be a (i, j) fuzzy γ-closed set, and it is verified in the following example.

Example 2.3 We take the FBTS given in the Example 2.1 and consider the fuzzy set $\mu = \{(x, 0.8), (y, 0.3)\}$. One can easily verify that μ is a (i, j) $gf\gamma$-closed set in X, but it is not a (i, j) fuzzy γ-closed set there in.

Remark 2.4 Both the ideas of (i, j) $gf\gamma$-closed and (i, j) γ-gf closed sets are completely independent of each other.

Example 2.4 Let us consider the FBTS given in the Example 2.1, and we consider the fuzzy set $\mu = \{(x, 0.1), (y, 0.1)\}$. Clearly, μ is a (i, j) $gf\gamma$-closed set in X. Now, $\mu < \{(x, 0.2), (y, 0.3)\}$, which is (i, j) fuzzy γ-open set in X. But $(i, j) cl(\mu) \nleq \{(x, 0.2), (y, 0.3)\}$, and accordingly, μ is not a (i, j) γ-gf closed set. Thus, every (i, j) $gf\gamma$-closed set is not a (i, j) γ-gf closed set.

Example 2.5 We suppose (X, τ_i, τ_j) be a FBTS with $i, j = 1, 2, i \neq j$ such that $X = \{x, y\}, \tau_i = \{0_X, 1_X, \{(x, 0.6), (y, 0.3)\}, \{(x, 0.3), (y, 0.3)\}\}, \tau_j = \{0_X, 1_X, \{(x, 0.3), (y, 0.7)\}, \{(x, 0.6), (y, 0.7)\}\}$. Then, (i, j) $FO(X) = \{0_X, 1_X, \{(x, 0.6),$

$(y, 0.3)\}, \{(x, 0.3), (y, 0.3)\}, \{(x, 0.3), (y, 0.7)\}, \{(x, 0.6), (y, 0.7)\}\}$ and (i, j) $FC(X) = \{0_X, 1_X, \{(x, 0.4), (y, 0.7)\}, \{(x, 0.7), (y, 0.7)\}, \{(x, 0.7), (y, 0.3)\},$ $\{(x, 0.4), (y, 0.3)\}\}$. Thus, (i, j) $F\gamma$-$O(X) = \{0_X, 1_X, \{(x, \alpha), (y, \beta)\} : \alpha > 0.7,$ $\beta > 0.7\}$ and (i, j) $F\gamma$-$C(X) = \{0_X, 1_X, \{(x, \alpha), (y, \beta)\} : \alpha < 0.3, \beta < 0.3\}$. We consider the fuzzy set $\mu = \{(x, 0.5), (y, 0.5)\}$. Evidently, μ is a (i, j) γ-gf closed set in X, but it is not a (i, j) $gf\gamma$-closed set, since $\mu < \{(x, 0.6), (y, 0.7)\}$, whereas (i, j) $cl_\gamma(\mu) = 1_X \not< \{(x, 0.6), (y, 0.7)\}$ and implies that every (i, j) γ-gf closed set is not a (i, j) $gf\gamma$-closed set.

Theorem 2.4 *If λ and μ be any two (i, j) fuzzy γ-closed sets in a FBTS X, then $\lambda \vee \mu$ is a (i, j) $gf\gamma$-closed set.*

Proof Since λ and μ are (i, j) fuzzy γ-closed sets, $\lambda \vee \mu$ is also a (i, j) fuzzy γ-closed set. Again, we have every (i, j) fuzzy γ-closed set is (i, j) $gf\gamma$-closed. Therefore, $\lambda \vee \mu$ is a (i, j) $gf\gamma$-closed set in X.

Remark 2.5 Both the notions of (i, j) $\gamma gf\gamma$-closed set and (i, j) γ-gf closed set are independent of each other in a FBTS.

Example 2.6 Let us take a singleton set $X = \{x\}$ and define two fuzzy topologies τ_i and τ_j with $i, j = 1, 2, i \neq j$ on it with $\tau_i = \{0_X, 1_X, \{(x, 0.2)\}, \{(x, 0.7)\}\}, \tau_j = \{0_X, 1_X, \{(x, 0.8)\}\}$. So (i, j) $FO(X) = \{0_X, 1_X, \{(x, 0.2)\}, \{(x, 0.7)\}, \{(x, 0.8)\}\}$ and (i, j) $FC(X) = \{0_X, 1_X, \{(x, 0.2)\}, \{(x, 0.3)\}, \{(x, 0.8)\}\}$. Then (i, j) $F\gamma$-$O(X)$ $= \{0_X, 1_X, \{(x, \alpha) : \alpha \leq 0.2 \text{ or } \alpha > 0.3\}\}$ and (i, j) $F\gamma$-$C(X) = \{0_X, 1_X, \{(x, \alpha) : \alpha < 0.7 \text{ or } \alpha \geq 0.8\}\}$. We consider the fuzzy set $\xi = \{(x, 0.4)\}$, which is itself a (i, j) fuzzy γ-open set. Again, (i, j) $cl(\xi) = \{(x, 0.8)\} \not\leq \xi$ and (i, j) $cl_\gamma(\xi) = \xi \leq \xi$. Therefore, ξ is a (i, j) $\gamma gf\gamma$-closed set, but it is not a (i, j) γ-gf closed set therein.

Example 2.7 We consider a FBTS (X, τ_i, τ_j) for $i, j = 1, 2$ and $i \neq j$ with $X = \{x, y\}$ and $\tau_i = \{0_X, 1_X, \{(x, 0.2), (y, 0.3)\}, \{(x, 0.4), (y, 0.4)\}\}$ and $\tau_j = \{0_X, 1_X, \{(x, 0.2), (y, 0.3)\}, \{(x, 0.6), (y, 0.7)\}\}$. Then, we have (i, j) $FO(X) = \{0_X, 1_X, \{(x, 0.2), (y, 0.3)\}, \{(x, 0.4), (y, 0.4)\}, \{(x, 0.6), (y, 0.7)\}\}$. Here, (i, j) $FC(X) = \{0_X, 1_X, \{(x, 0.4), (y, 0.3)\}, \{(x, 0.6), (y, 0.6)\}, \{(x, 0.8), (y, 0.7)\}\}$. Thus, we get (i, j) $F\gamma$-$O(X) = \{0_X, 1_X, \{\{(x, \alpha), (y, \beta)\} : \alpha > 0.2, \beta > 0.7\}\}$. Now, we take a fuzzy set $\lambda = \{(x, 0.4), (y, 0.4\}$. Clearly λ is a (i, j) γgf-closed set in X. Again, λ is itself a (i, j) fuzzy γ-open set. But (i, j) $cl_\gamma(\lambda) = 1_X \not\leq \lambda$. Thus, λ is not a (i, j) $\gamma gf\gamma$-closed set.

Theorem 2.5 *Every (i, j) fuzzy γ-closed set is necessarily a (i, j) $\gamma gf\gamma$-closed set in a FBTS (X, τ_i, τ_j), but the converse is not true always.*

Proof The proof is straightforward from the respective definitions and hence ignored.

Remark 2.6 The converse of the above proposition is not true in general. We verify this remark in the following example.

Example 2.8 Suppose $X = \{x\}$ be any non-empty set. (X, τ_i, τ_j) with $i, j = 1, 2$, $i \neq j$ be a fuzzy bitopological space such that $\tau_i = \{0_X, 1_X, \{(x, 0.2)\}, \{(x, 0.4)\}\}$, $\tau_j = \{0_X, 1_X, \{(x, 0.7)\}, \{(x, 0.9)\}\}$. Then, (i, j) $FO(X) = \{0_X, 1_X, \{(x, 0.2)\}, \{(x, 0.4)\}, \{(x, 0.7)\}, \{(x, 0.9)\}\}$ and (i, j) $FC(X) = \{0_X, 1_X, \{(x, 0.1)\}, \{(x, 0.3)\}, \{(x, 0.6)\}, \{(x, 0.8)\}\}$. So (i, j) $F\gamma\text{-}O(X) = \{0_X, 1_X, \{(x, \alpha) : 0.1 < \alpha \leq 0.2$ or $0.3 < \alpha \leq 0.4$ or $0.6 < \alpha \leq 0.7$ or $\alpha > 0.8\}\}$ and (i, j) $F\gamma\text{-}C(X) = \{0_X, 1_X, \{(x, \alpha) : \alpha < 0.2$ or $0.3 \leq \alpha < 0.4$ or $0.6 \leq \alpha < 0.7$ or $0.8 \leq \alpha < 0.9\}\}$. Let $\mu = \{(x, 0.21)\}$. Here, μ is a (i, j) $\gamma g f \gamma$-closed set, but it is not a (i, j) fuzzy γ-closed set in X.

Nevertheless, their equivalence is indicated in the following proposition.

Proposition 2.1 *If λ is a (i, j) gf closed set in a FBTS X and if (i, j) $cl(\lambda) \leq (i, j)$ $cl_\gamma(\lambda) \leq \mu$, where μ is a (i, j) fuzzy open set in X, then λ is a (i, j) $gf\gamma$-closed set therein.*

Proof The proof is obvious, hence ignored.

Theorem 2.6 *If λ is a (i, j) $gf\gamma$-closed set in a FBTS X and $\lambda \leq \delta \leq (i, j)$ $cl_\gamma(\lambda)$, then δ is a (i, j) $gf\gamma$-closed set in X.*

Proof Let us suppose that $\delta \leq \mu$, where μ is (i, j) fuzzy open set in X. Now, as $\lambda \leq \delta$ and λ is a (i, j) $gf\gamma$-closed set in X, so (i, j) $cl_\gamma(\lambda) \leq \mu$. But (i, j) $cl_\gamma(\delta) \leq (i, j)$ $cl_\gamma(\lambda)$ since $\delta \leq (i, j)$ $cl_\gamma(\lambda)$. Therefore, (i, j) $cl_\gamma(\delta) \leq \mu$, and hence, δ is a (i, j) $gf\gamma$-closed set in X.

Remark 2.7 Every (i, j) $gf\gamma$-closed set may not be a (i, j) $\gamma g f \gamma$-closed set. We establish this claim in the following example.

Example 2.9 Let us consider a FBTS (X, τ_i, τ_j) with $i, j = 1, 2, i \neq j$ such that $X = \{x\}$, $\tau_i = \{0_X, 1_X, \{(x, 0.2)\}\}$, $\tau_j = \{0_X, 1_X, \{(x, 0.7)\}, \{(x, 0.8)\}\}$. Then, (i, j) $FO(X) = \{0_X, 1_X, \{(x, 0.2)\}, \{(x, 0.7)\}, \{(x, 0.8)\}\}$ and (i, j) $FC(X) = \{0_X, 1_X, \{(x, 0.2)\}, \{(x, 0.3)\}, \{(x, 0.8)\}\}$. Thus, (i, j) $F\gamma\text{-}O(X) = \{0_X, 1_X, \{(x, \alpha) : \alpha \leq 0.2$ or $\alpha > 0.3\}\}$ and so (i, j) $F\gamma\text{-}C(X) = \{0_X, 1_X, \{(x, \alpha) : \alpha < 0.7$ or $\alpha \geq 0.8\}\}$. Now, we observe that the fuzzy set $\xi = \{(x, 0.75)\}$ is a (i, j) $gf\gamma$-closed set in X. Again, ξ itself is a (i, j) fuzzy γ-open set but (i, j) $cl_\gamma(\xi) = \{(x, 0.8)\} \nleq \xi$. Therefore, ξ is not a (i, j) $\gamma g f \gamma$-closed set in X.

In this context, we claim that both the concepts (i, j) $gf\gamma$-closed set and (i, j) $\gamma g f \gamma$-closed set are totally independent. But we could not find a situation to show that a (i, j) $\gamma g f \gamma$-closed set may not be a (i, j) $gf\gamma$-closed set.

Question Is there any fuzzy bitopological space where every (i, j) $\gamma g f \gamma$-closed set is not necessarily a (i, j) $g f \gamma$-closed set?

Note: We suppose
(1) (i, j) fuzzy closed set (2) (i, j) fuzzy γ-closed set
(3) (i, j) $g f$ closed set (4) (i, j) γ-$g f$ closed set
(5) (i, j) $g f \gamma$-closed set. (6) (i, j) $\gamma g f \gamma$-closed set.

The interrelationship between the above-mentioned different types of sets can be observed from the following diagram:

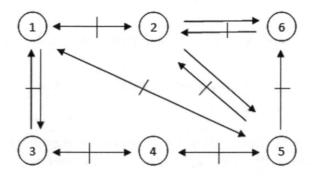

Definition 2.5 A fuzzy set λ in a FBTS (X, τ_i, τ_j) for all $i, j = 1, 2$ and $i \neq j$ is said to be (i, j) generalized fuzzy γ-open (in short, (i, j) $g f \gamma$-open) set iff its complement is a (i, j) $g f \gamma$-closed set.

Theorem 2.7 *In a FBTS X, the following statements are equivalent:*

(i) A fuzzy set λ in X is a (i, j) $g f \gamma$-open set.
(ii) If δ is a (i, j) fuzzy closed and $\delta \leq \lambda$ then, $\delta \leq (i, j)$ $int_\gamma(\lambda)$.

Proof We suppose λ is a (i, j) $g f \gamma$-open set in X and δ is a (i, j) fuzzy closed set such that $\delta \leq \lambda$. Then, $1_X - \lambda \leq 1_X - \delta$, where $1_X - \delta$ is a (i, j) fuzzy open set. Now, $1_X - (i, j)$ $int_\gamma(\lambda) = (i, j)$ $cl_\gamma(1_X - \lambda) \leq 1_X - \delta$, which implies, $\delta \leq (i, j)$ $int_\gamma(\lambda)$.
Conversely, Let λ is a fuzzy set in X such that $\delta \leq (i, j)$ $int_\gamma(\lambda)$, whenever δ is (i, j) fuzzy closed and $\delta \leq \lambda$. Now, if $1_X - \lambda \leq \mu$, where μ is (i, j) fuzzy open set in X, then $1_X - \lambda \leq \mu$, which means $1_X - \mu \leq \lambda$. Hence by assumption, $1_X - \mu \leq (i, j)$ $int_\gamma(\lambda)$. Thus, $1_X - (i, j)$ $int_\gamma(\lambda) \leq \mu$. Therefore, (i, j) $cl_\gamma(1_X - \lambda) \leq \mu$, and hence, $1_X - \lambda$ is a (i, j) $g f \gamma$-closed set. Therefore, λ is a (i, j) $g f \gamma$-open set in X.

Theorem 2.8 *If (i, j) $int_\gamma(\lambda) \leq \mu \leq \lambda$ and λ is a (i, j) $g f \gamma$-open set, then μ is also a $(i, j) g f \gamma$-open set.*

Proof Here, $(i, j) int_\gamma(\lambda) \leq \mu \leq \lambda$ implies that $1_X - \lambda \leq 1_X - \mu \leq 1_X - (i, j)$ $int_\gamma(\lambda)$. Then, $1_X - \lambda \leq 1_X - \mu \leq (i, j) cl_\gamma(1_X - \lambda)$. So, $1_X - \mu$ is $(i, j) gf\gamma$-closed, and thus, μ is a $(i, j) gf\gamma$-open set.

Remark 2.8 The union of two $(i, j) gf\gamma$-open sets is not a $(i, j) gf\gamma$-open set.

Example 2.10 We consider the FBTS taken in Example 2.1 and suppose $\mu_1 = \{(x, 0.2), (y, 0.7)\}$, $\mu_2 = \{(x, 0.8), (y, 0.5)\}$. Obviously, both μ_1 and μ_2 are $(i, j) gf\gamma$-open sets. Now, $\mu_1 \vee \mu_2 = \{(x, 0.8), (y, 0.7)\}$, which is not a (i, j) $gf\gamma$-open set.

Theorem 2.9 *The intersection of any two $(i, j) gf\gamma$-open sets is a $(i, j) gf\gamma$-open set.*

Proof It can be easily proved and hence omitted.

3 (i, j) Generalized Fuzzy γ-Continuous Functions and (i, j) Generalized Fuzzy γ-Irresolute Functions

Definition 3.1 A function $f : X \to Y$ is called (i, j) generalized fuzzy continuous (in short, $(i, j) gf$ continuous) if the inverse image of every (i, j) fuzzy closed set in Y is $(i, j) gf$ closed set in X.

Definition 3.2 A function $f : X \to Y$ is called (i, j) generalized fuzzy γ-continuous (in short, $(i, j) gf\gamma$-continuous) if the inverse image of every (i, j) fuzzy closed set in Y is $(i, j) gf\gamma$-closed set in X.

Remark 3.1 Both the notions of $(i, j) gf\gamma$-continuous function and $(i, j) gf$ continuous function are independent of each other. It is demonstrated in the following two examples.

Example 3.1 We consider a FBTS (X, τ_i, τ_j) with $i, j = 1, 2, i \neq j$ such that $X = \{x\}$, $\tau_i = \{0_X, 1_X, \{(x, 0.1)\}, \{(x, 0.3)\}\}$, $\tau_j = \{0_X, 1_X, \{(x, 0.7)\}, \{(x, 0.8)\}\}$. Then, we have $(i, j) FO(X) = \{0_X, 1_X, \{(x, 0.1)\}, \{(x, 0.3)\}, \{(x, 0.7)\}, \{(x, 0.8)\}\}$. Thus, $(i, j) FC(X) = \{0_X, 1_X, \{(x, 0.9)\}, \{(x, 0.7)\}, \{(x, 0.3)\}, \{(x, 0.2)\}\}$. Also, $(i, j) F\gamma\text{-}O(X) = \{0_X, 1_X, \{(x, \alpha) : \alpha \leq 0.1 \text{ or } 0.2 < \alpha \leq 0.8 \text{ or } \alpha > 0.9\}\}$ and $(i, j) F\gamma\text{-}C(X) = \{0_X, 1_X, \{(x, \alpha) : \alpha \geq 0.9 \text{ or } 0.2 \leq \alpha < 0.8 \text{ or } \alpha < 0.1\}\}$. Again, let (Y, σ_i, σ_j) be another FBTS with $Y = \{y\}$, $\sigma_i = \{0_Y, 1_Y, \{(y, 0.5)\}\}$, $\sigma_j = \{0_Y, 1_Y\}$. Then, $(i, j) FO(Y) = \{0_Y, 1_Y, \{(y, 0.5)\}$ and so $(i, j) FC(Y) = \{0_Y, 1_Y, \{(y, 0.5)\}\}$. We consider a function $f : X \to Y$ such that $f(x) = y$. Clearly, f is a $(i, j) gf\gamma$-continuous function. But, $f^{-1}\{(y, 0.5)\} = \{(x, 0.5)\} \notin (i, j) FC(X)$. Then, f is not a $(i, j) gf$ continuous function.

Example 3.2 Let us consider two FBTS (X, τ_i, τ_j), (Y, σ_i, σ_j) with $i, j = 1, 2, i \neq j$ such that $X = Y = \{a, b\}$, $\tau_i = \{0_X, 1_X, \{(a, 0.2), (b, 0)\}, \{(a, 0.2), (b, 0.3)\}\}$, $\tau_j = \{0_X, 1_X, \{(a, 0), (b, 0.3)\}\}$, $\sigma_i = \{0_Y, 1_Y, \{(a, 0.29), (b, 0.19)\}\}$, $\sigma_j = \{0_Y, 1_Y\}$. We define a function $f : X \to Y$ such that $f(a) = b$, $f(b) = a$. Here, f is a (i, j) gf continuous function. Again, the inverse of the (i, j) fuzzy closed set $\{(a, 0.71), (b, 0.81)\}$ in Y is $\{(a, 0.81), (b, 0.71)\}$, which is a (i, j) gf closed set in X, but not (i, j) $gf\gamma$-closed set. Therefore, f is not a (i, j) $gf\gamma$-continuous mapping.

Definition 3.3 A function $f : X \to Y$ is called (i, j) γ-generalized fuzzy continuous (in short, (i, j) γ-gf continuous) function if the inverse image of every (i, j) fuzzy closed set in Y is a (i, j) $gf\gamma$-closed set in X.

Definition 3.4 A function $f : X \to Y$ is called (i, j) γ-generalized fuzzy γ-continuous (in short, (i, j) $\gamma gf\gamma$-continuous) function if the inverse image of every (i, j) fuzzy closed set in Y is a (i, j) $\gamma gf\gamma$-closed set in X.

Remark 3.2 The notion of (i, j) $gf\gamma$-continuous function and the notion of (i, j) γ-gf continuous function are completely independent of each other. The claim is illustrated in the following two consecutive examples.

Example 3.3 We consider two FBTS (X, τ_i, τ_j) and (Y, σ_i, σ_j) with $i, j = 1, 2, i \neq j$ such that $X = \{x, y\}$, $Y = \{a, b\}$, $\tau_i = \{0_X, 1_X, \{(x, 0.2), (y, 0)\}, \{(x, 0)(y, 0.3)\}\}$, $\tau_j = \{0_X, 1_X, \{(x, 0.2), (y, 0.3)\}\}$, $\sigma_i = \{0_Y, 1_Y, \{(a, 0.9), (b, 0.9)\}\}$, $\sigma_j = \{0_Y, 1_Y\}$. We define a function $f : X \to Y$ such that $f(x) = a$, $f(y) = b$. Accordingly, we have (i, j) $FO(X) = \{0_X, 1_X, \{(x, 0.2), (y, 0)\}, \{(x, 0), (y, 0.3)\}, \{(x, 0.2), (y, 0.3)\}\}$ and $(i, j) FC(X) = \{0_X, 1_X, \{(x, 0.8), (y, 1)\}, \{(x, 1), (y, 0.7)\}, \{(x, 0.8), (y, 0.7)\}\}$. Now, (i, j) $F\gamma$-$O(X) = \{0_X, 1_X, \{(x, \alpha), (y, \beta)\} : \alpha < 0.2, \beta < 0.3$ or $\alpha \geq 0.8, \beta \geq 0.7\}$ and (i, j) $F\gamma$-$C(X) = \{0_X, 1_X, \{(x, \alpha), (y, \beta)\} : \alpha > 0.8, \beta > 0.7$ or $\alpha \leq 0.2, \beta \leq 0.3\}$. Also, (i, j) $FO(Y) = \{0_Y, 1_Y, \{(a, 0.9), (b, 0.9)\}\}$ such that $(i, j) FC(Y) = \{0_Y, 1_Y, \{(a, 0.1), (b, 0.1)\}\}$. Clearly, $f^{-1}\{(a, 0.1), (b, 0.1)\} = \{(x, 0.1), (y, 0.1)\}$ is a (i, j) $gf\gamma$-closed set but not a (i, j) γ-gf closed set therein. Hence, f is (i, j) $gf\gamma$-continuous function but not a (i, j) γ-gf continuous function.

Example 3.4 We consider two FBTS (X, τ_i, τ_j) and (Y, σ_i, σ_j) with $i, j = 1, 2, i \neq j$ such that $X = \{x, y\}$, $Y = \{a, b\}$, $\tau_i = \{0_X, 1_X, \{(x, 0.6), (y, 0.3)\}, \{(x, 0.3)(y, 0.3)\}\}$, $\tau_j = \{0_X, 1_X, \{(x, 0.3), (y, 0.7)\}, \{(x, 0.6), (y, 0.7)\}\}$, $\sigma_i = \{0_Y, 1_Y, \{(a, 0.5), (b, 0.5)\}\}$, $\sigma_j = \{0_Y, 1_Y\}$. We define a function $f : X \to Y$ such that $f(x) = a$, $f(y) = b$. Then, we have (i, j) $FO(X) = \{0_X, 1_X, \{(x, 0.2), (y, 0)\}, \{(x, 0), (y, 0.3)\}, \{(x, 0.2), (y, 0.3)\}\}$ and $(i, j) FC(X) = \{0_X, 1_X, \{(x, 0.8), (y, 1)\}, \{(x, 1), (y, 0.7)\}, \{(x, 0.8), (y, 0.7)\}\}$. Then, (i, j) $F\gamma$-$O(X) = \{0_X, 1_X, \{(x, \alpha), (y, \beta)\} : \alpha > 0.7, \beta > 0.7\}$ and (i, j) $F\gamma$-$C(X) = \{0_X, 1_X, \{(x, \alpha), (y, \beta)\} : \alpha < 0.3, \beta < 0.3\}$. Also, $(i, j) FO(Y) = \{0_Y, 1_Y, \{(a, 0.9), (b, 0.9)\}\}$. So, we have (i, j) $FC(Y) = \{0_Y, 1_Y, \{(a, 0.1), (b, 0.1)\}\}$. Here, $f^{-1}\{(a, 0.5), (b, 0.5)\} = \{(x, 0.5), (y, 0.5)\}$, which is a (i, j) γ-gf closed set but not a (i, j) $gf\gamma$-closed set. Consequently, f is a (i, j) γ-gf continuous function but not a (i, j) $gf\gamma$-continuous function.

Remark 3.3 Any (i, j) $gf\gamma$-continuous function may not be a (i, j) $\gamma gf\gamma$-continuous function.

Example 3.5 We consider a FBTS (X, τ_i, τ_j) with $i, j = 1, 2, i \neq j$ such that $X = \{x\}$, $\tau_i = \{0_X, 1_X, \{(x, 0.2)\}\}$, $\tau_j = \{0_X, 1_X, \{(x, 0.7)\}, \{(x, 0.8)\}\}$. Then (i, j) $FO(X) = \{0_X, 1_X, \{(x, 0.2)\}, \{(x, 0.7)\}, \{(x, 0.8)\}\}$ and (i, j) $FC(X) = \{0_X, 1_X, \{(x, 0.2)\}, \{(x, 0.3)\}, \{(x, 0.8)\}\}$. Thus, (i, j) $F\gamma\text{-}O(X) = \{0_X, 1_X, \{(x, \alpha) : \alpha \leq 0.2 \text{ or } \alpha > 0.3\}\}$ and so (i, j) $F\gamma\text{-}C(X) = \{0_X, 1_X, \{(x, \alpha) : \alpha < 0.7 \text{ or } \alpha \geq 0.8\}\}$. Now, we define a function $f : (X, \tau_i, \tau_j) \rightarrow (Y, \sigma_i, \sigma_j)$, where $Y = \{y\}$, $\sigma_i = \{0_Y, 1_Y, \{(y, 0.75)\}\}$ and $\sigma_j = \{0_Y, 1_Y\}$. Here obviously, f is a (i, j) $gf\gamma$-continuous mapping. But $f^{-1}\{(y, 0.75)\} = \{(x, 0.75)\}$, which is not a (i, j) $\gamma gf\gamma$-closed set. Thus, f is (i, j) $\gamma gf\gamma$-continuous.

Theorem 3.1 *Let* $f : (X, \tau_i, \tau_j) \rightarrow (Y, \sigma_i, \sigma_j)$, *with* $i, j = 1, 2$ *and* $i \neq j$, *be a function. Then, the following statements are equivalent:*

(i) *The function* f *is a* (i, j) $gf\gamma$*-continuous function.*
(ii) *The inverse image of every* (i, j) *fuzzy open set in* Y *is a* (i, j) $gf\gamma$*-open set in* X.

Proof The proof is obvious.

Theorem 3.2 *Let* $f : (X, \tau_i, \tau_j) \rightarrow (Y, \sigma_i, \sigma_j)$ *and* $g : (Y, \sigma_i, \sigma_j) \rightarrow (Z, \eta_i, \eta_j)$, *for* $i \neq j$ *be any two functions. If* f *is a* (i, j) $gf\gamma$*-continuous function and* g *is any* (i, j) *fuzzy continuous function, then their composition* $g \circ f$ *is a* (i, j) $gf\gamma$*-continuous function.*

Proof It can be easily proved and hence omitted.

Remark 3.4 The composition of any two (i, j) $gf\gamma$-continuous functions may not be a (i, j) $gf\gamma$-continuous function.

Example 3.6 Let us suppose that (X, τ_i, τ_j), (Y, σ_i, σ_j) and (Z, χ_i, χ_j) are three FBTS with $i, j = 1, 2, i \neq j$ such that $X = \{x\}$, $Y = \{y\}$, $Z = \{z\}$, $\tau_i = \{0_X, 1_X, \{(x, 0.2)\}, \{(x, 0.4)\}\}$, $\tau_j = \{0_X, 1_X, \{(x, 0.7)\}\}$, $\sigma_i = \{0_Y, 1_Y, \{(y, 0.7)\}\}$, $\sigma_j = \{0_Y, 1_Y, \{(y, 0.1)\}\}$, $\chi_i = \{0_Z, 1_Z, \{(z, 0.6)\}\}$, $\chi_j = \{0_Z, 1_Z, \{(z, 0.8)\}\}$. Also, we consider two fuzzy mappings $f : X \rightarrow Y$ and $g : Y \rightarrow Z$ such that $f(x) = y$ and $g(y) = z$. Obviously, f and g are (i, j) $gf\gamma$-continuous functions. Now, $(g \circ f)^{-1}\{(z, 0.4)\} = f^{-1}(g^{-1}(\{(z, 0.4)\})) = f^{-1}(\{(y, 0.4)\}) = \{(x, 0.4)\}$. Here, $\{(z, 0.4)\}$ is a (i, j) fuzzy closed set in Z, but $(g \circ f)^{-1}\{(z, 0.4)\} = \{(x, 0.4)\}$ is not a (i, j) $gf\gamma$-closed set in X. Therefore, $g \circ f$ is not a (i, j) $gf\gamma$-continuous function.

Following the definition of pairwise γ-continuity due to Tripathy et al. [10], we introduce the following definition and the related result.

Definition 3.5 A function $f :\rightarrow Y$ is said to be (i, j) fuzzy γ-continuous function if the inverse of any (i, j) fuzzy open set in Y is (i, j) fuzzy γ-open set in X.

Proposition 3.1 *A function $f : X \to Y$ is (i, j) $gf\gamma$-continuous as well as (i, j) fuzzy γ-continuous if the inverse image of every (i, j) fuzzy closed set a (i, j) fuzzy γ-closed set.*

Definition 3.6 A function $f : (X, \tau_i, \tau_j) \to (Y, \sigma_i, \sigma_j)$ with $i, j = 1, 2$ and $i \neq j$, is called (i, j) fuzzy γ-irresolute function if $f^{-1}(\lambda)$ is a (i, j) fuzzy γ-open (resp. (i, j) fuzzy γ-closed) set in X for every (i, j) fuzzy γ-open (resp. (i, j) fuzzy γ-closed) set λ in Y.

Definition 3.7 A function $f : (X, \tau_i, \tau_j) \to (Y, \sigma_i, \sigma_j)$ with $i, j = 1, 2$ and $i \neq j$, is called (i, j) generalized fuzzy irresolute (in short, (i, j) gf irresolute) function if $f^{-1}(\lambda)$ is a (i, j) gf closed set in X for every (i, j) gf closed set in Y.

Definition 3.8 A function $f : (X, \tau_i, \tau_j) \to (Y, \sigma_i, \sigma_j)$, for $i, j = 1, 2, i \neq j$, is called a (i, j) generalized fuzzy γ-irresolute (in short, (i, j) $gf\gamma$-irresolute) function if $f^{-1}(\lambda)$ is a (i, j) $gf\gamma$-closed set in X for every (i, j) $gf\gamma$-closed set λ in Y.

Remark 3.5 Both the notions of (i, j) $gf\gamma$-irresolute function and (i, j) $gf\gamma$-continuous function are independent of each other. It is verified by illustrating the following two successive examples.

Example 3.7 Let us take two FBTS (X, τ_i, τ_j) and (Y, σ_i, σ_j) with $i, j = 1, 2, i \neq j$ such that $X = \{x\}, Y = \{y\}, \tau_i = \{0_X, 1_X, \{(x, 0.2)\}\}, \tau_j = \{0_X, 1_X\}, \sigma_i = \{0_Y, 1_Y, \{(y, 0.2)\}, \{(y, 0.75)\}\}, \sigma_j = \{0_Y, 1_Y, \{(y, 0.8)\}\}$. We consider a fuzzy function $f : X \to Y$ such that $f(x) = y$. Here, f is a (i, j) $gf\gamma$-irresolute function. Here, $\{(y, 0.2)\}$ is (i, j) fuzzy closed in Y, but the inverse of this set is $\{(x, 0.2)\}$, which is not a (i, j) $gf\gamma$-closed set in X. Thus, f is not a (i, j) $gf\gamma$-continuous function.

Example 3.8 We consider two FBTS (X, τ_i, τ_j) and (Y, σ_i, σ_j) with $i, j = 1, 2, i \neq \tau_j$ such that $X = \{x\}, Y = \{y\}, \tau_i = \{0_X, 1_X, \{(x, 0.2)\}, \{(x, 0.4)\}\}, \tau_j = \{0_X, 1_X\}, \{(x, 0.8)\}, \sigma_i = \{0_Y, 1_Y, \{(y, 0.2), \{(y, 0.5)\}\}, \{(y, 0.75)\}\}, \sigma_j = \{0_Y, 1_Y, \{(y, 0.8)\}\}$. Let $f : X \to Y$ be a fuzzy function such that $f(x) = y$. Here, f is a (i, j) $gf\gamma$-continuous function. Now, $\{(y, 0.4)\}$ is a (i, j) $gf\gamma$-closed set in Y, but the inverse of this set is a $\{(x, 0.4)\}$, which is not a (i, j) $gf\gamma$-closed set in X. Evidently, f is not a (i, j) $gf\gamma$-irresolute function.

Remark 3.6 Both the notions of (i, j) $gf\gamma$-irresolute function and (i, j) gf irresolute function are independent of each other.

Example 3.9 We consider the Example 3.6. There, the function $f : X \to Y$ is a (i, j) $gf\gamma$-irresolute function. Now, $\{(y, 0.1)\}$ is a (i, j) gf closed set in Y. But the inverse of this set is $\{(x, 0.1)\}$, and it is not a (i, j) gf closed set in X. So, f is not a (i, j) gf irresolute function.

Example 3.10 Let (X, τ_i, τ_j) and (Y, σ_i, σ_j) be any two FBTS with $i, j = 1, 2, i \neq j$ such that $X = \{x\}, Y = \{y\}, \tau_i = \{0_X, 1_X, \{(x, 0.2)\}, \{(x, 0.6)\}\}, \tau_j = \{0_X, 1_X, \{(x, 0.7)\}\}, \sigma_i = \{0_Y, 1_Y, \{(y, 0.2)\}\}, \sigma_j = \{0_Y, 1_Y, \{(y, 0.7)\}\}$. Let $f : X \to Y$ be a fuzzy function such that $f(x) = y$. Here, f is a (i, j) gf irresolute function. Now, $\{(y, 0.6)\}$ is (i, j) $gf\gamma$-closed set in Y. But $f^{-1}\{(y, 0.6)\} = \{(x, 0.6)\}$, which is not a (i, j) $gf\gamma$-closed set in X. Hence, f is not a (i, j) $gf\gamma$-irresolute function.

Analogous to the Theorem 3.1, we state the following result but without any proof.

Theorem 3.3 *Let* $f : (X, \tau_i, \tau_j) \to (Y, \sigma_i, \sigma_j)$ *where* $i, j = 1, 2, i \neq j$ *be any function. Then, the following statements are equivalent:*

(i) *The function* f *is a* (i, j) $gf\gamma$-*irresolute function.*
(ii) *The inverse image of every* (i, j) $gf\gamma$-*open set in* Y *is a* (i, j) $gf\gamma$-*open set.*

Proof The proof is obvious.

Theorem 3.4: Let $f : X \to Y$ and $g : Y \to Z$ be two functions. If both f and g are (i, j) $gf\gamma$-irresolute function, then the composition $g \circ f$ is also a (i, j) $gf\gamma$-irresolute function.

Proof The proof can be obtained easily and hence omitted.

4 Conclusion

To the best of our knowledge, it is observed that maximum of the generalizations of fuzzy closed sets in the study of fuzzy topological space are linearly dependent among themselves. In our work, we have reported three generalizations of (i, j) fuzzy γ-closed sets, namely (i, j) generalized fuzzy γ-closed set, (i, j) γ-generalized fuzzy closed set and (i, j) γ-generalized fuzzy γ-closed set, and we have established interrelationships among them. The significance of these relationships may be applicable in the graphical representation of electric networks and parallel topologies of the same and other related problems.

References

1. Balasubramanian, G., Sundaram, P.: On some generalizations of fuzzy continuous functions. Fuzzy Sets Syst. **86**, 93–100 (1997)
2. Bhattacharya, B., Chakaraborty, J.: Generalized regular fuzzy closed sets and their applications. J. Fuzzy Math. **23**(1), pp. 227–239 (2015)
3. Bhattacharya, S.: On generalized regular closed sets. Int. J. Contemp. Math. Sci. **6**(3), 145–152 (2011)
4. Fukutake, T.: On generalized closed sets in bitopological spaces. Bull. Fukuoka. Univ. Ed. **35**(III), pp. 1928 (1986)
5. Kandil, A., Nouh, A.A., El-Sheikh, S.A.: On fuzzy bitopological spaces. Fuzzy Sets Syst. **74**, 353–363 (1995)
6. Levine, N.: Generalized closed sets in topology. Rend. Circ. Math. Palermo **19**, 89–96 (1970)
7. Palaniappan, N., Rao, K.C.: Regular generalized closed sets. Kyungpook Math. J. **33**(2), 211–219 (1993)
8. Park, J.H., Park, J.K.: On regular generalized fuzzy closed sets and generalization of fuzzy continuous functions. Indian J. Pure Appl. Math. **34**(7), 1013–1024 (2003)

9. Paul, A., Bhattacharya, B., Chakraborty, J.: On Λ^γ-sets in Fuzzy Bitopological Space. Bol. Soc. Paran. Mat. **35**(3), pp. 285–299 (2017)
10. Tripathy, B.C., Debnath, S.: γ-Open Sets and γ-continuous mappings in fuzzy bitopological spaces. J. Intell. Fuzzy Syst. **24**, 631–635 (2013)

ANN Application for Medical Image Denoising

M. Laxmi Prasanna Rani, G. Sasibhushana Rao and B. Prabhakara Rao

Abstract Nowadays, medical image denoising is crucial for accurate diagnosis of the critical diseases. For denoising these images, conventional wavelet technique (universal threshold) uses a fixed value of threshold which is non-adaptive. The main aim of this paper is to develop a steepest descent (SD)-based learning algorithm, which is used in Artificial Neural Networks (ANN), to reduce the noise in images adaptively. A new soft thresholding function is proposed as the activation function of the ANN. From the results, it is found that proposed algorithm performed well when compared with conventional wavelet technique in terms of mean squared error (MSE), peak signal-to-noise ratio (PSNR).

Keywords Artificial neural network (ANN) · Medical image denoising · Mean squared error (MSE) · Peak signal-to-noise ratio (PSNR) · Wavelet thresholding

1 Introduction

The images get distorted by noise while being acquired or transmitted. Denoising reduces the noise in images without reducing the quality of the image. The conventional techniques of reducing noise in an image are filtering. The nonlinear techniques of image denoising have been developed after a lot of research and mainly based on thresholding discrete wavelet transform (DWT) coefficients of the image [1]. The

M. L. P. Rani (✉)
Department of ECE, MVGR College of Engineering, Vizianagaram, Andhra Pradesh, India
e-mail: prassugowtham@gmail.com

G. Sasibhushana Rao
Department of ECE, AUCE(A), Andhra University, Visakhapatnam, Andhra Pradesh, India
e-mail: sasigps@gmail.com

B. Prabhakara Rao
Department of ECE, JNTUK, Kakinada 533003, Andhra Pradesh, India
e-mail: drbprjntu@gmail.com

© Springer Nature Singapore Pte Ltd. 2019
J. C. Bansal et al. (eds.), *Soft Computing for Problem Solving*,
Advances in Intelligent Systems and Computing 816,
https://doi.org/10.1007/978-981-13-1592-3_53

denoising preprocessing toolkit has to be applied to the high noise density images so that quality of the image is retained.

A new technique called steepest decent (SD)-based artificial neural network (ANN) is developed in this paper for denoising of images. A new soft thresholding function is proposed as the activation function of the ANN. This activation function is infinitely differentiable and is used in many learning algorithms based on gradients. The results obtained by using the soft thresholding [2] as activation function are presented in terms mean squared error (MSE) and peak signal-to-noise ratio (PSNR) in this paper.

2 Image Denoising

2.1 Denoising of an Image

The image can be degraded during acquisition process, by use of imperfect instruments and during transmission. Removing noise from original image without disturbing the image quality is called image denoising. The noisy image is expressed by the equation.

$$q(x, y) = p(x, y) + r(x, y) \tag{1}$$

where $p(x, y)$ is the original image of dimensions $M \times N$ which is noise free, $q(x, y)$ is the noisy image, and $r(x, y)$ is an additive white Gaussian noise with a zero mean and variance (σ^2).

The aim of image denoising [3] is to obtain a denoised or restored image $s(x, y)$ which is similar to original image $p(x, y)$ by reducing the noise in $r(x, y)$. This can be achieved by using the performance parameters of MSE and PSNR. The expressions for MSE and PSNR are given in equations.

$$J(s, p) = \text{MSE} = \frac{1}{MN} \sum_{x=0}^{M-1} \sum_{y=0}^{N-1} [p(x, y) - s(x, y)]^2 \tag{2}$$

$$\text{PSNR} = 10 \log\left(\frac{\text{MAX}^2}{\text{MSE}}\right) \tag{3}$$

where $M \times N$ is the size of the image and MAX is the maximum possible pixel values of the image.

2.2 Conventional Techniques

Conventional techniques like histogram equalization and specification, Gaussian filters, mean filters, and median filters for low frequency, unsharp masking for high frequency and also morphological filters [4] are used for image denoising, which are based on statistical behavior of noise in spatial domain [5].

2.3 Wavelet Transform (WT)

In the frequency domain, wavelet transform is a powerful tool to denoise an image. The wavelet transform of a noisy image is a combination of the wavelet transform of the image and noise, and this noise is spread among all the wavelet coefficients. The aim of noise reduction is to approximate the original image by reducing the value of MSE. In image decomposition using DWT, the image is decomposed into four sub-bands; HH, HL, LH, and LL [5]. The diagonal coefficients of the image are given by the sub-band HH, the horizontal details are given by sub-band HL, and the vertical coefficients are given by the sub-band LH. The LL sub-band is the approximation coefficients of low-frequency components, and this sub-band is further divided into different sub-bands in higher levels of decomposition.

2.4 Wavelet Thresholding

A thresholding technique for denoising of images has been proposed [6, 7] in the wavelet domain by Donoho and Johnstone. The thresholding techniques are of two types, i.e., soft thresholding and hard thresholding techniques.

2.4.1 Soft Thresholding

The amplitude of wavelet coefficients of an image below threshold value thr are zero and the wavelet coefficients [8] greater than threshold are the difference between the detailed values of image and threshold value thr.

$$\varphi_s(q, \text{thr}) = \left\{ \begin{array}{l} \text{sgn}(q)(|q| - \text{thr}), |q| \geq \text{thr} \\ 0, q < \text{thr} \end{array} \right\} \tag{4}$$

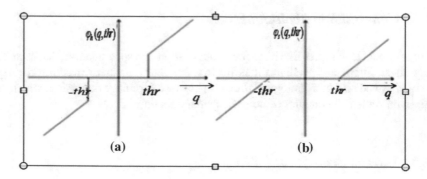

Fig. 1 a Hard thresholding function. b Soft thresholding functions

2.4.2   Hard Thresholding

This will be set to zero if its amplitude is less than a pre-assigned value of threshold; otherwise, it remains the same value.

$$\varphi_h(q, \text{thr}) = \begin{cases} q, q > \text{thr} \\ 0, |q| \le \text{thr} \end{cases} \tag{5}$$

These two hard and soft thresholding functions are shown in Fig. 1.

3   Image Denoising Using Universal Thresholding

In denoising, using the technique of thresholding, the important decision is to select a particular threshold value. The reconstructed image will remain noisy, if the value of threshold is too less. On the other hand, details of image will be smoothed out, if the threshold value is too large. So the selection of threshold value is important to get proper results. Three steps are involved in the method of universal threshold [1] for image denoising.

- Apply DWT to the noisy input image which is already added with white Gaussian noise and get the approximation, vertical, horizontal, and diagonal coefficients [7].
- Universal threshold is applied to all diagonal coefficients [8], where universal threshold is $t_{\text{universal}} = \sigma \sqrt{2 \log(M)}$ where M is the number of pixels in an image; σ is standard deviation of noisy image. $\sigma = \frac{\text{median}(|HH|)}{0.6745}$ where HH is the diagonal coefficients of an image.
- Apply the IDWT to the threshold coefficients and get an approximate image.

The steps of image denoising with universal thresholding are shown in Fig. 2.

In the wavelet domain, only few coefficients contain the energy of an image and all the coefficients have the noise energy. This is the fundamental idea of the wavelet

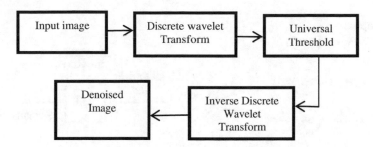

Fig. 2 Block diagram of image denoising using universal thresholding

thresholding. Therefore, nonlinear wavelet soft thresholding is used to reduce the noise coefficients to zero and get larger coefficients representing the image.

4 Steepest Decent-Based Artificial Neural Network

4.1 Artificial Neural Networks (ANN)

ANNs which involve elementary components operate parallelly. By using activation function and tuning, the weights between internal elements and neural network execute a required function. The output produced by the ANN is compared with the targeted output and the difference is the error of the network and this network is tuned until the output of the network matches the target output. By repeating the process of giving different inputs to the network by upgrading the weights of neural network is an epoch. In general, more number of iterations are required to understand the concept of the neural network. The block diagram of ANN with transfer function and threshold as activation function is shown in Fig. 3.

4.2 Steepest Decent-Based Artificial Neural Network

Steepest descent (SD)-based ANN consists of elements with interconnection of different inputs, nonlinear adaptive activation function, and optimum threshold value to get better PSNR and MSE. A fixed nonlinear adaptive activation function [9, 10] and linear wavelet transform are used in this SD-based neural network. The learning algorithms of this SD-based ANN consist of gradients and the activation function with higher order derivatives [10].

The block diagram of image denoising of SD-based ANN with wavelet transform, threshold as activation function, and error function is shown in Fig. 4.

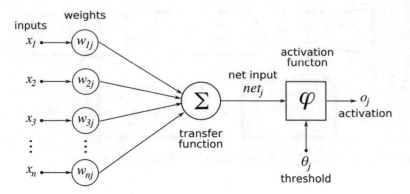

Fig. 3 Block diagram of artificial neural network with transfer function and threshold as activation function

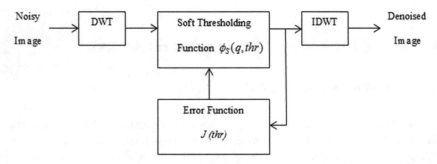

Fig. 4 Block diagram of image denoising using proposed SD-based artificial neural network with wavelet transform, threshold as activation function, and error function

The linear wavelet transform is applied to the noisy image $q(x, y)$ of dimensions $N \times N$. The wavelet coefficients are thresholded by the soft thresholding function $\phi_S(\text{thr})$. Then these thresholded values are varied with a gradient descent or steepest decent of the error function in every step and are again applied to the inverse wavelet transform to get denoised image $s(x, y)$.

The higher order derivatives of the network activation function provide good learning algorithm of a neural network. The best numerical properties are maintained by employing high-order variable activation functions. The standard universal thresholding function is a differentiated function but it does not have high-order derivatives. An infinitely differentiable soft thresholding activation function is used in this paper. Many gradient-based learning algorithms are operated easily by using this function.

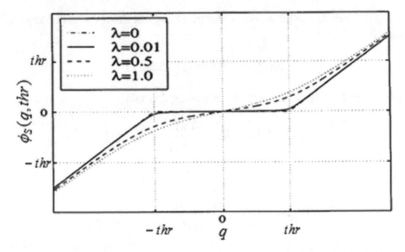

Fig. 5 New soft thresholding function $\varphi_S(q, \text{thr})$

4.3 Soft Thresholding Function

A soft thresholding function with higher order derivatives is used in this paper. This thresholding function is infinitely differentiable.

$$\phi_S(q, \text{thr}) = q + \frac{1}{2}\left(\sqrt{(q - \text{thr})^2 + \lambda} - \sqrt{(q + \text{thr})^2 + \lambda}\right) \qquad (6)$$

where thr is the universal threshold value, λ is a user-defined function parameter. When $\lambda = 0$, this function becomes the standard soft-thresholding function and when $\lambda > 0$, the function has higher order derivatives. When $\lambda \to \infty$ the new function becomes linear, and it does not have thresholding ability. For smaller values of λ, this thresholding function is a good approximation of the standard soft thresholding function. The new thresholding function with different values of λ is shown in Fig. 5.

4.4 Steepest Descent-Based Supervised ANN Learning Algorithm

In supervised learning, a reference signal is known to calculate MSE in the learning process. In this process, using neural networks concept and activation function of thresholding, the threshold value is varied or updated with a gradient descent or steepest decent of the error function in every step. As the general soft-thresholding function is weakly differentiable [5, 10], a steepest decent (SD) gradient-based optimization algorithm using new soft thresholding function is used to calculate the optimum value of threshold. In every learning step, threshold thr can be adjusted by the equation

$$\mathrm{thr}(j+1) = \mathrm{thr}(j) - \Delta\mathrm{thr}(j) \tag{7}$$

$\Delta th(j)$ can be computed using the following formula

$$\Delta\mathrm{thr}(j) = \beta(j) \cdot \frac{\partial J(\mathrm{thr})}{\partial\mathrm{thr}}\bigg|\mathrm{thr} = \mathrm{thr}(j) \tag{8}$$

where

$$J(\mathrm{thr}) = \sum_{k=0}^{N-1} e_j \cdot \frac{\partial\varphi_S(q, \mathrm{thr})}{\partial\mathrm{thr}}\bigg|\mathrm{thr} = \mathrm{thr}(j) \tag{9}$$

where e_j is the estimation error.

$$\phi_S(q, \mathrm{thr}) = q + \frac{1}{2}\left(\sqrt{(q - \mathrm{thr})^2 + \lambda} - \sqrt{(q + \mathrm{thr})^2 + \lambda}\right) \tag{10}$$

$\beta(j) = \mathrm{diag}[\beta_1(j), \beta_2(j), \beta_3(j), \ldots \beta_{N-1}(j)]$ is learning rate matrix at each step and β_j is the learning rate parameter for threshold value $\mathrm{thr}(j)$. This threshold parameter $\mathrm{thr}(j)$ depends on the wavelet detail coefficients.

5 Results and Discussions

For denoising of images, initially universal thresholding technique is implemented in MATLAB. Then, the coding is done for method of proposed SD-based ANN. The new function of soft thresholding is used here. This algorithm is tested with medical images consisting of white Gaussian noise. To compare perfectly, medical images (MRI of brain 1 and brain 2) taken as a reference in this paper. And also this algorithm is applied to another medical image but of X-Ray (hand) using Daubechies wavelet. The results of PSNR and MSE of universal threshold and SD-based ANN methods are shown in Table 1. Figure 6 shows the original, noisy, universal thresholded denoised, and SD-based ANN denoised images of MRI of brain 1, brain 2, and X-Ray of hand, respectively. The original PSNRs of noisy images are compared with the results of the non-adaptive, i.e., universal thresholding wavelet technique and SD-based neural network of adaptive method. The best results are obtained by using SD-based supervised ANN learning algorithm of adaptive denoising method for brain images. But the value of MSE is more, and PSNR is less for the SD-based supervised ANN method for X-Ray images.

Table 1 MSE and PSNR of universal threshold and SD-based supervised ANN

Reference image	PSNR of noisy image (dB)	Universal threshold		SD-based supervised ANN	
		MSE	PSNR (dB)	MSE	PSNR (dB)
MRI (Brain1)	26.6006	224.5354	27.8250	8.8455	31.7460
MRI (Brain2)	26.6006	207.5272	27.0581	33.4037	30.3420
X-ray	26.0286	45.3773	29.9834	102.6288	28.9520

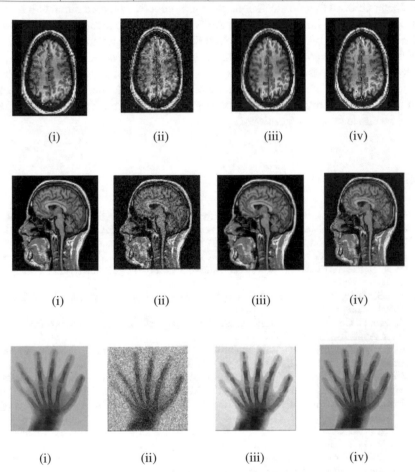

Fig. 6 (i) Original. (ii) Noisy. (iii) Denoised (Universal threshold), and (iv) denoised (SD-based supervised ANN) images of MRI of brain 1, brain 2, and X-ray of hand, respectively

6 Conclusion

This paper progressed with steepest decent-based ANN type to reduce adaptive noise, which involves methods of thresholding. The proposed soft thresholding function behaves as activation function for these ANNs. In contrast to the universal thresholding, the thresholding function which is designed is countlessly differentiable. By performing this new function of thresholding, a new algorithm which is based on SD method becomes possible and the process of learning becomes more serviceable. It has been exhibited that this technique of noise reduction is more efficient than universal thresholding. It is observed that ANN along with the algorithms of stochastic learning can be used as nonlinear adaptive filter of a real-time novel. It is concluded that this algorithm is an iterative process in some statistical sense. The results are presented for the methods of noise reduction which includes the algorithms of already existing wavelet thresholding and the adaptive noise reduction. It has been shown that maximum adaptive noise reduction algorithms which are based on ANN gives better results than other algorithms.

References

1. Al Jumah, A.: Denoising of an image using discrete stationary wavelet transform and various thresholding techniques. Published Online February 2013 J. Sig. Inf. Process., 33–41(2013)
2. Donoho, D.: De-noising by softthresholding. EEE Trans. Inform Theory 41(3), 612–627 (1995)
3. Portilla, J., Strela, V., Wainwright, M., Wainwright, M., Simoncelli, E.: Image denoising using Gaussian scale mixtures in the wavelet domain. IEEE Trans. Image Process. 12(11), 1338–1351 (2003)
4. Roy, S., Sinha, N., Sen, A.K.: A new hybrid image denoising method. Int. J. Inf. Technol. Knowl. Manage. 2(2), 491–497 (2010)
5. Hedaoo, P., Godbole, S.S.: Wavelet thresholding approach for image denoising. Int. J. Netw. Secur. Appl. 3(4), 16–21 (2011)
6. Martin, V., Chang, S.G.: Yu, B.: Adaptive wavelet thresholding for image denoising and compression. IEEE Trans. Image Process. 9(9) (2000)
7. Bahendwar, Y.S., Sinha, G.R.: A comparative performance analysis of discrete wavelet transforms for denoising of medical images. In: Mandal, D.K., Syan, C.S. (eds.) CAD/CAM, Robotics and Factories of the Future. Lecture Notes in Mechanical Engineering. Springer, New Delhi (2016)
8. Shang, H.-q., Gao, R.-p., Wang, C.-y.: An improvement of wavelet shrinkage denoising via wavelet coefficient transformation. J. Vibr. Shock, pp 165–168(2011)
9. Nezamabadi-Pour, M., Nasri, H.: Image denoising in the wavelet domain using a new adaptive thresholding function. Neuro Comput. 72, 1012–1025 (2009)
10. Zhang, X.-P.: Thresholding neural network for adaptive noise reduction. IEEE Trans. Neural Netw. 12(3), 261–270 (2001)

Unsupervised Machine Learning Algorithm for MRI Brain Image Processing

S. Saradha Rani, G. Sasibhushana Rao and B. Prabhakara Rao

Abstract Denoising of an image is the first and primary pre-processing step in image processing. In this paper, an algorithm is implemented using machine learning in conjunction with wavelet-based denoising method. Most learning algorithms use activation function that is continuously differentiable. Since standard threshold functions are weakly differentiable, a new type of thresholding function was proposed. *Stein's unbiased risk estimate* (*SURE*)-based updating algorithm is used for estimation. The proposed method is compared with conventional filtering and wavelet-based denoising methods, using performance evaluators like *PSNR* and *MSE*. Results indicate there is a significant reduction in MSE and increase in PSNR for the proposed method.

Keywords Image denoising · SURE · Thresholding · Unsupervised learning

1 Introduction

A digital image is fundamentally an arithmetical depiction of an entity. While acquiring, transmitting, and retrieving of images from the database, images are often degraded with noise. In order to enhance the quality of an image, which was degraded by noise, many linear and nonlinear algorithms [1] are developed. Two approaches exist to denoise the image, filtering in spatial domain and in transform domain.

S. S. Rani (✉)
Department of Electronics and Communication Engineering,
GITAM University, Visakhapatnam 530045, Andhra Pradesh, India
e-mail: ssaradarani@gmail.com

G. S. Rao
AU College of Engineering, Visakhapatnam 530003, Andhra University, India
e-mail: sasigps@gmail.com

B. P. Rao
JNTUK, Kakinada 533003, Andhra Pradesh, India
e-mail: drbprjntuk@gmail.com

© Springer Nature Singapore Pte Ltd. 2019   685
J. C. Bansal et al. (eds.), *Soft Computing for Problem Solving*,
Advances in Intelligent Systems and Computing 816,
https://doi.org/10.1007/978-981-13-1592-3_54

Image degradation or restoration can be modelled as,

$$\text{Spatial domain:} g(a, b) = h(a, b) * f(a, b) + \eta(a, b) \tag{1}$$

$$\text{Frequency domain:} G(c, d) = H(c, d)F(c, d) + N(c, d) \tag{2}$$

where f is the original image, h is the degraded function, η is the noise added, and g is the corrupted image.

2 Spatial Filtering

2.1 Noise Problem

The noise reduction technique that is applied to a degraded image must possess edge and detail features preservation up to the maximum extent. The model of an image formation is given as

$$\mathbf{v}(l) = \mathbf{u}(l) + \mathbf{n}(l) \tag{3}$$

where u indicates the original image, n represents the corrupted image, and l indicates the pixel location in spatial domain.

The noise is additive which has standard deviation, σ, and mean as zero. The denoising objective is to lessen the noise in u and guesstimate $\hat{u}$ as close to u as possible. Mean-squared error (MSE) is the commonly used metric to measure the closeness. Here, the signal may be a set of a limited samples or an infinite samples generated by a time-varying process. A real-time adaptive assessment technique is needed for an unbounded data sample set.

Consider the signal is a set of finite samples, denoted by a vector $u = [u_0, u_1, \ldots, u_{N-1}]^T$ and the signal that is observed is

$$V = u + n \tag{4}$$

i.e.,

$$v_i = u_i + n_i \tag{5}$$

where $i = 0, 1, \ldots N - 1$.

2.2 Spatial Filters

Spatial filtering is the regular technique to expel noise from image. The spatial filters make use of low-pass filtering on collection of pixels assuming that noise exists for higher frequencies. Spatial filters are classified as linear and nonlinear filters.

The easy linear spatial filter is the *arithmetic mean* filter and considered as:

$$\hat{f}(k, l) = \frac{1}{mn} \sum_{(s,t) \in S_{xy}} g(s, t) \tag{6}$$

This is a simple smoothing filter. In the process of removing noise, it blurs the image.

Median filtering is a nonlinear operation employed in image processing to scale back "salt-and-pepper" noise. When the noise is to be reduced while preserving the edges, median filter is more effective.

$$\hat{f}(k, l) = \text{median}\{g(s, t)\} \tag{7}$$

3 Wavelet Domain

Wavelet provides investigation of the signal, localized in both time and frequency. DWT is a sampled description of CWT that converts the discrete signals into discrete coefficients. The DWT of a signal x is generated by transmitting it all the way through a series of filters [2]. The samples are first allowed through a low-pass filter that gives the output as the convolution of filter impulse response g and the samples of x.

Approximation coefficient and detail coefficient are the outputs of low-pass and high-pass filters, respectively [3–5].

3.1 Wavelet Thresholding

Denoising using wavelet removes noise in the image while preserving the characteristics of the image. Despite its frequency content, thresholding is applied to the DWT coefficients that are affected by additive white Gaussian noise. Thresholding is of two types.

- Hard thresholding: In hard thresholding, a tolerance is chosen. If the numerical value of the wavelet is below the tolerance value, then it is changed to zero; that is, many zeros are introduced with no loss of great amount of detail [6]. The coefficients that are greater than the tolerance are left unchanged and are given in Eq. (8)

$$\mathcal{I}_h\left(u, t\right) = \left\{ \begin{array}{ll} u, & |u| > t \\ 0, & |u| \leq t \end{array} \right. \tag{8}$$

- Soft thresholding: In soft thresholding, coefficients which are less than the tolerance are made zero [6]. The coefficients that are greater in magnitude than threshold are shrunk towards zero by subtracting the coefficient value from threshold value and is given in Eq. (9)

$$\mathcal{I}_s\left(u, t\right) = sgn\left(u\right)(|u| - t) = \left\{ \begin{array}{ll} u + t, & u < -t \\ 0, & |u| \leq t \\ u - t, & u > t \end{array} \right. \tag{9}$$

Steps involved in wavelet thresholding are:

(i) Get the noisy image as input.

(ii) Apply discrete wavelet transform on noisy image and acquire noisy coefficients. DWT outputs two signals, approximation and details, when the signal is allowed throughout complimentary filters. This task is called decomposition or analysis.

(iii) To the coefficients, apply soft thresholding. Universal threshold is given as [11]:

$$t = \sigma \sqrt{2 \log N} \tag{10}$$

(iv) Apply inverse DWT on the coefficients to attain denoised image.

4 Unsupervised Learning

4.1 Thresholding Functions

For a method that deals with the problem of reducing noise, it is desired to choose the method that gives the best performance, and it is the objective that is aimed in the paper.

Machine learning is a form of artificial intelligence, providing the system with the facility to learn without any prior programming. To lessen the noise in wavelet domain and to perform the thresholding function, a new artificial neural network (ANN) is proposed. The transform may be any orthogonal transform that is linear [7]. In the proposed method, linear transform is fixed and nonlinear activation function is adjustable.

The network activation function gradients and higher derivatives are utilized by most of the learning algorithms. To develop a gradient-based algorithm, high-order derivatives are desired for the activation function [6, 7].

The standard hard-thresholding function cannot be differentiated whatsoever, since it is discontinuous. The function of standard soft thresholding is weakly differentiable and has no higher order derivatives.

The new soft thresholding is formulated as (Fig. 1c)

$$\eta_s\,(u,t) = u + \frac{1}{2}\,\left(\sqrt{(u-t)^2 + \alpha} \;-\; \sqrt{(u+t)^2 + \alpha}\right) \qquad (11)$$

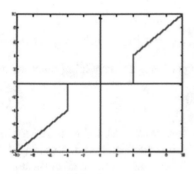

(a) Hard Thresholding Function

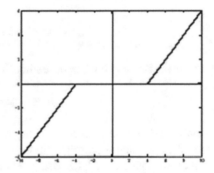

(b) Soft Thresholding Function

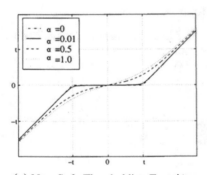

(c) New Soft Thresholding Function

Fig. 1 Thresholding functions

4.2 SURE-Based Learning Algorithm

In reality if only received signal which is noisy is known, then obtaining the reference signal is a difficult task. Because the reference signal is not there, the estimation error must be calculated in unsupervised fashion. Hence, unsupervised learning algorithm is proposed to reduce the error.

Practically, the variance of the noise σ^2 is generally pre-known or can be assessed easily. In those cases, for additive Gaussian noise, to assess the estimation error, Stein's unbiased risk estimate (*SURE*) is a good method. It is an unbiased estimator of MSE. To suppress the noise problem as in Eq. (3), let the variance of noise is normalized to unity, $\sigma^2 = 1$.

Stein's unbiased risk estimate (*SURE*) is formulated as

$$J_{\text{sure}}(t) = N + \|g(v)\|^2 + \sum \frac{\partial g_i}{\partial v_i} \tag{12}$$

where t is the threshold, when the estimation operator is a thresholding function, and

$$g_i = \Box_s(u, t) - u_i \tag{13}$$

Hence, the objective function of the neural network is SURE risk, and to minimize this, an adaptive learning algorithm which is gradient based is used [7–10]. However, to obtain the SURE risk gradient, the estimation operator must have no less than second-order derivative. Neither of the standard thresholding functions has second-order derivatives. A new soft-thresholding function Eq. (9) which is infinitely differentiable is chosen.

The projected method can be briefed as follows:

(i) Apply discrete wavelet transform on noisy image and acquire wavelet coefficients, i.e., approximation and detail components.
(ii) Calculate the threshold value from Eq. (11).
(iii) Gradient SURE risk is calculated.

$$\frac{\partial J_{\text{sure}}(t)}{\partial t} = 2 \sum_{i=0}^{N-1} g_i \frac{\partial g_i}{\partial t} + 2 \sum_{i=0}^{N-1} g_i \frac{\partial^2 g_i}{\partial v_i \partial t} \tag{14}$$

(vi) Then, the gradient-based adaptive learning steps are employed, i.e.,

$$t(k+1) = t(k) - \Delta t(k) \tag{15}$$

and

$$\Delta t(k) = \beta \frac{\partial J_{\text{sure}}(t)}{\partial t} \tag{16}$$

By using the above equations, find the estimation error and thus optimal threshold value.

(iv)   Reconstruct the image by applying inverse DWT to the threshold image.

5 Results and Discussion

Investigation of the performance of spatial filters and wavelet domain filters is done on MRI brain image. Performance of spatial filters such as average and median filters with different kernel sizes is shown in Fig. 2c–f. In wavelet domain filtering, images are denoised using standard threshold method and a method where the threshold is updated using unsupervised learning algorithm. Figure 2a is the original brain image, to which a Gaussian noise of density 20% is added and is given in Fig. 2b. Figure 2g is the denoised image using a standard threshold and Fig. 2h is the denoised image using SURE risk. Table 1 shows performance metrics such as MSE and PSNR for spatial filters. From the table, it is clear that median filter performs better than average filter. Table 2 compares the performance of standard thresholding and SURE estimate method in terms of MSE, PSNR, and structural similarity index measure ($SSIM$). On observing Table 2, it is seen that SURE performs better than standard threshold with MSE $= 3.3581$, PSNR $= 32.6430$, and SSIM $= 0.9311$ for the MRI brain image.

6 Conclusion

In this paper, a SURE-based learning technique has been implemented for adaptive noise reduction. It provides a combination of linear filtering and thresholding. Median filtering performs better in spatial domain filtering for MRI brain image. In thresholding method of denoising, SURE risk performs better than the standard threshold. Among spatial filters (average and median filters) and wavelet threshold-

Table 1 Performance metrics of image denoising using spatial filters

Filter	Mean 3×3	Mean 5×5	Median 3×3	Median 5×5
MSE	883.8539	1090.6227	831.7706	918.2003
PSNR	18.667	17.7541	18.9308	18.5014

Table 2 Performance metrics of image denoising using thresholding

Thresholding technique	Standard threshold	SURE method
MSE	979.1981	3.3581
PSNR	18.2221	32.643
SSIM	0.5988	0.9311

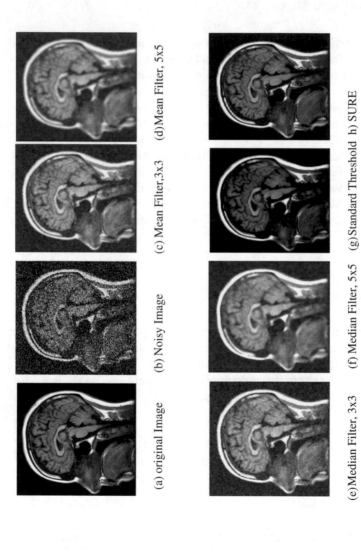

(a) original Image   (b) Noisy Image   (c) Mean Filter,3x3   (d) Mean Filter, 5x5

(e) Median Filter, 3x3   (f) Median Filter, 5x5   (g) Standard Threshold   h) SURE

Fig. 2 Denoising of MRI brain using different techniques

ing methods (standard thresholding and SURE method), the performance of SURE method is good subjectively and objectively, for the application of denoising MRI brain image.

References

1. Hosur, S., Tewfik, A.H.: Wavelet transform domain LMS algorithm. In: IEEE International Conference on Acoustics Speech Signal Processing, vol. III, pp. 508–510 (1993)
2. Heil, C.E., Walnut, D.F.: Continuous and discrete wavelet transforms. SIAM Rev. **32**, pp. 628–666 (1989)
3. Erdol, N., Basbug, F.: Performance of wavelet transform based adaptive filters. In: IEEE International Conference Acoustics Speech Signal Processing, vol. III, pp. 500–503 (1993)
4. Doroslovacki, M., Fan, H.: Wavelet-based adaptive filtering. In: IEEE International Conference Acoustics Speech Signal Processing, vol. III, pp. 488–491 (1993)
5. Haykin, S.: Neural Networks: A Comprehensive Foundation. Prentice-Hall, Englewood Cliffs (1994)
6. Donoho, D.L.: De-noising by soft-thresholding. IEEE Trans. Inform. Theory **41**, 613–627 (1995)
7. Zhang, X.-P.: Thresholding neural network for adaptive noise reduction. IEEE Trans. Neural Netw. **12**(3) (2001)
8. Zhang, X.P., Desai, M.: Adaptive denoising based on SURE risk. IEEE Sig. Process. Lett. **10**(5), 265–267 (1998)
9. Erdol, N., Basbug, F.: Wavelet transform based adaptive filtering. In: IEEE International Conference Acoustics Speech Signal Processing, vol. III, pp. 500–503 (1993)
10. Marshall, D.F., Jenkins, W.K., Murphy, J.J.: The use of orthogonal transforms for improving performance of adaptive filters. IEEE Trans. Circ. Syst. **36**(4), 474–483 (1989)
11. Abramovich, F., Sapatinas, T., Silverman, B.W.: Wavelet thresholding via a Bayesian approach. J. Roy. Stat. Soc. B **60** (1998)

Design of Semi-chaotic Integration-Based Particle Swarm Optimization Algorithm and Also Solving Travelling Salesman Problem Using It

Akanksha Samar and R. S. Sharma

Abstract Strategy is used to solve nonlinear swarm intelligence optimization problems where social sharing of information helps individual to get benefit from before experience of other companions in search for food. In PSO, solution depends on inertia weight (W), which is an element of acceleration that decides the step size of the solutions. Due to this component, sometimes the global search process may skip the global optima. It suffers from the global minima trap, stagnation and convergence problem. So to avoid this situation and improve the exploration, exploitation and rate of convergence, a new phase is added, named as semi-chaotic integrated PSO (SCIPSO). In SCIPSO, the inertia weight is integrated over a range to enhance the search efficiency. Due to this, solutions are motivated to exploit more desirable in search space. The modified strategy is successfully tested over 25 benchmark functions with other nature-inspired algorithms, and results are best comparatively, which can be further used for the real optimization problem. Along with this, in this paper, the SCIPSO is used to solve one of the widely studied NP-Hard problem named as Travelling Salesman Problem (TSP) to calculate the best minimal cost for maximum of 50 cities whose results are comparatively better than the basic PSO algorithm given under TSPlib.

Keywords Particle swarm optimization · Nature-inspired algorithm · Inertia weight · Travelling Salesman Problem · Cost · Route

A. Samar (✉) · R. S. Sharma
Rajasthan Technical University, Kota, Rajasthan, India
e-mail: akankshasamar19@gmail.com

R. S. Sharma
e-mail: rssharma@rtu.ac.in

© Springer Nature Singapore Pte Ltd. 2019
J. C. Bansal et al. (eds.), *Soft Computing for Problem Solving*,
Advances in Intelligent Systems and Computing 816,
https://doi.org/10.1007/978-981-13-1592-3_55

695

1 Introduction

Particle swarm optimization (PSO) is an algorithm which finds a solution to an optimization problem in a search space or model in the presence of objectives. The movements of the particles are conducted by their own best-known position in the search space and entire swarm's best-known position. Shi and Eberahart introduced a parameter called inertia weight (W) for original PSO algorithm. The W is used to balance global and local search abilities. Large W is more appropriate for global search, and small W facilitates local search. A linearly decreasing W over the entire dimension of search achieves good performance [10]. Researchers are steadily working in this field to refine the fulfilment of the PSO [4, 9].

Thus, this theory of PSO is transformed as a new modified algorithm and used to solve Travelling Salesperson Problem (TSP) and finds an optimal cost. The use of particle swarm optimization (PSO) in one of the recent theories is solving TSP to gain much efficiency. To find an optimal route has been a major problem in transportation. This type of problem in TSP has been solved through many methods, but the problem becomes more complex when having a large number of delivery and receiving locations [13].

In the paper, a modified integrated algorithm is proposed for PSO named as semi-chaotic integrated PSO (SCIPSO), which is based on the integration factor of W that helps to improve the local as well as global best search, while improving the convergence speed. This algorithm is used to solve the NP-Hard problem, TSP, to find out the best cost for at least 50 cities. The rest of the paper is classified as follows: Sect. 2 gives a brief introduction of PSO. Section 3 describes the proposed approach named as SCIPSO, and Sect. 4 discusses the application based on PSO that is TSP solved using modified PSO. Test benchmark functions, parameter settings, experimental results and discussions are presented in Sect. 5. In the end, conclusions and future work are given in Sect. 6.

2 Basic Particle Swarm Optimization Algorithm

In a PSO system, swarms fly through the search space. Each particle represents a best solution to the optimization problem. The best position obtained is referred to as the global best particle [3]. Each particle traverses the X-Y coordinate within a two-dimensional search space. Its velocity is expressed by v[S] and v[S + 1](the velocity along the X-axis and Y-axis, respectively). Modification of the particles position is realized by the position and velocity information [6]. Each swarm knows its best value obtained so far in the search space (pbest) and its X-Y position. Individual particles also have knowledge about the best value achieved by the group (gbest) among pbest. Each swarm uses information related to its current position (x[S], x[S + 1]), its current velocities (v[S], v[S + 1]), distance between its current position pbest, distance between its current position and the groups gbest to modify its position [12].

The velocity and position of each agent are changed according to Eqs. 1 and 2:

$$v_i(S+1) = v_i(S) + c_1 * rand * (p_{best} - x_i(S)) + c_2 * rand * (g_{best} - x_i(S)) \tag{1}$$

$$x_i(S+1) = x_i(S) + v_i(S+1) \tag{2}$$

where v(S + 1), v(S) are velocities of swarms for (S + 1) and (S) iterations, respectively. $rand$ is random variable whose range is between [0,1]. c_1, c_2 are coefficients of social cognitive and personal parameter whose values are under the range of [0,2]. x(S + 1), x(S) are positions for (S + 1) and (S) iterations, respectively. pbest and gbest are local best and global best fitness values.

The inertial weight (W) is used to balance the global and local search abilities. This W parameter is introduced in 1998 by Kennedy, which changes the velocity equation as [1]:

$$v_i(S+1) = W * v_i(S) + c_1 * rand * (p_{best} - x_i(S)) + c_2 * rand * (g_{best} - x_i(S)) \tag{3}$$

The pseudo-code of PSO is shown in Algorithm 1.

Initialization: Initialize the set of swarms with arbitrarily initial velocity and position in S-dimension by randomly assigning each particle to it.
while Stopping criteria is not filled **do**
   Evaluate fitness for each individual swarm.
   Update the personal best for each individual.
   Update the global best.
   Update velocities and positions of individuals by equation (1) and (2) respectively.
end while
Set the particle to new best position.

Algorithm 1: Basic PSO Algorithm

3   A Semi-chaotic Integration-Based Inertia Weight Particle Swarm Optimization Algorithm (SCIPSO)

In PSO process, distribution of swarms feature varies not only with the generation number but also with the development state. Basic PSO algorithm can easily get trapped in the local optima when solving complex multimodal problems. This weakness has restricted wider applications of PSO. As compared to the other algorithms, modified PSO has an advantage of faster convergence which increases the convergence speed and gives the utilization of better use of exploration as well as exploitation and avoiding the local optima trap.

A number of variant PSO algorithms have been proposed to achieve these goals. In this paper, SCIPSO is formulated by developing the semi-chaotic inertial weight parameter and integrating it to a range which on increasing the number of iterations gives faster convergence speed. This improves the velocity and position factors.

- New inertia weight chaotic parameter named as C is introduced here which helps to improve the convergence rate, given as:

$$C = (w1 - w2) * (1 - iter)/maxiter + w2 \qquad (4)$$

Here, w1 = 2 and w2 = 0.8 are taken randomly. iter is iteration from 1 to 30, and maxiter is maximum iteration which is taken as 250.

- Integration range lies between [0,2] to obtain best outcomes and to improve the convergence rate as:

$$A = integral(C, 0, 2) \qquad (5)$$

- Now, the velocity equation has to be calculated as:

$$v_i'(S + 1) = W * v_i'(S) + c_1 * rand * (p_{best} \\ - x_i'(S)) + c_2 * rand * (g_{best} - x_i'(S)) * A \qquad (6)$$

- New position of swarm is calculated by putting Eq. 6 in Eq. 2.

Fig. 1 SCIPSO flowchart

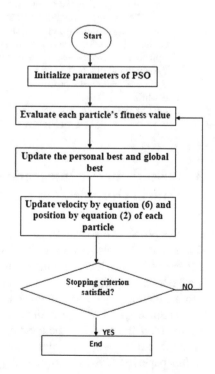

- The final velocity $v'(S + 1)$ and position $x'(S + 1)$ give the best optimal solution as compared to basic PSO and other algorithms.

The flowchart of SCIPSO is given in Fig. 1.

4   SCIPSO to Solve Travelling Salesperson Problem(TSP)

TSP is a NP-Hard problem and is used to find smallest route and cost for different number of cities. The approach inside the paper is—the SCIPSO algorithm—implemented for solving the TSP problem, whose optimal cost is found to be the minimum cost considering all other possible routes in the network. The result of PSO on TSP is compared with the result of SCIPSO on TSP, which formulates best cost. A graph is used to represent a network of routes with points (cities) connected to each other.

The SCIPSO-TSP algorithm is proposed below for finding the shortest path. Here, in the below algorithm, 'n' is total number of cities. $x(S + 1)$, $y(S + 1)$, $x(S)$ and $y(S)$ are new and old coordinates of positions, respectively, obtained by putting new velocity of Eq. 6 in Eq. 2.

The pseudo-code of SCIPSO-TSP is shown in Algorithm 2.

Get new velocity as $v(S + 1)$ and position as $x(S + 1)$ from Eqs. 6 and 2 respectively.
Consider city 1 as the starting and ending point the particles position using different points on the graph.
Generate all $(n-1)!$ permutations of cities.
Calculate distance of every permutation and keep track of minimum distance permutation of each graph as: $Dis(i, j) = \sqrt{((x(S + 1) - x(S))^2 + (y(S + 1) - y(S))^2)}$; $Dis(j, i) = Dis(i, j)$;
while Initialize the number of edges(NFE) is equal to zero, then do NFE = NFE+1 **do**
    Calculate minimum cost, MC: If NFE is 2, then NFE must be 1, i, MC(NFE, i) = dist(1, i)
    Else if size of NFE is greater than 2, C(NFE, i) = min MC(NFE-i, j) + dis(j, i) where j belongs to NFE, NFE != i and NFE != 1.
    Return the permutation with minimum cost(MC) which is new gbest position.
end while
Stop the algorithm by using target values as the stopping criterion given.

Algorithm 2: SCIPSO-TSP Algorithm

Let us consider a network of routes from TSPlib benchmark function with 14 points(burma14), the points connected to each other in all manner of combination (i.e. 14! ways of combination which means thousands of connections) using fourteen digit strings. The routes involved in this network are all explored in order to find the minimum cost and distance that does not form a cycle. Lets assume these points to be 0 to 13 and their locations on a graph to be: x = [16.47 16.47 20.09 22.39 25.23

Fig. 2 City1 (route of burma14 cities)

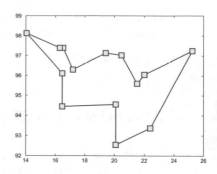

22.00 20.47 17.20 16.30 14.05 16.53 21.52 19.41 20.09]; y = [96.10 94.44 92.54 93.37 97.24 96.05 97.02 96.29 97.38 98.12 97.38 95.59 97.13 94.55].

Representing these points on a Cartesian graph (Fig. 2), we choose some route from any source to a destination and encode them as particles and named as city 1.

5 Experimental Result and Discussion

5.1 Test Problems

The paper consists of 25 benchmark functions f1 to f25 which are taken to inspect the execution of SCIPSO as appeared in Tables 1 and 2.

5.2 Experimental Setting

Comparison between the proposed algorithm SCIPSO and with other relative estimation is done among SCIPSO, PSO [11], GSA [2] and ABC [5], and SCIPSO generates best success rate and lowest mean error. Comparison of SCIPSO-TSP with PSO-TSP to minimize the cost of route for number of function evaluation, NFE = 12,550. The following experimental setting is used:

- Total number of runs = 30;
- Size of population = 250;
- Maximum number of iterations (maxiter) for PSO and to minimize cost (MC) = 250;
- Integrated range is = [0,2];

Table 1 Test problems, S: dimension, AE: acceptable error

Test problem	Objective function	Search range	Optimum value	S	AE
Sphere	$f_1(u) = \sum_{i=1}^{S} u_i^2$	[−5.12, 5.12]	$f(0) = 0$	30	1.0E−04
De Jong f4	$f_2(u) = \sum_{i=1}^{D} i.(u_i)^4$	[−5.12, 5.12]	$f(\mathbf{0}) = 0$	30	1.0E−04
Ackley	$f_3(u) = -20 + e + \exp(-\frac{0.2}{S}\sqrt{\sum_{i=1}^{S} u_i^3}) - \exp(\frac{1}{S}\sum_{i=1}^{S}\cos(2\pi.u_i)u_i)$	[−30, 30]	$f(0) = 0$	30	1.0E−04
Alpine	$f_4(u) = \sum_{i=1}^{S}|u_i\sin u_i + 0.1u_i|$	[−10, 10]	$f(0) = 0$	30	1.0E−04
Exponential	$f_5(u) = -(\exp(-0.5\sum_{i=1}^{S} u_i^2)) + 1$	[−1, 1]	$f(0) = -1$	30	1.0E−04
Zakharov	$f_6(u) = \sum_{i=1}^{D} u_i^2 + (\sum_{i=1}^{D} \frac{iu_i}{2})^2 + (\sum_{i=1}^{D} \frac{iu_i}{2})^4$	[−5.12 5.12]	$f(0) = 0$	30	1.0E−02
Cigar	$f_7(u) = u_0^2 + 100{,}000\sum_{i=1}^{D} u_i^2$	[−10, 10]	$f(\mathbf{0}) = 04$	30	1.0E−04
brown3	$f_8(u) = \sum_{i=1}^{S-1}(u_i^{2(u_{i+1})^2+1} + u_{i+1}^{2u_i^2+1})$	[−1, 4]	$f(0) = 0$	30	1.0E−04
Schwefel	$f_9(u) = \sum_{i=1}^{S}|u_i| + \prod_{i=1}^{S}|u_i|$	[−10, 10]	$f(0) = 0$	30	1.0E−04
Axis parallel hyper-ellipsoid	$f_{10}(u) = \sum_{i=1}^{S} i.u_i^2$	[−5.12, 5.12]	$f(0) = 0$	30	1.0E−04
Sum of different powers	$f_{11}(u) = \sum_{i=1}^{D}|u_i|^{i+1}$	[−1 1]	$f(\mathbf{0}) = 0$	30	1.0E−04
Neumaier 3 Problem (NF3)	$f_{12}(u) = \sum_{i=1}^{S}(u_i - 1)^2 - \sum_{i=2}^{S} u_i u_{i-1}$	[−900, 900]	$f(0) = -(S*(S+4)*(S-1))/6.0)$	10	1.0E−04
Rotated hyper-ellipsoid	$f_{13}(u) = \sum_{i=1}^{S}\sum_{j=1}^{i} u_j^2$	[−65.536, 65.536]	$f(0) = 0$	30	1.0E−04

Table 2 Test problems, S: dimension, AE: acceptable error

Test problem	Objective function	Search range	Optimum value	S	AE
Levy Montalvo 2	$f_{14}(u) = 0.1(\sin^2(3\Pi u_1) + \sum_{i=1}^{S-1}(u_i - 1)^2 \times (1 + \sin^2(3\Pi u_{i+1})) + (u_S - 1)^2(1 + \sin^2(2\Pi u_S))$	[−5, 5]	$f(1) = 0$	30	1.0E−04
Ellipsoidal	$f_{15}(u) = \sum_{i=1}^{D}(u_i - i)^2$	[−30 30]	$f(1, 2, 3, \ldots, D) = 0$	30	1.0E−04
Beale function	$f_{16}(u) = [1.5 - u_1(1 - u_2)]^2 + [2.25 - u_1(1 - u_2^2)]^2 + [2.625 - u_1(1 - u_2^3)]^2$	[−4.5,4.5]	$f(3, 0.5) = 0$	2	1.0E−04
Colville	$f_{17}(u) = 100[u_2 - u_1^2]^2 + (1 - u_1)^2 + 90(u_4 - u_3^2)^2 + (1 - u_3)^2 + 10.1[(u_2 - 1)^2 + (u_4 - 1)^2] + 19.8(u_2 - 1)(u_4 - 1)$	[−10 10]	$f(1) = 0$	4	1.0E−04
Kowalik	$f_{18}(u) = \sum_{i=1}^{11}[a_i - \frac{u_1(b_i^2 + b_i u_2)}{b_i^2 + b_i u_3 + u_4}]^2$	[−5, 5]	$f(0.192833, 0.190836, 0.12311, 0.135766) = 0.000307486$	4	1.0E−04
2D tripod function	$f_{19}(u) = p(u_2)(1 + p(u_1)) + |(u_1 + 50p(u_2)(1 - 2p(u_1)))| + |(u_2 + 50(1 - 2p(u_2)))|$	[−100, 100]	$f(0, -50) = 0$	2	1.0E−04
Shifted sphere	$f_{20}(u) = \sum_{i=1}^{D} z_i^2 + f_{bias}, z = u - o, u = [u_1, u_2, \ldots u_D], o = [o_1, o_2, \ldots o_D]$	[−100,100]	$f(o) = f_{bias} = -450$	10	1.0E−04
Goldstein–Price	$f_{21}(u) = (1 + (u_1 + u_2 + 1)^2(19 - 14u_1 + 3u_1^2 - 14u_2 + 6u_1u_2 + 3u_2^2)) \cdot (30 + (2u_1 - 3u_2)^2 \cdot (18 - 32u_1 + 12u_1^2 + 48u_2 - 36u_1u_2 + 27u_2^2))$	[−2, 2]	$f(0, -1) = 3$	2	1.0E−14
Easom's function	$f_{22}(u) = -\cos u_1 \cos u_2 e^{((-(u_1-\pi)^2 - (u_2-\pi)^2))}$	[−100, 100]	$f(\pi, \pi) = -1$	2	1.0E−13
Dekkers and Aarts	$f_{23}(u) = 10^5 u_1^2 + u_2^2 - (u_1^2 + u_2^2)^2 + 10^{-5}(u_1^2 + u_2^2)^4$	[−20 20]	$f(0, 15) = f(0, -15) = -24777$	2	5.0E−01
McCormick	$f_{24}(u) = \sin(u_1 + u_2) + (u_1 - u_2)^2 - \frac{3}{2}u_1 + \frac{5}{2}u_2 + 1$	$-1.5 \leq u_1 \leq 4, -3 \leq u_2 \leq 3$	$f(-0.547, -1.547) = -1.9133$	30	1.0E−04
Moved axis parallel hyper-ellipsoid	$f_{25}(u) = \sum_{i=1}^{S} 5i \times u_i^2$	[−5.12, 5.12]	$f(u) = 0; x(i) = 5 \times i, i = 1 : S$	30	1.0E−12

5.3 Results Comparison of Modified PSO with Other Variants

In Tables 3 and 4, SCIPSO is compared with other algorithms on same 25 benchmark functions in nature-inspired algorithms, under MATLAB 13 simulating environment which gives highest success rate (SR) and lowest average function evaluation (AFE).

Table 3 Comparison of the results of test function (TF), standard deviation (SD), mean error (ME), average number of function evaluations (AFE) for SCIPSO

TF	Algorithm	SD	ME	AFE	SR
f_1	SCIPSO	4.91E−07	9.48E−06	11, 683.333	30
	PSO	6.64E−07	9.36E−06	38, 386.666	30
	ABC	1.96E−06	8.30E−06	20, 508.333	30
	GSA	8.02E−07	8.99E−06	76, 470.000	30
f_2	SCIPSO	7.44E−07	9.22E−06	9190.000	30
	PSO	7.36E−07	9.21E−06	33, 253.333	30
	ABC	3.40E−06	5.20E−06	9931.666	30
	GSA	1.32E−06	8.40E−06	50, 305.000	30
f_3	SCIPSO	2.94E−07	9.68E−06	32, 598.333	30
	PSO	3.52E−07	9.63E−06	77, 275.000	30
	ABC	1.67E−06	8.10E−06	48, 431.666	30
	GSA	4.35E−07	9.44E−06	128, 835.000	30
f_4	SCIPSO	2.30E−07	9.69E−06	24, 146.666	30
	PSO	4.14E−07	9.43E−06	91, 338.333	30
	ABC	1.10E−06	8.75E−06	75, 470.000	30
	GSA	5.36E−07	9.25E−06	123, 951.666	30
f_5	SCIPSO	5.95E−07	9.16E−06	8156.666	30
	PSO	7.78E−07	9.14E−06	28, 241.666	30
	ABC	2.14E−06	7.19E−06	16, 625.000	30
	GSA	8.88E−07	8.97E−06	73, 110.000	30
f_6	SCIPSO	4.85E−04	9.66E−03	57, 695.000	30
	PSO	1.34E−02	1.98E−02	196, 405.000	11
	ABC	1.44E+01	1.00E+02	200, 000.000	0
	GSA	2.59748	5.06851	200, 000.000	0
f_7	SCIPSO	4.23E−07	9.50E−06	23, 666.667	30
	PSO	7.13E−07	9.29E−06	69, 461.667	30
	ABC	1.92E−06	7.99E−06	34, 711.667	30
	GSA	9.788438	8.780742	200, 000.000	0
f_8	SCIPSO	4.17E−07	9.44E−06	13, 606.667	30
	PSO	6.37E−07	9.34E−06	35, 061.667	30
	ABC	1.91E−06	8.12E−06	20, 918.333	30
	GSA	1.02E−06	8.74E−06	79, 390.000	30

Table 3 (continued)

TF	Algorithm	SD	ME	AFE	SR
f_9	SCIPSO	3.60E−07	9.66E−06	24,955.000	30
	PSO	4.59E−07	9.64E−06	71,468.333	30
	ABC	8.66E−07	9.11E−06	41,466.667	30
	GSA	3.95E−07	9.53E−06	145,243.333	30
f_{10}	SCIPSO	4.11E−07	9.41E−06	13,965.000	30
	PSO	6.99E−07	9.18E−06	44,131.667	30
	ABC	2.42E−06	7.53E−06	22,623.333	30
	GSA	7.29E−07	9.16E−06	88,296.667	30
f_{11}	SCIPSO	1.72E−06	7.46E−06	3333.333	30
	PSO	1.52E−06	8.01E−06	9828.333	30
	ABC	2.94E−06	5.69E−06	15,368.333	30
	GSA	2.52E−06	6.96E−06	37,470.000	30
f_{12}	SCIPSO	7.39E−07	9.32E−06	20,141.667	30
	PSO	4.04E−07	9.65E−06	66,488.333	30
	ABC	1.14E+00	1.22E+00	200,004.267	0
	GSA	1.49E−06	7.72E−06	69,508.333	30
f_{13}	SCIPSO	3.77E−07	9.44E−06	18,346.667	30
	PSO	5.37E−07	9.24E−06	56,290.000	30
	ABC	2.23E−06	7.81E−06	27,993.333	30
	GSA	9.58E−07	8.47E−06	76,086.667	30
f_{14}	SCIPSO	2.69E−07	9.63E−06	9725.00	30
	PSO	6.39E−07	9.27E−06	35,576.666	30
	ABC	2.43E−06	6.71E−06	19,723.333	30
	GSA	7.04E−07	9.16E−06	73,245.000	30
f_{15}	SCIPSO	4.83E−07	9.30E−06	16,346.666	30
	PSO	8.11E−07	9.01E−06	44,593.333	30
	ABC	2.36E−06	7.28E−06	24,511.666	30
	GSA	4.79E−05	0.0005253	106,043.330	28
f_{16}	SCIPSO	3.23E−06	4.25E−06	2268.333	30
	PSO	2.23E−06	6.60E−06	2748.333	30
	ABC	1.76E−06	8.47E−06	17,794.533	30
	GSA	2.96E−06	4.82E−06	55,576.666	30
f_{17}	SCIPSO	2.07E−04	8.68E−04	25,421.666	30
	PSO	1.66E−04	8.45E−04	40,583.333	30
	ABC	9.69E−02	1.49E−01	200,025.030	0
	GSA	0.1562894	0.0313651	128,590.000	27
f_{18}	SCIPSO	7.17E−06	9.48E−05	21,611.666	30
	PSO	1.94E−05	8.54E−05	34,856.666	30
	ABC	7.95E−05	1.75E−04	174,952.330	8
	GSA	7.97E−05	0.0002391	78.333	30

Table 3 (continued)

TF	Algorithm	SD	ME	AFE	SR
f_{19}	SCIPSO	2.20E−05	6.86E−05	10,156.666	30
	PSO	3.00E−01	1.00E−01	38,825.000	27
	ABC	2.29E−05	6.48E−05	7355.200	30
	GSA	2.34E−07	6.86E−07	120,410.000	30
f_{20}	SCIPSO	1.54E−06	8.26E−06	6975.00	30
	PSO	1.33E−06	8.47E−06	15,646.666	30
	ABC	1.90E−06	7.79E−06	9041.666	30
	GSA	1.89E−06	7.93E−06	70,280.000	30
f_{21}	SCIPSO	2.71E−15	4.75E−15	7673.333	30
	PSO	2.42E−15	4.59E−15	9760.000	30
	ABC	1.40E−06	4.43E−07	136,819.360	17
	GSA	1.22E−11	4.50E−12	115,734.760	23
f_{22}	SCIPSO	2.54E−14	5.10E−14	8083.333	30
	PSO	2.74E−14	4.18E−14	9873.333	30
	ABC	9.74E−05	2.90E−05	20,0025.933	0
	GSA	0.2084472	0.100	137,431.666	27
f_{23}	SCIPSO	6.15E−03	4.91E−01	3738.333	30
	PSO	6.05E−03	4.92E−01	5313.333	30
	ABC	5.50E−03	4.89E−01	1518.333	30
	GSA	6115.5683	8014.52713	533.333	30
f_{24}	SCIPSO	6.04E−06	9.00E−05	1273.333	30
	PSO	8.38E−06	8.73E−05	1475.000	30
	ABC	6.12E−06	9.03E−05	1346.733	30
	GSA	7.08E−06	8.84E−05	38,666.666	30
f_{25}	SCIPSO	4.65E−17	9.28E−16	34,216.666	30
	PSO	6.02E−17	9.22E−16	104,743.333	30
	ABC	1.27E−16	8.66E−16	59,833.3333	30
	GSA	3.82E−15	8.92E−16	78,652.666	30

Table 4 Comparison results of test problem, TP

No. of cities	Best cost for PSO-TSP	Best cost for SCIPSO-TSP
20	387.6095	361.5041
30	376.6954	344.4581
Burma14	30.8785	30.8785
Att48	50,625.4691	42,880.187

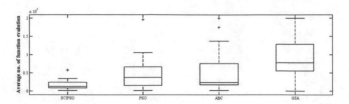

Fig. 3 Boxplot graph (average number of function evaluation)

Boxplot analysis of average number of function evaluations (AFE) can also be designed for the comparison of considered algorithms SCIPSO, PSO, GSA and ABC [7]. In Fig. 3, it is demonstrated as graphical representation of data for SCIPSO, PSO, GSA and ABC. Here results depict that SCIPSO is better at accuracy and efficiency level in comparison of PSO, GSA and ABC [8].

5.4 Results for Comparison of SCIPSO-TSP with PSO-TSP

On forming the graph with library datasets solution, if there are intersecting lines on route then we can conclude that solution is not optimal (best).

In Table 4, there is comparison between best minimum cost for SCIPSO-TSP algorithm and PSO-TSP algorithm; we can conclude on that note that the modified algorithm gives best minimum cost then basic algorithm.

Results of the routes and graph shown for TSPlib benchmark functions are as follows:

(a) for random 20 cities: route is generated by following dataset: x = [15 65 8 55 21 32 5 88 38 61 44 51 31 30 0 56 11 65 11 44];

y = [5 99 65 59 95 50 63 74 65 74 60 22 55 95 73 9 59 64 47 57]

Dataset of x and y coordinates gives the route of 20 cities. The comparison graph below shows the two graphs as shown in Fig. 4; green line represents PSO-TSP, and blue line represents SCIPSO-TSP. Both the lines represent the plot between number of edges (NFE) and best cost generated by using algorithm 2. As shown in Fig. 4, when NFE is 1 then best cost is approximately 600; when NFE is increased to 2, best cost value minimized to approximately 400. Hence, it is clearly displayed that as the rate of NFE increases, the cost rate decreases. The values of PSO-TSP are higher as compared to SCIPSO-TSP. It shows that by using algorithm 2, best cost values are minimized. Similarly, the value of best cost is minimized for all the four graphs as shown in Figs. 5, 6 and 7, respectively. All simulations are performed in MATLAB 13; test cases of TSP were chosen from TSPlib on github (https://github.com/pdrozdowski/TSPLib.Net/tree/master/TSPLIB95/tsp).

(b) for random 30 cities, route is generated by following dataset: x = [42 16 8 7 27 30 43 58 58 37 38 46 61 62 63 32 45 59 5 10 21 5 30 39 32 25 25 48 56 30];

Fig. 4 Comparison between
SCIPSO-TSP and PSO-TSP
random 20 cities

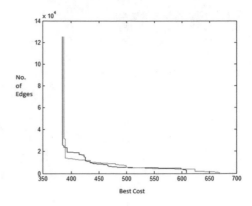

Fig. 5 Comparison between
SCIPSO-TSP and PSO-TSP
random 30 cities

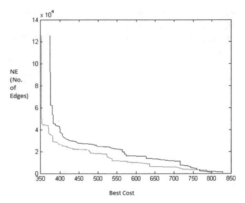

y = [57 57 52 38 68 48 67 48 27 69 46 10 33 63 69 22 35 15 6 17 10 64 15 10 39 32 55 28 37 40];

(c) for att48.tsp, route is generated by following dataset: x = [6734 2233 5530 401 3082 7608 7573 7265 6898 1112 5468 5989 4706 4612 6347 6107 7611 7462 7732 5900 4483 6101 5199 1633 4307 675 7555 7541 3177 7352 7545 3245 6426 4608 23 7248 7762 7392 3484 6271 4985 1916 7280 7509 10 6807 5185 3023];
y = [1453 10 1424 841 1644 4458 3716 1268 1885 2049 2606 2873 2674 2035 2683 669 5184 3590 4723 3561 3369 1110 2182 2809 2322 1006 4819 3981 756 4506 2801 3305 3173 1198 2216 3779 4595 2244 2829 2135 140 1569 4899 3239 2676 2993 3258 1942];

(d) for burma14.tsp: DataSet is: x = [16.47 16.47 20.09 22.39 25.23 22.00 20.47 17.20 16.30 14.05 16.53 21.52 19.41 20.09];
y = [96.10 94.44 92.54 93.37 97.24 96.05 97.02 96.29 97.38 98.12 97.38 95.59 97.13 94.55];

Fig. 6 Comparison between
SCIPSO-TSP and PSO-TSP
att48 cities

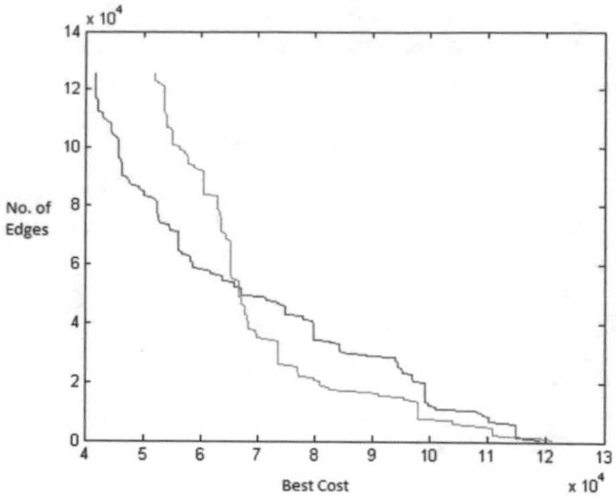

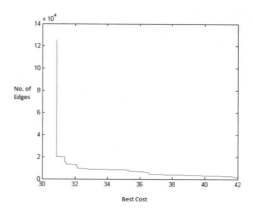

Fig. 7 Comparison between
SCIPSO-TSP and PSO-TSP
burma14 cities

6 Conclusion

PSO mainly focuses on to reduce the chances of premature convergence rate and
fall into the local minima trap. The paper introduces a new variant of PSO named as
semi-chaotic integrated particle swarm optimization (SCIPSO). SCIPSO algorithm
is comprised with the basic PSO to improve the functionality of this nature-inspired
algorithm. It is used a partial chaotic parameter which helps to increase the acceleration rate. W is an important parameter of PSO, which significantly affects the convergence, exploration and exploitation in PSO process. The integration of decreasing W
can improve the convergence speed. SCIPSO is compared with PSO, GSA and ABC
over different benchmark functions and gives best outcomes. The acquired outcomes
express that SCIPSO is a competitive variant of PSO.

Thus further, this SCIPSO is used to solve Travelling Salesman Problem (TSP) which
gives minimal cost result for a range of small dataset cities. On as far as by increasing
number of dataset above 50 cities in this paper, the quality of result degrades. So we
must balance between runtime and the solution cost.

In future, SCIPSO can be applied to more optimization problems as well as for different applications.

References

1. Bansal, J.C., Singh, P.K., Saraswat, M., Verma, A., Jadon, S.S., Abraham, A.: Inertia weight strategies in particle swarm optimization. In: 2011 Third World Congress on Nature and Biologically Inspired Computing (NaBIC), pp. 633–640. IEEE (2011)
2. Dowlatshahi, M.B., Nezamabadi-Pour, H., Mashinchi, M.: A discrete gravitational search algorithm for solving combinatorial optimization problems. Inf. Sci. **258**, 94–107 (2014)
3. Holliday, D., Resnick, R., Walker, J.: Fundamentals of physics (1993)
4. Jadon, S.S., Sharma, H., Bansal, J.C., Tiwari, R.: Self adaptive acceleration factor in particle swarm optimization. In: Proceedings of Seventh International Conference on Bio-Inspired Computing: Theories and Applications (BIC-TA 2012), pp. 325–340. Springer (2013)
5. Karaboga, D., Gorkemli, B., Ozturk, C., Karaboga, N.: A comprehensive survey: artificial bee colony (abc) algorithm and applications. Artif. Intell. Rev. **42**(1), 21–57 (2014)
6. Kennedy, J., Eberhart, R.: Particle swarm optimization. In: Proceedings of IEEE International Conference on Neural Networks IV, pages, vol. 1000 (1995)
7. Mazidi, A., Damghanijazi, E.: Meta-heuristic approaches for solving travelling salesman problem. Int. J. Adv. Res. Comput. Sci. **8**(5) (2017)
8. Odili, J.: Kahar, Mohmad, M.N., Noraziah, A., Kamarulzaman, S.F: A comparative evaluation of swarm intelligence techniques for solving combinatorial optimization problems. Int. J. Adv. Robot. Syst. **14**(3), 17298814–17705969 (2017)
9. Sedighizadeh, D., Masehian, E.: Particle swarm optimization methods, taxonomy and applications. Int. J. Comput. Theory Eng. **1**(5), 486 (2009)
10. Shi, Y., Eberhart, R.: A modified particle swarm optimizer. In: The 1998 IEEE International Conference on Evolutionary Computation Proceedings, 1998. IEEE World Congress on Computational Intelligence, pp. 69–73. IEEE (1998)
11. Yang, Q., He, G., Li, L.: Modifications of particle swarm optimization for global optimization. In: 2010 3rd International Conference on Biomedical Engineering and Informatics (BMEI), vol. 7, pp. 2923–2926. IEEE (2010)
12. Zhang, Y.-H., Lin, Y., Gong, Y.-J., Zhang, J.: Particle swarm optimization with minimum spanning tree topology for multimodal optimization. In: 2015 IEEE Symposium Series on Computational Intelligence, pp. 234–241. IEEE (2015)
13. Zhong, W.-h., Zhang, J., Chen, W.-n.: A novel discrete particle swarm optimization to solve traveling salesman problem. In: IEEE Congress on Evolutionary Computation, 2007. CEC 2007, pp. 3283–3287. IEEE (2007)

Keyword-Based Journal Categorization Using Deep Learning

T. Revathi and T. M. Rajalaxmi

Abstract Journal searching for particular context is nowadays a very challenging task because of the large availability of journals and also the finding reputed journals with impact factors also one of the tedious job. Our aim is to reduce the long procedure for journal searching using modern techniques. Applications in machine learning have witnessed a booming interest from last decade. The proposed work uses machine learning algorithm with NoSQL database for keyword-based journal retrieval. The complete work is divided into three subworks: (i) keywords modeling, (ii) journal categorization, and (iii) information retrieval of journal. The journal is categorized based on the keyword. The journal and their keywords are trained using deep neural network. For the given keyword, the similar keywords are extracted. The information retrieval gives the details of the journals for the appropriate keywords. The journal details are maintained in the NoSQL database, and the details are retrieved.

Keywords Word embedding · Deep learning · Word2vec · NoSQL · Machine learning

1 Introduction

Nowadays, there is a wide availability of journals in the electronic forms, which requires an automatic technique to identify the journals with the help of predefined keywords. The past decades contain many machine learning algorithms for text categorization. The main aim of journal categorization is the classification of journal into a fixed number of pre-determined keyword-based categories. So it is easy to search the journal based on the categories. Every journal will be either in multiple, or single

T. Revathi (✉)
Department of Computer Science, SSN College of Engineering, Chennai, India
e-mail: revathit@ssn.edu.in; revvlr@yahoo.com

T. M. Rajalaxmi
Department of Mathematics, SSN College of Engineering, Chennai, India
e-mail: laxmi.raji18@gmail.com

© Springer Nature Singapore Pte Ltd. 2019
J. C. Bansal et al. (eds.), *Soft Computing for Problem Solving*,
Advances in Intelligent Systems and Computing 816,
https://doi.org/10.1007/978-981-13-1592-3_56

711

category utilizing machine learning; the main purpose is to learn keyword–journal pair through instances which perform the category assignments automatically. Many Web sites give the journal list according to the keywords. But they did not concentrate on the journal list based on the impact factor and category.

Deep learning is one of the machine learning methods. The learning methods can be supervised or unsupervised in deep learning. There are many deep learning architecture like deep neural network, recurrent neural network, and deep belief networks. It can be applied to various application fields, such as computer vision, speech recognition. In this work, we have used the deep neural network architecture. The deep neural networks (DNNs) are an artificial neural network which has multiple layers between the input and output layers. DNN uses the feed forwarding networks in which there is no looping back from output layer to the input layer. The dataflow is from input layer to the output layer.

In this paper, we have used deep neural networks for both keyword modeling and journal categorization. The NoSQL database, MongoDB, is used to store the journal details. In the next section, the related works for this paper are discussed. Sect. 3 explains our proposed system with the methods, like word embedding and MongoDB. The experimental result shows our implementation of our work. Finally, we conclude with the future enhancement.

2 Related Works

Text categorization is the classification of documents or words into predefined categories. Tang et al. [1] proposed Bayesian classification for automatic text categorization using class-specific features. Baggenstoss's PDF projection theorem for class-specific feature classification. Wang et al. [2] classified the short text based on word embedding and convolution neural network. In word embedding, the semantically related words is chosen. Schmidhuber [3] discussed about deep neural networks with that he reviewed the deep supervised, unsupervised learning, and reinforcement learning. Dasari and Venu Gopala Rao [4] examined the main approaches for text categorization and compared the machine learning algorithm with the state of the art. Zamani and Bruce Croft [5] proposed two learning models, one learns the relevance distribution over the vocabulary set and other does the classification based on relevant or non-relevant class for each query. Sharma et al. [6] discussed about the first-level classification of some modern leading NoSQL representatives. They compared the inter-class and intra-class classification. Sharma et al. [7] reviewed the latest data model to process big data and NoSQL.

3 Proposed System

The proposed system uses word embedding package word2vec for training and testing the keyword–journal pair. Store the journal details in the NoSQL database. The following are the steps to be followed to find the journals for their appropriate keywords

(1) The keyword for the journal is given as the input to the system.
(2) The system checks if the keyword is trained in the keyword—journal pair.

    (a) If it is present, it produces the list of journals with their details for the appropriate keyword.
    (b) Else the keyword is sent to the trained word embedding vectors, to search the similar keyword to find the journal. The output gives the similarity-based keyword. Then do step 2.

(3) End.

3.1 Flow Chart

See Fig. 1.

3.2 Keyword Modeling and Journal Categorization

In this paper, the keyword–keyword pair and journal–keyword pair are modeled using word embedding method. The word embedding is a method to map a word to many similar words. Using this technique, the keywords for the journal trained for example: A machine learning keyword is mapped into many similar words: They are big data, fuzzy set, and neural network. Word embedding does not produce the synonyms for the word; rather it gives related words which have some similarity between them.

Word Embedding. Word embedding is one of the most interesting areas of research in deep learning. A word embedding W: words $\rightarrow$ R^n is a parameterized function mapping words in some language to high-dimensional vectors (perhaps 200–500 dimensions). Word embeddings is nothing but the conversion of the word to numbers, and the same word may have different number representations. There are two types of word embedding. They are (1) frequency-based and (2) prediction based. The frequency-based word embedding includes word count, term frequency–inverse document frequency (TF–IDF) vector and co-occurrence vector. The prediction-based model has two types which are continuous bag of words (CBOW) and skip-gram.

For example,

W(rat) = (0.2, −0.4, 0.7, …)
W(cat) = (0.0, 0.6, −0.1, …).

W is initialized to have random vectors for each word. It learns to have meaningful vectors in order to perform some task. The word embedding kind of 'map' of words is used in this application. Similar words are close together. This is not the synonym for the given word; rather the word has similarity between them.

Word2vec [8] is a model which used to generate word embeddings. This model uses two-layer neural networks which are trained to reconstruct linguistic contexts of words. There are two basic neural network models for word2vec which are CBOW and Skip-Gram. CBOW uses a window of word to predict the middle word. Skip-Gram uses a word to predict the surrounding ones in the window. In the proposed work, the Skip-Gram is used (Fig. 2).

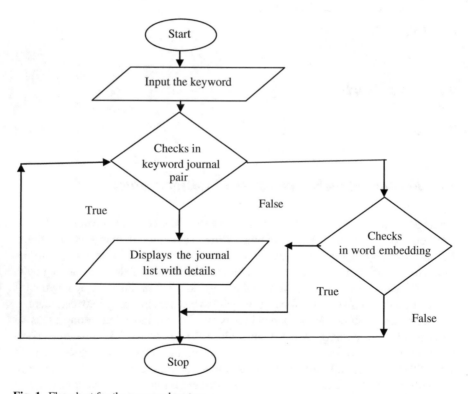

Fig. 1 Flowchart for the proposed system

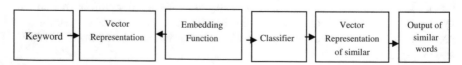

Fig. 2 Steps in Word2vec

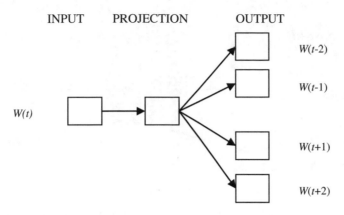

INPUT   PROJECTION   OUTPUT

$W(t)$

$W(t-2)$

$W(t-1)$

$W(t+1)$

$W(t+2)$

Fig. 3 Skip-Gram model

The advantages of using Skip-Gram model are it can have two vector representations of a single word (Fig. 3).

Skip-Gram uses negative sampling training method. Skip-Gram is quite opposite to the CBOW methods. It takes target word in the input layer and the context of words in the output layer.

The algorithm of the Skip-Gram model:

(1) It generates the hot vector for every word.
(2) Getting the embedding vectors for the context.
(3) Generate the score vectors.
(4) Turn each score into probabilities.
(5) The probability vector generated to match the true probability which is the hot vectors of the actual output.

3.3 Information Retrieval

Nowadays, the data generations are multiplying rapidly, So NoSQL database is preferred compared to RDBMS. It can effectively handle the fastest change than RDBMS. NoSQL stands for **Not Only SQL**, and it belongs to a class of non-relational data storage systems. It do neither require a fixed table schema nor do they use the concept of joins. The three properties of a system are consistency, availability, and partitions.

There are different types of storage in the NoSQL database: They are (1) column database (2) key-value database (3) document database and (4) graph database. It can handle both structured and semi-structure data format. The NoSQL database is open source, and some of them are Apache Cassandra, CouchDB, HBase, hypertable, MongoDB, Redis, and Riak.

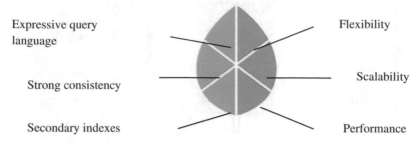

Expressive query language   Flexibility

Strong consistency   Scalability

Secondary indexes   Performance

Fig. 4 MongoDB nexus architecture

In this work, MongoDB database is used. MongoDB is an open source, and it is published under the GNUGPL. It is based on document database in NoSQL and leading NoSQL database. MongoDB is written in C++. MongoDB stores data in flexible format, and the data structure can be changed over time. The one major advantage of MongoDB is embedding documents and arrays, which lessen the need for joins statement. Added to it, dynamic schema helps in polymorphism (Fig. 4).

The proposed system uses MongoDB for the various features like flexibility, scalability, consistency. The proposed works store the data of various journals. The journal with the impact factors is considered to be stored in the database. The journal and its details are taken from IEEE, Springer, Elsevier, Science direct databases. Now for modeling the application, we have used few journals for training and testing the proposed system. The database contains journal ID, name, and impact factor, and publisher's name. The journal can be retrieved based on the journal ID.

4 Experimental Results

The proposed system is implemented in python as a front end, and for database storage, MongoDB is used. We have here taken Computer Science- and Mathematics-oriented keywords, for training and testing. The trained keyword-based journals are taken from IEEE, Springer, Elseveir, and Science direct database. For each category of keyword, at least five impact factors journals are trained. The same journal can fall under any number of keyword.

The result of our application is shown in Fig. 5. At first, the keyword is entered into the system, and it checks for the similar keywords. The similar keywords are trained using word2vec. The cosine similarity of the two keywords is considered for the selection of the similarity between words. If the cosine value is small, then the keywords are related to each other, and if the cosine value is high, then the keywords are not related to each other. The result shows the cosine value of the related keywords. The keywords and the journal ID are trained. The journal IDs

Fig. 5 Result in python

for the related keyword are displayed, and the corresponding journal information is retrieved from the backend.

5 Conclusions and Future Enhancement

In this paper, we used deep neural networks for a keyword-based journal categorization. The keyword–journal is paired based on the similarity of the words concept. When compared to the traditional learning methods, deep neural networks can generate more features in the training phase. With minimum effect, the approach can be scaled for huge amount of keywords and journals. In real time, the data are growing rapidly, so to handle big data the application has to be modeled accordingly. In future to handle large amount of data efficiently in less time, the application can be ported to GPU-based system.

References

1. Tang, B., He, H., Baggenstoss, P.M., Kay, S.: A Bayesian classification approach using class-specific features for text categorization. IEEE Trans. Knowl. Data Eng. **11**, 1602—1606 (2016)
2. Wang, P., Xu, B., Heng, C., Hongwei, F.: Semantic expansion using word embedding clustering and convolution neural network for improving short text classification. Neuro Comput. **74**, 806–814 (2016)
3. Schmidhuber, J.: Deep learning in neural networks: an overview. Neural Netw. **61**, 85–117 (2015)

4. Dasari, D.B., Venu Gopala Rao, K.: Text categorization and machine learning methods: current state of the art. Glob. J. Comput. Sci. Technol. Softw. Data Eng. **1** 806–814 (2012)
5. Zamani, H., Bruce Croft, W.: Relevance-based Word Embedding. arXiv preprint arXiv:1705.0 3556 (2017)
6. Sharma, S., Tim, U.S., Gadia, S., Wong, J., Shandilya, R., Peddoju, S.K.: Classification and comparison of NoSQL big data models. Int. J. Big Data Intell. **2**(3), 201–221 (2015)
7. Sharma, S., Shandilya, R., Patnaik, S., Mahapatra, A.: Leading NoSQL models for handling Big Data: a brief review. Int. J. Bus. Inf. Syst. **22**(1), 1–25 (2016)
8. Cerisara, C., Kral, P., Lenc, L.: On the effects of using word2vec representations in neural networks for dialogue act recognition. Comput. Speech Lang. **47**, 175–1193 (2017)

Hopf Real Hypersurfaces in the Complex Quadric Q^m with Recurrent Jacobi Operator

Pooja Bansal

Abstract In this article, we first study the generalised notion of parallel Jacobi opera-tor known as recurrent Jacobi operator in complex quadric $Q^m = SO_{m+2}/SO_m SO_2$. Using this generalised notion, we describe recurrent normal Jacobi operator and recurrent structure Jacobi operator in complex quadric Q^m and we investigate Hopf real hypersurfaces of complex quadric Q^m with recurrent normal Jacobi operator and recurrent structure Jacobi operator. Consequently, we prove the non-existence results with these geometric conditions.

Keywords Complex quadric · Hopf hypersurface · Recurrent normal Jacobi operator · Recurrent structure Jacobi operator

1 Introduction

In [1], Suh considered the idea of parallel normal Jacobi operator, that is, $\nabla_X \overline{R}_N = 0$, for tangent vector field X on M, where ∇ denotes the induced connection from Levi Civita connection $\overline{\nabla}$ of Q^m. Suh, in his paper [1], proved a theorem of non-existence for real hypersurfaces in complex quadric Q^m, $(m \geq 3)$ with parallel normal Jacobi operator. The geometrical meaning of this notion is that the eigenspaces of the normal Jacobi operator $\overline{R}_N$ are parallel along any curve in M. Here, the eigenspaces of the normal Jacobi operator are said to be parallel if they are invariant under any parallel displacement along any curves on M in Q^m. On the other hand, it implies if E is the eigenspace of normal Jacobi operator $\overline{R}_N$, then for any $Y \in E$, $\nabla_X Y \in E$ along any direction X on M in Q^m.

Perez, Santosh and Suh used the concept of structure Jacobi operator and recurrent structure Jacobi operator and found some geometric results [10, 11]. Further, Jeong et al. have used the recurrent Jacobi operator of real hypersurfaces in complex two-plane

P. Bansal (✉)
Department of Mathematics, Faculty of Natural Sciences, Jamia Millia Islamia,
New Delhi 110025, India
e-mail: poojabansal811@gmail.com

© Springer Nature Singapore Pte Ltd. 2019
J. C. Bansal et al. (eds.), *Soft Computing for Problem Solving*,
Advances in Intelligent Systems and Computing 816,
https://doi.org/10.1007/978-981-13-1592-3_57

Grassmannians [2]. Moreover, he has focused on real hypersurfaces M in Q^m with parallel structure Jacobi operator. Motivated by his results, we wish to generalise this notion to recurrent normal Jacobi operator and recurrent structure Jacobi operator.

An operator T is said to be *recurrent* for real hypersurface in complex quadric Q^m if it satisfies the following condition

$$(\nabla_X T)Y = \omega(X)T(Y),$$

where ω is 1-form on M, and X, Y are vector fields tangent to M.

Thus, we define recurrent normal Jacobi operator and recurrent structure Jacobi operator for real hypersurfaces in complex quadric Q^m which is the generalisation of parallel normal Jacobi operator and parallel structure Jacobi operator, respectively.

Now, *recurrent normal Jacobi operator* for a real hypersurface M in Q^m is defined by

$$(\nabla_X \overline{R}_N)Y = \omega(X)\overline{R}_N(Y),$$

and *recurrent structure Jacobi operator* for a real hypersurface M in Q^m is defined by

$$(\nabla_X R_\xi)Y = \omega(X)R_\xi(Y),$$

where ω is 1-form on M, and X, Y are vector fields tangent to M.

In this paper, we prove that there do not exist any Hopf real hypersurface M in complex quadric Q^m, $(m \geq 3)$ with recurrent normal Jacobi operator. Further, we have shown that there do not exist any Hopf real hypersurface M in complex quadric Q^m, $(m \geq 3)$ with recurrent structure Jacobi operator.

2 Geometry of Q^m

The geometry of complex quadric can be seen in detail from [3–5]. The complex hypersurface of CP^{m+1} is known as complex quadric Q^m defined by the equation $z_1^2 + \cdots + z_{m+1}^2 = 0$, where $z_1, \ldots, z_{m+1}$ are homogeneous coordinates on CP^{m+1} equipped with the induced Riemannian metric g. Then, the canonical Kaehler structure (J, g) on Q^m is induced by Kaehler structure on CP^{m+1}. The *one*-dimensional quadric Q^1 is congruent to the round 2-sphere S^2. The *two*-dimensional quadric Q^2 is congruent to the Riemannian product $S^2 \times S^2$. For this, we will assume $m \geq 3$ throughout the paper.

Apart from J, Q^m admits one more different important geometric structure known as complex conjugation A on the tangent spaces of Q^m satisfying $AJ + JA = 0$. This complex conjugation A is a bundle of two vectors $\mathcal{U}$ containing S^1-bundle of real structures [4, 6]. For each unit normal vector $\overline{z}$ at $z \in Q^m$, let $A_{\overline{z}}$ be the shape operator of Q^m in CP^{m+1}. Then, A is an involution, i.e., $A_{\overline{z}}W = W$ for $W \in T_z Q^m$, or $A_{\bar{z}}$ is complex conjugation restricted to $T_z Q^m$. Now, $T_z Q^m$ is decomposed as [7]:

$$T_z Q^m = \mathcal{V}(A_{\bar{z}}) \oplus J\mathcal{V}(A_{\bar{z}}),$$

such that $\mathcal{V}(A_{\bar{z}}) := (+1)$-eigenspace and $J\mathcal{V}(A_{\bar{z}}) := (-1)$-eigenspace of $A_{\bar{z}}$.

The expression for Riemannian curvature tensor $\overline{R}$ of Q^m is given as:

$$\overline{R}(X, Y)Z = g(Y, Z)X - g(X, Z)Y + g(JY, Z)JX - g(JX, Z)JY$$
$$-2g(JX, Y)JZ + g(JAY, Z)JAX - g(JAX, Z)JAY,$$

where X, Y and Z are vector fields in $T_z Q^m$.

A nonzero tangent vector $W \in T_z Q^m$ is called *singular* if it is tangent to more than one maximal flat in Q^m. There are two types of singular tangent vectors for the complex quadric Q^m:

1. If $\exists A \in \mathcal{U}$: W is an eigenvector corresponding to an eigenvalue $(+1)$, then the singular tangent vector W is known as $\mathcal{U}$-*principal*.

2. If $\exists A \in \mathcal{U}$ and orthonormal vectors $U, V \in \mathcal{V}(A)$ such that $W/\|W\| = (U + JV)/\sqrt{2}$, then the singular tangent vector W is known as $\mathcal{U}$-*isotropic*.

3 Some General Fundamental Formulas

Consider a real hypersurface M in Q^m. Then, a transform JX of the Kaehler structure J on Q^m for any tangent vector field X on M in Q^m defined by $JX = \phi X + \eta(X)N$ where N is the unit normal vector to M and ϕX denotes the tangential component of JX. Also, $\xi = -JN$ and $\eta(X) = g(\xi, X)$. Here, M associates an induced *almost contact metric structure* (ϕ, ξ, η, g) satisfying the following relations [8]:

$$\begin{cases} \phi^2 X = -X + \eta(X)\xi, \\ \phi\xi = 0, \eta(\phi X) = 0, \eta(\xi) = 1 \\ g(\phi X, \phi Y) = g(X, Y) - \eta(X)\eta(Y). \end{cases}$$

Moreover, we have

$$\nabla_X \xi = \phi SX$$

where S is the shape operator of M.

On the other hand, the Gauss and Weingarten formulas for M are given as respectively,

$$\overline{\nabla}_X Y = \nabla_X Y + \sigma(X, Y) \ \ and \ \ \overline{\nabla}_X N = -SX,$$

where σ is second fundamental form of M, X, Y are tangent vector fields in M and N is a normal vector field on M.

Now, $TM = \mathcal{C} \oplus R\xi$, where $\mathcal{C} = ker(\eta)$ is the maximal complex subbundle of TM. The structure tensor field ϕ restricted to $\mathcal{C}$ coincides with the complex structure J restricted to $\mathcal{C}$ and $\phi\xi = 0$.

Here, a maximal $\mathcal{U}$ invariant subspace of $T_z M$ is defined as:

$$\mathcal{Q}_z = \{X \in T_z M \mid AX \in T_z M \ for \ all \ A \in \mathcal{U}\}.$$

Now, we give the following important lemmas:

Lemma 1 ([9]) *For each $z \in M$, we have*

(i) *If N_z is $\mathcal{U}$-principal, then $\mathcal{Q}_z = \mathcal{C}_z$.*

(ii) *If N_z is not $\mathcal{U}$-principal, there exist a conjugation $A \in \mathcal{U}$ and orthonormal vectors $X, Y \in \mathcal{V}(A)$ such that $N_z = \cos(t)X + \sin(t)JY$ for some $t \in (0, \pi/4]$. Then, we have $\mathcal{Q}_z = \mathcal{C} \ominus \mathbb{C}(JX + Y)$.*

Lemma 2 ([7]) *Let M be a Hopf hypersurface in Q^m such that the normal vector field N is $\mathcal{U}$-principal everywhere. Then, α is constant. Moreover, if $X \in \mathcal{C}$ is a principal curvature vector of M with principal curvature λ, then $2\lambda = \alpha$ and ϕX is a principal curvature vector of M with principal curvature $\frac{\alpha\lambda + 2}{2\lambda - \alpha}$.*

Take M to be a Hopf hypersurface, i.e.,

$$S\xi = \alpha\xi,$$

where α is a smooth function defined by $\alpha = g(S\xi, \xi)$ on M.

Now, let $A \in \mathcal{U}_z$ such that $N = \cos(t)Z_1 + \sin(t)JZ_2$ where Z_1, Z_2 are orthonormal vectors in $V(A)$ and $0 \le t \le \pi/4$ which is a function on M. Since we know that $\xi = -JN$, we have

$$N = \cos(t)Z_1 + \sin(t)JZ_2,$$
$$AN = \cos(t)Z_1 - \sin(t)JZ_2,$$
$$\xi = \sin(t)Z_2 - \cos(t)JZ_1,$$
$$A\xi = \sin(t)Z_2 + \cos(t)JZ_1$$

which gives $g(\xi, \ AN) = 0$.

4 Some Consequences on Recurrent Normal Jacobi Operator

This section is devoted to two subsections in which we will consider $\mathcal{U}$-principal and $\mathcal{U}$-isotropic unit normal vector field. Let M be Hopf real hypersurface in Q^m and N be a unit normal vector field to M in Q^m. Now, the expression for normal Jacobi operator $\overline{R}_N$ of complex quadric Q^m by the virtue of Riemannian curvature tensor $\overline{R}$ of Q^m is given by

$$\overline{R}_N(X) = \overline{R}(X, N)N$$
$$= X + 3\eta(X)\xi + g(AN, N)AX - g(AX, N)AN - g(AX, \xi)A\xi.$$

where on $X = \xi$ using with $g(A\xi, N) = 0$ follows

$$\overline{R}_N(\xi) = 4\xi + g(AN, N)A\xi - g(A\xi, \xi)A\xi$$
$$= 4\xi + 2g(AN, N)A\xi.$$

4.1 $\mathcal{U}$-Principal Normal Vector Field

This subsection proves the non-existence of Hopf real hypersurface in Q^m satisfying recurrent normal Jacobi operator with unit normal vector field N is $\mathcal{U}$-principal. Assume that M has $\mathcal{U}$-principal unit normal vector field N. Then, the unit normal vector field satisfies $AN = N$ or $A\xi = -\xi$ for a complex conjugation $A \in \mathcal{U}$. We put

$$AY = BY + \rho(Y)N,$$

where BY denotes the tangential component of AY and

$$\rho(Y) = g(AY, N) = g(Y, N) = 0.$$

Hence, $AY = BY$ for any vector field Y on M with $\mathcal{U}$-principal unit normal vector field N in Q^m.

We have the following theorem:

Theorem 1 *There do not exist any Hopf real hypersurface in Q^m, $(m \geq 3)$ with recurrent normal Jacobi operator such that the unit normal vector field N is $\mathcal{U}$-principal.*

Proof Assuming the condition of N, the normal Jacobi operator has the form

$$\overline{R}_N(X) = \overline{R}(X, N)N$$
$$= X + 2\eta(X)\xi + AX. \tag{1}$$

By assumption of recurrent normal Jacobi operator, we have

$$(\nabla_X \overline{R}_N)(Y) = \omega(X)\overline{R}_N(Y).$$

From this with the formula $(\nabla_X A)Y = \nabla_X(AY) - A\nabla_X Y$, above relation reduces to

$$2g(\phi SX, Y)\xi + 2\eta(Y)\phi SX + (\nabla_X A)Y = \omega(X)(Y + 2\eta(Y)\xi + AY)$$

which on $Y = \xi$ gives

$$2\phi SX + (\nabla_X A)\xi = 2\omega(X)\xi. \tag{2}$$

Since we have

$$\begin{aligned}(\nabla_X A)\xi &= \nabla_X(A\xi) - A(\nabla_X \xi)\\ &= [-q(X) + 2\alpha\eta(X)]N\end{aligned}$$

Inserting above formula in (2) and on comparing tangent part, we obtain

$$\phi SX = \omega(X)\xi.$$

Operating ϕ on both sides, we get $\phi^2 SX = 0$, which gives

$$SX = \alpha\eta(X)\xi$$

that is, M is totally η-umbilical, which further says that S commutes with the structure tensor ϕ. Then, M is locally congruent to a tube over a totally geodesic CP^k in a Q^{2k}. But, they are never totally η-umbilical. So, we assert that there do not exist any Hopf real hypersurfaces in Q^m with recurrent normal Jacobi operator such that the unit normal vector field is $\mathcal{U}$-principal.

4.2 $\mathcal{U}$-Isotropic Normal Vector Field

Here, we will show the non-existence of Hopf real hypersurface in Q^m satisfying recurrent normal Jacobi operator when the unit normal vector field N is $\mathcal{U}$-isotropic. Let us assume that M has $\mathcal{U}$-isotropic unit normal vector field N. Then, N has the expression

$$N = \frac{1}{\sqrt{2}}(Z_1 + JZ_2)$$

for any $Z_1, Z_2 \in \mathcal{V}(A)$ which yields

$$g(\xi, A\xi) = 0,$$
$$g(\xi, AN) = 0,$$
$$g(AN, N) = 0,$$
$$g(AN, A\xi) = 0, \quad \left.\right\}\quad(3)$$

Thus, we have the following theorem.

Theorem 2 *There do not exist any Hopf real hypersurface in Q^m, $(m \geq 3)$ with recurrent normal Jacobi operator such that the unit normal vector field N is $\mathcal{U}$-isotropic.*

Proof Here, we are assuming that the unit normal vector field N is $\mathcal{U}$-isotropic, then the normal Jacobi operator can be rewritten as

$$\overline{R}_N(X) = X + 3\eta(X)\xi - g(AX, N)AN - g(AX, \xi)A\xi. \quad (4)$$

By taking covariant derivative along in any direction X and using the above relation, we have

$$(\nabla_X \overline{R}_N)(Y) = 3g(\phi SX, Y)\xi + 3\eta(Y)\phi SX - g(Y, \nabla_X AN)AN$$
$$-g(AN, Y)\nabla_X AN - g(\nabla_X A\xi, Y)A\xi - g(AY, \xi)\nabla_X A\xi$$

Let M has recurrent normal Jacobi operator, we see

$$3g(\phi SX, Y)\xi + 3\eta(Y)\phi SX - g(Y, \nabla_X AN)AN - g(AN, Y)\nabla_X AN$$
$$-g(\nabla_X A\xi, Y)A\xi - g(AY, \xi)\nabla_X A\xi$$
$$= \omega(X)[Y + 3\eta(Y)\xi - g(AY, N)AN - g(AY, \xi)A\xi]$$

On taking scalar product with ξ and using the assumption of $\mathcal{U}$-isotropic normal vector field, we obtain

$$4\omega(X)g(Y, \xi) = 3g(\phi SX, Y) - g(\nabla_X AN, \xi)g(AN, Y) - g(A\xi, Y)g(\nabla_X A\xi, \xi)$$

which gives

$$4\omega(X)g(Y, \xi) = 3g(\phi SX, Y) + g(AN, \phi SX)g(AN, Y) + g(A\xi, Y)g(A\xi, \phi SX)$$
$$(5)$$

Since $g(AN, \xi) = 0$ and $g(A\xi, \xi) = 0$. On taking covariant derivative with respect to X, we arrive

$$g(\nabla_X AN, \xi) = -g(AN, \phi SX) \text{ and } g(\nabla_X A\xi, \xi) = -g(A\xi, \phi SX),$$

Take $Y = AN$ in (5), one can get

$$g(\phi SX, AN) = 0. \tag{6}$$

Similarly, if we take $Y = A\xi$ in (5), we deduce that

$$g(\phi SX, A\xi) = 0. \tag{7}$$

Now taking account (6) and (7) into (5), we have

$$\phi SX = \frac{4}{3}\omega(X)\xi.$$

This implies that $SX = \alpha\eta(X)\xi$; that is, M is totally η-umbilical. Then, the shape operator S commutes with the structure tensor ϕ. Then, M is locally congruent to a tube over a totally geodesic CP^k in a Q^{2k}. But, they can never be a totally η-umbilical. This assert that there do not exist any Hopf real hypersurfaces in Q^m with recurrent normal Jacobi operator if the unit normal vector field is $\mathcal{U}$-isotropic.

5 Some Consequences on Recurrent Structure Jacobi Operator

From the Gauss equation for $Q^m \subset CP^{m+1}$, the Riemannian curvature tensor R of M is given by

$$\begin{aligned}
R(X, Y)Z = {} &g(Y, Z)X - g(X, Z)Y + g(\phi Y, Z)\phi X - g(\phi X, Z)\phi Y \\
&-2g(\phi X, Y)\phi Z + g(AY, Z)AX - g(AX, Z)AY \\
&+g(JAY, Z)JAX - g(JAX, Z)JAY.
\end{aligned}$$

where X, Y and Z are vector fields in $T_z M$.

First, we define *structure Jacobi operator* for a Hopf real hypersurface M in Q^m as follows:

$$\begin{aligned}
R_\xi(X) = {} &R(X, \xi)\xi \\
= {} &X - \eta(X)\xi + g(A\xi, \xi)AX - g(AX, \xi)A\xi - g(JAX, \xi)JA\xi \\
&+\alpha SX - \alpha^2\eta(X)\xi \tag{8}
\end{aligned}$$

where we have used the condition of Hopf hypersurface.

Further, assume that M has recurrent structure Jacobi operator

$$(\nabla_X R_\xi)Y = \omega(X)R_\xi Y.$$

5.1 $\mathcal{U}$-Principal Normal Vector Field

In this section, we will discuss non-existence of Hopf real hypersurfaces in Q^m with recurrent structure Jacobi operator such that the unit normal vector field N is $\mathcal{U}$-principal. Now, since we are assuming that M has $\mathcal{U}$-principal unit normal vector field N. Then, the unit normal vector field N satisfies $AN = N$ for a complex conjugation $A \in \mathcal{U}$, and hence, structure Jacobi operator becomes

$$
\begin{aligned}
R_\xi(X) &= R(X, \xi)\xi \\
&= X - 2\eta(X)\xi - AX + \alpha SX - \alpha^2\eta(X)\xi. \tag{9}
\end{aligned}
$$

Theorem 3 *There do not exist any Hopf real hypersurfaces in Q^m, $(m \geq 3)$ with recurrent structure Jacobi operator such that the unit normal vector field N is $\mathcal{U}$-principal.*

Proof Let the structure Jacobi operator be recurrent. Then, we have

$$
\begin{aligned}
&-(2 + \alpha^2)[g(\phi SX, Y)\xi + \eta(Y)\phi SX] - [q(X)JAY + g(SX, Y)N \\
&-g(SX, AY)N] + (X\alpha)SY + \alpha(\nabla_X S)Y - (X\alpha^2)\eta(Y)\xi = \\
&\omega(X)[Y - 2\eta(Y)\xi - AY + \alpha SY - \alpha^2\eta(Y)\xi]. \tag{10}
\end{aligned}
$$

Since

$$
\begin{aligned}
(\nabla_X S)\xi &= \nabla_X(S\xi) - S\phi SX \\
&= (X\alpha)\xi + \alpha\phi SX - S\phi SX
\end{aligned}
$$

Taking $Y = \xi$ in (10) and using above calculation, it follows that

$$
-2\phi SX - [-q(X) + 2\alpha\eta(X)]N + 2\alpha(X\alpha)\xi - \alpha S\phi SX - (X\alpha^2)\xi = 0
$$

whose tangential part gives

$$
-2\phi SX + 2\alpha(X\alpha)\xi - \alpha S\phi SX - (X\alpha^2)\xi = 0
$$

We know that, by taking inner product with ξ, coefficient of Reeb vector field vanishes and thus we are left with

$$
\phi SX = -\left(\frac{\alpha}{2}\right)S\phi SX \tag{11}
$$

Since we know that for a real hypersurface with $\mathcal{U}$-principal normal vector field, we have

$$2S\phi SX = \alpha(S\phi + \phi S)X + 2\phi X.$$

This with $SX = \lambda X$ gives

$$S\phi X = \mu\phi X, \ where \ \mu = \frac{\alpha\lambda + 2}{2\lambda - \alpha} \qquad (12)$$

Using $SX = \lambda X$ and (12) in (11), we see

$$\lambda\phi X = -\left(\frac{\alpha\lambda\mu}{2}\right)\phi X$$

or

$$\lambda = -\left(\frac{\alpha\lambda}{2}\right)\left(\frac{\alpha\lambda + 2}{2\lambda - \alpha}\right)$$

which further with $\lambda \neq 0$ yields $\alpha^2 + 4 = 0$, again which is a contradiction. This concludes that there do not exist any Hopf real hypersurface in Q^m satisfying recurrent structure Jacobi operator with $\mathcal{U}$-principal normal vector field.

5.2 $\mathcal{U}$-Isotropic Normal Vector Field

Here, we will discuss non-existence of Hopf real hypersurface in Q^m satisfying recurrent structure Jacobi operator with the unit normal vector field N is $\mathcal{U}$-isotropic. With this assumption of $\mathcal{U}$-isotropic normal vector field, structure Jacobi operator can be written as

$$R_\xi(X) = R(X, \xi)\xi$$
$$= X - \eta(X)\xi - g(AX, \xi)A\xi - g(X, AN)AN + \alpha SX - \alpha^2\eta(X)\xi. \quad (13)$$

Theorem 4 *There do not exist any Hopf real hypersurfaces in Q^m, $(m \geq 3)$ with recurrent structure Jacobi operator such that the unit normal vector field N is $\mathcal{U}$-isotropic.*

Proof Using $Y = \xi$ and (13), condition of recurrent structure Jacobi operator reduces to

$$R_\xi(\phi SX) = 0$$

which is further equivalent to

$$\phi SX - g(\phi SX, A\xi)A\xi + g(\phi SX, AN)AN + \alpha S\phi SX = 0 \qquad (14)$$

Case 1: If $\alpha = 0$, we have

$$\phi SX = g(\phi SX, A\xi)A\xi + g(\phi SX, AN)AN \tag{15}$$

On taking inner product with $A\xi$ and AN respectively, above equation gives

$$g(\phi SX, AN) = 0 \; and \; g(\phi SX, A\xi) = 0.$$

Using this in (15), we have $\phi SX = 0$ which directly conclude that $SX = \alpha \eta(X)\xi$, which means that M is totally η-umbilical. Then, the shape operator S commutes with the structure tensor ϕ. Thus, M is locally congruent to a tube over a totally geodesic CP^k in a Q^{2k}. But, they are never totally η-umbilical.

Case 2: If $\alpha \neq 0$, then applying AN to (14) gives

$$g(\phi SX, SAN) = 0$$

Let us suppose that $\lambda \neq 0$ be an eigenvalue of the shape operator corresponding to distinct eigenvectors AN, $A\xi$, i.e., $SAN = \lambda AN \; and \; SA\xi = \lambda A\xi$. Then,

$$g(\phi SX, AN) = 0 \tag{16}$$

Moreover, if we take inner product with $A\xi$, we have

$$g(\phi SX, SA\xi) = 0$$

or,

$$g(\phi SX, A\xi) = 0 \tag{17}$$

Using (16), (17) in (14), we obtain

$$\phi SX = -\alpha S\phi SX \tag{18}$$

Moreover, we have from Lemma 4.2 [7]

$$2S\phi SX = \alpha(\phi S + S\phi)X + 2\phi X - 2g(X, AN)A\xi + 2g(X, A\xi)AN. \tag{19}$$

Now, let us consider the distribution $Q^\perp$, which is an orthogonal complement of the maximal $\mathcal{U}$-invariant subspace Q in the complex subbundle C of $T_z M$, $z \in M$ in Q^m. Then, by Lemma 1 given in Sect. 3, the orthogonal complement $Q^\perp = C \ominus Q$ becomes $C \ominus Q = Span\{AN, A\xi\}$.

Then, on the distribution Q we know that $AX \in T_z M$, $z \in M$, because $AN \in Q$. So (19), together with the fact that $g(X, A\xi) = 0$ and $g(X, AN) = 0$ for any $X \in Q$, imply that

$$2S\phi SX = \alpha(S\phi + \phi S)X + 2\phi X.$$

We may put $SX = \lambda X$ in above one which results

$$S\phi X = \left(\frac{\alpha\lambda + 2}{2\lambda - \alpha}\right)\phi X. \tag{20}$$

On taking account (20) in (18) together with $SX = \lambda X$, it follows that

$$\lambda\phi X = -\alpha\lambda\left(\frac{\alpha\lambda + 2}{2\lambda - \alpha}\right)\phi X$$

Thus, we have two possibilities on λ

$$\lambda = 0 \; or \; \lambda = -\frac{\alpha}{\alpha^2 + 2}$$

If we take λ to be nonzero, then $\gamma = -\frac{\alpha}{\alpha^2+2}$ is the other principal curvature. Also, by the virtue of $SX = \gamma X$ in (18), we arrive at $S\phi X = -\frac{1}{\alpha}\phi X$. On the other hand, $S\phi X = \gamma\phi X$.

On simplifying gives, $\frac{\alpha^2}{\alpha^2+2} = 1$, which is a contradiction. This completes the theorem.

References

1. Suh, Y.J.: Real hypersurfaces in the complex quadric with parallel normal Jacobi operator. Math. Nachr. **290**(2–3), 442–451 (2017)
2. Jeong, I., Pérez, J.D., Suh, Y.J.: Recurrent Jacobi Operator of real hypersurfaces in complex two-plane Grassmannians. Bull. Korean Math. Soc. **50**(2), 525-536 (2013)
3. Smyth, B.: Differential geometry of complex hypersurfaces. Ann. Math. **85**, 246–266 (1967)
4. Reckziegel,H.: On the geometry of the complex quadric. In: Geometry and Topology of Submanifolds VIII, pp. 302–315. World Scientific Publishing, Brussels/Nordfjordeid, River Edge, NJ (1995)
5. Berndt, J., Suh, Y.J.: Contact hypersurfaces in Kähler manifold. Proc. Amer. Math. Soc. **23**, 2637–2649 (2015)
6. Klein, S.: Totally geodesic submanifolds in the complex quadric. Differential Geom. Appl. **26**, 79–96 (2008)
7. Suh, Y.J.: Real hypersurfaces in the complex quadric with Reeb parallel shape operator. Int. J. Math. **25**, 1450059, 17 pp. (2014)
8. Blair, D. E.: Contact Manifolds in Riemannian Geometry. Lecture Notes in Mathematics, vol. 509. Springer, Berlin (1976)
9. Berndt, J., Suh, Y.J.: Real hypersurfaces with isometric Reeb flow in complex quadrics. Int. J. Math. **24**, 1350050, 18 pp. (2013)
10. Perez, J.D., Santos, F.G.: Real hypersurfaces in complex projective space with recurrent structure Jacobi operator. Diff. Geom. Appl. **26**, 218–223 (2008)
11. Pérez,J.D., Santos,F.G., Suh,Y. J.: Real hypersurfaces in complex projective space whose structure Jacobi operator is D-parallel. Bull. Belgian Math. Soc. Simon Stevan **13**, 459-469 (2006)
12. Bansal, P., Shahid, M.H.: Optimization Approach for Bounds Involving Generalized Normalized δ-Casorati Curvatures, Accepted in AISC series, Springer, (2018)

Validation of Well-Known Population-Based Stochastic Optimization Algorithms Using Benchmark Functions

Byamakesh Nayak, Srikanta Kumar Dash and Jiban Ballav Sahu

Abstract Test functions are used for validation of reliability and the implementation of optimization process. There are a lot of benchmarked test functions reported in the literature. The classifications of benchmarked functions are based on steady, differentiable, unable to be separated and scaled or having one or several modes or maxima and dimensions. In this paper, the test functions are properly selected as objective functions with various properties in terms of modality, dimension, valleys, etc., for validation of well-known population-based stochastic optimization algorithms. The novelty of this paper is to strengthen the validation of the grey wolf optimization (GWO) and the particle swarm optimization (PSO) algorithms using ten well-known benchmark functions.

Keywords Validation · Grey wolf optimization (GWO) · Particle swarm optimization (PSO) · Benchmark functions

1 Introduction

There are various areas in the fields of engineering, medicines, science, economics, etc., where optimization is needed by proper selection of objective functions. The objective functions may be unitary, multiple, unconstrained, constrained with different dimensions. It may also be unimodal, multimodal, continuous differentiable and valley landscape. The selection of optimization algorithm is also one of the major tasks in research. Wrong selection of optimization theorem may give poor result by falling local maxima or minima instead of global one. This demands the test of relia-

B. Nayak (✉) · S. K. Dash · J. B. Sahu
School of Electrical Engineering, KIIT University, Bhubaneswar, India
e-mail: electricbkn11@gmail.com

S. K. Dash
e-mail: srikantdash12@gmail.com

J. B. Sahu
e-mail: jibanballav.sahu@gmail.com

© Springer Nature Singapore Pte Ltd. 2019
J. C. Bansal et al. (eds.), *Soft Computing for Problem Solving*,
Advances in Intelligent Systems and Computing 816,
https://doi.org/10.1007/978-981-13-1592-3_58

bility, efficiency and validation of proposed optimization algorithm by test functions which must be diverse and unbiased. Jamil and Yang presented 175 test functions in review literature [1]. The grouping of 175 test functions is based on modality, basin, valley, dimensions, and flat surface, area of global minima or maxima and separability.

An objective function with more than one local maximum or minimum point is called multimodal function. Multimodal test functions are used to test optimization algorithms to escape from any local minimum. For highly nonlinear problems and with increase in dimension, the search space increases exponentially. This creates significant barrier for poorly designed exploration process of optimization algorithms [2, 3]. A steep decline of objective function surrounded by large area is called basin. Plateaus correspond to steep incline for searching of global maxima point. The poorly designed exploitation principle of optimization algorithms can be easily attracted to such regions due to need of information and gives poor results [4]. A valley is defined as a constricted region of small alteration which is bounded by regions of sharp decline. Like basin, the floor of the valley slows down the search process of an algorithm and initially attracted towards this region [4]. Separability is another issue for failure of optimization algorithms. If all the parameters or variables of objective functions are independent, then solving of optimization algorithms are easy which results error free global minimum. Dependency of variables may prone to fall in local minimum [5]. The increase of dimensions of the search space of non-separable function further creates detrimental effect of searching procedures of algorithms. Objective functions with plane surface render a complexity of the algorithms for its search process towards minima [6].

Population-based stochastic optimization algorithms searched the global minima or maxima by assuming the problem as black box. It monitors the output of the black box by feeding the input. Due to this, no derivation is required for determining the global optimum point. Therefore, this can be applied to any field with high flexibility [7]. Population-based meta-heuristics follow certain rules. The rules are based on exploration and exploitation [8–11]. The advantages compared to single-based solution are sharing the information to just best positions and assisting each other to avoid falling in local minima. Meta-heuristics may be classified into three main categories: evolutionary, physics based and SI algorithms. One of interesting branches is the swarm intelligence (SI) originates from natural colonies. Some of the most popular SI techniques are ant colony optimization (ACO) [12], particle swarm intelligence (PSO) [13] and artificial bee colony (ABC) [14]. Besides SI techniques, grey wolf optimization and Whale Optimization Algorithms are found in the literature for different optimization problems [15–17]. This paper presents the validation of reliabilities and efficiencies of PSO and GWO by selections of recent benchmark objective functions with diverse effects. Section 2 presents the mathematical expression of benchmark functions which are used for validation of algorithms. Section 2 explains the brief explanation of algorithms taken for validation. The outcome and discussion are explained in Sect. 3 followed by Conclusion in Sect. 4.

2 Benchmark Functions

The test functions used for validation of PSO and GWO algorithms are collected from the literature [1]. They are briefly reviewed.

2.1. Continuous, two-dimensional, differentiable, non-scalable and non-separable

a. **Ackley 3 function (Unimodal):**

$$F_1(x) = -200e^{-0.2\sqrt{x_1^2 + x_2^2}} + 5e^{\cos(3x_1) + \sin(3x_2)}$$

The variables are subjected to $-32 \leq x_i \leq 32$. The location of global minimum point $x_{min} = (0, 0.4)$. The minimum value of objective function is at $F_1(x_{min}) = -219.1418$

b. **Beale function (Unimodal):**

$$F_2(X) = (1.5 - X_1 + X_1 X_2)^2 + (2.25 - X_1 + X_1 X_2^2)^2 + (2.625 - X_1 + X_1 X_2^3)^2$$

subject matter to $-4.5 \leq X_i \leq 4.5$. The overall lowest amount is located at $X_{min} = (3, 0.5)$ and minimum value $F_2(X_{min}) = 0$

c. **Booth function (Unimodal):**
$F_3(X) = (X_1 + 2X_2 - 7)^2 + (2X_1 + X_2 - 5)^2$ Subject to $-10 \leq X_i \leq 10$. The global minimum is located at $X_{min} = (1, 3)$ and minimum value $F_3(X_{min}) = 0$

d. **Adjiman function (Multimodal):**
$F_4(X) = \cos X_1 . \sin X_2 - \frac{X_1}{X_2^2 + 1}$ With limit of $-1 \leq X_1 \leq 2$ and $-1 \leq X_2 \leq 1$. The global minimum is located at $X_{min} = (2, 0.010578)$ and minimum value $F_4(X_{min}) = -2.02181$.

e. **Bartels function (Multimodal):** The equation can be formed as:
$F_5(x) = \left| x_1^2 + \frac{2}{-2} + x_1 x_2 \right| + |\sin(x_1)| + |\cos(x_2)|$ with a limit of $-500 \leq x_i \leq 500$. The overall lowest amount is located at $x_{min} = (0, 0)$ and lowest value $F_5(x_{min}) = 1$.

f. **Bohachevsky 2 function (Multimodal):** The multimodal equation can be written as
$F_6(x) = x_1^2 + 2x_2^2 - 0.3\cos(3.pi.x_1).0.4\cos(4.pi.x_2) + 0.3$ with matter to limit $-100 \leq x_i \leq 100$. The overall lowest amount is located at $x_{min} = (0, 0)$ and lowest value $F_6(x_{min}) = 0$.

g. **Branin Rcos function (Multimodal):** This function possesses three global minima and can be formed as:
$F_7(X) = \left(X_2 - \frac{5.1 X_1^2}{4.\pi^2} + \frac{5 X_1}{\pi} - 6 \right)^2 + 10\left(1 - \frac{1}{8\pi}\right)\cos(X_1) + 10$ With matter to limit $5 \leq X_1 \leq 10$ and $0 \leq X_2 \leq 15$. The global minimums are located at $X_{min} = (-\pi, 12.275), (\pi, 2.275)$ and $(3\pi, 2.245)$ and minimum value $F_7(X_{min}) = 0.3978873$.

h. **Bird function (Multimodal):** This function has two global minima and can be expressed as:

$F_8(x) = \sin x_1 e^{(1-\cos x_2)^2} + \cos x_2 e^{(1-\sin x_1)^2} + (x_1 - x_2)^2$ with matter to limit $-2\pi \leq x_i \leq 2\pi$. The overall lowest amount is located at $x_{min} = (4.7010, 3.1529)$ and $(-1.5821, -3.1302)$ and minimum value $F_8(x_{min}) = -106.764537$.

i. **Camel function six-hump function (Multimodal)**: This function has two global minima and can be expressed as:
$F_9(x) = \left(4 - 2.1x_1^2 + \frac{x_1^4}{3}\right)x_1^2 + x_1 x_2 + \left(4x_2^2 - 4\right)x_2^2$ with subject to *constrain* $-5 \leq x_i \leq 5$. *The overall lowest amount are located at* $x_{min} = (-0.0898, 0.7126)$ and $(0.0898, -0.7126)$ *and minimum value* $F_9(x_{min}) = -1.0316$.

j. **Chung Reynold function:** This function is continuous, multidimensional, differentiable, scalable, partially separable and unimodal. It is represented as:

$$F_{10}(x) = \{\sum_{i=1}^{D} x_i^2\}^2$$

Subjected to $-100 \leq x_i \leq 100$. *The global minimum* $(0, 0, 0 \ldots)$ *with* $F_{10}(x_{min}) = 0$.

2.1 Grey Wolf Optimizer (GWO)

Mirjalili et al. [16, 17] proposed an algorithm based on the hunting method of grey wolves who lives in a group (pack), containing of 5–12 members on average. Their social hierarchy has different levels. The first levels are the leaders called the alpha wolves. They are the decision-makers of the family. The second levels are called beta (β), which can be either male or female, helps in decision-making for the family. The beta reinforces the alpha's commands and gives the feedback to alpha. There different steps adopted by grey wolves for hunting are to locate the prey followed by run after the prey. These steps are followed by encircling and hunting of the prey. The mathematical expressions are

$$\vec{D} = \left|\vec{C}\vec{X_p(t)} - X(t)\right| \tag{1}$$

$$X(t+1) = X_P(t) - \vec{A}.\vec{D} \tag{2}$$

Here X_P is the location vector of the prey, X denote the location vectors of the grey wolf and t denote the recent iteration and $\vec{A}$, $\vec{C}$ are the coefficient vectors represented as:

$$\vec{A} = 2\vec{ar_1} - \vec{a} \tag{3}$$

$$\vec{C} = 2\vec{r_2} \tag{4}$$

here $\vec{a}$ decreases linearly from 2 to 0 during iteration and r_1, r_2 are arbitrary vectors in between [0, 1] and allows to reach any point a. It is assumed that α, β and δ have better information about the location of prey and are the best search agents. The hunting optimization is given by the following equations are formulated for updating the position

$$D_\alpha = \left| C_1 \vec{X}_\alpha - \vec{X} \right| \tag{5}$$

$$D_\beta = \left| C_2 \vec{X}_\beta - \vec{X} \right| \tag{6}$$

$$D_\delta = \left| C_3 \vec{X}_\delta - \vec{X} \right| \tag{7}$$

$$\vec{X_1} = \vec{X_\alpha} - \vec{A_1} \vec{D_\alpha} \tag{8}$$

$$\vec{X_2} = \vec{X_\beta} - \vec{A_2} \vec{D_\beta} \tag{9}$$

$$\vec{X_3} = \vec{X_\delta} - \vec{A_3} \vec{D_\delta} \tag{10}$$

$$\vec{X_{t+1}} = \frac{X_1 + X_2 + X_3}{3} \tag{11}$$

The mathematical modelling of optimization is carried out by linearly decline in each iteration. The $\vec{A}$ will take random variables between $[-2a, 2a]$ which is utilized by the search agent to move towards prey. When $|A| < 1$, convergence occurs and when $|A| > 1$ divergence occur.

2.2   Particle Swarm Optimization (PSO)

PSO is an efficient stochastic population-based method-based optimization proposed by Kennedy and Eberhart [18]. This algorithm works by maintaining several candidate solutions in the search space at the same time. In comparison with evolutionary-based methods, this method remembers the best solutions and the individual solutions and it can escape local maxima. The best solution achieved among all particles is the best solutions. The PSO algorithm works in three simple steps, i.e. fitness estimation by giving an objective function, updating of values of individual and global best fitness solution and position, updating velocity and position of each particle. These steps are repeated until a best solution is achieved. The velocity of each particle is given by the following equation:

$$V_i(t+1) = w\, V_i(t) + c_1 r_1 [x_1(t) - x_i(t)] + c_2 r_2 \big[g(t) - x_i(t)\big] \tag{12}$$

$$x_i(t+1) = x_i(t) + V_i(t+1) \tag{13}$$

The value of w, c_1 and c_2 in Eq. (12) is given by the user. The values of w lie between [0, 1.2]. r_1,r_2 are random values generated by each iteration update, and its values lie between [0, 1] and c_1, c_2 lie between [0,2]. If i is the index of each particle $x_i(t)$ and $V_i(t)$ is the location and velocity of the element at time t.

The first term w $V_i(t)$ of Eq. (12) keeps the particle moving in the direction of search and is known as inertia component 'w' is the inertia coefficient. The inertia component is responsible for increasing or decreasing the speed of search. The term $c_1r_1[x_1(t) - x_i(t)]$ is responsible for memory, and it is known as the cognitive coefficient. The term $c_2r_2[g(t) - x_i(t)]$ causes the move towards the best region found so far, and it is known as social component. In order to define the search space, velocity clamping is done where a range of $[-V_{max}, +V_{max},]$ and a search space $[-x_{max}, +x_{max}]$ is provided where $V_{max} = kx_{max}$ where k is the velocity clamping factor and lies between 0 and 1. Once the pace of each element is considered, each element location is updated by applying Eq. (13). This process is continued until the best solution is achieved. The entire process of GWO and PSO is represented in flow charts.

2.2.1 Flow Chart of GWO

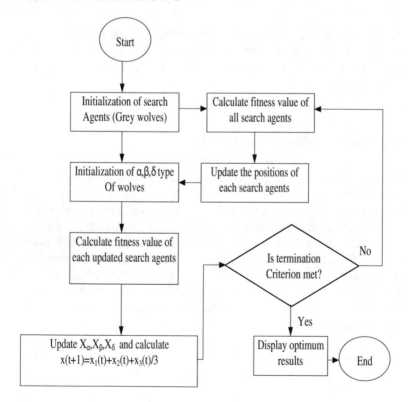

2.2.2   Flow Chart of PSO

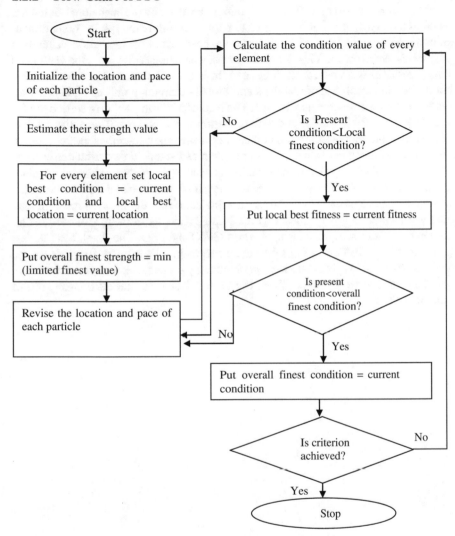

3   Result and Discussion

The GWO algorithm was benchmarked by using 29 benchmark functions [16] In this paper, ten different benchmark functions are considered for strengthening the validation of PSO and GWO and to their efficiency and strength are compared by taking unconstrained benchmark function into consideration. This paper uses the extra ten different benchmark functions with various characteristics in terms of modality, dimensions and valleys for strengthening the validation of PSO and GWO. The population size considered here is 30, and the numbers of iteration is 500. The first nine functions, i.e. $F_1(x)$ to $F_9(x)$ considered above are two-dimensional in nature, whereas the $F_{10}(x)$ is multidimensional in nature. In simulation results, eight dimensions are considered, but the behaviour of two positions of all population as iteration progresses, are shown in Figs. 1 and 2. For each benchmark function, the GWO and PSO algorithms were run for ten times and the best results were taken into consideration. The statistical data of positions and global minimum points of all benchmark functions are reported given in Table 1. The tests of performance were carried out for both GWO and PSO, and the standard deviation was found out. The statistical data of positions and global minimum points of all benchmark functions are reported in Table 1. Low statistical values in term of mean and standard deviation confirm the accurate optimization value if the function is converged. A lower value of standard deviation in the table indicates that the results tend to be close to the expected value, while a higher value of standard deviation indicates that the information is spread out over a wider range of values. Figures 1 and 2 show the convergence curve of all the ten optimization algorithms applied on GWO and PSO. The graph has the best performance of GWO, and PSO for the corresponding benchmark function can be analysed from tables. The tests were performed by proper selection of test function, and the results were found out. From Fig. 2, it can be seen that the convergence in case of GWO is faster.

Table 1 Comparative analysis of GWO and PSO

Function	Function and position value	Values given in [1]	GWO	PSO	Average (PSO)	Average (GWO)	Standard deviation (PSO)	Standard deviation (GWO)
F1	F1	−219.1418	−194.392	−194.3924	−194.3883	−194.3761	0.0407	0.3504
	X(1)	0	0.7758	0.7756	−0.7739	−0.7757	0.0137	0.1026
	X(2)	0.4	−0.7755	−0.7750	−0.7631	−0.7693	0.1479	0.1311
F2	F1	0	3.9354*e−08	10−30	0.0074	0.0029	0.1089	0.0742
	X(1)	3	2.9996	3	2.9872	2.9928	0.1045	0.0524
	X(2)	0.5	0.49994	0.5	0.4970	0.4983	0.0371	0.0096
F3	F1	0	$6 * 10^{-7}$	00	0.0123	0.0160	0.1392	0.4328
	X(1)	1	1.001	1.0	0.9917	0.9891	0.0373	0.0725
	X(2)	−3	3.003	3.0	3.0082	3.0092	0.0544	0.0315
F4	F1	−2.02181	−2.0218	−1.7504	−1866	−2	1066	0
	X(1)	2	2	1.7512	1869	2	1065	0
	X(2)	0.010578	0.10578	0.0211	0	0.1	0	0
F5	F1	1	1	1	251.6905	13.4495	1.7109e3	0.3258e3
	X(1)	0	3.5963e−17	0	−2.2220	−0.0256	0.0141e3	0.0021e3
	X(2)	0	1.2025e−08	0	2.3941	−0.0902	0.0163e3	0.0022e3
F6	F1	0	−0.12	−0.12	1.5517	0.1452	27.3720	7.5261
	X(1)	0	$2 * 10^{-8}$	0	−0.0362	−0.0198	0.6493	0.4559
	X(2)	0	$-3 * 10^{-10}$	0	0.0288	0.0072	0.7887	0.1674
F7	F1	0.397833	0.39789	0.3979	0.4049	0.4053	0.0888	0.1189
	X(1)		3.1418	−3.1416	3.1482	3.1441	0.1793	0.1969
	X(2)	12.275	2.2758	12.2750	2.2775	2.2807	0.0549	0.0407
F8	F1	−106.7645	−106.7645	−106.7645	−106.5942	−106.5969	1.1786	0.8571
	X(1)	−1.58214	−1.5817	4.7010	4.7009	4.6990	0.0201	0.2002
	X(2)	−3.13024	−3.1305	3.1529	3.1600	3.1549	0.0293	0.0241
F9	F1	−1.0316	−1.0316	−1.0316	−1.0316	−1.0283	0.0014	0.0724
	X(1)	0.7126	−0.08981	0.0898	−0.0891	0.0908	0.0097	0.0210
	X(2)	−0.7126	0.71262	−0.7127	0.7087	−0.7119	0.0748	0.0465
F10	F1	0	0	0	7.4644e5	1.7107e5	9.8256e6	3.7800e6
	X(1)	0	NA	NA	NA	NA	NA	NA
	X(2)	0	NA	NA	NA	NA	NA	NA

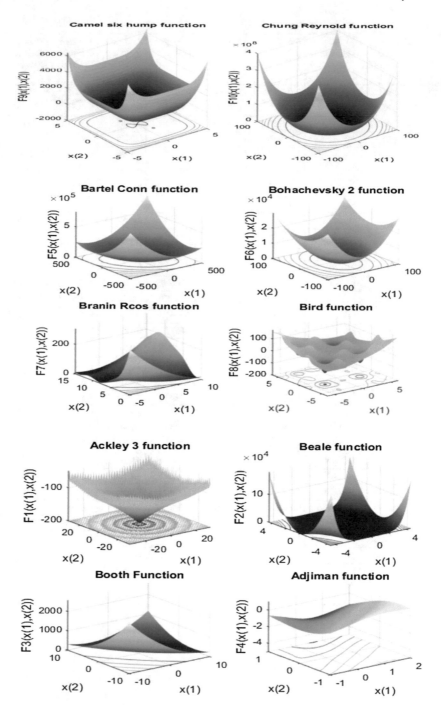

Fig. 1 Behavior of benchmark functions

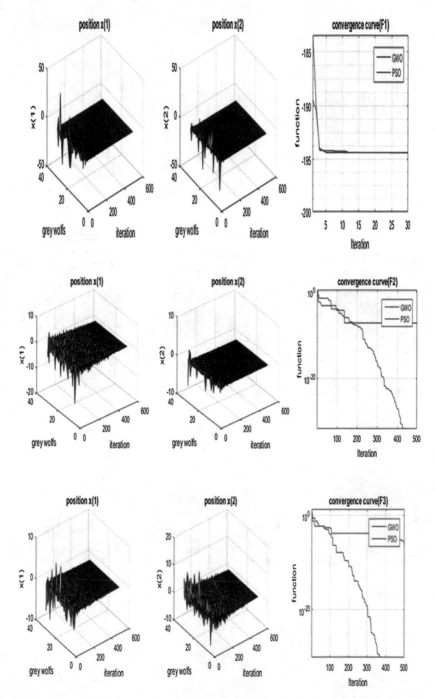

Fig. 2 Position of GWO and convergence curve of PSO and GWO of benchmark functions

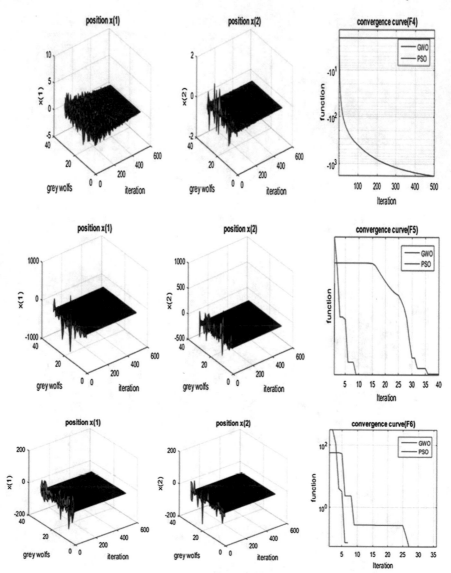

Fig. 2 (continued)

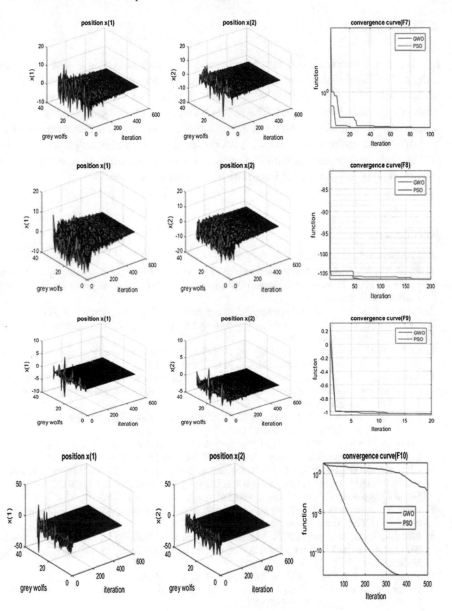

Fig. 2 (continued)

4 Conclusion

In this paper, a set of ten standard benchmarks functions are taken to test the reliability and efficiency of optimization algorithms. We have studied the validation and performance of GWO and PSO with a population size of 30, and the numbers of iterations taken is five hundred. The average values and the standard deviation for each benchmark functions shown in Table 1 and the performance curve are plotted. The time of computation in case of GWO is less and approximately around 3 s in all functions in comparison with PSO where the computation time is 4.5 s. GWO converges with less number of iterations in comparison with PSO.

References

1. Jamil, M., Yang, X.-S.: A literature survey of benchmark functions for global optimization problems. Int. J. Math. Model. Numer. Optim. **4**(2), 150–194 (2013). https://doi.org/10.1504/ijmmno.2013.055204
2. Winston, P.H.: Artificial Intelligence, 3rd edn. Addison-Wesley, Wokingham (1992)
3. Yao, X., Liu, Y.: Fast evolutionary programming. In: Proceedings of 5th Conference on Evolutionary Programming (1996)
4. Chung, C.J., Reynolds, R.G.: CAEP: an evolution-based tool for real-valued function optimization using cultural algorithms. Int. J. Artif. Intell. Tool **7**(3), 239–291 (1998)
5. Salomon, R.: Re-evaluating genetic algorithm performance under coordinate rotation of benchmark functions: a survey of some theoretical and practical aspects of genetic algorithms. Bio Syst. **39**(3), 263–278 (1996)
6. Boyer, D.O., Martfnez, C.H., Pedrajas, N.G.: Crossover operator for evolutionary algorithms based on population features. J. Artif. Intell. Res. **24**, 1–48 (2005)
7. Droste, S., Jansen, T., Wegener, I.: Upper and lower bounds for randomized search heuristics in black-box optimization. Theory Comput. Syst. **39**, 525–544 (2006)
8. Olorunda, O., Engelbrecht, AP.: Measuring exploration/exploitation in particle swarms using swarm diversity. In: IEEE Congress on Evolutionary Computation, 2008. CEC 2008 (IEEE World Congress on Computational Intelligence). pp. 1128–1134 (2008)
9. Alba, E., Dorronsoro, B.: The exploration/exploitation tradeoff in dynamic cellular genetic algorithms. IEEE Trans. Evol. Comput. **9**, 126–142 (2005)
10. Mirjalili, S., Mohd Hashim, S.Z., Moradian, Sardroudi H.: Training feed forward neural networks using hybrid particle swarm optimization and gravitational search algorithm. Appl. Math. Comput. **218**, 11125–11137 (2012)
11. Holland, J.H.: Genetic algorithms. Sci. Am. **267**, 66–72 (1992)
12. Dorigo, M., Birattari, M., Stutzle T.: Ant colony optimization. In: Computational Intelligence Magazine, vol. 1, pp. 28–39. IEEE (2006)
13. Kennedy, J., Eberhart, R.: Particle swarm optimization. In: Proceedings, IEEE International Conference On Neural Networks. pp. 1942–1948 (1995)
14. Basturk, B., Karaboga, D.: An artificial bee colony (ABC) algorithm for numeric function optimization. In: IEEE Swarm Intelligence Symposium, pp. 12–14 (2006)
15. Mech, L.D.: Alpha status, dominance, and division of labor in wolf packs. Can. J. Zool. **77**, 1196–1203 (1999)
16. Mirjalili, S., Mirjalili, S.M., Lewis, A.: Grey wolf optimizer. Adv. Eng. Softw. **69**, 46–61 (2014)
17. Mirjalili, S., Lewis, A.: The whale optimization algorithm. Adv. Eng. Softw. **95**, 51–67 (2016)
18. Kennedy, J., Eberhart, R.: Particle swarm optimization. In: Proceedings of the IEEE International Conference on Neural Networks, Perth, Australia, December 1995, pp. 1942–1948

Regularized Artificial Neural Network for Financial Data

Rajat Gupta, Shrikant Gupta, Muneendra Ojha and Krishna Pratap Singh

Abstract The paper deals with the application of artificial neural network on financial data. We applied different activation functions in hidden layer with regularization to overcome the problem of overfitting. We present a comparative analysis of all combinations of activation functions and regularizations applied on BSE Sensex and Nifty 50 dataset containing the stock indices of last 7 years.

Keywords Artificial neural network · Time series data · Regularization
Optimization · Gradient descent · Backpropagation

1 Introduction

Artificial neural network (ANN) achieved massive success in the field of machine learning and is applied to wide range of domains such as medical, financial, image processing, recognition, speech processing, engineering applications. A time series is a series of data points indexed (or listed or graphed) in time order. Most commonly, a time series is a sequence taken at successive equally spaced points in time. Thus, it is a sequence of discrete time data. Financial data are the type of time series data.

R. Gupta · S. Gupta · M. Ojha
International Institute of Information Technology, Naya Raipur, Naya Raipur, India
e-mail: rajat15101@iiitnr.edu.in

S. Gupta
e-mail: shrikant15101@iiitnr.edu.in

M. Ojha
e-mail: muneendra@iiitnr.edu.in

K. P. Singh (✉)
Indian Institute of Information Technology, Allahabad, Allahabad, India
e-mail: kpsingh@iiita.ac.in

© Springer Nature Singapore Pte Ltd. 2019
J. C. Bansal et al. (eds.), *Soft Computing for Problem Solving*,
Advances in Intelligent Systems and Computing 816,
https://doi.org/10.1007/978-981-13-1592-3_59

1.1   Related Work

Many researchers have worked with ANNs on different sets of problems that can be found in the literature. Li et al. [1] performed an empirical study on predictive time series modeling using ANN for Linac beam symmetry. In this work, artificial neural network and autoregressive moving average (ARMA) time series prediction modeling techniques were applied to 5 year daily Linac QA, and it was found that ANN modeling has advantages over ARMA technique. Sen et al. [2] in their paper discussed decomposition of time series data. The obtained results demonstrated the accuracy of decomposition results and efficiency of the proposed forecasting techniques. Ticknor [3] suggested Bayesian regularization on a simple feed-forward neural network where weights are considered random variables that need to be regularized using density function. Yan [4] in his paper described Bayesian regularization neural network for AI optimization.

Wang et al. [5] proposed wavelet de-noising-based backpropagation (WDBP) neural network for stock market predictions. They decomposed original data into several layers via wavelet transform and then established a backpropagation neural network model using the low-frequency signal of every layer for prediction. Guresen et al. [6] in their paper evaluated the effectiveness of neural network models which were known to be dynamic and effective in stock market predictions. The models analyzed are multilayer perceptron (MLP), dynamic artificial neural network (DAN2), and the hybrid neural networks which use generalized autoregressive conditional heteroscedasticity (GARCH) to extract new input variables. The results show that classical ANN model MLP outperforms DAN2 and GARCHMLP with a little difference. Chen et al. [7] proposed two-layer neural network for HFF. It hierarchically explores momentum signals from selected technical indicators. They showed that using the deep neural network is better than using single-layer neural network or any other machine learning algorithm. Patel et al. [2] in their work compared four machine learning algorithms in stock prediction: random forest, SVM, single-layer neural network, and Naïve Bayes, and suggested that random forest algorithm was the best performer and Naïve Bayes was worst.

In this paper, we have done a complete comparative analysis of multilayer artificial neural network with L1 and L2 regularization. Additionally, a different combination of activation functions such as tan hyperbolic, rectified linear unit, and linear activation is applied in the ANN for financial data of last 7 years stock indices of BSE Sensex and Nifty 50 and suggested the better model.

2   Artificial Neural Network

Artificial neural network or neural network is computing system which is inspired by the biological neural network in human brain where each neuron is capable to

Fig. 1 Schematic diagram of a feed-forward backpropagation artificial neural network

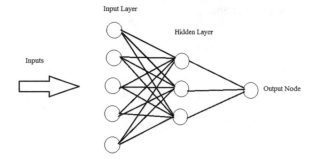

process incoming data (performing matrix operations) and is interconnected with each other forming a complete network capable of implementing complex functions.

2.1 Neural Network

A sequence of neurons connected with other neurons model a complex network which is capable of approximating a function is known as an artificial neural network. Artificial neurons mimic the biological neurons. Each neuron is able to process incoming data and when input exceeds a certain threshold, the neuron fires up depending on the activation function applied to them. An artificial neuron can accept a weighted sum of data having N-dimensions. Mathematically, consider a weight vector w, an input data vector x, and let b be the scalar unit representing bias in the system, then,

$$o = \sum_{i=0}^{n} w_i x_i + b_i \tag{1}$$

A neural net now composes an arbitrary number of layers of artificial neurons with each layer consisting of an arbitrary number of artificial neurons and data flow between them.

In Fig. 1, you can see the structure of feed-forward neural network with single hidden layer and one output layer with single node indication. This network will perform regression.

2.2 Activation Function

An activation function is a function which gives output if the input value is above the threshold. There are several activation functions suggested in the literature based on the type of data it is applied onto. The ones used in this paper are listed below.

- Tan hyperbolic

$$f(x) = \frac{1 - e^{-2x}}{1 + e^{-2x}} \tag{2}$$

It gives output in the range $(-1, 1)$.

- Rectified linear unit (ReLU)

$$f(x) = f(x) = \begin{cases} 0, & x < 0 \\ x, & x \geq 0 \end{cases} \tag{3}$$

Range of output is $[0, \infty)$.

- Linear activation

$$f(x) = \begin{cases} -x & x < 0 \\ 0 & x = 0 \\ x & x > 0 \end{cases} \tag{4}$$

Linear activation function has output in the range $(-\infty, \infty)$.

Depending upon the type of data, statement of problem, and normalization of dataset, we need to apply activation function.

2.3 Loss Function

Loss function is calculated between predicted output and actual output. We then try to minimize this cost by adjusting weights and biases in the network. Selection of the cost function is also an important step as it differs from problem point of view. If performing regression, then most suitable to use is mean squared error, and for classification, generally cross-entropy is used. In this paper, we have used mean squared error (MSE). In mean squared error, we find the difference between predicted value and actual value and then square the term and find mean of total loss. Mathematically for neural network, it is written as:

$$\text{loss}(Y, Y) = \frac{1}{2n} \sum_{i=1}^{N} (Y_i - Y_i)^2 \tag{5}$$

2.4 Overfitting and Regularization

Overfitting occurs when model fits the training data so accurately that it is not able to classify unknown situations. Loss of testing will be more than that of training. In order to overcome the problem of overfitting, we apply regularization. In regularization, we penalize the cost function by adding a regularization term to it and form a new cost function which is to be minimized by the optimizer. There are two type of regularizations considered in this paper:

- L1 Regularization

In L1 regularization, we add sum of absolute value of weights to penalize cost function along with regularization term λ,

$$\text{loss}(Y, \text{Y}) = \frac{1}{2n} \sum_{i=1}^{N} (Y_i - Y_i)^2 + \lambda \sum |W_i| \tag{6}$$

- L2 Regularization

In L2 regularization, we add sum of square of weights to penalize cost function along with regularization term λ,

$$\text{loss}(Y, \text{Y}) = \frac{1}{2n} \sum_{i=1}^{N} (Y_i - Y_i)^2 + \lambda \sum (W_i)^2 \tag{7}$$

Ng [8] in his paper compared L1 and L2 regularization and showed L1 regularization of parameters grows logarithmically in number of irrelevant features. While for L2 regularization with SVM and neural networks, worst-case sample grows linearly in number of irrelevant features.

2.5 Dataset

The dataset obtained from Yahoo Finance [9] contains 7 years of record of BSE Sensex and Nifty 50 stock indices. It is necessary before training that data are pre-processed and normalized. There are 1732 and 1789 records present in BSE Sensex and Nifty 50 dataset, respectively. Daily trading details are available out of which 14 entries were null and removed from BSE Sensex dataset and 158 entries were null and removed from Nifty 50 dataset. Features were normalized such that they have zero mean and unit variance. Target values are normalized according to activation function used in the neural network. In our case, we have used ReLU, Tanh, and linear activation. For target, the values are normalized in the range [0, 1].

Fig. 2 MSE versus epoch

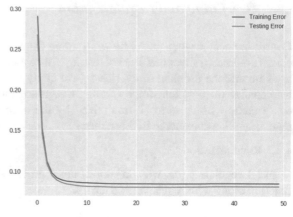

Fig. 3 Performance of ANN
with one hidden layer on
BSE Sensex data

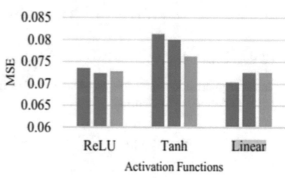

3   Results and Discussion

We ran our model on 60% of data and validated it on 40% of test data. We performed all combination of activation functions and applied both regularizations L1 and L2. We conducted experiments in two stages. In first stage, we trained our model without applying any regularization technique and tested results with our test data. While in second stage we applied different regularization while training our model and again tested with our test data.

Figure 2 presents the change in mean squared error vs. number of epochs for training and testing datasets. In Fig. 3, we observe that in the case of ReLU and Tanh activation functions regularization works better. However, with linear activation function, performance was better when no regularization was applied. No conclusive statement can be made regarding performance superiority between L1 and L2. L1 performs better in case of ReLU while L2 performs better for Tanh. Performance is almost similar for the linear function.

Fig. 4 Performance of ANN with one hidden layer on Nifty 50 data

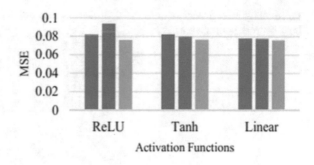

Fig. 5 BSE Sensex performance with two hidden layers

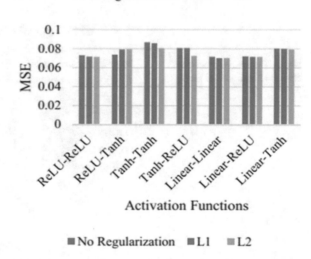

Figure 4 presents a comparative analysis of activation function and their corresponding regularization results with Nifty 50 dataset. The absence of regularization is found to be better than L1 regularization for ReLU function while L1 performs slightly better in Tanh and linear. However, L2 regularization consistently performs better in all three cases. Linear activation function, however, outperforms other activation functions in this case.

Extending our analysis of the results, we analyze ANN with two hidden layers wherein each layer has ten hidden nodes. As can be observed from Fig. 5, linear activation function in both hidden layers outperforms all other combinations of activation functions. Additionally, from the regularization point of view, L2 regularization obtained less mean squared error with linear activation closely followed by ReLU–ReLU and linear–ReLU.

For Nifty 50 dataset, we observe in Fig. 6 that neural network having linear activation function in both layers has less mean squared error than other activation functions. The linear–ReLU activation function combination closely follows behind.

Fig. 6 Nifty 50 two hidden layer network performance

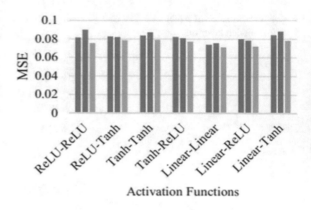

Fig. 7 Comparison between best combinations for Sensex

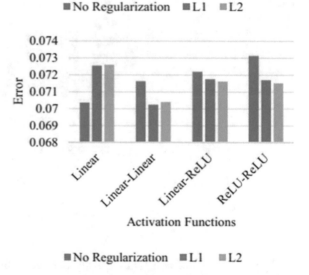

While considering regularization, we find that L2 regularization has done a better job and we can see lesser error in case of L2-regularized network having linear activation.

Now that we have seen performance of both architectures of network separately when applied to Sensex and Nifty dataset, we now take best results from both configurations and plot them together for better comparison and conclusions. We can then suggest which is better and which can be used in general. By comparing best prediction models on Sensex dataset from Fig. 7, we can say that linear activation in both hidden layers with L2 regularization outperforms every other combination followed by linear–ReLU and ReLU–ReLU in multiple hidden layer criteria. We can observe very less mean squared error. For Nifty 50, from Fig. 8, we can see that linear activation in both the hidden layers with L2 regularization has more advantage than other combination.

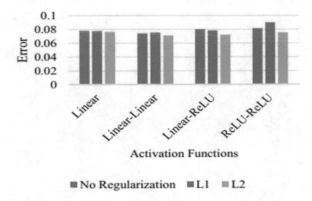

Fig. 8 Comparison between best combinations for Nifty 50

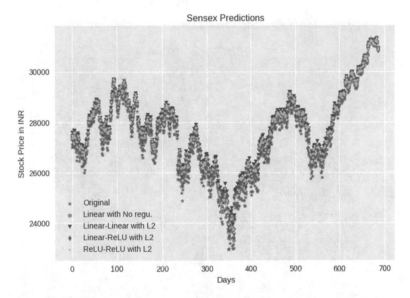

Fig. 9 BSE Sensex index prediction with all combinations of activation function and regularization techniques

From the above predictions of both Nifty 50 and BSE Sensex, we can see that our predictions are pretty good and having very less error. We can see Fig. 9 scatter plot prediction for Sensex dataset encompassing all variations of experiment and similarly for Fig. 10 which is scatter plot of prediction for Nifty 50 dataset. Figure 11 is best prediction so far with linear activation in both layers and having L2 regularization for BSE Sensex dataset. In Fig. 12, we can observe that prediction of our model for Nifty 50 is best. Overall best activation function we observed is linear activation and regularization method is L2 if overfitting occurs in case of time series data like stock market.

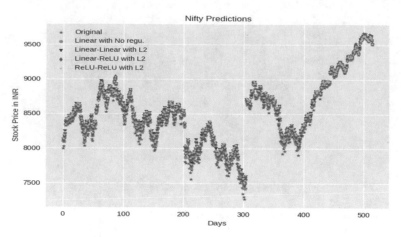

Fig. 10 Nifty 50 predictions with different activation

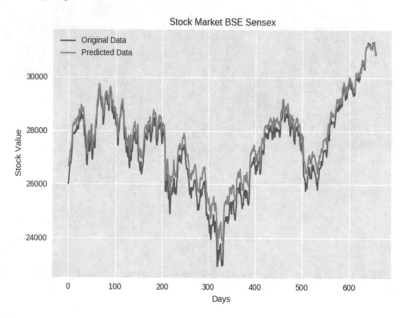

Fig. 11 BSE Sensex with linear activation in both hidden layer and L2 regularization

4 Conclusion

We described the theory behind artificial neural network, activation functions applied in layers, and regularization techniques for preventing overfitting. In our work, we find L2-regularized artificial neural network having linear activation functions with two hidden layers performs better in financial datasets used. Future work includes

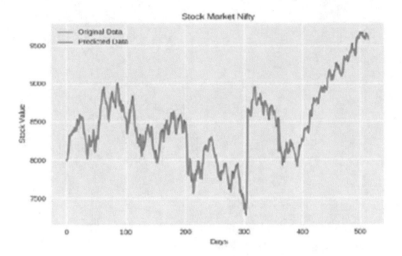

Fig. 12 Nifty 50 with linear activation in both layers and L2 regularization

a comparison of other regularization methods like Bayesian regularization, dropout, and drop connect for time series data.

References

1. Li, Q., Chan, M.F.: Predictive time-series modeling using artificial neural networks for Linac beam symmetry: an empirical study. Ann. N. Y. Acad. Sci. **1387**(1), 84–94 (2017)
2. Sen, J., Chaudhuri, T.D.: Decomposition of time series data of stock markets and its implications for prediction: an application for the indian auto sector (2016)
3. Ticknor, J.L.: A Bayesian regularized artificial neural network for stock market forecasting. Expert Syst. Appl. **40**(14), 5501–5506 (2013)
4. Yan, D., Zhou, Q., Wang, J., Zhang, N.: Bayesian regularisation neural network based on artificial intelligence optimisation. Int. J. Prod. Res. **55**(8), 2266–2287 (2017)
5. Wang, J.-Z., Wang, J.-J., Zhang, Z.-G., Guo, S.-P.: Forecasting stock indices with back propagation neural network. Expert Syst. Appl. (2011)
6. Guresen, E., Kayakutlu, G., Daim, T.U.: Using artificial neural network models in stock market index prediction. Expert Syst. Appl. **38**(8), 10389–10397 (2011)
7. Chen, H., Xiao, K., Sun, J., Wu, S.: A double-layer neural network framework for high-frequency forecasting. ACM Trans. Manage. Inf. Syst. **7**(4), 1–17 (2017)
8. Ng, A.Y.: Feature selection, L1 versus L2 regularization, and rotational invariance. In: Twenty-First International Conference on Machine learning—ICML'04, p. 78 (2004)
9. Yahoo Finance World Indices. https://in.finance.yahoo.com/world-indices/

A Multi-scale Convolutional Neural Network Architecture for Music Auto-Tagging

Tanmaya Shekhar Dabral, Amala Sanjay Deshmukh and Aruna Malapati

Abstract The application of deep neural networks, particularly convolutional neural networks, in the field of music auto-tagging has been gaining traction in recent times. These deep networks relieve the engineers from the burden of handcrafting domain-specific features. However, musical features often show great temporal diversity which traditional deep networks are unable to capture. Keeping this in mind, we propose a convolutional neural network architecture which attempts to learn features over multiple timescales. The architecture runs multiple convolutions over various subsampled versions of the original audio spectrogram. These convolution streams are then concatenated to make the tag predictions. We evaluate the architecture on the MagnaTagATune dataset, and we show that the proposed architecture yields results close to the state of the art and comprehensively beats shallow classifiers trained on handcrafted features.

1   Introduction

Traditionally, content-based information retrieval and classification poses the difficult challenge of extracting domain-specific features, which is often the first step in the pipeline. These extracted features are then fed into a classifier which performs the required task. Such features are typically handcraft; that is, the algorithm to derive these features from the raw data is designed by humans using some domain-specific knowledge. In the context of music information retrieval (MIR), mel-frequency cepstral coefficients [1] (MFCCs) provide a good example of such features. MFCCs, designed for speech processing, try to accommodate the nonlinear response of the human auditory system on the frequency scale. Engineering these features is often a tedious and time-consuming process. Moreover, these features are often specific to a particular task and do not easily generalize to other problems.

T. S. Dabral (✉) · A. S. Deshmukh · A. Malapati
Birla Institute of Technology and Science, Pilani, Hyderabad Campus, Hyderabad, India
e-mail: tanmaya.dabral@gmail.com

© Springer Nature Singapore Pte Ltd. 2019   757
J. C. Bansal et al. (eds.), *Soft Computing for Problem Solving*,
Advances in Intelligent Systems and Computing 816,
https://doi.org/10.1007/978-981-13-1592-3_60

In recent years, a paradigm of automatic feature learning is gaining traction as a concomitant of deep learning [2]. Deep learning involves learning a stack of features of increasing abstraction levels, with the lowest abstraction level being the raw, largely unprocessed data, and the highest being the desired classification. Each one of these levels is automatically learned as a piece-wise continuous function of the previous level using a gradient descent technique [3]. Such a shift in paradigm is especially evident in multimedia processing fields like computer vision and speech processing, where deep learning techniques such as recurrent neural networks (RNNs) and convolutional neural networks (CNNs) have shown great results [4, 5]. Similar approaches have been advocated in the field of MIR as well [6]. RNNs, CNNs or an amalgamation of the two [7] are becoming increasingly popular architectures for music classification or tagging tasks. In fact, RNNs have also been used to generate aesthetically pleasing music with moderate success [8].

Another paradigm generating interest, specifically in the field of MIR, is that of multi-scale feature extraction [9]. The features extracted over different temporal scales are often able to capture nuances that are not evident over one particular timescale. It is not unreasonable to assume, for example, that the time signature of a song is a feature of a relatively small timescale, while the overall mood of the song pertains to a relatively larger timescale.

In this paper, we present a novel deep architecture which, apart from learning, features in a supervised manner also exploits the multi-scale nature of musical data. We have tested the proposed architecture on the MagnaTagATune dataset [10], a popular dataset for testing automatic music tagging models. The rest of the paper follows the following structure: In Sect. 2, we present an overview of related past work. In Sect. 3, we present the motivation for using CNNs for MIR. In Sect. 4, we introduce our proposed architecture. In Sect. 5, we describe the experiments performed and the results obtained. Finally, we conclude our work in Sect. 6.

2 Related Work

Although automatic music tagging is a relatively young field, a substantial amount of work has been done in this direction. This section aims to review the literature relevant to automatic feature learning and multi-scale approaches, especially in the field of music auto-tagging.

Unsupervised feature learning using techniques like the k-means clustering algorithm has yielded good results in spite of the simplicity. Wulfing et al. [11] used the k-means clustering algorithm to learn a dictionary of features in an unsupervised manner. Nam et al. [12] compared various unsupervised feature-learning algorithms including k-means, sparse restricted Boltzmann machine and sparse coding and achieved results comparable to the state of the art on the CAL500 dataset. In a later publication, Nam et al. [13] extended this approach by using the bag of features generated by the unsupervised algorithms to initialize a deep neural network (DNN) in a layer-by-layer fashion and finally fine-tuning it in a supervised manner using the annotations and achieve results close to the supervised variants.

Supervised techniques, which aim to directly tag the audio using DNNs, have also been explored thoroughly. Using CNNs for MIR has been a popular approach [14, 15]. Dieleman et al. [16] showed that using spectrograms with CNNs is more effective than using raw audio directly. However, Lee et al. [17] developed an end-to-end approach which uses sample-level filters instead of the traditional frame level and achieved state-of-the-art results. A trend to be noted here is that deep architectures typically perform better than shallow ones although they come at a greater computational cost.

The MIR community has also shown an interest in exploiting multiple scales for generating features. Hamel et al. [18] used different sizes of the frames while computing the discrete Fourier transform to effect different timescales. Dieleman et al. [9], on the other hand, proposed the use of Laplacian or Gaussian pyramids to generate different timescales. Mesgarani et al. [19] described another multi-scale feature extraction algorithm based on a model of the human auditory cortex and used it for discriminating speech from ambient noise. Lee et al. [20] proposed a deep multi-scale architecture which used multiple CNNs pre-trained over various timescales as feature extractors. These features were then combined to predict the final tags.

3   CNNs for Music Tagging

The connectivity pattern of neurons in CNNs was inspired by the behaviour of the visual cortex; individual neurons collect information from a local receptive field, and the receptive fields of different neurons overlap to cover the entire visual field. When visual input is passed through multiple layers of such neurons, the local information obtained from the neurons in each layer is combined by the neurons in the next layer to capture higher-level features. In simple words, high-level features are derived from low-level features. CNNs thus learn a hierarchy of features where the degree of abstraction of the features corresponds to the depth of the layer from which they are obtained. Moreover, they do so while ensuring a certain degree of shift and distortion invariance [21].

Following their success in computer vision [4], CNNs have also been used for audio processing tasks like speech recognition [22], music information retrieval [15] and automatic music tagging by making use of the image-like mid-level representations of audio signals such as mel-spectrograms [14, 16]. Mel-spectrograms are time-frequency representation of sounds that reduce the resolution of frequencies at the higher end, thus matching the human auditory perception and at the same time reducing the size of the data. These characteristics of mel-spectrograms, supported by the knowledge that the perceptible characteristics of audio are closely related to the energies in different frequency bands, make them suitable for extracting high-level music audio features using CNNs. The use of CNNs for this task is further justified by their ability to ensure temporal and frequency invariance which makes them fit well with music data, where the events of interest can occur at any time or frequency.

4 Proposed Architecture

In the preprocessing step, we converted the raw audio signals to log-amplitude mel-spectrograms, which have been shown to work well for audio processing using CNNs [23]. To achieve feature extraction over multiple timescales, we adopted a method similar to Dieleman et al. in [9]. We generated multiple versions of the spectrograms with different temporal resolutions by mean pooling the original with different factors. We then ran parallel streams of alternating convolutions and max pools on each of these subsampled spectrograms. The outputs of these streams were concatenated depth-wise, generating a single tensor. This tensor was then convolved and pooled further. Finally, it was flattened and went through two fully connected layers to give the output tags.

In the architecture used for the experiments, we used three different convolution streams corresponding to three different resolutions. The first stream took an input of size (128, 628). This size was reduced to (64, 157) and (32, 39) for the next two streams by mean pooling. The mean pools used had the sizes 2 and 4 in the frequency and time axes, respectively, with the stride equal to the size on each axis. The streams corresponding to larger resolutions underwent more number of convolutions and pools to process the greater amount of information present. Finally, each stream reduced the input to the size (70, 16, 19), resulting in a combined tensor of shape (210, 16, 19). This tensor was further convolved twice and pooled once, leading to a final size of (70, 8, 9). Note that the convolutions used were 2D in nature instead of the typical 1D along the time axis. This is because we believe that pitch invariance on the mel-scale is also an integral feature of musical genres. All the convolutions were half-padded. The final tensor was then flattened to yield a vector of size (5040,). This vector passed through a fully connected layer consisting of 500 neurons. Finally, the outputs of this layer were passed through a fully connected sigmoidal layer with 50 units representing the tags. The complete architecture, along with the filter shapes used, is depicted in Fig. 1.

5 Experiments and Results

We used the MagnaTagATune dataset [10] to evaluate the proposed model. The dataset is hosted by the Music Informatics Group of City University London. Since its release, it has been used for various MIR tasks such as genre classification, mood classification, instrument identification and auto-tagging, and it serves as a basis for testing and comparing different models. It consists of over 25,000 sound clips

Fig. 1 The architecture used in the experiments. Conv (x, y, z) means a convolution layer with x feature maps, with the filter size (y, z). Fully connected (x) means a fully connected layer with x neurons. All neurons are ReLu activated, except for the output layer which uses the sigmoid activation

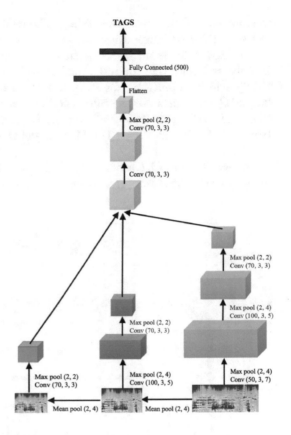

of length 29 s each (sampled at 16 kHz, 32 kbps, mono mp3) along with human annotations with a combination of 188 tags. It also consists of a description of the tracks' timbre, rhythm and harmonic-content features, and the codes required for generating the dataset distribution. The tags include, but are not limited to, genres, instruments, moods and the genders of the vocalists. Due to the skewed distribution of the tag, we used only the top 50 (frequency wise).

In our experiments, we trained our multi-scale CNN (MSCNN) model on log-amplitude mel-spectrograms of the audio clips, as proposed in Sects. 3 and 4. We evaluated the predictions made by the model by computing the average of the area under the ROC curve (AUC) [24] across all the 50 tags. Furthermore, we compared the performance of our model with that of a multilayer perceptron (MLP) with one hidden layer trained on handcrafted features of the audio clips. The MLP was trained on 30 handcrafted features extracted from each audio clip—20 MFCCs, spectral centroid, spectral contrast, spectral roll-off and zero crossings rate. The hidden layer of the MLP consisted of 20 neurons. For both the MSCNN and the MLP, all the neurons were activated with rectified linear units (ReLus) [25] except for the output layer, which had sigmoidal units. In the CNN, we also applied batch normalization [26] and dropout [27] to prevent over-fitting. The multi-scale CNN model was trained with the ADAM optimizer using the parameter values from the original paper [28].

The log-amplitude mel-spectrograms of the audio clips were generated using the Librosa library [29]. Both the models were implemented using Theano [30], a deep learning framework for Python. For the purposes of the experiment, the MagnaTagATune dataset was shuffled and divided into a training set (20,000 audio clips) and a testing set (5800 audio clips). The MSCNN set-up was run on a GTX 1080 GPU for 200 iterations. The MLP model was run on the same GPU for 300 iterations.

The average ROC AUC score for MSCNN and MLP is shown in Table 1 and Fig. 2 along with published results from previous works on the same dataset. It can be seen

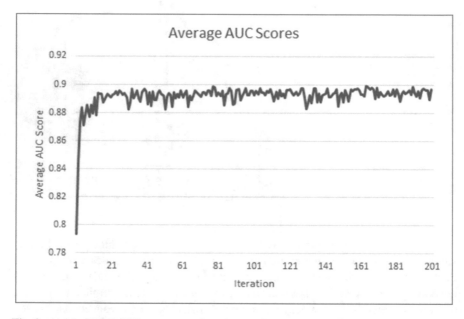

Fig. 2 A plot of ROC AUC score versus iteration number on the test set

Table 1 A comparison of the average ROC AUC scores for MSCNN, MLP and previous models on the MagnaTagATune dataset

Models	Average AUC scores
MLP	0.804
MSCNN	0.899
4-Layer FCN [14]	0.894
Multi-scale K-means features [9]	0.898
Transfer learning with pre-trained multi-scale CNN [20]	0.902

that our architecture performs very close to the state of the art and yields much better results than simple handcrafted features.

6 Conclusion

In this paper, we present a novel multi-scale CNN architecture for automatic music tagging. The architecture was designed to run different convolution streams on subsampled versions of audio spectrograms in an attempt to learn salient features over various timescales. We tested this architecture on the MagnaTagATune dataset and showed that it performs close to the state-of-the-art architectures and comprehensively beats a shallow classifier trained on handcrafted features. In future work, we intend to explore dilated convolutions with different dilation factors to learn features over different timescales instead of subsampling the input itself.

References

1. Davis, S.B., Mermelstein, P.: Comparison of parametric representations for monosyllabic word recognition in continuously spoken sentences. IEEE Trans. Acoust. Speech Sig. Process. **28**(4), 357–366 (1980)
2. Hinton, G.E., Osindero, S., Teh, Y.-W.: A fast learning algorithm for deep belief nets. Neural Comput. **18**, 1527–1554 (2006)
3. Schmidhuber, J.: Deep learning in neural networks: an overview (2014). arXiv:1404.7828
4. Krizhevsky, A., Sutskever, I., Hinton, G.E.: ImageNet classification with deep convolutional neural networks. In: Proceedings of the Neural Information Processing Systems Conference (2012)
5. Hinton, G., Deng, L., Dong, Y., Dahl, G., Mohamed, A.-R., Jaitly, N., Senior, A., Vanhoucke, V., Nguyen, P., Sainath, T., Kingsbury, B.: Deep neural networks for acoustic modeling in speech recognition. IEEE Sig. Process. Mag. **29**, 82–97 (2012)
6. Humphrey, E.J., Bello, J.P., LeCun, Y.: Moving beyond feature design: deep architecture and automatic feature learning in music informatics. In: Proceedings of the 13th International Society for Music Information Retrieval Conference (2012)
7. Choi, K., Fazekas, G., Sandler, M., Cho, K.: Convolutional recurrent neural networks for music classification (2016). arXiv:1609.04243
8. Briot, J.-P., Hadjeres, G., Pachet, F.: Deep learning techniques for music generation—A survey (2017). arXiv:1709.01620
9. Multiscale approaches to music audio feature learning. In: Proceedings of the 14th International Society for Music Information Retrieval Conference (2013)
10. Law, E., West, K., Mandel, M., Bay, M., Downie, J.S.: Evaluation of algorithms using games: the case of music annotation. In: Proceedings of the 10th International Conference on Music Information Retrieval (ISMIR) (2009)
11. Wulfing, J., Riedmiller, M.: Unsupervised learning of local features for music classification. In: Proceedings of the 13th International Society for Music Information Retrieval Conference (2012)
12. Nam, J., Herrera, J., Slaney, M., Smith, J.: Learning sparse feature representations for music annotation and retrieval. In: Proceedings of the 13th International Society for Music Information Retrieval Conference (2012)

13. Nam, J., Herrera, J., Lee, K.: A deep bag-of-features model for music auto-tagging (2015). arXiv:1508.04999
14. Choi, K., Fazekas, G., Sandler, M.: Automatic tagging using deep convolutional neural networks (2016). arXiv:1606.00298
15. van den Oord, A., Dieleman, S., Schrauwen, B.: Deep content-based music recommendation. In: Proceedings of the Neural Information Processing Systems Conference (2013)
16. Dieleman, S., Schrauwen, B.: End-to-end learning for music audio. In: Proceedings of the IEEE International Conference on Acoustic, Speech and Signal Processing (ICASSP) (2014)
17. Lee, J., Park, J., Kim, K.L., Nam, J.: Sample-level deep convolutional neural networks for music auto-tagging using raw waveforms (2017). arXiv:1703.01789
18. Hamel, P., Bengio, Y., Eck, D.: Building musically-relevant audio features through multiple timescale representations. In: Proceedings of the 13th International Society for Music Information Retrieval Conference (2012)
19. Mesgarani, N., Shamma, S., Slaney, M.: Speech discrimination based on multiscale spectro-temporal modulations. In: Proceedings of the IEEE International Conference on Acoustics, Speech, and Signal Processing (ICASSP) (2004)
20. Lee, J., Nam, J.: Multi-level and multi-scale feature aggregation using pre-trained convolutional neural networks for music auto-tagging. arXiv:1703.01793 (2017)
21. LeCun, Y., Boser, B.E., Denker, J.S., Henderson, D., Howard, R.E., Hubbard, W.E., Jackel, L.D.: Handwritten digit recognition with a back-propagation network. In: Proceedings of the Neural Information Processing Systems Conference (1989)
22. Sainath, T.N., Mohamed, A.-R., Kingsbury, B., Ramabhadran, B.: Deep convolutional neural networks For LVCSR. In: Proceedings of the IEEE International Conference on Acoustics, Speech and Signal Processing (ICASSP) (2013)
23. Dorfler, M., Bammer, R., Grill, T.: Inside the spectrogram: convolutional neural networks in audio processing. In: Proceedings of the International Conference on Sampling Theory and Applications (SampTA) (2017)
24. Fawcett, T.: An introduction to ROC analysis. Pattern Recogn. Lett. **27**, 861–874 (2006)
25. Nair, V., Hinton, G.E.: Rectified linear units improve restricted Boltzmann machines. In: Proceedings of the 27th International Conference on Machine Learning (ICML) (2010)
26. Ioffe, S., Szegedy, C.: Batch normalization: accelerating deep network training by reducing internal covariate shift (2015). arXiv:1502.03167
27. Srivastava, N., Hinton, G., Krizhevsky, A., Sutskever, I., Salakhutdinov, R.: Dropout: a simple way to prevent neural networks from overfitting. J. Mach. Learn. Res. **15**, 1929–1958 (2014)
28. Kingma, D.P., Adam, J.B.: A method for stochastic optimization (2014). arXiv:1412.6980
29. McFee, B., Raffel, C., Liang, D., Ellis, D.P.W., McVicar, M., Battenberg, E., Nieto, O.: librosa: audio and music signal analysis in python. In: Proceedings of the 14th Python in Science Conference, pp. 18–25 (2015)
30. Theano Development Team, Theano: A python framework for fast computation of mathematical expressions (2016). arXiv:1605.02688

Machine Learning Approaches for the Estimation of Particulate Matter (PM$_{2.5}$) Concentration Levels: A Case Study in the Hyderabad City, India

Latha Krishnappa and C. P. Devatha

Abstract Particulate matter concentration is one among several variables monitored at regular intervals to calculate air quality indices (AQI) which are intended to help understand the acute and chronic effects of air quality on human health. The fine particulate (PM$_{2.5}$) samplers installed at pollution monitoring stations continuously monitor the concentration of pollutant in air over time. The specific time-averaged concentration is then estimated from the continuous records. Missing data records in the PM$_{2.5}$ time series is quite normal, which is attributed by faulty equipment, routine maintenance schedules, or replacement of equipment. When one or more point observations in a time series are missing, it is very essential to estimate or predict the missing values. This study presents the application of machine learning techniques such as support vector regression (SVR), group method of data handling (GMDH) network, and evolutionary adaptive neuro fuzzy inference system to estimate the 24-h average PM$_{2.5}$ concentration levels at a particular station using PM$_{2.5}$ concentration levels observed at neighborhood stations as inputs. The performance of these models are evaluated in terms of widely used statistical metrics such as centered root mean square difference (CRMSD), normalized Nash–Sutcliffe efficiency (NNSE), and correlation coefficient (R). The findings of the study reveal that the GMDH model provided reasonably accurate estimates of daily PM$_{2.5}$ levels.

Keywords Air quality index (AQI) · Particulate matter (PM$_{2.5}$) · SVR · GMDH GA-ANFIS · PSO-ANFIS

L. Krishnappa (✉) · C. P. Devatha
Department of Civil Engineering, National Institute of Technology Karnataka,
P.O Srinivasnagar, Surathkal, Mangalore 575025, India
e-mail: lalisujay@gmail.com

C. P. Devatha
e-mail: revacp@gmail.com

© Springer Nature Singapore Pte Ltd. 2019
J. C. Bansal et al. (eds.), *Soft Computing for Problem Solving*,
Advances in Intelligent Systems and Computing 816,
https://doi.org/10.1007/978-981-13-1592-3_61

1 Introduction

A mixture of fine solid particles and fluid droplets found in the atmosphere is generally discussed as particulate matter (PM). $PM_{2.5}$ refers to the fine particles that have mean aerometric diameter of 2.5 μm (micro meter) or less, and the particles of this group can only be identified with an electron microscope. The coarse fraction particles which are greater than 2.5 μm and within 10 μm diameter are referred to as PM_{10}. These PM can arise from various sources which include emission from power plants, motor vehicles, airplanes, residential wood burning, forest fires, combustion of agricultural residues, volcanic eruptions, and dust storms. Some are released directly into the air, while others are produced when gases and particles react with one another in the atmosphere [1]. The PM from combustion sources emitted directly into the air are regarded as "primary" particles such as black carbon (soot), and the particles that are formed by exothermic reactions in the atmosphere from primary particles are referred to as "secondary" particles such as nitrates and sulfates. In general, the primary particles yield the "coarse PM" and most of the "fine PM" reap from the secondary particles [2].

Generally, ground-based $PM_{2.5}$ observations often contain missing data values due to defective equipment or unintentional human error. Such missing $PM_{2.5}$ data is estimated or predicted using a variety of methods. Dong et al. [3] developed a framework and methodology based on hidden semi-Markov models to predict high $PM_{2.5}$ concentration levels using past-meteorological measurements and $PM_{2.5}$ observation levels as model inputs. de Mattos Neto et al. [4] developed an evolutionary hybrid system, TAEF method, for time-series prediction of particulate matter concentrations using multilayer perceptron neural networks as predictor. Lary et al. [5] used a combination of meteorological and remote sensing data products together with ground-based $PM_{2.5}$ observations to train the machine learning algorithms that estimate the daily $PM_{2.5}$ levels. Feng et al. [6] developed a hybrid model integrating wavelet transformation and air mass trajectory analysis to improve the accuracy of artificial neural network (ANN) in forecasting daily average concentrations of $PM_{2.5}$. They also state that the wavelet transformation and trajectory-based geographic models as reliable tools to enhance the accuracy of the $PM_{2.5}$ forecasts. In the literature, there are only few studies that use $PM_{2.5}$ time series of exogenous station to predict a time series of interest station.

Nowadays, $PM_{2.5}$ levels in Indian metropolitan urban areas are around four to five times higher than in the European cities. $PM_{2.5}$ levels is of special concern since it is comprised of a huge proportion of distinct toxic compounds and acids, which can aerodynamically diffuse deeper into the respiratory tract. An air quality index (AQI) scheme transforms the weighted values of individual air pollutant concentrations into a single number or set of numbers. In India, based on short-term (up to 24-h averaging period) observed data of eight variables (PM_{10}, $PM_{2.5}$, SO_2, NO_2, CO, NH_3, O_3, and Pb), the AQI is depicted by a single number (index value), nomenclature, and color [7]. The missing records in the data of any of the above variables due to routine maintenance schedules may hinder the near real-time dissemination of AQI. When one or more point observations in a time series are missing, it is very essential to estimate or predict the missing values. This study presents the application of machine learning techniques such as support vector regression (SVR), group method of data handling (GMDH) network, evolutionary adaptive neuro fuzzy inference system to estimate the 24-h average $PM_{2.5}$ concentration levels at a particular station using $PM_{2.5}$ time-series data observed at neighborhood stations as model inputs.

2 Study Area and Data Analysis

The 24-h average data of $PM_{2.5}$ recorded at four air quality monitoring stations—Sanathnagar, IDA Pashamylaram, Bollaram Industrial Area, ICRISAT Patancheru of Hyderabad, Telangana state, India, were used in the present study. The location of air pollutant sampling laboratories of Hyderabad is shown in Fig. 1. Higher levels of pollution at Hyderabad is obvious due to heavy vehicular traffic, and the city owns well-established large-scale industrial landscape, including government-owned electronics manufacturing and defense industries in addition to a plethora of private industrial undertakings.

The short-term data from May 20, 2017, to August 30, 2017, were used to develop and test the machine learning models. The data were taken from the Central Pollution Control Board (CPCB) Web site [8]. About 70% of total data was used as a training

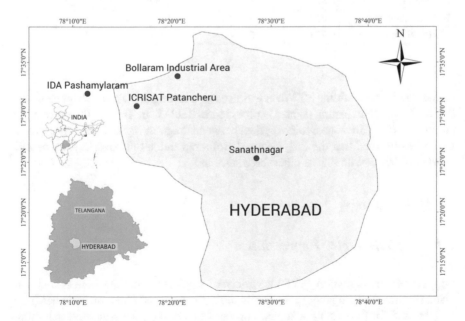

Fig. 1 Study area—location of air quality monitoring stations

Table 1 Descriptive statistics of PM$_{2.5}$ (μg/m^3) data series

Statistic	Sanathnagar		Bollaram inds. area		IDA Pashamylaram		ICRISAT Patancheru	
	Train	Test	Train	Test	Train	Test	Train	Test
Min	6.44	6.09	1.35	5.28	12.24	10.09	13.06	12.53
Max	59.83	35.2	55.56	25.35	56.7	38.75	62.94	59.99
Mean	19.21	18.16	13.09	16.40	24.03	25.51	29.60	35.61
STD	10.39	7.55	9.72	5.29	8.93	6.99	9.86	10.09
VAR	108.03	57.00	94.40	28.01	79.71	48.84	97.24	101.85

STD: Standard deviation; VAR: Variance

Table 2 Correlation coefficient (R) with respect to PM$_{2.5}$ data between neighboring stations

R	Train data			
	Sanathnagar	Bollaram inds. area	IDA Pashamylaram	ICRISAT Patancheru
Sanathnagar	1	0.938	0.839	0.695
Bollaram inds. area	0.938	1	0.866	0.702
IDA Pashamylaram	0.839	0.866	1	0.628
ICRISAT Patancheru	0.695	0.702	0.628	1
R	Test data			
	Sanathnagar	Bollaram inds. area	IDA Pashamylaram	ICRISAT Patancheru
Sanathnagar	1	0.805	0.669	0.723
Bollaram inds. area	0.805	1	0.760	0.767
IDA Pashamylaram	0.669	0.760	1	0.676
ICRISAT Patancheru	0.723	0.767	0.676	1

dataset, and the remaining 30% data was used as a testing set. The descriptive statistics of training and testing dataset is provided in Table 1. In Table 2, the correlation coefficient (R) with respect to PM$_{2.5}$ data between neighboring stations is presented. It can be observed that the PM$_{2.5}$ data of Bollaram Industrial Area had reasonable good correlation with that of other three stations.

3 Methodology

3.1 Support Vector Regression

Support vector regression (SVR) a variety of support vector machine is a supervised machine learning algorithm developed by Vapnik and Chervonenkis has been widely used for forecasting and prediction tasks. Based on the structural risk minimization principle, SVR minimizes the errors in the model using special nonlinear functions called kernels which transform/maps the input data in the feature space to a higher-dimensional feature space (where the linear regression is feasible) [9]. SVR minimizes both the empirical risk and lower or upper bound of a confidence interval. For additional information regarding SVR, its architecture, parameter tuning, model building procedure, one may refer to Cortes and Vapnik [10]; Raghavendra and Deka [11].

3.2 Group Method of Data Handling (GMDH) Network

Group Method of Data Handling (GMDH)—a method of inductive statistical learning developed by Alexey Grigorevich Ivakhnenko is widely used for nonlinear regression problems [12]. GMDH network generates a relatively simple and numerically stable polynomial neural network of quadratic neurons that are configured in a special structure to map a given set of training vectors to desired outputs. The number of layers and neurons in hidden layers, model structure, and other optimal neural network parameters are all self-organizing in nature. For additional information regarding GMDH, its architecture, model building procedure, one may refer to Anastasakis and Mort [13] and http://www.gmdh.net/GMDH_his.htm.

3.3 Evolutionary ANFIS

The fuzzy inference system (FIS) based on fuzzy c-means (FCM) clustering is implemented in the adaptive neuro fuzzy inference system (ANFIS). The metaheuristics or evolutionary algorithms such as genetic algorithm (GA) or particle swarm optimization (PSO) is applied to tune the parameters of ANFIS structure. The evolutionary-trained ANFIS is specially useful for nonlinear regression problems. In the FIS based on FCM clustering, the input membership function type is 'gaussmf' by default and the output membership function type is 'linear'. The evolutionary algorithm (GA or PSO) assists the ANFIS to adjust the parameters of the membership function. For additional information regarding GA, PSO, and ANFIS, their architecture, parameters, model building procedure, one may refer to Yu and Gen [14] and Jang [15].

3.4 Model Development

In the present study, since, the $PM_{2.5}$ data of Bollaram Industrial Area had reasonable good correlation with that of other three stations, the $PM_{2.5}$ time-series data of Sanathnagar, IDA Pashamylaram, and ICRISAT Patancheru were used as input variables to estimate the $PM_{2.5}$ levels at Bollaram Industrial Area. The time-series data from 20th May 2017 to 31st July 2017 was considered for training the models, and the data from 1st August 2017 to 30th August 2017 were used for testing. The SVR model based on radial basis kernel function was developed using LIBSVM toolbox [16] in MATLAB. A three-dimensional grid search was used for finding the optimal parameters (C, ϵ and γ) of SVR. The GMDH network, GA-ANFIS and PSO-ANFIS models were developed using the source codes provided by Yarpiz [17, 18] in MATLAB.

4 Performance Evaluation Metrics

To evaluate the performance of SVR, GMDH, GA-ANFIS, and PSO-ANFIS models, measures such as centered root mean square difference (CRMSD), normalized Nash–Sutcliffe efficiency (NNSE), and correlation coefficient (R) were employed. The NSE measures the efficiency of a model by relating the goodness of fit of the model to the variance of the observed data [19].

$$CRMSD = \sqrt{\frac{1}{N} \sum_{i=1}^{N} \left[(O_i - \overline{O}) - (E_i - \overline{E}) \right]^2} \tag{1}$$

$$NNSE = \frac{1}{(2 - NSE)} \quad \text{where} \quad NSE = 1 - \left[\frac{\sum_{i=1}^{N} (O_i - E_i)^2}{\sum_{i=1}^{N} (O_i - \overline{O})^2} \right] \tag{2}$$

$$R = \frac{\sum_{i=1}^{N} (O_i - \overline{O}) \cdot (E_i - \overline{E})}{\sqrt{\sum_{i=1}^{N} (O_i - \overline{O})^2 \cdot \sum_{i=1}^{N} (E_i - \overline{E})^2}} \tag{3}$$

where O and E refer to the observed $PM_{2.5}$ and estimated $PM_{2.5}$ by the model. $\overline{O}$ is the mean of observed values, and $\overline{E}$ represents the mean of estimated values. N represents the total number of dataset samples.

5 Results and Discussion

The optimal SVR parameters searched via three-dimensional grid search were C = 42.15; ϵ = 0.0625, and γ = 9.0125 × 10^{-4}. The optimal GMDH network was obtained for maximum number of layers = 3, maximum number of layer neurons = 6, and alpha = 0.05. The evolutionary ANFIS algorithm of either GA or PSO was run for 1000 iterations in order to arrive at optimized models. The performance statistics of SVR, GMDH, GA-ANFIS, and PSO-ANFIS models are provided in Table 3. It

Table 3 Performance evaluation by statistical indices

Models	CRMSD (μg/m)		NNSE		R	
	Train	Test	Train	Test	Train	Test
GA-ANFIS	2.396	2.617	0.941	0.572	0.968	0.874
PSO-ANFIS	1.999	2.595	0.958	0.602	0.978	0.905
GMDH	2.459	2.688	0.938	0.658	0.967	0.871
SVM	2.577	2.717	0.932	0.561	0.964	0.860

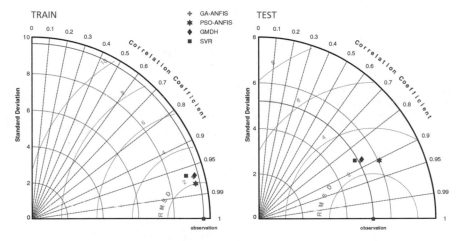

Fig. 2 Taylor diagram showing correlation coefficient, standard deviation, and CRMSD

can be observed that the PSO-ANFIS model has least CRMSD and highest R among the models. However, the NNSE of GMDH model is slightly better than that of PSO-ANFIS model. Also, from the Taylor diagram (as shown in Fig. 2), one can observe that during test phase, the standard deviation of the PSO-ANFIS estimates is slightly more when compared to that of the standard deviation of the observed time series. The GMDH model is considered to provide better estimates of PM$_{2.5}$ during both train and test phases based on the evaluation using different statistical indices. Figure 3 presents the time-series plot of PM$_{2.5}$ estimates of SVR, GMDH, GA-ANFIS, and PSO-ANFIS models with respect to test period, wherein it can be visualized that the GMDH model estimates are closely following the observed PM$_{2.5}$ time series. In addition, the scatter plots, analyzing the relation between the observed and the model estimates, are presented in Fig. 4. PM$_{2.5}$ time series are difficult to estimate or predict, as they tend to be noisy, highly nonlinear, non-stationary, and chaotic. GMDH network is advantageous in the case of small data samples since their network structures are flexible to obtain the optimal solutions by automatically adapting to the model complexity. One more advantage of the GMDH network is that it cannot over train.

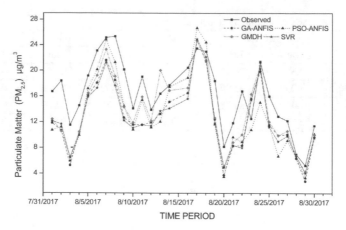

Fig. 3 Time-series plots of observed versus estimated PM$_{2.5}$

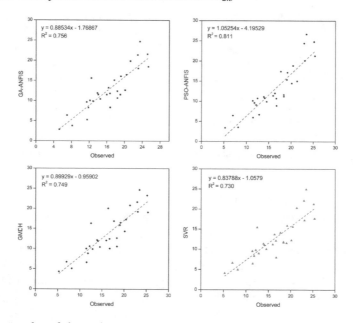

Fig. 4 Scatter plots of observed versus estimated PM$_{2.5}$

6 Conclusions

In this study, an attempt was made to evaluate the performance of SVR, GMDH, GA-ANFIS, and PSO-ANFIS models to estimate the PM$_{2.5}$ data of Bollaram Industrial Area using the same PM$_{2.5}$ data of exogenous stations as inputs. This kind of model development is particularly useful for filling missing data records in the time series. The GMDH network performed relatively better when compared to that of other models in the estimation of $PM_{2.5}$. The Taylor diagram provides a concise evaluation of individual model performance and summarizes how close a set of predictions/estimates match the real observations. The suggested strategy can be adopted to model other air pollutants of similar statistical behavior

Acknowledgements The authors would like to thank the Central Pollution Control Board, India, for hosting the necessary data on its Web site which is used in this study.

References

1. Tan, Z.: Properties of aerosol particles. In: Air Pollution and Greenhouse Gases. Green Energy and Technology Series, pp. 91–116. Springer, Singapore (2014). https://doi.org/10.1007/978-981-287-212-8_4
2. Vallero, D.: The science of air pollution, In: Fundamentals of Air Pollution, 5th edn., Chap. 3, pp. 43–81. Academic Press, Boston (2014). https://doi.org/10.1016/B978-0-12-401733-7.00003-7
3. Dong, M., Yang, D., Kuang, Y., He, D., Erdal, S., Kenski, D.: $PM_{2.5}$ concentration prediction using hidden semi-Markov model-based times series data mining. Expert Syst. Appl. **36**(5), 9046-9055 (2009). https://doi.org/10.1016/j.eswa.2008.12.017
4. de Mattos Neto, P.S.G., Madeiro, F., Ferreira, T.A.E., Cavalcanti, G.D.C.: Hybrid intelligent system for air quality forecasting using phase adjustment. Eng. Appl. Artif. Intell. **32**, 185–191 (2014). https://doi.org/10.1016/j.engappai.2014.03.010
5. Lary, D.J., Lary, T., Sattler, B.: Using machine learning to estimate global $PM_{2.5}$ for environmental health studies. Environ. Health Insights **9**(Suppl 1), 41–52 (2015). https://doi.org/10.4137/EHI.S15664
6. Feng, X., Li, Q., Zhu, Y., Hou, J., Jin, L., Wang, J.: Artificial neural networks forecasting of $PM_{2.5}$ pollution using air mass trajectory based geographic model and wavelet transformation. Atmos. Environ. **107**, 118–128 (2015). https://doi.org/10.1016/j.atmosenv.2015.02.030
7. CPCB: National Air Quality Index. CPCB, New Delhi, 42 pp. (2014). http://www.indiaenvironmentportal.org.in/files/file/Air%20Quality%20Index.pdf
8. CPCB: Ambient Air Quality Data at Various Locations in the Country. http://cpcb.nic.in/RealTimeAirQualityData.php (2017)
9. Vapnik, V.N.: The Nature of Statistical Learning Theory. Springer, New York (1995). https://doi.org/10.1007/978-1-4757-2440-0
10. Cortes, C., Vapnik, V.: Support-vector networks. Mach. Learn. **20**(3), 273–297 (1995). https://doi.org/10.1007/BF00994018
11. Raghavendra, N.S., Deka, P.C.: Support vector machine applications in the field of hydrology: a review. Appl. Soft Comput. **19**, 372–386 (2014). https://doi.org/10.1016/j.asoc.2014.02.002
12. Ivakhnenko, A.G.: The group method of data handling—A rival of the method of stochastic approximation. Sov. Autom. Control **13**(3), 43–55 (1968)
13. Anastasakis, L., Mort, N.: The development of self-organization techniques in modelling: a review of the group method of data handling (GMDH). ACSE Research Report 813, University of Sheffield, UK (2001)
14. Yu, X., Gen, M.: Introduction to Evolutionary Algorithms. Springer, London (2010). https://doi.org/10.1007/978-1-84996-129-5
15. Jang, J.S.: ANFIS: adaptive-network-based fuzzy inference system. IEEE Trans. Syst. Man Cybern. **23**(3), 665–685 (1993)
16. Chang, C., Lin, C.: LIBSVM : a library for support vector machines. ACM Trans. Intell. Syst. Technol. **2**(3), 1–27. http://www.csie.ntu.edu.tw/~cjlin/libsvm

17. Heris, S.M.K.: Implementation of Group Method of Data Handling in MATLAB. Project Code: YPML113, Yarpiz (2015). http://www.yarpiz.com
18. Heris, S.M.K.: Evolutionary ANFIS Training in MATLAB. Project Code: YPFZ104, Yarpiz (2015). http://www.yarpiz.com
19. Nash, J.E., Sutcliffe, J.V.: River flow forecasting through conceptual models. Part I : a discussion of principles. J. Hydrol. **10**(3), 282–290 (1970). https://doi.org/10.1016/0022-1694(70)90255-6

Theoretical Estimation of the Microalgal Potential for Biofuel Production and Carbon Dioxide Sequestration in India

Bunushree Behera, Nazimdhine Aly, M. Asok Rajkumar and P. Balasubramanian

Abstract The unprecedented decline in petroleum reserves along with the rising concerns of global warming and environmental pollution has resulted in the search for alternative energy. India receives an abundant amount of solar insolation that can be easily transformed into other bioenergy sources. In current years, microalgal biofuels have gained attention owing to the presence of substantial amount of lipids and ease of cultivation in the presence of light energy, wastewater, and carbon dioxide (CO_2). In spite of the theoretical knowledge, the lack of convincing technical data and economic hindrances limit their field-scale application. The current study utilizes the global horizontal solar irradiance data (from the year 2002 to 2008) of India as input into the photon energy balance equations, which were solved in MATLAB to predict the theoretical microalgal biomass, lipid productivity, and CO_2 sequestration potential. The maximum biomass productivity was predicted as 90.1 g m^{-2} d^{-1}, corresponding to the lipid productivity of 31.3 ml m^{-2} d^{-1} and CO_2 sequestration potential of 23.6 g m^{-2} d^{-1} in the southern peninsular regions and Western Ghats. Since the solar irradiance varies from 3.25 to 6.08 kWh m^{-2} d^{-1} for the entire Indian subcontinent, most parts of India were projected to be suitable for growing microalgae. Decline in biomass productivity by 32.5% was evident accounting for photoinhibition effects such preliminary estimates would help in assessing the real-time potential of microalgae before going for cost-intensive field-scale analysis.

Keywords Microalgae · Solar insolation · Photoinhibition · Biomass productivity · Lipid productivity · CO_2 sequestration

B. Behera · N. Aly · P. Balasubramanian (✉)
Agricultural & Environmental Biotechnology Group, Department of Biotechnology & Medical Engineering, National Institute of Technology Rourkela, Rourkela, Odisha, India
e-mail: biobala@nitrkl.ac.in

N. Aly
Groupe-SMTP, Antananarivo, Madagascar

M. Asok Rajkumar
Department of Mechanical Engineering, Gnanamani College of Technology, Namakkal, Tamilnadu, India

© Springer Nature Singapore Pte Ltd. 2019
J. C. Bansal et al. (eds.), *Soft Computing for Problem Solving*,
Advances in Intelligent Systems and Computing 816,
https://doi.org/10.1007/978-981-13-1592-3_62

1   Introduction

The substantial utilization of fossil fuels in the present era is the major contributor to environmental pollution and global warming. Further, the depletion of these reserves is highly alarming and demanded the search for alternative fuels [9]. Indian subcontinent holds vast reserves of renewable resources that harbor tremendous potential for replacing the conventional fossil fuels. However, most of these resources are unharnessed due to the lack of appropriate technical knowledge and economic hindrances [8, 9]. In recent years, microalgae have attracted lots of attention as the most promising source of bioenergy since they contain up to 60% lipids along with substantial amount of carbohydrates and proteins [5]. Further, their ease of cultivation compared to the first- and second-generation energy crops as the microalgae could be cultivated in either marine or brackish water in the presence of light energy increases their desirability in terms of energy requirements and cost economics [5, 10, 11]. The integrated symbiotic approach combining wastewater treatment with biofuel production utilizing the flue gas from industries is acknowledged as the strategy for curbing environmental pollution and global climate change [19]. In spite of the very well-known advantages of microalgae, still their real-time, large-scale application is highly restricted, especially in developing countries. Lack of convincing data on the productivity and economics limits their scale-up. As microalgae are autotrophic microorganisms that use the light energy to convert CO_2 into biomass, the algal productivity is a direct function of the horizontally received solar insolation [16]. Hence, site-specific studies via mathematical modeling approaches are essential to select the desired locations with appropriate solar insolation and climatic condition to facilitate easy decision making by policy makers and stakeholders.

Various researchers around the globe have theoretically predicted the microalgal productivity. **Slegers et al.** (**2011**, **2013**) developed a mathematical model to estimate the biomass productivity of microalgae using the open pond and flat panel photobioreactors considering the influence of climatic conditions in Algeria, Netherlands, and France [12, 13]. **Sudhakar et al.** (**2012**) had predicted the biomass productivity and CO_2 sequestration potential of different regions of India taking into account the horizontal solar insolation data from RETScreen database [16]. **Asmare et al.** (**2013**) studied the biomass and oil productivity of microalgae for five different regions of Ethiopia concerning the variation in spatiotemporal conditions [4]. **Aly and Balasubramanian** (**2017**) evaluated the effects of various global geographical coordinates on microalgal growth rate and biomass productivity [1]. **Aly et al.** (**2017**) evaluated the growth of microalgae in a trackable solar photobioreactor for the state of Odisha, India [3]. Most of the earlier studies have taken into account the influence of solar insolation directly without considering the photoinhibition effects which cause irreparable damage to the photosynthetic system, thus declining the algal productivity. Another study by **Aly and Balasubramanian** (**2016**) calculated the microalgal productivity in hypothetical open ponds of National Institute of Technology (NIT) Rourkela, Odisha, including the effects of photoinhibition [2].

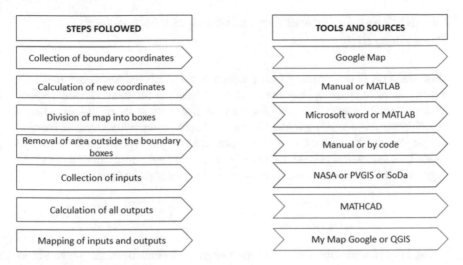

Fig. 1 Flowchart of the detailed methodology for map generation

The current study evaluated the potential of microalgal and lipid productivity along with the CO_2 sequestration potential for the entire Indian subcontinent based on global horizontal solar irradiance (GHI). The solar insolation received from different regions of India for the year 2002–2008 was collected, averaged, and fed as input into the photon energy balance equations to obtain the desired theoretical yields. Such kind of studies acts as a benchmark for the stakeholders and decision makers working in the area of large-scale microalgal cultivation to facilitate decision making, thus making algal biofuels a commercial reality.

2 Methodology

2.1 Data Collection and GHI Map Generation

Indian subcontinent lies toward the north of the equator between 6°44′ and 37°30′ north latitude and 68°7′ and 97°25′ east longitude. The entire Indian country was divided into 30,243 spatial coordinates based on the available GHI data from the Ministry of New and Renewable Energy (MNRE), Solar Energy Center database, Government of India. Latitude and longitude of all these spatial coordinates were considered for the study. The available GHI data were retrieved from MNRE databases for the year 2002–2008 and utilized further for calculations. The flowchart starting from the collection of the data until map generation has been detailed as presented in Fig. 1. The coordinates of the selected site and the map plotting were done as outlined in detail by the earlier studies of the authors **Aly et al. (2017)** [3].

2.2 Modeling Equations for Estimating the Microalgal Productivity

Solar radiation is the primary factor affecting the photosynthesis and metabolic growth rate of microalgae. However, the solar radiation incident over the raceway pond or the total energy available for microalgal growth is much less than that of the full spectrum solar energy. The biomass and lipid productivity along with carbon dioxide sequestration capacity of autotrophic microalgae depend on the photosynthetically active radiation (PAR). PAR is the amount of solar insolation in the visible range from 400 to 700 nm available to the algae for photosynthesis as given in Eq. (1).

$$\%PAR = \frac{PAR\ energy}{Full\ spectrum\ energy} * 100 = \frac{\int_{\lambda=400}^{700\ nm} Esolar\,(\lambda)d\lambda}{\int_{\lambda=0}^{4000\ nm} Esolar\,(\lambda)d\lambda} * 100 \qquad (1)$$

The daily biomass productivity of microalgae is dependent on the photon transmission efficiency, photon conversion efficiency, and photon utilization efficiency of microalgae. Photon transmission efficiency of microalgae takes into account the absorption efficiency of algal cells which is dependent on the light distribution efficiency of the microalgae ($\eta_{light\ distribution}$) considering them as particulates distributed in the media, land use efficiency ($\eta_{land\ use}$), and %PAR. Equation (2) has been used to calculate the photon transmission efficiency.

$$\eta_{transmission} = \eta_{light\ distribution} * \eta_{land\ use} * \alpha * \%PAR \qquad (2)$$

Photon conversion efficiency (PCE_{PAR}) as given by Eq. (3) denotes the number of photons converted into useful energy by algae.

$$PCE_{PAR} = \frac{HV\ of\ CH_2O\left(\frac{KJ}{mol\ CH_2O}\right)}{n_{photons}\left(\frac{mol\ photons}{mol\ of\ CH_2O}\right) X\ ME\ of\ PAR\left(\frac{KJ}{mol\ photon}\right)} \qquad (3)$$

The capture efficiency of solar insolation by microalgae depends on the metabolic process efficiency of photosynthesis, absorption, and respiration, and thus can be represented by Eq. (4).

$$\eta_{capture} = \eta_{photosynthesis} * \eta_{photo\text{-}utilization} * (1 - r) \qquad (4)$$

To have a more realistic estimation of the photosynthetic ability, the photon utilization efficiency ($\eta_{photo\text{-}utilization}$) of microalgae as given in Eq. (5) incorporates the Bush equation with photoinhibition effect.

$$\eta_{photo\text{-}utilization} = \frac{I_s}{I_l}[\ln\left(\frac{I_s}{I_l}\right) + 1] \qquad (5)$$

Apart from the solar irradiation, the inherent biochemical content of microalgae which constitutes the biomass energy of microalgae ($E_{microalgae}$) is given by Eq. (6).

$$E_{microalgae}\left(\frac{MJ}{Kg}\right) = f_L * E_L + f_P * E_P + f_C * E_C \tag{6}$$

where f_L, f_P, and f_C represent the fraction of lipids (30%), proteins (35%), and carbohydrates (35%), respectively. E_L, E_P, and E_C represent the energy content of lipids, proteins, and carbohydrates which were presumed as 16.7, respectively.

The daily microalgal biomass productivity ($MB_{(daily)}$) is dependent on incident solar radiation and can be calculated by Eq. (7).

$$MP_{(daily)}\left(\frac{g}{m^2 * day}\right) = \frac{\eta_{transmission} * \eta_{capture} * SI\left(\frac{KWh}{m^2 * day}\right)}{E_{microalgae}\left(\frac{MJ}{Kg}\right)} \tag{7}$$

2.3 Evaluation of Lipid Production Potential and CO_2 Sequestration Capacity

The lipid productivity of microalgae $\left(ML_{(daily)}\right)$ is directly proportional to the biomass productivity ($MB_{(daily)}$) and is given by Eq. (8).

$$ML_{(daily)}\left(\frac{ml}{m^2 * day}\right) = \frac{f_L * MB_{(daily)}\left(\frac{g}{m^2 * day}\right)}{\rho_L\left(\frac{Kg}{l}\right)} \tag{8}$$

where f_L represents the fraction of lipids in microalgae that is assumed to be 35% and ρ_L represents the density of microalgal lipids.

The CO_2 sequestration potential of microalgae is largely strain-specific and also depends on the cultivation conditions. Assuming the algal biomass molecular formula as $CO_{0.48}H_{1.83}N_{0.11}P_{0.01}$ and applying the law of conservation of mass to microalgae result in Eq. (9).

$$4CO_2 + H_2O + Nutrients + Sunlight \rightarrow 4CO_{0.48}H_{1.83}N_{0.11}P_{0.01} + 3.5O_2 \tag{9}$$

Rate constant (K) for the reaction is calculated considering the ratio of the molar mass of CO_2 (M_{CO_2}) to that of microalgal biomass ($M_{Biomass}$) as given in Eq. (10)

$$\frac{M_{CO_2}}{M_{Biomass}} = \frac{44}{23.2} = 1.89 \tag{10}$$

Total carbon dioxide (total CO_2) fixed is given by Eq. (11)

$$\text{Total } CO_2 = K * MB * \text{fixation efficiency} \tag{11}$$

The fixation efficiency has been assumed to be 15% to guarantee the minimum amount of CO_2 captured by microalgae.

2.4 Assumptions and Limitations of the Proposed Model

The mathematical model encompasses a set of empirical equations as described in the previous section which was written and solved using Mathcad. As the productivity is a function of solar insolation received, all the calculations were carried out considering the energy of one photon. The equations used for energy balances are inclusive of the losses due to atmospheric conditions and metabolic maintenance of microalgae. The biochemical composition of microalgae, i.e., the carbohydrate: protein: lipid ratio, has been assumed to be 35:35:30. The energy content of proteins, carbohydrates, and lipids is taken as 16.7, 15.7, and 37.6 kJ g^{-1}, respectively. The density of lipids (ρ) has been considered as 0.85 kg l^{-1}. The list of constants used in the modeling along with their values as given in the previous literature is given in Table 1. Even though the model takes into account the effects of photoinhibition, still the inclusion of the effects of other parameters like the water temperature, initial microalgal concentration, reactor geometry could be done in future for improvising the current model.

3   Results and Discussion

3.1   Biomass Productivity of Microalgae as a Function of Solar Insolation

India being a tropical country receives a good amount of sunlight throughout the year. The average daily solar insolation varies from 3.25 to 6.08 kWh m^{-2} d^{-1} as illustrated in GHI map (Fig. 2a). Indian subcontinent can be divided into three different solar hotspot boundaries, i.e., regions receiving less than 4 kWh m^{-2} d^{-1}, regions receiving solar insolation between 4 and 5.5 kWh m^{-2} d^{-1}, and the regions receiving more than 5.5 kWh m^{-2} d^{-1}. The daily average solar insolation received is maximum in the regions of southern peninsula and the Western Ghats which are more than 5.5 kWh m^{-2} d^{-1}. The Eastern Ghats and central plateau regions receive a moderate range of global solar insolation daily that varies from 4.25 to 5.52 kWh m^{-2} d^{-1}. The Himalayan region receives the lowest range of daily average global insolation along with northern plains and the northeastern parts in the range of 3.25–4.29 kWh m^{-2} d^{-1}. **Sudhakar et al. (2012)** had reported that an average solar insolation received by most parts of India varies from 4 to 7 kWh m^{-2} d^{-1} [16].

Table 1 List of constants used in the empirical equations for modeling

Parameters	Values	References
Full spectrum solar energy (E_{solar}) (MJ m^{-2} year)	5623–7349	[18]
Photon transmission efficiency[a] ($\eta_{Transmission}$)	0.43–0.44	[15]
Photosynthetically active radiation (PAR) (%)	0.45–0.47	[6]
Energy of photons (E_{photon}) (kJ mol^{-1})	225.3	[6]
Photon conversion efficiency[a] ($PCE_{PAR}/\eta_{Photosynthesis}$)	0.26–0.27	[15]
Light distribution efficiency[a] ($\eta_{light\ distribution}$)	0.96	[15]
Photon utilization efficiency[a] ($\eta_{Photon-utilisation}$)	1	[15]
Land use efficiency[a] ($\eta_{land\ use}$)	0.96	[15]
Fraction of energy used in respiration[a] (r)	0.2	[15]
Light absorption efficiency of algae[a] (α)	1	[15]
Saturation light used in photosynthesis (I_S) (μmol m^{-2} s^{-1})	200	[17]
Incident light used in photosynthesis (I_1) (μmol m^{-2} s^{-1})	200	[17]
Mean energy of 1 mol PAR (M.E of PAR) (kJ mol^{-1})	217.4	[6]
Higher heating value of microalgae (HV of CH_2O) (MJ kg^{-1})	14.21	[15]

[a]Represents dimensionless value

Since microalgae are autotrophic photosynthetic microorganisms, it is expected that the biomass productivity must be directly proportional to the amount of solar insolation received. As evident from Fig. 2b, the pattern of biomass productivity also follows a similar trend as that of the GHI map. The predicted daily average biomass productivity for the Indian subcontinent varies from 48.10 to 90.12 g m^{-2} d^{-1}. The maximum range of biomass productivity varying from 81.48 to 90.12 g m^{-2} d^{-1} was predicted in the regions of southern peninsula and Western Ghats which are receiving the maximum amount of horizontal solar insolation. Biomass productivity ranging from 66.33 to 81.48 12 g m^{-2} d^{-1} has been predicted for the central plateau regions, the Gangetic Plains, and the Western Ghats which receive a moderate range of solar

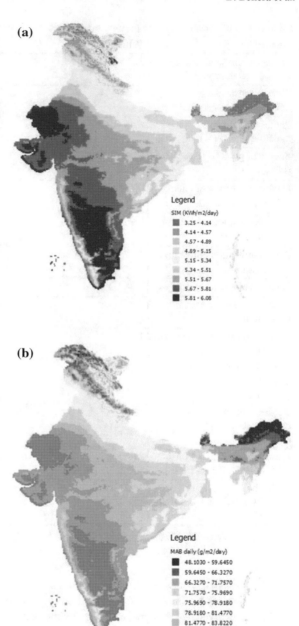

Fig. 2 a Daily average global horizontal solar insolation (GHI) map of India. **b** Predicted microalgal biomass productivity as a function of the received solar radiation

insolation. The northeastern regions as well as the northern Himalayan regions with less than 4 kWh m^{-2} d^{-1} are expected to have the biomass productivity in the range of 48.10–66.32 12 g m^{-2} d^{-1}. Most of the biomass productivities are consistent

Table 2 Comparison of the model output with the previous studies in the literature

Study sites	GHI (kWh m^{-2} d^{-1})	Biomass prod. (g m^{-2} d^{-1})	Lipid prod. (ml m^{-2} d^{-1})	CO$_2$ capture (g m^{-2} d^{-1})	References
Without the inclusion of photoinhibition effect					
Parts of India	4.5	80	18.82	–	[16]
Parts of India	5.5	97	22.82	–	[16]
Parts of India	6.5	115	27.05	–	[16]
Una, Himachal Pradesh, India	5.3	75.68	35.61	–	[14]
Chennai, Tamil Nadu, India	5.1	75.35	34.51	–	[14]
NIT Rourkela, Odisha, India	4.9	72.59	25.21	20.65	[2]
Entire Indian subcontinent	6.0	90.10	31.3	23.64	Present study
With the inclusion of photoinhibition effect					
NIT Rourkela, Odisha, India	4.9	54.1	18.8	15.4	[2]
Indore, Madhya Pradesh, India	5.1	55.9	–	15.3	[1]
Entire Indian subcontinent	6.0	60.6	21.0	17.2	Present study

with those of the previous studies as given in Table 2. Variations obtained might be attributed to the differences in climatological parameters and differences in strain of microalgae taken under consideration. The study clearly depicts the unleashed potential of different locations in Indian subcontinent that can be used for growing algae. Thus, the presented data provides estimates to be used by different policy makers for making the large-scale microalgal productivity a reality in near future.

3.2 Lipid Productivity and Carbon Dioxide Sequestration Capacity of Microalgae

Figures 3 and 4 illustrate the pattern of predicted lipid productivity profile and CO$_2$ sequestration capacity of microalgae for different regions of India, respectively. The lipid productivity of microalgae is determined by the biomass productivity as well as the strain-specific biochemical content of microalgae [5, 16]. The predicted lipid productivity profile follows a trend similar to that of the predicted biomass produc-

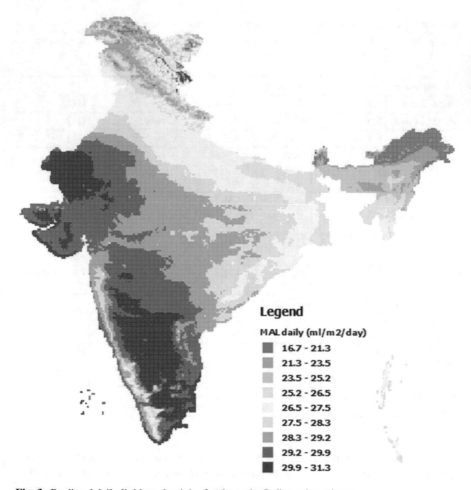

Fig. 3 Predicted daily lipid productivity for the entire Indian subcontinent

tivities with the maximum productivity achieved in the regions of southern peninsula and Western Ghats (28.3–31.3 ml m^{-2} d^{-1}). It is followed by the central plateaus (25.2–28.3 ml m^{-2} d^{-1}) and the minimum in the regions of northern Himalayan foothills and northeastern parts (16.7–25.2 ml m^{-2} d^{-1}). **Aly and Balasubramanian (2016)** have predicted a lipid productivity of 25.21 ml m^{-2} d^{-1} for NIT Rourkela, Odisha, India [2]. The lipid productivity potential of microalgae over various regions of the Indian subcontinent is evidently higher than that of the other energy crops (**Asmare et al. 2013**), making it suitable for large-scale generation of algal biofuel [4].

The CO$_2$ sequestration potential of microalgae is dependent largely on the biomass productivity as well as the carbon capture efficiency of the strain under study. It is noteworthy to mention that 15% CO$_2$ capture efficiency has been assumed for the

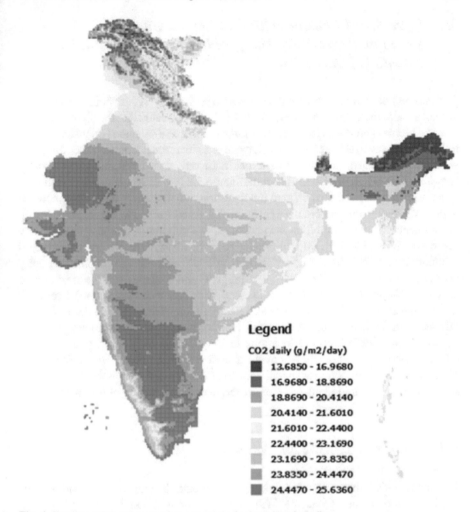

Fig. 4 Predicted CO_2 sequestration potential of microalgae in India

present study to provide the minimum possible values for CO_2 sequestration potential. The average daily CO_2 capture potential of possible large-scale microalgal cultivation ranges from 13.69 to 23.64 g m^{-2} d^{-1}. Maximum average CO_2 sequestered has been predicted for the southern peninsular India and the Western Ghats. The predicted profile for daily CO_2 captured is similar to the biomass productivity mapping. The CO_2 sequestration potential reported in the current study is several times higher than that of the capture potential of terrestrial plants [4]. It emphasizes the suitability of large-scale microalgal cultivation as a potential biological CO_2 sequestration strategy in future.

3.3 Influence of Photoinhibition Effects on Biomass and Lipid Productivity Along with Carbon Sequestration Potential of Microalgae

The photon flux density of microalgae has a direct impact on the photosynthetic rate as well as with the microalgal productivity [16]. It is expected that the photosynthetic rate and biomass productivity would be increased with the enhanced intensity of solar radiation. However, at the cellular level, there is an increase in photosynthetic activity with the increase in the amount of solar insolation only up to a certain threshold value beyond which the photosystem II becomes saturated [7]. This phenomenon is known as photoinhibition, and thus beyond the threshold level, there is no further increase in photosynthetic rate and biomass productivity. As illustrated in Fig. 5, considering the influence of photoinhibition into account, the maximum biomass productivity decreases by 32.5% from 90.1 to 60.6 g m^{-2} d^{-1}. Similarly, the maximum lipid productivity declines from 31.3 to 21.0 ml m^{-2} d^{-1} and the CO_2 capture capacity decreases from 25.6 to 17.2 g m^{-2} d^{-1} (Figs. 6 and 7). However, the generalized profile of biomass, lipid productivities, and CO_2 sequestration potential remains fairly same as that of the previous case. Thus photoinhibition shows a more realistic difference in the microalgal productivity. Similar kind of results were also reported **by Aly et al. (2017)** for 1142 locations in the entire state of Odisha using solar-trackable photobioreactor [3]. **Aly and Balasubramanian (2017)** also reported the effects of photoinhibition over spatial coordinates at ten different global locations on algal productivity, lipid yield, and carbon sequestration capacity of microalgae [1].

3.4 Model Comparison and Validation

The mathematical model was built taking into account the empirical equations relating to the photon energy balance. With the range of solar insolation varying from 3.25 to 6.08 kWh m^{-2} d^{-1} for the Indian subcontinent, the microalgal biomass productivity has been predicted to range from 66.3 to 90.10 g m^{-2} d^{-1}. The corresponding lipid productivity of 16.7–31.3 ml m^{-2} d^{-1} and CO_2 sequestration capacity of 13.69–23.64 g m^{-2} d^{-1} were also postulated in the current study. As given in Table 2, the data obtained in the present study bears resemblance and is consistent with that of data reported in the literature for the Indian scenario. Variation in the values of yields obtained in different case studies is due to the difference in geographical coordinates as well as the other parameters taken into account. Further, most of the previous studies did not consider the effect of photoinhibition. Photoinhibition causing irrevocable damage to the photosynthetic system reduces the efficiency of absorption of photons by 30–40%, thus declining the net productivity [7]. In the present study, biomass productivity declines by 32.5% while considering the effects of photoinhibition. **Aly and Balasubramanian (2016, 2017)** have also reported a decrease in

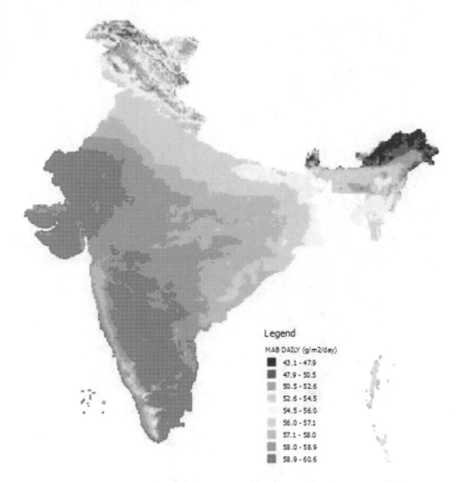

Fig. 5 Predicted microalgae productivity including the effects of photoinhibition

biomass and lipid productivity along with CO_2 sequestration potential of microalgae with the inclusion of photoinhibition effects on the model [2, 3].

4 Conclusion

The present study aimed to predict the microalgal biomass, lipid productivity and CO_2 capture potential for the entire Indian subcontinent. It is very well evident from the predicted ranges that algal technology is a potentially viable option for substituting the fossil fuels in the Indian scenario. The maximum biomass productivity predicted was 90.1 g m^{-2} d^{-1}, corresponding to the lipid productivity of 31.3 ml m^{-2} d^{-1} and CO_2 sequestration potential of 23.6 g m^{-2} d^{-1}. Microalgal biomass productivity

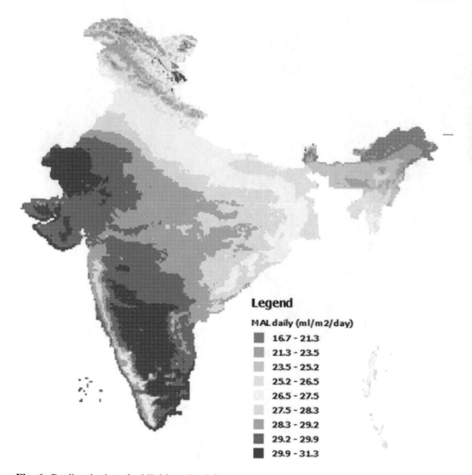

Fig. 6 Predicted microalgal lipid productivity along with the photoinhibition effect

has been found to decline by 32.5% due to photoinhibition, which also subsequently decreases the lipid productivity and CO_2 sequestration potential. However, the data obtained is limited only by solar insolation received, but the net microalgal productivity is an annexation of several other environmental and operational parameters. The inclusion of other influencing parameters like effects of hydrodynamic variation of water temperature, air temperature, initial microalgal concentration, and reactor geometry could further improvise the accuracy level of the model. Such theoretical studies would be helpful to the researchers, decision makers, and other stakeholders in accessing the algal biofuel potential in the arena of alternative energy and biological carbon sequestration.

Acknowledgements The authors are thankful to the Department of Biotechnology and Medical Engineering, NIT Rourkela, for providing the necessary facilities to carry out this research work.

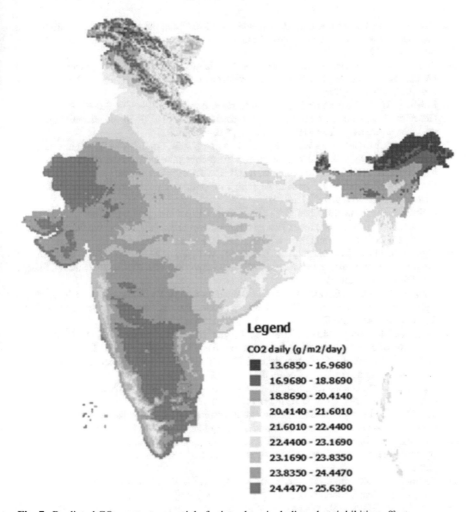

Fig. 7 Predicted CO_2 capture potential of microalgae including photoinhibition effect

The authors greatly acknowledge the Ministry of Human Resources Development, Government of India, for sponsoring the Ph.D. program of the first author.

References

1. Aly, N., Balasubramanian, P.: Effect of geographical coordinates on carbon dioxide sequestration potential by microalgae. Int. J. Environ. Sci. Dev. **8**(2), 147–152 (2017)
2. Aly, N., Balasubramanian, P.: Effect of photoinhibition on microalgal growth in open ponds of NIT Rourkela, India. J. Biochem. Technol. **6**(3), 1034–1039 (2016)

3. Aly, N., Tarai, R.K., Kale, P.G., Paramasivan, B.: Modelling the effect of photoinhibition on microalgal production potential in fixed and trackable photobioreactors in Odisha, India. Curr. Sci. **113**(2), 272–283 (2017)
4. Asmare, A.M., Demessie, B.A., Murthy, G.S.: Theoretical estimation the potential of algal biomass for biofuel production and carbon sequestration in Ethiopia. Int. J. Renew Energy Res. **3**(3), 560–570 (2013)
5. Chisti, Y.: Biodiesel from microalgae. Biotechnol. Adv. **25**(3), 294–306 (2007)
6. González, J.A., Calbó, J.: Modelled and measured ratio of PAR to global radiation under cloudless skies. Agric. For. Meteorol. **110**(4), 319–325 (2002)
7. Huesemann, M., Crowe, B., Waller, P., Chavis, A., Hobbs, S., Edmundson, S., Wigmosta, M.: A validated model to predict microalgae growth in outdoor pond cultures subjected to fluctuating light intensities and water temperatures. Algal Res. **13**, 195–206 (2016)
8. Kandpal, T.C., Broman, L.: Renewable energy education: a global status review. Renew and Sust. Energ. Rev. **34**, 300–324 (2014)
9. Maity, J.P., Bundschuh, J., Chen, C.Y., Bhattacharya, P.: Microalgae for third generation biofuel production, mitigation of greenhouse gas emissions and wastewater treatment: present and future perspectives—A mini review. Energy **78**, 104–113 (2014)
10. Milano, J., Ong, H.C., Masjuki, H.H., Chong, W.T., Lam, M.K., Loh, P.K., Vellayan, V.: Microalgae biofuels as an alternative to fossil fuel for power generation. Renew Sustain. Energy Rev. **58**, 180–197 (2016)
11. Rangabhashiyam, S., Behera, B., Aly, N., Balasubramanian, P.: Biodiesel from microalgae as a promising strategy for renewable bioenergy production—A review. J Environ. Biotechnol. Res. **6**(4), 260–269 (2017)
12. Slegers, P.M., Lösing, M.B., Wijffels, R.H., Van Straten, G., Van Boxtel, A.J.B.: Scenario evaluation of open pond microalgae production. Algal Res. **2**(4), 358–368 (2013)
13. Slegers, P.M., Wijffels, R.H., Van Straten, G., Van Boxtel, A.J.B.: Design scenarios for flat panel photobioreactors. Appl. Energy **88**(10), 3342–3353 (2011)
14. Sudhakar, K., Premalatha, M., Rajesh, M.: Large-scale open pond algae biomass yield analysis in India: a case study. Int. J. Sustain. Energy **33**(2), 304–315 (2014)
15. Sudhakar, K., Premalatha, M.: Theoretical assessment of algal biomass potential for carbon mitigation and biofuel production. Iranica J. Energy Environ. **3**(3), 232–240 (2012)
16. Sudhakar, K., Rajesh, M., Premalatha, M.: A mathematical model to assess the potential of algal bio-fuels in India. Energy Sour. A: Recovery Util. Environ. Effects **34**(12), 1114–1120 (2012)
17. Torzillo, G., Pushparaj, B., Masojidek, J., Vonshak, A.: Biological constraints in algal biotechnology. Biotechnol. Bioprocess Eng. **8**(6), 338–348 (2003)
18. Weyer, K.M., Bush, D.R., Darzins, A., Willson, B.D.: Theoretical maximum algal oil production. Bioenerg Res. **3**(2), 204–213 (2010)
19. Zhou, Y., Schideman, L., Yu, G., Zhang, Y.: A synergistic combination of algal wastewater treatment and hydrothermal biofuel production maximized by nutrient and carbon recycling. Energy Environ. Sci. **6**(12), 3765–3779 (2013)

Investigation of Comparison Approach for Optimal Location of STATCOM Based Transient Stability Improvement Using Computational Algorithms

P. K. Dhal

Abstract In practical approach of power system scenario, the voltage instability and voltage collapse have been very important consideration due to their stress operation. So the location of flexible AC transmission system (FACTS) helps to improve power system performance. The FACTS is a power electronic based system. It is used to enhance the power system controllability and transfer capability. In this paper investigate the comparison analysis of transient stability is achieved using particle swarm optimization, Genetic algorithm and Biogeography based optimization. These algorithms are used to optimally locate STATCOM device. The various operating condition like without STATCOM, with STATCOM, with STATCOM tuned by PSO, BBO and GA are used to evaluate the performance of the proposed system. The simulation results have been obtained by PSAT software. The main aim is to identify best computational algorithm and to improve voltage stability in modern power system.

Keywords STATCOM · Transient stability · Particle swarm optimization
Genetic algorithm and biogeography based optimization

1 Introduction

The modern environment of power system scenario is being operated under highly stress condition due to increase in continuous power demand. It is directly threats to society for maintain the required steady state bus voltage [1, 2]. It is necessary to preserve the voltage at all buses close to nominal value at all the times. The ability of a power system is maintained in an acceptable range called as voltage stability. If the voltage moves out too much from their nominal value, it is called as voltage collapse. The main factor causing voltage instability is an ability of the power

P. K. Dhal (✉)
Department of Electrical and Electronics Engineering, Veltech Dr. RR & Dr. SR University,
Chennai, India
e-mail: pradyumna.dhal@rediffmail.com

© Springer Nature Singapore Pte Ltd. 2019
J. C. Bansal et al. (eds.), *Soft Computing for Problem Solving*,
Advances in Intelligent Systems and Computing 816,
https://doi.org/10.1007/978-981-13-1592-3_63

system to meet the demand for particularly reactive power [3, 4]. Therefore in the field of power system, new technologies are established. It is called as Flexible AC transmission system (FACTS). The flexible AC transmission systems are the modern devices which are used in many aspects like power stability, power quality etc. The FACTS device i.e. STATCOM is a device used to inject or absorb the reactive power. It is also used to regulate the system voltage by absorbing and injecting reactive power [5, 6]. The optimal location and tuning of STATCOM has an essential role to enhance the stability. This paper presents the investigation of best location and tuning of STATCOM to enhance the transient stability after clearing three phase fault [7–9].

2 Proposed of Particle Swarm Optimization Algorithm with Flow Chart

The particle swarm optimization is an evolutionary computation technique. It is utilized a population of individuals. It is called as particle. It flies through the problem hyper space with initial velocities. There are two best fitness value called as particle best and global best.

Step 1: To initialize every particle's velocity rate and position.
Step 2: To Calculate the fitness value and to determine P_{best}
Step 3: To determine G_{best} from the P_{best}
Step 4: To Update the velocity rate and position.
Step 5: To Check the solution is feasible or not.
Step 6: If the solution is feasible, check the iteration count.
Step 7: If the iteration count reaches the maximum, stop the process.
Step 8: If the iteration count does not reach the maximum, then continue the process from 2–7.
Step 9: if iteration reaches the maximum value, then its stop (Fig. 1).

3 Proposed of Genetic Algorithm with Flow Chart

The genetic algorithm is an optimization technique which works based on the principle of natural selection i.e. genetic mutation and survival of the fittest. It is subclass of evolutionary computing. This algorithm uses natural selection, crossover and mutation operators. By using the operators, the genetic algorithm locates the optimal solution. The procedure of the algorithm is

Step 1: To make a random population from the solution area.
Step 2: To Evaluate the fitness of each solution.
Step 3: To selection or reproduction based on the fitness.

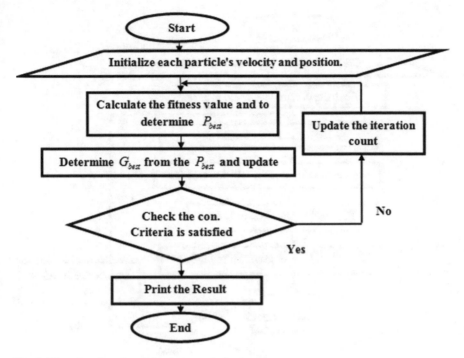

Fig. 1 Flow chart based particle swarm optimization

Step 4: To Crossover the selected solutions.
Step 5: To change the selected solutions.
Step 6: To Replace the old solution with new one.
Step 7: To take test problem criterion.
Step 8: continue steps 2–7 until the criterion is satisfied.
Step 9: Its complete, then its stop (Fig. 2).

4 Proposed of Biogeography Based Optimization Algorithm with Flow Chart

The geographical distribution of biological organism is the base of the biogeography based optimization. The biogeography mathematical model explains that the species migration from one island to another island. The island is a habitat. It is isolated from other habitats. The habitat is high suitability index (HSI). The algorithm procedure is

Step 1: To Initialize the BBO parameters.
Step 2: To Initialize a random set of habitats. (Each habitats have solution to the given problem).

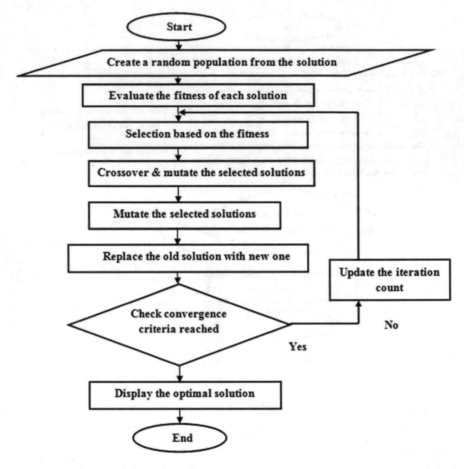

Fig. 2 Flow chart based genetic algorithm

Sep 3: Calculate the HSI to the number of the species S, the immigration rate λ and emigration rate μ.

Step 4: To Modify non elite habitats and recomputed each HSI.

Step 5: Update the probability of its species count and change every non elite habitat based on the probability.

Step 6: Recomputed each HSI.

Step 7: Go to step 3 for next iteration. If acceptable solution is found, then terminate the process (Fig. 3).

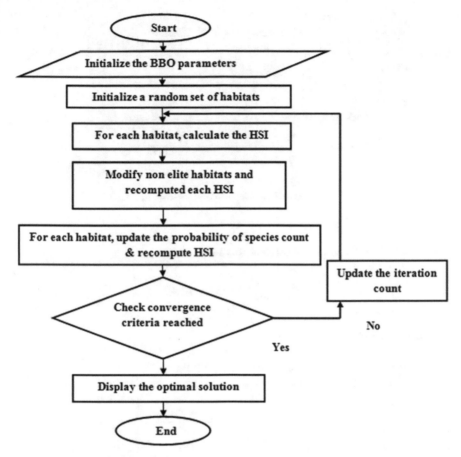

Fig. 3 Flow chart of biogeography based optimization

5 Analysis of 9-Bus System

In this system have 6 transmission lines, 3 generators, 3 loads and a local load D is considered to analyze by PSAT software. The performance of the STATCOM has been evaluated through the system. The system performance has been studied by applying a 3 phase fault. It is assumed that three phase fault time at 1.05 s and clearing time 1.15 s. It is identified that the bus 5 and bus 6 are weak buses. It is assumed that the fault is occurred either bus 5 or bus 6. Now it is considered fault is occurred at bus 6 as shown in Fig. 4.

The complete voltage profile is shown in the Table 1. That not only bus-5 and bus-6 are severely affected but real power and reactive power also affected in load side without STATCOM.

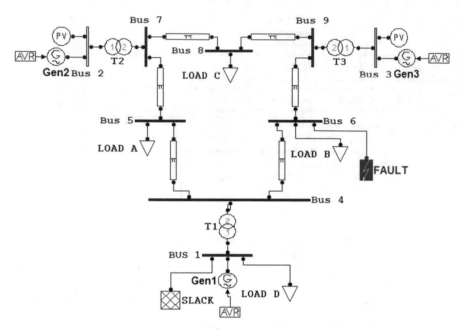

Fig. 4 Bus-9 system with fault location

Table 1 Voltage profile parameter in p.u without STATCOM

Bus	V (p.u)	Phase angle [p.u]	Input power (p.u)	Input reactive power (p.u)	Power (load) (p.u)	Reactive power (load) (p.u)
Bus 1	1.04	12.03	4.62	2.02	1.60	0.65
Bus 2	1.03	11.87	1.63	0.62	0.00	0.00
Bus 3	1.03	11.80	0.85	0.43	0.00	0.00
Bus 4	0.98	11.86	0.00	0.00	0.00	0.00
Bus 5	**0.92**	**11.71**	**0.00**	**0.00**	**2.00**	**0.90**
Bus 6	**0.93**	**11.72**	**0.00**	**0.00**	**1.60**	**0.65**
Bus 7	0.99	11.77	0.00	0.00	0.00	0.00
Bus 8	0.97	11.69	0.00	0.00	1.80	0.60
Bus 9	1.00	11.75	0.00	0.00	0.00	0.00

In Figs. 5, 6 and 7, it is identified that at bus 5, voltage value is 0.92 p.u and bus 6, voltage value is 0.93 p.u. it is severely affected due to three phase fault (Figs. 8, 9, 10 and 11).

Now the STATCOM device is connected to bus-6 as shown in Fig. 12 to improve voltage profile (Figs. 13, 14, 15, 16, 17, 18, 19, 20 and 21; Tables 2, 3, 4 and 5).

Table 2 Voltage profile in p.u with STATCOM tuned by PSO

Bus	V in (p.u)	Phase in (p.u)	Power input (p.u)	Reactive power input (p.u)	Real power load (p.u)	Reactive power load (p.u)
Bus 1	1.04	−91.75	4.61	1.56	1.60	0.65
Bus 2	1.03	−91.89	1.63	0.49	0.00	0.00
Bus 3	1.03	−91.97	0.85	0.16	0.00	0.00
Bus 4	1.01	−91.92	0.00	0.00	0.00	0.00
Bus 5	**0.94**	**−92.05**	**0.00**	**0.00**	**2.00**	**0.90**
Bus 6	**1.02**	**−92.05**	**0.00**	**0.66**	**1.60**	**0.65**
Bus 7	1.00	−91.99	0.00	0.00	0.00	0.00
Bus 8	0.98	−92.07	0.00	0.00	1.80	0.60
Bus 9	1.02	−92.02	0.00	0.00	0.00	0.00

Table 3 Voltage profile in p.u with STATCOM tuned by GA

Bus	V in (p.u)	Phase in (p.u)	Power input (p.u)	Reactive power input (p.u)	Real power load (p.u)	Reactive power load (p.u)
Bus 1	1.04	−54.03	4.61	1.56	1.6	0.65
Bus 2	1.03	−54.17	1.63	0.49	0.00	0.00
Bus 3	1.03	−54.25	0.85	0.16	0.00	0.00
Bus 4	1.01	−54.20	0.00	0.00	0.00	0.00
Bus 5	**0.94**	**−54.34**	**0.00**	**0.00**	**2.00**	**0.90**
Bus 6	**1.01**	**−54.33**	**0.00**	**0.66**	**1.60**	**0.65**
Bus 7	1.00	−54.27	0.00	0.00	0.00	0.00
Bus 8	0.98	−54.36	0.00	0.00	1.80	0.60
Bus 9	1.02	−54.30	0.00	0.00	0.00	0.00

Table 4 Voltage profile in p.u with STATCOM tuned by BBO

Bus	V in (p.u)	Phase in (p.u)	Power input (p.u)	Reactive power input (p.u)	Real power load (p.u)	Reactive load (p.u)
Bus 1	1.04	−165.77	4.61	1.56	1.60	0.65
Bus 2	1.03	−165.92	1.63	0.49	0.00	0.00
Bus 3	1.03	−166.00	0.85	0.16	0.00	0.00
Bus 4	1.02	−165.95	0.00	0.00	0.00	0.00
Bus 5	**0.95**	**−166.08**	**0.00**	**0.00**	**2.00**	**0.90**
Bus 6	**1.04**	**−166.09**	**0.00**	**0.66**	**1.60**	**0.65**
Bus 7	1.01	−166.02	0.00	0.00	0.00	0.00
Bus 8	0.99	−166.10	0.00	0.00	1.80	0.60
Bus 9	1.03	−166.05	0.00	0.00	0.00	0.00

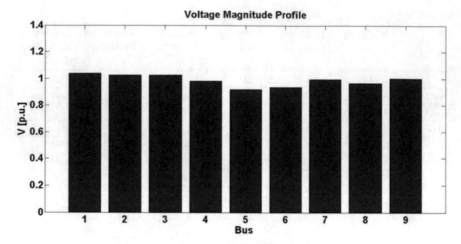

Fig. 5 Voltage profile in p.u without STATCOM

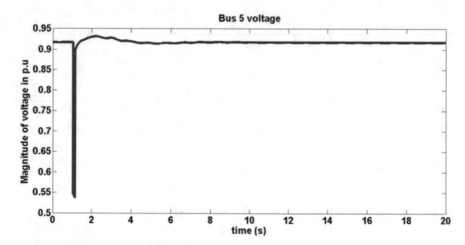

Fig. 6 Bus 5 voltage in p.u without STATCOM

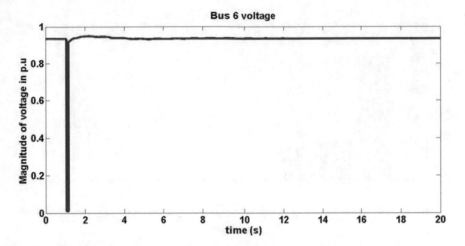

Fig. 7 Bus 6 voltage in p.u without STATCOM

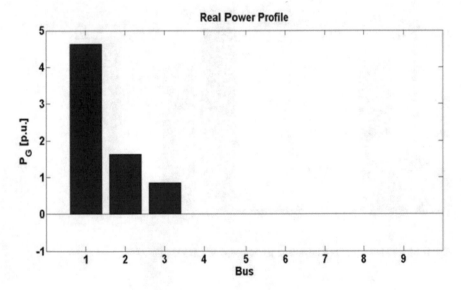

Fig. 8 Real power profile generator side in p.u without STATCOM

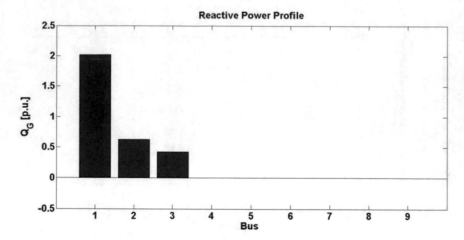

Fig. 9 Reactive power profile generator side in p.u without STATCOM

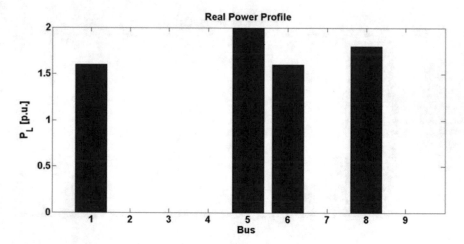

Fig. 10 Real power profile load side in p.u without STATCOM

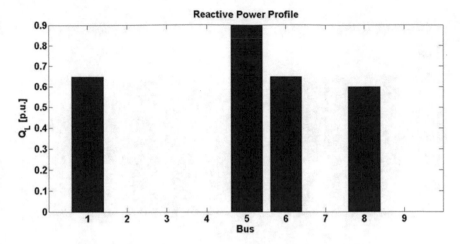

Fig. 11 Reactive power profile Load side in p.u without STATCOM

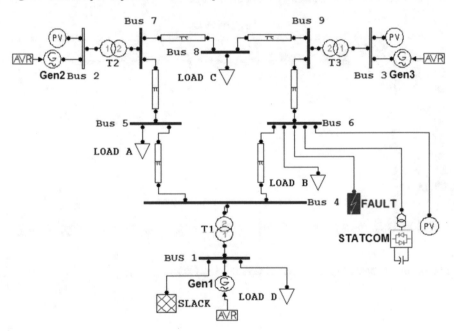

Fig. 12 Bus-9 system with STATCOM at bus 6 position

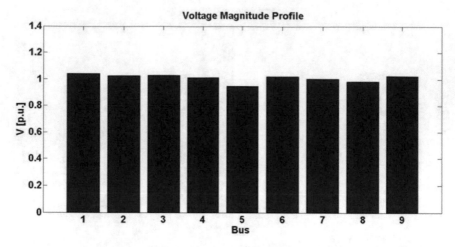

Fig. 13 Voltage profile in p.u with STATCOM tuned by PSO

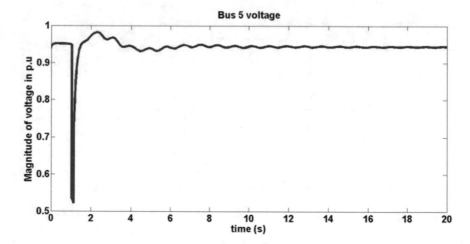

Fig. 14 Bus 5 voltage in p.u with STATCOM tuned by PSO

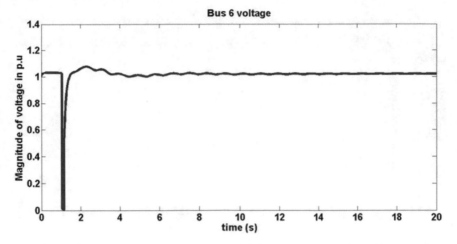

Fig. 15 Bus 6 voltage in p.u with STATCOM tuned by PSO

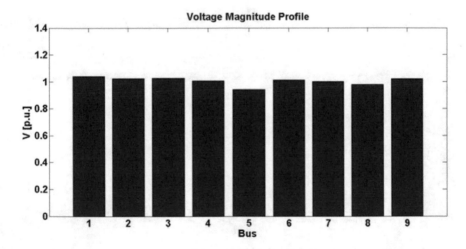

Fig. 16 Voltage profile in p.u with STATCOM tuned by GA

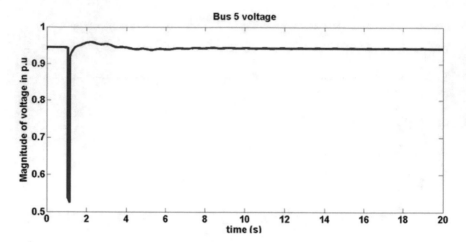

Fig. 17 Bus 5 voltage in p.u with STATCOM tuned by GA

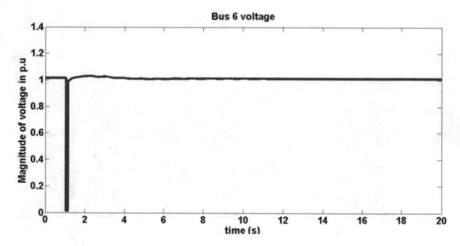

Fig. 18 Bus 6 voltage in p.u with STATCOM tuned by GA

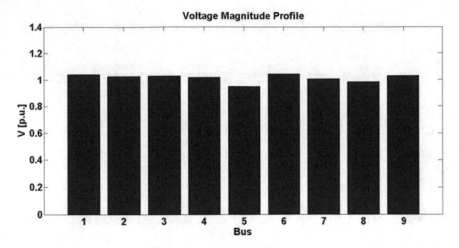

Fig. 19 Voltage profile in p.u with STATCOM tuned by BBO

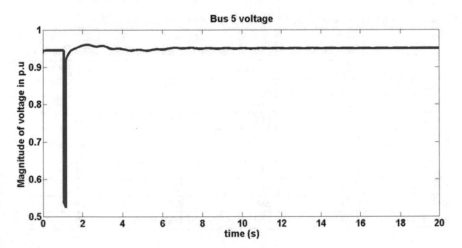

Fig. 20 Bus 5 voltage in p.u with STATCOM tuned by BBO

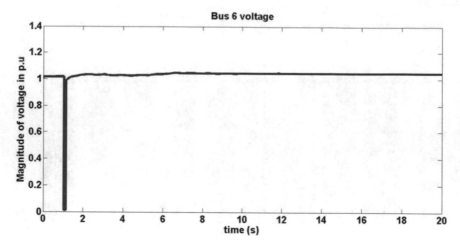

Fig. 21 Bus 6 voltage in p.u with STATCOM tuned by BBO

Table 5 Comparison of voltage profile in p.u using computational algorithms

Bus	Without STATCOM	With STATCOM (tuned by GA)	With STATCOM (tuned by PSO)	With STATCOM (tuned by BBO)
	V in p.u	V in p.u	V in p.u	V in p.u
Bus 1	1.04	1.04	1.04	1.04
Bus 2	1.03	1.03	1.03	1.03
Bus 3	1.03	1.03	1.03	1.03
Bus 4	0.98	1.01	1.01	1.02
Bus 5	**0.92**	**0.94**	**0.94**	**0.95**
Bus 6	**0.93**	**1.01**	**1.02**	**1.04**
Bus 7	0.99	1.00	1.00	1.01
Bus 8	0.97	0.98	0.98	0.99
Bus 9	1.00	1.02	1.02	1.03

6 Conclusion

The voltage improvement in 9-bus system is being discussed. The optimal location at which, the value of objective function is minimum. It can be found at bus 6 that means a STATCOM device gives best optimum value for the objective function. The voltage stability of the system using STATCOM is investigated through the computational algorithms. The proposed algorithms are used to find best method to get voltage stability improvement. By proper tuning of the STATCOM using PSO, GA and BBO algorithm, voltage stability of the system is achieved. The results are compared with the PSO tuned by STATCOM. It is an improved to 0.94 pu and 1.02 pu value. The voltage value is improved by genetic algorithm which is 0.94 and 1.01 pu value. Similarly by using biogeography algorithm, voltage value is improved to 0.95 and 1.04 pu. It shows the effectiveness of the tuned system.

References

1. Karthikeyan, K., Dhal, P.K.: Multi verse optimization (MVO) technique based voltage stability analysis through continuation power flow in IEEE57 bus. Energy Proc. **117**, 583–591 (2017). Elsevier
2. Venkateswara Rao B., Nagesh Kumar, G.V.: Firefly algorithm based optimal power flow for sizing of thyristor controlled series capacitor to enhance line based voltage stability. J. Electr. Eng., **16** (2016)
3. Mirjalili, S., Mirjalili, S.M., Hatamlou, A.: Multi-verse optimizer: a nature-inspired algorithm for global optimization. Nat. Comput. Appl. Forum (2015)
4. Aysha, P.A., Baby A.: Transient stability assessment and enhancement in power system. Int. J. Mod. Eng. Res. **4**(9), 61–65 (2014)
5. Elsheikh, A., Helmy, Y., Abouelseoud, Y., Elsherif, A.: Optimal power flow and reactive compensation using a particle swarm optimization algorithm. J. Electr. Syst. **10-1**, 63–77 (2014)
6. Welhazi, Y., Guesmi, T., Jaoued, I., Abdallah, H.H.: Power system stability enhancement using FACTS controllers in multi machine power systems. J. Electr. Syst. **10-3**, 276–291 (2014)
7. Sambariya, D.K., Prasad, R.: Robust tuning of power system stabilizer for small signal stability enhancement using meta heuristic bat algorithm. Electr. Power Energy Syst. **61**, 229–238 (2014)
8. Fughar, A., Nwohu, M.N.: Optimal location of STATCOM in Nigerian 330 kv network using ant colony optimization meta-heuristic. Glob. J. Res. Eng.: F Electr. Electron. Eng. **14**(3) 2014 (version 1.0)
9. Marefatjou, H., Soltani, I.: optimal placement of STATCOM to voltage stability improvement and reduce power losses by using QPSO algorithm. J. Sci. Eng. **2**(2), 105–119 (2013)

Genetic Algorithm for Multi-choice Integer Linear Programming Problems

D. K. Mohanty, R. K. Jana and M. P. Biswal

Abstract Genetic algorithms (GAs) are very powerful techniques to solve difficult combinatorial optimization problems. Multi-choice programming (MCP) belongs to a class of combinatorial optimization problems where the decision maker (DM) has to choose a value from a number of alternative choices and to find a combination which optimizes an objective function subject to a given set of constraints. In many of the real-life optimization problems, the solution of some problems cannot be fractions and must be specified as integers, and solving of such types of problems is called integer programming (IP). In this paper, a genetic algorithm (GA) for nonlinear integer programming (NLIP) to solve a multi-choice integer linear programming (MCILP) problem is proposed. Here, we consider an MCILP problem where some or all right-hand side parameters of the constraints may have multiple choices. The multi-choice parameters in the problem are handled using the interpolating polynomials. After constituting interpolating polynomials corresponding to all multi-choice parameters, an NLIP model is formulated. Using our proposed GA, optimal integer solution of the NLIP model is obtained. Lastly, some illustrative examples are presented to support the solution procedure.

D. K. Mohanty (✉) · M. P. Biswal
Department of Mathematics, Indian Institute of Technology Kharagpur,
Kharagpur 721302, India
e-mail: dkmohanty.iitkgp@gmail.com

M. P. Biswal
e-mail: mpbiswal@maths.iitkgp.ernet.in

R. K. Jana
Indian Institute of Management Raipur, GEC Campus, Sejbahar 492015, India
e-mail: rkjana1@gmail.com

© Springer Nature Singapore Pte Ltd. 2019
J. C. Bansal et al. (eds.), *Soft Computing for Problem Solving*,
Advances in Intelligent Systems and Computing 816,
https://doi.org/10.1007/978-981-13-1592-3_64

1 Introduction

In some real-life decision-making problems, in which the DM is allowed to set multiple numbers of values for a parameter. Such types of optimization problems are known as multi-choice programming problem. Multi-choice programming (MCP) problem is first developed by Healey [1]. Multi-choiceness occurs for any parameter in an optimization problem where decision maker has to choose one value from a set of values of a parameter. Lin [2] presented a survey of MCP problems and their different solution procedures. Chang [3] has developed a method to solve multi-choice goal programming(MCGP) in which aspiration levels are multi-choice types, and Paksoy and Chang [4] have proposed a binary mixed-integer programming model and used the revised MCGP approach of Chang [5] to tackle multi-choice parameters for solving a supply chain model in guerrilla marketing. Liao [6] applied the method of Chang [3] to solve the multi-segment goal programming problem with multi-segment aspiration levels. Then, Biswal and Acharya [7]introduced multi-choice linear programming (MCLP) problems having two or more goals in the right-hand side constraints. They developed a transformation technique using binary variables to obtain an equivalent mixed-integer nonlinear programming (MINLP) problem. Further, Biswal and Acharya [8] proposed interpolating polynomials for each multi-choice type parameters to transform the model to a MINLP problem. Chang et al. [9] used Chang's [3] transformation technique to solve multi-coefficient goal programming for group pricing discrimination problems. Pradhan and Biswal [10] proposed a solution procedure to solve stated multi-level programming problem where some (or all) of the parameters of the objective functions and the constraints are multiple alternative choices. Recently, they [11] presented a multi-choice stochastic programming problem where alternative choices of multi-choice parameters are considered to be random variables.

In computer science and operations research, genetic algorithms (GAs) are general-purpose population-based stochastic search techniques inspired by the process of natural selection and evolution. GA was first introduced by John Holland [12], at the University of Michigan, and GAs were initially designed as computer-based models which were used to solve problems based on adaptive processes associated with natural genetics. The GA is a suitable optimization method, especially for solving problems, which involve nonsmooth and multimodal search spaces. GAs have been successfully applied to solve many optimization problems by Dejong [13], Goldberg and Richardson [14], and Abdullah [15]. Homaifar et al. [16] presented an application of genetic algorithms (GAs) to nonlinear constrained in which penalty function was utilized in the objective function to account for violation optimization. Deb [17] developed a constraint handling method for GAs which does not require any penalty parameter. Jana and Biswal [18] presented a stochastic simulation-based GA for solving chance constraint programming problems, where the random variables present in the problem follow some specific continuous distributions. Today GAs are used to solve the difficult problems such as mixed-integer nonlinear programming (MINLP) problems. Nazario and Ruggiero [19] proposed a heuristic algorithm to solve mixed-integer programming problem. Yokota et al. [20] proposed a penalty function approach for solving MINLP problems using genetic algorithm (GA). Wasanapradit et al. [21] solved mixed-integer nonlinear problem by using modified genetic algorithms. He improved the work of Yokota et al. [20] by using secant method and bisection method to convert infeasible chromosomes to feasible chromosomes.

In some real-life mathematical programming problems, the fractional solutions of decision variables are not realistic; therefore, it must be integer values. So to solve this kind of problem, integer programming (IP) is required. IP is a basic approach to real-life application problems such as multi-commodity flow network design, production planning, vehicle routing, manpower planning, job scheduling, and plant location. It usually depends on Branch and Bound technique, but in case of some complicated problems more developed method such as genetic algorithm is required. In this study, we consider an integer linear programming problem in which some parameters in the right-hand side of constraints are having multiple values. We use interpolating polynomial technique to deal with the multi-choice parameters and obtain a nonlinear integer programming (NLIP) problem; then, we propose an integer GA to solve the transformed NLIP model to obtain integer solution.

2 Multi-choice Integer Linear Programming Problems

In a multi-choice linear programming (MCLP) problem, the DM has to choose one value from a set of values for a particular parameter to optimize the objective function. In many real-life applications of MCLP model, some or all the variables are restricted to be integers; such problems are called multi-choice integer linear programming (MCILP) problems. In practice, MCILP model can be extended to applications like multi-item scheduling, transportation problem, sales resource allocation, timetabling, supply chain management, multi-choice knapsack problems. The mathematical model of the MCILP problem where right-hand side parameters of the constraints are of multi-choice types is given by:

$$\min : z = \sum_{j=1}^{n} c_j y_j \tag{1}$$

subject to

$$\sum_{j=1}^{n} a_{ij} y_j \geq \{b_i^{(1)}, b_i^{(2)}, \ldots, b_i^{(t_i)}\}, \quad i = 1, 2, 3, \ldots, m \tag{2}$$

$$y_j = 0, 1, 2, \ldots, \quad j = 1, 2, 3, \ldots, n. \tag{3}$$

where parameter b_i of the ith constraint is a multi-choice type having t_i number of alternative choices. To solve the above MCILP problem (1)–(3) directly, we need to solve $\prod_{i=1}^{m} t_i$ number of linear programming problems. Since this procedure is lengthy, some new techniques are required to solve the problem more efficiently. There are three different transformation approaches available in the literature to solve an MCP problem; these are auxiliary variable approach, binary variable approach, and interpolating polynomial approach. Using any one of these techniques, MCILP problem can be transformed into a nonlinear integer problem(NLIP). By using the integer programming technique, the optimal solution of the transformed model can be obtained and hence the optimal solution of the MCILP problem can be obtained. These three techniques are discussed below.

2.1 Transformation Techniques

Here, we discuss the aforementioned transformation approaches used to transform an MCILP problem into an NLIP problem.

Auxiliary Variable Approach:
In this approach (Phillip et al. [22], Sarker and Newton [23]), the number of binary variables needed for a particular constraint equals to the total number of alternative values of the multi-choice parameter of that constraint. There will be also one extra constraint required for each of the multi-choice parameter. Using this technique, the ith constraint of the model (1)–(3) is transformed as:

$$\sum_{j=1}^{n} a_{ij} y_j \geq v_i^{(1)} b_i^{(1)} + v_i^{(2)} b_i^{(2)} + \cdots + v_i^{(t_i)} b_i^{(t_i)}, \quad i = 1, 2, 3, \ldots, m \qquad (4)$$

$$v_i^{(1)} + v_i^{(2)} + \cdots + v_i^{(t_i)} = 1, \quad i = 1, 2, 3, \ldots, m \qquad (5)$$

where $v_i^{(t)}$, $(i = 1, 2, \ldots, m; t = 1, 2, \ldots, t_i)$ are required binary variables. Here, m number of additional constraints and $\sum_{i=1}^{m} t_i$ number of additional binary variables are required to transform the MCILP problem.

Binary Variable Approach:
This method was proposed by Biswal and Acharya [7]. Here, $\prod_{i=1}^{m} \lceil (\frac{\ln(t_i)}{\ln 2}) \rceil$ number of binary variables are needed for the problem (1)–(3) and the number of binary variables required depends on the number of alternative values of the multi-choice parameter. Some new constraints are also required to avoid the repetition of alternative choices of the parameters. In this approach, we need less number of additional constraints and less number of binary variables than the previous technique.

Interpolating Polynomial Approach:
Some major difficulties were faced when transforming an MCLP problem into a standard mathematical model in above two methods. These difficulties arise while choosing binary variables, selecting bounds for binary code and restricting binary

Table 1 Table for multi-choice coefficient b_i

v_i	0	1	2	$\cdots$	$t_i - 1$
$f_{b_i}(v_i)$	$b_i^{(1)}$	$b_i^{(2)}$	$b_i^{(3)}$	$\cdots$	$b_i^{(t_i)}$

codes using auxiliary constraints. To overcome such difficulties, interpolating polynomial technique for MCLP problem is developed by Biswal and Acharya [8]. In this technique, an integer variable v_i is introduced for each multi-choice parameter b_i for the ith constraint of the model (1)–(3). Let $0, 1, 2, ..., (t_i - 1)$ be t_i number of node points, where $b_i^{(1)}, b_i^{(2)}, ..., b_i^{(t_i)}$ are the associated functional values of the interpolating polynomial at t_i different node points. A polynomial $f_{b_i}(v_i)$ of degree $(t_i - 1)$ is derived which interpolates the given data: $f_{(t_i-1)}(b^j) = b^(j + 1)$, $j = 0, 1, 2, ..., (t_i - 1), i = 1, 2, 3, ..., m$.

We construct an interpolating polynomial $f_{b_i}(v_i)$ of degree $t_i - 1$ which interpolates the given data (Table 1). Using Lagrange's interpolation formula (Atkinson [24]), the interpolating polynomial can be established as:

$$
\begin{aligned}
f_{b_i}(v_i) = &\ \frac{(v_i - 1)(v_i - 2) \cdots (v_i - t_i + 1)}{(-1)^{(t_i-1)}(t_i - 1)!} b_i^{(1)} + \frac{v_i(v_i - 2) \cdots (v_i - t_i + 1)}{(-1)^{(t_i-2)}(t_i - 2)!} b_i^{(2)} \\
&+ \frac{v_i(v_i - 1)(v_i - 3) \cdots (v_i - t_i + 1)}{(-1)^{(t_i-3)} 2!(t_i - 3)!} b_i^{(3)} + \cdots \\
&+ \frac{v_i(v_i - 1)(v_i - 2) \cdots (v_i - t_i + 2)}{(t_i - 1)!} b_i^{(t_i)}, \quad i = 1, 2, ..., m.
\end{aligned}
\tag{6}
$$

Then replacing the multi-choice parameters of the problem (1)–(3) with their corresponding interpolating polynomials, equivalent transformed model is established as:

$$
\min : z = \sum_{j=1}^{n} c_j y_j
\tag{7}
$$

subject to

$$
\begin{aligned}
\sum_{j=1}^{n} a_{ij} y_j \geq &\ \frac{(v_i - 1)(v_i - 2) \cdots (v_i - t_i + 1)}{(-1)^{(t_i-1)}(t_i - 1)!} b_i^{(1)} + \frac{v_i(v_i - 2) \cdots (v_i - t_i + 1)}{(-1)^{(t_i-2)}(t_i - 2)!} b_i^{(2)} \\
&+ \frac{v_i(v_i - 1)(v_i - 3) \cdots (v_i - t_i + 1)}{(-1)^{(t_i-3)} 2!(t_i - 3)!} b_i^{(3)} + \cdots \\
&+ \frac{v_i(v_i - 1)(v_i - 2) \cdots (v_i - t_i + 2)}{(t_i - 1)!} b_i^{(t_i)}, \quad i = 1, 2, ..., m.
\end{aligned}
\tag{8}
$$

$$
y_j = 0, 1, 2, ..., j = 1, 2, 3, ..., n; \ 0 \leq v_i \leq t_i - 1; \ v_i \in \mathbb{N}_0 \ \ i = 1, 2, 3, ..., m.
\tag{9}
$$

Instead of Lagrange's interpolating polynomial, other interpolating polynomials can be used for the transformation . In comparison with the previous approaches, this approach needs less number of new variables, and no extra constraints are required for the transformation. But the model becomes a nonlinear integer programming (NLIP) problem after using transformation techniques. Using any NLIP technique, the above model (7)–(9) can be solved. We have used integer GA to solve this model, which has been discussed in the following section.

3 GA for Nonlinear Integer Programming

GAs belong to the class of direct method approach and are more robust than indirect methods. GA being a direct method has a greater chance of success when nonlinear programming (NLP) problem is solved. It does not require any derivative and also avoids getting stuck in local minima. Since after using transformation technique, MCILP problem is transformed into an NLIP problem, GA can be a better option to solve this NLIP problem.

3.1 Step-by-Step Procedure

Our proposed GA is designed to solve NLIP problems. The steps of the algorithm are described in details:

1. **Representation and initialization**
   We define an integer P_{size} as the number of chromosomes or size of the population. Let $y_1, y_2, ..., y_n$ be the n number of decision variables, all the variables are declared as integer types. Then, each chromosome can be represented as

$$Y_P = (y_1, y_2, ..., y_n)_P$$

   where $P = 1, 2, ..., P_{size}$; chromosomes' P_{size} are randomly initialized. The values of $y_j (j = 1, 2, ..., n)$ are chosen uniformly between 0 and the upper limits of jth decision variable y_j, $(j = 1, 2, ..., n)$.

2. **Constraint Checking**
   In this step, constraint checking is done for both linear and nonlinear constraints present in NLIP problem (7)–(9)

3. **Fitness**
   Here, objective function values which satisfy the given constraints are taken as the fitness function. After all the constraints conditions are satisfied, the fitness value of each chromosome is calculated from the objective function value.

4. **Selection**

   In the selection process, chromosomes are stochastically selected from one generation to take part in the next generation. Fitter chromosome will have a better survival rate and will take part to create a mating pool in the next generation. To select fittest chromosome for the next generation, we have used roulette-wheel selection strategy [12].

5. **Crossover**

   The crossover operator combines the genes of two or more parents to generate better offspring. First, we define the probability of crossover as a parameter P_c. Then, a random number r is generated within $(0, 1)$ for each pair of chromosomes in the current population; we select the given chromosome for crossover if $r <$ P_c. This operation is repeated for P_{size} times and produces an expected number $P_c \times P_{size}$ of parents.

6. **Mutation**

   Mutation is a process of modifying the genetic material of a chromosome from its initial state. In this step, a parameter p_m of a genetic system as the probability of mutation is defined. An expected number $P_m \times P_{size}$ of chromosomes will take part in mutation operation. For every bit in the population, we generate a random real number r from the interval $(0, 1)$; if $r < P_m$, then the corresponding chromosome is selected for mutation. The bitwise changing of genes is called mutation. This mutation operation is repeated for P_{size} number of times.

7. **Termination**

   The algorithm terminates after a given number of cyclic repetitions (generations) of the above steps or a suitable solution has been found. The elite chromosome is returned as the best solution found so far.

3.2 Summary of the Proposed GA for NLIP

Step 1: Initialize P_{size} number of chromosomes at random according to the initialization process described above.

Step 2: Check the system constraints by constraint handling method. ·

Step 3: Apply crossover operation as discussed above.

Step 4: Apply mutation operation as described above to update the chromosomes.

Step 5: Compute the fitness value of each chromosome via corresponding objective value.

Step 6: Select the chromosomes according to the selection process described above.

Step 7: Repeat the Steps 2–6 until termination criteria are met.

Step 8: Report the best chromosome as the optimal solution.

4 Numerical Example and Result Discussion

In this section, we have presented two numerical examples to illustrate our methodology.

Example 1 Here, we consider an example stated by Biswal and Acharya [7]. In this study, the MCILP problem is formulated as:

$$min : z = 160y_1 + 400y_2 + 300y_3 \tag{10}$$

subject to

$$3y_1 + 6y_4 + 5y_3 \geq \{33, 36, 39\} \tag{11}$$

$$4y_1 + 6y_2 + 3y_3 \geq \{15, 17, 20, 24, 28\} \tag{12}$$

$$2y_1 + 8y_2 + 2y_3 \geq \{30, 33, 35, 37, 40, 45\} \tag{13}$$

$$y_j = 0, 1, 2, \ldots; \quad j = 1, 2, 3. \tag{14}$$

Instead of binary variable approach, we have used Lagrange's interpolating polynomial technique (2.1), model (7)–(9) to establish the transformed model as:

$$min : z = 160y_1 + 400y_2 + 300y_3 \tag{15}$$

subject to

$$3y_1 + 6y_4 + 5y_3 \geq 3v_1 + 33 \tag{16}$$

$$4y_1 + 6y_2 + 3y_3 \geq \frac{v_2^4}{24} + \frac{v_2^3}{4} + \frac{v_2^2}{24} + \frac{7}{4}v_2 + +15 \tag{17}$$

$$2y_1 + 8y_2 + 2y_3 \geq \frac{37}{12}v_3^5 - 37v_3^4 + \frac{605}{4}v_3^3 - \frac{483}{2}v_3^2 + \frac{763}{6}v_3 + 30 \tag{18}$$

$$v_1 = 0, 1, 2; \ v_2 = 0, 1, 2, 3, 4; \ v_3 = 0, 1, 2, 3, 4, 5 \tag{19}$$

$$y_j = 0, 1, 2, \ldots; \quad j = 1, 2, 3. \tag{20}$$

The above nonlinear integer program is solved using GA. The proposed GA is coded in C programming language with population size = 100, total number of generations = 100,000, crossover probability = 0.2, mutation probability = 0.01. The solution to the above mathematical model is obtained as: $y_1 = 7$, $y_2 = 2$, $y_3 = 0$ and minimum value of $z = 1920$.

Our proposed procedure gives same value for the objective function with slight changes in values of the decision variables as compared to the result given by Biswal and Acharya [7].

Example 2 Here, we consider an example stated by Biswal and Acharya [8].

By considering all the information given in the example, an MCILP problem can be established as:

$$min : z = \sum_{j=1}^{7} y_j \tag{21}$$

subject to

$$y_1 + y_4 + y_5 + y_6 + y_7 \geq \{14, 16, 17\} \tag{22}$$
$$y_1 + y_2 + y_5 + y_6 + y_7 \geq \{12, 13\} \tag{23}$$
$$y_1 + y_2 + y_3 + y_6 + y_7 \geq \{14, 15, 17, 19\} \tag{24}$$
$$y_1 + y_2 + y_3 + y_4 + y_7 \geq \{15, 16, 18, 19, 21\} \tag{25}$$
$$y_1 + y_2 + y_3 + y_4 + y_5 \geq 14 \tag{26}$$
$$y_1 + y_3 + y_4 + y_5 + y_6 \geq \{15, 17\} \tag{27}$$
$$y_3 + y_4 + y_5 + y_6 + y_7 \geq \{11, 14, 15\} \tag{28}$$
$$y_j = 0, 1, 2, \ldots; \quad j = 1, 2, 3, \ldots, 7. \tag{29}$$

Using Lagrange's interpolating polynomial technique (2.1), model (7)–(9) the MCILP model (21)–(29) can be transformed into a NLIP problem as

$$min : z = \sum_{j=1}^{7} y_j \tag{30}$$

subject to

$$y_1 + y_4 + y_5 + y_6 + y_7 \geq -\frac{1}{2}v_1^2 + \frac{5}{2}v_1 + 14 \tag{31}$$
$$y_1 + y_2 + y_5 + y_6 + y_7 \geq v_2 + 12 \tag{32}$$
$$y_1 + y_2 + y_3 + y_6 + y_7 \geq -\frac{1}{6}v_3^3 + v_3^2 + \frac{7}{6}v_3 + 14 \tag{33}$$
$$y_1 + y_2 + y_3 + y_4 + y_7 \geq \frac{1}{6}v_4^4 - \frac{4}{3}v_4^3 + \frac{20}{6}v_4^2 - \frac{7}{6}v_4 + 15 \tag{34}$$
$$y_1 + y_2 + y_3 + y_4 + y_5 \geq 14 \tag{35}$$
$$y_1 + y_3 + y_4 + y_5 + y_6 \geq 2v_5 + 15 \tag{36}$$
$$y_3 + y_4 + y_5 + y_6 + y_7 \geq -v_6^2 + 4v_6 + 11 \tag{37}$$
$$v_1, v_6 = 0, 1, 2; \quad v_2, v_5 = 0, 1; \quad v_3 = 0, 1, 2, 3; \quad v_4 = 0, 1, 2, 3, 4 \tag{38}$$
$$y_j = 0, 1, 2, \ldots; \quad j = 1, 2, 3, \ldots, 7. \tag{39}$$

The above nonlinear integer program is solved using GA. The proposed GA is coded in C programming language with population size = 100, total number of generations $= 100{,}000$, crossover probability $= 0.2$, mutation probability $= 0.01$. The optimal solution obtained is as follows: $y_1 = 4$, $y_2 = 3$, $y_3 = 2$, $y_4 = 5$, $y_5 = 0$, $y_6 = 5$, $y_7 = 1$, $v_1 = 0$, $v_2 = 0$, $v_3 = 0$, $v_4 = 0$, $v_5 = 0$, $v_6 = 0$ and minimum value of $z = 20$.

Our proposed procedure gives same value for the objective function with slight changes in values of the decision variables as compared to the result given by Biswal and Acharya [8] where the values of the decision variables are: $y_1 = 4$, $y_2 = 5$, $y_3 = 0$, $y_4 = 5$, $y_5 = 0$, $y_6 = 5$, $y_7 = 1$, $v_1 = 0$, $v_2 = 0$, $v_3 = 0$, $v_4 = 0$, $v_5 = 0$, $v_6 = 0$ and minimum value of $z = 20$.

5 Conclusions

The stated MCILP problem is first transformed to a NLIP programming using interpolating polynomial approach. As genetic algorithms (GAs) are powerful tools for solving integer linear or nonlinear programming problems, our proposed GA for NLIP is used to solve the transformed problem. Our proposed GA for NLIP can be used to solve MCLP problem where technological coefficients and cost coefficients are of multi-choice type. GA can be used to solve MCILP problem directly without transforming to its deterministic form (NLIP). The proposed study can be extended to multi-objective and/or multi-level framework for multi-choice programming. Instead of GA, other evolutionary computing techniques such as particle swarm optimization, ant colony optimization, artificial bee colony optimization can be used to solve MCP problems, and better result may be obtained.

References

1. Healy Jr., W.C.: Multiple choice programming (a procedure for linear programming with zero-one variables). Oper. Res. **12**(1), 122–138 (1964)
2. Lin, E.Y.H.: Multiple choice programming: a state-of-the-art review. Int. Trans. Oper. Res. **1**(4), 409–421 (1994)
3. Chang, C.T.: Multi-choice goal programming. Omega **35**(4), 389–396 (2007)
4. Paksoy, T., Chang, C.T.: Revised multi-choice goal programming for multi-period, multi-stage inventory controlled supply chain model with popup stores in Guerrilla marketing. Appl. Math. Model. **34**(11), 3586–3598 (2010)
5. Chang, C.T.: Revised multi-choice goal programming. Appl. Math. Model. **32**(12), 2587–2595 (2008)
6. Liao, C.N.: Formulating the multi-segment goal programming. Comput. Ind. Eng. **56**(1), 138–141 (2009)
7. Biswal, M.P., Acharya, S.: Transformation of a multi-choice linear programming problem. Appl. Math. Comput. **210**(1), 182–188 (2009)
8. Biswal, M.P., Acharya, S.: Solving multi-choice linear programming problems by interpolating polynomials. Math. Comput. Model. **54**(5), 1405–1412 (2011)
9. Chang, C.T., Chen, H.M., Zhuang, Z.Y.: Multicoefficients goal programming. Comput. Ind. Eng. **62**, 616–623 (2012)

10. Pradhan, A., Biswal, M.P.: Multi-level nonlinear programming problem with some multi-choice parameter. In: Mathematics and Computing 2013, pp. 91–101. Springer, New Delhi (2014)
11. Pradhan, A., Biswal, M.P.: Multi-choice probabilistic linear programming problem. OPSEARCH **54**(1), 122–142 (2017)
12. Holland, J.H.: Adaptation in Natural and Artificial Systems: An Introductory Analysis with Applications to Biology, Control, and Artificial Intelligence. MIT Press, Cambridge (1992)
13. De Jong, K.: Adaptive system design: a genetic approach. IEEE Trans. Syst. Man Cybern. **10**(9), 566–574 (1980)
14. Goldberg, D.E., Richardson, J.: Genetic algorithms with sharing for multimodal function optimization. In: Genetic Algorithms and Their Applications: Proceedings of the Second International Conference on Genetic Algorithms, pp. 41–49. Lawrence Erlbaum, Hillsdale, NJ (1987)
15. Abdullah, A.R.: A robust method for linear and nonlinear optimization based on genetic algorithm. Cybernetica **34**(4), 279–287 (1991)
16. Homaifar, A., Qi, C.X., Lai, S.H.: Constrained optimization via genetic algorithms. Simulation **62**(4), 242–253 (1994)
17. Deb, K.: An efficient constraint handling method for genetic algorithms. Comput. Methods Appl. Mech. Eng. **186**(2), 311–338 (2000)
18. Jana, R.K., Biswal, M.P.: Stochastic simulation-based genetic algorithm for chance constraint programming problems with continuous random variables. Int. J. Comput. Math. **81**(9), 1069–1076 (2004)
19. Ramrez-Beltrn, N.D., Aguilar-Ruggiero, K.: Application of an geuristic procedure to solve mixed-integer programming problems. Comput. Ind. Eng. **33**(1–2), 43–46 (1997)
20. Yokota, T., Gen, M., Li, Y.X.: Genetic algorithm for non-linear mixed integer programming problems and its applications. Comput. Ind. Eng. **30**(4), 905–917 (1996)
21. Wasanapradit, T., Mukdasanit, N., Chaiyaratana, N., Srinophakun, T.: Solving mixed-integer nonlinear programming problems using improved genetic algorithms. Korean J. Chem. Eng. **28**(1), 32–40 (2011)
22. Phillip, D., Revindran, A., Solberg, J.: Operations Research: Principles and Practice (1975)
23. Sarker, R.A., Newton, C.S.: Optimization Nodelling: A Practical Approach. CRC Press, New York (2007)
24. Micha, D.A.: An Introduction to Numerical Analysis. Atkinson, Kendall E, New York (1980)

Multi-channel, Multi-slice, and Multi-contrast Compressed Sensing MRI Using Weighted Forest Sparsity and Joint TV Regularization Priors

Sumit Datta and Bhabesh Deka

Abstract In Compressed Sensing based Magnetic Resonance Imaging (CS-MRI) reconstruction, wavelet transform finds the widest application as a sparsifying transform. It has been reported that along with the *standard wavelet sparsity*, group sparsity terms, like, the *tree sparsity*, the *joint sparsity* and the *forest sparsity* exist in the wavelet decomposition of multi-channel, multi-slice, and multi-contrast MR images. To develop an efficient unified CS-MRI reconstruction model for these images, different wavelet based sparsity priors together plays a pivotal role. In this paper, we propose a simultaneous weighted forest sparsity and joint total variation based efficient CS-MRI reconstruction model for multi-channel, multi-slice, and multi-contrast MR images. The proposed technique shows significant improvements in terms of quality of reconstruction over existing methods for different datasets.

Keywords Compressed sensing · Magnetic resonance imaging · Tree sparsity Forest sparsity · Joint total variation

1 Introduction

Magnetic resonance imaging (MRI) is one of the most commonly used soft-tissue imaging techniques in medical diagnostics mainly because it is non-invasive and provides good contrast without using any ionizing radiation. But, it comes with slow imaging speed due to some instrumental and physiological limitations.

Lustig et al. [11] introduce the idea of compressed sensing (CS) in MRI (CS-MRI). As MRI is compressible in the wavelet domain and acquired in the frequency (Fourier) domain sufficiently incoherent to the wavelet domain, so naturally it is a good candidate for compressed sensing [2]. Application of CS in MRI, makes

S. Datta · B. Deka (✉)
Computer Vision and Image Processing Laboratory, Department of Electronics
and Communication Engineering, Tezpur University, Tezpur, Assam 784028, India
e-mail: bdeka@tezu.ernet.in

© Springer Nature Singapore Pte Ltd. 2019   821
J. C. Bansal et al. (eds.), *Soft Computing for Problem Solving*,
Advances in Intelligent Systems and Computing 816,
https://doi.org/10.1007/978-981-13-1592-3_65

it possible to reconstruct good quality MR images from just 25 to 30% of the total k-space data which is a significant step towards the development of rapid MRI.

Let $\mathbf{x} \in \mathbb{R}^n$ be an underlying MR image and $\mathbf{y} \in \mathbb{R}^m$ the measured k-space data i.e. $\mathbf{y} = \mathbf{F}_u \mathbf{x}$, where $\mathbf{F}_u \in \mathbb{R}^{m \times n}$ is the partial Fourier operator and $m \ll n$. To reconstruct the MR image, $\mathbf{x}$ from the undersampled data, $\mathbf{y}$ authors in [11] solve the following minimization problem-

$$\hat{x} = \arg \min_{\mathbf{x}} \left\{ \tfrac{1}{2} \|\mathbf{F}_u \mathbf{x} - \mathbf{y}\|_2^2 + \lambda_1 \|\mathbf{x}\|_{\text{TV}} + \lambda_2 \|\boldsymbol{\Psi} \mathbf{x}\|_1 \right\} \tag{1}$$

The first term $\|\mathbf{F}_u \mathbf{x} - \mathbf{y}\|_2^2$ in the right side of Eq. 1 is used to impose data consistency, the second term $\|\mathbf{x}\|_{\text{TV}}$, called the total variation (TV) norm of $\mathbf{x}$: $\|\mathbf{x}\|_{\text{TV}} = \sum_{i,j} \sqrt{\left(\nabla_x x_{ij}\right)^2 + \left(\nabla_y x_{ij}\right)^2}$ where ∇_x and ∇_y denotes the first order forward finite difference operator in horizontal and vertical directions, respectively, is used to enhance the spatial domain gradient sparsity. Third term i.e. $\|\boldsymbol{\Psi} \mathbf{x}\|_1$ is the ℓ_1-norm of wavelet coefficients, sparsifies the input data in the wavelet domain.

Lustig et al. [11] solve the above problem using the nonlinear conjugate gradient (NCG) algorithm. But it is quite slow due to high per iteration cost. Huang et al. [10] proposed a new algorithm for solving the above CS-MRI reconstruction problem which is not only faster but also efficient, calling it as the fast composite splitting algorithm (FCSA). It is quite fast as compared to the previous techniques as they divide the whole problem into two independent sub-problems which are efficiently solved by iterative soft-thresholding, and finally, linearly combine their individual solutions to obtain the solution of Eq. 1. As MR images are compressible in transform domain one can still improve the quality of reconstruction by incorporating regularization terms which would exploit the natural structure of the wavelet coefficients better for sparsification.

For analysis of anatomical structures of any organ in the human body, 3D MRI is definitely a good choice. However, due to the long scan time there are only a few applications, like, contrast angiography where 3D MRI data is obtained directly. In most of the applications, 3D data are reconstructed from 2D multi-slice images [6, 7]. In case of 2D multi-slice MRI neighboring slices are highly correlated because the inter-slice gap is either zero or very small (1–3 mm) hence variation of anatomical structures across different slices is very less or negligible as shown in first row in Fig. 1. Due to this high similarity in adjacent slices there is a tendency that both the spatial domain gradients and the inter-slice wavelet coefficients are very similar. So, during the CS based MR image reconstruction if one can utilize these information then better reconstruction quality may be achieved without increasing the measured data. Similarly, in multi-channel or parallel MRI (pMRI), although the coil is more sensitive towards the portion nearer to the field of view (FoV), images corresponding to different coils are highly correlated as the underlying anatomical object is the same for all coils as shown in in Fig. 1. In the same vein, in multi-contrast MRI, although contrast is varying for different images but the underlying FoV is same. Therefore, in multi-contrast MRI different images are highly correlated as shown in Fig. 1.

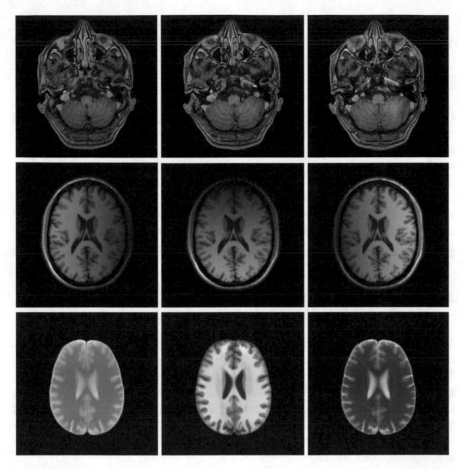

Fig. 1 Different types of MR image data. First row: multi-slice, second row: multi-channel, and third row: multi-contrast

In order to exploit the redundancy present in multi-slice/multi-channel/multi-contrast MR images as discussed above, the idea of group sparsity is used on both wavelet and gradient domain based regularization terms given in Eq. 1. Wavelet coefficients of an MR image follows the quadtree structure [3]. Since, multi-slice MR images are highly correlated, we can group the wavelet domain tree structures (overlapping parent-child pairs) of the same position from adjacent slices together. Similarly, in gradient domain, magnitude of gradients of the same position from different adjacent slices can also be grouped together. This particular arrangement penalizes all members of a particular group uniformly irrespective of the slice from which they originally come from, hence, enforces a particular group either to zero or non-zero after thresholding. Thus, we can form distinct groups from multi-slice image data both in the wavelet and gradient domains, which are either sparse or non-sparse and thereby successfully obtaining *group sparsity* in each group.

Therefore, the proposed new CS-MRI reconstruction model consists of two regularization terms; one is the group sparsity from the knowledge of wavelet domain tree structure and other is the group sparsity from the spatial domain gradient magnitudes. We use the wavelet domain forest sparsity and spatial domain joint total variation (JTV) norms to define the above two types of regularization, respectively. The proposed model is solved using an approach based on the ideas of fast composite splitting algorithm (FCSA) [10]. Simulation results outperforms existing forest sparsity [4] and the Joint TV [9] based CS-MRI algorithms. To the best of the authors' knowledge this work for the first demonstrates the application of forest sparsity and JTV in a single reconstruction model to reconstruct multi-slice/multi-channel/multi-contrast MRI image at one go using the CS.

The rest of the paper is organized as follows, Sect. 2 details about the forest sparsity in multi-sice MRI. Next, we discuses the proposed work with mathematical details in Sect. 3. Then performance of the proposed work with other reconstruction techniques are compared in Sect. 4. Finally, Sect. 5 concludes the proposed work.

2 Forest Sparsity in Multi-slice MRI

Wavelet coefficients of MR image follow a quadtree structure i.e. if a parent coefficient is large (or small) then its children coefficients also tend to be large (or small). To utilize this property of wavelet coefficients, Chen and Huang [3] added a separate regularization term with optimization problem given in in Eq. 1. They call it as the wavelet domain tree sparsity regularization term. Here, each pair of parent-child coefficients are kept in a group so that sparsity can be enforced amongst different overlapping groups depending on their magnitudes. With this addition, the new reconstruction problem can be expressed as-

$$\hat{x} = \arg\min_{\mathbf{x}} \left\{ \frac{1}{2} \|\mathbf{F}_u \mathbf{x} - \mathbf{y}\|_2^2 + \lambda_1 \|\mathbf{x}\|_{\mathrm{TV}} + \lambda_2 \|\boldsymbol{\Psi}\mathbf{x}\|_1 + \lambda_3 \sum_{g \in \mathbf{G}} \|\boldsymbol{\Psi}\mathbf{x}\|_2 \right\} \quad (2)$$

where g represents indices of each overlapping parent-chid pair and $\mathbf{G}$ indices of all such parent-child pairs in the wavelet decomposition.

Huang et al. [9] introduce the concept of joint sparsity in multi-contrast CS based MR image reconstruction to exploit the similarity of non-zero wavelet transform coefficients and magnitudes of gradients in different images. For better performance, group sparsity is applied on both wavelet and gradient domains separately. In both the wavelet and the gradient domains, they keep coefficients of the same position from different images in a single group. This grouping arrangement uniformly penalizes all coefficients within a group irrespective of their origins.

Chen et al. [4] proposed a new sparsity model, namely, the forest sparsity. . According to them, multi-channel MRI data follow joint sparsity across different channels whereas the image in each channel follow the wavelet domain tree sparsity.

This is an extension of the tree sparsity model where data can be represented by forest i.e. mutually connected trees. Here, forest structure is approximated by overlapping groups constructed from different channels. If any coefficient is large (or small) then its children and neighboring coefficients from other channels also tend to be large (or small). To explore this *a priori* information in multi-channel MRI all parent-child pairs of same position from different channels are kept in a single group. This forest sparsity based reconstruction model gives *state-of-the-art* performance in CS-MRI reconstruction because it removes both intra and inter channel similarities in multi-channel MRI data. A pictorial representation of different wavelet domain sparsity terms is given in Fig. 2 for better understanding the concept of forest sparsity.

3 Proposed Method

Motivated from the work in [4], we propose a forest sparsity based novel CS-MRI reconstruction technique. To give an idea of the work done in [4], we will reproduce their multi-channel CS-MRI reconstruction model here. Let $\mathbf{X} = [\mathbf{x}_1, \mathbf{x}_2, \ldots, \mathbf{x}_T] \in \mathbb{R}^{Tn}$, where T denotes the number of channels or coils in pMRI, be the multi-channel MR image data and $\mathbf{Y} = [\mathbf{y}_1, \mathbf{y}_2, \ldots, \mathbf{y}_T] \in \mathbb{R}^{Tm}$ the acquired multi-channel data i.e. $\mathbf{F}_u \in \mathbb{R}^{Tm \times Tn}$ in the k-space. The reconstruction model is defined as follows:

$$\hat{X} = \arg\min_{\mathbf{x}} \left\{ \frac{1}{2} \|\mathbf{F}_u \mathbf{X} - \mathbf{Y}\|_2^2 + \lambda_1 \|\mathbf{X}\|_{\text{TV}} + \lambda_2 \sum_{g \in \mathbf{G}} \|\boldsymbol{\Psi} \mathbf{X}\|_2 \right\} \tag{3}$$

The first regularization term representing the TV-norm for multi-channel images is expressed as $\|\mathbf{x}\|_{\text{TV}} = \sum_{i,j}^{T\sqrt{n},T\sqrt{n}} \sqrt{\left(\nabla_x x_{ij}\right)^2 + \left(\nabla_y x_{ij}\right)^2}$. The third term representing the forest sparsity regularization is mathematically expressed by $\ell_{2,1}$-norm. Here, g represents index of a group containing elements as overlapping parent-child pairs of the same position across different channels and $\mathbf{G}$ contains indices of all such groups in the multi-channel data. First, ℓ_2-norm is applied on all individual groups at the same position of different channels followed by ℓ_1-norm applied on the results of ℓ_2-norm.

According to Candes et al. [1] ℓ_1-norm has a fundamental limitation i.e. it penalizes larger coefficients more as compared to smaller coefficients. To solve this imbalance they proposed the idea of weighted ℓ_1-norm minimization where coefficients of sparse vector are individually weighted and the weights are inversely proportional to the magnitudes of coefficients [5, 8]. Further, images to be reconstructed in multi-channel, multi-contrast, and multi-slice MRI are highly correlated. So, magnitudes of gradient coefficients at the same position in different images tend to be similar. Therefore, one can also include this prior information in the standard TV-regularization term to further improve the reconstruction quality.

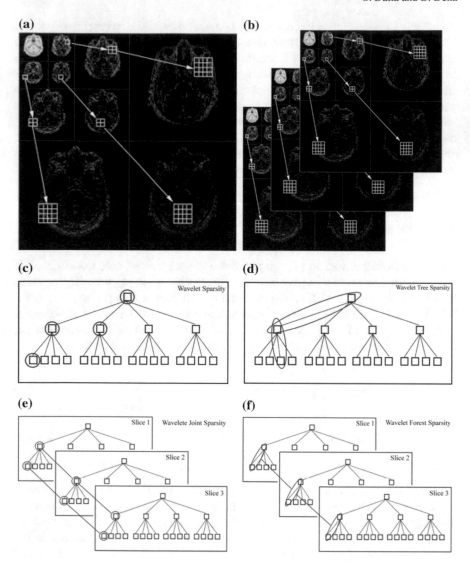

Fig. 2 **a** Wavelet quadtree structure, **b** wavelet quadtree structure in multi-dimensional orientation and (**c–f**), respectively represent grouping arrangements of wavelet domain sparsity, tree sparsity, joint sparsity and forest sparsity regularization terms. In (**c–f**) red line indicates grouping arrangements

Keeping in mind the above issues of the two regularization terms, we propose a new unified CS-MRI reconstruction model based on weighted forest sparsity and joint total variation (JTV) norm for multi-slice, multi-channel, multi-contrast MR image data as follows:

$$\widehat{\mathbf{X}} = \arg\min_{\mathbf{x}} \left\{ \tfrac{1}{2} \|\mathbf{F}_u \mathbf{X} - \mathbf{y}\|_2^2 + \lambda_1 \|\mathbf{X}\|_{\mathrm{JTV}} + \lambda_2 \sum_{g \in \mathbf{G}} \mathbf{W}_g \|\mathbf{Z}_g\|_2 \right\} \tag{4}$$

where $\mathbf{Z} = \boldsymbol{\Psi} \mathbf{X}$ and joint TV norm $\|\mathbf{X}\|_{\mathrm{JTV}} = \sum_{i,j}^{\sqrt{n},\sqrt{n}} \sqrt{\sum_{t=1}^{T} \left(\left(\nabla_x x_{t,ij} \right)^2 + \left(\nabla_y x_{t,ij} \right)^2 \right)}.$

The above problem can be rewritten as an unconstrained problem as-

$$\widehat{\mathbf{X}} =$$

$$\arg\min_{\mathbf{x}} \left\{ \tfrac{1}{2} \|\mathbf{F}_u \mathbf{X} - \mathbf{y}\|_2^2 + \lambda_1 \|\mathbf{X}\|_{\mathrm{JTV}} + \lambda_2 \sum_{g \in \mathbf{G}} \mathbf{W}_g \|\mathbf{Z}_g\|_2 + \tfrac{\beta}{2} \|\mathbf{Z}_g - (\mathbf{G}(\boldsymbol{\Psi}\mathbf{X}))_g\|_2^2 \right\} \tag{5}$$

where $\mathbf{W}_\mathbf{G}$ is a weight vector and its elements are inversely proportional to the ℓ_2-norm of all individual overlapping groups, can be defined as-

$$w_g = \frac{1}{\|(\boldsymbol{\Psi}\mathbf{X})_g\| + \varepsilon}, \quad g \in \mathbf{G} \tag{6}$$

where ε is a small positive parameter. Now, we decompose the proposed composite problem in Eq. 5 into two simpler subproblems-

$$\widehat{\mathbf{Z}} = \arg\min_{\mathbf{Z}} \left\{ \lambda_2 \sum_{g \in \mathbf{G}} \mathbf{W}_g \|\mathbf{Z}_g\|_2 + \tfrac{\beta}{2} \|\mathbf{Z}_g - (\mathbf{G}(\boldsymbol{\Psi}\mathbf{X}))_g\|_2^2 \right\} \tag{7}$$

$$\widehat{\mathbf{X}} = \arg\min_{\mathbf{x}} \left\{ \tfrac{1}{2} \|\mathbf{F}_u \mathbf{X} - \mathbf{y}\|_2^2 + \lambda_1 \|\mathbf{X}\|_{\mathrm{JTV}} + \tfrac{\beta}{2} \|\mathbf{Z} - (\mathbf{G}(\boldsymbol{\Psi}\mathbf{X}))\|_2^2 \right\} \tag{8}$$

Groupwise first subproblem has a closed form solution, given by-

$$\widehat{\mathbf{Z}}_{g_i} = \max\left(\|(\mathbf{G}(\boldsymbol{\Psi}\mathbf{X}))_g\|_2 - \frac{\lambda_2 \mathbf{W}_g}{\beta}, 0 \right) \frac{(\mathbf{G}(\boldsymbol{\Psi}\mathbf{X}))_g}{\|(\mathbf{G}(\boldsymbol{\Psi}\mathbf{X}))_g\|_2} \tag{9}$$

which is commonly known as $shrinkgroup(.)$ operator [4]. On the other hand for solving second subproblem we adopt the technique proposed in [9, Algorithm 2].

The proposed algorithm is summarized in Algorithm-1. In algorithm the *project* function is given by:

$$\mathbf{x}_i = project\,(\mathbf{x}_i, [l, \, u]) = \begin{cases} \mathbf{x}_i, & if\, l \leq \mathbf{x}_i \leq u \\ l, & if\, \mathbf{x}_i < l \\ u, & if\, \mathbf{x}_i > u \end{cases}, \qquad (10)$$

where i represents pixel location and l and u are denotes the intensity range MR image. This new reconstruction model is suitable for all the three types of MR image data i.e. multi-channel MRI, multi-slice MRI, and multi-contrast MRI.

Algorithm 1 Proposed Algorithm

Input: $\mathbf{\Psi}$, $\mathbf{F}_u$, $\mathbf{Y}$, λ_1, λ_2

Initialization: $\left\{ \mathbf{X}^{(0)}, \mathbf{r}^{(1)} \right\} \leftarrow \mathbf{F}_u^T \mathbf{Y}$, $\mathbf{W}_{\mathrm{G}}^{(1)} \leftarrow \mathbf{I}$, $\left\{ t^1, k \right\} \leftarrow 1$

1: **while** not converged **do**

2: $\quad \mathbf{Z}^k \leftarrow shrinkgroup\left(\mathbf{G\Psi X}^{k-1}, \frac{\lambda_2 \mathbf{W}_{\mathrm{G}}^k}{\beta} \right)$

3: $\quad \mathbf{X}^k \leftarrow \arg\min_{\mathbf{x}} \left\{ \frac{1}{2} \|\mathbf{F}_u \mathbf{X} - \mathbf{Y}\|_2^2 + \lambda_1 \|\mathbf{X}\|_{\mathrm{JTV}} + \frac{\beta}{2} \|\mathbf{Z} - (\mathbf{G}\,(\mathbf{\Psi X}))\|_2^2 \right\}$

4: $\quad \mathbf{X}^k \leftarrow project\left(\mathbf{X}^k, [l, \, u] \right)$

5: $\quad$ Update $\left(\mathbf{W}_{\mathrm{G}} \right)^{k+1}$ using Eq. 6

6: $\quad t^{k+1} \leftarrow \left(1 + \sqrt{1 + 4\left(t^k \right)^2} \right) \Big/ 2$

7: $\quad \mathbf{r}^{k+1} \leftarrow \mathbf{X}^k + \frac{t^k - 1}{t^{k+1}} \left(\mathbf{X}^k - \mathbf{X}^{k-1} \right)$

8: $\quad k \leftarrow k + 1$

9: **end while**

Output: $\hat{\mathbf{X}} \leftarrow \mathbf{X}^{(k)}$

---

4 Results

All simulations are performed in MATLAB 8.0(R2012b) on a HP desktop having 3.40GHz Intel(R) Core(TM) i7-2600 CPU, 64-bit OS with 16GB memory. We have performed a number of experiments on multi-slice, multi-channel and multi-contrast MR images to evaluate the performance of the proposed reconstruction technique. Multi-slice images are collected from a local hospital,[1] multi-channel images are obtained by multiplying simulated coil sensitivity profiles with a randomly selected multi-slice image, and finally multi-contrast images are collected from SRI24b brain Atlas Dataset [12].

For evaluation of reconstruction techniques we considered signal-to-noise ratio (SNR) for mean square error (MSE) with ground truth, mean structural similarity (MSSIM) index for structural similarity which is more sensitive for human visual system and CPU time for measuring the computational complexity. For comparison we consider some well known recently developed fast CS MR image reconstruction methods, namely, the FCSA [10], the WaTMRI [3], the FCSA Joint [9], and the FCSA Forest [4].

---

[1]GNRC Hospital, Sixmile, Guwahati, India; www.gnrchospitals.com.

For fair comparisons, we made a common simulation setup for all reconstruction techniques. A fixed common stoping criterion for convergence i.e. if relative change of objective function is less than 10^{-4} is taken. In case of multi-slice and multi-contrast MR images we consider a group of three MR images at a time for reconstruction whereas in case of multichannel MR a group of four images are considered.

Reconstruction quality in terms of SNR and MSSIM are shown in Fig. 3. From figure, it is clear that irrespective of MR images the proposed method shows better SNR and MSSIM values as compared to other methods. On an average, the proposed method achieves 1 dB improvement in SNR compared to the FCSA Forest. Computational complexity in terms of CPU Time and number of iterations are shown in Table 1. From table, we observe that the CPU Time of the proposed method is somewhat less than the FCSA Forest. However, it is sometimes higher compared to the FCSA, the WaTMRI, and the FCSA Joint as they do not consider forest sparsity regularization term in reconstruction. From table, we also observe that the number of iterations is less as compared to other methods. Finally, for better visualization a $3\times$ zoomed cropped portion of a multi-slice MR image and a multi-contrast (T1 weighted) MR image are shown in Figs. 4 and 5, respectively. From these figures, it is clearly observed that the proposed reconstruction technique outperforms all other compared techniques in terms of preservation of edges and less visual aliasing artifacts.

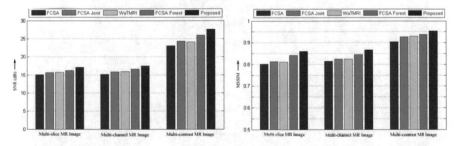

Fig. 3 Comparison of SNR (in dB) and MSSIM, respectively, using different reconstruction techniques for multi-slice, multi-channel, and multi-contrast MR images

Table 1 Comparison of CPU Time and number of iterations using various reconstruction techniques for different types of MR images

MR images	Parameters	FCSA	FCSA joint	WaTMRI	FCSA forest	Proposed
Multi-slice	Iteration	56	58	43	42	37
	CPU time	3.39	3.44	3.35	4.18	3.92
Multi-channel	Iteration	66	63	44	46	42
	CPU time	4.82	5.27	5.07	6.03	6.12
Multi-contrast	Iteration	51	82	47	52	39
	CPU time	3.21	4.88	4.10	5.18	4.24

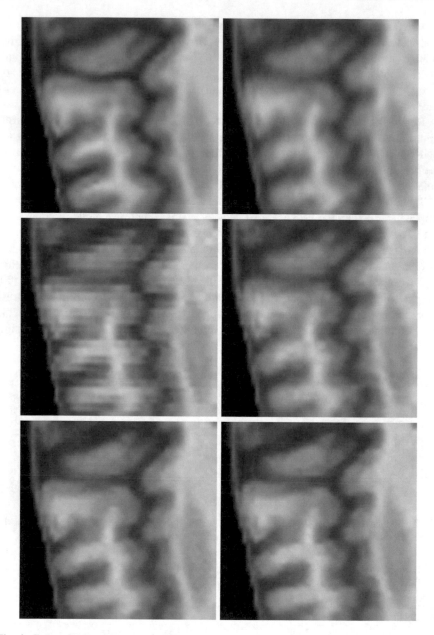

Fig. 4 Comparison of 3× zoomed cropped portion on a multi-contrast (T1 weighted) MR image. First row left to right: Original and reconstructed image using the FCSA, second row left to right: reconstructed images using FCSA Joint and WaTMRI, and third row left to right: reconstructed images using FCSA Forest and proposed

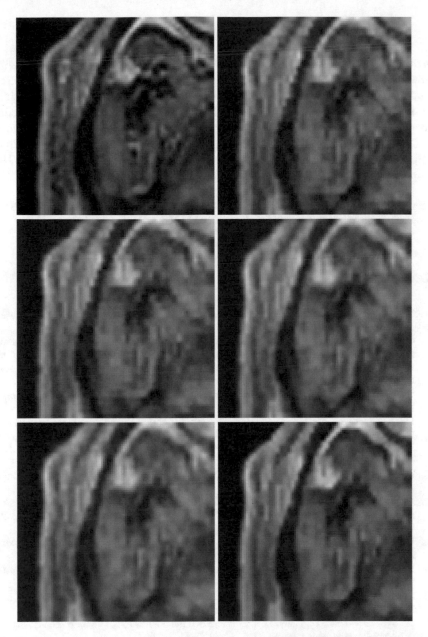

Fig. 5 Comparison of 3× zoomed cropped portion on a multi-slice MR image. First row left to right: Original and reconstructed image using the FCSA, second row left to right: reconstructed images using FCSA Joint and WaTMRI, and third row left to right: reconstructed images using FCSA Forest and proposed

5 Conclusions

In this paper, first we briefly discuss the evolution of different wavelet sparsity regularization priors in CS-MRI reconstruction. Next, we propose a novel reconstruction model based on wavelet forest sparsity and joint total variation sparsity for efficient reconstruction of multi-slice, multichannel, multi-contrast MR images. We have conducted a set of experiments to evaluate the performance of the proposed technique in CS reconstruction. Experimental results demonstrate that the proposed technique outperforms some of the very recent techniques both objectively and visually. The proposed technique shows high potential for rapid 3D MRI.

Acknowledgement Authors would like to thank Dr. S. K. Handique, MD (Radiodiagnosis), GNRC Hospitals, Guwahati, India for providing different 3D MRI data sets for simulation.

References

1. Candes, E., Wakin, M., Boyd, S.: Enhancing sparsity by reweighted L1 minimization. J. Fourier Anal. Appl. **14**(5), 877–905 (2008)
2. Candes, E.J., Romberg, J.K., Tao, T.: Robust uncertainty principles: exact signal reconstruction from highly incomplete frequency information. IEEE Trans. Inf. Theory **52**(2), 489–509 (2006)
3. Chen, C., Huang, J.: Exploiting the wavelet structure in compressed sensing MRI. Magn. Reson. Imaging **32**, 1377–1389 (2014)
4. Chen, C., Li, Y., Huang, J.: Forest sparsity for multi-channel compressive sensing. IEEE Trans. Sig. Process. **62**(11), 2803–2813 (2014)
5. Datta, S., Deka, B.: Efficient adaptive weighted minimization for compressed sensing magnetic resonance image reconstruction. In: Proceedings of the Tenth Indian Conference on Computer Vision, Graphics and Image Processing, ICVGIP'16, pp. 95:1–95:8. ACM, New York, NY, USA (2016)
6. Datta, S., Deka, B.: Magnetic resonance image reconstruction using fast interpolated compressed sensing. J. Opt. (2017)
7. Datta, S., Deka, B., Mullah, H.U., Kumar, S.: An efficient interpolated compressed sensing method for highly correlated 2D multi-slice MRI. In: 2016 International Conference on Accessibility to Digital World (ICADW), pp. 187–192 (2016)
8. Deka, B., Datta, S.: Weighted wavelet tree sparsity regularization for compressed sensing magnetic resonance image reconstruction, pp. 449–457. Springer, Singapore (2017)
9. Huang, J., Chen, C., Axel, L.: Fast multi-contrast MRI reconstruction. Magn. Reson. Imaging **32**(10), 1344–1352 (2014)
10. Huang, J., Zhang, S., Metaxas, D.N.: Efficient MR image reconstruction for compressed MR imaging. Med. Image Anal. **15**(5), 670–679 (2011)
11. Lustig, M., Donoho, D., Pauly, J.M.: Sparse MRI: the application of compressed sensing for rapid MR imaging. Magn. Reson. Med. **58**, 1182–1195 (2007)
12. Rohlfing, T., Zahr, N., Sullivan, E., Pfefferbaum, A.: The SRI24 multi-channel atlas of normal adult human brain structure. Hum. Brain Mapp. **31**(5), 798–819 (2010)

Understanding Single Image Super-Resolution Techniques with Generative Adversarial Networks

Amit Adate and B. K. Tripathy

Abstract Single Image Super-Resolution techniques have the function of retrieving a high resolution image from a single low resolution input. They implement deep learning heuristics which perform the techniques to form pixel-accourate reproductions. In this paper we have experimented upon various neural architectures with unique approaches towards the task of super-resolution. We have especially elaborated upon adversarial training networks which are yielding progressive results in both conditional and quantifiable benchmarks.

Keywords Single image super-resolution · Generative adversarial networks
Adversarial training · Handwritten digits

1 Introduction

The objective of Super-Resolution (SR) is to retrieve a high resolution image one or more input images of lower resolution. The techniques implemented for SR can be widely classified into the following, Multi-Image Super-Resolution [1] and Example-Based Super-Resolution [2]. In the time-honoured Multi Image Super-Resolution, a well defined collection of sample images are taken in low resolution. Every one of the sample images are then transformed into a set of linear constraints. The operations on these constraints leads to the recovery of the high resolution image. But essentially this technique is limited only to small increases in resolution with the thresholds smaller than 2.

These shortcomings led to the evolution of Example-Based Super-Resolution introduced by Elad and Datsenko [3], and also further development has been done by others [4]. In this technique, a database is generated that has the low-high resolution

A. Adate (✉) · B. K. Tripathy
SCOPE, VIT University, Vellore 632014, Tamil Nadu, India
e-mail: adateamit.sanjay2014@vit.ac.in

B. K. Tripathy
e-mail: tripathybk@vit.ac.in

© Springer Nature Singapore Pte Ltd. 2019
J. C. Bansal et al. (eds.), *Soft Computing for Problem Solving*,
Advances in Intelligent Systems and Computing 816,
https://doi.org/10.1007/978-981-13-1592-3_66

image pairs, the correspondence between these pairs is learned. Upon learning from the database, the model is applied to a new low resolution image to retrieve its most likely high resolution image. Even after applying the aforementioned techniques, there was a very wide room for optimization. In recent times, state-of-the-art methods for SR are almost all example-based or its variant.

Known for its monumental success in image classification, Convolutional Neural Networks (CNN) have shown most raging popularity in tackling SR. The key factors for its success are, dynamic training on modern GPU's, the introduction of the Rectified Linear Units [5] which decreases the computational time while maintaining the image quality and the availability of abundant data-sets for training bulkier models.

2 Background: CNN

When it comes to tackling SR problems for contemporary real world problems, the augmentation criteria is mean square error (MSE) between the retrieved high resolution image and the ground truth. The secondary metric to be optimized is peak signal-to-noise-ratio (PSNR). But the competence of both MSE and PSNR to obtain intuitively pertinent differences is very limited as they are based on pixel data based differences throughout the image. There have been many models which are based on CNN's, namely SRCNN (Super-Resolution-CNN) [4], PixelCNN [6], SRResNet (Super-Resolution Residual CNN) [6] are among the ones most discussed.

All of them try to tackle the same problem but use different approaches. SRCNN's implements an end-to-end optimization of the image. It tries to surpass the bicubic benchmarks with a few iterations, also it significantly outperforms the sparse coding(SC) based methods [7]. The PixelCNN architecture is built upon the basis that MSE and PSNR fail to measure the extent of supervised SR. The architecture is a probabilistic generative model that inflicts an order on each pixel of the image by interpreting them as a sequence of pixels [2]. The PixelCNN obtained a very high log-likelihood upon experimenting with MNIST and CIFAR-10 datasets. The SRResNet is built upon Microsoft's Residual Network(Resnet), it uses more than 100 layers of the ResNet blocks stacked together to obtain state-of-the-art PSNR [8]. But the architecture is deep and requires a hefty training time, but the changes in the time complexity and the model size led the SRResNet to beat the SR benchmarks set by the PixelCNN (Fig. 1).

3 Generative Adversarial Networks

Generative adversarial networks (GAN's) were introduced in 2014 [9] and have been experimented upon various image datasets for the task of hybrid processing techniques and more commonly for generating images from visualizations. Their main aim is to artificially generate realistic data. The GAN framework is the most

Fig. 1 The PixelCNN
architecture for super
resolution

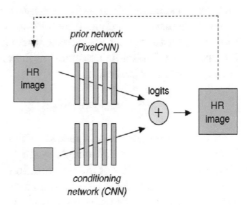

popular architecture with successes in the line of research on adversarial training in deep learning [10] where the two player min-max game is crafted carefully so that the convergence of the networks attains the optimal criteria. Among all the architectures mentioned in Sect. 2 there is one commonly observed obstacle, the recovery of finer texture details while implementing SR at large up-scaling factors.

3.1 Super Resolution Generative Adversarial Network

To increase the benchmark on up-scaling factors, a GAN was designed to perform SR. The architecture is termed as Super Resolution Generative Adversarial Network (SRGAN)[10]. The SRGAN is built upon microsoft's ResNet with skip-connection and diverge from MSE as the only optimization metric. The loss function used by them cannot be MSE as it struggles in retrieving the high frequency feature details. A similar usage neural network was introduced, VGG net [11]. Upon conducting classification upon ImageNet [12], the VGG net model, its validation and training set are both publically available. The optimization function on the SRGAN is to minimize the MSE in the feature spaces of VGG net. As the pretrained VGG model is available, SRGAN weight updation mechanism is heuristically worked up for extracting features instead of low-level pixel by pixel computed error measures.

3.2 Wasserstein Generative Adversarial Network

Every GAN consists of a generator and a discriminator, the generator does the ask of mustering data. The probability distributions upon the pre-trained data provide a usable gradient for the generator. To optimize the distance between two distributions, Wasserstein Distance [13] was implemented in the generator. The architecture introduced was called as Wasserstein Generative Adversarial Networks (WGAN's). The wasserstein distance is defined by the authors as the minimum cost of transporting mass in order to transfer mass from distribution f to distribution g. The measure is continous and differentiable almost in the entire domain, after genralizing wasserstein distance. Its equation:

$$W(_{r,\theta}) = \sup\nolimits_{\|f\|_L \le 1} x_{\sim r}[f(x)] - x_{\sim \theta}[f(x)]$$

WGAN introduces a loss metric, which compared to standard GAN's has an improved convergence and sample quality. There is also a significant improvement observed in the stability of the optimization process. The WGAN algorithm intends to train the discriminator well before hand than the generator. For the task of SR, WGAN's have to avoid using momentum based optimizer such as Adam optimizer. For implementing superresolution in contention with other networks mentioned above, there is room for optimization in a WGAN. Its the norm of penalizing the gradient of the discriminator with respect to its input [14]. The improved network is called as a WGAN-GP (WGAN with Gradient Penalty). Well, in short, the WGAN-GP stabilizes GAN training over a wide range of applications by adding a gradient penalty to the discriminator loss, hence reducing weight clipping. This is the equation govering the gradient penalty:

$$\max\nolimits_{w \in \mathcal{W}} x_{\sim r}[f_w(x)] - z_{\sim p(z)}[f_w(g_\theta(z))]$$

4 Experiments

The deep learning networks benefit heavily from big data training, for comparison we use the MNIST Handwritten digits database. The factor of $2\times$ between low- and high-resolution is performed on all the experiments performed. This corresponds to a $8\times$ reduction in image pixels. The PSNR and Structural Similarity(SSIM) measures are reported for comparison and were calculated on the y-channel of center-cropped after the daala package was used for the removal of a 4-pixel wide strip from each border. While training model for the MNIST database, we compared the model with 4 modes, MSE, GAN, WGAN and SRGAN. We also tried to evaluate the performance of WGAN's for 2 losses (*lrG 1e-2*) and (*lrG 1e-4*). While training the SRGAN, we added an additional total variation loss with weight 2×10^{-6}. The comparative visual examples are provided in Fig. 2.

Even accounting for adversarial loss, MSE provides solutions with the highest PSNR values, but perceptually a rather straightforward and convincing results were observed by the SRGAN. We further focused on the trait of minor artifacts and their reconstruction. This is caused by the competition between MSE based pixel data loss and the observed adversarial loss. Figure 2 observes a clear distinction between SRGAN based minor reconstruction of the attribute artifacts (Fig. 3).

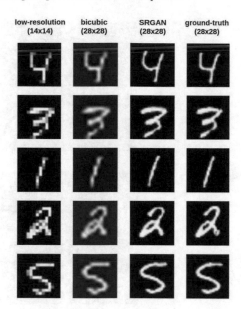

Fig. 2 The SRGAN comparative examples

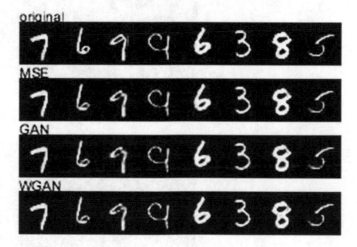

Fig. 3 The WGAN comparative examples

Figure 6 observes the difference between the ground truth and the WGAN model for the first loss function. For the same loss function, we applied to the MSE to gain results in Fig. 5 and to the standard GAN model to obtain Fig. 4. We also implemented the second loss function proposed to generate the images in Fig. 7. We could not determine a decidedly optimal loss function for the WGAN and the SRGAN. In all the figures below, the image on the upper half is generated by a network and the other half is the corresponding original image.

Fig. 4 Images generated by
GAN (lrG 1e-2)

Fig. 5 Images generated by
GAN (lrG 1e-4)

Fig. 6 Images generated by
WGAN (lrG 1e-2)

Fig. 7 Images generated by
WGAN (lrG 1e-4)

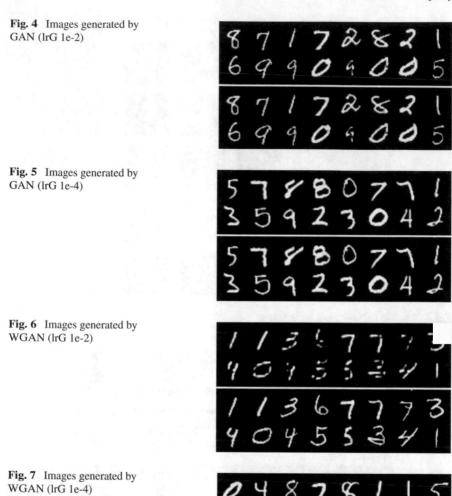

5   Results

In our experiments we have observed decreased content loss compared to the architectures mentioned in Sect. 2. This is observed in Fig. 3, reason being that we are able to represent features of higher abstraction by going deeper in the neural networks. Hence the feature-map deals with the content loss while the adversarial loss focuses mainly on the texture details.

To view our experiments and the images generated:
https://github.com/amitadate/gan-sr.

6 Conclusion

We observed a trend after applying various loss functions and training on the same validation set for all the aforementioned models, the observation being that higher level feature maps yield to a very better texture detail in higher resolution. As the SRGAN and WGAN act on different adversarial training techniques, we could not provide a common quantitative comparison between them. But we have presented enough data to provide a perceptual difference in post super-resolution for the same image dataset. Our findings conclude that SRGAN significantly outperformed the other models, the closest being WGAN. Our experiments conclude that the implementation of generative adversarial neural networks has created a higher benchmark for the task of single image super resolution.

References

1. Park, S.C., Park, M.K., Kang, M.G.: Super-resolution image reconstruction: a technical overview. IEEE Sig. Process. Mag. **20**(3), 21–36 (2003)
2. Glasner, D., Bagon, S., Irani, M.: Super-resolution from a single image. In: 2009 IEEE 12th International Conference on Computer Vision, pp. 349–356. IEEE (2009)
3. Elad, M., Datsenko, D.: Example-based regularization deployed to super-resolution reconstruction of a single image. Comput. J. **52**(1), 15–30 (2007)
4. Dong, C., Loy, C.C., He, K., Tang, X.: Learning a deep convolutional network for image super-resolution. In: European Conference on Computer Vision, pp. 184–199. Springer (2014)
5. Nair, V., Hinton, G.E.: Rectified linear units improve restricted Boltzmann machines. In: Proceedings of the 27th International Conference on Machine Learning (ICML-10), pp. 807–814 (2010)
6. Dahl, R., Norouzi, M., Shlens, J.: Pixel recursive super resolution (2017). arXiv:1702.00783
7. Yang, J., Wright, J., Huang, T.S., Ma, Y.: Image super-resolution via sparse representation. IEEE Trans. Image Process. **19**(11), 2861–2873 (2010)
8. He, K,, Zhang, X., Ren, S., Sun, J.: Deep residual learning for image recognition. CoRR (2015). arXiv:1512.03385
9. Goodfellow, I., Pouget-Abadie, J., Mirza, M., Xu, B., Warde-Farley, D., Ozair, S., Courville, A., Bengio, Y.: Generative adversarial nets. In: Advances in Neural Information Processing Systems, pp. 2672–2680 (2014)
10. Ledig, C., Theis, L., Huszar, F., Caballero, J., Aitken, A.P., Tejani, A., Totz, J., Wang, Z., Shi, W.: Photo-realistic single image super-resolution using a generative adversarial network. CoRR (2016). arXiv:1609.04802
11. Simonyan, K., Zisserman, A.: Very deep convolutional networks for large-scale image recognition. CoRR (2014). arXiv:1409.1556
12. Krizhevsky, A., Sutskever, I., Hinton, G.E.: Imagenet classification with deep convolutional neural networks. In: Advances in Neural Information Processing Systems, pp. 1097–1105 (2012)

13. Vallender, S.S.: Calculation of the wasserstein distance between probability distributions on the line. Theory Prob. Appl. **18**(4), 784–786 (1974)
14. Gulrajani, I., Ahmed, F., Arjovsky, M., Dumoulin, V., Courville, A.C.: Improved training of Wasserstein GANs. CoRR (2017). arXiv:1704.00028

Detecting Image Forgery
in Single-Sensor Multispectral Images

Mridul Gupta and Puneet Goyal

Abstract With the advancements in digital technology, multispectral images have found use in fields like forensics, remote sensing due to their ability to perceive things which were otherwise non-existent. They are used to obtain more information about terrains, land cover and in forensics as certain things like blood stains are not visible in visible spectrum. But with newly developed photo-editing softwares, they can be easily manipulated without leaving any visible clue of manipulation, but will destroy the underlying correlation between different bands. Newly developed digital cameras employ a single sensor along with multispectral filter array (MSFA) and then interpolate the data at other locations, hence introducing a correlation between bands. In this paper, we have proposed an algorithm that can identify the lack of correlation at tampered locations in a multispectral image and can thus help in establishing the authenticity of the given multispectral image. We show the efficiency of our approach with respect to the size of tampered regions in images interpolated with one the most common demosaicking algorithm—binary tree-based edge sensing (BTES).

Keywords Multispectral image forgery · Multispectral filter array (MSFA)
MSFA demosaicking · Interpolation · EM algorithm

1 Introduction

Multispectral images since they were brought into use have found use in numerous fields such as military planning, urban planning, forensics. If it becomes possible to forge these images then due to the severity of their uses, it will also become increasingly important to have the methods to authenticate the image. Several algorithms

M. Gupta (✉)
Indian Institute of Technology Roorkee, Roorkee, India
e-mail: mridulgupta9@gmail.com

P. Goyal
Indian Institute of Technology Ropar, Rupnagar, India
e-mail: puneet@iitrpr.ac.in

© Springer Nature Singapore Pte Ltd. 2019
J. C. Bansal et al. (eds.), *Soft Computing for Problem Solving*,
Advances in Intelligent Systems and Computing 816,
https://doi.org/10.1007/978-981-13-1592-3_67

Fig. 1 A typical Bayer pattern

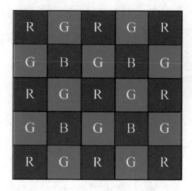

have been proposed for detecting the forged 3-band images [1–8]. Li et al. recently proposed an improved method of image forgery localization via integrating tampering possibility maps [1]. Shen et al. described a novel passive splicing image forgery detection approach using textural features based on the gray-level co-occurrence matrices [2]. Pun et al. proposed a copy-move forgery detection scheme that first integrates both block-based and keypoint-based methods and then performs feature point matching [4]. Ferrarra et al. [5] and Popescu et al. [6] proposed the image forgery detection techniques based on color filter array (CFA) artifacts. These techniques generally assume that although digital forgeries may leave no visual clues of having been tampered with, they may, nevertheless, alter the underlying statistics of an image. A typical digital camera will capture the value of one color at one pixel location, and its value at the rest of the locations will be determined by interpolating the available data. The patterns in which the value of colors at different pixels is recorded are CFA (color filter array) patterns, and the most prominent among them is Bayer pattern as shown in Fig. 1. Missing values at different pixels locations are then interpolated from available data, and this process is called color filter array (CFA) interpolation or demosaicking. Interpolation introduces some correlations between the samples of a color image. Aberrations from these correlations can be quantified to determine whether the image is genuine or not. Our proposed method utilizes an iterative way as inspired by Popescu's and Farid's work [6] to determine the authenticity of a multispectral image. As CFA involves only three colors so the pattern is not complex and it is easier to determine the relationship, but as the number of bands increases, based on the probability of appearance of each band, we get a profusion of possible patterns. These patterns are called multispectral filter array (MSFA). We use BTES [5] demosaicking algorithm to interpolate data from MSFA (4-band and 5-band images), edit them and then implement our algorithm on these images to determine its effectiveness in detecting digital tampering and analyze its accuracy with increase in size of forgeries.

1.1 MSFA Interpolation Algorithm

There are many algorithms that have been proposed for demosaicking [9–14] a multispectral image, but we primarily focus here on binary tree-based edge sensing (BTES) demosaicking method as it is the most referred work, and unlike many other methods, BTES is a generic algorithm that can handle more diversified MSFA patterns. On the basis of the number of spectral bands and probability of appearance (POA) of each band, BTES suggests an approach to generate an MSFA. To prepare an MSFA for a 5-band image, {A, B, C, D, E} with POA as {1/4, 1/4, 1/4, 1/8, 1/8}, a tree is generated where each leaf node represents a band and its depth represents its POA as shown in Fig. 2a. After generating the tree, checkerboard separation is carried out (Fig. 2b) and the leaf nodes are combined to obtain the final MSFA as in Fig. 2c.

To interpolate the MSFA thus obtained using BTES, a band is selected at random having highest POA (contains more information). To generate complete band D, values at locations of band E are determined using the available data for band D (Fig. 3b), and then the values at locations of band A are computed using newly computed data as shown in Fig. 3c. Final band D is computed using its all available data (Fig. 3d). The method to obtain the value at required location is explained below.

For patterns at odd level (lets say, k) of tree, only down-sampling by $(k - 1)$, $k \geq 1$ is sufficient to convert the pattern to basic pattern, and for even levels, it is down-sampled by 2 level/2 − 1, level > 1 and then rotated by 45° For instance in Fig. 4a, A can be converted to basic pattern just by rotation whereas in Fig. 4b D is down-sampled to get the basic pattern.

Now, for interpolation, weights of four neighboring pixels are calculated. For vertical,

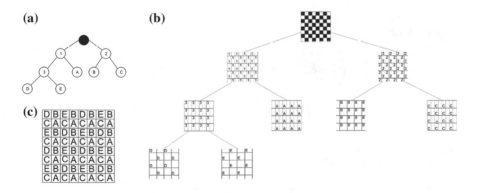

Fig. 2 MSFA generation **a** binary tree, **b** checkerboard separation, **c** 5-band MSFA

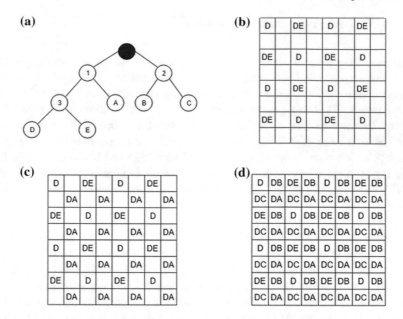

Fig. 3 Demosaicking process **a** binary tree, **b** interpolating D at locations of E, **c** interpolating D at locations of A, **d** interpolating and obtaining D at all locations

$$V : W_{m,n} = \left(1 + \left|R_{m+2,n} - R_{m,n} + \left|R_{m-2,n} - R_{m,n}\right| + \frac{1}{2}R_{m-1,n-1}\right.\right.$$
$$\left.\left. - R_{m+1,n-1} + \frac{1}{2}R_{m-1,n+1} - R_{m+1,n+1}\right|\right)^{-1} \tag{1}$$

$m \in \{i - 1, i + 1\}, n = j.$
For horizontal,

$$H : W_{m,n} = \left(1 + R_{m,n+2} - R_{m,n} + \left|R_{m,n-2} - R_{m,n}\right| + \frac{1}{2}R_{m+1,n-1}\right.$$
$$\left. - R_{m+1,n+1} + \frac{1}{2}R_{m-1,n-1} - R_{m-1,n+1}\right)^{-1} \tag{2}$$

where $n \in \{j - 1, j + 1\}, m = i.$
So, the interpolated value is:

$$\hat{R}_{i,j} = \frac{\sum_{s,t(|s+t|=1)} W_{i+s,j+t} R_{i+s,j+t}}{\sum_{s,t(|s+t|=1)} W_{i+s,j+t}} \tag{3}$$

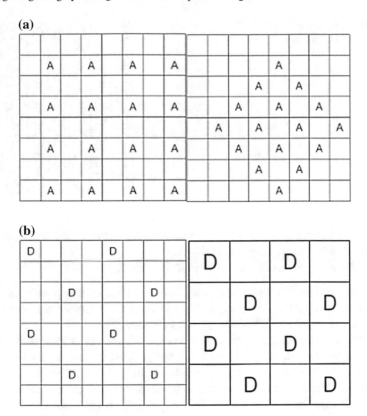

Fig. 4 Transforming the pixels **a** pixels have been rotated by 45° to get a basic pattern, **b** simple down-sampling gives basic pattern

2 Proposed Algorithm

In an MSFA-interpolated image, all pixels of a particular band are likely to have the correlation with each other. If the image is tampered with, it introduces some aberrations in those correlations and identifying them would help us in establishing the originality of an image. We assume that each pixel has a linear relationship with its neighboring pixels as it can reasonably estimate the correlations and is faster.

Inspired by [6], we use the expectation/maximization algorithm to determine the most probable correlation between the pixels, and the pixels that deviate from the expected values more will have higher probability of not belonging to the MSFA.

Let $f(x, y)$ denote a color channel of an MSFA-interpolated image. We begin by assuming that each sample in $f(x, y)$ belongs to one of two models:

(1) M1 if the sample is linearly correlated to its neighbors, satisfying: $f(x, y) = \sum \alpha_{u,v} f(x + u, y + v) + n(x, y)$

Fig. 5 MSFA for 5-band
images (using BTES [6])

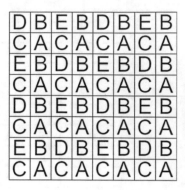

$\alpha_{0,0} = 0$ and $n(x, y)$ denotes random samples drawn from a Gaussian distribution with zero mean and unknown variance.

(2) M2 if the sample is not correlated to its neighbors.

Here, u, v are the distances of the pixels from central pixel which were considered during interpolation; e.g., Assuming five bands: A, B, C, D, E, with probabilities 1/4, 1/4, 1/4, we generate an MSFA as depicted in Fig. 5.

To find the contributing neighbors for band A at locations of band B, the nearest location where A is known is identified. In Fig. 5, immediate vertical neighbors of B hold known values of A so $(u, v) = \{(-1, 0), (+1, 0), (0, -1), (0, +1)\}$. To accomplish this task, a grid of 8×8 size is taken and 1 is stored at locations of band A, 2 at locations of band B, and so on. To solve for band A at location of other bands, nearest occurrence of "1" is searched for and its distance from central pixel is stored as (x, y). From these values, (u, v) are obtained as either $\{(-x, x), (x, x), (x, -x), (-x, -x)\}$ for diagonal nearest neighbor or $\{(x, 0), (-x, 0), (0, x), (0, -x)\}$ for the other case.

From Fig. 5, interpolation of band A at location of band D and E requires the four corner neighbors so $(u, v) = \{(-1, -1), (-1, +1), (+1, -1), (+1, +1)\}$. Interpolation of band D at location of B and E requires values of band D available at locations next to immediate neighbors in horizontal and vertical direction, so $(u, v) = \{(-2, 0), (+2, 0), (0, -2), (0, +2)\}$. The term α represents the relationship between central pixel and its neighbors, so alpha and probability maps for all the cases are calculated separately to ensure higher accuracy and better probability maps.

The EM algorithm is a two-step iterative algorithm: First, we calculate the probability of each sample belonging to each model, and then in the second step, the value of α is estimated, i.e., the correlation. The probability of $f(x, y)$ belonging to M_1 is calculated using Bayes' rule:

$$A = \Pr\{f(x, y) | f(x, y) \in M_1\} \Pr\{f(x, y) \in M_1\}$$

$$\Pr\{f(x, y) \in M_1\} \text{ and } \Pr\{f(x, y) \in M_2\} = 1/2$$

The probability of observing a sample $f(x, y)$ generated from model M1 is given by (probability map):

$$\Pr\{f(x, y)|f(x, y) \in M_1\} = \frac{1}{\sigma\sqrt{2\pi}}\exp\left[-\frac{1}{2\sigma^2}\left(f(x, y) - \sum_{u,v=-N}^{N} \alpha_{u,v} f(x + u, y = v)\right)^2\right]$$

(4)

Since first step requires value of α, it is randomly initialized. A new α is obtained by minimizing the squared error function:

$$E(\vec{\alpha}) = \sum_{x,y} w(x, y)\left(f(x, y) - \sum_{u,v=-N}^{N} \alpha_{u,v} f(x + u, y + v)\right)^2$$

(5)

Differentiating it and putting it equal to zero yields:

$$\sum_{u,v=-N}^{N} \alpha_{u,v}\left(\sum_{x,y} w(x, y)f(x + s, y + t)f(x + u, y + v)\right)$$
$$= \sum_{x,y} w(x, y)f(x + s, y + t)f(x, y)$$

(6)

This is comparatively easier and faster way to get the result.
Detailed Algorithm:

```
/*initialize*/
Solving for band A
Choose { α⁰ᵤ,ᵥ } randomly
Choose σ₀
Set p₀ as 1 over the size of range of possible values of f(x,y).
for each possible MSFA
    n=0
    /*expectation step*/
    for every other band than A(say B)
        //(x, y) represents locations of band B
        //(xₐ, yₐ) represents locations of band A
        Calculate u,v:-
            grid=zeros(8,8)
            grid(xₐ,yₐ)=1
            grid(x,y)=2
            neigh=nearest(grid,1,2)
            //neigh=(x,x) for nearest neighbor at distance x in
                            //diagonal direction
```

```
    if neigh(0)~=neigh(1)
        (u,v)={(-x,0),(x,0),(0,-x),(0,x)}
    else
        (u,v)={(-x,-x),(x,x),(x,-x),(-x,x)}
    end
    for each location of band B in current MSFA
```
$$R(x,y) = f(x,y) - \sum_{u,v=-N}^{N} \alpha_{u,v} f(x+u, y+v)$$
```
    end
    for each location of band B in current MSFA
```

$$P(x,y) = \frac{1}{\sigma\sqrt{2\pi}} exp\left[-\frac{R(x,y)^2}{2\sigma^2}\right]$$

$$W(x,y) = \frac{P(x,y)}{P(x,y)+P_0}$$

```
    end
end
/* maximization step*/
Compute α_{u,v}^{n+1} by solving linear equation.
```
$$\sigma_{n+1} = \left(\frac{\sum_{x,y} W(x,y) R^2(x,y)}{\sum_{x,y} W(x,y)}\right)^{1/2}$$
```
n=n+1
        Until (α_{u,v}^{n+1} - α_{u,v}^n <ε)
end
```

A probability map $(P(x, y))$ is generated by running this algorithm which helps in determining if the image is genuine.

3 Experimental Results

We tested our algorithm on cave dataset of 31 images (512×512) [15] provided by Columbia University. We generated 31 4-band and 31 5-band images and stored first 3 bands of 4-band images and 1, 3, 5 bands of the 5-band image. All of these images were edited in varying proportions of 1–16% hence giving five groups of 31 images each.

To quantify the results, we prepared a synthetic map and compared it with the obtained probability maps of original images and (manually tampered) edited images.
Synthetic map:

$$\text{if } (S(x, y) = r_{x,y}) \quad S_r(x, y) = 0 \quad \text{else } S_r(x, y) = 1$$

where r represents any channel and S represents the MSFA and we already have the probability map p_r obtained from channel r of an image. P_r and S_r represent their respective Fourier transforms.

The measure of similarity M is: $M = \sum |P_r(x, y)|.|S_r(x, y)|$

If the value of M is above a specified threshold, then it is assumed that a correlation is present and the image is genuine. When all three channels of an image

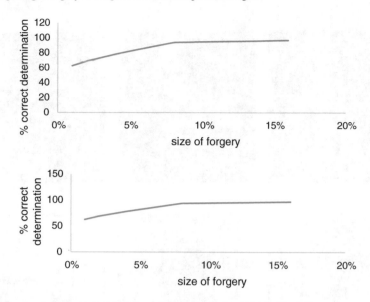

Fig. 6 % correct determination in 4-band images (upper) and 5-band images

for all possible MSFA patterns is below threshold, then the image is considered a forged image. Accuracy of the algorithm was determined on the basis of its correct determination of forged images out of 31 images, and results have been plotted in Fig. 6. The accuracy increased with increase in size of forgeries.

To determine the authenticity of an image with an accuracy of about 80%, image should at least be 7% forged for 4- and 5-band images. Since the algorithm determines originality on the basis of correlations between the pixels produced due to demosaicking, it can be used to determine authenticity of an image with any number of bands(including 3-band images), with the only constraint on their probability of appearances.

We compared the probability maps obtained from our algorithm with the probability maps as described in [6] for five channel images by applying the algorithm on only one channel without any changes. Figure 7 shows the obtained probability maps. The proposed algorithm gives much better results for multispectral images.

The image shown in (Fig. 8a) was reproduced using 4-band BTES (IR, R, G, B) and saved as NIR-G-B image, and the blood stains on the T-shirt were removed using Photoshop as in Fig. 8b. The probability map of image was then prepared using the suggested algorithm, and it clearly demarcates the tampered portion; Fig. 8c.

(a)

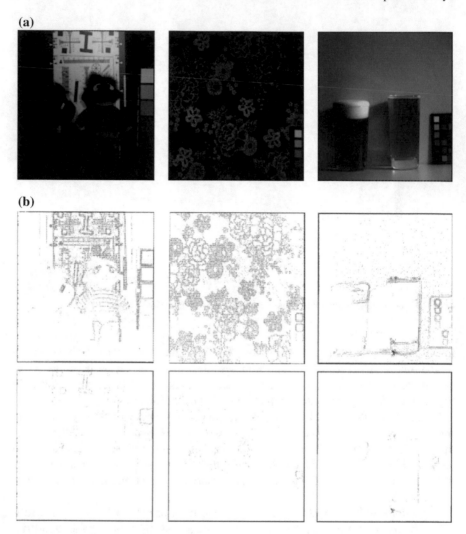

(b)

Fig. 7 **a** Shows the test images and **b** their probability maps obtained from the proposed algorithm (down) and Popescu et al. [6]

(a)   (b)

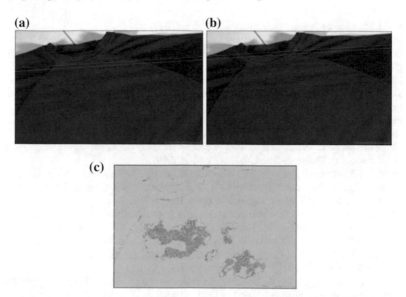

(c)

Fig. 8 **a** Blood-stained T-shirt (NIR-G-B), **b** photoshopped T-shirt, **c** probability map of (**b**)

4 Conclusion

In this paper, we studied the potential of expectation/maximization algorithm in determining the authenticity of a single-sensor multispectral image reconstructed using some demosaicking method. We discussed in relation to the binary tree-driven demosaicking method BTES, but approach seems applicable to other demosaicking methods in general. The demosaicking methods lead to correlations which would get affected if the original multispectral image is tampered. By using the expectation/maximization algorithm, these correlations can be estimated and it can then be used to determine whether the image is forged or not. With the use of multispectral images increasing with time, it was imperative that an algorithm to determine an image's authenticity be developed. This algorithm also highlights the importance of using an MSFA. It not only reduces the cost of capturing these high-content images but also provides the means to detect digital manipulations/any tampering in the multispectral images.

References

1. Li, H., Luo, W., Qiu, X., Huang, J.: Image forgery localization via integrating tampering possibility maps. IEEE Trans. Inf. Forensics Secur. **12**(5), 1240–1252 (2017). https://doi.org/10.1109/TIFS.2017.2656823
2. Shen, X., Shi, Z., Chen, H.: Splicing image forgery detection using textural features based on the grey level co-occurrence matrices. IET Image Process. **11**(1), 44–53 (2017). https://doi.org/10.1049/iet-ipr.2016.0238

3. Farid, H.: How to Detect Faked Photos. Am. Sci. (2017)
4. Pun, C.M., Yuan, X.C., Bi, X.L.: Image forgery detection using adaptive over-segmentation and feature point matching. IEEE Trans. Inf. Forensics Secur. **10**(8), 1705–1716 (2015). https://doi.org/10.1109/TIFS.2015.2423261
5. Ferrara, P., Bianchi, T., Rosa, A.D., Piva, A.: Image forgery localization via fine-grained analysis of CFA artifacts. IEEE Trans. Inf. Forensics Secur. **7**(5), 1566–1577 (2012). https://doi.org/10.1109/TIFS.2012.2202227
6. Popescu, A.C., Farid, H.: Exposing digital forgeries in color filter array interpolated images. IEEE Trans. Signal Processing **53**(10), 3948–3959 (2005)
7. Cao, H., Kot, A.C.: Accurate Detection of demosaicing regularity for digital image forensics. IEEE Trans. Inf. Forensics Secur. **4**(4), 899–910 (2009)
8. Farid, H.: A Survey of image forgery detection. IEEE Sig. Process. Mag. **26**(2), 16–25 (2009)
9. Miao, L., Qi, H., Ramanath, R., Snyder, W.E.: Binary tree-based generic demosaicking algorithm for multispectral filter arrays. IEEE Trans. Image Process. **15**(11), 3550–3558 (2006)
10. Miao, L., Qi, H.: The design and evaluation of a generic method for generating mosaicked multispectral filter arrays. IEEE Trans. Image Process. **15**(9), 2780–2791 (2006)
11. Jaiswal, S.P., Fang, L., Jakhetiya, V., Pang, J., Mueller, K., Au, O.C.: Adaptive multispectral demosaicking based on frequency-domain analysis of spectral correlation. IEEE Trans. Image Process. **26**, 953–968 (2017)
12. Aggarwal, H.K., Majumdar, A., Ward, R.: A reconstruction al-gorithm for multi-spectral image demosaicing. In: Proceedings of the IASTED International Conference on Signal and Image, Banff, AL, Canada, vol. 1719 (2013)
13. Monno, Y., Kikuchi, S., Tanaka, M., Okutomi, M.: A practical one shot multispectral imaging system using a single image sensor. IEEE Trans. Image Process. **24**(10), 3048–3059 (2015)
14. Mihoubi, S., Losson, O., Mathon, B., Macaire, L.: Multispectral demosaicing using pseudo-panchromatic image. IEEE Trans. Comput. Imaging **3**(4), 982–995 (2017). https://doi.org/10.1109/TCI.2017.2691553
15. Yasuma, F., Mitsunaga, T., Iso, D., Nayar, S.K.: Generalized assorted pixel camera: post-capture control of resolution, dynamic range and spectrum. IEEE Trans. Image Process. **19**(9), 2241–2253 (2010). http://www.cs.columbia.edu/CAVE/projects/gap_camera/

Neuro-Fuzzy Analysis of Demonetization on NSE

Rashmi Bhardwaj and Aashima Bangia

Abstract This paper studies the impact of demonetization on National Stock Exchange (NSE) for the daily basis data of opening prices of NSE Nifty India consumption from April 1, 2016, to July 31, 2017. The two time periods, before and after demonetization, are taken into consideration. Artificial neural network (ANN) and adaptive neuro-fuzzy inference system (ANFIS) are used to study the time series analysis of opening price of stock market. This study shows that the optimized number of neurons in the hidden layer cannot be always determined by using a particular formula but trial-and-error method. The results show a rather temporary effect on the stock market as investors tend to look beyond the existing crisis.

Keywords ANFIS · Demonetization · Fuzzy · ANN · NSE

1 Introduction

According to Black's Law Dictionary, demonetization means the disuse of a particular metal for purpose of coinage or, in general, the withdrawal of the value of a metal as money. As per the RBI Act, Section 26(2), the central government, on the sanction of the Central Board of RBI, may, by the notification in the Gazette of India, declare that, with effect from such date as may be specified in the notification, any series of bank notes of any denomination shall cease to be legal tender. On November 8, 2016, the government of India took the step to demonetize Rs.500 and Rs.1000 currency, which meant that the legal tender of currency units was declared invalid from November 9, 2016. This resulted in 86% of the circulated money being removed from the economy overnight.

R. Bhardwaj (✉) · A. Bangia
Non-Linear Dynamics Research Lab, University School
of Basic and Applied Sciences, Guru Gobind Singh
Indraprastha University, Dwarka, New Delhi, India
e-mail: rashmib22@gmail.com

© Springer Nature Singapore Pte Ltd. 2019
J. C. Bansal et al. (eds.), *Soft Computing for Problem Solving*,
Advances in Intelligent Systems and Computing 816,
https://doi.org/10.1007/978-981-13-1592-3_68

To predict long-term trends in financial markets, different methods of the mathematical model like time series modelling, statistical modelling, stochastic modelling can be used. For nonlinear characteristics, many models including artificial neural network (ANN), fuzzy inference system (FIS), etc., are used.

Neuro-fuzzy model is the complexity of decision-making for fuzzy operator and may improve the short-term forecasting. Neural networks are important from a statistical point of view [1, 8–11, 22]. Wavelet–neuro-fuzzy conjunction model is used to develop the model for improved forecasting [16–20]. Statistical analysis is the basic tool used for the analysis of time series [2–7]. Time series analysis using wavelet, neuro-fuzzy techniques provides better approximation with minimum error in forecasting [21]. Long-range variability analysis is very useful for future prediction [12–15].

Fuzzy reasoning was developed to predict NSE stock market prices for before and after demonetization in India. The data is divided into training and testing phases. The model results are compared with measured data. The present study is used to study the effect of demonetization on NSE for stock market prediction. In this study, the NSE Nifty India Consumption Opening Prices data on a daily basis is trained and modelled using ANFIS to study the scenario of NSE when there was unhindered cash flow in the market.

2 Methodology

2.1 Fuzzy Inference System (FIS)

The fuzzy inference system (Takagi and Sugeno 1985) is the standard computing framework based on the theories of fuzzy set theory, fuzzy if-then rules and fuzzy reasoning. It is used in an automatic control, data classification, decision analysis, expert systems, robotics and pattern recognition.

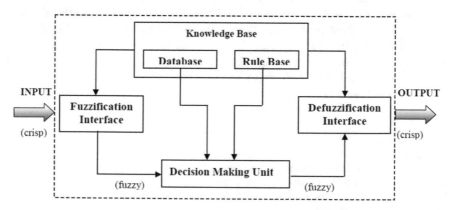

The steps of fuzzy reasoning, i.e. inference operations on fuzzy if-then rules, performed by fuzzy inference system are as follows:

(i)   The input variables with the membership functions on the antecedent part to obtain the membership values of every semantic label are compared (fuzzification step).

(ii) The membership values on the premise part to get weight of each rule are combined.

(iii) The qualified consequents either fuzzy or crisp of each rule depending on the weights are generated.

(iv) The qualified consequents to produce a crisp output are finally aggregated (defuzzification step).

2.2   *Artificial Neural Networks (ANNs)*

Artificial Neural networks are used for proficiently modelling large and complex problems. Neural networks are useful for: (i) classification of problems where the output is a categorical variable or (ii) for regressions where the output variable is continuous. It detects and learns the interrelated patterns between input datasets and aimed values. Neural network described a structure of simple processing nodes, i.e. neurons interconnected with each other in a definite order performing simple numerical manipulations. These networks are made up of an input layer consisting of nodes which represent different input variables; the hidden layer consists of many hidden nodes, and an output layer consists of output variables. Learning is the training of updating the connecting weights in response to external stimuli presented at the input barrier. The network "learns" in arrangement with a learning rule that governs the fine-tuning of connecting weights in response to learning examples applied at the input and output buffers. Recall is the practice of accepting an input and producing a response determined by the geometry and synaptic weights of the network.

Training of ANN
The learning techniques can be classified as:

(i)   Supervised learning or associative learning: In this, the structure is trained by providing it with input and matching output pattern. These input–output pairs can be provided by an external teacher or by the system which contains the structure.

(ii) Unsupervised learning or self-organization: In this, an output unit is trained to respond to clusters of pattern within the input. The salient features of the input dataset are considered by the system for statistical analysis. There is no a priori set of categories into which the patterns are to be classified rather the system must develop its own representation of the input stimuli.

2.3 Adaptive Neuro-Fuzzy Inference System (ANFIS)

The basic concepts and motivation of integrating fuzzy logic and neural networks into a working functional system are used to enhance the existing procedure. The amalgamation of the two techniques: fuzzy logic system and neural networks, suggest the unusual idea of transforming. ANFIS is a well-known estimator which is perhaps able to approximate any real continuous function on a compact set. The basic structure of the fuzzy inference system is to map input characteristics to their membership functions. Then, input membership function to rules and the rules to a set of output characteristics are connected. Further, it maps output characteristics to output membership functions and output membership function to a single output. Each fuzzy system contains three main parts: fuzzifier, fuzzy databank and defuzzifier. Fuzzy databank includes two main parts: fuzzy rule base and inference engine.

ANFIS Architecture
According to the Takagi and Sugeno type, the fuzzy inference system has two inputs x and y and one output f.

Rule 1: If x is A_1 and y is B_1 then $f_1 = p_1x+q_1y+r_1$ ... (1)
Rule 2: If x is A_2 and y is B_2 then $f_2 = p_2x+q_2y+r_2$ ... (2)

Layer 1: Each node i in this layer is a square node with a node function

$$O_i^1 = \mu A_i(x)$$

where x is the input to the node i, A_i is the semantic label associated with this node, and μ is the membership function of A_i, and it defines the degree to which the given x gratifies the quantifier A_i. Let $\mu A_i(x)$ assumed to be bell-shaped with a maximum equal to 1 and minimum equal to 0 such as the generalized bell function or Gaussian function.

Layer 2: Each node in this layer is a circular node which multiplies the incoming signals and sends the product out.

$$O_i^2 = w_i = \mu A_i(x) \times \mu B_i(x), \qquad i = 1, 2, 3 \ldots$$

Each node output represents the firing strength of a rule.
Layer 3: Each node in the layer is a circular node and is referred as N. The ith node calculates the ratio of the ith rule's firing strength to the sum of all rule's firing strengths

$$O_i^3 = w_i = \frac{w_i}{w_1 + w_2}, \qquad i = 1, 2, 3 \ldots$$

The outputs of this layer will be called normalized firing strength.

Layer 4: Each node i in this layer is a square node with a node function

$$O_i^4 = \overline{w_i} f_i = \overline{w_i}(p_i x + q_i y + r_i)$$

where $\overline{w_i}$ is the output of layer 3 and (p_i, q_i, n_i) is the set of parameters.
Layer 5: The single node in this layer is a circular node labelled that calculates the overall output as the summation of all such incoming signals, i.e.

$$O_i^5 = \sum \overline{w_i} f_i = \frac{\sum\limits_i w_i f_i}{\sum\limits_i w_i}$$

Thus, it constructs an adaptive network which is functionally equivalent to a type 3 fuzzy inference system (Fig. 1).

3 Results and Discussions

The neuro-fuzzy designer is trained with backpropagation method for generalization and Sugeno inference for specialization. The training of data is done for 2000 epochs. The conjecture model uses ANFIS system to estimate the succeeding values. Neuro-fuzzy model of the NSE Nifty India Consumption Opening Prices for the time period before demonetization, i.e. from April 1, 2016, to November 8, 2016, on a daily basis is used for the analysis. Also, neuro-fuzzy model of the NSE Nifty India Consumption Opening Prices for the time period after demonetization, i.e. from November 9, 2016, to July 31, 2017, on a daily basis is used for the analysis. Figures 2 and 3 show the ANFIS trained data of opening prices of NSE training results and model output surface view before demonetization and after demonetization, respectively. Using neuro-fuzzy model, actual and forecast for opening price of NSE before demonetization and after demonetization are shown in Figs. 4 and 5, respectively.

It is observed that ANN model has 9% error (approximately) in forecast future; neuro-fuzzy coupled model results in 2% error (approximately). The results show a rather temporary effect on the stock market as investors tend to look beyond the existing crisis.

It is observed that the predicted values of opening prices and actual values did not match at all after demonetization. Actual opening prices were different from the forecasted/expected outcomes. Also, as compared to before demonetization, the opening values were higher after demonetization.

(a)

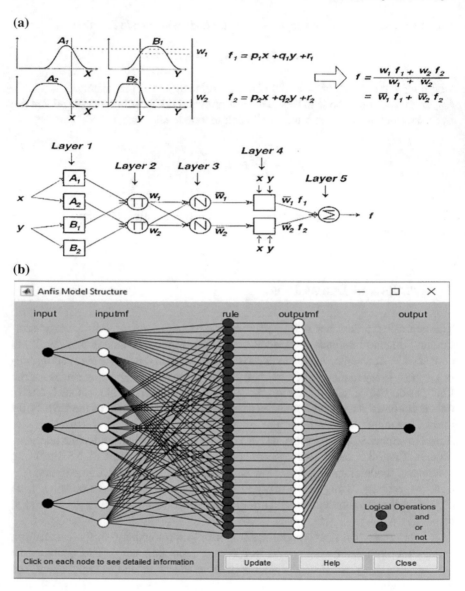

(b)

Fig. 1 Graphical representation of neuro-fuzzy and ANFIS

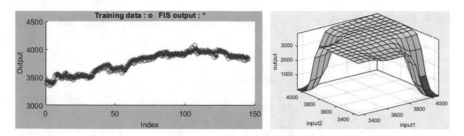

Fig. 2 ANFIS trained data, training results, model output surface view before demonetization

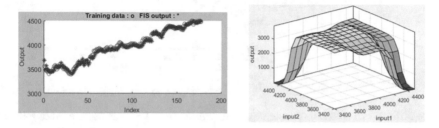

Fig. 3 ANFIS trained data, training results, model output surface view after demonetization

Fig. 4 Actual and forecast for opening price of NSE before demonetization by neuro-fuzzy model

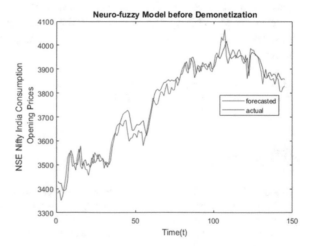

4 Conclusion

The RBI Act of demonetization in India came into effect at the midnight of November 8, 2017, was a one of its kind of experience for the world to witness. It had a great impact on the cash flow in the market as the currency notes of Rs. 500 and Rs. 1000 had lost their value and declared invalid. The stock markets were bound to face its repercussions. After demonetization, it is observed that the predicted values of opening prices and actual values did not match at all. Actual opening prices

Fig. 5 Actual and forecast for opening price of NSE after demonetization by neuro-fuzzy model

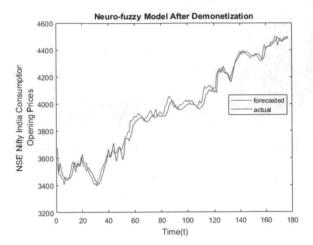

were different from the forecasted/expected outcomes. Also, as compared to before demonetization, the opening values were higher after demonetization. The negative impact shown after demonetization is gradually rebounding back to normalcy. Now, masses wanted to convert their currency notes to electronic money. The impact was, however, seen to be temporary as gradually the money came back in the market and so did investors. This trend is also witnessed in the forecasted model using neuro-fuzzy model.

Acknowledgements Authors are thankful to Guru Gobind Singh Indraprastha University for providing financial support and research facilities.

References

1. Aqil, M., Kita, I., Yano, A., Nishiyama, A.: Analysis and prediction of flow from local source in a river basin using a Neuro-fuzzy modeling tool. J. Environ. Manage. **85**, 215 (2007)
2. Bhardwaj, R.: Wavelets and Fractal methods with environmental applications. In: Siddiqi, A.H., Manchanda, P., Bhardwaj, R. (eds.) Mathematical Models, Methods and Applications, pp. 173–195 (2016)
3. Bhardwaj, R., Bangia, A.: Complex dynamics of meditating body. Indian J. Ind. Appl. Math. **7**(2), 106–116 (2016)
4. Bhardwaj, R.: Wavelet and correlation analysis of weather data. Int. J. Curr. Eng. Technol. **2**(1), 178–183 (2012)
5. Bhardwaj, R.: Wavelet & correlation analysis of air pollution parameters using Haar wavelet (Level 3). Int. J. Therm. Technol. **2**(2), 160–164 (2012)
6. Bhardwaj, R., Kumar, A., Maini, P., Kar, S.C., Rathore, L.S.: Bias free rainfall forecast and temperature trend based temperature forecast based upon T-170 Model during monsoon season. Meteorol. Appl. **14**(4), 351–360 (2007)
7. Bhardwaj, R., Srivastava, K.: Real time nowcast of a cloudburst and a thunderstorm event with assimilation of doppler weather radar data. Nat. Hazards **70**(2), 1357–1383 (2014)

8. Chen, H.W., Chang, N.B.: Using fuzzy operators to address the complexity in decision making of water resources redistribution in two neighboring river basins. Adv. Water Resour. **33**, 652 (2010)
9. Chaturvedi, D.K., Singh, M.M., Kalra, P.K.: Improved generalized neuron model for short term load forecasting. Int. J. Soft Comput.—Fusion Found. Methodologies Appl. **8**, 10 (2004)
10. Cheng, B., Titterington, D.M.: Neural networks: a review from a statistical perspective. Stat. Sci. **9**, 2–54 (1994)
11. Chang, F.J., Chang, Y.T.: Adaptive neuro-fuzzy inference system for prediction of water level in reservoir. Adv. Water Resour. **29**, 1 (2006)
12. Durai, V.R., Bahrdwaj, R.: Location specific forecasting of maximum and minimum temperature over india by using the statistical bias corrected output of global forecasting system. J. Earth Syst. Sci. **123**(5), 1171–1195 (2014)
13. Durai, V.R., Bahrdwaj, R.: Evaluation of statistical bias correction methods for numerical weather prediction model (NWP) forecasts of maximum and minimum temperatures. Nat. Hazards **73**(3), 1229–1254 (2014)
14. Durai, V.R., Bahrdwaj, R.: Forecasting quantitative rainfall over india using multi-model ensemble technique. Meteorol. Atmos. Phys. **126**, 31–48 (2014)
15. Elmitwally, A., Farghal, S., Kandil, M., Abdelkader, S., Elkateb, M.: Proposed wavelet–Neurofuzzy combined system for power quality violation detection and diagnosis. IEE Proc.-Gener Transm. Distrib. **148**, 15 (2001)
16. Mittal, A., Bhardwaj, R.: Prediction of daily air pollution using wavelet decomposing and adaptive-network-based fuzzy inference system. Int. J. Environ. Sci. **2**(1), 174–184 (2011)
17. Mittal, A., Bhardwaj, R.: Index, fractal dimension and hurst exponent estimation of air pollution parameters. Int. J. Adv. Sci. Tech. Res. **2**(1), 363–375 (2011)
18. Moosavi, V. et al.: A Wavelet-ANFIS Hybrid Model for Groundwater Level Forecasting for Different Prediction Periods, Water Resources Management (2013)
19. Moosavi, V. et al.: Optimization of Wavelet-ANFIS and Wavelet-ANN Hybrid Models by Taguchi Method for Groundwater Level Forecasting, Arabian Journal for Science and Engineering (2014)
20. Parmar, K.S., Bhardwaj, R.: Water quality index and fractal dimension analysis of water parameters. Int. J. Environ. Sci. Technol. **10**, 151–164 (2013)
21. Parmar, K.S., Bhardwaj, R.: Statistical, time series and fractal analysis of full stretch of river Yamuna (India) for water quality management. Environ. Sci. Pollut. Res. **22**(1), 397–414 (2015)
22. Zadeh, L.A.: Fuzzy sets. Inf. Control **8**, 338 (1965)

A Hardware Architecture Based on Genetic Clustering for Color Image Segmentation

Rahul Ratnakumar and Satyasai Jagannath Nanda

Abstract Color image segmentation finds several real-life applications on hyperspectral image processing, brain tumor detection (Biomedical), facial recognition (Biometric), object tracking (Video analysis), etc. In this manuscript, the color image segmentation is dealt as a clustering problem. A genetic algorithm (GA)-based hardware architecture is proposed to perform the segmentation task in a fast manner. Testing of the proposed architecture is carried out on four standard RGB color images like Pepper, Baboon, Lenna, and Colorbars. Comparison with three other benchmark architectures of genetic algorithm reveals that the proposed architecture provides satisfactory results in terms of complexity, system clock frequency, and resource utilization. The three other architectures used for comparison are compact implementation of GA, used for simple optimization tasks, whereas the proposed one is used for clustering huge number of pixels within an image, for executing the task of segmentation.

Keywords Genetic algorithm · Clustering · Finite state machine
Linear-feedback shift register · Image segmentation

1 Introduction

Clustering is referred to as an unsupervised classification problem [1]. The input patterns (vectors) in the many dimensional spaces are assigned into different unique clusters, such that the vectors within an individual cluster are similar in some characteristics and those of different clusters are dissimilar. Usually for this comparison is based on a measure of distance like Euclidian, Manhattan or cosine is used [2]. The

R. Ratnakumar (✉) · S. J. Nanda
Department of Electronics and Communication Engineering,
Malaviya National Institute of Technology, Jaipur 302017, Rajasthan, India
e-mail: rahul.ratnakumar@gmail.com

S. J. Nanda
e-mail: nanda.satyasai@gmail.com

© Springer Nature Singapore Pte Ltd. 2019
J. C. Bansal et al. (eds.), *Soft Computing for Problem Solving*,
Advances in Intelligent Systems and Computing 816,
https://doi.org/10.1007/978-981-13-1592-3_69

traditional K-means, K-medoids, K-nearest neighborhood, and similar statistical-based approaches fail to produce effective clusters in case of overlapping datasets due to their limitation of hill climbing approach to determine solutions [1]. Under such scenarios, the evolutionary approaches like genetic algorithm (GA) and their variants play a significant role to determine effective cluster [3].

The color image segmentation finds extensive applications in identification of specific targets in images and determines tumors in medical brain images, in space application performing hyperspectral image analysis, etc. [4]. Under such complex applications, accuracy of clusters plays a crucial role, as each cluster reflect a specific portion of the image. Based on the pixel values, the decision to find out effective categories can be managed in a better way in evolutionary approach compared to conventional clustering [4]. The present research paper aims to develop hardware architecture to implement the GA-based clustering. The main task of this architecture is to segment color images.

The first FPGA implementation of a general-purpose GA engine was reported by Scott et al. in 1995 [5]. The features of this architecture were modular hardware implementation of a simple miniaturized GA in VHDL. A roulette wheel selection and one-point crossover were employed.

It employs a fixed population size of 16 and member width of 3 bits. The architecture was dumped onto multiple Xilinx FPGA BORG Boards. Later on, Tommiska and Vuori in [6] implemented a general-purpose GA system, where round robin parent selection, one-point crossover was employed. A fixed population size of 32 gave it very good evolutionary characteristics. This architecture was dumped onto Altera FPGAs mounted on PCI cards. Yoshida et al. in [7] implemented a GA processor with a steady-state architecture supporting efficient pipelining applying a simplified tournament selection. Shackleford et al. in [8] implemented a survival-based, steady-state GA, Coded in VHDL, and it was optimized for higher performance. The same architecture was tested on set-covering and protein-folding problems. This was later implemented on an Aptix AXB-MP3 field programmable populated with six FPGAs. Tang and Yip in [9] in implemented a PCI-based hardware GA system using two Altera FPGAs mounted on a PCI board, which had multiple crossover and mutation operators implemented with programmable crossover and mutation thresholds. This architecture focused on the parallel implementations of the PCI-based GA system.

This paper focuses on development of a hardware architecture based on GA clustering to segment benchmark RGB color images. Manhattan distance matrix is used as similarity measure for clustering. Testing is carried out on four color images of Peppers, Baboon, Leena, and Color bars. The performance of the proposed approach is compared with three other architectures implementing genetic algorithm, for simple optimization tasks.

2 Problem Formulation

2.1 Image Segmentation as a Clustering Problem

Clustering is the process of dividing the given input data into different groups such that the data elements within a cluster are more similar to each other, and the elements of different cluster are dissimilar to each other, based on some of their characteristics known as features. For example in the case of K-means clustering, we randomly generate 'K' centroids and update them by computing the mean of the data elements nearest to each centroids (hence the name K-means).

This process is repeated until the centroid values converge to an optimal point in terms of minimizing the average distance between data point and centroid or maximizing the average distance between two centroids.

Image segmentation is mathematically expressed as a process that groups an input image into n classes [10]. Each class is treated as a cluster, and sum of all the cluster results in the total image.

$$\bigcup_{i=1}^{n}(C_i) = 1 \tag{1}$$

where C_i is a connected set with $i = 1, 2, \ldots n$. This reflects that data (pixel) points within a region have to be connected in some way. Different regions in an image should be disjoint in properties reflected by

$$(C_i \cap C_j) = \phi \tag{2}$$

where $i, j \in 1, 2, \ldots n$.

2.2 Genetic Algorithm for Clustering

Genetic algorithms (GA) are randomized meta-heuristic optimization or search techniques inspired by the natural principles of genetics and evolution. One of the intrinsic features of GA is its large parallelism. Even in large complicated multidimensional solution spaces, GA is competent to provide us at least with non-optimal solutions, for the fitness or objective function of any given search or optimization problem. Like any other evolutionary/nature-inspired algorithm, the strength of GA lies in the fact that it is robust, versatile, simple but stochastic in nature and has the ability to jump over local minima, applies the mechanism of survival of the fittest, focused on the strategy—'first explore and then exploit.'

In GA, we use encoded strings called chromosomes as the parameter for searching the solution space. A group of such strings are termed as a population. Initially, we fill

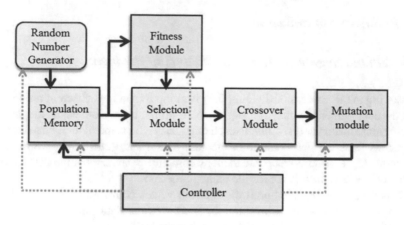

Fig. 1 Block diagram for implementation of genetic clustering

the population table with random strings. The quality of the chromosome is judged by computing the fitness (goodness) function associated with each and every string. The better the fitness function of the chromosome, the better are the chances for its survival. The best chromosomes are then selected after sorting based on their fitness, and the selected ones enter into the reproduction process (mating pool). Nature-inspired evolutionary mechanisms like crossover and mutation are executed on these parent chromosomes to yield offspring, which are nothing but the next generation parameters (chromosomes). This cycle continues till a fixed number of generations is emerged or condition for termination is fulfilled.

The block diagram to implement genetic algorithm for clustering [11, 12] is shown in Fig. 1. A random number generator initializes the centroid memory, which acts as the chromosome for the genetic algorithm. Once all the required chromosomes are randomly assigned in a parallel fashion, their fitness is found by the fitness module. Based on the ranking of the chromosomes according to their fitness, the best ones are selected from the population memory as parents for the upcoming generation of centroids. These are passed on to undergo natural selection process involving crossover and mutation paradigm. These evolutionary processes of crossover and mutation also require generation of random numbers; thus, the random number generator module is also utilized for the same. During the reproduction process, more offsprings are generated which in turn undergo further fitness evaluation, before their selection to population memory for the upcoming evolutionary processes. This cycle has to continue till a sufficiently good solution is produced. The control unit coordinates all the activities of the hardware architecture.

3 Proposed Hardware Architecture for Genetic Clustering

The detailed architecture of proposed genetic clustering hardware is shown in Fig. 2. It constitutes of a linear-feedback shift register (LFSR) random number generator initializing a chromosome memory [13], whose values are compared with input data memory values, consolidated to find the fitness of the chromosomes which is the sum of the distances of all data points to its nearest cluster centroid, utilizing a summing and sorting module. The distance metric used is Manhattan owing to its reduced complexity, as the conventional Euclidian metric requires a squaring unit. The best-fit chromosomes (centroids) are selected through a sorting mechanism to form the parents for the new generation. The crossover and mutation modules act upon the selected parents to produce offspring, which are tested for their fitness function. The best ones among the offspring are matched with their parent community and refined further for better solutions in the future generations.

String Representation: Each string is a sequence of numbers used to represent the K clusters centers. For example, for a N-dimensional space, the length of chromosome will be $N * k$ words. In this architecture, a 48-bit string denotes a single chromosome. ($N = 2$, $k = 24$, for 3 clusters each represented by 8 bits). For the present problem, we have the input consisting of 8-bit pixels elements; therefore, the

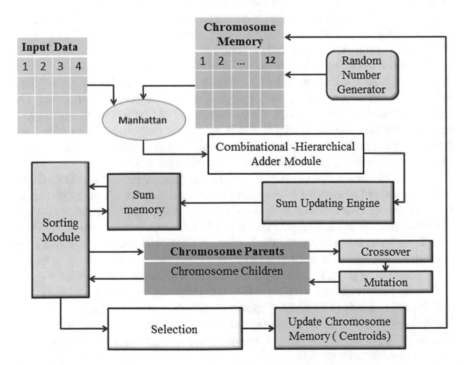

Fig. 2 Proposed architectural diagram for genetic clustering hardware

Fig. 3 Mutation module of
the architecture for genetic
clustering

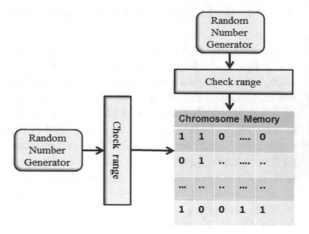

input domain is 0 (Black colored pixels) to 255 (white colored pixels). These pixels
numbering around 512×512, arranged as two matrices (dimensions).

Population Initialization: The K cluster centroids are initialized to k random
points in the starting of the process. Total population size is 10. So the chromosome
memory will have 480 bits. The population size is determined by taking into account
many factors like complexity of the problem given, presence and absence of multiple
maxima and minima within the solution space, the size, dimension of the input and
the accuracy, precision of the solutions required.

Crossover Module: It is used to perform the crossover operation on the two
parent individuals, which are selected because of their better fitness. This design
unit offers only single-type crossover operation, i.e., single-point crossover. The
input to the module constitutes the whole population of size 10 chromosomes, with
each chromosome represented by 8 bits. Single-point crossover is used with fixed
crossover point assigned as the third bit from MSB, with a crossover rate of 100%.
The output chromosomes are then sent to the mutation module.

Mutation Module: Figure 3 shows the hardware structure of the mutation module.
It is used to avoid converging of the chromosomal solutions to a local optimum and
instead explore newer and better solutions [9]. The random number generator assigns
dynamic positions within the children chromosomes for the mutation to occur. Here,
a fixed mutation rate-setting scheme is used and mutation rate of 1%. Finally, the
generated chromosome is checked for its fitness using a tournament selection method
and if found to be good, is fed into the population memory, as shown in the Fig. 1.

Linear-Feedback Shift Register: Random number generator is designed from
a 16-bit shift register (SR) which is multiplexed with a seed initialized as shown in
Fig. 4. It regulates the initial point of the counting within the range $0–32,767$ ($2^{15}-1$).
Specially chosen points in the shift register are XOR-ed to form the new input which
is given as a feedback to SR.

FSM Mealy machine: The state machine shown in Fig. 5 starts with a Go signal,
then the machine waits for the input N_ld, which gives the number of data points

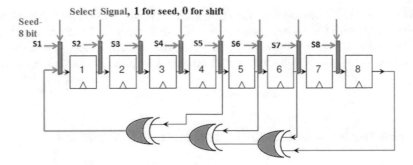

Fig. 4 8-bit linear-feedback shift register (LFSR)

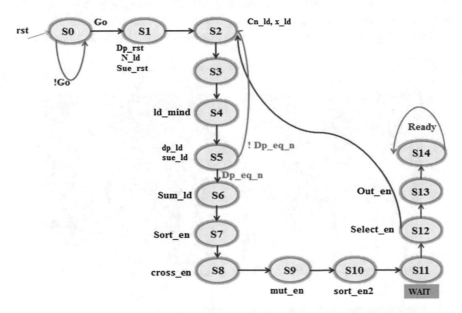

Fig. 5 State diagram of the FSM Mealy machine in the genetic clustering architecture

to the machine. Dp_rst and Sue_rst reset the data pointer and sum updating engine, respectively. In the second state, Cn_ld loads the centroid memory with the random values generated by the LFSR-RNG. S3 is a wait state for completing the combinational hierarchical adder calculation. Ld_mind loads the minimum distance to the sum updating engine to find the fitness function. After this routine, the data pointer is updated (dp_ld). The sum of the distance of the data with the nearest centroid is also added and stored within the accumulator by the signal sue_ld. The control unit checks the data pointer and if it is equal to N, the number of data points Dp_eq_n is triggered, which enables the state machine to exit from the loop. Then in the next two states (s6 and s7), the fitness (sum) is loaded into the sum memory and sorting machine is activated. After that, crossover and mutation modules are executed in the states (s8 and s9). Then the machine again goes for final selection for the best chromosomes (parent) for the next generation (s10). In parallel to the selection process, the termination criteria are checked; if it is true, the best chromosomes are given to the output (s13–s14) with a ready signal, else next iteration is initiated in the state (s2).

4 Hardware Implementation and Results Discussions

Behavioral modeling of the evolutionary clustering using genetic algorithm has been conducted in MATLAB 2015a. Convergence is found to be satisfactory. The above architecture is modeled in hardware using Mentor Graphics tool ModelSim 10.4a in Verilog. The histogram analysis of these images revealed the possibility of three

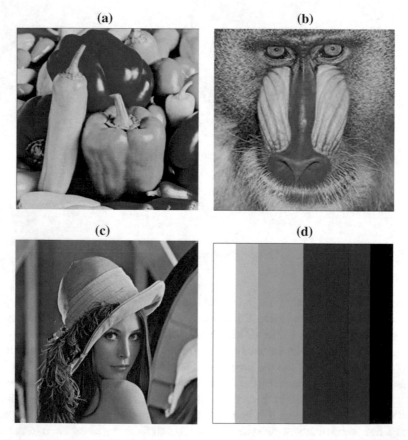

Fig. 6 Benchmark standard color images used for testing the hardware architecture for genetic clustering **a** Peppers.png, **b** Baboon.png, **c** Lenna.png, **d** Colorbars.png

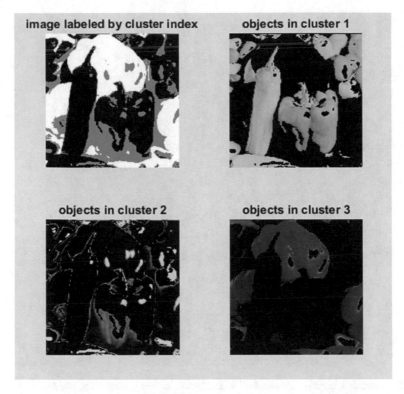

Fig. 7 Segmented results of the standard benchmark color image Peppers.png with $K = 3$

clusters ($k = 3$) of independent chromaticity within the pixel population. Thus, the hardware was designed to divide the pixel into three different cluster heads. The results were matching the system simulation results of MATLAB. The Verilog design was further synthesized in ISE Project Navigator 14.7. The target device used was XC7K70T-3FBG676, of Kintex-7 family. Test case study is conducted for testing the designed architecture for genetic clustering using the standard benchmark Portable Network Graphics (PNG) color images like Peppers, Baboon, Lenna, and Colorbars as shown in Fig. 6.

Cluster analysis of Peppers.png: The Peppers.png image is grouped into three clusters as shown in Fig. 7, each capturing the red (R), green (G), and blue (B) pixel (pair) elements within the image. It can be easily verified that cluster-1 consists of pixels pertaining to Green color ranging from lime green to emerald green. Cluster-2 has pixel (pairs) elements of higher magnitudes (above 200), ranging from lighter (whiter) pixels gradually decreasing to pixel magnitudes reaching till sienna brown. The final group, cluster-3, consists of all pixel pairs which are emitting Red shades (Fig. 7).

Cluster Analysis of Baboon.png: As shown in Fig. 8, cluster-1 groups all the pixel pairs denoting the reddish components of the image, i.e., the bright-red nose

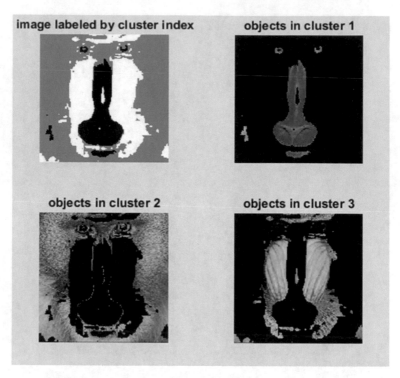

Fig. 8 Segmented results of the standard benchmark color image Baboon.png with $K = 3$

of the Baboon. The orange-red iris of the Baboon is also captured within the same cluster. Cluster-2 consists of greenish components of the pixel pairs showcasing the hair as well as the greenish-yellow mane of its face. The cluster-3 groups all the pixels covering the blue snout of the Baboon.

Cluster analysis of Lenna.png: Unlike the other color images, the result of clustering of Lenna.png yields a slightly different result, as compared to the results of other benchmark images. The cluster-1 consists of the brighter parts of the image which has a very high illumination. Cluster-2 is consisting of pixel pairs making up the darker parts of the image, like hair, eyes, and the dark door wall. Cluster-3 group areas are having illumination in between the bright and dark areas, as shown in Fig. 9.

Cluster Analysis of Colorbars.png: The objects of the image Colorbars.png pertaining to the Blue region, get grouped into the cluster-1. Cluster-2 consists of all pixel pairs emitting the shades of Green color. In the same way, cluster-3 groups all pixel pairs falling within the Red region of the image. See Fig. 10.

The implementation details of the proposed genetic algorithm with other benchmark methods are shown in Table 1. The comparative hardware specification of the proposed architecture with three other architectures of [14, 15] are presented in Table 2. From Table 2, it is observed that the proposed architecture having satisfac-

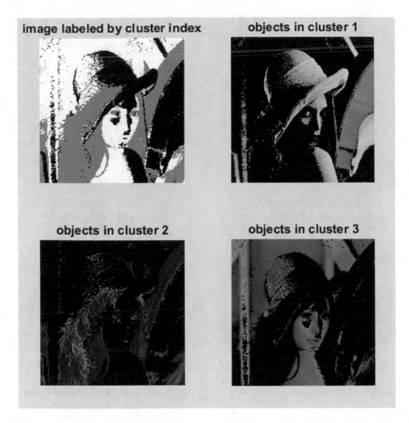

Fig. 9 Segmented results of the standard benchmark color image Lenna.png with $K = 3$

Table 1 Comparison of proposed GA with other designs

References	Selection	RNG	Fitness module	Mutation rate	Population size/individual & fitness length
[5]	Roulette	CA	Redesign	None	Fixed/Fixed/Fixed
[6]	Simplified tournament	CA	Reprogram	None	Fixed/Fixed/Fixed
[7]	Roulette/elitist	CA	Reprogram	Dynamic	(64 or 128)/Fixed/Fixed
[16]	Tournament	CA	Reprogram	Multipoint	Fixed/Fixed/Fixed
This work	Tournament	CA	Redesign	Multipoint	10/Fixed/Fixed

tory clock frequency compared to the other compact genetic implementations. The present implementation has more requirement of area over other architectures due to the implementation of image segmentation. The other architectures are meant for simple optimization problems (Fig. 11).

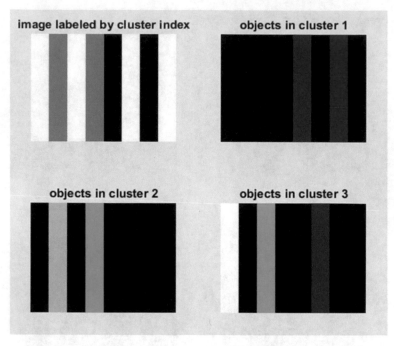

Fig. 10 Segmented results of the standard benchmark color image Colorbars.png with $K = 3$

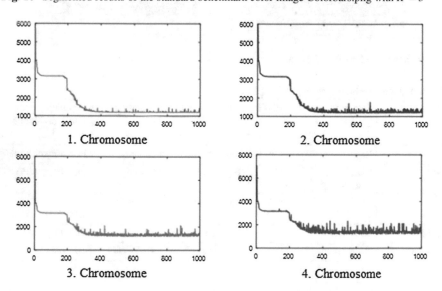

Fig. 11 Convergence plot (mean distance between data points and centroid, against iterations) for the best four chromosomes in the segmentation of the image Colorbars.png

Table 2 Comparison of hardware specification of the proposed architecture with other architectures

	Implementation-1 [14]	Implementation-2 [14]	Implementation-3 [15]	This work
Family	XC4000 series	Virtex II	Virtex	Kintex 7
Device	XC4010D/BORG	Xc2v1000	V1000FG680	XC7K70T-3FBG676
Slice flip flops	689	712	615	1223
Total slices	403 out of 6192	310 out of 5120	813 out of 12,288	1132 of 10,250 (11%)
Total gate count Max frequency Max net delay	17,469 25.03 MHz 11.49 ns	18,732 87.187 MHz 13.312 ns	15,210 23.572 MHz 10.537 ns	23,191 14.32 MHz 69.8 ns

5 Conclusion

In this manuscript, hardware architecture is developed for clustering using genetic algorithm. The architecture is used for color image segmentation. The proposed architecture has lower computational complexity due to the reduction of hardware units for range checking and data normalization. Hardware reduction is further achieved with the use of distance metric Manhattan distance instead of the classical Euclidean distance. Test case study is conducted for testing the designed architecture for genetic clustering using the standard benchmark Portable Network Graphics (PNG) color images like Lenna, Peppers, Baboon, and Colorbars. Behavioral modeling of the evolutionary clustering using genetic algorithm has been conducted in MATLAB 2015a. Convergence is found to be satisfactory. The above architecture is modeled in hardware using Verilog HDL in Modelsim 10.4a. The simulation result is studied for further improvements. Satisfactory results are obtained after synthesis in Xilinx FPGA, Kintex7.

References

1. Jain, A.K.: Data clustering: 50 years beyond K-means. Pattern Recogn. Lett. **31**(8), 651–666 (2010)
2. Ujjwal, M., Bandyopadhyay, S.: Genetic algorithm-based clustering technique. Pattern Recogn. **33**(9), 1455–1465 (2000). ISSN 0031-3203
3. Nanda, S.J., Panda, G.: A survey on nature inspired metaheuristic algorithms for partitional clustering. Swarm Evol. Comput. **16**, 1–18 (2014)
4. Saegusa, T., Maruyama, T.: An FPGA implementation of real-time K-means clustering for color images. J. Real-Time Image Process. **2**(4), 309–318 (2007)

5. Scott, S.D., Samal, A., Seth, S.: HGA: a hardware-based genetic algorithm. In: Proceedings of the 1995 ACM Third International Symposium on Field-Programmable Gate Arrays, pp. 53–59 (1995)
6. Tommiska, M., Vuori, J.: Implementation of genetic algorithms with programmable logic devices. In: Proceedimgs of 2nd Nordic Workshop Genetic Algorithm, pp. 71–78 (1996)
7. Yoshida, N., Yasuoka, T.: Multi-gap: parallel and distributed genetic algorithms in VLSI. In: IEEE SMC'99 Conference Proceedings on Systems, Man, and Cybernetics, vol. 5, pp. 571–576 (1999)
8. Shackleford, B., et al.: A high-performance, pipelined, FPGA-based genetic algorithm machine. Genet. Algorithms Evolvable Mach. 2(1), 33–60 (2001)
9. Tang, W., Yip, L.: Hardware implementation of genetic algorithms using FPGA. In: The 2004 47th Midwest Symposium on Circuits and Systems MWSCAS'04, vol. 1, pp. I-549 (2004)
10. Xu, R., Wunsch, D.: Survey of clustering algorithms. IEEE Trans. Neural Netw. 16(3), 645–678 (2005)
11. Chen, T.W., Chien, S.Y.: Bandwidth adaptive hardware architecture of K-means clustering for video analysis. IEEE Trans. Very Large Scale Integr. VLSI Syst. 18(6), 957–966 (2010)
12. Fernando, P.R., et al.: Customizable FPGA IP Core implementation of a general-purpose genetic algorithm engine. IEEE Trans. Evol. Comput. 14(1), 133–149 (2010)
13. Ratnakumar, R, Nanda, S.J.: A FSM based approach for efficient implementation of K-means algorithm. In: 20th International Symposium on VLSI Design and Test (VDAT) 2017, Guwahati, India (2016)
14. Gallagher, J.C., Vigraham, S., Kramer, G.: A family of compact genetic algorithms for intrinsic evolvable hardware. IEEE Trans. Evol. Comput. 8(2), 111–126 (2004)
15. Aporntewan, C., Chongstitvatana, P.: A hardware implementation of the compact genetic algorithm. In: Proceedings of the 2001 Congress on Evolutionary Computation, vol. 1, pp. 624–629 (2001)
16. Kim, J.J., Chung, D.J.: Implementation of genetic algorithm based on hardware optimization. In: Proceedings of the IEEE Region 10 Conference TENCON 99, vol. 2, pp. 1490–1493 (1999)

An in-silico Approach for Enhancing the Lipid Productivity in Microalgae by Manipulating the Fatty Acid Biosynthesis

Bunushree Behera, S. Selvanayaki, R. Jayabalan and P. Balasubramanian

Abstract To fulfill the impetus of demands on alternative energy, microalgal biofuels have attracted significant attention due to the ease of cultivation, higher photosynthetic rate, as well as, the presence of significant quantity of lipids. However, from an energy perspective, the polyunsaturated fatty acids (PUFA) (substrate for transesterification to biodiesel) constitute only 10–20% of the total lipids. Approaches for increasing lipids include coercing the algal cells under nutrient depletion which also declines their growth rate. Improving the lipid accumulation without compromising growth requires strain modification via genomic or metabolic engineering which necessitates the core understanding of the critical regulators of *denovo* lipid biogenesis. Increase in activity of the enzyme acetyl-CoA carboxylase (ACCase) has been postulated to improve the lipid synthesis. Thus, the current study utilized the *Chlamydomonas reinhardtii* as the model organism for understanding the lipid metabolism. In-silico computational approach was used to design the 3D structure of ACCase, the key enzyme that catalyzes the rate-limiting step of lipid synthesis. The accuracy of the predicted structure was validated by the presence of 94% of amino acid residues in the favorable region of Ramachandran plot. The docking studies with four selected ligands (ACP, AMP, Biotin, and Glycine) showed biotin as the suitable ligand with a lowest binding affinity (−5.5 kcal/mol). The ligand–protein complex is expected to increase the enzyme activity driving lipid accumulation in vivo. Such in-silico studies are essential to design and decipher the role of different regulatory enzymes in improving the quantity and quality of microalgal biodiesel.

Keywords Microalgae · ACCase · Lipid production · Homology modeling
In-silico · Docking

B. Behera · S. Selvanayaki · P. Balasubramanian (✉)
Department of Biotechnology and Medical Engineering, National Institute of Technology
Rourkela, Rourkela, Odisha, India
e-mail: biobala@nitrkl.ac.in

S. Selvanayaki
Department of Bioinformatics, Karunya University, Coimbatore, Tamil Nadu, India

R. Jayabalan
Department of Life Science, National Institute of Technology Rourkela, Rourkela, Odisha, India

© Springer Nature Singapore Pte Ltd. 2019   877
J. C. Bansal et al. (eds.), *Soft Computing for Problem Solving*,
Advances in Intelligent Systems and Computing 816,
https://doi.org/10.1007/978-981-13-1592-3_70

1 Introduction

Depleting fossil fuel reserves, increase in crude oil prices and growing environmental concern about greenhouse gas emissions have raised the stimulus for alternative fuels [1]. Biofuels derived from plant biomass popularly termed as energy crops are increasingly gaining attention. However, the food/feed versus fuel dilemma with the first-generation biofuels and the increased processing costs of second-generation biofuels delimit their commercialization on a large scale [2]. To sort out the problems mentioned above, microalgae have recently emerged as an attractive option for replacing the conventional fossil fuels due to the higher photosynthetic rate, faster growth rate, ease of cultivation of microalgae in wastewater or marine water and their ability to sequester as well as use carbon dioxide from the atmosphere as a nutrient [1, 3]. Biodiesel is one of the popular alternative fuels obtained from the polyunsaturated fatty acids (PUFAs) or triacylglycerols (TAGs) stored in microalgal biomass [3, 4].

In current decades, there has been an intense research in the arena of microalgal lipids. The lipids in microalgae are synthesized mainly via *denovo* fatty acid biosynthesis pathway. The primary step includes carboxylation of acetyl-CoA to malonyl-CoA and leads to the formation of palmitic and stearic acids, which undergo further saturation and elongation giving rise to oleic acid. All steps of fatty acid conversion and esterification occur in both plastids and the endoplasmic reticulum [4–6]. The lipids in algae exist in the form of lipid droplets and are used to provide structural membrane support, signaling, and as an energy source. Lipids utilized for a specialized function differ structurally from each other. From an energy perspective, most of the microalgal lipids are stored at the levels of 10–20% of total lipids in the form of TAGs which are formed by esterification of omega three fatty acids or PUFA [4]. The major drawback to produce PUFA rich lipid for use as biodiesel at field scale or industrial level is the low desirable lipid content in microalgae and the low biomass productivity in a photobioreactor that increases the harvesting cost considerably [7, 8].

Strategies for enhancing the lipid productivity include subjecting the microalgae to stress under environmental or operating conditions like nitrogen and phosphate limitations. Consequently, most of these approaches are time and energy consuming. Further, increase in the lipid productivity inhibits the growth and biomass productivity [8]. Strain selection is also an essential aspect of improving the lipid productivity. There is a need to understand and analyze the metabolism of lipid biosynthetic pathway to improvise the desirable lipid productivity in selected strains. Till recently, the biochemical understanding of the lipid metabolism in plants and unicellular algae is still lacking. Recently, Bellou et al. [5] have analyzed the biochemistry underlying the lipid metabolism in microalgae. Yu et al. [4] have discussed the modifications in the rate-limiting steps of TAGs formation to increase the quantum of lipids. A comprehensive review of the various databases and bioinformatics tools along with the omics approaches to improve the algal lipid metabolism has been discussed by different researchers [6, 8–10]. The molecular dynamic simulations have been used by Kumar et al. [11] to predict the effect of nitrogen and phosphorous on the activity of acetyl-CoA carboxylase (ACCase) activity in cyanobacteria. Studies have also postulated that an increase in the cytosolic concentration of ACCase enhances the oil content [12, 13]. Nevertheless, there still lies a knowledge gap between the available omics data and the metabolic pathways underlying the lipid biosynthesis.

The current study utilizes the in-silico computational modeling approach to design a 3D model of ACCase enzyme/protein which is the key enzyme of the lipid biogenesis pathway. The ligands responsible for enhancing lipid production are then identified through docking with the protein/enzyme. Computational biology approaches like homology modeling, annotating the function of modeled protein, and docking were used to study the ligand–protein binding effects. The best and stable protein–ligand docked molecule could be taken into system biology studies for network analysis and some high-throughput experiments to realise the microalgal biofuel in the market.

2 Computational Methodology

2.1 Selection of Enzyme from the Lipid Biosynthesis Pathway

The model organism selected for study is *Chlamydomonas reinhardtii* as it is one of the algae whose genome has been completely sequenced and is the most common algae for studying the metabolic pathways [14]. The lipid synthesis pathway in microalgae (*C. reinhardtii*) consists of two essential enzymes, i.e., type II fatty acid synthase (FAS) and acetyl-CoA carboxylase (ACCase), occurring in chloroplasts [4, 6]. The enzyme ACCase determines the rate-limiting pathway for fatty acid biosynthesis, converting acetyl-CoA to malonyl-CoA which then elongate in endoplasmic reticulum giving rise to PUFAs or TAGs [6]. Since ACCase is the enzyme that determines the fatty acid pools in microalgae, it was selected for homology modeling to predict the 3D structure that is unavailable in the literature.

2.2 Design of the Three-Dimensional Structure of ACCase Protein

The target protein sequence of enzyme ACCase from *E. coli* in FASTA format was retrieved from NCBI database (http://www.ncbi.nlm.nih.gov/). The template structure of the retrieved sequence was identified using BLASTP (https://blast.ncbi.nl m.nih.gov/Blast.cgi?PAGE=Proteins). The sequence of the protein and the template structure was fed into MODELER software (http://www.salilab.org/modeller/) using python scripts for obtaining the 3D structure of the protein. Discrete Optimized Protein Structure (DOPE) score and GA341 values were considered for finding the best structure from the various modeled structures. Further structural analysis of the protein chain, folds, family, and domain was done using PHYRE software (http://www.sbg.bio.ic.ac.uk/phyre2/). Sequence-level annotations (position-specific annotations) for finding the binding sites in the protein were carried out using amiGO gene ontology tool (http://www.amigo.geneontology.org/) using the BLAST link. The UniProtKB ID was used to search the unique gene product. Structural validation of the model in PDB format was done using the RAMPAGE tool (http://mordred.bi oc.cam.ac.uk/~rapper/rampage.php) to obtain the phi–psi torsion angles which were then analyzed using the Ramachandran plot.

2.3 Docking of Ligands to the Protein and Analysis of Ligand–Protein Interaction

Four different ligands like ACP (phosphomethyl phosphonic acid adenylate ester), AMP (adenosine monophosphate), biotin and Gly (glycine) as mentioned in the literature were selected for docking with the protein modeled using homology modeling. The structure of the ligands was obtained from PubChem (https://pubchem.ncbi.nlm.nih.gov/). Docking was done using the AutoDock Vina (http://vina.scripps.edu/) after finding the docking site of protein using metaPocket (http://metapocket.eml.org/) tool. The interaction was viewed using PYMOL software (https://pymol.org/). The ligand–protein stability was studied in terms of the binding affinity of the docked structures.

3 Results and Discussion

3.1 Identification of Target Amino Acid Sequence and Template Structure

The target sequence of the ACCase enzyme (acetyl-CoA carboxylase, EC 6.4.1.2, Alpha subunit) (https://www.brenda-enzymes.org/) obtained from *Escherichia coli* strain. K-12 substrain MG1655 is given in FASTA format as in Table 1. Submitting the target sequence to BLASTP (protein–protein BLAST) resulted in a series of template structures as shown in Fig. 1. Color indications obtained were as follows: <40% black, 40–60% blue, 60–80% green, 80–200% pink, and ≥200% red. The identity should be more than 30%, and the *e*-value must be minimum for the structure to be chosen as the target structure [15]. The list of different PDB structures with their description, total scores, *e*-value, and similarity index retrieved from BLASTP is illustrated in Fig. 2. Four PDB structures were taken as a template which are having scores greater than 30%. The structures selected for modeling were 2F9Y, 2F9I, 27AS, and 2BZR.

Table 1 Sequence of amino acid for the enzyme ACCase in FASTA format

Source	FASTA sequence:
Escherichia coli str. K-12 substr. MG1655	gi|16128178|ref|NP_414727.1| acetyl-CoA carboxylase, carboxytransferase, alpha subunit [Escherichia coli str. K-12 substr. MG1655]
Accession no. NP_414727 **Sequence length** 319 aa	MSLNFLDFEQPIAELEAKIDSLTAVSRQD EKLDINIDEEVHRLREKSVELTRKIFADL GAWQIAQLARHPQRPYTLDYVRLAFD EFDELAGDRAYADDKAIVGGIARLDG RPVMIIGHQKGRETKEKIRRNFGMPAPE GYRKALRLMQMAERFKMPIITFIDTPG AYPGVGAEERGQSEAIARNLREMSRL GVPVVCTVIGEGGSGGALAIGVGDKV NMLQYSTYSVISPEGCASILWKSADKAPL AAEAMGIIAPRLKELKLIDSIIPEPLGGA HRNPEAMAASLKAQLLADLADLDVLSTED LKNRRYQRLMSYGYA

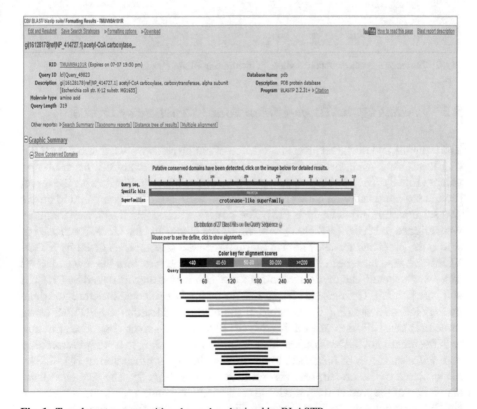

Fig. 1 Template structures with color codes obtained by BLASTP

⊖ Descriptions

Sequences producing significant alignments:

Select: All None Selected:0

‖ Alignments Download ▾ GenBank Graphics Distance tree of results Multiple alignment ⚙

Description	Max score	Total score	Query cover	E value	Ident	Accession
☐ Chain A, The Crystal Structure Of The Carboxyltransferase Subunit Of Acc From Escherichia Coli	643	643	100%	0.0	99%	2F9Y_A
☐ Chain A, Crystal Structure Of The Carboxyltransferase Subunit Of Acc From Staphylococcus Aureus	305	305	98%	2e-101	52%	2F9I_A
☐ Chain A, Crystal Structure Of Propionyl-coa Carboxylase, Beta Subunit (tm0716) From Thermotoga Maritima At 2.30 A Resolution	62.4	62.4	54%	1e-10	28%	1VRG_A
☐ Chain A, Crystal Structure Of The Acyl-Coa Carboxylase, Accd5, From Mycobacterium Tuberculosis	57.4	57.4	44%	6e-09	33%	2A7S_A
☐ Chain A, Crystal Structure Of Accd5 (Rv3280), An Acyl-Coa Carboxylase Beta-Subunit From Mycobacterium Tuberculosis	57.4	57.4	44%	7e-09	33%	2BZR_A
☐ Chain A, Propionyl-Coa Carboxylase Beta Subunit, D422a	53.9	53.9	43%	8e-08	29%	3IB6_A
☐ Chain A, Propionyl-Coa Carboxylase Beta Subunit, D422I	53.1	53.1	43%	2e-07	27%	3IB9_A
☐ Chain A, Crystal Structure Of The Carboxyl Transferase Subunit Of Putative Pcc Of Sulfolobus Tokodaii	52.4	52.4	43%	2e-07	26%	1X0U_A
☐ Chain A, Propionyl-Coa Carboxylase Beta Subunit, D422v	52.4	52.4	43%	3e-07	28%	3IAV_A
☐ Chain C, Crystal Structures And Mutational Analyses Of Acyl-Coa Carboxylase Subunit Of Streptomyces Coelicolor	52.0	52.0	43%	3e-07	28%	3MFM_C
☐ Chain A, Acyl-Coa Carboxylase Beta Subunit From S. Coelicolor (Pccb), Apo Form #2, Mutant D422i	52.0	52.0	43%	3e-07	26%	1XNW_A
☐ Chain A, Acyl-Coa Carboxylase Beta Subunit From S. Coelicolor (Pccb), Apo Form #1	51.6	51.6	43%	4e-07	28%	1XNV_A
☐ Chain B, Crystal Structure Of The Holoenzyme Of Propionyl-coa Carboxylase (pcc)	51.6	51.6	44%	5e-07	30%	3N6R_B
☐ Chain A, Transcarboxylase 12s Crystal Structure: Hexamer Assembly And Substrate Binding To A Multienzyme Core (With Methylmalonyl-Coe	48.1	48.1	54%	6e-06	26%	1ON3_A

Fig. 2　Template sequences obtained via alignment using BLASTP

3.2　Predicting the Three-Dimensional Structure of Protein

The template and the target protein sequence of amino acids were fed into the MOD-ELER software using python scripts. The templates were analyzed for resolution values (Fig. 3). The template structure 2F9I showed 52% similarity with the target protein structure and with a lower resolution value of 2.0, while compared to others was selected for further analysis [15]. The target sequence and the chosen template were aligned in MODELER along with the DOPE score, and the GA341 was used to screen the predicted model. The five predicted model structures as given by MOD-ELER software have been illustrated in Fig. 4. The structure with the lowest DOPE score was selected, and the 3D structure of the protein was then analyzed by PHYRE software [16, 17]. The analysis of the folds, families of the selected structure was done by PHYRE software (Fig. 5). The fold of the protein is identified by PHYRE using Structural Classification of Protein (SCOP) 3D-BLAST server [18]. The structure with the identity of 99% was selected with fold corresponding to c14:clpP/crotonase core with four turns of (beta beta-alpha in superhelix) as illustrated in Fig. 6. The 3D structure of the protein was viewed in RASMOL (Fig. 7). The sequence-level annotation of the predicted 3D model structure with 319 amino acids was done using amiGO gene ontology tool (as shown in Fig. 8) which was further used in docking studies [19].

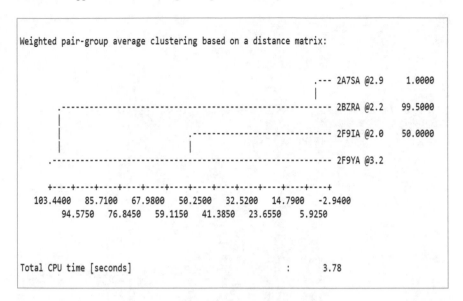

Fig. 3 Screening of templates using MODELER software and python script

```
>> Summary of successfully produced models:
Filename                        molpdf     DOPE score     GA341 score
------------------------------------------------------------------------
tar.B99990001.pdb             1763.94641   -33101.62500       1.00000
tar.B99990002.pdb             1608.43323   -33803.69531       1.00000
tar.B99990003.pdb             1690.56812   -33480.96875       1.00000
tar.B99990004.pdb             1738.59448   -33630.75000       1.00000
tar.B99990005.pdb             1828.81238   -33629.04688       1.00000

Total CPU time [seconds]                                :     440.19
```

Fig. 4 Results of DOPE score for selecting the desired templates

3.3 Validation of Protein Structure Using Ramachandran Plot

The 3D predicted protein structure from MODELER was validated using RAM-PAGE. The stereochemical properties of protein and the accuracy of the predicted model were analyzed via Ramachandran plot using RAMPAGE tool (Fig. 9). The analysis of the Ramachandran plot (Fig. 10) showed that 94% of the residues of

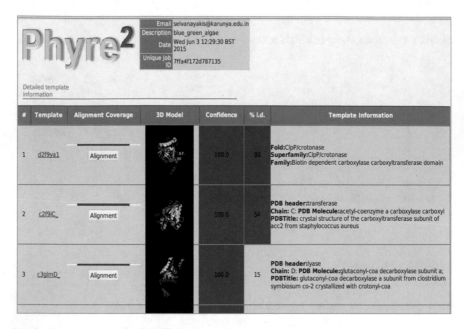

Fig. 5 Structural analysis of the 3D structure of protein in PHYRE

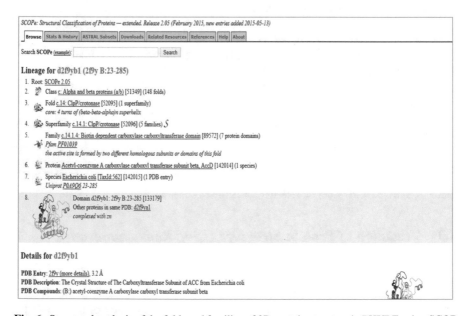

Fig. 6 Structural analysis of the folds and families of 3D protein structure in PHYRE using SCOP

Fig. 7 3D structure of the protein as viewed in RASMOL software

```
>UNIPROTKB|P0ABD5 [details] [associations]
            symbol:accA species:83333 "Escherichia coli K-12"
            [GO:0006633 "fatty acid biosynthetic process" evidence=IEA;IGI;IMP]
            [GO:0042759 "long-chain fatty acid biosynthetic process"
            evidence=NAS] [GO:0005737 "cytoplasm" evidence=IC] [GO:2001295
            "malonyl-CoA biosynthetic process" evidence=IEA] [GO:0005524 "ATP
            binding" evidence=IEA] [GO:0000166 "nucleotide binding"
            evidence=IEA] [GO:0009329 "acetate CoA-transferase complex"
            evidence=IDA] [GO:0016874 "ligase activity" evidence=IEA]
            [GO:0005515 "protein binding" evidence=IPI] [GO:0003989 "acetyl-CoA
            carboxylase activity" evidence=IEA] UniPathway:UPA00655
            HAMAP:MF_00823 InterPro:IPR001095 InterPro:IPR011763 Pfam:PF03255
            PRINTS:PR01069 PROSITE:PS50989 GO:GO:0005524 EMBL:U00096
            EMBL:AP009048 GenomeReviews:AP009048_GR GenomeReviews:U00096_GR
            GO:GO:0003989 GO:GO:2001295 GO:GO:0042759 eggNOG:COG0825
            HOGENOM:HOG000273832 KO:K01962 PANTHER:PTHR22855:SF3
            TIGRFAMs:TIGR00513 OMA:QLTKDIY ProtClustDB:PRK05724 EMBL:M96394
            EMBL:D49445 EMBL:U70214 EMBL:D87518 EMBL:M19334 PIR:A43452
            RefSeq:NP_414727.1 RefSeq:YP_488487.1 PDB:2F9Y PDBsum:2F9Y
            ProteinModelPortal:P0ABD5 SMR:P0ABD5 DIP:DIP-35897N IntAct:P0ABD5
            MINT:MINT-1228651 PaxDb:P0ABD5 PRIDE:P0ABD5
            EnsemblBacteria:EBESCT00000001533 EnsemblBacteria:EBESCT00000014303
            GeneID:12930759 GeneID:944895 KEGG:ecj:Y75_p0181 KEGG:eco:b0185
            PATRIC:32115481 EchoBASE:EB1600 EcoGene:EG11647
            BioCyc:EcoCyc:CARBOXYL-TRANSFERASE-ALPHA-MONOMER
            BioCyc:ECOL316407:JW0180-MONOMER
            BioCyc:MetaCyc:CARBOXYL-TRANSFERASE-ALPHA-MONOMER SABIO-RK:P0ABD5
            EvolutionaryTrace:P0ABD5 Genevestigator:P0ABD5 GO:GO:0009329
            Uniprot:P0ABD5
      Length = 319
```

Fig. 8 Results of sequence annotation of the 3D structure of protein using amiGO

the predicted 3D structure from MODELER is in favored regions. It signifies the accuracy of the predicted model and stereospecific stability of the 3D protein [20].

Fig. 9 Analysis of protein residues in the favored regions of Ramachandran plot

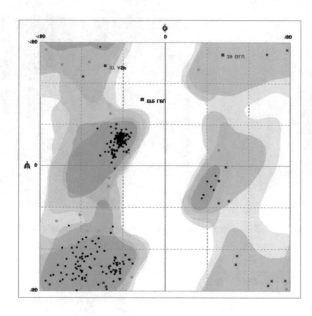

Fig. 10 Evaluation of the model accuracy based on the number of favored regions

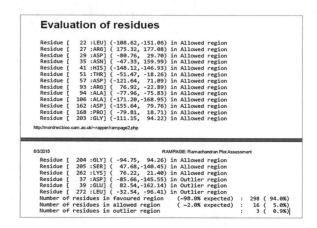

3.4 Analysis of Stability of Ligand-Protein Interaction via Docking

The selected ligands were successfully docked into 3D protein structure. Docking of the ligands with protein resulted in significant ligand–protein interaction. The recognition surface for ligand–protein interaction was studied using the PYMOL software (Fig. 11). The stability of these interactions depends on the interaction energy which is usually denoted in terms of the binding affinity (Table 2). Since the interaction of biotin had the lowest binding affinity, binding of biotin is expected to

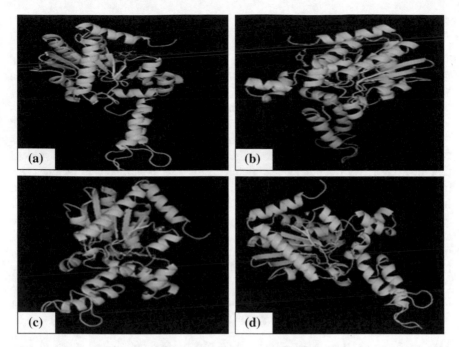

Fig. 11 Structure of the protein docked with **a** ACP; **b** BIO; **c** AMP; **d** GLY

Table 2 Binding affinities of various ligands used in docking studies

Ligand		Binding affinity (kcal mol^{-1})
Phosphomethyl phosphonic acid adenylate ester	ACP	−5.2
Biotin	BIO	−5.5
Adenosine monophosphate	AMP	−4.8
Glycine	GLY	−2.6

remodel the configuration of substrate–enzyme complex resulted in increasing the activity of ACCase. Further increase in ACCase activity is expected to increase the accumulation of PUFA/TAGs in microalgae. Interaction of biotin with ACCase could transfer the carboxyl group to acetyl-CoA to form malonyl-CoA, which further got esterified and elongated to form PUFAs and then stored as TAGs [21]. Blatti et al. [22] had also highlighted the significance of protein–protein interaction in increasing the lipid biogenesis in algae. These principles provide a fundamental understanding of the role of ACCase in algal fatty acid biosynthesis, paving the way for future metabolic engineering. Nevertheless, further system biology-based network analysis along with the experimental studies is required to validate the results mentioned above.

4 Conclusion

Though microalgae are considered to be the promising feedstocks for alternative fuels, relatively, the lower lipid accumulation necessitates the strain modification via genomics and metabolic engineering. It is essential to understand the primary biochemical pathways of their lipid accumulation in vivo in microalgae. Improvising the activity of the critical enzyme ACCase via ligand–protein interaction could enhance the accumulation of lipids. Homology modeling was used to design the 3D structure of enzyme ACCase successfully. The Ramachandran plot analysis showed that the accuracy of the predictive structure as 94% of the amino acid residues was in the favored regions. The docking studies confirmed the stability of interaction of the selected ligands (ACP, AMP, biotin, and glycine). The interaction of biotin with the enzyme having the least binding affinity (-5.5 kcal/mol) showed a stable enzyme–substrate complex formation, which could favorably reconfigure the structure, and thus enhances the enzyme activity for lipid accumulation via *denovo* lipid biosynthesis pathway. However, the experimental analysis is essential to validate the predicted results further. Such in-silico studies are essential for increasing the fundamental understanding of in vivo lipid regulation and biosynthesis before proceeding for metabolic or genetic engineering experiments at laboratory scale.

Acknowledgements The authors are thankful to the DBT sponsored Bioinformatics Infrastructure Facility (BIF) at Department of Biotechnology and Medical Engineering of NIT Rourkela for their support during the research work. The authors are grateful to Ministry of Human Resources and Development of Government of India (MHRD, GoI) for sponsoring the first author's Ph.D. program.

References

1. Rangabhashiyam, S., Behera, B., Aly, N., Balasubramanian, P.: Biodiesel from microalgae as a promising strategy for renewable bioenergy production—a review. J Environ. Biotechnol. Res. **6**(4), 260–269 (2017)
2. Maity, S.K.: Opportunities, recent trends and challenges of integrated biorefinery: Part I. Renew. Sust. Energ. Rev. **43**, 1427–1445 (2015)
3. Chisti, Y.: Biodiesel from microalgae. Biotechnol. Adv. **25**(3), 294–306 (2007)
4. Yu, W.L., Ansari, W., Schoepp, N.G., Hannon, M.J., Mayfield, S.P., Burkart, M.D.: Modifications of the metabolic pathways of lipid and triacylglycerol production in microalgae. Microb. Cell Fact. **10**(1), 1–11, 91 (2011)
5. Bellou, S., Baeshen, M.N., Elazzazy, A.M., Aggeli, D., Sayegh, F., Aggelis, G.: Microalgal lipids biochemistry and biotechnological perspectives. Biotechnol. Adv. **32**(8), 1476–1493 (2014)
6. Banerjee, C., Dubey, K.K., Shukla, P.: Metabolic engineering of microalgal based biofuel production: prospects and challenges. Front Microbiol. **7**, 1–8 (2016)
7. Tan, K.W.M., Lee, Y.K.: The dilemma for lipid productivity in green microalgae: importance of substrate provision in improving oil yield without sacrificing growth. Biotechnol. Biofuels **9**(1), 255 (2016)
8. Reijnders, M.J., van Heck, R.G., Lam, C.M., Scaife, M.A., dos Santos, V.A.M., Smith, A.G., Schaap, P.J.: Green genes: bioinformatics and systems-biology innovations drive algal biotechnology. Trends Biotechnol. **32**(12), 617–626 (2014)

9. Misra, N., Panda, P.K., Parida, B.K.: Agrigenomics for microalgal biofuel production: an overview of various bioinformatics resources and recent studies to link OMICS to bioenergy and bioeconomy. OMICS **17**(11), 537–549 (2013)

10. Ramakrishnan, G.S., Kamath, M.M., Niranjan, V.: Increasing Microbial Biofuel Production by In-silico Comparative Genomic Studies. Int. J. Biosci. Biochem. Bioinform. **4**(5), 386–390 (2014)

11. Kumar, R., Biswas, K., Singh, P.K., Singh, P.K., Elumalai, S., Shukla, P., Pabbi, S.: Lipid production and molecular dynamics simulation for regulation of acc D gene in cyanobacteria under different N and P regimes. Biotechnol. Biofuels **10**(1), 1–14, 94 (2017)

12. Bao, X., Ohlrogge, J.: Supply of fatty acid is one limiting factor in the accumulation of triacylglycerol in developing embryos. Plant Physiol. **120**(4), 1057–1062 (1999)

13. Roesler, K., Shintani, D., Savage, L., Boddupalli, S., Ohlrogge, J.: Targeting of the Arabidopsis homomeric acetyl-coenzyme a carboxylase to plastids of rapeseeds. Plant Physiol. **113**(1), 75–81 (1997)

14. Merchant, S.S., Prochnik, S.E., Vallon, O., Harris, E.H., Karpowicz, S.J., Witman, G.B., Terry, A., Salamov, A., Fritz-Laylin, L.K., Maréchal-Drouard, L., Marshall, W.F.: The Chlamydomonas genome reveals the evolution of key animal and plant functions. Science **318**(5848), 245–250 (2007)

15. Altschul, S.F., Gish, W., Miller, W., Myers, E.W., Lipman, D.J.: Basic local alignment search tool. J. Mol. Biol. **215**(3), 403–410 (1990)

16. Shi, J., Blundell, T.L., Mizuguchi, K.: FUGUE: sequence-structure homology recognition using environment-specific substitution tables and structure-dependent gap penalties. J. Mol. Biol. **310**(1), 243–257 (2001)

17. Holm, L., Rosenstrom, P.: Dali server: conservation mapping in 3D. Nucleic Acids Res. **38**, W545–W549 (2010)

18. Andreeva, A., Howorth, D., Chothia, C., Kulesha, E., Murzin, A.G.: Investigating protein structure and evolution with SCOP2. Curr. Protoc. Bioinform. **49**, 1–21 (2015)

19. Carbon, S., Ireland, A., Mungall, C.J., Shu, S., Marshall, B., Lewis, S.: AmiGO hub & web presence working group. AmiGO: online access to ontology and annotation data. Bioinformatics **25**(2), 288–289 (2008)

20. Hollingsworth, S.A., Karplus, P.A.: A fresh look at the Ramachandran plot and the occurrence of standard structures in proteins. Biomol. Concepts. **1**(3–4), 271–283 (2010)

21. Beld, J., Lee, D.J., Burkart, M.D.: Fatty acid biosynthesis revisited: structure elucidation and metabolic engineering. Mol. BioSyst. **11**(1), 38–59 (2015)

22. Blatti, J.L., Beld, J., Behnke, C.A., Mendez, M., Mayfield, S.P., Burkart, M.D.: Manipulating fatty acid biosynthesis in microalgae for biofuel through protein-protein interactions. PLoS ONE **7**(9), 1–12 (2012)

Application of the Relevance Vector Machine to Drought Monitoring

Alok Kumar Samantaray, Gurjeet Singh and Meenu Ramadas

Abstract The study demonstrates the application of relevance vector machines (RVMs) to drought monitoring, specifically, agricultural drought classification. The model is based on a crop water stress function that serves as an indicator of agricultural drought in the study area. The RVM framework performs a multi-class classification on the crop stress feature vector and yields probabilistic classification of drought classes. The results indicate that the uncertainty involved in classification is known with the help of the RVM-based classification model.

Keywords Relevance vector machines · Drought monitoring · Agricultural droughts · Crop water stress

1 Introduction

Statistical learning algorithms and soft computing techniques are extensively used in solving regression and classification problems in different fields of engineering [3, 4, 6, 13, 15, 20, 25] and in health sector [10, 27, 28] nowadays. Recent advances in soft computing techniques such as support vector machines (SVMs), relevance vector machines (RVMs), artificial neural networks (ANNs), and Bayesian approaches have made it possible to work with complex time series and spatial information datasets. In the field of hydrometeorology and agricultural water management, the data-driven models are gaining popularity and are able to cater to applications such as hydrologic prediction, extreme event modeling, flood and drought studies, climate change and water quality studies [1, 7, 11, 23].

In this study, we present the application of relevance vector machine (RVM) to drought monitoring in the Mid-Mahanadi River Basin, Odisha, India, specifically, in the context of agricultural drought classification. The functional form of RVM developed by Tipping [22] is identical to that of the support vector machine (SVM)

A. K. Samantaray · G. Singh · M. Ramadas (✉)
School of Infrastructure, IIT Bhubaneswar, Bhubaneswar, Odisha, India
e-mail: meenu@iitbbs.ac.in

© Springer Nature Singapore Pte Ltd. 2019
J. C. Bansal et al. (eds.), *Soft Computing for Problem Solving*,
Advances in Intelligent Systems and Computing 816,
https://doi.org/10.1007/978-981-13-1592-3_71

approach of Vapnik [24]. In the field of hydrologic studies, the applications of RVMs are manifold. It has been widely used in climate change impact assessment studies for the purpose of downscaling hydrologic variables to basin scale [5, 12]. Downscaling is the process by which coarse-resolution data (for instance, 25 km spatial resolution variables) are scaled down to finer resolution data (for instance, 1 km spatial resolution data or station measurements). Ghosh and Mujumdar [5] had utilized both SVM and RVM to downscale streamflow from large-scale climatic information obtained from the general circulation models (GCMs). The authors pointed out the demerits of SVMs: computational complexity due to large number of support vectors and that only point estimates are available as outputs from the model. Compared to SVM, the Bayesian formulation of RVM yielded sparse regression model and the model uncertainty (or posterior probability) is known. Regression models based on RVM, SVM, and ANN were compared by Samui and Dixon [18] for predicting evaporation losses from reservoirs. The ANN model has capabilities to explore the relationships and dependencies between hydrological and climatic variables and has served as a useful tool for several hydrologic applications [6]. The authors found RVM and SVM models to be better performing and sparse. Chen et al. [2] used RVMs to predict future runoff of Danjiang Kou reservoir in China. Surface soil moisture estimation using RVM and SVM models with satellite data and meteorological variables as inputs was successfully conducted by Zaman et al. [26]. Another study by Srivastava et al. [19] investigated the downscaling capabilities of RVM for developing high spatial resolution soil moisture datasets from satellite observations. The model downscaled ~40 km resolution soil moisture and ocean salinity (SMOS) satellite data to local scale soil moisture deficit values.

Previous hydrologic studies have formulated various indices to represent hydrological, meteorological, and agricultural droughts. Few examples of the existing hydrometeorological drought indices are the standardized precipitation index (SPI), Palmer drought severity index (PDSI), vegetation condition index (VCI), soil moisture deficit index (SMDI), and the standardized precipitation evapotranspiration index (SPEI). However, these indices yield point estimates of drought condition and do not reflect the uncertainty involved in classification. The RVM is therefore incorporated in the drought monitoring framework to aid in probabilistic drought classification. The drought status is indicated by using crop water stress function calculated using both soil and crop data of the location and is more suitable for agricultural droughts in comparison with SPI, PDSI, and SPEI. The drought class definitions for the proposed index would be set by the RVM model and varies from location to location. Such a non-unique classification scheme across space was previously adopted by Mallya et al. [8, 9] and Ramadas and Govindaraju [14] utilizing mixture models and graphical models. For instance, from the drought indicator data, the model essentially learns what is a severe drought class for the particular location, and what is a near normal drought class. The present study highlights the use of RVM in agricultural drought classification. The study is built on the probabilistic drought classification model developed by Ramadas and Govindaraju [14] using hidden Markov models (HMM). The HMM-based model yields probability of a particular drought belonging to different drought categories but required more computational time and effort, estimation of parameters and also assumes Markovian dependence in time. The proposed approach could save computational time and costs. The methodology and results of the drought monitoring study over Mid-Mahanadi River Basin using RVM are presented in subsequent sections of the paper.

2 Methodology

2.1 Relevance Vector Machines (RVM)

Relevance vector machines (RVMs) are a powerful technique in machine learning that is often used in classification and regression problems. Unlike the previously developed support vector machines (SVMs) models that yield point estimates, RVM offers predictive distributions of classes. Supervised classification using RVM is founded on the concept of sparse Bayesian learning.

In the present context of drought classification, the input vector (crop water stress) and the corresponding target (drought category) are defined as x_n and t_n, respectively. Therefore, the data set for this model is $D = [X_{N \times 1}, y_{N \times 1}]$, where $X_{N \times 1}$ represents N input samples with one feature, i.e., the crop water stress, and the associated drought category is in $y_{N \times 1}$. Using RVM, we seek posterior probabilities of class membership. Suppose the dataset of input-target pairs have a length N, and their model incorporates noise, as shown in Eq. (1):

$$t_n = y(x_n, w) + \varepsilon_n \tag{1}$$

where the modeled error term ε_n is Gaussian with mean zero and variance. Here, $y(x_n, w)$ is a function defined over the input space specific to the problem, and parameters of this function are popularly known as weights w, as shown in Eq. (2):

$$y(x_n, w) = \sum_{i=1}^{N} w_i \phi_i(x) \tag{2}$$

Through the learning process, the best estimates of these weights are determined. The basis functions $\phi_i(x)$ are kernel functions. Kernels are mathematical functions that take input and transform it into a required form. Few examples of kernels used in machine learning models are linear, nonlinear, polynomial, Gaussian, radial basis function (RBF), and sigmoid. Sparse RVM framework utilizes fewer kernel functions, and hence, fewer nonzero parameters that are relevant for the model. The details of RVM are available in Tipping [22]. It was initially developed as a binary classification tool.

The binary RVM classification scheme [22] was modified into a multi-class relevance vector machine classification algorithm by Thayananthan [21]. The code for multi-class RVM classification written by Thayananthan [21] in MATLAB software is available free of cost from http://mi.eng.cam.ac.uk/~at315/MVRVM.

2.2 Agricultural Drought Monitoring Framework

The input to the RVM model, the crop water stress function is estimated from variables such as surface soil moisture, and parameters such as soil moisture values at wilting point (minimum value of soil moisture below which the plant wilts and fails to recover upon rewetting of soil) and at incipient stomatal closure (the soil moisture value at which the plant leaf openings begin to restrict transpiration). The crop stress function is based on the formulation of static water stress by Rodriguez-Iturbe et al. [16, 17]. The authors define the stress function Z as follows: It is zero for values of soil moisture beyond level of incipient stomatal closure (S^*, lower limit), it is at its maximum value (1) when soil moisture falls below the wilting point (S_w, upper limit), and it varies nonlinearly between 0 and 1 for values between S^* and S_w. This formulation utilized by Ramadas and Govindaraju [14] for the crop water stress-based drought index in HMM framework was adopted in this study for agricultural drought. This is given by Eq. (3).

$$Z(t) = \left[\begin{array}{ll} 0; & S(t) \geq S^* \\ 1; & S(t) \leq S_w \\ \left(\frac{S^* - S(t)}{S^* - S_w} \right)^m; & S_w < S(t) < S^* \end{array} \right] \tag{3}$$

where $S(t)$ is the surface soil moisture data that is obtained from the essential climate variable-soil moisture (ECV-SM) data under the European Space Agency-Climate Change Initiative (ESA-CCI).

The ECV-SM global soil moisture dataset provides volumetric soil moisture from 1979 to 2014, at daily time step, at 0.25° spatial resolution. This is a merged product of different sensor-specific data sets at the level of the retrieved surface soil moisture data (Level 2) and utilizes the different satellites and sensors of the ESA, EUMETSAT, NASA, JAXA, to name a few. The ECV-SM on an average represents soil moisture conditions over a few centimeters, i.e., the surface soil moisture.

The crop water stress function incorporates crop response to deficit in soil moisture content. In this study, the crop distribution pattern and soil types over the Mid-Mahanadi River Basin region were used to compute the lower and upper limits of soil moisture content in order to evaluate the crop stress at each grid location. The spatial variability in crop water stress values is a major indicator of agricultural drought susceptibility of the region especially under periods of no precipitation input and excessive cultivation. The spatial variability of crop water stress and hence of the monitored agricultural drought is well preserved by this model formulation.

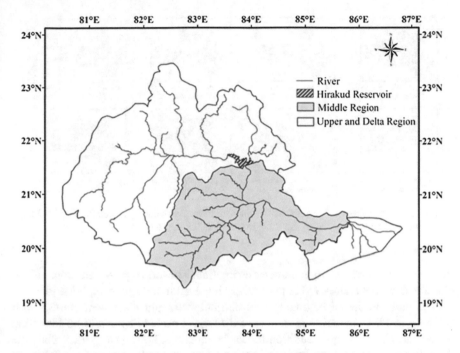

Fig. 1 Map of the Mahanadi River Basin showing the mid-Mahanadi region, the study area

3 Results of Drought Classification Using RVMs

3.1 Drought Outlook for Mid-Mahanadi River Basin

The crop and soil type data available for the region were utilized for evaluating the crop water stress. The crop water stress was calculated for all the grid locations in the study region (Middle Region shown in Fig. 1). As mentioned previously, a lower value corresponds to less severe condition and value close to 1 indicates extreme agricultural drought. Figure 2 shows the crop water stress maps for the study region during the period December 2000 to March 2001 in the first row, and from December 2004 to March 2005, in the second row. These represent drought and non-drought years for the region. According to the statistics of Odisha Agriculture Department, 2000–2001 was a drought year for the state of Odisha that resulted in crop losses and reduced crop yields.

The RVM model-based classification was then run using the multi-class RVM code in MATLAB software. A four-class model was adopted, such that the categories are near normal, moderate drought, severe drought, and extreme drought. Among the various RVM kernels available, the Gaussian radial basis kernel was used. The width of the kernel was kept as 0.5. The code was run for all the grid locations at a lesser computational cost compared to the HMM-based classification model, as the classi-

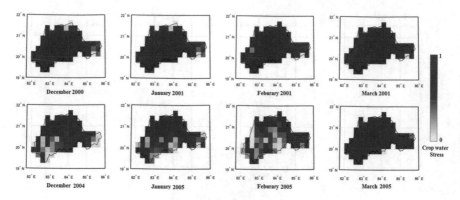

Fig. 2 Crop water stress status in the Mid-Mahanadi River Basin during December 2000 to March 2001 (first row) and during December 2004 to March 2005 (second row)

fication did not involve any parameter estimation. The model provided probabilistic classification that addresses the uncertainty involved in drought categorization.

The probabilities of belonging to a particular drought class were known with the help of RVM model, and the one with largest probability could be identified as the prevailing drought condition for further analysis. For instance, in a moderate drought month, the results indicated that the crop water stress value was 0.24 and the probabilities of the drought states were 77% probability of being in moderate drought category and 18, 2, and 3% probabilities of being in near normal, severe and extreme categories, respectively.

A consolidated analysis of results from all grid locations in the region was used for identifying drought-prone areas. The results of RVM-based drought monitoring have been used to determine the prevailing drought class. The drought events in region during the period December 2000 to March 2001, and from December 2004 to March 2005 are shown in Fig. 3 in first and second rows, respectively. The comparison of spatial extent and severity of droughts during the period December 2000–March 2001 (drought year) with that of the non-drought year agree with the drought statistics for the Odisha State. This shows that the proposed index and classification model are helpful tools for drought monitoring in the region. For policymakers, the knowledge of probabilities associated with different drought classes could be of help in deciding allocation of resources to assist with mitigation and in overall drought management.

4   Conclusions

The study demonstrates that data-driven techniques can be useful for learning patterns in drought occurrences that aid in agricultural drought monitoring. The proposed crop stress function is a better indicator of agricultural drought as it accounts for crop water stress. The main highlight of the work is that probabilistic classification of droughts

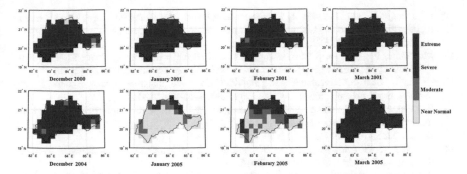

Fig. 3 Agricultural drought outlook for the Mid-Mahanadi River Basin during December 2000 to March 2001 (first row) and during December 2004 to March 2005 (second row)

was facilitated by RVMs thereby saving time and computational effort. From the results, it is evident that the drought-affected periods are correctly identified by the formulated model.

It is observed that the spatial resolution of soil moisture data to be utilized for drought studies need improvement. The soil moisture dataset derived from remote sensing satellites could be developed into a reliable proxy for field data with proper ground validation. In the proposed framework, they can be utilized to perform drought monitoring with uncertainty assessment.

References

1. Bai, Y., Wang, P., Li, C., Xie, J., Wang, Y.: A multi-scale relevance vector regression approach for daily urban water demand forecasting. J. Hydrol. **517**, 236–245 (2014)
2. Chen, H., Xiong, W., Guo, J.: Application of relevance vector machine to downscale GCMs to runoff in hydrology. In: Fifth International Conference on Fuzzy Systems and Knowledge Discovery, 2008, FSKD'08, vol. 5, pp. 598–601. IEEE, Piscataway (2008)
3. Deka, P.C.: Support vector machine applications in the field of hydrology: a review. Appl. Soft Comput. **19**, 372–386 (2014)
4. Fei, S.W., He, Y.: Wind speed prediction using the hybrid model of wavelet decomposition and artificial bee colony algorithm-based relevance vector machine. Int. J. Electr. Power Energy Syst. **73**, 625–631 (2015)
5. Ghosh, S., Mujumdar, P.P.: Statistical downscaling of GCM simulations to streamflow using relevance vector machine. Adv. Water Resour. **31**(1), 132–146 (2008). https://doi.org/10.101 6/j.advwatres.2007.07.005
6. Govindaraju, R.S., Rao, A.R. (eds.): Artificial Neural Networks in Hydrology, vol. 36. Springer Science & Business Media (2013)
7. Khalil, A., McKee, M., Kemblowski, M., Asefa, T.: Sparse Bayesian learning machine for real-time management of reservoir releases. Water Resour. Res. **41**(11) (2005)
8. Mallya, G., Tripathi, S., Govindaraju, R.S.: Probabilistic drought classification using gamma mixture models. J. Hydrol. **526**, 116–126 (2015)
9. Mallya, G., Tripathi, S., Kirshner, S., Govindaraju, R.S.: Probabilistic assessment of drought characteristics using hidden Markov model. J. Hydrol. Eng. **18**(7), 834–845 (2012)

10. Månsson, K.N., et al.: Predicting long-term outcome of Internet-delivered cognitive behavior therapy for social anxiety disorder using fMRI and support vector machine learning. Transl. Psychiatry **5**(3), e530 (2015)
11. Mujumdar, P.P., Ghosh, S.: Modeling GCM and scenario uncertainty using a possibilistic approach: application to the Mahanadi River, India. Water Resour. Res. **44**(6) (2008)
12. Okkan, U., Inan, G.: Bayesian learning and relevance vector machines approach for downscaling of monthly precipitation. J. Hydrol. Eng. **20**(4), 04014051 (2014)
13. Pradhan, B.: A comparative study on the predictive ability of the decision tree, support vector machine and neuro-fuzzy models in landslide susceptibility mapping using GIS. Comput. Geosci. **51**, 350–365 (2013)
14. Ramadas, M., Govindaraju, R.S.: Probabilistic assessment of agricultural droughts using graphical models. J. Hydrol. **526**, 151–163 (2015)
15. Rebentrost, P., Mohseni, M., Lloyd, S.: Quantum support vector machine for big data classification. Phys. Rev. Lett. **113**(13), 130503 (2014)
16. Rodriguez-Iturbe, I., D'Odorico, P., Porporato, A., Ridolfi, L.: On the spatial and temporal links between vegetation, climate, and soil moisture. Water Resour. Res. **35**, 3709–3722 (1999)
17. Rodríguez-Iturbe, I., D'Odorico, P., Porporato, A., Ridolfi, L.: Tree-grass coexistence in Savannas: the role of spatial dynamics and climate fluctuations. Geophys. Res. Lett. **26**, 247–250 (1999)
18. Samui, P., Dixon, B.: Application of support vector machine and relevance vector machine to determine evaporative losses in reservoirs. Hydrol. Process. **26**(9), 1361–1369 (2012)
19. Srivastava, P.K., Han, D., Ramirez, M.R., Islam, T.: Machine learning techniques for downscaling SMOS satellite soil moisture using MODIS land surface temperature for hydrological application. Water Resour. Manage. **27**(8), 3127–3144 (2013)
20. Tehrany, M.S., Pradhan, B., Mansor, S., Ahmad, N.: Flood susceptibility assessment using GIS-based support vector machine model with different kernel types. CATENA **125**, 91–101 (2015)
21. Thayananthan, A.: Relevance Vector Machine based Mixture of Experts. Technical Report, Department of Engineering, University of Cambridge, England (2005)
22. Tipping, M.E.: Sparse Bayesian learning and the relevance vector machine. J. Mach. Learn. Res. **1**, 211–244 (2001)
23. Tripathi, S., Govindaraju, R.S.: On selection of kernel parameters in relevance vector machines for hydrologic applications. Stoch. Env. Res. Risk Assess. **21**(6), 747–764 (2007)
24. Vapnik, V.: The Nature of Statistical Learning Theory. Springer, New York (1995)
25. Yan, J., Liu, Y., Han, S., Qiu, M.: Wind power grouping forecasts and its uncertainty analysis using optimized relevance vector machine. Renew. Sustain. Energy Rev. **27**, 613–621 (2013)
26. Zaman, B., McKee, M., Neale, C.M.: Fusion of remotely sensed data for soil moisture estimation using relevance vector and support vector machines. Int. J. Remote Sens. **33**(20), 6516–6552 (2012)
27. Zhang, Y., Wang, S., Dong, Z.: Classification of Alzheimer disease based on structural magnetic resonance imaging by kernel support vector machine decision tree. Prog. Electromagnet. Res. **144**, 171–184 (2014)
28. Zheng, B., Yoon, S.W., Lam, S.S.: Breast cancer diagnosis based on feature extraction using a hybrid of K-means and support vector machine algorithms. Expert Syst. Appl. **41**(4), 1476–1482 (2014)

Fuzzy-Based Integration of Security and Trust in Distributed Computing

P. Suresh Kumar and S. Ramachandram

Abstract To protect the resources from various vulnerability factors, resources should have various routine security mechanisms such as antivirus capability, firewall capability, usage of secure network connections, provision of execution sandbox, invoking dynamic checkpointing, and intrusion detection system-related capabilities. Security is one of the key issues in distributed computing systems like grid and cloud. Whole system is secured when resources have self-defense capability. Adapting security measures in grid environment is an expensive mechanism and leads to delays in service provisioning whereas trust can be, relatively, a simple and fast solution. In view of the interest of the users and quick delivery of the services by the provider, integration of different combinations of trust levels and security mechanisms can reduce the costs and delays involved in adapting security measures. This paper proposes a new approach for integrating security levels along with trust in general for distributed computing systems and in particular for grid computing systems. Our previous work proposed a T-grid computational model suitable for grid computing systems, which will be used for experimenting and testing the proposed idea of the integration. Results of the studies are produced.

Keywords Distributed computing · Cloud computing · Grid computing · Trust
Indirect trust · Fuzzy · Recommendations · Security · Integration · T-grid

P. Suresh Kumar (✉)
Department of CSE, KITS, Warangal, Telangana, India
e-mail: peddojusuresh@gmail.com

S. Ramachandram
Department of CSE, University College of Engineering, Osmania University,
Hyderabad, Telanagana, India

© Springer Nature Singapore Pte Ltd. 2019   899
J. C. Bansal et al. (eds.), *Soft Computing for Problem Solving*,
Advances in Intelligent Systems and Computing 816,
https://doi.org/10.1007/978-981-13-1592-3_72

1 Introduction

Distributed computing systems like cloud and grid computing [1] are such an environment with ability to share the resources dynamically, flexibly, and fast. Moreover, it must have the ability to scalable. What it lacks is the level of assurance or confidence given to the potential end-users due to weaknesses in the traditional methods [2] adapted in security mechanisms. Security and trust [3] are so closely related that majority of the times, people get confused in considering each other. Some researchers feel that without trust, security does not exist whereas others feel that security builds trust. Few researchers feel that there is only security in two different flavors, hard security, and soft security. Implementing various strict security mechanisms like authentication, access controls, and other cryptographic techniques considered as hard security mechanisms, whereas trust is considered as soft security mechanism. Trust is implied inherently in hard security measures [4].

New approaches are definitely needed to improve the quality of assurance given to the end-users. Particularly, the existing systems [5] lack in providing the security requirements dynamically, flexibly, and fast. The major drawback is in providing the scalability. Hence, there is a need for the system that is highly scalable, dynamic, and flexible. The motivation for this paper relies on this fact that both trust and security must be flexibly and dynamically used in such a way that the entire system is not expensive and further scalable. The additional advantage is in reducing the delays and computational complexities involved in implementing hard security mechanisms. The focus of this paper is to integrate security and trust in an efficient manner as per the requirements of the end-user to provide faster and scalable services.

Organization of the paper is as follows: In Sect. 2, related work is discussed. The proposed enhanced trust evaluation approach for T-Grid computational model is presented in Sect. 3. Experimental setup for simulating the model is given in Sect. 4. In Sect. 5, obtained simulation results were analyzed and evaluated the performance of proposed approach with different parameters. Finally, Sect. 6 concludes the paper.

2 Related Work

Very less work was carried out in integration of security and trust area. Viduto et al. [6] addressed trust and reputation in ubiquitous networks based on graph theory. Authors in this paper defined trust as a function of quality of a task and proposed a new authentication scheme zero-common knowledge (ZCK) through which one agent can authenticate another. Initially, a graph theory concept is applied to formulate a trust metric. Das et al. [7] proposed a comprehensive quantitative model called SecuredTrust, which addresses varying behavior of fraudulent agents and balance the workload among providers. Vijay Kumar et al. [8] proposed a model wherein resource selection can be done with a trust and reputation based on calculated trust factor (TF). This model considered the security features in evaluation of trust factor of entities allocated in grid environment.

Song et al. [9] proposed a novel scheme called security-binding scheme in which reputation of grid sites is assessed by considering trust value of site by applying fuzzy methods. Patni et al. [10] proposed some major methods and mechanisms of security in different levels and presented related issues. Authors in this paper discuss load factors affected with different security mechanisms and corresponding solutions were also presented. The authors in paper [11] demonstrated a randomized algorithm in evaluating trust value between service provider and user with consisting rating among them. The domains participating in grid will communicate each other with quantified trust value with rating. Ennahbaoui et al. [12] proposed an intrusion detection system based on recording behavior and actions of nodes in grid with the help of mobile agents to detect malicious intruders. Another trust model [13] based on evaluation of trust for data-intensive applications with proposed task scheduling algorithm.

Authors in paper [14] were proposed a model for trust management in cloud computing environment. In this paper, authors stated that privacy and security are important aspects as well as trust but it only considers feedback of user. In another paper [15], authors presented vast survey on trust and security issues in cloud computing environment. They have mentioned major factors that are affecting trust on cloud and security of data. Although in paper [16] authors have examined different risk factors and trends in adopting trust and security in distributed environment like cloud in mathematical equations. In paper [17], authors identified different trust management techniques to increase the trustworthiness of results in cloud environment and they want to improve performance of trusted service with optimization techniques.

3 Proposed Model

The trust evaluation discussed in paper [18] by the same authors is very specific to direct trust in its implementation. One possible improvement identified can be the way of eliminating unreliable transactions from the system, and other improvement can be to eliminate the inactive transactions over a period of time which is presented in elimination trust model in [19]. In view of the importance of security, it also plays a major role in shielding distributed systems like cloud and grid from different attacks from users. Various defense mechanisms such as antivirus capability, firewall capability, usage of secure network connections, provision of execution sandbox, invoking dynamic checkpointing, and IDS-related capabilities can be adapted to make the entire system more robust and trusted.

In the proposed framework, the security levels offered by individual resources are expected to be available in the trust database and client along with various other parameters provides required security levels. According to this model, trust evaluation module sits inside the resource broker module and is responsible for evaluation

Table 1 Proposed security levels with their typical security attributes

Security level	Security attributes
LOW	Antivirus capability, firewall capability
MEDIUM	Antivirus capability, firewall capability, usage of secure network connections, provision of execution sandbox
HIGH	Antivirus capability, firewall capability, usage of secure network connections, provision of execution sandbox, Invoking dynamic checkpointing, IDS-related capability

of final trust of the resource providers based on trust and security values due to fuzzification process. This final trust is, yet, updated in trust database for future use.

3.1 Security Levels

This paper proposes a new approach, further extending the technique proposed in paper [19], by integrating additional security features along with the trust to make the entire model highly robust. In this proposed, new approach, various security levels are defined for tuning according to the needs of the user such as LOW, MEDIUM, and HIGH. Number of security levels can be increased or decreased based on the design of the model and its corresponding attributes. The mapping depends on the availability of these features (or attributes) with the corresponding provider. We defined each security level according to security attribute's priority. For example, if user requirements prefer a security level of LOW, may be in order to reduce the execution cost, the broker will look for a resource which is having security level LOW (e.g., it may be capable of providing only antivirus capability, firewall capability or both). Table 1 defines the proposed security levels with their typical security attributes considered in this model.

3.2 Framework

In order to integrate the security levels in trust evaluation module, the general framework discussed in paper [19] is revised and is shown in Fig. 1.

3.3 Methodology

Authors of this paper proposed the T-Grid model, discussed in [18, 19], is enhanced by integrating the security features. The idea behind the new approach proposed in this

Fig. 1 T-Grid system framework with security levels inclusive

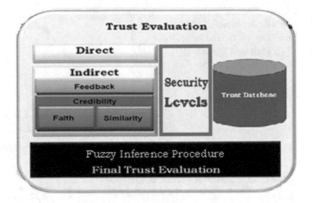

paper is to enhance the robustness and give the flexibility to the client in choosing the security levels depending on the requirement of the stringent security of the information stored. Further, this approach will help the end customer to concentrate on costs and delays involved in taking the security impositions. Relying just alone on trust on one hand helps the customer in reducing the costs and delays involved in adapting the services; on the other hand, it helps in flexibly and dynamically choosing the resources.

Hence, the proposed trust model evaluates the final trust (FT) in extension to the previous model [19] by considering security features of the resources apart from elimination of unreliable transactions from the grid environment which can be applicable to cloud or any distributed computing environment. In this proposed approach, security features are also fuzzified and defuzzified [20] along with the FT to make the system more robust.

3.4 Trust with Security

Similar to elimination approach discussed in [19], membership functions are defined and used during the course of fuzzification and defuzzification process of the security features along with FT. However, there is a substantial change in choosing the fuzzy rules.

3.5 Fuzzy Inference

Figure 2 shows membership functions used to represent both final trust and security level during fuzzification process. As it is the integration approach unlike in elimination approach, there is a substantial change in choosing the operator for fuzzy

Fig. 2 Example
membership functions for
parameters like final trust
(FT), security levels (SLs),
and secured final trust (ST)

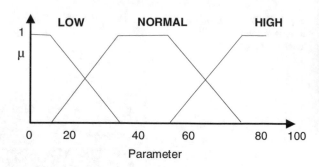

Table 2 Fuzzy rules for secured final trust

Rule #	Final trust (FT)	Security level (SL)	Secured final trust (ST) (St = Ft + Sl)
1	LOW	LOW	LOW
2	LOW	NORMAL	NORMAL
3	LOW	HIGH	HIGH
4	NORMAL	LOW	NORMAL
5	NORMAL	NORMAL	NORMAL
6	NORMAL	HIGH	HIGH
7	HIGH	LOW	HIGH
8	HIGH	NORMAL	HIGH
9	HIGH	HIGH	HIGH

rule. In view of the integration of both the trust and security levels, OR (+) operator is used in place of AND operator. Table 2 lists fuzzy rules adapted for integrating security features.

3.6   *Final Trust*

Final trust can be, at the end, defuzzified using center of gravity to arrive at crisp value in order to store in trust database (TD). In fact, this secured final trust value becomes the reputation value for future use. The TD stores a mapping of this reputation value with the user who has given the feedback for corresponding resource provider.

Table 3 User job submissions when applied security approach

User	PEs	Memory	Cost	Required trust	Required security level
U1	309	75,044	1191.13	0.68	0.30
U2	303	75,021	1123.06	0.76	0.65
U3	304	75,021	1528.81	0.76	0.75
U4	302	75,008	978.49	0.04	0.85
U5	305	75,010	1319.99	0.87	0.8

Table 4 Resource availability when applied security approach

Resource	PEs	Memory	Cost	Security level
R1	308	75,033	1051.66	0.27
R2	302	75,035	1983.84	0.74
R3	301	75,036	932.86	0.92
R4	305	75,044	257.83	0.62
R5	309	75,021	1364.96	0.53

4 Experimental Evaluation

The experimental setup used for evaluation of T-Grid model with security levels considered remains same as discussed in [19].

4.1 Sample Data

The sample data used for experiments are shown in Tables 3 and 4. The tables are shown with the sample data having five users (U1, U2, U3, U4 and U5) and five resources (R1, R2, R3, R4, and R5).

4.2 Simulation Parameters

The simulation parameters used in experimental setup carried out for the approach used in this paper, i.e., integration of trust as well as security level in T-Grid model are listed in Table 5.

Table 5 Simulation parameters used in experiments when applied security approach

Security level	Security attributes
Number of users (or Nodes or Recommenders)	5–50
Number of jobs submitted	5–00
Number of resources	5–0
Processing elements (PEs)/Machine	5–1000
Cost ($/PE)	0–2000 Units
Memory	500–100,000 MB
Trust levels	0.0–1.0
Security levels	0.0–1.0
Feedback levels	0.0–1.0
Membership functions	Low (0.0–0.4) Normal (0.3–0.7) High (0.6–1.0)

4.3 Evaluation Metrics

The evaluation on the proposed security approach is done using different metrics like effect of security levels, apart from the average trustworthiness and utilization of the resources.

5 Results and Analysis

The results obtained due to experiments conducted during the simulation of T-Grid model by integrating security features apart from eliminating unreliable transactions are discussed in this section. The evaluation metrics are considered for analysis purpose.

5.1 Trustworthiness of the Resources

Figure 3 shows trust value of five resources, i.e., {R1, R2, …, R5} averaged over a period of one year. The results indicate that there is a considerable increase in the average trust value of each resource based on their security levels when compared with approach discussed in [19]. In this approach, R3 is evolved as the highest trusted and secured resource compared to other resources because its security value is 0.92. Now with this method, average trust values of each resource are revised based on their security levels. The average trust values of R2, R4, R5, and R1 are increased due to security levels they have. Figure 4 shows the comparative graph of average trust value

Fig. 3 Average trust value of the resources over a period of one year when applied security approach

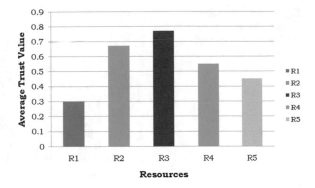

Fig. 4 Average trust value of the resources over a period of one year when applied security approach comparative graph of average trust value of the resources before elimination, after elimination, and by integrating security levels over a period of one year

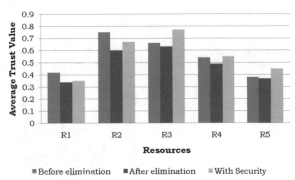

of the resources before elimination, after elimination, and by integrating security levels over a period of one year. Compared to other approaches, this approach is giving more efficient results means more trusted and secured resources participating and hence giving more better average trust value.

5.2 Resource Utilization

To estimate the utilization of resources for this approach, experiment was conducted considering the same five resources {R1, R2, R3, R4, R5} but by integrating security levels, corresponding security levels of individual resources including {0.27, 0.74, 0.92, 0.62, 0.53}, respectively, with a total of 100 job requests. Results achieved in case of resource utilization are shown in Fig. 5. Experiment was carried out by considered 100 job requests submitted by a user and estimated utilization percentage of each resource. The results show that 26% of the submitted job requests were served by resource R3 due to its trustworthiness and security levels. Because the order in which resources are used with trust levels is same as the order in which resources are used with security levels. Next in the sequence, the resources R2, R4, R5, and R1 are

Fig. 5 Resource utilization
of each resource among 100
jobs requests when applied
security approach

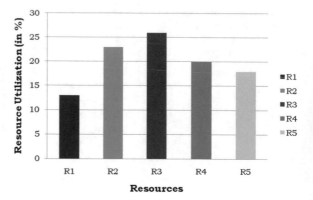

Fig. 6 Comparative graph
of resource utilization before
elimination, after
elimination, and by applying
security levels among 100
job requests

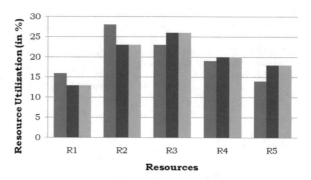

highly trustable and secured, respectively. According to the results shown in Fig. 5, the resources are chosen based on their trustworthiness as well as security levels.

Figure 6 depicts the comparative graph of resource utilization before elimination, after elimination, and by applying security levels among 100 job requests.

Further experiments were conducted by varying the number of job requests submitted by the user. The results achieved are shown in Fig. 7. This experiment ensures the scalability of the system sustained with inclusion of stringent security levels. Highly trusted and secured resources are selected to serve more number of job requests. Further, the linearity in the graph indicates that the entire system is scalable even if there is an increase in the number of job requests with stringent security levels.

Fig. 7 Resource utilization with varying number of job requests when applied security approach

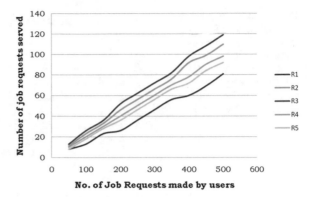

5.3   Security Effect on Final Trust Value

In this experiment, the main motive is to evaluate final trust when security is integrated with elimination approach. The resource R3 is considered for the experiments as it is proved to be a highly trusted resource during the elimination approach discussed in [19]. For the resource R3, various possibilities are explored at different levels of its existing trust due to the elimination approach such as 0.2, 0.5, and 0.9 represented with LOW, NORMAL, and HIGH categories, respectively. Similarly, various levels of security are integrated with the elimination model trust levels for the resource R3. In this integrated approach, existing trust of R3 and various security levels provided by the RP are fuzzified with membership functions of security.

Figure 8 shows the effect of integrating security on final trust value for the resource R3 with different values chosen in the range of LOW, NORMAL, and HIGH. Figure 8a depicts that resource R3 existing trust value of 0.2 is getting improved when applied with different security levels. The same is represented as new trust value in the figure. There is no change in new trust value without applying security. When LOW-scaled security levels are applied on R3 with existing trust value of 0.5, new trust value is decreased. Whenever security levels are increasing toward NORMAL and HIGH, then new trust value is also increased as shown in Fig. 8b. Figure 8c shows the case where existing trust value of resource R3 is considered as 0.9, and whenever security levels with LOW and NORMAL are applied, new trust value is decreased and it reached to HIGH because of HIGH-security levels.

The results in Fig. 8 show that trust of the resource R3 is drastically increasing when HIGH-security levels are integrated with existing trust levels. This result is quite a strong indication for robustness of the proposed T-Grid model when integrating security levels along with elimination approach. It is also indicating that trust is highly dependent on security features.

Fig. 8 Impact on final trust
value of resource R3 when
applied different security
levels with **a** R3 with
existing trust value = 0.2 **b**
R3 with existing trust
value = 0.5 **c** R3 with
existing trust value = 0.9

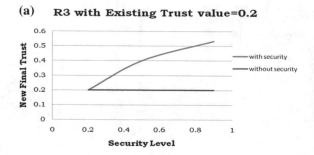

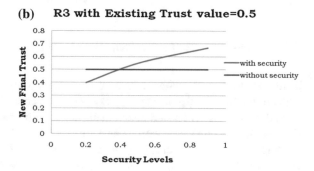

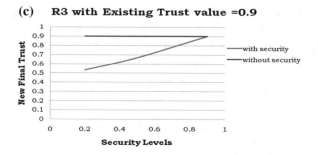

6 Conclusion

The approach discussed in this paper is unique in its nature such that it considers both
trust and security features in protecting the distributed systems like grid and cloud
systems. However, different levels of security are considered in view of the user
affordability. Based on the user's affordability for security levels and trust requirements, selection of resource provider is done. At its best possible conditions, i.e.,
higher trust and higher security levels, the whole system can be treated as highly
trusted as well as robust.

Results indicate that robustness of the system increases as the model adapts security features and trust combined together. This is demonstrated through various results

analysis done across different parameters like trustworthiness of the recommendations, utilization of the resources, and effect of security levels over the trust.

References

1. IBM Red Book, G.: Fundamentals of Grid Computing, REDP-3613-00 (2004)
2. Foster, I., Kesselman, C.: The anatomy of the grid—enabling scalable virtual organizations. Int. J. Supercomput. Appl. (2002)
3. Merriam-Webster's Collegiate Dictionary.: (2005)
4. Kavecky, S.: Grid security and trust management overview. Int. J. Comput. Sci. Issues **10**(3), 225–233 (2013). (No. 2)
5. Suresh Kumar, P., Sateesh Kumar, P., Ramachandram, S.: Recent trust models in grid. Int. J. Theor. Appl. Inf. Technol. **26**(1), 64–68 (2011)
6. Viduto, V., Townend, P., Xu, J., Djemame, K., Bochenkov, A.: A graph-based approach to address trust and reputation in ubiquitous networks. In: 2013 IEEE 7th International Symposium on Service Oriented System Engineering (SOSE), pp. 397, 402, 25–28 May 2013
7. Das, A., Islam, M.M.: SecuredTrust: a dynamic trust computation model for secured communication in multiagent systems. IEEE Trans. Dependable Secure Comput. **9**(2), 261–274 (2012)
8. Vijayakumar, V., Wahidha Banu, R.S.D.: Trust and reputation aware security for resource selection in grid computing. In: International Conference on Security Technology, pp. 121–124 (2008)
9. Song, S., Hwang, K., Kwok, Y.-K.: Trusted grid computing with security binding and trust integration. J. Grid Comput. (2005)
10. Patni, J.C., Rastogi, P., Jayant, V.K., Aswal, M.S.: Methods and mechanisms of security in grid computing. In: 2015 2nd International Conference on Computing for Sustainable Global Development (INDIACom), New Delhi, pp. 1040–1043 (2015)
11. Kumar, G.M., Ramachandran, S., Gyani, J.: Randomized algorithm for trust model in grid computing using GridSim components. In: 2015 IEEE International Advance Computing Conference (IACC), Bangalore, pp. 936–941 (2015)
12. Ennahbaoui, M., Idrissi, H., Hajji, S.E.: Secure and flexible grid computing based intrusion detection system using mobile agents and cryptographic traces. In: 2015 11th International Conference on Innovations in Information Technology (IIT), Dubai, pp. 314–319 (2015)
13. Xu, Y., Qu, W.: A trust model-based task scheduling algorithm for data-intensive application. In: 2011 Sixth Annual Chinagrid Conference (ChinaGrid), pp. 227–233 (2011)
14. Noor, T.H., Sheng, Q.Z., Yao, L., Dustdar, S., Ngu, A.H.H.: CloudArmor: supporting reputation-based trust management for cloud services. IEEE Trans. Parallel Distrib. Syst. (2015)
15. Harbajanka, S., Saxena, P.: Survey paper on trust management and security issues in cloud computing. In: 2016 Symposium on Colossal Data Analysis and Networking (CDAN), Indore, pp. 1–3 (2016)
16. JaganRaja, V., Sathish Kumar, P., Venkatachalam, V.: Survey on trust management in distributed cloud environment. Int. J. Innov. Sci. Eng. Technol. **3**(1) (2016)
17. Nalavade, U., Lomte, V.M.: Survey on various trust management issues in cloud environment. Int. J. Recent Innov. Trends Comput. Commun. **5**(1), 15–18 (2017)
18. Suresh Kumar, P., Ramachandram, S.: User satisfaction based quantification of direct trust in T-grid computation model. In: Proceedings of IEEE International Conference on Computer, Communication and Control Technology (I4CT'2014), Langkawi, Kedah, Malaysia (2014)
19. Suresh Kumar, P., Ramachandram, S.: A study on impact of untrustworthy and older recommendations over T-grid computational model. Springer Int. J. Syst. Assur. Eng. Manag. (2014). Available Online

20. Liao, H., Wang, Q., Li, G.: A fuzzy logic-based trust model in grid. In: International Conference on Networks Security, Wireless Communications and Trusted Computing, 2009. NSWCTC '09, vol. 1, pp. 608–614 (2009)

Offline Handwritten Malayalam Word Recognition Using a Deep Architecture

P. J. Jino, Kannan Balakrishnan and Ujjwal Bhattacharya

Abstract Handwriting recognition is an important application of pattern recognition subject. Although some research studies of handwriting recognition of a few major Indian scripts can be found in the literature, the same is not true for many of the Indian scripts. Malayalam is one such script, and automatic recognition issues of this script remain largely unexplored till date. On the other hand, there are nearly 40 million people mainly living in the southern part of India whose native language is Malayalam. In the present article, we present our recent study of Malayalam offline handwritten word recognition. The main contributions of the present study are (a) pioneering development of a database for offline handwritten word samples of Malayalam script and (b) its benchmark recognition results based on a transfer learning strategy which involves a deep convolutional neural network (CNN) architecture for feature extraction and a support vector machine (SVM) for classification. Recognition result of the proposed architecture on the writer independent test set of Malayalam handwritten word sample database is quite satisfactory. Moreover, the same architecture has been found to improve the existing state of the art of offline handwriting recognition of several major Indian scripts.

Keywords Handwritten word recognition · Malayalam handwriting recognition
Convolutional neural network · Offline handwriting recognition

P. J. Jino
Artificial Intelligence Lab, Department of Computer Applications,
Cochin University of Science and Technology, Kerala, India
e-mail: jino@cusat.ac.in

K. Balakrishnan
Department of Computer Applications, Cochin University of Science
and Technology, Kerala, India
e-mail: mullayilkannan@gmail.com

U. Bhattacharya (✉)
CVPR Unit, Indian Statistical Institute, 203 B.T. Road, Kolkata, India
e-mail: ujjwal@isical.ac.in

© Springer Nature Singapore Pte Ltd. 2019
J. C. Bansal et al. (eds.), *Soft Computing for Problem Solving*,
Advances in Intelligent Systems and Computing 816,
https://doi.org/10.1007/978-981-13-1592-3_73

913

1 Introduction

Offline or online handwriting recognition was an open research problem for decades. Some extensive study [1, 24, 29] of the problem has led to remarkable success in recent past on a number of scripts of developed countries. Also, some handwriting recognition studies of a few Indian scripts such as Devanagari [3], Bangla [5], Tamil [4], Oriya [6], Malayalam [12] are found in the literature [21]. Although a majority of offline handwriting recognition studies of Indian scripts considered only isolated characters, a few others [22, 26, 27] considered offline handwritten words as well. In these studies, various handcrafted feature representations of input sample image were used for their recognition. On the other hand, the recent success of convolutional neural networks (CNN) in image recognition tasks has opened up new direction of handwriting recognition research to improve the state of the art and similar approaches do not require selection of an efficient handcrafted feature representation of input samples. Recently, Maitra et al. [18] have improved the state of the art of handwritten Bangla character recognition using a CNN architecture.

Usually, a deep CNN architecture involves a large number of connection weights, and thus, its proper training requires a significantly high volume of training samples which often stands as a bottleneck of using such an architecture in handwriting recognition tasks. A solution to this problem is the use of transfer learning strategy [13], where a moderately trained CNN is used for feature extraction, and the values computed at its last convolution or subsampling layer are fed as features to another classifier such as a support vector machine (SVM) which has the capacity of being sufficiently trained based on a relatively smaller number of samples. Recently, similar transfer learning strategy has been used [18] to recognize handwritten numerals of several Indian scripts such as Devanagari, Bangla, English, Telugu and Oriya.

Usually, handwriting recognition studies are performed in two different paradigms—analytic and holistic. In an analytic approach, explicit segmentation or segmentation at the time of recognition of input word sample is considered, whereas holistic recognition does not require such a segmentation module. In [17], advantages and requirements of holistic recognition have been discussed. The goal of the present study is holistic recognition of offline handwritten Malayalam words. Here, a deep CNN architecture is used for computation of feature of the input image sample, and an SVM is used for its holistic recognition based on this feature. Also, since there is no benchmark database of handwritten Malayalam word samples, a moderately large database of such samples has been generated as a part of the present study.

Rest of this article is organized as follows. Details of the sample database have been described in Sect. 2 while the network architecture, its training and the SVM used have been detailed in Sect. 3. Simulation results are provided in Sect. 4. Finally, Sect. 5 concludes the present article.

2 Handwritten Malayalam Word Samples: The New Database

2.1 Malayalam Script

Malayalam is one of the twenty-two scheduled languages of India and was declared as a classical language by the Government of India in 2013. Spoken by around 35 million people, it is the official language of Kerala province, union territories of Lakshadweep and Puducherry. The script in which the language is written is also called Malayalam, and it is one among the ten major official Indian scripts. It is derived from Grantha, an inheritor of ancient Brahmi script. Malayalam script is alpha-syllabic and non-cursive in nature. Its basic character set consists of vowels and consonants. Each vowel has an independent form and another dependent form except the first vowel (/a/) which has no corresponding dependent form. Dependent vowel signs do not appear on their own. Such a vowel appears as a diacritic attached to a consonant. These are composed of one or more glyph pieces which appear either to the left, right or both sides of a consonant.

A new reformed Malayalam script was introduced in the year 1971 by an order of Kerala Government on the basis of recommendations made by committees formed for this purpose. In this reformed script, the number of glyphs had been reduced substantially for the ease of printing and typewriting. The basic reforms include (i) substitution of irregular ligatures by corresponding sequences of basic glyphs and (ii) replacement of severely complex shaped conjuncts (combinations of multiple characters) by the corresponding sequences of separated basic characters and diacritic [20]. Although textbooks recommended by Govt. sponsored schools follow this new script, in practice, this is followed only partially in daily writing in Malayalam. In the present scenario, both printed and handwritten texts appear to have a mixture of characters from both old and new scripts. It makes their automatic recognition a more difficult task.

2.2 Database Lexicon

Offline handwritten Malayalam word sample database developed as a part of the present study is based on a lexicon consisting of 241 city (Panchayath) names and another 73 words selected from Sabdatharavali Malayalam Dictionary. The present lexicon set of 314 distinct words covers nearly 20% of the entire Malayalam corpus of [2]. Lengths of words of the present lexicon vary widely—its shortest words consist of 2 characters, while its longest word has 14 characters. Frequencies of words versus their lengths in this lexicon are shown in Fig. 1. According to a treatise of Malayalam grammar [30], the new script has 97 symbols: 36 consonants, 5 pure consonants, 13 vowels, 13 dependent vowels, 4 consonant signs and 26 compound characters. The present lexicon set includes all of these symbols except eleven of them: "ഌ

Fig. 1 Distribution of word length in Malayalam handwritten word database

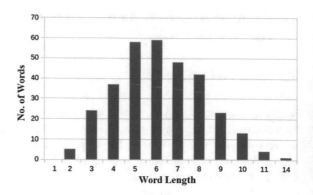

/Malayalam Consonant Letter NGA", "ഃ /Malayalam Consonant Sign Visarga" and eight compound characters. According to a recent study of Malayalam corpus [25], occurrence percentage of "ങ" and "ഃ" are only 0.043 and 0.002, respectively, whereas the eight left out compound characters did not appear in this report. On the other hand, words of the present lexicon set include 13 compound characters of the traditional (old) Malayalam script. The occurrence frequencies of these 100 symbols in the present lexicon set are shown in Table 1.

Table 1 Frequency of various characters in the lexicon of the present database

Type	Char.	Freq.	Char.	Freq.	Char.	Freq.	Char.	Freq.	Char.	Freq.
Vowels	അ	15	ആ	13	ഇ	4	ഈ	5	ഉ	7
	ഊ	2	ഋ	3	എ	6	ഏ	6	ഐ	2
	ഒ	5	ഓ	1	ഔ	2				
Dependent vowels♣	ാ	109	ി	130	ീ	9	ു	114	ൂ	61
	ൃ	5	െ	24	േ	29	ൈ	5	ൊ	88
	ൌ	12	്	31						
Consonants	ക	68	ഖ	7	ഗ	16	ഘ	6	ച	18
	ഛ	3	ജ	6	ഝ	4	ഞ	3	ട	58
	ഠ	1	ഡ	2	ഢ	5	ണ	11	ത	32
	ഥ	4	ദ	13	ധ	6	ന	43	പ	46
	ഫ	4	ബ	6	ഭ	5	മ	56	യ	37
	ര	88	റ	23	ല	47	ള	21	ഴ	22
	വ	48	ശ	15	ഷ	3	സ	13	ഹ	5
Consonant Signs	്	7	ൗ	3	ൃ	11	ഃ	45		
Pure Consonants	ൻ	4	ൺ	6	ർ	61	ൽ	16	ൾ	3
Compound Characters	ക്ക	44	ങ്ക	7	ങ്ങ	24	ച്ച	8	ഞ്ച	5
	ട്ട	15	ഞ്ഞ	3	ണ്ണ	6	ത്ത	16	ന്ത	5
	ന്ന	12	ന്റ	26	മ്പ	16	മ്മ	3	യ്യ	4
	ല്ല	13	ള്ള	2	ണ്ട	6				
Compound Characters of Old Script	ള്ള	14	ക്ഷ	4	ൂ	14	ഭ	2	ഡ	4
	ഭ	3	ഝ	1	ൗ	4	ജ്ജ	1	സ്ഥ	1
	ശ്ല	8	സ്ന	26	ൠ	1				

♣Light gray circles show the position of a consonant character.

The set of first 20 characters occurring most frequently in our lexicon is in agreement with similar results presented in the study of [19].

2.3 Samples of Present Database

Samples of our database were collected from a group of 49 natives belonging to different sections of the population with respect to age, sex, education, profession and income. Writers of its samples had age in the range 10–60. The set of writers consists of both left-handed and right-handed ones. They were asked to write the words of a given lexicon set on a specific form printed on A4 size paper. Header part of this form was used to collect information about the writer such as name, age, qualification, signature. So the present database can be used for several other applications of handwriting analysis. There are 45 writers who wrote the entire set of samples at two different points of time, and the remaining four writers wrote it only once. The form was so designed that automatic extraction of individual word samples should be easy. Writers used their own pens. Filled-in forms were scanned using a flatbed scanner at 300 dpi. Automatically extracted samples were manually checked for necessary corrections. A few (20) word samples from the present database are shown in Fig. 2. Each word sample of this figure belongs to a distinct word class. It consists of pairs of classes of similar shapes. As for example, the pair of words belonging to (first row, first column) and (second row, first column) have similar shapes. Similarly, the pair of words belonging to (third row, first column) and (fourth row, first column) look similar.

This database consists of 29,516 handwritten Malayalam word samples. The entire database is divided into training and test sets. Samples provided by 30 writers form its training set, while the samples of remaining 19 writers form the test set. The training and test sets consist of 18,840 and 10,676 samples, respectively.

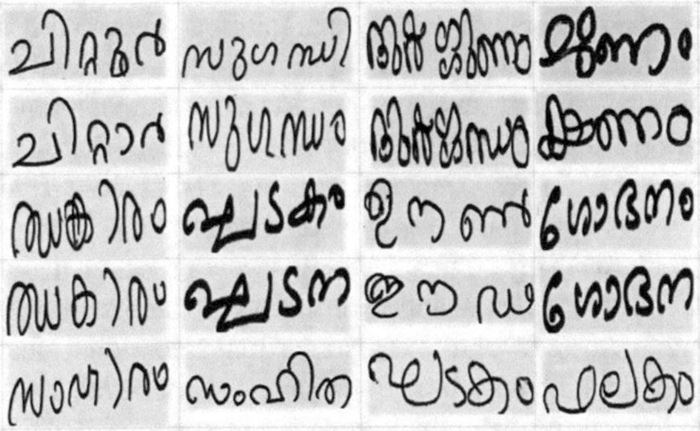

Fig. 2 A few samples (each belonging to a distinct word class) from the present database—these provide a broad idea of the interclass similarity present in our database.

3 Proposed Approach

The work flow of the proposed recognition approach includes a brief preprocessing stage followed by feature extraction using a convolutional neural network (CNN) and finally classification with the help of a support vector machine (SVM). Details of the approach are presented below. Further details of CNN and SVM can be found in [9, 16], respectively.

3.1 Preprocessing

In traditional approaches of offline handwriting recognition, several preprocessing operations are performed on input samples to reduce the variations in their images. However, similar preprocessing modules do not have many roles in CNN-based recognition approaches because the design of a convolutional neural network architecture has the inherent capacity to handle various sources of variations in input samples. The only preprocessing operation required to be performed before feeding the samples to a CNN architecture is size normalization because the input layer of such a neural network has certain fixed size. In the present study, we experimented with several choices of the size of the input layer and observed 64×100 input layer as the optimal size. Thus, in the present approach, the preprocessing stage involves (i) cropping the minimum bounding rectangle of the input word image and (ii) size normalization of the cropped image to the size 64×100 using bicubic transformation.

3.2 Convolutional Neural Network

A fully connected multilayer perceptron (MLP) can successfully recognize handwritten samples when these are fed with efficient discriminative feature vectors. Such feature vectors are usually handcrafted ones, and it is obviously hard to decide on the feature vector which should lead to sufficiently successful recognition. As an alternative, raw image of handwritten samples may also be fed at the input layer of an MLP and the network may be allowed to learn to discriminate samples of different classes. However, the number of connection weights of such a network must be huge, particularly when the number of classes of the underlying recognition problem is large. Thus, training of such a heavy network requires a large number of labelled samples which is difficult to arrange. An alternative to the use of MLP for handwriting recognition disregarding selection of feature vector is the use of a convolutional neural network. It has now been well established that CNNs are capable of learning efficient discriminative features needed for a classification task. Since the lower part of a CNN architecture is not fully connected, these involve smaller number of connection weights than its MLP counterpart. Moreover, if a CNN is used only as a feature vector extractor leaving the classification task to another suitable tool such as the support vector machine (SVM), then the CNN need not to be sufficiently trained and it helps to get the job done even with the availability of a limited number of training samples. Such a strategy is known as "Transfer Learning" [23]. On the other hand, it has now been established that the larger depth of a CNN architecture translates into its performance [28]. However, there is a limit of the depth of a CNN beyond which the network fails to be trained successfully due to the well-known problem of vanishing/exploding gradients [10].

3.2.1 CNN Architecture

The deep architecture of the CNN used in the present task of offline handwritten Malayalam word recognition consists of ten layers: four convolution layers, four subsampling layers and a fully connected part which includes one hidden layer and an output layer. A diagram of this architecture is provided in Fig. 3. The size of the input image to this network is 64×100 on which 5×5 kernel with stride 1 is used to generate 30 feature maps each of size 60×96 at the first convolutional layer C1. Maxpooling based on non-overlapping 2×2 kernel on these feature maps produces 30×48 feature maps at the subsampling layer S2. Next, we apply convolution operation using 5×5 kernel with stride 1 followed by maxpooling with non-overlapping 2×2 kernel to obtain 35 feature maps of size 26×44 at second convolutional layer C3 and the same number of feature maps of size 13×22 at the second subsampling layer S4 successively. Output of the third convolutional layer C5 consists of 40 feature maps of size 10×18, and these are obtained by using 4×5 kernel at stride 1. These are next reduced to the size 13×22 at the subsampling layer S6 with the help of maxpooling based on the same non-overlapping 2×2 kernel. Finally, 4×4 kernel at stride 1 is used to obtain 45 feature maps of size 2×6 at the convolution layer C7, and maxpooling as before is used to obtain 1×3 feature maps at the last subsampling layer S8. There are 135 values at S8 which are fed as input to the fully connected part of the network. This part has a hidden layer of 128 nodes and an output layer 314 nodes which is the number of underlying word classes.

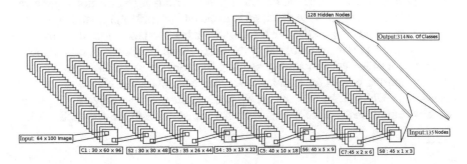

Fig. 3 Proposed architecture of the convolutional neural network

We arrived at the above architecture of the CNN based on simulations of a large number of various architectures, and the present architecture provided an optimal performance on the database described in Sect. 2. Activation used in all the convolutional layers and fully connected layer is rectified linear unit (ReLU) [14] while the output layer uses softmax [7] as the activation.

3.2.2 Training of the CNN

Weights of the CNN were initialized randomly with values in the range 0–0.5. Gradient descent is the most popularly used method for training of neural networks. In the literature, there exist different algorithms for optimization of gradient descent training of neural networks. These include (i) batch gradient descent which updates the connection weights after each epoch, i.e., after presentation of the whole set of training samples to the network, (ii) mini-batch gradient descent which update the network weights after each presentation of a batch of n training samples and (iii) stochastic gradient descent, popularly known as SGD [8], which randomly selects a training sample at each iteration for presentation to the network, and each time, the weights are updated. In the present implementation, we used mini-batch gradient descent variant with batch size equals to 50.

Since mini-batch gradient descent algorithm cannot guarantee convergence of the training at a good local minimum, different additional strategies are adopted by the practitioners for the required effective training of the network. In the present implementation, we used a particular gradient descent optimization strategy, called ADADELTA [31] for this purpose. In this strategy, a distinct learning rate is used for each connection weight and this is dynamically automatically updated using first order information.

A deep convolutional network involves a large number of parameters. Thus, the problem of overfitting occurs in situations when there are only a limited number of training samples. Although the deep learning strategy had shown its promising performance in various applications, the two major issues of this strategy are (i) overfitting and (ii) computational burden of its training algorithm. The effect of computational burden has, in the meantime, become manageable due to the availability of high-speed GPUs. On the other hand, various regularization techniques have been experimented in the literature to avoid the overfitting problem. Dropout is a comparatively new regularization technique that has been more recently employed in deep learning [15] to get rid of this problem. The term 'dropout' refers to dropping out units in a neural network during training. By dropping a unit out, we mean temporarily ignoring it from the learning task along with all its incoming and outgoing connections. Units of the network for dropout are chosen randomly. In the present implementation, each unit of the network is dropped out with the probability 0.5.

Data augmentation [15] is another strategy which is used to prevent overfitting of deep neural network architecture. Here, we randomly apply one of the transformations from (i) rotation ($-5°$ to $+5°$), (ii) Gaussian noise with variance 0.2 and (iii) width and height shifting by one-tenth of the corresponding dimension before feeding a training sample to the network.

The stopping of training epochs of a neural network is another crucial issue for its successful application. We used a validation set of samples for this purpose. In fact, 20% of the training samples of each class is selected randomly to form the validation set, and the remaining training samples were actually used for adjustment of connection weights. Initially, the network error on validation set remains high, and this gets decreased as the training progresses, but after a certain number of iterations, the validation error starts increasing and this is the instant when we stopped training and obtained the network performance on the test set.

3.3 Support Vector Machine

In the literature, support vector machines (SVMs) have been established as an efficient classification model even in the presence of a limited number of training samples. It maps input samples into a higher-dimensional space where an optimal separating hyperplane is constructed. Since its computation involves solving a quadratic programming problem, SVM does not have the difficulty of the existence of multiple local minima unlike the gradient descent-based learning method of CNN architecture.

Traditionally, SVM is a two-class classifier. It solves multiclass classification problem using one of the two strategies: "one-versus-all" or "one-versus-one". In the present task, we used the latter strategy. In this strategy, $n(n-1)/2$ classifiers are constructed for an n-class classification problem and each one is trained by using samples from the two classes. Finally, results of these all these two-class classifiers are combined to reach the decision.

SVM uses the well-known kernel trick. A kernel provides the similarity of two inputs to it as the output. Usually, an implementation of SVM provides various options for the kernel function such as (i) linear, (ii) polynomial, (iii) RBF. In the present task, we used RBF kernel. This kernel involves two parameters which are often denoted by C and γ. We obtained the suitable values of these two parameters by a grid search strategy.

An SVM as described above has been trained using the 135 feature values computed at the last subsampling layer S8 of our CNN architecture. Since the training of SVM does not need any validation sample set, we have used the entire set of 18,840 training samples for training of the SVM. In the recognition phase, feature vector of an unknown (test) sample is first generated by the CNN at S8 layer of the CNN which is next fed to the SVM to get its classification output.

4 Results of Experimentations

We have simulated the proposed holistic approach of handwritten word recognition on (i) our database of 314 class handwritten Malayalam words, (ii) two databases of 114 class handwritten Marathi legal amount words and (iii) another database of 106 class Hindi legal amount words.

4.1 Results on Malayalam Word Database

Here, we obtained recognition accuracy percentage of the proposed approach on the test set of 10,676 handwritten samples corresponding to 314 classes of Malayalam words. These samples were written by 19 writers, and 15 of them wrote all the 314 words at two different instances, while another 4 writers wrote all the words only once.

The hybrid architecture consisting of CNN and SVM provided 96.90% recognition accuracy on this test set, while the CNN alone provided 95.74% accuracy. An analysis of misclassified words shows that when a word is misclassified as another word in the lexicon, these two words have necessarily a common character string. Examples of a few such word pairs are shown in Table 2.

4.2 Hindi Legal Amount Word Database

The model implemented for the Malayalam Word recognition is also simulated on a database of handwritten Hindi legal amount words [11]. The lexicon size of this database is 106, and it consists of 8480 samples provided by 80 writers. It contains all the possible words that one needs to use to write a legal amount in Hindi language. The training set consists of 6360 word samples written by 60 writers and the remaining 2120 samples written by 20 writers form the test set. A hybrid architecture similar to

Table 2 Misclassification scenarios of Malayalam words

Actual word		Recognized Word	
Malayalam	Transliteration	Malayalam	Transliteration
ആലപ്പാട്	Alappad	ആലപ്പുഴ	Alappuzha
അരിക്കുളം	Arikkulam	കരുംകുളം	Karumkulam
ചെങ്ങന്നൂർ	Chengannur	പന്ന്യന്നൂർ	Pannyannur
ചിറ്റൂർ	Chittoor	ഒറ്റൂർ	Ottoor
ഇടമുളയ്ക്കൽ	Idamulakkal	പുഴയ്ക്കൽ	Puzhakkal
കൊടംതുരുത്ത്	Kodamthuruth	കടുതുരുത്തി	kaduthuruthi
കൊടംതുരുത്ത്	Kodamthuruth	മുളന്തുരുത്തി	Mulamthuruthi
ധകാരം	Dhakaram	ധങ്കാരം	Dhamkaram
ഋണം	Wranam	ക്ഷണം	Kshanam

the one used for Malayalam word database is trained for the present 106 class word recognition problem, and the recognition performance of the same is verified on its test samples. We obtained 94.15% accuracy on the test set of Hindi legal amount word database which improves the existing state-of-the-art accuracy value of 83.07% published in [11].

4.3 Marathi Legal Amount Word Database

The hybrid deep neural network model presented in this article has been simulated on two handwritten word databases DB1 and DB2 [11] of Marathi legal amount. The lexicon size of both of these two Marathi word databases DB1 and DB2 is 114. Samples of DB1 were written by 90 writers, while the same of DB2 were written by another 70 writers. Compared to DB2, samples of DB1 are neat and legible. Word databases DB1 and DB2 consist of 10260 and 7980 image samples, respectively. The proposed approach provided 92.60 and 92.19% recognition accuracies on DB1 and DB2, respectively, which improved the existing respective state-of-the-art accuracy values of 85.78 and 78.79% [11].

5 Conclusions

The present study explores a deep hybrid neural network architecture for offline handwritten Malayalam word recognition. The same architecture has also been simulated on existing handwritten word databases of Hindi and Marathi legal amounts. The later simulation results show that the proposed approach improves the state of the art on these two databases.

References

1. Arica, N., Yarman-Vural, F.T.: An overview of character recognition focused on off-line handwriting. IEEE Trans. Syst. Man Cybern. Part C (Applications and Reviews) 31(2), 216–233 (2001)
2. Bharati, A., Rao, K.P., Sangal, R., Bendre, S.: Basic statistical analysis of corpus and cross comparison among corpora. Technical Report of the Indian Institute of Information Technology, pp. 1–11 (2000)
3. Bhattacharya, U., Chaudhuri, B.B.: Handwritten numeral databases of Indian scripts and multistage recognition of mixed numerals. IEEE Trans. Pattern Anal. Mach. Intell. 31(3), 444–457 (2009)
4. Bhattacharya, U., Ghosh, S.K., Parui, S.: A two stage recognition scheme for handwritten Tamil characters. In: 9th International Conference on Document Analysis and Recognition (ICDAR 2007), vol. 1, pp. 511–515 (2007)

5. Bhattacharya, U., Shridhar, M., Parui, S.K., Sen, P.K., Chaudhuri, B.B.: Offline recognition of handwritten Bangla characters: an efficient two-stage approach. Pattern Anal. Appl. **15**(4), 445–458 (2012)
6. Bhowmik, T.K., Parui, S.K., Bhattacharya, U., Shaw, B.: An HMM based recognition scheme for handwritten Oriya numerals. In: 9th International Conference on Information Technology (ICIT '06), pp. 105–110 (2006)
7. Bishop, C.M.: Pattern Recognition and Machine Learning. Springer, Berlin (2006)
8. Bottou, L.: Stochastic Gradient Descent Tricks, pp. 421–436. Springer, Berlin, Heidelberg (2012)
9. Burges, C.J.C.: A tutorial on support vector machines for pattern recognition. Data Min. Knowl. Discov. **2**(2), 121–167 (1998)
10. Glorot, X., Bengio, Y.: Understanding the difficulty of training deep feedforward neural networks. In: International Conference on Artificial Intelligence and Statistics (2010)
11. Jayadevan, R., Kolhe, S.R., Patil, P.M., Pal, U.: Database development and recognition of handwritten Devanagari legal amount words. In: International Conference on Document Analysis and Recognition, pp. 304–308 (2011)
12. John, J., Pramod, K.V., Balakrishnan, K., Chaudhuri, B.B.: A two stage approach for handwritten Malayalam character recognition. In: 14th International Conference on Frontiers in Handwriting Recognition, pp. 199–204 (2014)
13. Kang, L., Kumar, J., Ye, P., Li, Y., Doermann, D.: Convolutional neural networks for document image classification. In: 22nd International Conference on Pattern Recognition (2014)
14. Krizhevsky, A., Sutskever, I., Hinton, G.E.: Imagenet classification with deep convolutional neural networks. In: Advances in Neural Information Processing Systems, pp. 1097–1105 (2012)
15. Krizhevsky, A., Sutskever, I., Hinton, G.E.: Imagenet classification with deep convolutional neural networks. In: Pereira, F., Burges, C.J.C., Bottou, L., Weinberger, K.Q. (eds.) Advances in Neural Information Processing Systems 25, pp. 1097–1105 (2012)
16. LeCun, Y., Bottou, L., Bengio, Y., Haffner, P.: Gradient-based learning applied to document recognition. Proc. IEEE **86**(11), 2278–2324 (1998)
17. Madhvanath, S., Govindaraju, V.: The role of holistic paradigms in handwritten word recognition. IEEE Trans. Pattern Anal. Mach. Intell. **23**(2), 149–164 (2001)
18. Maitra, D.S., Bhattacharya, U., Parui, S.K.: CNN based common approach to handwritten character recognition of multiple scripts. In: 13th International Conference on Document Analysis and Recognition (ICDAR), pp. 1021–1025 (2015)
19. Neeba, N.: Large Scale Character Classification. Ph.D. thesis, International Institute of Information Technology Hyderabad, India (2010)
20. Neeba, N., Namboodiri, A., Jawahar, C., Narayanan, P.: Recognition of Malayalam documents. In: Guide to OCR for Indic Scripts, pp. 125–146. Springer, Berlin (2009)
21. Pal, U., Chaudhuri, B.: Indian script character recognition: a survey. Pattern Recognit. **37**(9), 1887–1899 (2004)
22. Pal, U., Roy, K., Kimura, F.: A lexicon driven method for unconstrained Bangla handwritten word recognition. In: 10th International Workshop on Frontiers in Handwriting Recognition (2006)
23. Pan, S.J., Yang, Q.: A survey on transfer learning. IEEE Trans. Knowl. Data Eng. **22**(10), 1345–1359 (2010)
24. Plamondon, R., Srihari, S.N.: On-line and off-line handwriting recognition: a comprehensive survey. IEEE Trans. Pattern Anal. Mach. Intell. **22**(1), 63–84 (2000)
25. Prema, S., Joseph, M.: Malayalam Frequency Count Study Report. Technical Report, Department of Linguistics, University of Kerala (2001)
26. Shaw, B., Bhattacharya, U., Parui, S.K.: Combination of features for efficient recognition of offline handwritten Devanagari words. In: 14th International Conference on Frontiers in Handwriting Recognition, pp. 240–245 (2014)
27. Shaw, B., Bhattacharya, U., Parui, S.K.: Offline handwritten Devanagari word recognition: information fusion at feature and classifier levels. In: 3rd IAPR Asian Conference on Pattern Recognition (ACPR), pp. 720–724 (2015)

28. Szegedy, C., Liu, W., Jia, Y., Sermanet, P., Reed, S., Anguelov, D., Erhan, D., Vanhoucke, V., Rabinovich, A.: Going deeper with convolutions. In: IEEE Conference on Computer Vision and Pattern Recognition (CVPR), pp. 1–9 (2015)
29. Trier, Ø., Jain, A.K., Taxt, T.: Feature extraction methods for character recognition–a survey. Pattern Recogn **29**, 641–662 (1996)
30. Varma, A.R.R.R.: Kerala Panineeyam. Kozhikode Poorna, 3rd ed (1997)
31. Zeiler, M.D.: Adadelta: An Adaptive Learning Rate Method. arXiv:1212.5701 (2012)

Bat Algorithm-Based Traffic Signal Optimization Problem

Sweta Srivastava and Sudip Kumar Sahana

Abstract The need for the transport services and road network development came into existence with the development of civilization. In the present urban transport scenario with ever-mounting vehicles on the road network, it is very much essential to tackle network congestion and to minimize the overall travel time. This work is based on determining the optimal wait time at traffic signals for the microscopic discrete model. The problem is formulated as bi-level models based on Stackelberg game. The upper layer optimizes time spent in waiting at the traffic signals, and the lower layer solves stochastic user equilibrium. Soft computing techniques like genetic algorithms, ant colony optimization and many other biologically inspired techniques are proven to give good results for bi-level problems. Here, this work uses bat intelligence to solve the problem. The results are compared with the existing techniques.

Keywords Bat algorithm · Genetic algorithm · Ant colony optimization

1 Introduction

Nowadays, the eternally growing number of vehicles creates a challenge to avoid the traffic jam in the modern urban transportation scenario. The most profitable way to put up with it can be optimizing the wait time at traffic signals. In order to save the priceless time of vehicle users and for the reduction the traffic congestion on road intersections optimized traffic signals play a vital role. Along with improving the road safety, they also smooth the progress of medical emergencies and industrial needs.

The need for the transport and road network planning came on track with the expansion of civilization, but in the literature, work in this field can be seen from

S. Srivastava (✉) · S. K. Sahana
Department of Computer Science, Birla Institute of Technology, Mesra,
Ranchi, India
e-mail: ssrivastava.rnc09@gmail.com

© Springer Nature Singapore Pte Ltd. 2019
J. C. Bansal et al. (eds.), *Soft Computing for Problem Solving*,
Advances in Intelligent Systems and Computing 816,
https://doi.org/10.1007/978-981-13-1592-3_74

the last five decades. Abdullaal [1] designed an innovative method for vehicular equilibrium network design problem using Hooke–Jeeves' technique with continuous variables. A mutually consistent (MC) traffic assignment and an innovative setting for the traffic signal for implementation on an average size of the road network were developed by Allsop [2]. A linear constraint approximation model was suggested by Heydecker [3] for solving the bi-level optimization problem with certain constraints.

Biologically inspired techniques present practical advantages to the researchers for getting to the bottom of difficult computational and optimization problems with its prime features like robustness, self-adaptability, flexibility and many other facets.

Ceylan and Bell [4] integrated GA traffic assignment using the TRANSYT software. The traffic control and minimization were solved using the path flow estimator (PFE). He integrated the model to develop GATRANSPFE to solve the network design problem. The applicability of GATRANSPFE was put side by side with mutually consistent (MC) solution using some numerical examples. The performance index (PI) was enhanced by 34% over the mutually consistent solution of the problem in the 75th generation. Koh [5] used differential evolution algorithm for a bi-level continuous network design problem (CNDP). Basken and Haldenbilen [6] developed an ACO reduced search space (ACORSS) to find better signal timing in the signal setting problem of a bi-level model for traffic optimization problem. Hu [7] solved urban transportation equilibrium network design problem using bi-level programming and was solved using PSO and Frank Wolfe. The algorithm was found effective and took less iteration to give a better solution in comparison to simulated annealing. Srivastava and Sahana [8, 9] designed a discrete evolutionary model to reduce the waiting time of vehicles at traffic signals within the urban transportation system using bi-level Stackelberg game model. Five test networks with 12, 16, 20, 24, 28 nodes were designed using Petri net. The proposed hybrid technique was solved for optimizing wait time at traffic signals and for SUE. Hybrid algorithm outperformed ACO and GA. Discrete model for traffic optimization problem was previously proposed by Canteralla et al. [10, 11] and many other researchers. These models were developed for macroscopic simulation. The projected model works on a level of sections inside the road network, and thence, it will pay attention on a variety of microscopic tribulations.

The echolocation of microbats inspired Yang [12–14] to develop an innovative metaheuristic technique, the bat algorithm (BA), for solving complex constrained optimization problems. This proposed technique was found to be furthermore efficient than particle swarm optimization, genetic algorithms and harmony search because of its robust and innovative parameter control features and frequency tuning abilities. BA is proven to give good results for many optimization problems. Kiełkowicz and Grela [15] used BA for nonlinear optimization problems. Abatari et al. [16] used a BA-inspired method to solve the optimal power flow (OPF) problem. Yassine Saji et al. [17] applied BA to solve discrete travelling salesperson problem.

This paper is structured as follows: Sect. 2 defines the problem. Section 3 presents the solution technique for the specified problem. Section 4 displays the experimental setup for the bi-level model that consists of five test cases. The results and discussions over the findings are presented in Sect. 5. The conclusions drawn from this work are presented wrapping the work at the end.

2 Problem Definition

The problem is designed as a bi-level model inspired by the Stackelberg game where the traffic optimization model is grouped into two layers. The "traffic signals" and the "stochastic user equilibrium" are represented by the "Upper Layer" and "Lower Layer", respectively, as shown in Fig. 1 [9]. The output of the upper layer depends on the input of lower layer, and the lower layer analyses the situation from external input and output value of the upper layer.

The violation of convexity of the possible region is the major attraction of this type of problems. The problems are well recognized for presenting multiple local optimum values. The prime idea is to find a method that optimizes both the layers give optimal travel time by reducing total wait time. An illustration of the road intersection [8] for the proposed model is projected in Fig. 2.

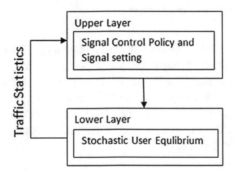

Fig. 1 Proposed traffic optimization model using bi-level technique

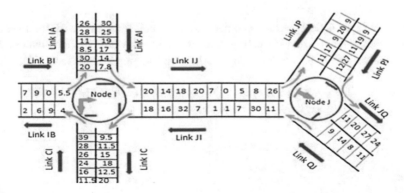

Fig. 2 A snapshot of road intersection displaying two junctions

The road network consists of two ways path connecting the nodal junctions. This can be taken analogical to a bi-directional digraph. Every link in the network consists of a several small "time-sections", which are most of the time dissimilar for every link. The time-sections decide the vehicle load to be cleared at a particular green signal value. These time-sections are intercepted by dividing the length of each link in the road network with the average allowed speed of vehicles on that link. As shown in Fig. 2, every time-section encompasses a number written to it. These digits denote the quantity of vehicles (or load) available at a particular time in that section of the link. The values are converted into "per car unit" in order to maintain the uniformity [18]. That is, the load is weighed against the space that a car will take and accordingly their relative values are computed. The time quantum is considered to be consistent for all the intersections; however, the signal duration at the junction may vary. The overall wait time is computed as the sum of wait time of the vehicle at every intersection in that road network. The traffic signal adjusts its value synchronously over all intersections. The calculation of the total wait time is done in several numbers of phases. Each phase consists of the present signal value taken from all the intersections of the network.

2.1 Mathematical Formulation

A mathematical model for the bi-level problem is given in this section. The model is analogous to the leader and follower model of the Stackelberg game where both the layers depend on each other. The prime objective of the upper layer is to trim down the wait time of the vehicles on every road junction, thus proving with an overall reduction of wait time on the road network. The task of assigning the load based on initial source–destination requests so as to minimize the "total travel time" of vehicle users is the task of the lower layer.

2.1.1 Upper Layer

The upper layer is designed to optimize the total wait time TW at all the intersections I encountered while traversing from the origin towards the destination. The mathematical model for the upper layer objective function is as follows [9]:

$$\text{Minimize TW} \, (\Psi, q * (\Psi)) = \sum_{a \in I} \sum_{b \in L(a)} q_{ab}(\Psi, q * (\Psi)) \times t_{ab}(\Psi, q * (\Psi)) \quad (1)$$

$$\text{Subject to: } \Psi_{\min} \leq \Psi \leq \Psi_{\max}$$

where

Ψ is the signal variable.

q^* is the optimized load distribution function taken from the lower layer of the problem.

l_a is the set of all the links in the network, attached to the intersection 'a'.

q_{ab} is the total load which is waiting on link 'ab' at intersection 'a'.

t_{ab} is the time for which the load q_{ab} waited.

2.1.2 Lower Layer

The stochastic user equilibrium is denoted by the lower layer [8, 9]. This gives the total load assigned to the links in the network. Depending upon the signal value taken from the upper layer, it clears the desired load. The model can be represented as follows:

$$\text{Minimize } D(\Psi p, C) \tag{2}$$

where D is the distance cost function for travelling on a link determined by previous signal value (Ψ_p) and C is the length of the link connecting junctions.

3 Research Methodology

Solution technique for the traffic signal optimization model is presented in Fig. 3. The proposed scheme for optimizing the overall wait time at traffic signals is based on bi-level model.

The upper layer objective function is solved using bat algorithm, and the obtained solution is further used to optimize the lower layer. Bat algorithm is a technique that uses the echolocation behaviour of the bats for its hunt and survives. In order to identify and keep away from obstacles, the bats use sonar echoes which reflect back from the obstacle and are then transformed into the frequency. There is a time delay between the rate of emission and reflection. The bats utilize this delay for navigation purpose. The generalized bat algorithm [11–13] is stated below.

Step 1: *Definition of the objective function F(x)*

Step 2: *Initialization of the bat population xi and bat's velocity vi*
           where i = 1 to n
           Initialize the pulse rates ri and loudness Ai

Step 3: *Set the pulse frequency value Qi [Qmin, Qmax]*

Step 4: *Initialize the number of bat iterations*

Step 5: *Initialize the loop*
           while (t < Tmax)

Step 6: *Adjusting the frequency update the velocities and the location of bats for calculating the new solutions*
           if(rand (0, 1) > ri)

Fig. 3 A representation of
the solution method
for traffic signal optimization
problem using BA

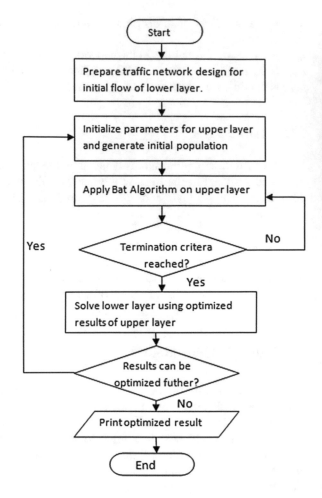

Choose the best solution among the set of the solutions and generate the
local solution in the region of the global best solution
End if

Step 7: Generate the new position xi of bats by flying randomly in the solution space
if(rand(01) < Ai and f(xi) < f(x))
Update new xi
Increase the value of pulse rate ri and reduce the loudness Ai
End if

Step 8: Find the current best solution by ranking the bats

Step 9: End the loop
End while

Algorithm 1. Bat Algorithm for traffic signal optimization problem.

4 Experimental Setup

For testing the applicability of the described model, five different test cases were adapted from [9] corresponding to five different networks. The specifications are shown in Table 1.

Figure 4 shows a 12-node network showing signal values. Figure 5 shows 16, 20, 24 and 28 node networks.

The termination criterion for BA is taken as 100 iterations after a repetitive number of executions on MATLAB. The frequency Q lies in a range between [0, n], where n denotes the number of nodes in the network. Pulse rate was taken as [0, 1]. Intersections also entertained additional attributes like the value of the signal and related positional information on the links attached along with the time spent in waiting at the signal.

Certain assumptions were taken into account to simplify the execution while maintaining the reliability of the problem. Initially, the lower layer statistics were generated in an arbitrarily manner to simulate the upper layer. All intersections were taken to be three or four ways where every pair of origin–destination (O–D) assigned in the networks is supposed to be connected with at least one of the intersections.

Table 1 Specification of test networks

	Number of nodes	Number of intersections
Test case 1	8	4
Test case 2	12	6
Test case 3	16	8
Test case 4	20	10
Test case 5	24	12

Fig. 4 A 12-node and 4-intersection network model showing one of the signal values

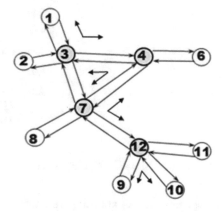

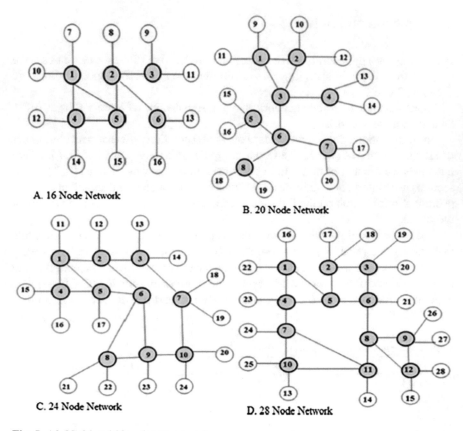

Fig. 5 16, 20, 24 and 28 node test network

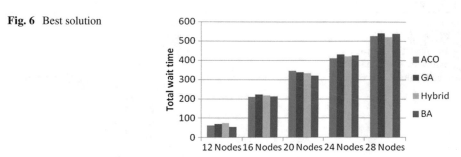

Fig. 6 Best solution

5 Results and Discussion

The results for BA are compared with ACO, GA and Hybrid ACO GA [8, 9]. Figure 6 shows the best solution for the algorithms. BA is found to give better results than other three for 12, 16 and 20 node, but for 24 and 28 nods, ACO and Hybrid algorithm outperforms BA.

Fig. 7 Worst solution

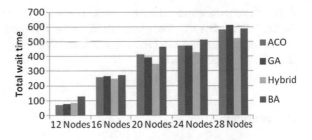

Fig. 8 Average wait time

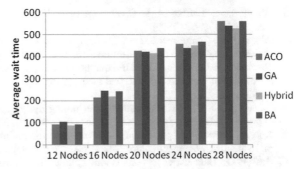

Fig. 9 Range of solution

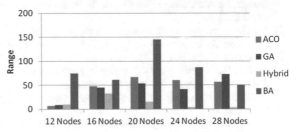

Figure 7 gives the worst performance of all the four algorithms. The hybrid algorithm achieves a better value than the rest of the algorithm. In most of the cases, BA gives the worse solution.

Figure 8 gives average solution for ACO, GA, hybrid ACO GA and BA. The hybrid ACO GA outperforms ACO, GA and BA. The average wait time for BA is comparatively high for the other techniques.

The range of all the algorithms is compared in Fig. 9. It can be observed that BA explores the highest range of solution sets.

6 Conclusion

The simulation work was carried out for various sizes of networks for multiple cases and a number of times. It can be concluded that the BA explores a wide range of solution set and gives better results than ACO, GA and Hybrid ACO GA algorithm as per the results obtained in this work. However, the hybrid ACO GA outperforms ACO, GA and BA for average solution. In the future, a lot of enhancements can be carried out on the proposed BA algorithm and that might enforce to give an excellent deal of improved solution.

References

1. Abdulaal, M., LeBlanc, L.J.: Continuous equilibrium network design models. Transporting Res. **13B**(1), 19–32 (1979)
2. Allsop, R.E., Charlesworth, J.A.: Traffic in a signal-controlled road network: an example of different signal timings including different routings. Traffic Eng. Control **18**(5), 262–264 (1977)
3. Heydecker, B.G., Khoo, T.K.: The equilibrium network design problem. In: Proceeding(s) of AIRO'90 conference on Models and methods for Decision Support, Sorrento, pp. 587–602 (1990)
4. Ceylan, H., Bell, M.G.H.: Traffic signal timing optimization based on genetic algorithm approach, including drivers' routing. Transp. Res. **38B**(4), 329–342 (2004)
5. Koh, A.: Solving transportation Bi-level problem with differential evaluation. IEEE Congress on Evolutionary Computing, pp. 2243–2250 (2007) ISBN-978-1-4244-1340-9
6. Baskan, O., Haldenbilen, S.: Ant colony optimization approach for optimizing traffic signal timings. In: Ant Colony Optimization-Methods and Application, InTech, ISBN 978-953-307-157-2-2011
7. Hu, H.: A particle swarm optimization algorithm for Bi-level programming models in Urban traffic equilibrium network design. ICCTP **2009**, 1–7 (2009)
8. Srivastava, S., Sahana, S.: Nested hybrid evolutionary model for traffic signal optimization. Applied Intelligence, pp. 1–11. Springer, Berlin (2016)
9. Srivastava, S., Sahana, S., Pant, D., Mahanti, P.: Hybrid synchronous discrete distance, time model for traffic signal optimization. J. Next Gener. Inf. Technol. **6**, 1–8 (2015)
10. Cantarella, G.E., Pavone, G., Vitetta, A.: Heuristics for urban road network design: Lane layout and signal settings'. Eur. J. Oper. Res. **175**(3), 1682–1695 (2006)
11. Cantarella, G.E., Vitetta, A.: The multi-criteria road network design problem in an Urban area. Transportation **3**(6), 567–588 (2006)
12. Yang X.S.: A new metaheuristic bat-inspired algorithm. In: Gonzalez, J.R. et al. (eds.) Nature Inspired Cooperative Strategies for Optimization (NISCO 2010) Studies in Computational Intelligence, vol. 284, pp. 65–74. Springer, Berlin (2010)
13. Yang, X.S.: Nature-Inspired Metaheuristic Algorithms, 2nd edn. Luniver Press, Frome, UK (2010)
14. Yang, X.S., Gandomi, A.H.: Bat algorithm: a novel approach for global engineering optimization. Eng. Comput. **29**(5), 464–483 (2012)
15. Kiełkowicz, K., Grela, D.: Modified Bat algorithm for nonlinear optimization. IJCSNS Int. J. Comput. Sci. Netw. Secur. **16**(10) (2016)
16. Delkhosh Abatari, H., Seydali Seyf Abad, M., Seifi, H.: Application of Bat optimization algorithm in optimal power flow. In: 24th Iranian Conference on Electrical Engineering (ICEE) 2016
17. Saji, Y., Riffi, M.E., Ahiod, B.: Discrete bat-inspired algorithm for travelling salesman problem. Complex Systems (WCCS) (2014)
18. IRC: 64 – 1990, Guidelines for Road Capacity in Rural Area, Indian Road congress (1990)

Application of Greedy and Heuristic Algorithm-Based Optimisation Methods Towards Aerodynamic Shape Optimisation

Shuvayan Brahmachary, Ganesh Natarajan, Vinayak Kulkarni, Niranjan Sahoo and Soumya Ranjan Nanda

Abstract In the present work, application of evolutionary algorithm and gradient-based optimisation techniques are extended towards obtaining minimum drag axisymmetric bodies in hypersonic flows. An attempt has been made to study the comparative performance of greedy and heuristic algorithm-based optimisation algorithm of interest with its application towards generating optimal shape configurations. We compare the performance of memetic meta-heuristic-based shuffled frog-leaping algorithm (SFLA), biological evolution-based genetic algorithm (GA), stochastic method-based simulated annealing (SA) and gradient-based steepest descent (SD) method. The suitability of each optimisation algorithm is analysed for a common test case of minimum drag axisymmetric body with the use of theoretical correlation as its flow solver. This is then followed by the implementation of a computationally expensive but accurate in-house Euler flow solver based on Immersed Boundary (IB) method, which results in a discrete solution space. This naturally results in greater computational cost per function evaluation. Results indicate that evolutionary algorithm-based optimisation technique requires much greater number of function evaluations as compared to gradient-based optimisation technique. Moreover, for a uni-modal problem considered in this work, the choice of gradient-based optimisation method proves to be quite robust and computationally efficient.

1 Introduction

Ever since the introduction of meta-heuristic and efficient stochastic algorithms, they have been subjected to many hard combinatorial optimisation problems. Some of the typical meta-heuristic-based optimisation methods include genetic algorithm (GA) [1], Ant colony optimisation (ACO) [2], particle swarm optimisation (PSO) [3], Artificial Bee Colony (ABC) [4]. While these algorithms have been able to demonstrate

S. Brahmachary (✉) · G. Natarajan · V. Kulkarni · N. Sahoo · S. R. Nanda
Department of Mechanical Engineering, Indian Institute of Technology Guwahati,
Guwahati, India
e-mail: b.shuvayan@iitg.ernet.in

© Springer Nature Singapore Pte Ltd. 2019
J. C. Bansal et al. (eds.), *Soft Computing for Problem Solving*,
Advances in Intelligent Systems and Computing 816,
https://doi.org/10.1007/978-981-13-1592-3_75

its robustness in terms of its ability to ensure feasible solution, these methods often prove its advantageous when the design space is limited with discrete solutions. In such scenario, exact methods such as branch and bound, dynamic programming are often plagued with greater computational cost. As a result, researchers have often resorted towards more robust methods like GA, PSO, ACO with the intention of obtaining feasible optimal solution for hard combinatorial problems [5].

One such active area of research is the one concerning optimal shape configuration based on some aerodynamic flow model, otherwise known as aerodynamic shape optimisation or ASO. ASO primarily aims at achieving optimal shape contour for the object under consideration, by maximising or minimising some aerodynamic parameter/quantity of interest, within the feasible design space. It typically comprises of three units, viz. a flow solver, an optimiser and a mathematical model for shape parametrisation. The conjunction of the optimiser and the flow solver primarily dictates the computational time required to arrive at a feasible solution, whereas the shape parametrisation influences the number of design variables used for the same. While one could argue that the use of greater number design/control variables leads to better shape representation, it is always advisable to use simpler models for mathematically representing the shapes, especially during the initial design phase [6].

Implementation of ASO has been significantly made for improving upon the design metrics of flight vehicles or its components. Some of the typical examples of ASO with the use of a conventional Computational Fluid Dynamics (CFD)-based flow solver are for airfoil [7, 8], wings [9, 10], aeroshell [11, 12], bullet [13, 14], scramjet/ramjet Engine [15, 16], waveriders [17], etc. However, the use of CFD-based flow solver as compared to empirical relation or theoretical correlation makes the process computationally expensive [18]. Thus, from the point of view of computationally cheap optimisation framework, it is always desired to have lesser number of function evaluations per optimisation cycles, so that the total turn-around time from initial guess to final optimal solution is within permissible limits [19]. As reported by Obayashi and Tsukahara [20], gradient-based method (GM) requires lesser number of function evaluation to reach the final solution, as compared to other optimisation techniques such as genetic algorithm (GA) and simulated annealing (SA). However, GM got stuck in a local optima, as is the case with greedy-based optimisation method, whereas the heuristic-based GA took greater number of function evaluations to reach global optimal solution. Thus, there is a compromise between feasible optimal solution and computational cost that one needs to pay while choosing between greedy method like gradient methods or heuristic or meta-heuristic optimisation methods like GA, SA. In this work, we thus make an attempt to address the issue of computational cost while ensuring that the final optimal solution is indeed feasible as well as optimal. We implement gradient-based steepest descent (SD) method, genetic algorithm (GA), simulated annealing (SA) and shuffled frog-leaping algorithm (SFLA) as optimiser and compare their merits for a common problem of obtaining minimum drag body.

2 Optimisation Framework

In this section, we describe the computational framework used in ASO, i.e. the flow solver, optimiser and shape representation (Fig. 1).

2.1 Shape Parametrisation

Shape parametrisation is the mathematical representation of the body whose configuration is to be altered iteratively. We use a simple power-law representation of the axisymmetric forebody as,

$$\frac{y}{R} = a\left(\frac{x}{L}\right)^n$$

where R and L represent the radius and length of the body, respectively, and n refers to the design variable such that $0 < n < 1$. As iterated earlier, for an initial design phase, choosing power-law over Bezier or NURBS saves significant amount of computational time by reducing the number of design variable.

2.2 Flow Solver

The present work makes use of two broad types of flow solver, i.e. a theoretical correlation based on Newtonian theory [21] as well as in-house developed continuum CFD Immersed Boundary (IB)-based Finite Volume solver [22]. Using the former,

Fig. 1 Aerodynamic shape optimisation framework

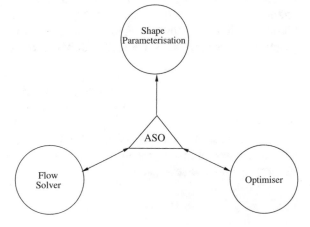

one can explicitly arrive at the following expression for objective function (i.e. minimum drag),

$$Min. \; C_d = \frac{1}{\pi R^2} \sum_{j=0}^{R} C_p \sin(tan^{-1} nax^{n-1}) \, 2\pi r_j dr_j \qquad (1)$$

$$n \in \{0, 1\}$$

$$0 < x < L; \; 0 < r < R$$

The above correlation enables quick estimation of coefficient of drag (C_D) and when coupled with an optimiser, yields near-optimal solution in quick turn-around time [18]. For accurate estimation of C_D, we implement Immersed Boundary-based FV Euler solver, which has been validated extensively for inviscid high-speed flows. Moreover, it has been shown that the use of the IB solver can be seamlessly made towards aerodynamic shape optimisation problems [22]. The FV solver makes use of van Leer scheme for second order convective flux discretisation, while using Euler explicit scheme for time discretisation. We make use of a fixed, non-uniform grid to properly resolve the leading edge of the forebody in a domain of 0.1×0.15 with 37,500 control volumes.

2.3 Optimiser

As discussed in earlier sections, we analyse the performance of two broad categories of optimisation techniques, i.e. greedy algorithm-based optimiser and heuristic-based optimiser. The flow chart depicting the algorithm followed for each of the optimisation techniques is shown in Figs. 2, 3, 4 and 5. The symbol α in Fig. 2 refers to adaptive step size, which is modified with iterations as,

$$\alpha^k = \begin{cases} A\alpha^{k-1} \; \text{if} \; \phi^k < \phi^{k-1} \\ B\alpha^{k-1} \; \text{otherwise} \end{cases}$$

where we choose $A = 1.1$ and $B = 0.5$. The tolerance limit here equals the relative difference between two consecutive function value being less than 10^{-10}. In Fig. 3, initial temperature is set equal as,

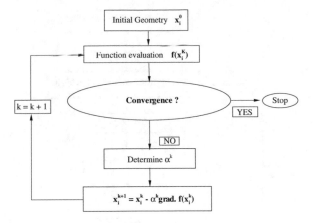

Fig. 2 Flow chart depicting steepest descent

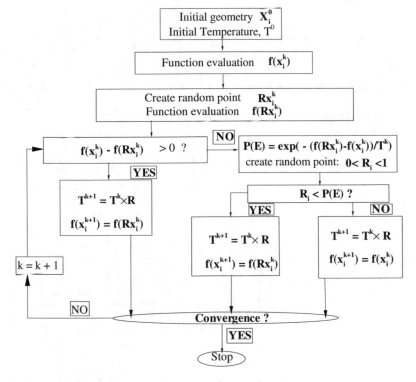

Fig. 3 Flow chart for simulated annealing

$$T^0 = \frac{\sum_{i=1}^{4} f(Rx_i)}{4}$$

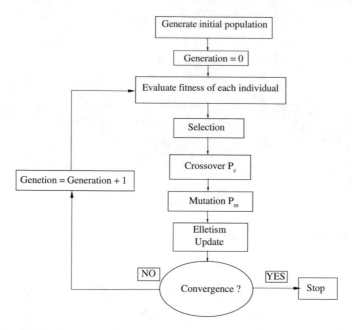

Fig. 4 Flow chart for genetic algorithm

Fig. 5 Flow chart depicting shuffled frog-leaping algorithm

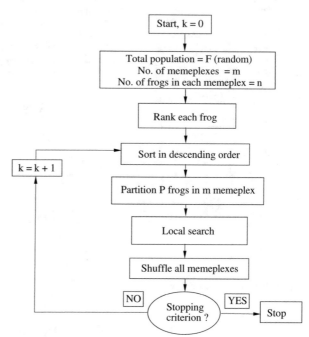

where Rx_i is the random number created for a sample size equal to 4. Also, the symbol R is equal to 0.9. In Fig. 4, the symbols P_c and P_m refer to probability of crossover and mutation, respectively. In particular, $P_c = 0.9$ whereas, $P_m = 0.1$, where the latter is kept low to create lesser randomness in the solution as they progress. These probability values are adaptively reduced by a factor equal to 0.99. Here, the population size is taken as 4, while the termination criterion is set equal to a very low value of P_m. Lastly, in Fig. 5, total number of frogs $F = 20$ is divided into five memeplexes, each containing four frogs.

3 Results and Discussion

We initiate our numerical experiments by comparing the performance of all optimisation techniques for a common problem of minimum drag body.

3.1 Test Problem 1

For this test problem, we use theoretical correlation [21] as flow solver. We impose a constraint of fixed fineness ratio $L/D = 2$, while the termination criterion for all the optimisation methods is set equal to 100, to ensure a fair comparison among them. To reflect upon the computation cost, we show the convergence of all optimisation techniques coupled with flow solver, in terms of the function evaluations. Since same flow solver is used for al the optimisation methods, the total number of function calls made to arrive at the final optimal solution reflects upon the computational cost.

Figure 6 shows the convergence obtained from all the optimisation techniques, in terms of the power-law index 'n' as the decision variable. All the optimisation methods generate an final optimal solution equal to 'n'=0.735, which agrees very well with Sahai et al. [23]. It can be seen that while shuffled frog-leaping algorithm (SFLA) converges faster, the total number of function evaluations required to reach 100 optimisation cycles is the maximum. Same observation can be seen for GA, with minor oscillations in the convergence. It must be mentioned that SFLA convergence corresponds to the best solution in the entire population for every optimisation cycle, whereas for genetic algorithm (GA), it is the best solution at the beginning of each optimisation cycle. The least amount of function evaluations are recorded by simulated annealing (SA), however, as is expected from a stochastic optimisation method, large oscillations are seen during the initial phase of the optimisation cycles. A smooth convergence history is seen for steepest descent (SD), which is typical of a gradient-based optimisation method. Moreover, the convergence can be greatly accelerated by opting for a larger value of step size α^0, as can be observed from Fig. 7. It must, however, be mentioned that for very large value of α^0, the solution diverges and leads to infeasible solutions.

Fig. 6 Convergence
obtained from all
optimisation methods

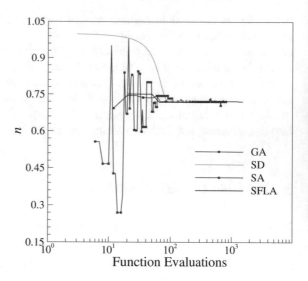

Fig. 7 Convergence
acceleration using larger step
size α^0

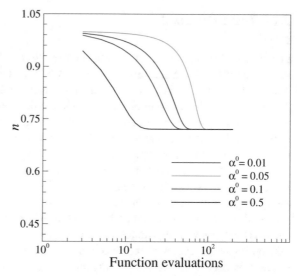

3.2   Test Problem 2

For this test case, we again use the theoretical correlation based on [21], to arrive
at the optimal solution. However, unlike the previous case, this time the termination
criterion set for each optimisation cycle is different. For SD method, the optimisation
cycles are terminated when the relative difference between two consecutive function
value is less than 10^{-10}. In the case of GA, the convergence criterion is set equal to a
low value for probability of mutation (i.e. relative difference between initial $P_{m,i}$ and
current P_m is greater than 0.99). For SA, the convergence is set to be obtained when
the cooling temperature T^k is less than 10^{-10}. Lastly, for SFLA, the optimisation is

Fig. 8 Convergence obtained from all optimisation methods

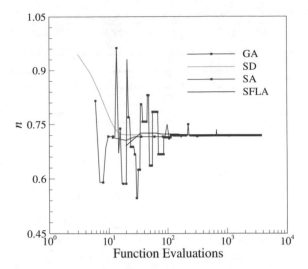

terminated when the difference between global best design variables for consecutive 30 iterations is same.

Figure 8 shows the convergence history from all the optimisation techniques. Once again it can be observed that meta-heuristic-based optimisation methods like GA and SFLA converge to near-optimal solution faster than other methods, however the do so by taking greater number of function evaluations and hence greater computational cost. SA, on the other hand, requires lesser number of function evaluations compared to GA and SFLA, but shows oscillatory nature of convergence. Lastly, a rapid convergence from greedy algorithm-based SD optimisation method is seen. It was seen that SD converged within 18 optimisation cycles, requiring just 37 function evaluations. This shows that for aerodynamic shape optimisation problems such as considered herein, the use of SD-based optimisation method can result in significant reduction in computational cost without any compromise with the nature of the final solution. This is in contrast to the observations of Obayashi [20], who reports that gradient-based optimisation method took lesser number of function calls to arrive at local optimal solution. Unlike his observations, for the present test case, we notice that all the optimisation methods generate same optimal solution, while requiring different number of function evaluations per optimisation cycles. The total computational cost in terms of number of optimisation cycles, function evaluations per optimisation cycles, etc., is shown in Table 1.

3.3 Test Problem 3

In this test case, we implement the in-house developed Finite Volume Immersed Boundary method-based Euler flow solver to obtain the optimal solution. Considering the test cases considered previously, we make use of steepest descent (SD)-based

Table 1 Computational cost for each optimisation method

Opt. method	Fun. eval. per opt. cycle	Total number of fun. eval.	Total opt. cycles
SD	2	37	18
SA	1	207	202
GA	8	3676	459
SFLA	15	605	39

Fig. 9 Convergence obtained from steepest descent method

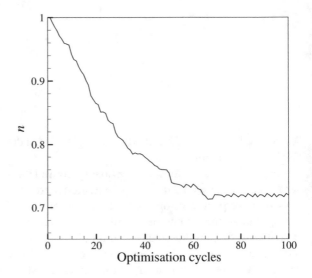

optimisation, as it is associated with the least computational cost. Initial guess corresponding to a right circular cone is made (i.e. 'n' = 1). It can be seen from Fig. 9 that a similar super-linear convergence, typical of a gradient-based optimisation method, is obtained. The final optimal corresponding to minimum drag axisymmetric body for $L/D = 2$ is shown in Fig. 10, which roughly corresponds to 'n' = 0.7 and agrees well with [22].

4 Conclusions

In the present work, we perform aerodynamic shape optimisation for obtaining minimum drag axisymmetric configurations using different types of optimisation algorithms. In particular, we use greedy algorithm-based steepest descent method and compare its performance against heuristic-based shuffled frog-leaping algorithm, genetic algorithm and simulated annealing optimisation methods. Three test cases are performed; it was observed that meta-heuristic-based optimisation methods like

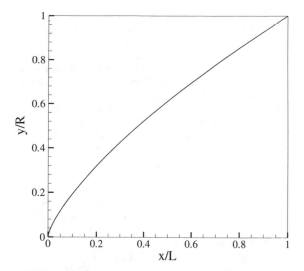

Fig. 10 Final optimal body obtained using IB solver in conjunction with steepest descent optimisation method

genetic algorithm and shuffled frog-leaping algorithm-based optimisation converges rapidly to final optimal solution, however they do so requiring greater number of function evaluations. Stochastic algorithm-based simulated annealing requires less number of function evaluations as compared to GA and SFLA; however, it has oscillatory nature of convergence. Lastly, we find that gradient-based steepest descent method serves as a robust method which can yield feasible optimal solution with lesser turn-around time from initial guess to final optimal solution. This is particularly important as the use of CFD-based flow solver in conjunction with an optimiser results in significant amount of computational time.

References

1. Deb, K., Agrawal, S., Pratap, A., et al.: A fast elitist non-dominated sorting genetic algorithm for multi-objective optimization: NSGA-I. In: Schoenauer, M., et al. (eds.) Parallel Problem Solving from Nature PPSN VI, pp. 849–858. Springer, Berlin (2000)
2. Dorigo, M., Maniezzo, V., Colorni, A.: Ant system: optimization by a colony of cooperating agents. IEEE Trans. Syst. Man Cybern. Part B **26**, 29–41 (1996)
3. Kennedy, J., Eberhart, R.C.: Particle swarm optimization. In: ROC IEEE International Conference on Neural Networks (Path. Australia), pp. 1942–1948. IEEE Service Center, Piscataway, NJ (1995)
4. Karaboga, D., Basturk, B.: A powerful and efficient algorithm for numerical function optimization: artificial bee colony (ABC) algorithm. J. Glob. Opt. **39**, 459–471 (2007)
5. Burke, E.K., Gendreau, M., Hyde, M., et al.: Hyper-heuristics: a survey of the state of the art. J. Oper. Res. Soc. **64**, 1695–1724 (2013)
6. Anderson, G.R., Aftosmis, M.J.: Adaptive shape parameterization for aerodynamic design. NAS Technical Report, NAS-2015-02 (2015)
7. Barrett, T.R., Bressloff, N.W., Keane, A.J.: Airfoil shape design and optimisation using multifidelity analysis and embedded inverse design. AIAA J. **44**, 2051–2060 (2006)

8. Vecchia, P.D., Daniele, E., D'Amato, E.: An airfoil shape optimization technique coupling PARSEC parameterization and evolutionary algorithm. Aerosp. Sci. Technol. **32**, 103–110 (2014)
9. Epstein, B., Peigin, S., Tsach, S.: A new efficient technology of aerodynamic design based on CFD driven optimization. Aerosp. Sci. Technol. **10**, 100–110 (2006)
10. Obayashi, S., Sasaki, D., Takeguchi, Y., et al.: Multiobjective evolutionary computation for supersonic wing-shape optimization. IEEE Trans. Evol. Comput. **4**, 182–187 (2000)
11. Theisinger, J.E., Braun, R.D.: Multi-objective hypersonic entry aeroshell shape optimization. J. Spacecraft Rockets **46**, 957–966 (2009)
12. Bopp, M.S., Ruffin, S.M., Braun, R.D., et al.: Multi-fidelity approach to estimate heating for three-dimensional hypersonic aeroshells. J. Spacecraft Rockets **50**, 754–762 (2013)
13. Ben-Dor, G., Dubinsky, A., Elperin, T.: Shape optimization of penetrator nose. Theo. Appl. Fract. Mech. **35**, 261–270 (2001)
14. Schinetsky, P.A., Brooker, B.T., Treadway, A., et al.: Numerical and experimental analysis of projectile nose geometry. J. Spacecraft Rockets **52**, 1515–1519 (2015)
15. Smart, M.K.: Optimization of two-dimensional scramjet inlets. J. Aircraft **36**, 430–433 (1999)
16. Raj, N.O.P., Venkatasubbaiah, K.: A new approach for the design of hypersonic scramjet inlets. Phys. Fluid **24**, 1–15 (2012)
17. Mangin, B., Chpoun, A.: Optimization of viscous waveriders derived from axisymmetric power-law blunt body flows. J. Spacecraft Rockets **43**, 990–998 (2006)
18. Brahmachary, S., Natarajan, G., Sahoo, N.: On maximum ballistic coefficient axisymmetric geometries in hypersonic flows. J. Spacecraft Rockets (2017). https://doi.org/10.2514/1.A33887
19. Thevenin, D., Janiga, G.: Optimization and Computational Fluid Dynamics, 1st edn. Springer, Berlin, Heidelberg (2008)
20. Obayashi, S., Tsukahara, T.: Comparison of optimization algorithms for aerodynamic shape design. AIAA J. **35**, 1413–1415 (1997)
21. Anderson, J.D.: Modern Compressible Flow with Historical Perspective, 1st edn. McgrawHill Publishing Company, New York (1990)
22. Brahmachary, S., Natarajan, G., Kulkarni, V., Sahoo, N.: A sharp interface immersed boundary framework for simulations of high speed inviscid compressible flows. Int. J. Numer. Methods Fluids (2017). https://doi.org/10.1002/fld.4479
23. Sahai, A., John, B., Natarajan, G.: Effect of fineness ratio on minimum-drag shapes in hypersonic flows. J. Spacecraft Rockets **51**, 900–907 (2014)

Performance Evaluation of Runner–Root Algorithm on CEC 2013 Benchmark Functions

A. J. Umbarkar, A. C. Adamuthe and S. M. Nale

Abstract From past four decades, many heuristic algorithms are proposed by researchers for solving complex engineering problems. The main source of inspiration of these algorithms is the intelligence present in different parts of nature. These algorithms are found to be better than traditional optimization algorithms for combinatorial optimization problems. F. Merrikh-Bayat proposed a new heuristic algorithm named runner–root, which is inspired by the function of runners and roots of strawberry and spider plants. F. Merrikh-Bayat had tested performance of the algorithm on standard CEC 2005 benchmark problems. Objective of this paper is to test performance of runner–root algorithm on benchmark problems reported in Congress on Evolutionary Computation (CEC-2013) Technical Report. Results show that RRA gives optimal values for all the test functions for dimension two. RRA performance is satisfactory on dimension five except composite functions. For dimension 10 and 30, RRA is better than GA and CMA-ES. But, RRA is suffering from the problem of curse of dimensionality. Its performance is degraded for dimensions 50 and 100.

Keywords Runner–root algorithm · Function optimization · CEC 2013
Nature-inspired algorithm · Evolutionary algorithm

A. J. Umbarkar (✉) · S. M. Nale
Department of Information Technology, Walchand College of Engineering,
Sangli, MS, India
e-mail: anantumbarkar@rediffmail.com

S. M. Nale
e-mail: snehalnale2506@gmail.com

A. C. Adamuthe
Department of Information Technology, Rajarambapu Institute of Technology,
Rajaramnagar, MS, India
e-mail: amol.admuthe@gmail.com

© Springer Nature Singapore Pte Ltd. 2019
J. C. Bansal et al. (eds.), *Soft Computing for Problem Solving*,
Advances in Intelligent Systems and Computing 816,
https://doi.org/10.1007/978-981-13-1592-3_76

1 Introduction

Evolutionary algorithms (EAs) are widely used for solving unconstrained and constrained optimization problems. Nature inherently has the optimization processes everywhere and in all live creatures. In the last two decades, researchers developed and investigated a wide range of algorithms based on intelligence in nature. Genetic algorithms, particle swarm optimization, harmony search, ant colony optimization, artificial bee colony, firefly algorithm, differential evolution are few popular algorithms [1, 2]. Nature-inspired heuristic algorithms are based on swarm intelligence, biological systems, physical, and chemical systems. Swarm intelligence-based algorithms such as cuckoo search and firefly algorithms are found to be very efficient [3]. Nature-inspired algorithms are very successful for solving optimization problems with good solutions in finite time [3].

Runner–root algorithm (RRA) is inspired by the function of runners and roots of some plants in nature such as strawberry and spider plants. F. Merrikh-Bayat proposed this algorithm for both unimodal and multimodal optimization problems. RRA explores the search space in all iterations, and exploitation is performed when exploration does not lead to a considerable improvement in the objective value.

According to no free lunch (NFL) theorem, there is no best algorithm. Hence, there is need to investigate the performance of any new heuristic with different benchmark problems. In this paper, we tested the performance of RRA on benchmark functions designed for CEC-2013 [4]. It comprises 28 minimization functions representing different types of response surface (e.g., unimodal, multimodal, composition functions, separable, non-separable, shifted, and rotated).

Section 2 gives brief information about runner–root algorithm. Section 3 describes datasets, experiential details, results, and discussion. Section 4 is about conclusions.

2 Runner–Root Algorithm

RRA is influenced by search methodology of plants like strawberry and spider plant for water resources and minerals. The runner is a creeping stalk produced in the leaf axils and grows out from the mother (parent) plant. The second node of runner a new plant is called daughter plant. A new runner arises on the daughter plant to generate another new daughter plant [5].

Randomly generate fixed number of points in the domain of problem is called as mother plant. Every iteration mother plant generates one root and one runner (daughter plant). Root is in its vicinity, and runner is relatively farther location. Computational agents, namely runners and roots, move with large and small random steps, respectively. Objective function is evaluated at the points referred to by runners and roots. Half of these points are selected by selection mechanism as mother plants for next iteration [6]. The selection mechanism is based on elite selection and roulette-wheel method.

Fig. 1 Flowchart of RRA
(Taken from [6])

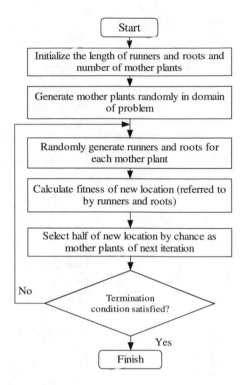

RRA is a population-based algorithm. It is not parameter-free algorithm. Parameters of RRA are population size, number of iterations, number of runners to create for each solution, and the distance for each runner.

Strawberry and spider plant with runners performs optimization with both global and local search. Exploration is global search with random large steps is performed at all iterations. Local search with random small steps (exploitation) is performed only if global search fails. The RRA is equipped with re-initialization strategy if eventually the algorithm traps in a local optimum point (Fig. 1).

3 Experimental Details, Results, and Discussion

In this section, the RRA algorithm's performance and ability to deal with unimodal, multimodal, and composite functions are evaluated using CEC 2013 benchmarks functions [4].

Table 1 gives the list of CEC problem set with optimal value [4]. The dataset consists of five unimodal, fifteen multimodal, and eight composite functions. All functions are single objective with search space ranging from 100 to −100.

Table 1 CEC 2013 problem set

Short name	Name of function	Optimum value (I)
Unimodal functions		
F01	Sphere function	−1400
F02	Rotated high conditioned elliptic function	−1300
F03	Rotated Bent Cigar function	−1200
F04	Rotated discus function	−1100
F05	Different power function	−1000
Basic multimodal functions		
F06	Rotated Rosenbrock's function	−900
F07	Rotated Schaffers F7 function	−800
F08	Rotated Ackley's function	−700
F09	Rotated Weierstrass function	−600
F10	Rotated Griewank's function	−500
F11	Rastrign's function	−400
F12	Rotated Rastrign's function	−300
F13	Non-continuous rotated Rastrign's function	−200
F14	Schwefel's function	−100
F15	Rotated Schwefel's function	100
F16	Rotated Katsuura function	200
F17	Lunacek Bi_Rastrigin function	300
F18	Rotated Lunacek Bi_Rastrigin function	400
F19	Expanded Griewank's plus Rosenbrock's function	500
F20	Expanded Scaffer's F6 function	600
Composition function		
F21	Composition function 1	700
F22	Composition function 2	800
F23	Composition function 3	900
F24	Composition function 4	1000
F25	Composition function 5	1100
F26	Composition function 6	1200
F27	Composition function 7	1300
F28	Composition function 8	1400

RRA implemented in Java and executed on personal computer with a 3.0 GHz QuadCore-Processor, 4 GB RAM, and Windows XP operating system. The algorithm is executed 10 times for each test function. To test the performance of RRA, best value (B), mean (M), median (Med), worst (W), and standard deviation (SD) measures are used. To test the performance of algorithm on different dimensions, experiments are carried out for 2, 10, 30, 50, and 100 design variables. Population size taken in experiments is 1000, and the number of iteration is 100. The algorithm-specific parameter, namely number of roots and number of runners, is set to 1000 and 100, respectively.

Tables 2 and 3 shows the results for RRA with dimensions 2 and 5, respectively. RRA gives optimal values for all the test functions for dimension 2. RRA performance is satisfactory on dimension 5 except composite functions.

Table 2 Experimentation of RRA with dimension $(D) = 2$

F	I	B	M	W	SD
1	−1.40E+03	−1.40E+03	−1.40E+03	−1.40E+03	8.93E−02
2	−1.30E+03	−1.13E+03	−1.23E+03	−1.22E+03	1.83E+00
3	−1.20E+03	−1.12E+03	−1.12E+03	−1.12E+03	3.27E+00
4	−1.10E+03	−1.02E+03	−1.03E+03	−1.03E+03	1.05E+00
5	−1.00E+03	−9.99E+02	−1.00E+03	−1.00E+03	3.50E−01
6	−9.00E+02	−8.99E+02	−9.00E+02	−9.00E+02	2.51E+00
7	−8.00E+02	−7.99E+02	−7.99E+02	−8.00E+02	1.52E+02
8	−7.00E+02	−6.99E+02	−7.00E+02	−7.00E+02	5.40E−01
9	−6.00E+02	−5.99E+02	−6.00E+02	−6.00E+02	1.04E+00
10	−5.00E+02	−4.99E+02	−5.00E+02	−5.00E+02	1.02E+00
11	−4.00E+02	−3.99E+02	−4.00E+02	−4.00E+02	1.20E+00
12	−3.00E+02	−2.99E+02	−3.00E+02	−3.00E+02	1.54E+00
13	−2.00E+02	−1.99E+02	−2.00E+02	−2.00E+02	1.02E+33
14	−1.00E+02	−9.95E+01	−9.98E+01	−9.99E+01	1.00E+01
15	1.00E+02	1.01E+02	1.02E+02	1.02E+02	1.34E+04
16	2.00E+02	2.00E+02	2.00E+02	2.00E+02	2.12E+00
17	3.00E+02	3.00E+02	3.00E+02	3.00E+02	1.01E+01
18	4.00E+02	4.00E+02	4.01E+02	4.01E+02	1.03E+00
19	5.00E+02	5.00E+02	5.00E+02	5.00E+02	0.00E+00
20	6.00E+02	6.00E+02	6.00E+02	6.00E+02	1.39E+00
21	7.00E+02	7.04E+02	7.04E+02	7.05E+02	0.00E+00
22	8.00E+02	8.01E+02	8.01E+02	8.01E+02	0.00E+00
23	9.00E+02	9.01E+02	9.02E+02	9.02E+02	0.00E+00
24	1.00E+03	1.00E+03	1.00E+03	1.00E+03	0.00E+00
25	1.10E+03	1.10E+03	1.10E+03	1.10E+03	0.00E+00
26	1.20E+03	1.20E+03	1.20E+03	1.20E+03	0.00E+00
27	1.30E+03	1.32E+03	1.32E+03	1.32E+03	0.00E+00
28	1.40E+03	1.40E+03	1.40E+03	1.40E+03	0.00E+00

Table 3 Experimentation of RRA with dimension $(D) = 5$

F	I	B	M	W	SD
1	−1.40E+03	−1.40E+03	−1.40E+03	−1.40E+03	0.00E+00
2	−1.30E+03	−1.30E+03	−1.30E+03	−1.30E+03	1.95E+00
3	−1.20E+03	−1.20E+03	−1.20E+03	−1.12E+03	2.01E+00
4	−1.10E+03	−1.09E+03	−1.09E+03	−1.09E+03	8.90E−01
5	−1.00E+03	−9.97E+02	−9.97E+02	−9.98E+02	0.00E+00
6	−9.00E+02	−8.99E+02	−9.00E+02	−9.00E+02	0.00E+00
7	−8.00E+02	−7.98E+02	−7.99E+02	−7.99E+02	8.99E+00
8	−7.00E+02	−6.80E+02	−6.81E+02	−6.81E+02	2.12E+00
9	−6.00E+02	−5.99E+02	−6.00E+02	−6.00E+02	2.02E+00
10	−5.00E+02	−4.99E+02	−4.99E+02	−4.99E+02	3.25E−01
11	−4.00E+02	−3.96E+02	−3.97E+02	−3.97E+02	3.12E−01
12	−3.00E+02	−2.98E+02	−2.99E+02	−2.99E+02	9.36E−01
13	−2.00E+02	−1.93E+02	−1.93E+02	−1.94E+02	1.61E+00
14	−1.00E+02	−9.91E+01	−9.95E+01	−1.00E+02	1.21E+00
15	1.00E+02	1.63E+02	1.64E+02	1.65E+02	9.99E+00
16	2.00E+02	2.00E+02	2.01E+02	2.02E+02	7.21E+00
17	3.00E+02	3.06E+02	3.07E+02	3.07E+02	4.15E−01
18	4.00E+02	4.06E+02	4.06E+02	4.06E+02	1.34E+00
19	5.00E+02	5.00E+02	5.01E+02	5.01E+02	2.03E+00
20	6.00E+02	6.01E+02	6.01E+02	6.01E+02	2.61E+00
21	7.00E+02	8.24E+02	8.24E+02	8.24E+02	1.78E+00
22	8.00E+02	9.61E+02	9.61E+02	9.61E+02	2.02E+00
23	9.00E+02	1.25E+03	1.24E+03	1.25E+03	1.05E+00
24	1.00E+03	1.11E+03	1.11E+03	1.11E+03	7.89E+00
25	1.10E+03	1.21E+03	1.21E+03	1.21E+03	1.21E+00
26	1.20E+03	1.30E+03	1.30E+03	1.30E+03	2.14E+00
27	1.30E+03	1.61E+03	1.61E+03	1.61E+03	1.10E+01
28	1.40E+03	1.51E+03	1.51E+03	1.51E+03	1.25E+00

Performance of RRA is compared with standard particle swarm optimization (PSO) [7], genetic algorithms (GA) [8], and BIPOP-based covariance matrix adaptation evolution strategy (BIPOP-CMA-ES) [8]. Tables 4 and 5 shows mean, median, and SD for $D = 10$ and $D = 30$, respectively. Table 4 shows that the performance of PSO and RRA is better than GA and BIPOP-CMA-ES. All the functions except functions 2 and 3 are not performing in both PSO and RRA. Table 5 shows that PSO is best performing with respect to other algorithms.

Table 4 Comparison between RRA, PSO [7], GA [8], and BIPOP-CMA-ES [9] for $D = 10$

F	PSO [7] M	Med	SD	GA [8] M	Med	SD	BIPOP-CMA-ES [9] M	Med	SD	RRA M	Med	SD
1	–	-1.400E+03	0.0000E+00	0.0000E+00	0.0000E+00	0.0000E+00	0.00E+00	0.00E+00	0.00E+00	-1.37E+03	–	3.18E-12
2	–	3.504E+04	7.356E+04	0.0000E+00	0.0000E+00	0.0000E+00	0.00E+00	0.00E+00	0.00E+00	4.74E+05	–	7.96E+03
3	–	2.670E+05	1.656E+07	0.0000E+00	0.0000E+00	0.0000E+00	0.00E+00	0.00E+00	0.00E+00	4.76E+07	–	3.56E+03
4	–	7.769E+03	4.556E+03	0.0000E+00	0.0000E+00	0.0000E+00	0.00E+00	0.00E+00	0.00E+00	1.04E+04	–	1.20E+02
5	–	-1.000E+03	3.142E-05	0.0000E+00	0.0000E+00	0.0000E+00	0.00E+00	0.00E+00	0.00E+00	-9.78E+02	–	0.00E+00
6	–	-8.902E+02	4.974E+00	0.0000E+00	0.0000E+00	0.0000E+00	0.00E+00	0.00E+00	0.00E+00	-8.94E+02	–	2.30E+01
7	–	-7.789E+02	1.327E+01	4.4413E-02	1.4248E-03	2.1039E-01	0.00E+00	0.00E+00	0.00E+00	-7.83E+02	–	5.61E+00
8	–	-6.797E+02	6.722E-02	2.0423E+01	2.0416E+01	8.6351E-02	2.03E+01	2.04E+01	8.20E-02	-6.80E+02	–	7.68E+00
9	–	-5.952E+02	1.499E+00	3.4278E+00	2.6034E+00	2.8956E+00	5.54E-01	1.04E-01	6.84E-01	-5.97E+02	–	1.52E+00
10	–	-4.997E+02	2.713E-01	4.0285E-02	3.6914E-02	2.8188E-02	0.00E+00	0.00E+00	0.00E+00	-4.93E+02	–	5.78E-02
11	–	-3.891E+02	5.658E+00	2.7313E-01	0.0000E+00	4.9060E-01	9.36E-01	9.95E-01	8.06E-01	-3.72E+02	–	5.35E+00
12	–	-2.861E+02	6.560E+00	6.3599E+00	5.9698E+00	2.2869E+00	5.85E-01	9.95E-01	6.03E-01	-2.71E+02	–	2.87E+00
13	–	-1.792E+02	9.822E+00	1.0131E+01	8.5235E+00	6.2965E+00	9.44E-01	9.95E-01	8.38E-01	-1.74E+02	–	1.08E+01
14	–	7.338E+02	2.335E+02	2.7406E+01	1.8597E+01	2.7503E+01	5.48E+01	3.35E+01	6.98E+01	8.40E+02	–	5.87E+01
15	–	8.743E+02	2.507E+02	8.3132E+02	8.5097E+02	2.5758E+02	5.28E+01	4.00E+01	5.78E+01	8.32E+02	–	2.64E+02
16	–	2.005E+02	2.457E-01	1.2820E+00	1.3367E+00	3.2578E-01	3.05E-01	9.00E-02	4.57E-01	2.01E+02	–	1.46E+00
17	–	3.189E+02	5.873E+00	1.1233E+01	1.1072E+01	7.7321E-01	1.16E+01	1.15E+01	5.59E-01	3.44E+02	–	5.38E+00
18	–	4.178E+02	4.534E+00	1.8553E+01	1.7396E+01	5.2164E+01	1.16E+01	1.17E+01	1.16E+00	4.39E+02	–	5.21E+00
19	–	5.009E+02	3.886E-01	5.3245E-01	5.1083E-01	1.4835E-01	6.04E-01	5.91E-01	1.20E-01	5.03E+02	–	3.66E+01
20	–	6.034E+02	4.194E-01	3.2119E+00	3.2079E+00	5.0495E-01	2.58E+00	2.62E+00	5.52E-01	6.03E+02	–	3.53E+00
21	–	1.100E+03	0.000E+00	2.5020E+02	3.0000E+02	5.0020E+01	2.84E+02	3.00E+02	1.22E+02	1.10E+03	–	0.00E+00
22	–	1.706E+03	3.431E+02	9.3574E+01	7.5099E+01	6.2780E+01	9.93E+01	7.82E+01	6.56E+01	1.64E+03	–	7.86E+01
23	–	1.810E+03	3.596E+02	9.3213E+02	9.0944E+02	3.3198E+02	1.17E+02	1.11E+02	6.51E+01	1.78E+03	–	3.67E+02
24	–	1.214E+03	9.166E+00	2.1487E+02	2.1357E+02	6.7898E+00	1.30E+02	1.11E+02	3.87E+01	1.21E+03	–	5.66E+01
25	–	1.309E+03	5.943E+00	2.1747E+02	2.1858E+02	6.7062E+00	1.93E+02	2.02E+02	2.84E+01	1.31E+03	–	6.27E+00
26	–	1.400E+03	5.513E+01	1.9641E+02	2.0002E+02	1.8056E+02	1.19E+02	1.08E+02	2.49E+02	1.40E+03	–	2.90E+01
27	–	1.636E+03	7.359E+01	4.3536E+02	4.1524E+02	7.4197E+01	3.46E+02	3.54E+02	4.40E+01	1.72E+03	–	1.68E+02
28	–	1.700E+03	8.362E+01	2.9216E+02	3.0000E+02	3.9208E+02	2.80E+02	3.00E+02	6.01E+01	1.56E+03	–	1.98E+02

Table 5 Comparison between RRA, PSO [7], GA [8], and BIPOP-CMA-ES [9] for $D = 30$

F	PSO [7]			GA [8]			BIPOP-CMA-ES [9]			RRA		
	M	Med	SD	M	Med	SD	M	Med	SD	M	Med	SD
1	–	−1.400E+03	1.875E−13	0.0000E+00	0.0000E+00	0.0000E+00	0.00E+00	0.00E+00	0.00E+00	2.02E+03	–	0.00E+00
2	–	3.075E+05	1.667E+05	1.5481E+05	1.1017E+05	1.3704E+05	0.00E+00	0.00E+00	0.00E+00	3.56E+07	–	1.28E+00
3	–	1.188E+08	5.243E+08	3.2787E+07	8.4130E+06	7.5503E+07	8.20E−02	0.00E+00	5.09E−01	1.30E+09	–	6.80E+07
4	–	3.804E+04	6.702E+03	9.0770E−01	2.7695E−01	1.2588E+00	0.00E+00	0.00E+00	0.00E+00	7.50E+04	–	1.64E+02
5	–	−1.000E+03	4.909E−05	0.0000E+00	0.0000E+00	0.0000E+00	0.00E+00	0.00E+00	0.00E+00	1.26E+02	–	0.00E+00
6	–	−8.717E+02	2.825E+01	2.0420E+01	1.9697E+01	7.9244E+00	0.00E+00	0.00E+00	0.00E+00	−5.99E+02	–	2.89E+03
7	–	−7.131E+02	2.107E+01	4.5781E+01	4.0441E+01	2.9705E+01	9.43E+00	1.16E+00	1.33E+01	−6.32E+02	–	2.04E+01
8	–	−6.791E+02	5.893E−02	2.1017E+01	2.1022E+01	5.3401E−02	2.09E+01	2.09E+01	5.10E−02	−6.79E+02	–	1.98E+01
9	–	−5.716E+02	4.426E+00	3.7016E+01	4.0294E+01	6.4427E+00	6.49E+00	6.42E+00	2.38E+00	−5.78E+02	–	2.50E+00
10	–	−4.997E+02	1.478E−01	8.3530E−02	7.3936E−02	4.6649E−02	0.00E+00	0.00E+00	0.00E+00	−4.05E+02	–	1.02E+01
11	–	−2.916E+02	2.740E+01	2.1337E+01	1.9899E+01	1.0698E+01	3.08E+00	2.99E+00	1.52E+00	−2.36E+02	–	1.73E+01
12	–	−2.055E+02	3.539E+01	3.7740E+01	3.6813E+01	9.5489E+00	2.41E+00	1.99E+00	1.43E+00	−1.56E+02	–	8.90E+00
13	–	−2.322E+00	3.862E+01	8.0977E+01	8.0217E+01	1.9538E+01	2.39E+00	1.99E+00	1.47E+00	−3.64E+01	–	2.01E+01
14	–	3.923E+03	6.194E+02	1.0096E+03	9.6605E+02	4.7359E+02	6.69E+02	5.14E+02	6.98E+02	4.81E+03	–	5.21E+02
15	–	3.904E+03	6.938E+02	4.0970E+03	4.1112E+03	6.9288E+02	6.10E+02	4.93E+02	4.51E+02	5.35E+03	–	1.22E+03
16	–	2.014E+02	3.588E−01	2.7216E+00	2.8320E+00	5.0533E−01	7.75E−01	4.20E−02	1.14E+00	2.05E+02	–	3.15E+01
17	–	4.152E+02	2.018E+01	6.0073E+01	5.7819E+01	1.1028E+01	3.63E+01	3.60E+01	1.77E+00	5.40E+02	–	4.12E+01
18	–	5.168E+02	2.460E+01	7.4491E+01	7.2342E+01	1.7987E+01	5.44E+01	4.14E+01	3.40E+01	6.39E+02	–	4.56E+01
19	–	5.090E+02	4.418E+00	4.1383E+00	3.6472E+00	1.9876E+00	2.40E+00	2.50E+00	4.18E−01	5.18E+02	–	8.15E+01
20	–	6.140E+02	1.109E+00	1.3690E+01	1.3885E+01	4.7759E−01	1.42E+01	1.43E+01	6.36E−01	6.12E+02	–	4.05E+01
21	–	1.000E+03	6.796E+01	2.7960E+02	3.0000E+02	7.8956E+01	2.00E+02	2.00E+02	2.83E+01	1.26E+03	–	4.12E+01
22	–	5.151E+03	7.670E+02	1.1460E+03	1.1253E+03	3.8599E+02	8.39E+02	7.06E+02	5.77E+02	5.86E+03	–	3.90E+03
23	–	5.663E+03	8.227E+02	4.2831E+03	4.3806E+03	6.7183E+02	7.17E+02	6.64E+02	4.52E+02	5.97E+03	–	7.89E+02
24	–	1.264E+03	1.246E+01	2.7876E+02	2.7688E+02	1.4120E+01	1.80E+02	1.62E+02	5.02E+01	1.38E+03	–	3.89E+01
25	–	1.400E+03	1.045E+01	2.9962E+02	3.0299E+02	8.9007E+00	2.31E+02	2.25E+02	2.12E+01	1.45E+03	–	2.89E+01
26	–	1.540E+03	8.240E+01	3.2065E+02	3.5035E+02	6.4888E+01	1.64E+02	1.49E+02	3.23E+01	1.50E+03	–	1.97E+02
27	–	2.326E+03	1.119E+02	1.0803E+03	1.1041E+03	1.4696E+02	5.04E+02	5.13E+02	7.13E+01	2.69E+03	–	4.83E+02
28	–	1.700E+03	4.761E+02	3.0000E+02	3.0000E+02	0.0000E+00	2.92E+02	3.00E+02	3.92E+01	3.43E+03	–	6.74E+01

Table 6 Experimentation of RRA with dimension $(D) = 50$

F	I	B	M	W	SD
1	−1.40E+03	1.06E+03	1.02E+03	1.39E+03	6.75E−02
2	−1.30E+03	3.75E+07	3.92E+07	4.22E+07	4.93E+05
3	−1.20E+03	5.28E+09	5.41E+09	6.65E+09	1.00E+00
4	−1.10E+03	8.84E+04	9.04E+04	1.16E+05	1.15E+03
5	−1.00E+03	−3.99E+02	−3.16E+02	−2.75E+02	5.10E−02
6	−9.00E+02	−7.61E+02	−7.33E+02	−7.23E+02	2.53E+01
7	−8.00E+02	−7.52E+02	−7.50E+02	−7.47E+02	5.23E+01
8	−7.00E+02	−6.80E+02	−6.79E+02	−6.78E+02	1.69E+01
9	−6.00E+02	−5.45E+02	−5.40E+02	−5.25E+02	1.03E+02
10	−5.00E+02	−1.54E+02	−1.13E+02	−1.12E+02	2.28E+01
11	−4.00E+02	−6.91E+01	−5.09E+01	−3.57E+01	5.17E+01
12	−3.00E+02	6.91E+01	2.44E+02	4.02E+02	2.42E+01
13	−2.00E+02	1.68E+02	1.64E+02	1.71E+02	6.31E+01
14	−1.00E+02	1.09E+04	1.10E+04	1.11E+04	1.85E+03
15	1.00E+02	1.17E+04	1.20E+04	1.26E+04	2.06E+04
16	2.00E+02	2.03E+02	2.04E+02	2.40E+02	4.46E−02
17	3.00E+02	8.16E+02	8.18E+02	9.40E+02	1.41E+02
18	4.00E+02	8.97E+02	9.01E+02	9.15E+02	9.91E+01
19	5.00E+02	5.51E+02	5.54E+02	5.59E+02	7.47E+01
20	6.00E+02	6.22E+02	6.22E+02	6.22E+02	4.15E+00
21	7.00E+02	1.55E+03	1.53E+03	1.56E+03	4.22E+01
22	8.00E+02	1.22E+04	1.24E+04	1.25E+04	6.81E+02
23	9.00E+02	1.21E+04	1.24E+04	1.25E+04	2.46E+04
24	1.00E+03	1.53E+03	1.62E+03	1.71E+03	1.25E+02
25	1.10E+03	1.65E+03	1.66E+03	1.67E+03	3.67E+01
26	1.20E+03	1.70E+03	1.70E+03	1.64E+03	1.60E+02
27	1.30E+03	3.82E+03	3.87E+03	4.13E+03	2.55E+02
28	1.40E+03	1.29E+04	1.30E+04	1.40E+04	1.71E+03

Tables 6 and 7 show the results for dimension 50 and 100, respectively. Performance of RRA is reduced with respect increase in dimensions is degraded.

Table 7 Experimentation of RRA with dimension $(D) = 100$

F	I	B	M	W	SD
1	−1.40E+03	1.06E+03	1.02E+03	1.39E+03	6.75E−02
2	−1.30E+03	3.75E+07	3.92E+07	4.22E+07	4.93E+05
3	−1.20E+03	5.28E+09	5.41E+09	6.65E+09	1.00E+00
4	−1.10E+03	8.84E+04	9.04E+04	1.16E+05	1.15E+03
5	−1.00E+03	−3.99E+02	−3.16E+02	−2.75E+02	5.10E−02
6	−9.00E+02	−7.61E+02	−7.33E+02	−7.23E+02	2.53E+01
7	−8.00E+02	−7.52E+02	−7.50E+02	−7.47E+02	5.23E+01
8	−7.00E+02	−6.80E+02	−6.79E+02	−6.78E+02	1.69E+01
9	−6.00E+02	−5.45E+02	−5.40E+02	−5.25E+02	1.03E+02
10	−5.00E+02	−1.54E+02	−1.13E+02	−1.12E+02	2.28E+01
11	−4.00E+02	−6.91E+01	−5.09E+01	−3.57E+01	5.17E+01
12	−3.00E+02	6.91E+01	2.44E+02	4.02E+02	2.42E+01
13	−2.00E+02	1.68E+02	1.64E+02	1.71E+02	6.31E+01
14	−1.00E+02	1.09E+04	1.10E+04	1.11E+04	1.85E+03
15	1.00E+02	1.17E+04	1.20E+04	1.26E+04	2.06E+04
16	2.00E+02	2.03E+02	2.04E+02	2.40E+02	4.46E−02
17	3.00E+02	8.16E+02	8.18E+02	9.40E+02	1.41E+02
18	4.00E+02	8.97E+02	9.01E+02	9.15E+02	9.91E+01
19	5.00E+02	5.51E+02	5.54E+02	5.59E+02	7.47E+01
20	6.00E+02	6.22E+02	6.22E+02	6.22E+02	4.15E+00
21	7.00E+02	1.55E+03	1.53E+03	1.56E+03	4.22E+01
22	8.00E+02	1.22E+04	1.24E+04	1.25E+04	6.81E+02
23	9.00E+02	1.21E+04	1.24E+04	1.25E+04	2.46E+04
24	1.00E+03	1.53E+03	1.62E+03	1.71E+03	1.25E+02
25	1.10E+03	1.65E+03	1.66E+03	1.67E+03	3.67E+01
26	1.20E+03	1.70E+03	1.70E+03	1.64E+03	1.60E+02
27	1.30E+03	3.82E+03	3.87E+03	4.13E+03	2.55E+02
28	1.40E+03	1.29E+04	1.30E+04	1.40E+04	1.71E+03

4 Conclusions

This work is test performance of runner–root algorithm to solve unimodal, multi-modal, and composite test functions of CEC 2013. The obtained solutions are compared with the best-known results. Obtained results are compared with PSO, GA, and BIPOP-CMA-ES from dimension 10 and 30. Performance of RRA is better than GA and CMA-ES. Performance of many evolutionary algorithms (EA) reduces with increasing dimensionality of problems. Results show that performance runner–root algorithm is also affected with increasing dimensions.

RRA has research potential to solve engineering optimization problems. RRA's performance can be improved by balancing exploration and exploitation operators.

References

1. Fister, I., X. Yang, Fister, I., Brest, J., Fister, D.: A Brief Review of Nature-Inspired Algorithms for Optimization. arXiv preprint arXiv:1307.4186 (2013)
2. Xing, B., Gao, W.: Innovative Computational Intelligence: A Rough Guide to 134 Clever Algorithms. Springer, Berlin (2014)
3. Yang, X.: Nature-Inspired Optimization Algorithms. Elsevier, New York City (2014)
4. Liang, J., Qu, B., Suganthan, P., Hernández-Díaz, A.: Problem Definitions and Evaluation Criteria for the CEC 2013 Special Session on Real-parameter Optimization, vol. 201212, pp. 3–18. Computational Intelligence Laboratory, Zhengzhou University, Zhengzhou, China and Nanyang Technological University, Singapore, Technical Report (2013)
5. Merrikh-Bayat, F.: The runner-root algorithm: A metaheuristic for solving unimodal and multimodal optimization problems inspired by runners and roots of plants in nature. Appl. Soft Comput. **33**, 292–303 (2015)
6. Merrikh-Bayat, F.: A Numerical Optimization Algorithm Inspired by the Strawberry Plant. arXiv preprint arXiv:1407.7399, Cornell University Library (2014)
7. Zambrano-Bigiarini, M., Clerc, M., Rojas, R.: Standard particle swarm optimization 2011 at CEC-2013: a baseline for future PSO improvements. In: 2013 IEEE Congress on Evolutionary Computation (CEC 2013), pp. 2337–2344 (2013)
8. Elsayed, S., Sarker, R., Essam, D.: A genetic algorithm for solving the CEC'2013 competition problems on real-parameter optimization. In: 2013 IEEE Congress on Evolutionary Computation (CEC 2013), pp. 356–360 (2013)
9. Loshchilov, I.: CMA-ES with restarts for solving CEC 2013 benchmark problems. In: 2013 IEEE Congress on Evolutionary Computation (CEC 2013), pp. 369–376 (2013)

Determination of DG Allocation for Minimizing Annual Grid Energy Transaction

Gulnar Niazi, Soniya Lalwani and Mahendra Lalwani

Abstract The integration of distributed generation (DG) units in power distribution network has become progressively imperative in recent years. The main goal of this paper is to decide the optimal number, size and location of DG units to reduce annual grid energy transaction. An optimal DG placement (ODGP) denotes obtaining the best possible location of the DG along with its size on the transmission line without causing any disturbance to the system. Optimal integration of DG in distribution grid is one of the important and effective options. Optimal allocation with a suitable sizing of DG units plays an efficient role in terms of improving voltage profile and power quality, reduce flows and system losses. The implemented technique is based on particle swarm optimization (PSO) for optimal allocation of DG unit in power systems. The proposed algorithm has been tested on IEEE 33 bus standard radial distribution system (DS) in MATLAB programming environment.

Keywords Distributed generation (DG) · Distribution system (DS) · Optimal DG placement (ODGP) · Particle swarm optimization (PSO)

1 Introduction

The traditional generation system which uses the conventional resources (like fossil fuels, hydro and nuclear) for electricity generation depends upon centralized generating station for its operation. The operation of such traditional generation systems relies on centralized control utility generators. In the current scenario, the support for large central station plants is diminishing, due to decreasing conventional resources, raised transmission cost, heightened environmental concerns and

G. Niazi · M. Lalwani (✉)
Department of Electrical Engineering, Rajasthan Technical University, Kota, India
e-mail: mlalwani.ee@gmail.com

S. Lalwani
Department of Computer Science & Engineering,
Rajasthan Technical University, Kota, India

© Springer Nature Singapore Pte Ltd. 2019
J. C. Bansal et al. (eds.), *Soft Computing for Problem Solving*,
Advances in Intelligent Systems and Computing 816,
https://doi.org/10.1007/978-981-13-1592-3_77

technological advancements. DG is a method which can be used to combat these challenges. The DG adopts distributed generation units (DGU) to generate electrical power from a local energy source that can either be renewable or non-renewable [1, 2]. There are several benefits of DG, and DGs are implemented because of the rational consumption of energy, heterogeneity of energy sources, easiness of finding sites for smaller generators, deregulation policy, briefer construction times, lesser capital costs of smaller plants and adjacency of the generation plant to heavy loads [3, 4].

In order to achieve these benefits, DG size has to be optimized. The main factors included in the optimization problem are system losses, active/reactive power costs, operating cost, network configuration and voltage profile. Many interesting methods have been adopted to solve this optimization problem. Some of the method mentioned in [5] are analytical approach, numerical method, heuristic method [6, 7]. All these methods have their own advantages and disadvantages which confide in data and system under consideration. This paper presents a variant of PSO technique for compensating the optimization problem enriched with ability of a global or near global optimum solution with small simulation time. The paper is organized as follows: Sect. 2 describes the methods for optimal DG allocation. Section 3 presents proposed technique in which PSO is used for finding out the optimal location of DGs for reduction in annual grid energy transaction. In Sect. 4, simulation and results are described.

2 Methodologies for Optimal DG Allocation

The multi-objective optimization problem of DG allocation uses various types of techniques. These optimization techniques can be divided into three categories: analytical method, numerical method, and heuristic method.

1. Analytical Method (AM): AM is easy to effectuate and fast to execute. Also known as the two-third rule to install two-third capacity of DG of the incoming generation at two-third of the line length [8]; however, this method is not efficacious on non-uniformly distributed loads.
2. Numerical Methods (NM): The main advantage of the numerical methods is that it is an exhaustive search method that assures the reporting of global optimum; even so, it is not pertinent for large-scale systems. The various numerical methods applied for ODGP are shown in Table 1.
3. Heuristic Method (HM): HM can be applied to solve large, complex ODGP problems owing to their robustness. These methods require high computational effort which does not prove as a drawback in critical DG placement applications. Table 2 depicts some HM to solve ODGP.

Table 1 Types of numerical methods

S. No.	NM type	Procedure
1	Gradient search (GS)	GS for optimal sizing in meshed networks [9]
2	Linear programming (LP)	Solving ODGP for maximum DG penetration and harvesting [10]
3	Sequential quadratic programming (SQP)	Solving ODGP without and with fault constraints [11]
4	Nonlinear programming (NLP)	ODGP formed as multi-period AC optimal power flow (OPF), solved by NLP [12]
5	Dynamic programming (DP)	Solving ODGP model with the objective to maximize the distribution operator with light, peak and nominal conditions [13]
6	Ordinal optimization (OO)	For specifying the location and sizes of multiple DGs [14]
7	Exhaustive search (ES)	For resolving ODGP in distribution network with variable power loads [15]

Table 2 Types of heuristic methods

S. No.	HM type	Procedure
1	Genetic algorithm (GA)	Used for achieving the exact or near-optimal solutions for multi-objective optimization problems. To solve ODGP with reliability constraints [16], for variable load models [17], distributed load [18] and constant power loads [19]
2	Tabu search (TS)	Solving ODGP for uniformly distributed loads [20]
3	Particle swarm optimization (PSO)	Solving an ODGP model in distribution network with no unity power factor variable power load models [21]
4	Ant colony optimization (ACO)	Proposed to solve ODGP [22]
5	Artificial bee colony (ABC)	Tuned two control parameters [23]

3　Proposed Technique

Optimal allocation of DGs on distribution networks is administrated by a standard test system (IEEE 33 bus test feeder). MATLAB/SIMULINK is employed for the network fabrication and simulations. The PSO algorithm is employed for determination of placement and sizing of DG in the chosen network with the objective of minimizing annual grid energy in MATLAB programming environment, with Newton-Raphson load flow program and some predetermined conditions of grid code. The algorithm was run for nominal, peak and light load conditions, and the results are obtained for given DG size and DG number constraint. The result hence obtained is used to study the reduction in annual grid energy transactions, power losses and improvement in voltage profiles.

3.1 Particle Swarm Optimization (PSO)

PSO is a population-based meta-heuristic technique inspired by social behaviour of bird flocking or fish schooling which was developed in 1995 by [24, 25]. For the initialization of this method, a definite number of particles are dispersed randomly in the problem search space. Each particle is randomly assigned a velocity during the first iteration. This velocity of each individual particle is updated in each iteration depending upon three factors: particles past velocity, particles best position up to now and swarms best position up to now. The particle gradually starts moving towards the optimal due to the shared and personal experiences.

PSO Implementation

In PSO algorithm, the number of particles in the search space represents the solution candidates. Each particle is an m dimension real vector denoting the m number of parameter that is to be optimized.

Steps of PSO technique:

Step 1. Process start: Set the iteration counter $k = 1$. Determine the population size and randomly generate an initial population of n particles. Initial velocity of each particle is randomly generated to evaluate the objective function.

Step 2. Objective function calculation: The objective function is calculated to determine the fitness value of each particle. This fitness value for the first iteration becomes *pbest*, and the best fitness values from all the particles become *gbest*.

Step 3. Velocity updating: The velocity of each particle is modified in each iteration with the help of Eq. (2). Then new values of particles are generated using Eq. (3).

Step 4. *gbest*, *pbest* updating: From the previous iterations, the individual best of each particle is assigned to *pbest* and the best from all the particles is taken as *gbest*.

Step 5. Updating iteration counter: The iteration counter is updated $k = k + 1$.

Step 6. Repeat steps 3–5, if the termination criterion is not met, otherwise go to step 7.

Step 7. Stop: The particles generating the latest *gbest* is the optimal solution of PSO.

Mathematical Representation of PSO

The minimization objective function is formulated as:

$$min \ f(x) \ x \in S \subseteq R^D \tag{1}$$

here, x is decision variables matrix, composed of m vectors with dimension D. S is the feasible solution space [26]. Particles rectify their position conferring their personal flying experience, i.e. personal best (*pbest*) along with the flying experience of the other particles in the swarm, i.e. global best (*gbest*):

$$v_i^{(k+1)} = wv_i^{(k)} + c_1r_1(pbest_i^{(k)} - x_i^{(k)}) + c_2r_2(gbest_i^{(k)} - x_i^{(k)}) \qquad (2)$$

The current position is updated by adding current velocities in the previous position:

$$x_i^{(k+1)} = x_i^{(k)} + v_i^{(k+1)} \qquad (3)$$

here, for ith particle at kth iteration: v_i^k represents the velocity; x_i^k is the position; $pbest_i^k$ and $gbest_i^k$ as personal best and global best positions, respectively; w as the inertia weight; c_1 and c_2 are cognitive and social acceleration coefficient, respectively; r_1 and r_2 are random number in range 0–1.

Exponentially decreasing weight strategy [27] is employed in proposed work for obtaining optimal solution and better convergence. w is formulated as:

$$w = w_{max} - \frac{(w_{max} - w_{min})}{\{-exp(k/t_k)\}} \qquad (4)$$

where w_{min} and w_{max} are minimum and maximum weights, respectively. Here, $w_{max} = 0.9$; $w_{min} = 0.4$; k = iteration number and t_k is the total number of iterations. Further, parameter setting in presented work is as follows:

- Number of particles (n) = 30
- Total number of iterations (t_k) = 1000
- Cognitive accceleration coefficient (c_1) = 1.4962
- Social acceleration coefficient (c_2) = 1.4962.

4 Simulation Results

The base case was run using Newton-Raphson load flow to obtain bus voltages, annual grid energy loss and annual grid energy transaction. By using PSO, the optimal location of three DGs was obtained and the reduction in annual grid energy transaction was observed. The comparisons of the radial 33 bus network with and without DG implementation for three different load levels are mention as follows:

4.1 Bus Voltage Profile for Nominal, Peak and Light Loads

Table 3 shows the voltage comparison for nominal, peak and light load levels without DG in the system and also with DGs located optimally, at buses 18, 25 and 28 and sized using PSO.

1. For nominal load: The minimum voltage improves from 0.9131 before DG installation at the 18th bus to 0.9817 after DG installation at the 29th bus.

Table 3 Bus voltages (pu) of IEEE 33 bus systems before and after installation of DG for nominal, peak and light load

Bus No.	Nominal		Peak		Light	
	Before DG	After DG	Before DG	After DG	Before DG	After DG
1	1.0000	1.0000	1.0000	1.0000	1.0000	1.0000
2	0.9970	0.9993	0.9951	0.9975	0.9986	1.0007
3	0.9829	0.9973	0.9715	0.9871	0.9917	1.0053
4	0.9755	0.9958	0.9588	0.9811	0.9881	1.0073
5	0.9681	0.9947	0.9463	0.9756	0.9846	1.0096
6	0.9497	0.9900	0.9151	0.9597	0.9757	1.0134
7	0.9462	0.9878	0.9092	0.9555	0.9741	1.0128
8	0.9414	0.9877	0.9010	0.9526	0.9718	1.0149
9	0.9351	0.9882	0.8904	0.9495	0.9688	1.0182
10	0.9293	0.9893	0.8804	0.9473	0.9660	1.0218
11	0.9284	0.9897	0.8790	0.9473	0.9656	1.0226
12	0.9269	0.9907	0.8764	0.9474	0.9648	1.0241
13	0.9208	0.9941	0.8660	0.9475	0.9619	1.0301
14	0.9185	0.9954	0.8621	0.9475	0.9608	1.0323
15	0.9171	0.9978	0.8597	0.9492	0.9602	1.0352
16	0.9158	1.0012	0.8573	0.9520	0.9595	1.0391
17	0.9137	1.0073	0.8539	0.9573	0.9585	1.0460
18	**0.9131**	1.0114	**0.8528**	0.9613	**0.9583**	1.0501
19	0.9965	0.9988	0.9942	0.9967	0.9983	1.0004
20	0.9929	0.9952	0.9885	0.9909	0.9965	0.9987
21	0.9922	0.9945	0.9873	0.9898	0.9962	0.9983
22	0.9916	0.9939	0.9863	0.9888	0.9958	**0.9980**
23	0.9794	0.9974	0.9656	0.9851	0.9900	1.0071
24	0.9727	0.9981	0.9547	0.9818	0.9867	1.0110
25	0.9694	1.0020	0.9493	0.9839	0.9850	1.0165
26	0.9477	0.9901	0.9119	0.9586	0.9748	1.0144
27	0.9452	0.9903	0.9075	0.9573	0.9736	1.0158
28	0.9338	0.9894	0.8880	0.9495	0.9681	1.0202
29	0.9255	**0.9817**	0.8740	0.9365	0.9642	1.0165
30	0.9220	0.9783	0.8680	0.9308	0.9625	1.0148
31	0.9178	0.9744	0.8609	0.9242	0.9605	1.0130
32	0.9169	0.9735	0.8593	0.9228	0.9601	1.0126
33	0.9166	0.9733	0.8588	**0.9223**	0.9599	1.0124

2. For peak load: The minimum voltage improves from 0.8528 before DG installation at the 18th bus to 0.9223 after DG installation at the 33rd bus.
3. For light load: The minimum voltage improves from 0.9583 before DG installation at the 18th bus to 0.9980 after DG installation at the 22nd bus.

Figure 1 shows the node voltage (pu) for IEEE 33 bus with and without DG
for nominal, peak and light loads. Figure 2 shows the rate of convergence of PSO
for maximizing the fitness, the graph gradually increases and then finally becomes
constant at a given maximum fitness value, i.e. the optimum solution. Table 4 presents
the DG placement for annual grid energy for IEEE 33 bus system with and without
DG for different loads. The following results are attained:

1. The net annual energy transacted from the grid is negligible (0.0003%) in comparison with the system's annual energy demand.
2. There is a 46.36% reduction in annual energy losses after the installation of DGs
   with energy losses increasing during light load by 26.4455% while decreasing
   during nominal loads and peak loads by 49.88% and 60.87%, respectively.
3. The maximum energy transaction from the grid after DG installation is 10.4063%
   and the minimum energy transacted −10.7184% of the annual energy transaction that occurred during peak load conditions and during light load conditions,
   respectively.
4. Figure 3 shows that the grid transaction remains negligible (31.27%) during
   nominal load conditions that prevail for the maximum duration of the year.
   However, the transaction remains almost same (about 11%), but with opposite
   signs, during peak and light load conditions. This shows the strength of the
   proposed method for the future distribution system with distributed storages.

Moreover, the performance of implemented PSO variant (referred EPSO in this
paragraph) is compared with standard PSO (SPSO) and genetic algorithm (GA) at
the parameters: standard deviation (SD), coefficient of variation (COV) and CPU
time. At SD criteria, EPSO shows better performance (SD = 5706.75), whereas
SPSO has worst performance (SD = 13,089.33) and GA is the next better performer
(SD = 7391.14). Similar trend is followed in the results of COV: 0.68 for EPSO; 0.93
for GA and 1.62 for SPSO. The CPU time (in seconds) taken by EPSO is 48.38;
by GA is 57.21 and by SPSO is 61.74. Hence, EPSO is a better performer than
compared algorithms. As an extension of the proposed work, developing a more
efficient version of PSO is planned with more comparisons and variants.

5 Conclusion

In this paper, an implementation of PSO technique to the optimal placement problem
of DG unit is explored. PSO is used for placement of DG in DS to reduce annual
energy loss and annual grid energy transaction. From the numerical analysis, it can be
observed that the annual grid energy with DGs is reduced than the case without DGs.

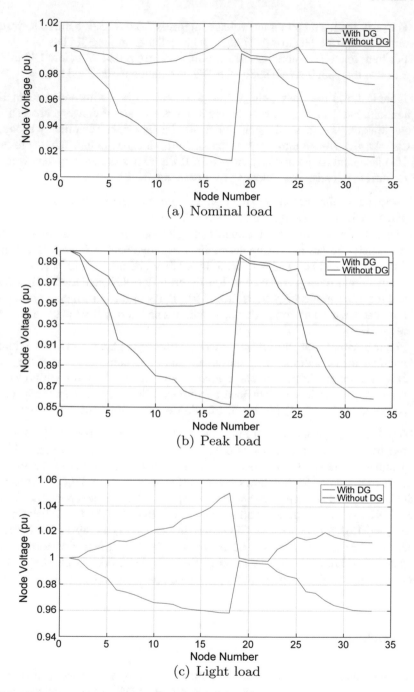

(a) Nominal load

(b) Peak load

(c) Light load

Fig. 1 Node voltage (pu) with and without DG at different loads

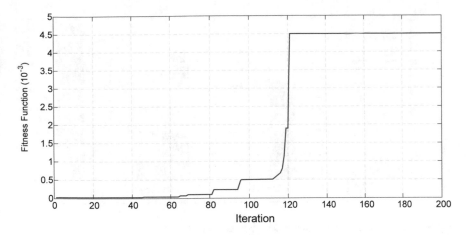

Fig. 2 Convergence analysis of algorithm

Table 4 DG placement for annual grid energy for IEEE 33 bus system

Scenario	Parameters	Load levels			Total
		Light (0.5)	Nominal (1.0)	Peak (1.6)	
Without DG	Energy loss (kWh)	9.4136×10^4	1.0652×10^6	8.63091×10^5	2.0224×10^6
	Annual energy transacted from grid (kWh)	3.8090×10^6	2.0601×10^7	9.7791×10^6	3.4189×10^7
With DG DG location (Bus No.: 18, 25, 28) DG size (kW:1018, 1287, 1491)	Energy loss (kWh)	2.1316×10^5	5.3391×10^5	3.3776×10^5	1.0848×10^6
	Annual energy transacted from grid (kWh)	-3.6645×10^6	1.0692×10^5	3.5578×10^6	125.2321
	% Energy loss reduction	-126.44	49.88	60.87	46.36
	% Grid energy transacted reduction	109.62	99.48	63.62	99.9996

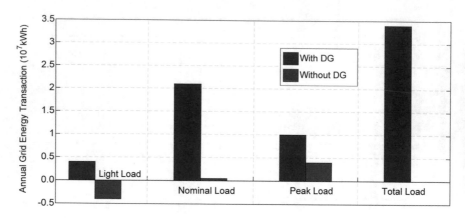

Fig. 3 Annual grid energy transaction with and without DG

By optimal location of DGs, annual grid transaction reduces 109.62% for nominal load, 99.48% for peak load and 63.62% for light load. The results show the efficiency of proposed (PSO) technique for reduction of kWh flows and kWh intake from the grid. Distribution network across the world generally has the same structure and operating principle, and thus, this methodology is applicable to such distribution networks.

Acknowledgements The second author (S.L.) gratefully acknowledges Science and Engineering Research Board, DST, Government of India, for the fellowship (PDF/2016/000008).

References

1. Rao, N.S., Wan, Y.-H.: Optimum location of resources in distributed planning. IEEE Trans. Power Syst. **9**(4), 2014–2020 (1994)
2. Lamarre, L.: The vision of distribution. EPRI J. **18**(3), 6–18 (1993)
3. Chiradeja, P., Ramakumar, R.: An approach to quantify the technical benefits of distributed generation. IEEE Trans. Energy Convers. **19**(4), 764–773 (2004)
4. Brown, R.E., Pan, J., Feng, X., Koutlev, K.: Sitting distributed generation to defer T & D expansion. In: IEEE Transmission and Distribution Conference and Exposition, vol. 12, Atlanta, USA, pp. 1151–1159 (2001)
5. Ng, H.N., Salama, M.M.A., Chikhani, A.Y.: Capacitor allocation by approximate reasoning: fuzzy capacitor placement. IEEE Trans. Power Delivery **15**(1), 393–398 (2000)
6. Gallego, R.A., Monticelli, A.J., Romero, R.: Optimal capacitor placement in radial distribution networks. IEEE Trans. Power Syst. **16**(4), 630–637 (2001)
7. Varilone, P., Carpinelli, G., Abur, A.: Capacitor placement in unbalanced power systems. In: Proceeding 14th PSCC, Sevilla, vol. 4(3), pp. 1–6 (2002)
8. Willis, H.L.: Analytical methods and rules of thumb for modeling DG-distribution interaction. In: IEEE Power Engineering Society Summer Meeting, vol. 3, WA, USA, pp. 1643–1644 (2000)
9. Vovos, P., Bialek, J.: Direct incorporation of fault level constraints in optimal power flow as a tool for network capacity analysis. IEEE Trans. Power Syst. **20**(4), 2125–2134 (2005)

10. Keane, A., O'Malley, M.: Optimal allocation of embedded generation on distribution networks. IEEE Trans. Power Syst. **20**(3), 1640–1646 (2001)
11. Al-Hajri, M.F., Al-Rashidi, M.R., El-Hawary, M.E.: Improved sequential quadratic programming approach for optimal distribution generation deployments via stability and sensitivity analyses. Electr. Power Compon. Syst. **38**(14), 1595–1614 (2010)
12. Ochoa, L.F., Gareth, H.P.: Minimizing energy losses: optimal accommodation and smart operation of renewable distributed generation. IEEE Trans. Power Syst. **26**(1), 198–205 (2011)
13. Singh, R.K., Goswami, S.K.: Optimum allocation of distributed generations based on nodal pricing for profit, loss reduction, and voltage improvement including voltage rise issue. Int. J. Electr. Power Energy Syst. **32**(6), 637–644 (2010)
14. Jabr, R., Pal, B.C.: Ordinal optimisation approach for locating and sizing of distributed generation. IET Gener. Transm. Distrib. **3**(8), 713–723 (2009)
15. Singh, D., Misra, R.K.: Effect of load models in distributed generation planning. IEEE Trans. Power Syst. **22**(4), 2204–2212 (2007)
16. Borges, C.L., Falcao, D.M.: Optimal distributed generation allocation for reliability, losses, and voltage improvement. Int. J. Electr. Power Energy Syst. **28**(6), 413–420 (2006)
17. Singh, D., Verma, K.: Multi-objective optimization for DG planning with load models. IEEE Trans. Power Syst. **24**(1), 427–436 (2009)
18. Singh, R.K., Goswami, S.K.: Optimal siting and sizing of distributed generations in radial and networked systems considering different voltage dependent static load models. In: IEEE 2nd International Conference on Power and Energy, Baharu, Malaysia, pp. 1535–1540 (2008)
19. Shukla, T., Singh, S., Srinivasarao, V.: Optimal sizing of distributed generation placed on radial distribution systems. Electr. Power Compon. Syst. **38**(3), 260–274 (2010)
20. Nara, K., Hayashi, Y., Ikeda, K., Ashizawa, T.: Application of tabu search to optimal placement of distributed generators. In: IEEE Power Engineering Society Winter Meeting, vol. 2, Ohio, USA, pp. 918–923 (2001)
21. El-Zonkoly, A.M.: Optimal placement of multi-distributed generation units including different load models using particle swarm optimization. Swarm Evol. Comput. **5**(7), 757–764 (2011)
22. Wang, L., Singh, C.: Reliability-constrained optimum placement of reclosers and distributed generators in distribution networks using an ant colony system algorithm. IEEE Trans. Syst. Man Cybern. Part C (Appl. Rev.) **38**(6), 757–764 (2008)
23. Abu-Mouti, F.S., El-Hawary, M.E.: Optimal distributed generation allocation and sizing in distribution systems via artificial bee colony algorithm. IEEE Trans. Power Delivery **26**(4), 2090–2101 (2011)
24. Aman, M.M., Jasmon, G.B., Bakar, A.H.A., Mokhlis, H.: A new approach for optimum DG placement and sizing based on voltage stability maximization and minimization of power losses. Energy Convers. Manage. **70**, 202–210 (2013)
25. Eberhart, R., Kennedy, J.: A new optimizer using particle swarm theory. In: Sixth International Symposium on Micro Machine and Human Science, Nogaya, Japan, pp. 39–43 (1995)
26. Lalwani, Soniya, Kumar, Rajesh, Gupta, Nilama: A novel two-level particle swarm optimization approach for efficient multiple sequence alignment. Memetic Comput. **7**(2), 119–133 (2015)
27. Lalwani, S., Kumar, R., Gupta, N.: A study on inertia weight schemes with modified particle swarm optimization algorithm for multiple sequence alignment. In: 6th IEEE International Conference on Contemporary Computing, JIIT, Noida, pp. 283–288 (2013)

Author Index

© Springer Nature Singapore Pte Ltd. 2019
J. C. Bansal et al. (eds.), *Soft Computing for Problem Solving*,
Advances in Intelligent Systems and Computing 816,
https://doi.org/10.1007/978-981-13-1592-3

Printed in the United States
By Bookmasters